立 心

——华中师范大学心理学院成立10年论文选辑（2015）

◇ 名誉主编　刘华山

◇ 主　　编　周宗奎

◇ 副 主 编　郭永玉　江光荣　马红宇

世界图书出版公司

华中师范大学出版社

图书在版编目（CIP）数据

立心：华中师范大学心理学院成立 10 年论文选辑：2015 / 周宗奎主编 . -- 广州：世界图书出版广东有限公司，2016.2

ISBN 978-7-5192-0784-7

Ⅰ . ①立… Ⅱ . ①周… Ⅲ . ①心理学—文集 Ⅳ . ① B84-53

中国版本图书馆 CIP 数据核字（2016）第 037901 号

立　心
——华中师范大学心理学院成立 10 年论文选辑（2015）

责任编辑　吕贤谷

封面设计　汤　丽

出版发行　世界图书出版广东有限公司

地　　址　广州市新港西路大江冲 25 号

印　　刷　虎彩印艺股份有限公司

规　　格　889mm×1194mm　1/16

印　　张　31.625

字　　数　836 千字

版　　次　2016 年 2 月第 1 版　2016 年 2 月第 1 次印刷

ISBN 978-7-5192-0784-7/B·0135

定　　价　110.00 元

立 心

——华中师范大学心理学院成立 10 年论文选辑（2015）

名誉主编：刘华山

主　　编：周宗奎

副 主 编：郭永玉　江光荣　马红宇

编　　委（按姓氏笔画排序）：

　　马红宇　刘华山　江光荣　佐　斌

　　周治金　周宗奎　范翠英　郭永玉　龚少英

心理咨询与健康

管理心理

序

　　十年光阴之于天地宇宙，转瞬即逝，不足一瞥；十年时间之于一所大学，以动辄百年计的校史烟云，亦不过弹指一挥，匆匆而过；而十年历史之于一个学院，则似乎可轻可重：其轻者，若学院书脉兴旺，学子繁盛，源远流长，则十年亦可谓白驹过隙，轻忽飘渺；其重者，譬如十年树木，塑其形，成其荫，十年足矣。

　　适逢心理学院立院十年，同事诸君，劳心费神，精选专文，终成此辑。文集选题广泛，恰如心理学大树之分枝繁多。似乎天马行空，各行其是，而审视渊源，则各项研究自有缘起，或横向联展，或纵向相依。细细读来，虽风格各异，方法有别，然而篇篇凝聚心血，探寻真知灼见，实乃文同此境，贯穿始终。求真务实之心路，历历在目。

　　文章千古事，得失寸心知。虽不免敝帚自珍，实向往珠玑之制。

　　北宋大家张载名言："为天地立心，为生民立命，为往圣继绝学，为万世开太平"。千余年来，多少名儒学子奉此为圭臬。吾辈可谓身居太平之世，治学历年，钻研心理，更当引此为鉴，磨砺心志，治顶天立地之学，求为天地立心之境！

　　是为序。

周宗奎

公元二〇一五年十一月三十日

于武昌桂子山

教育心理研究所团队成员

刘华山，教授，主要研究领域为学习与教学心理、学校心理辅导、中小学生心理健康教育、网络学习。

龚少英，博士，教授，教育心理研究所所长，主要研究领域为学习与教学心理、网络学习、教师心理。

张微：博士，副教授，主要研究领域为特殊儿童学习与促进、学校心理辅导、网络心理健康。

王福兴：博士，副教授，主要研究领域为网络学习、教学心理、媒体与儿童、眼动技术。

熊俊梅：博士，讲师，主要研究领域为学习与教学心理、网络学习、心理健康。

唐云：博士，讲师，主要研究领域为心理统计与测量、网络学习。

彭明：博士，讲师，主要研究领域为社会认知，包括情绪识别和调节的神经机制，进化角度下情绪的功能与发展，情绪对社会认知的影响，网络情境中的社会认知特点。

中国科学院院刊，2012，27（心理学理论体系与方法论专辑），156 - 163.

教育心理学：沟通心理学与教育的桥梁

刘华山，龚少英，熊俊梅

（华中师范大学心理学院）

摘 要 教育心理学是研究人的心理与行为改变规律的科学，是心理科学的一个重要应用分支，学校情境中学与教的基本心理学规律是其研究的主要对象。教育心理学是心理学与教育长期结合的产物，当其作为一门独立学科在 20 世纪初诞生后，曾给当时教育研究走向科学化以强有力的推动。近一二十年来教育心理学研究在学习与认知，动机、情绪与信念，教学与学科学习，学习的个体差异等主题上有了较大进展。研究发展趋势表现在：一般认知过程的研究转变为课堂里具体学科学习中的认知研究；教学心理研究受到重视；研究视角多元化；开始从对个体的关注转到对社会文化情境中个体的关注；计算机与网络等技术因素对学习的影响成为一个新的研究领域。最后，作者从发挥教育心理学作为沟通心理学与教育桥梁的功能出发，对我国教育心理学发展提出了若干建议。

关键词 教育心理学；发展脉络；研究进展；发展趋势

1 教育心理学的诞生

教育心理学是研究人的心理与行为改变规律的科学。人的心理与行为的改变，或曰人的学习，可以在自然和社会环境影响下自发地产生，也可以在学校教育影响下有目的、有计划地进行。由于后一种学习形式的重要性，当今大多数教育心理学家都认同把教育心理学定义为"研究学校情境中学与教的基本心理学规律的科学"。

以冯特 1879 年在莱比锡建立世界上第一个心理学实验室为标志，心理学从哲学母体中脱离出来，实现了从哲学心理学向科学心理学的转变；稍后，教育心理学也从教育哲学中脱离出来，成为一门独立的学科，其标志是 1903 年美国心理学家桑代克（E.L.Thorndike）《教育心理学》一书的问世。

从起源上说，教育心理学是心理学与教育结合的产物。心理学与教育的结合并非始自科学心理学诞生之后，而是经历了哲学取向下的结合与科学取向下的结合两个阶段的漫长历史过程。因此，虽然从学科体系来说，教育心理学是心理科学的一个分支学科，但这并不意味着教育心理学是从普通心理学中分化出来，它也不单纯是科学心理学的衍生物。

直到实验心理学成为一门独立的学科之后，欧洲的教育家和心理学家才开始运用实验、测量与统计的方法研究儿童心理发展及教育问题。其中为科学心理学与教育的结合作出早期努力的是冯特的学生莫伊曼（E.Meumann）。莫伊曼十分推崇实验研究对教育工作的重要性，他与教育家拉伊（A.Lay）所倡导的实验教育学运动，以及重视对儿童身心发展与改进教育方法进行实验研究的思想，深深地打动欧美许多教育家和心理学家。而在科学心理学与教育结合的历程中迈出决定性一步，并对教育心理学的创建作出突出贡献的则是美国心理学家桑代克。桑代克在 19 世纪末开始用实验、测量的方法研究学习及个别差异问题，于 1903 年出版了《教育心理学》。他所提出的"教育心理学的研究是以了解人性及改变人性而实现教育目的"的观点为本学科的性质及其与教育的关系做出了明确的定位。

桑代克《教育心理学》一书的问世，不仅使教育心理学走向科学化，而且也带动整个的教育学研究走向科学化。他认为教育将"依赖其领导者用科研结果而不是一般意见指导其方法选择的程度而得到改进"。许多过去视为不证自明的传统教育观点，现在都要经过实证研究的检验而决定存废。例如：

教育心理学对学习迁移现象（学习中的"举一反三"）有过各种理论解释。其中一种早期的迁移理论"形式训练说"认为：注意力、记忆力、思考力、意志力等，都是每个人的"心"所具有的官能（能力），它们可以通过训练而得到加强。学习的迁移则是在一个领域通过训练而得到加强的心理官能，在其它领域自动地发挥作用。特定的学科对于训练人的某种心理官能可能具有独特的作用。早已退出生活领域的拉丁语，据说就是因其有利于训练学生的观察、推理、记忆等官能，故仍值得学生花费大量的时间去学习。按照这一观点，学生所学知识、技能的内容并不重要，重要的是它能否使学生的思考力、记忆力、想象力得到训练。这种观点在欧美流行了 200 多年，对我国教育亦产生过一定的影响，可以说是长期误导了学校教育实践。正值教育心理学诞生之际，桑代克和伍德沃斯（R. S. Woodworth）于 1901 年以大学生为被试，在知觉领域进行了一系列实验，结果表明：形式训练说的许多假设都没有经得起科学检验的证据。该研究的结论影响深远。对于当今我国教育界一些人在倡导"素质教育"时忽视知识技能掌握的观点也有警示作用。它警示人们，轻视学科知识、技能、认知策略的学习和迁移，而一味地追求素质、能力的普遍提高，只能是南辕北辙。

2　教育心理学与教育关系发展的历史脉络

教育心理学诞生后，科学心理学与教育实践的关系，走过了百年曲折的发展道路。梅耶（R.E.Mayer）将这种关系的发展比喻为三条道路：单向道，死胡同，双向道。

2.1　单向道时期

20 世纪初至 30 年代一段时期，心理学家对科学心理学改进学校教学的作用普遍持有乐观态度。在这一精神鼓舞下，美国教育学界推动了一场以桑代克为主导的教育科学运动，开展了 4 项大型的教育心理学研究，包括贾德主持的儿童阅读心理研究；桑代克主持的智力测量研究；推孟主持的天才儿童研究；全国教育研究会负责的先天遗传与后天教养问题的研究。这场以教育心理学为主导的教育科学运动有力地促进了当时教育心理学的发展。这一时期的教育心理学家对教育现实问题都表示极大的关注；注重用科学精神和严密的科学方法指导教育问题的解决。但对教育问题的症结缺乏深层次的把握。

2.2　死胡同时期

由于前一时期教育科学运动的成就没有达到预期的理想，20 世纪 30-60 年代一段时间内，许多教育心理学家开始脱离学校教育实际，回到他们擅长的实验室工作，热心于根据动物学习实验中得到的资料去建立各种庞大的学习理论体系。行为主义学习理论在当时占有主导地位。由于这些理论都不是以学校情境中的学习活动为基础，所以很难在解决学校教育问题上发挥作用。其中行为主义者斯金纳（B.F.Skinner）在其强化原理基础上倡导的程序教学，由于符合学习的部分规律，曾在许多国家引起反响，并对教学技术现代化产生过积极影响。

2.3　双向道时期

20 世纪 60-90 年代，心理学与教育的关系进入双向道时期。心理学在适应教育实践需要方面取得了进展，同时教育也成了推动心理学发展的动力。

由于60年代初美苏国防竞赛的压力，美国人普遍增强了视教育为国防的观念，通过对教育的反省，认识到加强中小学知识教学，提高国民知识水平的重要性，在教育上掀起了"恢复基础运动"。教育心理学研究也开始由行为主义范式转向认知范式，心理学家从人为控制实验室重新回到面向教育实际的研究，学科学习心理和认知研究受到关注。布鲁纳以其结构主义学习观为依据所倡导的课程改革运动，对许多国家教育改革产生了强大的推动作用。奥苏伯尔对有意义言语学习的过程、条件、心理机制的研究，加涅对人类学习的分类及内外条件的研究，维特洛克（M.C.Wittrock）通过对阅读教学和自然科学教学的考察而开展的对生成过程（意义建构过程）的研究，以及信息加工理论的许多代表人物对学习中知识表征和内部加工过程的研究，在认知领域的学习和教学规律探讨方面都取得了可观的进展。

3 教育心理学的内容框架、研究方法与应用价值

3.1 教育心理学的内容框架

教育心理学研究学校教育情境中学生"学"的心理规律，同时也研究旨在有效地指导这种学习的教师的"教"的心理规律，而以学生学习的基本心理学规律为其研究主线。其基本内容框架包括：

学生学习的性质、特点和分类。研究涉及学习的实质，内部结构，学习中人的行为改变的心理机制，学习、记忆与脑的关系，各类学习（机械学习与有意义学习，陈述性知识学习与程序性知识学习，外显学习与内隐学习等）独特的过程和特点。

学生学习的过程。揭示学习的一般过程，也揭示各类学习的特殊过程。例如认知心理学家将阅读理解分为解码过程、字面理解过程、推理性理解过程和理解监控过程，使我们得以了解阅读过程中所发生的心理事实，从而为分析学生阅读能力差异提供了框架，也为对阅读障碍儿童实施有效干预引导了思路。

影响学生学习的因素。包括对影响学生学习的个体因素和外部因素及其复杂的交互作用的考察。在内部因素方面最为受到关注的是学生的认知结构特征（背景知识）和动机情感因素。外部因素则包括课堂里的社会心理因素、学校人际关系、家庭变量、社会文化背景、计算机与网络的技术环境因素等。

基于科学心理学的教学设计研究。本领域研究旨在将对学生学习的心理规律的了解转化为合理的教学原则、教学组织和教学设计。

3.2 教育心理学研究方法的特殊性

心理学研究使用的一般方法都被大量地用于教育心理学研究。计算机模拟、反应时实验、出声思考、作业展示以及眼动技术和某些认知神经科学的方法在对学生阅读、解题过程的研究中也都有广泛应用。例如：

四则运算是学生必须掌握的自动化技能，布朗和范莱恩（Brown & Vanlehn）提出了一种减法能力模型，据此编制了减法运算的计算机程序，并假定学生减法错误是由于使用了错误规则（称为"程序障碍"）。对所编制的程序作出各种改变，就可以模拟学生所犯的各类错误。利用这种模拟的程序可以对学生的错误进行分析，获得诊断信息，选择适合学生需要的补救教学措施。

由于教育心理学研究对象和研究目的的特殊性，在研究方法选用时，也有一些独特的问题需要考虑：

（1）强调真实教学环境中的研究。教育心理学研究特别注重实验室研究与现实课堂研究的结合、量的研究与质的研究的结合。在开展面向教育实际的问题研究时，重视自然实验法、各种准实验设计以及改进的观察法、深度访谈、学生作品分析等质性研究方法的运用。这也符合20世纪80年代以后出现的儿童与教育心理学研究中的"生态化运动"的基本趋势。

（2）注重包含多变量的综合研究。由于影响学生学习的因素众多，各种生理的、认知的、情感的、社会的因素交互作用，故在研究中不宜总是简化变量及变量间的关系，而是需要更多地采用多因素设计、更有弹性的理论模型和处理数据的多元统计方法，

提高研究的内部和外部效度。

3.3 教育心理学的应用价值

教育心理学主要服务于学校教育实际，宏观层面上，它能为课程改革、教学内容、方法的改革提供理论支持。20 世纪中期世界上影响较大的教育改革运动，如美国布鲁纳（J.S.Bruner）倡导的课程改革运动，苏联赞可夫（Л.В.Занков）主持的小学教育体制改革，都是受到教育心理学理论的推动而兴起的；在微观层面上它能为解决学校教育与课堂教学中的实际问题、为改善有特殊需要儿童的学习提供建议。

除此以外，教育心理学的基本原理和研究成果，特别是作为其核心部分的学习心理学，因其本身就带有基础性质，故在其他有关领域亦具有应用价值。例如，行为疗法的理论基础，就是学习联结理论中关于两种条件反射的形成、消退、强化、惩罚、接近学习、交互抑制以及观察学习、生物反馈等一系列原理；教育心理学的知识体系也部分适用于成人教育、员工培训、罪犯改造；美国心理学会成立学校心理学家分会时曾将学校心理学定义为应用临床与教育心理学（1945），反映了教育心理学在学校心理服务中的应用价值；班杜拉（A.Bandura）的观察学习论主要用来解释人的社会行为的习得过程，对于社会公众的态度改变和社会文明建设具有参考价值。

4 教育心理学的研究进展

NolenAL 对美国心理学与教育学杂志中 2007 年影响因子居高端的 6 种权威期刊（《教育心理学杂志》、《教育心理学家》、《学习科学杂志》、《学习和个体差异》、《教育心理学评论》和《当代教育心理学》）进行分析。这些期刊 2003 年 ~2007 年间总共发表了 758 篇论文。采用 SPSS TAS 组织和分析数据，从这些论文的主题词中形成 25 个类属。各类属在 758 篇论文中出现的比率反映了其内容的相对重要程度。结果显示，排在前 5 位的是：36% 的论文（n=279）聚焦于课堂成就，33.2% 的论文（n=251）

关注学习与记忆。情绪 / 动机 / 信念、认知 / 推理，教学的主题分别占总论文的 31%，21% 和 21%（一篇论文可聚焦于多个类属）。这说明近期教育心理学研究的内容重心集中在课堂成就、认知 / 推理、学习 / 记忆、情绪 / 动机 / 信念，以及教学几个方面。

Mitchell 和 McConnell Ⅲ 对 1995-2010 年在《当代教育心理学》杂志上发表的 440 篇专业论文进行了内容分析，发现教育心理学研究的几大主题是：学科、认知过程、个体差异、方法论及专业思考、教与学。以下仅就若干研究主题，对教育心理学近一、二十年来的研究进展做些说明。

4.1 认知与学习

涉及的主题有学习过程中的认知负荷、注意力、理解、记忆、推理、元认知和迁移；学生的各类知识（观念性理解、自动化技能和认知策略）在新知识获得和问题解决中的作用；阅读、数学学习、记忆、问题解决过程中的学习与大脑的关系等。不少研究探查了"基于计算机的协作和合作学习"这种新的学习形式的性质和条件，以及多媒体、超媒体和网络条件下的学习者的自我效能感、自我调节等问题。

4.2 学习动机与学业情绪

学习动机是研究得最多的主题之一，学习者个体认知因素，个体信念，如自我效能感、成就目标定向、学业成败的归因、自我价值等对动机的作用在近期研究中受到重视。研究内容有成就目标定向的发展、升学对目标定向的影响以及课堂目标结构对个体目标的影响、自我效能感与考试焦虑。

学业情绪是指发生于学习过程中与学生学业相关的情绪体验，如自豪，满足，焦虑、内疚，羞愧，无助和厌倦。德国心理学家 Pekrun 等按唤醒度和愉悦度的高低将学业情绪分为四类。并基于社会认知的视角，提出了社会认知控制—价值学业情绪理论（Social Cognitive Control-Value Theory of Achievement Emotions），引发了关于学业情绪的一系列实证研究。该理论认为，学业情绪受学习者个人的控制感（自我概念、自我效能感和归因）和成就价值的影响，也受父母期望、教学质量、同伴关系等外部环境的影响。

4.3　学科学习与教学

认知心理学为当代的教学研究提供了深厚的基础。研究的问题有：课堂环境不同方面，如班级目标结构、性别比例、教师特征、师生关系等对学生学习有何影响；计算机和网络条件下的学习对于不同知识经验、认识能力、学习动机的学生有何积极作用和消极作用；什么样的多媒体和网络学习的内容和形式能产生最佳效果。一些研究探查了基于计算机的合作学习以及网络学习对学生知识结构、问题解决的影响。此外，具体科目（阅读、写作、数学、科学等）的认知过程和个别差异的研究也占有很高的数量比例。

5　教育心理学的研究的发展趋势

5.1　研究内容的变化与拓展

近 20 多年来教育心理学研究内容显现出一些新特点，实验室中人为控制下对认知过程的研究转变为课堂里具体学科学习中的认知研究；教学心理研究受到重视，力图将"描述性"的学习理论与"处方式"的教学理论结合起来；教师的作用和训练的研究受到关注，通过新手和专家教师的比较研究，试图揭示专家型教师成长的途径；从 20 世纪 90 年代以后开始运用认知神经科学技术来探讨阅读、学习效率与脑部活动的关系，预计将会对深化教育心理学若干主题的研究（如能力的实质、复杂的学习过程、教学活动）产生积极影响。

5.2　研究视角的多元化

当代教育心理学研究主要采用认知、社会认知、元认知、信息处理、建构主义和行为主义 6 种视角。其中认知和社会认知视角采用得最为广泛。认知视角下的研究取得了令人瞩目的成就，对信息加工过程的精细描述，促进了学习和思维策略的研究以及认知策略教学的发展，多项研究显示了学习和思维策略教学成功地改善了学生在各科学习领域的表现。采用社会认知视角的研究者倾向于将个体特征与环境因素整合到一起考察。随着建构主义学习理论的兴起，教育心理学家开始从对个体的关注转到对社会文化情境中个体的关注，越来越倾向于探索基于真实生活和具体情境中的个体学习与合作学习，关注社会的、人际的和文化的环境对学习者的信念、态度和认知的影响。

5.3　环境变量和技术因素的影响

Berliner（2006）认为教与学通常是在"教师 × 学生 × 任务 × 情境"的交汇处发生。而近 20 年来情境的变化最为迅速。随着计算机辅助教学、多媒体教学、网上课程、视频会议实现的合作学习的日益增多，探讨环境如何激发学习者的认知和动机投入，网络课程设计如何符合学生的认知特点，避免学生认知负荷过重和信息迷航，使其成为有效地自主调节学习者，就成为必须研究的问题。专家认为，计算机、网络等技术作为一种环境因素，在学习中的作用应成为"真正重要的问题之一" 而列入 21 世纪教育心理学的研究日程。

6　发展我国教育心理学的建议

6.1　加强教育心理学研究与教育实践的联系

科学心理学独立以后，教育心理学承担着沟通心理学与教育理论、教育实践的桥梁作用。如何在坚持心理学科的方法规范和学术专长的同时关注教育现实问题的解决，则是教育心理学永远无法绕开的话题。如果在解决我国教育实际问题时，教育心理学研究严重缺位，教育者单凭权威人物意志、对西方某种流行理论的迷信、经验常识、理论思辨等去制定教育政策和教育措施，必定会使有关的教学理论和教育改革实践失却科学基础和实证依据。1996年台湾教育当局为改进小学数学教育，在缺乏充分准备的情况下全面推行建构式教学。实施 6 年造成200 万小学生数学能力普遍降低，最后在社会抨击声中于 2002 年全面喊停，即是一个深刻的教训。反之，如果像 20 世纪 20~50 年代的大多数教育心理学家那样坚持方法中心主义，一味地热心于在实验室情境中建立精密的学习理论体系，而置开发学生智力和创造力、完善学生人格、解决教育实际问题的基本目标于不顾，教育心理学必定会走入发展的"死胡同"，重蹈历史覆辙，丢失本学科的应用价值。

6.2 扩充教育心理学研究的视野

"活到老，学到老"是中国的古训，也是符合"终身学习"观点的先进理念。建设学习型组织，学习型社会已是建设中国和谐社会的重要内容。教育心理学主要研究中小学教育情境中学与教的心理学规律。在坚守这一重要研究领域的同时，可以适当将研究的视野扩展到成人，探讨各类组织中的员工学习、领导和管理人员学习以及教师培训与专业成长的心理学规律，以及高层次专业人才、创新人才成长的心理规律。在所要考察的学习的形式上，除了学校课堂学习外，社区的学习、基于博物馆、科技场馆的学习的特有规律也应该纳入教育心理学的研究范围。

6.3 提高人才培养质量

参照 Berliner 的有关观点，我们对我国教育心理学专业研究生培养提出如下建议：（1）使学生注重方法论思考和实证研究方法训练，能将量化研究和质性研究相结合，能运用多因素研究设计和多变量统计技术；（2）对我国教育现实问题有较深入的了解；（3）须从事为期一年的教育实践；（4）教师为研究生提供复杂环境下进行学术研究的实践机会；（5）确保学生对教育政策有相当程度的了解。

参考文献

胡锦涛 . (2007). *高举中国特色社会主义伟大旗帜，为夺取全面建设小康社会新胜利而奋斗——在中国共产党第十七次全国代表大会上的报告*，见《十七大报告》辅导读本，人民出版社，36-37.

胡谊 . (2007). 改良教育心理学：来自认知神经科学的影响，*心理学探新*，27（1），15-18.

李季湄 . (2001). 教育心理学的发展历程综述 - 梅耶的四隐喻说，*心理科学*，24（4），454-457.

林崇德 . (2005). 试论发展心理学与教育心理学研究中的十大关系，*心理发展与教育*，（1）：1-5.

皮连生 . (2004). *教育心理学*，第三版，上海教育出版社，10-13.

吴庆麟 . (2000). *认知教学心理学*，上海科技出版社，305-309.

张春兴 . (2005). 从思想演变看教育心理学发展宜采的取向，*北京大学教育评论*，3（1），77-93.

Azevdo. R Using hypermedia as a metacognitive tool for enhancing student learning? The role of self-regulated learning. *Educational Psychologist, 40*, 199‑209.

Berliner, D. C. (2006). Educational psychology: searching for essence throughout a century of influence. In Alexander，P A, Winne, P H (Eds.), Handbook of educational psychology (2nd ed). *Mahwah: Erlbaum,* 1-27.

Berliner, D. P. (2003). *Toward a future as rich as our past*. Carnegie Initiative on the Doctorate, Carnegie Foundation for the Advancement of Teaching, Stanford CA.

Calfee, R. (2006). Educational psychology in the 21st century. In Alexander，P A, Winne, P H （Eds.），Handbook of educational psychology (2nd ed.). *Mahwah: Erlbaum*, 29‑42.

Elliott, A. J. (2006). *A conceptual history of the achievement goal structure*. In Elliott，A J, Dweck, C S (Eds.), Handbook of Competence and Motivation. New York: Guilford Press.

Elliott, A. J., Murayama, K., Pekrun, R. A. (2011). 3 × 2 Achievement Goal Model. *Journal of Educational Psychology,* 103(3), 632-648.

Mandl, H., Ertl, B., Kopp, B. (2006). Computer support for collaborative learning environments. In Verschaffel, L, Dochy, F (Eds), Instructional psychology: Past, present and future trends. *Netherlands: Elsevier*, 223-237.

Miller, G., Reynolds, W. (2003). *Future Perspectives in Educational Psychology*. IN Reynold W M, Miller G E (Eds), Handbook of Psychology V7: Educational

Psychology. New Jersey: John Wiley & Sons, 609–628.

Mitchell, A. W., McConnell III, J. R. (2012). A historical review of Contemporary Educational Psychology from 1995 to 2010. *Contemporary Educational Psychology*. doi: 10.1016/j.cedpsych.2011.11.001.

Moos, D. C., Azevedo, R. (2009). Learning with computer–based learning environments: A literature review of computer self–efficacy. *Review of Educational Research, 79*, 2, 576–600.

Nolen, A. L. (2009). The Content of Educational Psychology: an Analysis of Top Ranked Journals from 2003 Through 2007. *Educ Psychol Rev, 21,* 279 – 289.

Pekrun, R., Goetz, T., Titz, W., et al. (2002). Academic emotions in students' self–regulated learning and achievement: A program of qualitative and quantitative research. *Educational Psychologist, 37,* 91 – 106.

Pekrun, R. (2006). The control–value theory of achievement emotions: Assumptions, corollaries, and implications for educational research and practice. *Educational Psychology Review, 18,* 315 – 341.

Simons, R., deLaat, M. (2006). E–pedagogies of networked learning. In Verschaffel, L, Dochy, F (Eds), Instructional psychology: Past, present and future trends. *Netherlands: Elsevier*, 239–255.

Educational psychology: A bridge between psychology and education

LIU Huashan; GONG Shaoying; XIONG Junmei

(School of Psychology, Central China Normal University 430079 Wuhan)

Abstract　Educational Psychology is an important applied discipline of psychology science investigating the psychology of man and laws for behavior modification. The primary laws for learning and instruction in school context are the foci of Educational Psychology. Educational Psychology has been the product of psychology and education. When it was founded as an independent discipline at the beginning of the last century, it enormously promoted the scientific study of education. In the last twenty years, there has been great advances for Educational Psychology in learning and cognition, motivation, emotions and belief, instruction and subject learning, and individual differences in learning, etc. The trends of research development are: The study of general cognitive process has shifted to the cognitive study of concrete subjects in classroom settings; The psychology of teaching has been more emphasized; Multiple perspectives for research are preferred; Educational psychologists have begun to pay attention to persons in social cultural context instead of only paying attention to individual characteristics; The impact of computer and internet technology on learning has been transformed into a new research field. Lastly, the authors put forward some suggestions as on how to further nurture the development of Educational Psychology, with its function as the communicative platform for psychology and education.

Keywords　educational psychology; developmental thread; research advancement; trends of development

心理学报，2014，46(8)，1192－1207.

道德推理与道德行为关系的元分析 *

吴 鹏[1]，刘华山[2,3]

（[1] 湖北大学教育学院心理学系，武汉，430062)

（[2] 华中师范大学心理学院）

（[3] 青少年网络心理与行为教育部重点实验室，武汉，430079 ）

摘 要 目前道德心理学中存在对道德推理作用的质疑，这一质疑源自哲学领域的著名争论。从经典道德心理学理论来说，道德推理应该是道德行为的重要决定因素，但新近的观点则否定这一重要作用。本研究采用元分析技术探讨道德推理与道德行为的关系。通过文献搜索与检查，获得了50项研究和83个独立效应量，共包含16738名被试。检验表明发表偏差不会影响元分析的结果，选择随机效应模型是准确的。通过随机效应模型的元分析表明，道德推理与道德的行为有显著的正相关，与不道德的行为有显著的负相关。调节效应分析表明，道德推理测量工具的类型会影响道德推理与道德行为的关系，被试年龄阶段会影响道德推理与不道德行为的关系。这些结果肯定了道德推理的作用，也强调了研究过程中要关注测量工具的类型，指出了需要开发更全面的道德推理测量工具。

关键词 道德推理；道德行为；元分析；调节效应

分类号 B849:C91

1 问题提出

在哲学领域，存在一个千百年来都未得到解决的问题：到底是情绪还是认知决定着人的道德？古今中外，有很多学者对这个问题发表过观点。在众多观点中，以休谟 (David Hume) 和康德 (Immanuel Kant) 两派的思想最具代表性。休谟认为情感驱动我们的道德判断，理性可以对道德判断发挥作用，但它必须依靠情感。他认为在人类的道德中，非理性因素是最为重要的。与之相反，康德的观点认为理性才

是影响道德判断的首要因素，推理决定着道德判断，非理性不能影响道德判断。与休谟和康德的截然相反的观点相对应，现代心理学的道德理论存在着两种相对立的视角。

1.1 道德心理学中的两个视角

现代心理学对于道德的早期研究持有一个普遍的观点——人类有意识的道德推理决定着后续的道德判断和道德行为。这一思想是源自皮亚杰 (Jean Piaget) 和科尔伯格 (Lawrence Kohlberg) 从心理学角度对道德发展的开创性研究，他们共同的理论假设就

---

* 湖北大学青年科学基金 (095200) 资助。

通讯作者：刘华山，E-mail:hsliupsycho@263.net

是个体的道德发展阶段是以其不同的道德推理水平来划分。显然，这一理论的哲学根源是康德的理性主义。康德的理性主义强调有意识的推理在道德判断过程中的重要作用，当需要我们判断行为或决定的对错时，个体是通过将外在的推理原则应用到一个具体情境中 (Murphy, Wilde, Ogden, Barnard, & Caldera, 2009)。据此哲学思想，皮亚杰从心理学的角度提出了道德发展理论，这也就成为心理学中道德研究的开创性、基石性理论。此后，另一位道德心理学家科尔伯格，在皮亚杰的基础上提出来自己的道德发展阶段理论。皮亚杰和科尔伯格都认为道德判断来自道德推理，他们只强调和承认认知因素对道德判断的作用，不承认非认知因素的作用。自此之后，广大道德心理研究者均秉承他们的道德判断的认知观而开展道德心理学的研究。因此，到目前为止的大量道德心理学的研究探讨了道德推理的重要作用。

另外一个道德心理学的理论视角，可以追溯到19世纪末。当时的心理学开始重视实证研究，道德心理学者认为推理并不重要。比如弗洛伊德 (Sigmund Freud) 就认为人们的判断来自于无意识的动机和感觉，然后才会用公众接受的理由进行辩解。而行为主义学家将道德推理看作道德行为的附带结果，认为道德行为只是社会奖励或惩罚的结果。伴随着上世纪60年代的认知革命，心理学家们开始一边倒的关注道德推理。但在上世纪末，受情感革命的影响，一些道德心理学家们开始认识到过去几十年的道德理论和研究过分夸大了有意识的理性因素（特别是道德推理）的作用而忽视无意识因素的影响，于是休谟的哲学思想又开始占据道德心理学理论。持这一观点的学者们认为至少有些道德判断是无意识心理过程的结果，这些无意识的心理过程也可以被认为是直觉性的。于是，道德心理学研究者又重新将研究焦点集中到非认知因素，其中受到更多关注的是道德直觉和道德情绪 (Hauser, Cushman, Young, Jin, & Mikhail, 2007)。

在目前的道德心理学研究中，研究的出发点几乎都是上述两种视角中的一个。而最近的研究者更多是以道德直觉和道德情绪来探讨道德心理，他们认为在道德心理和行为中推理并不会起重要作用。这其中以 Haidt 的社会直觉模型 (Social Intuitionist Model, SIM) 最为有名，Haidt 认为道德推理几乎不会直接影响道德判断，除非需要才会有慢速的、事后的道德推理来影响道德判断 (Haidt, 2001; 徐平，迟毓凯，2007;)。从 Haidt 的观点来看，个体的道德推理与其道德行为没有直接关系。但大量的研究结果均表明道德推理与道德行为是有关系的，这里就存在一个疑问：道德推理与道德行为有关系吗？

1.2 道德推理、道德判断

道德推理、道德判断是道德心理学家最早关注的因素之一，也是目前为止道德心理学理论中最为重要的构成因素。道德推理是指个人运用已有的道德概念和道德认识，对道德现象进行分析、评价、推断和选择的心理过程 (余宏波，刘桂珍，2006)。道德判断则是指个体（基于内心的道德原则）对哪些是道德的进行决策和判断的能力，以及能够根据这些判断付诸行为的能力 (杨韶刚，吴慧红，2006)。简言之，可以将道德判断定义为对一个人的行为或特性的评价（好或坏）(徐平，迟毓凯，2007)。从定义上可以清晰发现这两者之间的区别，但是大量的研究将这两者混为一体，很多的研究者将道德推理冠以道德判断 (Maeda, Thoma, & Bebeau, 2009; Narvaez & Gleason, 2007)。虽然这两者都是道德心理中的认知因素，但它们是不同的。道德推理侧重于推理与分析，它关注道德现象背后的理由与解释。而道德判断则关注评价与判断，不一定涉及判断的理由。因此，道德推理可以看作是道德判断的深入，是对判断结果的探讨。这两者是完全不同的，个体可能有相同的道德判断，但是其道德推理可以完全不一样。比如，小学生可能都会认为帮助同学是好的，但认为这样做可以得到老师的认可与可以得到他人的回报是两种完全不同的道德推理，也反映个体不同的道德水平。

1.3 道德行为

道德行为是一个涵盖面很广的道德因素，很多学者给出了自己的定义。比如道德行为是指在道德意志支配下表现出来的符合社会道德规范的行为（刘华山，2008）。彭蕾则认为道德行为是人们在道德方面有意识的行动，是指个体在一定道德意识的支配下所表现出来的有利或有害于他人与社会的实际行动，是个体道德品质的外在表现（彭蕾，2004）。心理学是从知、情、意、行来划分道德结构，这里面的行就是指道德行为。道德行为的另一面应该是非道德领域的行为（如学习、工作），即非道德行为。从上述定义来看，道德行为应该包含两部分——道德的行为与不道德的行为，而在目前的研究中学者们则会使用很多与这两者相近似的名词，比如助人行为、亲社会行为、利他行为、攻击行为、反社会行为等（Eisenberg, Zhou, & Koller, 2001; Janssens & Dekovi, 1997; Manning & Bear, 2011; Shumaker, 2006;Wyatt & Carlo, 2002）。但总的来说，道德行为是道德研究的最终点，所有的研究都是为了促进道德的行为、抑制不道德的行为。我们对于道德行为的探讨应该包含上述两方面的内容，这样才能完整地展现道德行为。

1.4 道德推理与道德行为的关系

针对道德的行为的研究中，Eisenberg 的团队成果最为丰富，他们探讨了青少年阶段、成年早期的亲社会行为发展，发现道德推理与其亲社会行为相关（Eisenberg , Carlo, & Murphy,1995; Eisenberg, Miller, Shell, McNalley, & Shea, 1991; Eisenberg et al., 2002）。此外，研究者也发现儿童、成人、护士、会计师、商业人士的道德推理水平与道德的行为有显著的正相关（Ketefian, 1981; Krebs & Rosenwald, 1977; Malti, Gasser, & Gutzwiller-Helfenfinger, 2010; Ryan, 2001）。同时，有一些研究结果发现道德推理并不会与道德的行为有关系（Lai, Siu, Chan, & Shek, 2012; Schonert-Reichl, 1999; Simmons & Zumpf, 1986; 朱丹 , 李丹 , 2005）。

另一方面，大量的研究发现个体的道德推理与攻击行为、青少年犯罪、逃学等有显著的负相关（Gasser & Malti, 2012; Guzman, 2006; Wyatt & Carlo, 2002），但也有很多的研究发现道德推理与不道德行为并没有联系（Lai et al., 2012; Richards, Bear, Stewart, & Norman, 1992）。

对于道德推理与道德行为的关系，以认知发展理论的角度来看，道德认知因素应该起着重要的作用（Blasi, 1980）。而道德推理应该是最为重要的认知因素，因此道德推理理应与道德行为有密切关系，这一推论也符合康德的思想。但近 10 年来，道德心理学研究中开始重视非认知因素的作用、减弱认知因素的作用（Eisenberg, 2000;Haidt, 2001; Tangney, Stuewig, & Mashek, 2007）。从这些学者的理论来看，非认知因素才是道德行为的重要影响因素，道德推理等认知因素可能与道德行为没有关系。针对这两种观点，本研究想通过元分析来探讨道德推理与道德行为（道德的行为与不道德的行为）的关系。依据皮亚杰与科尔伯格的开创性研究与理论，我们提出假设：道德推理与道德的行为之间有正相关，道德推理与不道德的行为之间有负相关。

1.5 道德推理与道德行为关系的调节变量

元分析技术不仅仅是得到一个合成效应量，它还可以就这个效应量的影响因素展开分析。由于本元分析是探讨道德推理与道德行为之间的关系，所以可以将效应量的影响因素看作是一个调节变量，而本研究想探讨测量工具和被试年龄阶段这两个调节变量。

1.5.1 道德推理测量工具的类型

在几十年的道德研究中，出现了很多的道德推理测量工具。我们认为可以分为两大类：非结构化测量和标准的结构化测量。第一类工具是以访谈的形式提供一些小故事，然后让被试判断行为的合理性或允许性，并指出其理由。在数据分析时，研究者则根据一定的标准进行计分，或将被试的道德推理分成不同的类型。最为常用的故事当属科尔伯格编制的列车困境和天桥困境（Cushman, Young, & Hauser, 2006; Greene et al., 2009）。此外，研究者也采用自编的故事，如哭泣的婴儿、背叛的妻子、严刑拷问等（Banerjee, Huebner, & Hauser, 2010; Tarrant, Branscombe, Warner, & Weston, 2012; Ugazio, Lamm, & Singer, 2012; Vandello, Michniewicz, & Goldschmid, 2011）。

第二类工具是按照标准的心理测量学程序

编制而成，同时在施测时也有严格的操作要求。这类工具中使用较多的有 MJT(the Moral Judgment Test)、DIT(the Defining Issues Test) 和 PROM(Prosocial Reasoning Objective Measure)。它们都依据科尔伯格的理论，向被试呈现一定数量的小故事，被试需要做出行为选择，最后还要对行为选择的理由进行评定。对于被试的作答，MJT、DIT 和 PROM 以自己独特的计分方式给每个被试一个道德推理分数。

大量的道德研究采用上述两类测量工具 (Banerjee et al., 2010; Hardy, 2006, Eisenberg et al., 2001; Maeda et al., 2009; Mouratidou, Barkoukis, & Rizos, 2012; Ugazio et al., 2012)，但我们认为第一类工具存在以下问题：①道德两难情景不统一。很多研究者会采用自编的故事来进行测量，但这些测量工具并没有进行严格的心理测量学检验。就算是采用列车困境或天桥困境的研究，对于故事的描述也存在不一致。比如有的研究中会要求被试回答是否应该救多数铁道工，而另一些研究则询问是否不该杀害那个铁道工。已有的研究早已表明，这一不同表述会影响个体的道德推理 (Broeders, van der Bos, Muller, & Ham, 2011; Christensen & Gomila, 2012)，因此这种测量方法有很大的研究特异性。②道德困境多涉及生死。这一主题的特殊性可能使其完全不同于其他道德话题（如违反规则等），对这一问题的推理是否可以完全反映被试的道德推理，值得商榷 (Graham et al., 2011)。③工具计分的主观性强。没有一个统一的严格计分方式，导致不同的研究结果不能进行比较。而主试的语言表达能力会极大影响被试的作答，被试的语言表达能力也会影响主试的记录 (Carlo, Eisenberg, & Knight, 1992; Rest, Cooper, Coder, Masanz, & Anderson, 1974)。因此，我们认为道德推理测量工具会影响元分析的结果，假设采用不同测量工具的研究中道德推理与道德行为的关系有显著差异。

1.5.2 被试年龄阶段

皮亚杰与科尔伯格的道德理论都认为个体的道德推理能力是发展的，不同发展阶段的个体有不同的道德推理水平或类型。已有研究指出不同道德推理水平导致的道德行为是不一样的 (Blasi, 1980; Comunian & Gielen, 1995; Eisenberg, Cumberland, Guthrie, Murphy, & Shepard, 2005)，而不同年龄阶段（儿童、青少年和成人）的个体在道德稳定性与认知能力上是有差异的。因此假设对不同年龄阶段的被试，其道德推理与道德行为之间的关系存在显著差异。

2 研究方法

2.1 文献搜索

本研究全面搜索了相关文献，包含了中文和英文文献搜索。中文文献的搜索过程如下：首先，在 CNKI 数据库、中国科技期刊数据库、万方数据库、中国优秀硕士学位论文全文数据库以及中国博士学位论文全文数据库中，以（*道德推理或道德判断*）或（*道德行为、亲社会行为、助人行为、利他行为、不道德行为、攻击行为、反社会行为或青少年犯罪*）为关键词进行搜索。此外，也在互联网 google 学术中以相应关键词进行搜索。英文文献的搜索过程：在 PsycARTICLES, PsycINFO, JSTOR, SAGE, Springer, Elsevier, ProQuest 博硕士论文全文数据库中，以 (moral reasoning 或 moral judgment) 或 (moral behavior、moral conduct、moral action、prosocial behavior、helping behavior、altruistic behavior、altruism、immoral behavior、immoral conduct、immoral action、aggression、aggressive behavior、antisocial behavior、delinquency、truancy) 为关键词进行搜索。同时，也在互联网 google 学术中进行搜索。

对于搜索到的、但没有结果内容的文献记录，我们尽量通过可以寻找到的联络方式给作者发送电子邮件以获取全文或结果。

2.2 文献纳入的标准

对于搜索到的相关研究，我们按照以下标准来决定是否将其纳入后面的元分析：①必须报告了数字结果的实证研究，而纯理论的、综述性的研究被排除。②如果仅仅只进行了道德判断测试，即只呈现了被试的判断结果，没有进一步测试判断理由的研究将

表 1　元分析中纳入的原始研究

研究	样本量	性别群体	年龄	结果变量类型	工具类型	相关系数
Aleixo&Norris, 2000	100	M	Ad	IB	U	−0.230
Barriga et al., 2001	193	B	A	IB	U	−0.200
Bear &Richards, 1981	91	B	C	IB	U	−0.290
Bear & Rys, 1994	133	B	C	IB	U	−0.228
Bear & Rys, 1994	60	F	C	IB	U	−0.040
Bear & Rys, 1994	73	M	C	IB	U	−0.380
Bear, 1999	77	B	C	IB	U	−0.290
Blair, Monson, & Frederickson, 2001	102	B	C	IB	U	−0.034
Blair, Monson, & Frederickson, 2001	102	B	C	IB	U	−0.005
Blair, Monson, & Frederickson, 2001	102	B	C	IB	U	−0.070
Blair, Monson, & Frederickson, 2001	102	B	C	IB	U	−0.138
Bredemeier, 1994	106	B	C	IB	U	−0.386
Bredemeier, 1994	106	B	C	IB	U	−0.375
Bredemeier, 1994	106	B	C	IB	U	−0.280
Bredemeier, 1994	106	B	C	IB	U	−0.464
Bruggeman & Hart, 1996	221	B	Ad	IB	S	−0.030
Carlo & Randall, 2002	249	B	A	MB	S	0.450
Carlo et al., 2003	80	B	A	MB	S	0.250
Carlo et al., 2003	58	B	A	MB	S	0.220
Carlo et al., 1996	130	B	A	MB	S	0.270
Carlo et al., 1996	55	M	A	MB	S	0.260
Carlo et al., 1996	75	F	A	MB	S	0.280
Cummings, 2001	145	B	Ad	IB	S	−0.202
Eisenberg−berg & Hand, 1979*	35	B	C	IB	U	0.000
Eisenberg et al., 1985	58	B	C	MB	U	0.310
Eisenberg et al., 1991	64	B	A	MB	U	0.300
Eisenberg et al., 1995	32	B	A	MB	S	0.510
Eisenberg et al., 2002	30	B	Ad	MB	S	0.420
Eisenberg, Zhou, & Koller, 2001	149	B	A	MB	S	0.300
Gasser & Malti, 2012	118	B	C	IB	U	−0.230
Gasser & Malti, 2012	118	B	C	IB	U	−0.180
Guzman, 2006	192	B	A	IB	U	−0.140
Guzman, 2006	195	B	A	IB	U	−0.200
Hardy, 2006	91	B	Ad	MB	S	0.040
Janssens & Dekovic, 1997	125	B	C	MB	S	0.170
Ketefian, 1981	79	B	Ad	MB	S	0.280
Krebs & Rosewald, 1977	31	B	Ad	MB	S	0.525
Kumru et al., 2012	330	B	A	MB	S	0.180
Kumru et al., 2012	1252	B	A	MB	S	0.170
Lai et al., 2012	566	B	A	IB	S	−0.050
Lai et al., 2012	566	B	A	MB	S	0.100
Malinowski & Smith, 1985	53	M	Ad	IB	S	−0.390
Malti, Gasser, & Gutzwiller, 2010	312	B	C	IB	U	−0.170
Malti, Gasser, & Gutzwiller, 2010	312	B	C	MB	U	0.060
Manning & Bear, 2011	216	B	C	IB	U	−0.270
Manning & Bear, 2011	132	B	A	IB	U	−0.210
Miller et al., 1996	74	B	C	MB	U	0.240
Palmer & Hollin, 1996	64	B	Ad	IB	U	−0.200
Palmer & Hollin, 2000	58	B	Ad	IB	U	−0.250
Palmer & Hollin, 2001	94	B	A	IB	U	−0.440
Raaijmakers, Engles, & Hoof, 2005	846	B	A	IB	S	−0.120
Raaijmakers, Engles, & Hoof, 2005	846	B	A	IB	S	−0.150
Raaijmakers, Engles, & Hoof, 2005	846	B	A	IB	S	−0.170

续表 1

研究	样本量	性别群体	年龄	结果变量类型	工具类型	相关系数
Richards et al., 1992	143	B	C	IB	U	−0.100
Rubin & Schneider, 1973	57	B	C	MB	U	0.310
Ryan, 2001	116	B	Ad	MB	S	0.210
Schonert−Reichl, 1999	54	F	A	MB	U	0.320
Schonert−Reichl, 1999	54	M	A	MB	U	0.160
Schonert−Reichl, 1999	54	F	A	IB	U	−0.080
Schonert−Reichl, 1999	54	M	A	IB	U	−0.220
Shumaker, 1993	64	B	A	MB	U	0.260
Simmons, 1996	428	B	Ad	IB	S	−0.010
Simmons, 1996	428	B	Ad	IB	S	−0.050
Simmons, 1996	429	B	Ad	IB	S	−0.020
Simmons, 1996	429	B	Ad	IB	S	−0.010
Tarry & Emler, 2007	475	M	A	IB	U	−0.080
Turner et al., 2007	74	B	As	MB	S	0.260
Wyatt & Carlo, 2002	80	B	A	MB	S	0.280
Wyatt & Carlo, 2002	76	B	A	MB	S	0.280
Wyatt & Carlo, 2002	58	B	A	MB	S	0.220
Wyatt & Carlo, 2002	80	B	A	IB	S	−0.420
Wyatt & Carlo, 2002	76	B	A	IB	S	−0.270
Wyatt & Carlo, 2002	58	B	A	IB	S	−0.100
Wyatt & Carlo, 2002	76	B	A	IB	S	−0.260
Wyatt & Carlo, 2002	58	B	A	IB	S	−0.430
洪丽 , 2005	481	B	A	MB	U	0.049
李炜 , 2012	203	B	Ad	MB	S	0.680
刘美辰 , 2012	580	B	A	IB	U	−0.353
刘志军 , 2001	286	B	A	MB	U	0.014
毛静思 , 2001	667	B	A	MB	U	0.083
彭蕾 , 2004	410	B	C	MB	U	0.202
朱丹 & 李丹 , 2005	217	B	C	MB	U	0.026

注：B 表示研究样本含男性与女性，M 表示研究样本为男性，F 表示研究样本为女性，C 表示样本为儿童，A 表示样本为青少年，Ad 表示样本为成人，MB 表示结果变量为道德的行为，IB 表示结果变量为不道德的行为，S 表示道德推理为标准化工具，U 表示道德推理为非标准化工具。

被排除。③研究探讨的行为必须是道德领域的，非道德领域行为的研究将被排除。④没有报告完整效应量的研究将被排除。最终，我们得到符合元分析要求的文献有 50 篇。其中公开发表的文献有 43 篇，中文文献有 7 篇。

2.3 文献编码

对纳入元分析的文献进行如下编码：文献信息（作者名 + 文献时间），样本性别群体（男性、女性或两者均有），样本年龄（儿童、青少年或成人），结果变量类型（道德的行为、不道德的行为），道德推理测量工具类型（标准化、非标准化），见表 1。

针对每一个独立样本，得到一个效应量。同时，考虑有的研究针对多种行为变量（道德的行为与不道德的行为），有的研究则报告了不同样本群体的结果（男性与女性），有的研究则报告了多次测量的结果（追踪研究）。我们分别呈现每一个研究文献中包含的多个独立效应量，于是有的研究文献会包含多个独立效应量。最后，我们一共得到 83 个独立的效应量。

2.4 元分析过程

2.4.1 效应量

最终纳入的 50 篇研究文献主要报告了道德推理与道德行为的相关系数，因此我们的元分析以相关系数作为效应量。其中，在研究文献 (Bear & Rys, 1994) 中，研究者分别报告了男性与女性被试道德推理与道德行为的相关系数和样本量，我们采用相关系数

合成的方法 (r-Fisher Z) 得到被试总样本道德推理与道德行为的相关系数 (张厚灿，徐建平，2004)。另一文献中 (Krebs & Rosenwald, 1977)，研究者没有报告相关系数，但呈现了不同道德推理水平被试是否实施助人行为的具体人数，我们通过计算得到道德推理与道德行为的相关系数。

2.4.2 模型的选定

目前的元分析主要采用固定效应模型或随机效应模型，这两者最主要的区别在于权重成分的不一样。固定效应模型假设元分析中所有研究背后只存在一个真效应量，而每个研究效应量的不同是由抽样误差引起的。随机效应模型则认为每个研究的真效应量都是不同的，每个研究效应量的不同是由真效应量的不同和抽样误差共同引起的 (Borenstein, Hedges, Higgins, & Rothstein, 2009)。两个模型的不同假设会导致元分析中平均效应量的显著性检验、区间估计以及调节变量的显著性检验方法不同 (Hunter & Schmidt, 2000)。在进行元分析之前，研究者就应该从理论与实际层面选定好模型。而不能先假设一个模型开始分析，结果发现与假设不符又换另一个模型进行分析 (Borenstein et al., 2009)。在模型的选定上，Borenstein 等建议主要考虑元分析的研究是否拥有一个共同的效应量以及元分析的目的。具体来说，如果认为元分析中的研究在功能上是相同的，而我们的元分析得到的总效应量只是针对包含的研究所涉及的总体，不推广到其它总体的话，我们应该使用固定效应模型。相反，如果元分析中包含的研究中被试群体、测量工具不同，并且有理由相信这种不一样会影响结果时，就不能假设存在一个真效应量，此时使用随机效应模型更加合理 (Borenstein et al., 2009)。

在我们最终确定的 50 篇研究文献中，被试包含儿童、青少年、成年人等，被试职业涵盖学生、商务人士、护士等。要进行元分析的研究文献中被试各异，元分析得到的效应量不能只局限于某一个研究所涉及的样本群体，因此不适合采用固定效应模型。此外，我们的元分析本来就想探讨测量工具的调节作用，

因此我们有理由相信随机效应模型更适合本元分析。在后面的元分析中，将采用异质性检验来验证我们的模型选择。

2.4.3 发表偏差

当被发表的研究文献系统性地不能代表该领域已经完成的研究总体时，就认为发生了发表偏差 (Rothstein, Sutton, & Borenstein, 2006)。发表偏差的结果就是某一领域的研究文献不完整，这会严重影响元分析的结果。任何一个元分析研究都应该关注发表偏差的问题，因为它会导致最终得到的效应高于真实值 (Kuppens, Laurent, Heyvaert, & Onghena, 2013)。针对发表偏差的问题，我们首先在文献搜索阶段尽可能获取了没有发表的文献。在后面的元分析过程中，我们还会采用漏斗图 (funnel plot)、Rosenthal's *Fail-safe N* 与 Egger's 检验等方法来评估本元分析的发表偏差。

2.4.4 元分析过程及软件

我们的元分析首先想探讨道德推理与道德行为之间的关系，考虑到道德行为分为道德的行为与不道德的行为，而道德推理与这两种行为之间的相关方向是相反的，于是我们分别针对道德的行为与不道德的行为来计算总效应量。在调节变量的检验过程中，我们同样是分开分析。本研究采用 CMA 2.2(Comprehensive Meta Analysis 2.2) 进行元分析。

表 2 效应量异质性检验结果

结果变量	Q	df	p	I^2	σ^2
道德的行为	168.567	34	<0.001	79.830	0.022
不道德的行为	147.342	47	<0.001	68.101	0.011

3 研究结果

3.1 异质性检验

针对道德的行为与不道德的行为分别进行异质性检验，结果见表 2。

从表 2 的结果来看，两个结果变量的 Q 检验均显著，表明元分析中各研究的效应量是异质的。另外，依据 Borenstein 等人对 I^2 的解释 (Borenstein et al.,

2009)，针对道德的行为的元分析的 I^2 为 79.830，说明在道德推理与道德的行为的关系研究中有 79.83% 的观察变异是由这一关系中真正差异所造成的。针对不道德的行为，I^2 为 68.101，说明在不道德推理与道德的行为的关系研究中有 68.10% 的观察变异是由这一关系中真正差异所造成的。σ^2 表示真效应量的方差，两个 σ^2 表明真效应量都有一定的变异。异质性检验的结果表明，我们选定以随机效应模型来进行元分析是准确的。

3.2 发表偏差检验

首先，通过漏斗图 (funnel plot) 来检查本元分析的发表偏差，两类道德行为的漏斗图见图 1 和图 2。从漏斗图来看，涉及道德的行为的研究文献并未均匀分布于总效应量两侧，多数研究位于总效应量右侧。而涉及不道德的行为的研究文献基本均匀分布于总效应量两侧。这一分布特点表明，针对道德的行为的研究可能存在发表偏差，而针对不道德的行为的研究不存在发表偏差。因为漏斗图仅仅只能从主观的角度初步检查发表偏差，为了更准确的检验发表偏差，我们紧接着进行了 Rosenthal's *Fail-safe N* 与 Egger's 检验，结果见表 3。

从 Egger's 检验的结果来看，涉及道德的行为的研究与涉及不道德的行为的研究均存在一定的发表偏差。从 Rosenthal's *N* 值来看，需要再纳入大量 (>2200) 涉及两个行为的研究文献才可能使两个总效应量不显著，这说明涉及两个行为的本研究并不存在严重的发表偏差。

上述 3 个发表偏差检验中，有两个结果 (漏斗图和 Rosenthal's *N*) 表明针对不道德的行为的元分析不存在发表偏差，一个结果 (Rosenthal's *N*) 表明针对道德的行为的元分析不存在发表偏差，都没有得到三

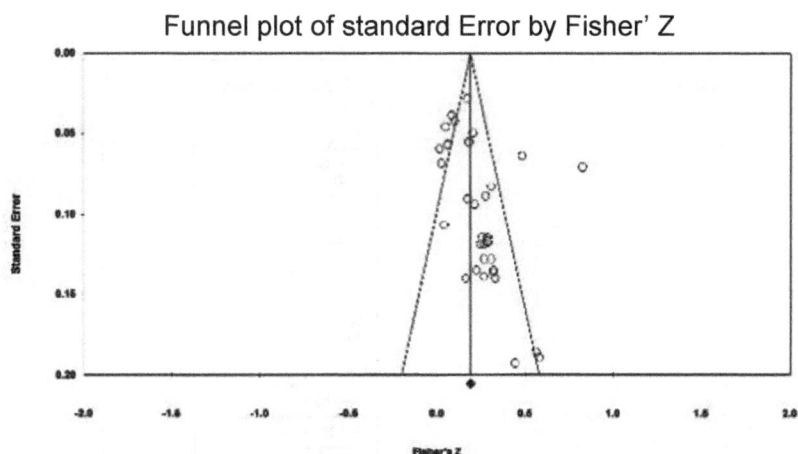

图 1　涉及道德的行为的研究的漏斗图

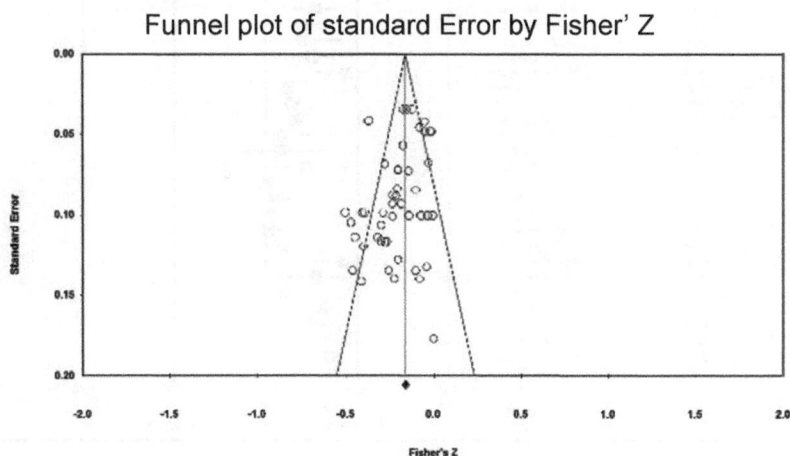

图 2　涉及不道德的行为的研究的漏斗图

表 3 发表偏差检验结果

结果变量	Rosenthal's N	Egger's Intercept	SE	LL	UL	p
道德的行为	2285	1.78240	0.72154	0.31443	3.25038	0.019
不道德的行为	3293	−1.52362	0.54727	−2.62521	−0.42203	0.008

注：LL、UL 表示 Egger's Intercept 的 95% 置信区间的下限与上限。

表 4 道德推理与道德行为关系的随机效应模型分析结果

结果变量	N	k	r	LL	UL	Z	p
道德的行为	6663	35	0.238	0.181	0.293	7.981	<0.001
不道德的行为	10065	48	−0.188	−0.225	−0.151	−9.734	<0.001

注：N 表示样本量，k 表示研究个数，LL、UL 表示 r 的 95% 置信区间的下限与上限。

个检验都认可的结果。但按照 Borenstein 等人的看法，发表偏差的检验目的应该是确定元分析结果属于以下哪种类型：①偏差的影响可以忽略不计；②偏差的影响不能忽略，但研究结果还是有效的；③研究结果可能存在问题 (Borenstein et al., 2009)。因此需要作进一步分析，我们采用 Duval 和 Tweedie 提出的剪粘法 (Trim and Fill) 来检验发表偏差对元分析结果造成的影响 (Duval & Tweedie, 2000)。结果发现，剪粘研究文献后，针对两种行为分别采用随机效应模型得到的总效应仍然都显著。此外，最终进行元分析的文献中，未发表的文献占 14%，这一比例已经很大。综合以上结果表明虽然本研究的两个元分析中可能存在轻

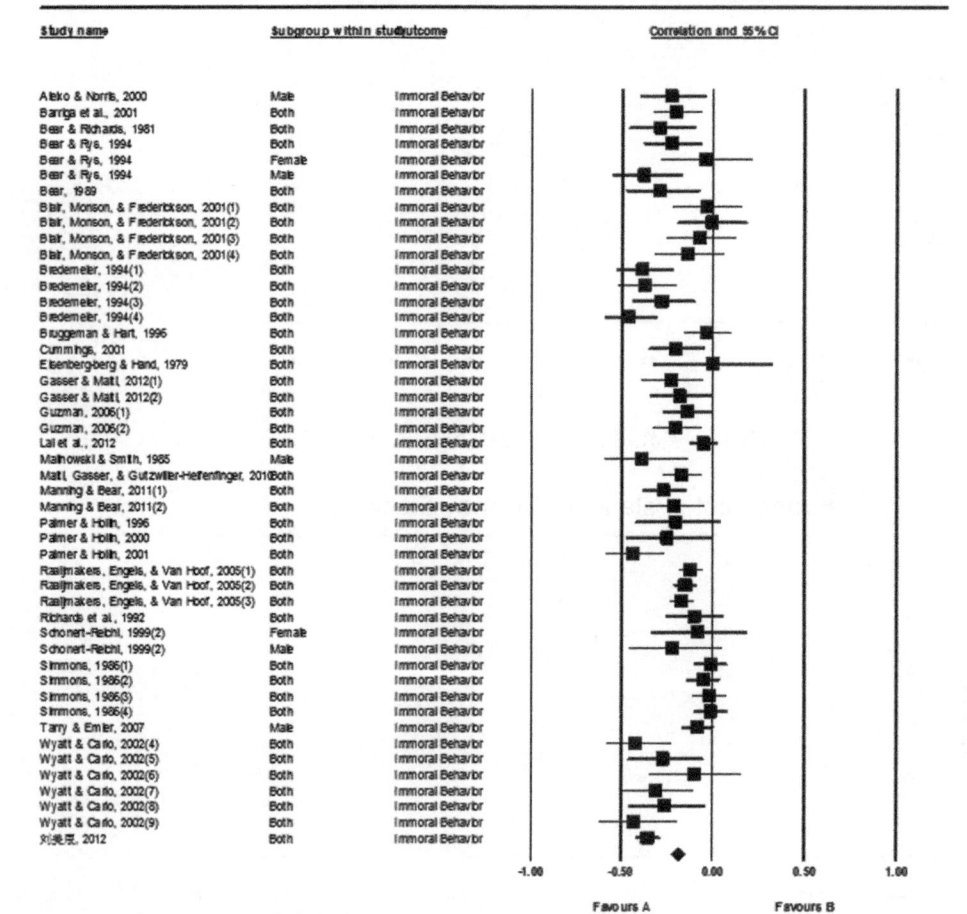

图 3 针对道德的行为的效应量的分布图

微的发表偏差，但是元分析的主要结论还是有效的。

3.3 主效应

从整体检验道德推理与道德行为的关系，结果见表4。结果表明，共有35项独立的道德推理与道德的行为的效应量，被试总数为6663，道德推理与道德的行为的整体相关系数为0.238，见图3。共有48项独立的道德推理与不道德的行为的效应量，被试总数为10065，道德推理与不道德的行为的整体相关系数为-0.188，见图4。

3.4 调节效应检验

从图3与图4来看，各个研究的效应量分布于总效应量（图中菱形）左右两侧，而且各研究的效应量之间存在很大的变异。为了分析这一变异，我们分别检验道德推理测量工具类型（标准化与非标准化）、被试年龄（儿童、青少年与成人）对道德推理与道德行为关系的调节作用，结果见表5。

从调节效应分析的结果来看，道德推理测量工具的类型可以影响道德推理与道德的行为间的关系（Q_b=9.577，p= 0.002），也可以影响道德推理与不道德的行为间的关系（Q_b= 5.550，p= 0.018）。被试的年龄只能影响道德推理与不道德的行为间的关系（Q_b= 10.183，p=0.006）。

4 讨论

本研究是道德心理学领域首次通过元分析技术

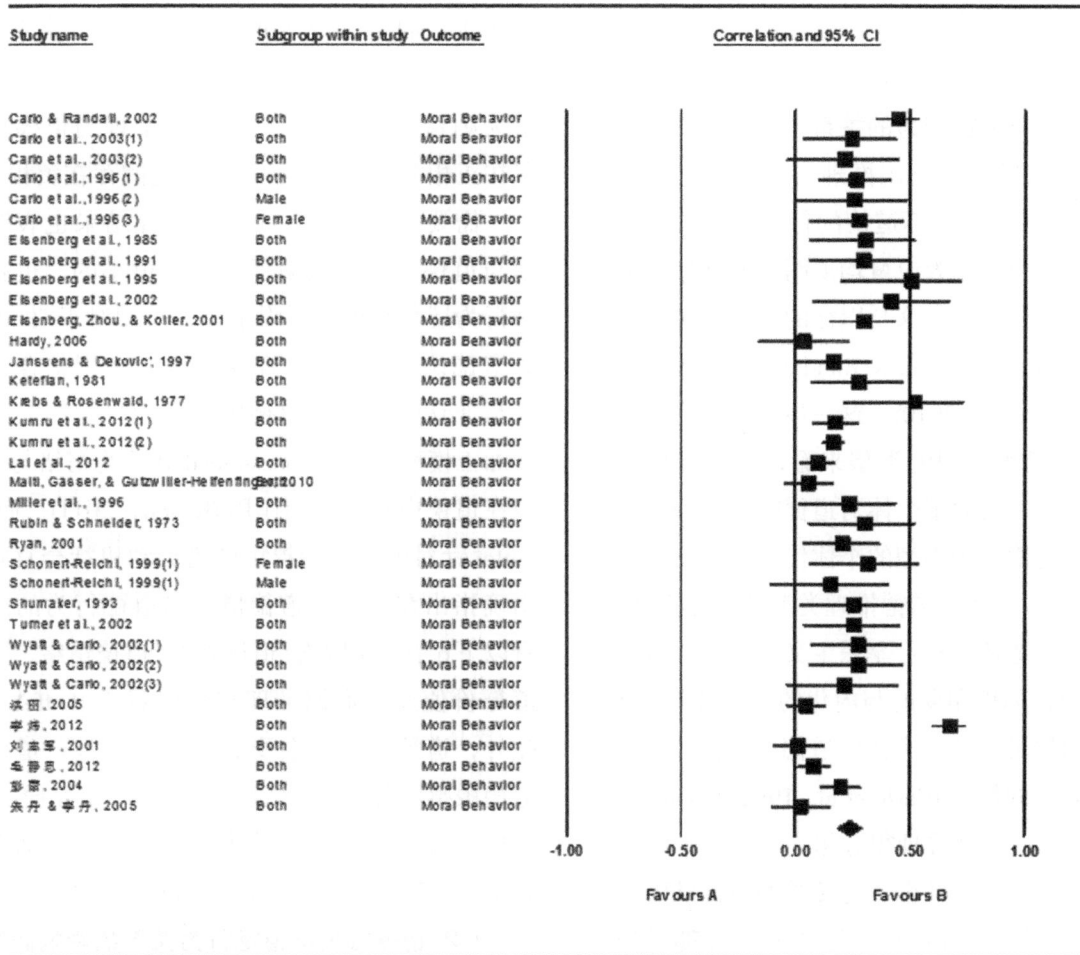

图4　针对不道德的行为的效应量的分布图

经常做出不道德行为 (Eisenberg et al., 2002)。因为不道德行为往往会带来一定的"好处"，比如金钱、权势等等，成人会有更强烈的动机去获得这些"好处"。另一方面，成人更能够为自己的不道德行为寻找理由、做出辩解 (Detert, Trevino, & Sweitzer, 2008)，这也就可以减少不道德行为可能带来的负性影响。于是成人间的不道德行为会有很大的差异，但其道德推理已经趋于稳定与成熟，不会有很大的差异 (Eisenberg et al., 2002; Reynolds & Ceranic, 2007)，两者之间不对称的变异就导致了成人的道德推理与不道德行为的相关很弱。

本研究没有发现被试年龄阶段对道德推理与道德的行为间关系的调节作用，原因可能是：①道德的行为是整个社会大力提倡的行为，每个年龄阶段的群体都会要求自己做出道德的行为。儿童和青少年会经常被要求或奖励去做出道德的行为，学校、家庭与社会也会教育他们进行正确的道德推理，这两方面的作用就使儿童和青少年的道德推理与道德的行为之间存在一定的正相关 (Eisenberg et al., 2002; Malti et al., 2010)。而相比于儿童，成人对自己做出道德的行为会有更强烈的要求 (Carlo, Crockett, Randall, & Roesch, 2007; Maeda et al., 2009)。成人又具有更强的体力、更好的能力来完成道德的行为，因此成人会有更多的道德行为。同时，随着年龄的增长，个体的道德推理水平会越来越高 (Eisenberg et al., 2002; Narvaez & Gleason, 2007)。于是，道德推理与道德的行为随着年龄共同增长，成人的道德推理与道德的行为会都强于儿童与青少年。但相关关系只是关注这两者之间的联系，随时间共同的增长可能不会显著影响两者之间的相关系数，于是年龄阶段不会影响道德推理与道德的行为之间的关系。②正如上述提到的，目前的道德推理测量工具主要针对于道德的行为，因此道德推理与道德的行为关系可能更为密切。而一些道德推理的测量工具针对不同年龄群体有不同的版本，比如 PROM 有儿童版、青少年版与成人版。测量工具针对各个年龄阶段进行修订，确保各个年龄阶段被试都能准确作答，也就提高了道德推理的测量准确性，这也就保证了在各个年龄阶段的

道德推理与道德的行为之间都能展现出紧密的联系。③本元分析涉及的原始研究中探讨青少年的道德推理与道德的行为之间关系的研究数量 (21) 远多于针对儿童或成人的研究 (7)，3 种年龄群体的研究数量的不均衡可能也会影响元分析中调节效应的分析结果 (Borenstein et al., 2009)。

4.3 不足与展望

本研究的不足：①没有考虑性别对道德推理与道德行为间关系的可能影响。道德心理研究领域中，性别通常都是一个重要的考虑因素。由于最终纳入元分析的原始研究基本都没有报告不同性别的效应量，也就无法分析性别可能的作用。②调节效应分析中的样本较少且分配不均衡，这都会影响分析结果。③本元分析所纳入的原始研究中，未发表的国外文献数量较少。作为重要的心理学研究主题，探讨道德行为的影响因素具有很大的理论与现实意义，未来的研究可以：①在探讨道德心理与行为中理性因素的作用时，同时考虑非理性因素的作用，从"道德双加工"角度全面探讨道德心理。②针对不道德行为编制道德推理测量工具，以准确测量不同道德行为的推理。③随着年龄的增长，个体的认知能力逐步成熟。在不同的年龄阶段，道德推理与道德行为之间的关系可能有一定的差异。甚至不同年龄阶段，理性因素与非理性因素的作用强度可能会有不同，今后的研究应该关注这一方面。

5　结论

本元分析发现道德推理与道德行为之间存在联系，道德推理测量工具的类型可以影响道德推理与道德行为（道德的行为或不道德的行为）的关系，被试的年龄阶段只能影响道德推理与不道德的行为间的关系。

参考文献

Aleixo, P. A., & Norris, C. E. (2000). Personality and

moral reasoning in young offenders. *Personality and Individual Differences, 28*(3), 609 - 623.

Banerjee, K., Huebner, B., & Hauser, M. (2010). Intuitive moral judgments are robust across variation in gender, education, politics and religion: A large-scale web-based study. *Journal of Cognition and Culture, 10*(3), 253 - 281.

Barriga, A. Q., Morrison, E. M., Liau, A. K., & Gibbs, J. C. (2001). Moral cognition: Explaining the gender difference in antisocial behavior. *Merrill-Palmer Quarterly, 47*(4), 532 - 562.

Bar-Tal, D., & Nissim, R. (1984). Helping behaviour and moral judgment among adolescents. B*ritish Journal of Developmental Psychology, 2*(4), 329 - 336.

Bear, G. G. (1989). Sociomoral reasoning and antisocial behaviors among normal sixth graders. *Merrill-Palmer Quarterly, 35*(2), 181 - 196.

Bear, G. G., & Richards, H. C. (1981). Moral reasoning and conduct problems in the classroom. *Journal of Educational Psychology, 73*(5), 664 - 670.

Bear, G. G., & Rys, G. S. (1994). Moral reasoning, classroom behavior, and sociometric status among elementary school children. *Developmental Psychology, 30*(5), 633 - 638.

Blair, R. J. R., Monson, J., & Frederickson, N. (2001). Moral reasoning and conduct problems in children with emotional and behavioural difficulties. *Personality and Individual Differences, 31*(5), 799 - 811.

Blasi, A. (1980). Bridging moral cognition and moral action: A critical review of the literature. *Psychological Bulletin, 88*(1), 1 - 45.

Borenstein, M., Hedges, L., Higgins, J., & Rothstein, H. (2009). *Introduction to meta-analysis*. West Sussex, UK: Wiley & Sons.

Bredemeier, B. J. L. (1994). Children's moral reasoning and their assertive, aggressive, and submissive tendencies in sport and daily life. *Journal of Sport & Exercise Psychology, 16*(1), 1 - 14.

Broeders, R., van der Bos, K., Muller, P. A., & Ham, J. (2011). Should I save or should I not kill? How people solve moral dilemmas depends on which rule is most accessible. *Journal of Experimental Social Psychology, 47*(5), 923 - 934.

Bruggeman, E. L., & Hart, K. J. (1996). Cheating, lying, and moral reasoning by religious and secular high school students. *The Journal of Educational Research, 89*(6), 340 - 344.

Carlo, G., Crockett, L. J., Randall, B. A., & Roesch, S. C. (2007). A latent growth curve analysis of prosocial behavior among rural Adolescents. *Journal of Research on Adolescence, 17*(2), 301 - 324.

Carlo, G., & Randall, B. A. (2002). The development of a measure of prosocial behaviors for late adolescents. *Journal of Youth and Adolescence, 31*(1), 31 - 44.

Carlo, G., Hausmann, A., Christiansen, S., & Randall, B. A. (2003). Sociocognitive and behavioral correlates of a measure of prosocial tendencies for adolescents. *The Journal of Early Adolescence, 23*(1), 107 - 134.

Carlo, G., Koller, S. H., Eisenberg, N., Da Silva, M. S., & Frohlich, C. B. (1996). A cross-national study on the relations among prosocial moral reasoning, gender role orientations, and prosocial behaviors. *Developmental Psychology, 32*(2), 231 - 240.

Carlo, G., Eisenberg, N., & Knight, G. P. (1992). An objective measure of adolescents' prosocial moral reasoning. *Journal of Research on Adolescence, 2*(4), 331 - 349.

Christensen, J. F., & Gomila, A. (2012). Moral dilemmas in cognitive neuroscience of moral decision-making: A principled review. *Neuroscience and Biobehavioral Reviews, 36*(4), 1249 - 1264.

Cummings, R., Dyas, L., Maddux, C. D., & Kochman, A. (2001). Principled moral reasoning and behavior of preservice teacher education students. *American Educational Research Journal, 38*(1), 143 – 158.

Comunian, A. L., & Gielen, U. P. (1995). Moral reasoning and prosocial action in Italian culture. *The Journal of Social Psychology, 135*(6), 699 – 706.

Cushman, F., Young, L., & Hauser, M. (2006). The role of conscious reasoning and intuition in moral judgment: Testing three principles of harm. *Psychological Science, 17*(12), 1082 – 1089.

Detert, J. R., Trevino, L. K., & Sweitzer, V. L. (2008). Moral disengagement in ethical decision making: A study of antecedents and outcomes. *Journal of Applied Psychology, 93*(2), 374 – 391.

Duval, S., & Tweedie, R. (2000). Trim and fill: A simple funnel - plot - based method of testing and adjusting for publication bias in meta - analysis. *Biometrics, 56*(2), 455 – 463.

Eisenberg, N. (2000). Emotion, regulation, and moral development. *Annual Review of Psychology, 51*(1), 665 – 697.

Eisenberg, N., Boehnke, K., Schuhler, P., & Silbereisen, R. K. (1985). The development of prosocial behavior and cognitions in German children. *Journal of Cross-Cultural Psychology, 16*(1), 69 – 82.

Eisenberg, N., Carlo, G., & Murphy, B. (1995). Prosocial development in late adolescence: *A longitudinal study. Child Development, 66*(4), 1179 – 1197.

Eisenberg, N., Cumberland, A., Guthrie, I. K., Murphy, B. C., & Shepard, S. A. (2005). Age changes in prosocial responding and moral reasoning in adolescence and early adulthood. *Journal of Research on Adolescence, 15*(3), 235 – 260.

Eisenberg, N., Guthrie, I. K., Cumberland, A., Murphy, B. C., Shepard, S. A., Zhou, Q., et al. (2002). Prosocial development in early adulthood: A longitudinal study. *Journal of Personality and Social Psychology, 82*(6), 993 – 1006.

Eisenberg, N., Miller, P. A., Shell, R., McNalley, S., & Shea, C. (1991). Prosocial development in adolescence: A longitudinal study. *Developmental Psychology, 27*(5), 849 – 857.

Eisenberg, N., Zhou, Q., & Koller, S. (2001). Brazilian adolescents' prosocial moral judgment and behavior: Relations to sympathy, perspective taking, gender - role orientation, and demographic characteristics. *Child Development, 72*(2), 518 – 534.

Eisenberg–Berg, N., & Hand, M. (1979). The relationship of preschoolers' reasoning about prosocial moral conflicts to prosocial behavior. *Child Development, 50*(2), 356 – 363.

Gasser, L., & Malti, T. (2012). Children's and their friends' moral reasoning Relations with aggressive behavior. *International Journal of Behavioral Development, 36*(5), 358 – 366.

Graham, J., Nosek, B. A., Haidt, J., Iyer, R., Koleva, S., & Ditto, P. H. (2011). Mapping the moral domain. *Journal of Personality and Social Psychology, 101*(2), 366 – 385.

Greene, J. D., Cushman, F. A., Stewart, L. E., Lowenberg, K., Nystrom, L. E., & Cohen, J. D. (2009). Pushing moral buttons: The interaction between personal force and intention in moral judgment. *Cognition, 111*(3), 364 – 371.

Guzman, C. M. (2006). *Moral reasoning and truancy in early adolescence*(Unpublished Doctorial Dissertation), Fordham University, New York.

Haidt, J. (2001). The emotional dog and its rational tail: A social intuitionist approach to moral judgment. *Psychological Review, 108*(4), 814 – 834.

Hardy, S. A. (2006). Identity, reasoning, and emotion: An empirical comparison of three sources of moral motivation. *Motivation and Emotion, 30*(3), 205 – 213.

Hauser, M., Cushmen, F., Young, L, Jin K–X., & Mikhail, J. (2007). A dissociation between moral judgments and justifications. *Mind & Language, 22*(1), 1‑21.

Hong, L. (2005). *A study on the relationship among altruistic behavior, empathy and moral judgment of senior middle school students.*（Unpublished Master thesis）, Fujian Normal University.

[洪丽. (2005). *高中生利他行为与移情，道德判断关系研究*（硕士学位论文）. 福建师范大学 .]

Hunter, J. E., & Schmidt, F. L. (2000). Fixed effects vs. random effects meta‑analysis models: Implications for cumulative research knowledge. *International Journal of Selection and Assessment, 8*(4), 275‑292.

Janssens, J. M., & Dekovi, M. (1997). Child rearing, prosocial moral reasoning, and prosocial behaviour. *International Journal of Behavioral Development, 20*(3), 509‑527.

Ketefian, S. (1981). Moral reasoning and moral behavior among selected groups of practicing nurses. *Nursing Research, 30*(3), 171‑176.

Koven, N. S. (2011). Specificity of meta‑emotion effects on moral decision‑making. *Emotion, 11*(5), 1255‑1261.

Krebs, D., & Rosenwald, A. (1977). Moral reasoning and moral behavior in conventional adults. *Merrill‑Palmer Quarterly of Behavior and Development, 23*(2), 77‑87.

Kumru, A., Carlo, G., Mestre, M. V., & Samper, P. (2012). Prosocial moral reasoning and prosocial behavior among Turkish and Spanish adolescents. *Social Behavior and Personality, 40*(2), 205‑214.

Kuppens, S., Laurent, L., Heyvaert, M., & Onghena, P. (2013). Associations between parental psychological control and relational aggression in children and adolescents: A multilevel and sequential meta‑analysis. *Developmental Psychology, 49*(9), 1697‑1712.

Lai, F. H. Y., Siu, A. M. H., Chan, C. C. H., & Shek, D. T. L. (*2012*). Measurement of prosocial reasoning among Chinese adolescents. *The Scientific World Journal, 2012,* Article ID 174845. doi:10.1100/2012/174845

Leffel, G. M. (2008). Who cares? Generativity and the moral emotions, part 2: A "Social Intuitionist Model" of moral motivation. *Journal of Psychology and Theology, 36*(3), 182‑201.

Li, W. (2012). *Research on the applicability of the internet altruistic behavior scale and the relationship between moral judgment and internet altruistic behavior of post graduates* (Unpublished Master thesis). East China Normal University, Shanghai.

[李炜. (2012). *硕士研究生网络利他行为的量表适用性及其与道德判断间关系* .（硕士学位论文）. 华东师范大学 , 上海].

Liu, H. S. (2008). Formation of morality. In L. Mo, (Eds.), *Educational Psychology* (pp. 228). Beijing: Educational Science Publishing House.

[刘华山. (2008). 品德的形成 . 见 莫雷（编）, *教育心理学* (pp. 228). 北京：教育科学出版社].

Liu, M. C. (2012). *Research on the relationship between moral judgment and aggression of middle school students: The effect of moderator of moral disengagement* (Uxnpublished Master thesis), Sichuan Normal University.

[刘美辰. (2012). *中学生道德判断与攻击行为的关系：道德推脱的调节效应*（硕士学位论文）. 四川师范大学 .]

Liu, Z. J. (2001). Research on the relationship between moral judgment ability, peer relationship and prosocial behavior of middle school students. *Psychological Science, 24*(5), 629‑630.

[刘志军. (2001). 中学生的道德判断推理水平，同伴关系和亲社会行为关系的研究 . *心理科学, 24*(5), 629‑630].

Maeda, Y., Thoma, S. J., & Bebeau, M. J. (2009). Understanding the relationship between moral judgment development and individual characteristics: The role of educational contexts. *Journal of Educational Psychology, 101*(1), 233‑247.

Malinowski, C. I., & Smith, C. P. (1985). Moral reasoning and moral conduct: An investigation prompted by Kohlberg's theory. *Journal of Personality and Social Psychology, 49*(4), 1016 – 1027.

Malti, T., Gasser, L., & Gutzwiller-Helfenfinger, E. (2010). Children's interpretive understanding, moral judgments, and emotion attributions: Relations to social behaviour. *British Journal of Developmental Psychology, 28*(2), 275 – 292.

Manning, M. A., & Bear, G. G. (2011). Moral reasoning and aggressive behavior: Concurrent and longitudinal relations. *Journal of School Violence, 10*(3), 258 – 280.

Mao, J. S. (2012). *Research on the relationship among moral judgment ability, guilt and pro-social behaviors of middle school students* (Unpublished Master thesis), Sichuan Normal University.

[毛静思 . (2012). *中学生道德判断能力，内疚与亲社会行为的关系研究* (硕士学位论文). 四川师范大学].

Miller, P. A., Eisenberg, N., Fabes, R. A., & Shell, R. (1996). Relations of moral reasoning and vicarious emotion to young children's prosocial behavior toward peers and adults. *Developmental Psychology, 32*(2), 210 – 219.

Mouratidou, K., Barkoukis, V., & Rizos, S. (2012). Achievement goals and moral competence in sport: Examining the moderating role of demographic characteristics. *European Psychologist, 17(1)*, 34 – 43.

Murphy, F. C., Wilde, G. Ogden, N. Barnard, P. J., & Caldera, A. J. (2009). Assessing the automaticity of moral processing: Efficient coding of moral information during narrative comprehension. *The Quarterly Journal of Experimental Psychology, 62(1)*, 41 – 49.

Narvaez, D., & Gleason, T. (2007). The relation of moral judgment development and educational experience to recall of moral narratives and expository texts. *The*

Journal of Genetic Psychology, 168(3), 251 – 276.

Palmer, E. J., & Hollin, C. R. (1996). Sociomoral reasoning, perceptions of own parenting and self-reported delinquency. *Personality and Individual Differences, 21*(2), 175 – 182.

Palmer, E. J., & Hollin, C. R. (2000). The interrelations of socio - moral reasoning, perceptions of own parenting and attributions of intent with self - reported delinquency. *Legal and Criminological Psychology, 5*(2), 201 – 218.

Palmer, E. J., & Hollin, C. R. (2001). Sociomoral reasoning, perceptions of parenting and self - reported delinquency in adolescents. *Applied Cognitive Psychology, 15*(1), 85 – 100.

Peng, L. (2004). *The development situation and the relativity research between moral judgment and moral behavior of primary and middle school students* (Unpublished Master thesis), Yunnan Normal University.

[彭蕾 . (2004). *中小学生道德判断与道德行为的发展现状及二者的相关研究* . 硕士学位论文 . 云南师范大学].

Raaijmakers, Q. A., Engels, R. C., & Van Hoof, A. (2005). Delinquency and moral reasoning in adolescence and young adulthood. *International Journal of Behavioral Development, 29*(3), 247 – 258.

Rest, J., Cooper, D., Coder, R., Masanz, J., & Anderson, D. (1974). Judging the important issues in moral dilemmas: An objective measure of development. *Developmental Psychology, 10*(4), 491 – 501.

Rest, J. R., Narvaez, D., Thoma, S. J., & Bebeau, M. J. (1999). DIT2: Devising and testing a revised instrument of moral judgment. *Journal of Educational Psychology, 91*(4), 644 – 659.

Reynolds, S. J., & Ceranic, T. L. (2007). The effects of moral judgment and moral identity on moral behavior: An empirical examination of the moral individual. *Journal of Applied Psychology, 92*(6), 1610 - 1624.

Richards, H. C., Bear, G. G., Stewart, A. L., & Norman, A. D. (1992). Moral reasoning and classroom conduct: Evidence of a curvilinear relationship. *Merrill−Palmer Quarterly, 38*(2), 176 - 190.

Rothstein, H. R., Sutton, A. J., & Borenstein, M. (Eds.). (2006). *Publication bias in meta−analysis: Prevention, assessment and adjustments*. Hoboken, NJ: Wiley.

Rubin, K. H., & Schneider, F. W. (1973). The relationship between moral judgment, egocentrism, and altruistic behavior. *Child Development, 44*(3), 661 - 665.

Ryan, J. J. (2001). Moral reasoning as a determinant of organizational citizenship behaviors: A study in the public accounting profession. *Journal of Business Ethics, 33*(3), 233 - 244.

Schonert−Reichl, K. A. (1999). Relations of peer acceptance, friendship adjustment, and social behavior to moral reasoning during early adolescence. *The Journal of Early Adolescence, 19*(2), 249 - 279.

Shumaker, D. M. (1993). *Altruism, empathy, and moral reasoning in high−risk children* (Unpublished Doctorial Dissertation). University of South Carolina.

Simmons, C. H., & Zumpf, C. (1986). The gifted child: Perceived competence, prosocial moral reasoning, and charitable donations. *The Journal of Genetic Psychology, 147*(1), 97 - 105.

Simmons, R. C. (1986). *Relationship between moral reasoning and participation in and acceptance of library theft−behaviors among undergraduates in a large academic library* (Unpublished Doctorial Dissertation). University of Illinois at Urbana-Champaign.

Tangney, J. P., Stuewig, J., & Mashek, D. J. (2007). Moral emotions and moral behavior. *Annual Review of Psychology, 58*, 345 - 372.

Tarrant, M., Branscombe, N. R., Warner, R. H., & Weaton, D. (2012). Social identity and perceptions of torture: It's moral when we do it. *Journal of Experimental Social Psychology, 48*(2), 513 - 518.

Tarry, H., & Emler, N. (2007). Attitudes, values and moral reasoning as predictors of delinquency. *British Journal of Developmental Psychology, 25*(2), 169 - 183.

Turner, N., Barling, J., Epitropaki, O., Butcher, V., & Milner, C. (2002). Transformational leadership and moral reasoning. *Journal of Applied Psychology, 87*(2), 304 - 311.

Ugazio, C., Lamm, C., & Singer, T. (2012). The role of emotions for moral judgments depends on the type of emotion and moral scenario. *Emotion, 12*(3), 579 - 590.

Vandello, J. A., Michniewicz, K. S., & Goldschmied, N. (2011). Moral judgments of the powerless and powerful in violent intergroup conflicts. *Journal of Experimental Social Psychology, 47*(6), 1173 - 1178.

Wu, H. H. (2005). *A new visual angle of moral research: The theory and empirical research of moral judgment test (Unpublished Master thesis)*, Nanjing Normal University.

[吴慧红 . (2005). *道德研究新视角 : 道德判断测验的理论和实证研究* (硕士学位论文). 南京师范大学].

Wu, P., Liu, H. S., Lu, L. J., & Tian, M. X. (2013). The relationship between adolescent immoral behaviors on internet and parenting style. *Psychological Science, 36*(2), 372 - 377.

[吴鹏 , 刘华山 , 鲁路捷 , 田梦潇 . (2013). 青少年网络不道德行为与父母教养方式的关系——道德脱离、责任心、道德同一性的中介作用 . *心理科学 , 36*(2), 372 - 377].

Wyatt, J. M., & Carlo, G. (2002). What will my parents think? Relations among adolescents' expected parental reactions, prosocial moral reasoning, and prosocial and antisocial behaviors. *Journal of*

Adolescent Research, 17(6), 646－666.

Xu, P., & Chi, Y. K. (2007). A review on the social intuitionist model of moral judgment. *Psychological Science, 30*(2), 403－405.

[徐平，迟毓凯 . (2007). 道德判断的社会直觉模型述评 . 心理科学 , *30*(2), 403－405].

Yang, S. G., & Wu, H. H. (2006). Research on the moral judgment abilities of the adolescents. *Psychological Exploration, 26*(2), 55－60.

[杨韶刚，吴慧红 . (2006). 青少年道德判断能力的研究 . 心理学探新 , *26*(2), 55－60].

Yu, H. B., & Liu, G. Z. (2006). Advance of research on the relationship between empathy, moral reasoning, perspective taking and prosocial behavior. *Psychological Development and Education, 22*(1), 113－116.

[余宏波，刘桂珍 . (2006). 移情，道德推理，观点采择与亲社会行为关系的研究进展 . 心理发展与教育 , *22*(1), 113－116].

Zhang, H. C., & Xu, J. P. (2004). *Modern Psychology and Education Statistics.* Beijing Normal University Press.

[张厚粲，徐建平 . (2004). 现代心理与教育统计学 . 北京：北京师范大学出版社]

Zhu, D., & Li, D. (2005). Moral reasoning, empathetic response, and prosocial behavior: Their interrelations in middle school students. *Psychological Science, 28*(5), 1231－1234.

[朱丹，李丹 . (2005). 初中学生道德推理，移情反应，亲社会行为及其相互关系的比较研究 . 心理科学 , *28*(5), 1231－1234].

Association between moral reasoning and moral behavior: A systematic review and meta–analysis

WU Peng[1] ; LIU Huashan[2,3]

([1]Faculty of Education, Hubei University, Wuhan 430062, China)

([2]School of Psychology, Central China Normal University, Wuhan 430079, China)

([3]Key Laboratory of Adolescent Cyberpsychology and Behavior, Ministry of Education, Wuhan 430079, China)

Abstract　According to the theory of Piaget and Kohlberg, whose focuses were the role of moral reasoning in morality, some researchers are inclined to identify moral reasoning as a key factor in predicting moral behavior, moral judgment or moral decision. Over the past decade, some moral psychologists put forward that cognitive factors played a trivial role in morality, while non–cognitive and unconscious factors having great impacts on morality. The controversy of whether moral reasoning can affect moral behavior remained in both theoretical researches and empirical studies. A systematic review was conducted to synthesize empirical results about relationship between moral reasoning and ethical behavior or immoral behavior. Through literature retrieval and selection, in terms of the criteria for inclusion in the meta–analysis, 83 independent effect sizes (50 studies, 16738 participants) were pick out as meta–analysis unit. After coding of data, independent effect sizes were analyzed by CMA 2.2 program. There are four analyses in this research, including heterogeneity test, publication bias test, main effect analysis and moderation effect analysis. In addition, in terms of tentative review analysis and research hypotheses, random effects model was used as meta–analysis model. The test for heterogeneity illustrated that there was significant heterogeneity in 83 independent effect sizes, and also random effects model was a appropriate model for subsequent meta–analysis. The publication bias test indicated that the impact of publication bias was modest but the major finding remained still valid. The research revealed that a positive association between moral reasoning and ethical behavior was found out (r = 0.238) and contrarily a negative relationship between moral reasoning and immoral behavior (r = −0.188) was disclosed. The moderator analysis revealed that the standardization of moral reasoning measurements affected the relationship between moral reasoning and moral behavior, and additionally participant's age could affect the relationship between moral reasoning and immoral behavior. Specifically, there was a stronger link between moral reasoning and ethical behavior in the process of using standardized instrument, while there being a weaker link between moral reasoning and immoral behavior. Meanwhile, the association between adult moral reasoning and immoral behavior was weaker than adolescent' or children' association. The results suggested that moral reasoning could play an important role in moral behavior (ethical behavior and immoral behavior). Researchers, what's more, are expected to pay much attention to measurement instruments of moral reasoning and standardized instruments in the research of moralities, and especially, the area of ethical behavior. Overall, the findings provided an evidence to prove moral reasoning being the key factor of morality and suggested that developing moral reasoning measurement in connection with immoral behavior is in urgent need.

Keywords moral reasoning; ethical behavior; immoral behavior; meta-analysis

心理学报，2010, 42(3), 415 – 422.

"热"执行对注意缺陷多动障碍和阅读障碍儿童言语工作记忆的影响 *

张　微[1]，刘翔平[2]，宋红艳[3]

([1] 华中师范大学心理学院暨湖北省人的发展与心理健康重点实验室，武汉，430079)
([2] 北京师范大学心理学院，北京，100875)
([3] 华中师范大学汉口分校心理咨询中心，武汉，430070)

摘　要　ADHD 儿童在与背外侧前额叶 (DLPFC) 相关的"冷"执行功能上的缺陷已大量证实，但在与眼眶和中前额叶皮层(OMPFC)相关的"热"执行功能上，ADHD 儿童是否存在缺陷则未可知。与儿童赌博任务实验范式（该任务中"热"启动对"冷"执行起到抑制作用）不同，本研究考察趣味言语 N-back 任务是否对 ADHD 和阅读障碍儿童的成绩有促进作用。枯燥 N-back 任务考察的是言语工作记忆的纯认知特征，而趣味任务则卷入了"热"执行对"冷"执行的影响。结果表明，在枯燥任务上，ADHD 和阅读障碍儿童的成绩均明显低于正常儿童，二者之间差异不显著，在趣味任务成绩上，ADHD 儿童与正常儿童的成绩没有显著差异，而阅读障碍儿童成绩落后于正常控制组和 ADHD 组，在成绩变化的趋势上，ADHD 儿童在趣味任务上成绩明显提高，而阅读障碍儿童则无明显改善。结果说明 ADHD 和阅读障碍儿童的言语工作记忆均存在明显的缺陷，但是机制不同，"热"执行对提高 ADHD 的言语工作记忆有明显的促进作用，ADHD 的"冷"执行缺陷能够通过"热"执行的调节得到改善。

关键词　注意缺陷多动障碍；阅读障碍；"热"执行；言语工作记忆

分类号　B844

1 前言

注意缺陷多动障碍（Attention Deficit Hyperactivity Disorder, ADHD）是一种临床常见的儿童行为问题，核心症状是注意缺陷、多动和冲动 (APA,1994)。通常发病于学龄前期，在小学阶段，症状表现尤为明显，随着年龄的增长，症状逐渐得到改善或者消失，国内调查显示，ADHD 的检出率在 3.1%–6.3% 之间（张微，刘翔平，廖冉，顾群，2007)。

以 Wilding（2005）为代表的研究者认为，执行功能（executive function, EF）缺陷是 ADHD 的核心缺陷，能够解释 ADHD 症状的全部特征，Wilding 对不少研究进行分析总结后，将持续性注意任务上的准确性差异归结于 ADHD 儿童的执行功能的基本缺损，执行功能的缺陷会影响 ADHD 儿童反应准备，反应

* 全国"十一五"教育规划国家青年基金项目 (CBA090119),湖北省人的发展与心理健康重点实验室项目 (200903)，"应用实验心理北京市重点实验室"资助项目。

通讯作者：刘翔平，E-mail：lxp599@163.com; 张微，E-mail：zhangwei2008@mail.ccnu.edu.cn

速度，刺激－反应组织，焦点的保持，目标特征的组合，目标的定义甚至更多的过程。

考虑到在前额叶皮层功能的差异，Zelazo 和 Müller（2002）区分了两类执行功能，与背外侧前额叶 (DLPFC) 相关的"冷"执行功能，更多涉及的是 EF 的纯认知方面，眼眶和中前额叶皮层 (orbital and medial prefrontal cortex，OMPFC）则与"热"执行功能有关，该区分得到了 Habib, Boulanger, Soubias, Delarbre 和 Joly-Pottuz（2003）的进一步确认。冷认知相对抽象，去情境化，目前多数用来测试 ADHD 执行功能的任务，如 Stroop, flanker, Go/No-Go, Stop, 持续性操作任务和工作记忆任务，均属此类。而在"热"执行的研究情境中，需要进行可能的冒险决策，或者需要对刺激的情感意义进行灵活性的估价，卷入情感和动机。Zelazo 和 Muller（2002）认为 ADHD 是一种"冷"执行功能障碍，而 Castellanos, Sonuga-Barke, 和 Tannock（2006）则认为，注意缺陷症状与"冷"执行功能有关，多动冲动症状与"热"执行有关，一些 ADHD 个体主要是"热"执行功能缺损，另一些 ADHD 个体的缺陷则主要表现在"冷"执行功能上，还有一些则表现出双重缺陷。

暂不论 ADHD 儿童的执行缺陷是以"热"执行还是"冷"执行缺陷为主，但"热"执行对"冷"执行加工能起到调节作用，且是以特定的神经回路为基础的。根据 Haber（2000）模型，皮质—纹状体—丘脑—皮质螺旋回路（spiraling cortico-striato-thalamo-cortical circuits）担当此任。此复杂的，非交互的通路包括纹状体—黑质—纹状体（SNS）和丘脑—皮质—丘脑网络。这是通过与 OMPFC 通路相关的情绪及动机影响"冷"执行的 DLPFC 通路的解剖学基础，而 DLPFC 接着影响了运动区。具体是，OMPFC 投射到壳区（shell），DLPFC 主要投射到中部纹状体，前运动皮层和运动皮层主要投射到背外侧纹状体。壳区投射到中脑两个区域：腹侧被盖区（ventral tegmental area，VTA）和黑质致密部（substantia nigra, pars compacta，SNc）。VTA 投射到壳区，形成 SNS 回路。这种通过这些螺旋回路的非交互成分形成的单向信息流通道的本质特征表明了动机和情感对认知加工的影响的层次性。认知加工能够调控动作输出并且与 ADHD 等相关的障碍明显有关系。

另有研究者 Sonuga-Barke（2002）虽然未明确提出 ADHD 在"冷""热"执行机制上的表现差异，但其双通道理论也强调了 ADHD 在奖励相关的动机加工上的缺陷，认为 ADHD 抑制缺陷（EDF）与厌恶延迟（DEL）遵循不同的神经通路，ADHD 是 EDF 和 DEL 双通道作用的结果。一方面，前额叶问题导致抑制缺陷，另一方面，与奖励环路（reward circuit）中央－边缘多巴胺分支（meso-limbic dopamine branch）相关的延迟满足能力的改变促成了 ADHD 的特殊的动机风格。ADHD 既表现出执行功能障碍，也表现出厌恶延迟满足的动机风格。枯燥的任务测查的是"冷"认知功能，而能够立刻得到满足的任务则与"热"认知功能相关，ADHD"冷"认知任务上存在缺陷，是因为其在前额叶多巴胺系统的神经通路上出现障碍，在"热"认知任务上，由于奖励回路被激活，ADHD 儿童能够通过情绪和动机的激活，提高任务卷入程度，改善了其兴奋、抑制等认知功能，从而弥补了其在 EDF 神经通路上的缺陷。也就是说，ADHD 儿童在"冷"认知任务上缺陷严重，但在"热"认知任务上的则有可能表现正常。

基于背外侧前额叶、扣带回以及纹状体等为基础的以抑制缺陷为代表的"冷"执行缺陷在 ADHD 身上已经被毫无争议地证实。但是，对 ADHD 儿童在"热"执行任务上的关注并不多，也没有形成明确的结论。来自强化依随（reinforcement contingencies）范式的总体研究结果表明，奖励和反应代价在 ADHD 和控制组的成绩和动机水平上有积极的效果，但 ADHD 儿童在奖励和反应代价条件下成绩比无强化依随条件下成绩提高的程度要比控制组大；另外，与控制组相比，ADHD 更多选择即刻满足（Luman, Oosterlaan, & Sergean, 2005）。但是来自生理记录方面的研究却发现，ADHD 在心理生理上（皮肤电和心律指标）对强化依随的敏感性低于控制组 (Crone, Jennings, & Van der Molen, 2003)。来自儿童赌博任

务的研究发现 (Habib, Boulanger, Soubias, Delarbre, & Joly-Pottuz, 2008)，ADHD 儿童的在此任务上的成绩落后于正常控制组，他们并不偏爱少的有利的选择，且 ADHD 儿童在此任务上的失败与 Stroop 测验上的成绩没有明显相关，Habib 等人认为 ADHD 在奖励机制上存在缺陷。国内朱昭红 (2006) 的研究发现，在赌博任务中儿童更倾向于不利选择，两种亚型儿童在赌博任务上的选择模式相似，均倾向于不利选择，表明儿童在情感性决策中的缺损源于对奖励的高敏感性，并且可能与 OFC 损伤者有同样的机制。

据上所述，尽管 Zelazo 和 Muller 认为 ADHD 可以看成是"冷"执行功能障碍，但是仍然有研究发现了 ADHD 在"热"执行上的缺陷，这有任务范式的原因，或可以假定在不同性质的实验任务中，"热"执行所发挥的协调作用并不相同，甚至截然相反。凡是情境化的，能够很好地卷入情感动机的任务，均涉及到"热"执行和"冷"执行的参与，但是情绪的高度卷入和动机的加强是否一定对有效地执行任务起到促进作用呢？在强化依随的任务中，每当个体作出有效的反应，都有可能得到积极的奖励，追求奖励的动机与能否有效地作出反应并不存在冲突，是一致的，为了获取高奖励，个体需要坚持有效地进行反应，动机越高，坚持性越强，成绩也越好，例如根据个体对计算机屏幕中出现信号的注视时间越久，就能得到越多的奖励。研究发现，奖励对 ADHD 儿童自我评定的动机水平有积极的效果，强化依随对 ADHD 儿童的动机促进作用较正常儿童更为明显 (Carlson, Mann, & Alexander, 2000；McInerny, & Kerns, 2003；Scheres, Oosterlaan, & Sergeant)。在儿童赌博任务中，研究者鼓励儿童尽量多赢得奖励，与反向择物任务相似，表面上的高奖励并不一定意味着高奖励，可能带来更高的损失，但是被试在每次试验中，首先感受到的是所作选择的有利面（例如笑脸，代表奖励），然后才有可能出现不利面（打开被遮盖的纸牌的下半区，可能露出数量不等的哭脸，意味着损失），此时涉及到一个对优势心理表征进行有效抑制的心理过程，动机和情绪的高度卷入

构成了不利因素，它会使不利的优势表征过度加强，从而提高了有效抑制的难度，对最终的任务结果产生负面影响。此时，动机和情感的过度卷入阻碍了任务的执行，由此带来的过"热"的冲动反应需要被有效抑制，这显然是与前面强化伴随的实验任务不同。

然而，遗憾的是有一类"热"执行任务没有引起重视。这类任务中，任务本身就很有趣，完成任务本身就是奖励，不刻意需要额外的人为的强化施加，个体的动机和情绪就得到加强，被试对任务本身的喜爱要甚于对完成任务的结果的奖励趋向，任务本身的积极特性激发了个体的动机和情感卷入，同时高度的动机和情感卷入并不阻碍任务本身的有效执行。此类任务可以认为是纯趣味任务。与强化依随任务相比，趣味任务所激发的动机指向于完成任务的过程，而强化依随任务本身是无趣的，动机指向于结果，并且要求个体人为地将反应和强化建立联系，当条件发生变化时，需要中断旧的反应－强化联系，建立新的反应－强化联系。根据刺激强化值的灵活表征假设 (Rolls, 2000)，眶额叶（OFC）卷入了根据情感或动机对刺激的再估价，以及对刺激与强化刺激联系的分离，OFC 的损伤则导致了刺激－奖励转换能力的损伤。但如果 OFC 存在缺陷，可能不会不会影响儿童在趣味任务上的表现，这是因为在此类趣味任务中，刺激奖励之间的联系并不重要，也不涉及到对刺激强化值的灵活表征和刺激－奖励转换。研究发现，ADHD 儿童在那些比较有意思的游戏任务中的工作记忆并不存在明显的缺陷，但是，在那些枯燥的记忆任务中，缺陷比较明显，研究中，任务是需要对连续的空间位置进行记忆，结果 ADHD 青少年却不存在明显缺损 (Barkley, Edwards, Laneri, , Fletcher, & Metevi, 2001)。

阅读障碍与 ADHD 关系极为密切，伴随发生率很高，语音加工缺陷被广泛认为是阅读障碍的核心问题，而阅读障碍的执行功能缺陷则并没有得到广泛认同。两种障碍在特定的神经心理加工缺陷上既有相似性又有差异，它们在"冷"言语工作记忆任务上的缺陷已得到大量研究的证实（Bowers, Steffey, & Tate,

1988；Xavier, Castellanos, & Tannock,2002）。但在"热"言语工作记忆任务上是否存在缺损则不甚明了。"热"执行对于儿童阅读障碍的言语工作记忆的成绩是否有影响呢？两种障碍是否表现出同样的模式呢？探索这些问题将有利于解释 ADHD 和阅读障碍言语工作记忆缺陷的核心问题，以及能够说明"热"执行对于"冷"执行的影响机制。

因此，本研究以阅读障碍被试和正常被试作为参照，采用 N-back 范式的两类性质言语工作记忆任务（趣味任务和枯燥任务）来要考察 ADHD 儿童的表现，以探索"热"执行对于 ADHD 和阅读障碍儿童言语工作记忆的影响。

2 方法

2.1 被试

共有 86 名来自某学习障碍儿童咨询矫治机构的障碍被试和 25 名来自学校样本的正常控制组被试参加了研究。ADHD 儿童最近半年内未服用过利他林等精神类药物治疗，无明显的器质性损伤，不伴随明显的品行障碍、情绪障碍、语言障碍和智力缺陷，通过 DSM-IV，Conners 家长和教师量表，自编 ADHD 问卷和 IVA-CPT 视听整合持续性操作测验来评估选择 ADHD 儿童，以瑞文标准推理测验的成绩作为智力衡量标准，排除得分在 25% 以下和 90% 以上儿童。

从北京市一所城区小学选取正常控制组被试 20 名，年龄、性别、智力以及教育水平和教育环境都与实验组匹配。发展性阅读障碍的选择主要工具为《小学生识字量测验》，以智力与识字水平的不匹配作为选择依据。

ADHD 儿童入选标准及甄选程序：在甄别 ADHD 时，我们对家长和教师（班主任）进行了访谈，访谈的工具有 DSM-IV，Conners 问卷，自编 ADHD 问卷（包括 ADHD 症状以及一些排除 ADHD 诊断的可能情况），另外，我们对儿童施测了 IVA — CPT 测验。诊断标准主要以 DSM-IV 家长评估为依据。在此量表上，至少有一个维度得分在 6 分或 6 分以上，

在 IVA-CPT 上被诊断为 ADHD（少数被试在 DSM-IV 家长评定上接近可诊断水平，CPT 得分较低，也可被视为 ADHD），同时，考虑到教师评定尺度较宽松，教师评定的 DSM-IV 得分不一定要求在 6 分或 6 分以上，但是需与家长评定一致，且相应维度得分不能在 3 分以下。另外，以 Conners 家长和教师评定量表以及自编的 ADHD 问卷作为辅助工具，凡符合或者边缘符合 DSM-IV 的 ADHD 诊断的儿童在 Conners 量表上得分较高（10 分或 10 分以上），且符合自编 ADHD 问卷的多数症状标准，可进一步确认，少数在不同诊断标准上有明显矛盾，且不能为访谈所证实为 ADHD 的儿童，将被排除在研究之外。正常儿童的取样来自于一所普通小学，其家长评定的 DSM-IV 两个维度得分均在 3 分以下，否则被排除。

确定阅读障碍的标准：我们采用了两个工具，分别是《小学生识字量测验》（上海教育出版社，信效度均为 0.98）和《标准化阅读理解测验》（本测验分低中高三个年级版本，分别适用于一二年级、三四年级和五六年级。项目构成如下：低年级阅读理解测验由 4 篇 600 左右的小短文及其与短文内容相关的问题构成，其中 3 篇记叙文，一篇说明文，共 19 个问题，均为四选一的选择题；中年级测验由 4 篇 800 字左右的短文和相应的题目构成，其中议论文 1 篇，说明文 1 篇，记叙文 1 篇，散文 1 篇，每篇各附 5 个选择题；高年级阅读理解测验由 5 篇约 1000 字左右的短文组成，其中说明文 2 篇，议论文 2 篇，记叙文一篇，各附 4 道选择题。低年级、中年级和高年级测验的分半信度分别为 0.83、0.82 和 0.86，效标效度分别为 0.71、0.87 和 0.77（通过计算阅读理解测验成绩与教师采用五点量表对学生阅读理解能力的主观评估的相关获得），信效度均符合要求。），纳入标准：识字量落后 0.5 年级，且阅读理解成绩在 50% 以下。非阅读障碍被试识字量必须在 –0.2 年级以上，且阅读理解成绩必须在 25% 以上。（识字量为主要的筛选标准，阅读理解测验仅为参考）

最后确定有效被试共 77 人，其中正常被试 17 人，ADHD 不伴随阅读障碍被试 21 名，ADHD 伴随阅读

障碍被试24名，阅读障碍（不伴随ADHD）被试15名，年龄在7到12岁之间的二到五年级儿童，男56人，女21人，平均年龄8.5岁。障碍被试均来自临床样本，而正常被试均来自某小学。四组被试年龄、智力相匹配，识字量存在明显差异，见表1：

从上表可以看出，四组被试的年龄和智力水平没有差异，但在识字量上存在明显差异。事后检验结果表明，识字水平上，控制组=ADHD，ADHD>ADHD+RD，ADHD+RD=RD（前后两两比较，

$P=0.315$，$P<0.001$，$P=0.695$）。这说明被试分组是合理的。

2.2 实验材料与设计

程序采用AB 6.0中文版编写。实验包括两个独立的任务：枯燥任务和趣味任务。两个任务的材料、难度、任务要求完全一致，只是在趣味性上有很大差异。为防止练习效应，枯燥任务和趣味任务并不在一个时间段完成（相隔1周）。且在完成任务的顺序上，进行了平衡。实验材料均为大写字母。

表1 四组被试瑞文成绩、识字量和年龄的比较

统计项	控制组（17）	ADHD（21）	ADHD+RD（24）	RD（15）	F	P
年龄（岁）	8.6（1.50）	8.42（1.01）	8.38（1.01）	8.87（1.30）	0.88	0.457
瑞文	91.81（8.22）	73.92（28.02）	79.89（16.21）	70.83（24.47）	2.29	0.087
识字量（相对年级）	0.61（.68）	0.37（.62）	−0.81（.48）	−0.73（.46）	25.53	0.000

注：识字量为相对年级水平，正表示在正常水平之上，负表示落后于正常水平。ADHD：儿童注意缺陷多动障碍不伴随阅读障碍，ADHD+RD：儿童注意缺陷多动障碍伴随阅读障碍，RD：阅读障碍不伴随儿童注意缺陷多动障碍。下同。

考虑到被试的特殊性，我们对经典N-back任务进行了改进，降低了难度，与Kiss, Pisio, Francois和Schopflocher（1998）改进后的N-back范式相似，有良好的效度。枯燥任务中所有trial形式相同，所有字母字体、颜色相同，没有反馈，没有积分，完成任务没有奖励，背景和前景色单调。趣味任务的趣味性明显增加。登陆时有非常漂亮的登录界面，并且伴随非常舒缓动听的背景音乐。登录后进入指导语界面，在非常漂亮的背景上呈现字体美观的指导语。指导语暗示被试任务将是一个有趣的小游戏。为了让被试明白指导语要求和奖励情况，主试需耐心给被试讲解。点击"进入游戏"按钮后，进入练习任务窗口，背景音乐消失。呈现大写字母序列，不同的trial的大写字母序列字体、颜色、大小各异，同一个trial内的字母大写、颜色、字体相同。练习共6题，至少答对5道题方可进入正式实验。正式实验中，屏幕背景色为红色，字母的颜色与屏幕背景色有区分。完成正式实验后，会有一个最后总分提示的窗口。实验结束后，根据被试的得分挑选奖品。具体任务操作见实验程序部分。

趣味游戏实验材料为大写字母，同一个trial中，目标刺激不能有相同的字母。共计60个trial，其中长度为4个字母、5个字母和6个字母的trial各20次，正负反应各半，随机分布。共有五种字体，分别是隶书，幼圆，Aribal，方正姚体，方正舒体，字体大小也有五种。根据情况挑选了15中字体类型、大小和颜色组合，即trial中字母的外观上有15种可能，均出现4次。

趣味任务中，被试对每个trial作出反应之后都有反馈，每次反馈的内容都不相同。用于反馈的材料为动画、声音、文字。动画为72项大小为150mm×150mm的动画，均取材于数码宝贝、三国等儿童熟悉的游戏动画人物，随机出现；声音72项，长度在3秒左右，均是非常动听的音乐；文字有180项，分为答对了的反馈、答对了且很快的反馈、答错了的反馈，例如："真是又快又准！加15分！"，这些反馈内容随机出现在各个trial的对应的反应条件下。反馈窗口持续时间为5s。练习实验同正式实验形式相同。各种类型的trial随机出现在实验中。只有在正确回答的条件下，才会有各种各样的动画和音乐的反馈，在答错的条件下，每次反馈的内容都相同，即一张苦脸，一个"哦哟"的声音和文字鼓励"不要灰心！"，这样做的目的是要让被试努力表现好才能看到各种各样的动画和听到美妙的音乐。

实验为一对一进行，主试坐在被试身边约 30 厘米处，斜对着屏幕，随时监控实验的进行。被试正坐在计算机屏幕前，眼睛与显示器中央成 15 度水平视角，事先熟悉键盘和鼠标。被试戴上立体声端坐于屏幕前，显示器分辨率为 1024×768。实验步骤如下：每个实验任务都是先出现注视点（屏幕中心的一个蓝色的"+"字）300ms，消失，然后呈现空屏 300ms，呈现字母序列的第一个字母 500ms，消失后出现空屏 500ms，随后呈现第二个字母 500ms，消失后再次呈现空屏 500ms，呈现第三个字母 500ms，以此类推，直到呈现最后字母序列的最后一个字母 500ms 和空屏 500ms，要求被试判断最后一个字母与倒数第三个字母是否相同，并按键反应，正反应按"P"键，负反应按"Q"键，探测字母消失，在枯燥任务中，trial 结束，之后是 800ms 的空屏刺激准备时间，然后进入下一个 trial；在趣味任务中，被试反应后，呈现反馈窗口，持续 5000ms，然后自动跳入下一个 trial。系统自动记录反应时和反应正确与否。如果被试在探测刺激出现后的 10s 内未能做出有效反应，将进入下一个 trial。枯燥任务大约持续 15 分钟，趣味任务大约持续 25 分钟。

3 结果

考虑到两类任务操作上的差异，比较反应时的差异意义不大，本研究中只是比较四组被试在各个水平上的正确率的差异。剔除反应时低于 200ms 和反应时明显过长（大于 3 个标准差）的数据，四组被试在不同性质任务下的正确率以及四组被试在枯燥和趣味记忆任务上成绩的变化的差异（趣味任务成绩减去枯燥任务成绩）见图 1 和图 2。

进行重复测量方差分析，枯燥任务，趣味任务，趣味任务和枯燥任务成绩的差异的组间效应均达到了显著性水平（$F(3,73)=15.67$, $p<0.01$；$F(3,73)=3.76$, $p<0.05$；$F(3,73)=3.42$, $p<0.05$）。进一步进行多重比较后发现，在枯燥记忆任务上，控制组的成绩（82.6%）明显高于 ADHD（70.7%）（$p<0.01$）、

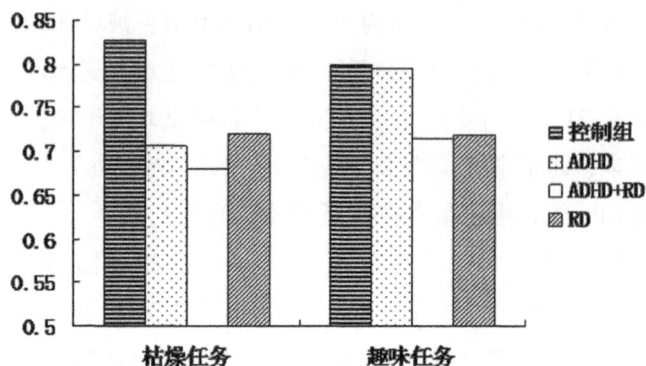

图 1　四组被试在枯燥和趣味言语工作记忆任务上的成绩（正确率）

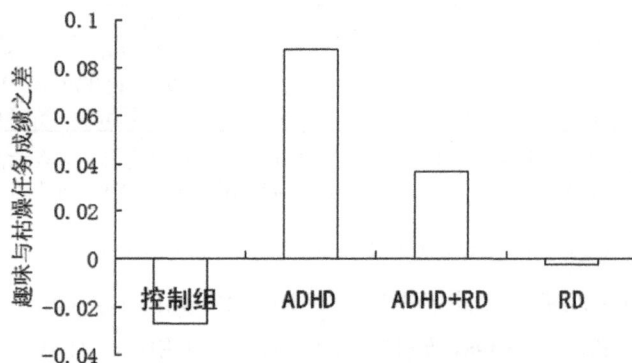

图 2　四组被试趣味与枯燥任务成绩（正确率）之差（趣味－枯燥）

ADHD 伴随阅读障碍（67.9%）（$p<0.01$) 和阅读障碍组（72.1%）（$p<0.01$），虽然 ADHD 伴随阅读障碍成绩最低，但是差异检验发现，ADHD、ADHD 伴随阅读障碍和阅读障碍三组的成绩差异并不显著 (ADHD, ADHD+RD：$p=0.565$；ADHD+RD，RD：$p=0.952$；ADHD,RD: $p=0.560$)。在趣味记忆任务上，四组被试的成绩依次为 79.9%, 79.5%, 71.5%, 71.9%, 结果是：正常控制组 =ADHD,（$p=0.912$);ADHD>ADHD+RD,($p<0.05$); ADHD+RD =RD($p=0.917$)。在趣味任务与枯燥任务的差异（趣味记忆任务的成绩减去枯燥记忆任务的成绩）上，正常控制组儿童由枯燥任务到趣味任务，成绩下降了 2.7%，ADHD 儿童提高了 8.8%，ADHD 伴随阅读障碍儿童提高了 3.7%，阅读障碍儿童下降了 0.2%，ADHD 儿童的成绩显著得到了提高 ($p<0.01$)，增幅最大（ADHD，ADHD+RD：$p=0.1$），其次是 ADHD 伴随阅读障碍儿童，成绩相

比正常组也有提高 (ADHD+RD,RD: $p<0.1$)，阅读障碍儿童在趣味记忆刷新任务上的成绩没有显著变化 ($p=0.753$)，反观正常儿童，成绩不但没有提高，反而下降 ($p<0.1$)。关于四组被试在枯燥趣味任务上的成绩变化趋势上，我们控制反应时后进行了方差分析得到了相同的结果 $F(54, 21)=4.079$，$p<0.01$。

从以上结果可以看出，相比"冷"工作记忆任务，"热"执行明显提高了 ADHD 儿童的成绩，但是对于阅读障碍和正常儿童的成绩影响不大。

4 讨论

本研究对比了 ADHD 和阅读障碍儿童在同等难度的趣味和枯燥工作记忆言语 N-back 任务上的表现，该任务范式被认为既很好的涉及到语音回路的功能，同时又卷入了中央执行（处理前摄抑制和记忆更新操作）的参与，结果发现 ADHD 儿童在趣味工作记忆任务上的成绩有明显提高，相反，阅读障碍儿童和正常儿童的成绩则几乎没有变化，甚至有小幅下降。这与来自儿童赌博任务的研究得到的 ADHD 儿童在"热"执行上的缺陷结果不同，本研究结果说明：1）ADHD 和阅读障碍儿童都表现出言语工作记忆能力的缺陷，但是缺损模式不同；2）ADHD 伴随阅读障碍儿童在"热"执行上成绩一定程度的补偿说明了 ADHD 伴随阅读障碍的缺陷既有 ADHD 的特征又有阅读障碍的特征；3）在趣味任务上，ADHD 的"热"执行功能很好的发挥了作用，ADHD 可能并不存在特异性的"热"执行缺陷。这与 Barkely（2001）的研究一致，Carlson, Booth, Misung 和 Canu (2002) 的研究也证实了 ADHD 儿童更多通过外部的反馈来评量他们行为的水平，表现出更低的掌握目标定向水平，内部动机缺失，成绩不稳定，容易受到趣味性奖励的影响。

可以认为"热"执行是否能够很好的发挥作用关键在于依据特定的任务情境。趣味任务中"热"执行和"冷"执行共同参与，"热"执行对"冷"执行起到一定的协调作用。与儿童赌博任务不同，

赌博任务中，过度的情感反应容易导致冲动的不利判断，虽然提高了动机，但是更需要加以抑制，进行冷却再加工，而相反，趣味任务卷入的"热"执行不仅提高了动机水平，而且对任务相关的"冷"执行功能起到促进作用，例如抑制能力和认知资源的利用效率的提高，这些都有利于任务执行，因此，对于"热"的情感反应，无须刻意对抗。然而，这种"热"执行的激发作用对 ADHD 的效果明显，而对阅读障碍和正常儿童没有明显的促进作用，这可能与不同的障碍的缺损模式的差异以及"热"执行的作用机制有关。对于正常儿童，在枯燥任务时，正常儿童成绩已经很高，其成绩并没有像 ADHD 一样，在枯燥任务时受动机缺失的影响而受损，故面临趣味任务时，虽然提高了动机水平，但是并不明显提高成绩；另外，与障碍儿童相比，正常儿童的成绩已经处于一个相对较高的水平，再提高可能有瓶颈效应；再次，趣味任务过多的声音变化、trial 变化和动画变化对正常儿童而言除了提高动机，也形成了干扰刺激，从而影响了成绩，尽管同样也对 ADHD 儿童造成了干扰，但后者受到动机和情绪的激发效果是最主要的。对于阅读障碍儿童，成绩在趣味任务条件下没有明显提高可能受到了任务本身的干扰外，另一个可能的原因是动机的加强对于阅读障碍儿童的言语工作记忆功能并没有实质性的促进作用，依据 Haber 模型，通过皮质—纹状体—丘脑—皮质螺旋回路，与"热"执行相关的 OMPFC 区调节了与"冷"执行相关的 DLPFC 区，并进一步调节了动作输出，即动机和情绪的积极卷入提高了执行功能，从而改善了任务操作。在记忆刷新操作过程中，主要的执行操作是与更新操作相关的处理前摄抑制作用，之前的研究发现 (张微，2008)，ADHD 儿童的处理前摄抑制能力较差，并导致了记忆刷新成绩的下降，而阅读障碍儿童并不存在这个问题，后者主要的缺陷表现在更为基础的语音回路的编码和复述功能上，而热工作记忆任务并不能改善阅读障碍儿童的语音编码和复述能力，故没有明显提高其成绩。

从另一个角度，Sonuga-Barke（2002）等人强调

了ADHD特殊的动机风格以及对强化的过度敏感性，本研究中两类任务的差异在于趣味任务能够获得即时奖励和满足，枯燥任务则不能。Sonuga-Barke认为厌恶延迟，强烈的追求即时满足的倾向是ADHD儿童的动机风格，并提出了厌恶延迟假说，认为ADHD是潜在的动机风格的功能特征而非自我调节系统失能。ADHD儿童被推动去逃避或者避免延迟。他们的注意缺陷、多动、冲动可以被认为是厌恶延迟的功能表现。当面临即时满足和延迟满足，ADHD会好不犹豫选择立刻满足。当没有选择时，他们会为了在延迟中减少时间知觉，与环境发生作用，创造或者参与非当时的环境事件。认知缺损被假定为由于厌恶延迟所导致的二级缺损。特殊的动机风格的基础是奖励通路。相反，在那些能够得到即时满足的情景中，奖励回路的激活，伴随情感动机的卷入，ADHD儿童的认知功能的缺陷得到了补偿。类似的，Douglas和Parry（1994）曾认为ADHD核心问题之一是他们是否对奖励具有异常的敏感性。Seageant（2000）的认知能量模型认为，ADHD儿童可能在不同的认知水平上表现出不同程度的认知过程缺陷，认知的加工机制中编码、中央加工和运动反应三个阶段与能量机制唤醒、努力和激活三种"能量库"联系紧密，同时这些认知加工过程和状态因素受到一个更高级的执行功能控制系统（也称管理评价机制）的监督和调控，ADHD儿童主要缺陷是与认知能量模型的高级管理机制相联系的三种能量库，但是动机的提高可以提高认知能量水平，从而影响了执行功能。同样的，Philip，Vaughan，Terry，Sally，Diane，和Belinda（1999）也认为ADHD儿童的成绩较容易受到动机和唤醒的影响，在有奖励的情况下，他们在那些需要意志努力的控制加工任务的表现会明显得到提高。

虽然研究者从不同的角度对ADHD在卷入热认知功能的趣味任务中的成绩的提高做出了解释，但有一点是共同的，即ADHD儿童在趣味任务中，都毫无疑问的表现出的强烈的情绪和动机卷入，"热"执行对"冷"执行进行了积极的调节。

参考文献

American Psychiatric Association. (1994). *Diagnostic and statistical manual of mental disorders (4th ed.)*.Washington, DC: American Psychiatric Press.

Barkley, R.W., Edwards, G., Laneri, M., Fletcher, K &Metevi, L.(2001). Executive functioning, temporal discounting, and sense of time in adolescents with Attention Deficit Hyperactivity Disorder (ADHD) and Oppositional Defiant Disorder (ODD). *Journal of Abnormal Child Psychology, 29*, 541－556.

Bowers, P., Steffey, R., & Tate, E.(1988). Comparison of the effects of IQ control methods on memory and naming speed predictors of reading disability. *Reading Research Quarterly, 23*,304-309

Carlson ,C., Booth, J. E., Misung, S., & Canu. W.H. (2002). Parent-teacher-,and self-related motivational styles in ADHD subtypes. *Journal of Learning Disabilities,35*,2 104-114.

Carlson, C., Mann, M., & Alexander, D. K.(2000) Effects of reward and response cost on the performance and motivation of children with AD/HD. *Cognitive Therapy and Research, 24*, 87－98.

Castellanos, E., Sonuga-Barke, E.J., & Tannock,M.R. (2006).Characterizing cognition in ADHD: beyond executive dysfunction. *TRENDS in Cognitive Sciences,10*（3）：117-124

Crone, E., Jennings, J. R., & Van der Molen, M. W.(2003). Sensitivity to interference and response contingencies in attention-deficit/hyperactivity disorder. *Journal of Child Psychology and Psychiatry, 44*, 224－226.

Douglas, V., & Parry, P.(1994). Effects of reward and nonreward on frustration and attention in attention deficit disorder. *Journal of Abnormal Child Psychology, 22*, 281－302.

Haber, S.(2000). Striatonigrostriatal pathways in primates

form an ascending spiral from the shell to the dorsolateral striatum. *Journal of Neuroscience, 20,* 2369 - 2382

Haber, S.(2003). The primate basal ganglia: parallel and integrative networks. *Journal of Chemical Neuroanatomy, 26,*317 - 330

Habib, M., Boulanger, C., Soubias, M., Delarbre, C., & Joly-Pottuz, B. (2008). The neuropsychology of the human reward system: Impaired gambling performance in ADHD children and adults with psychopathic tendencies . *Brain and Cognition, 67.* 11 - 47

Kiss, I., Pisio, C., Francois, A.,& Schopflocher, D.(1998). Central executive function in working memory: event-related brain potential studies. *Cognitive Brain Research*，*6*，235 - 247

Luman, M., Oosterlaan, J.,& Sergean, J. (2005). The impact of reinforcement contingencies on AD/HD: A review and theoretical appraisal. *Clinical Psychology Review, 25,*183 - 213

McInerny, R.,& Kerns, K.(2003). Time reproduction in children with ADHD: Motivation matters. *Child Neuropsychology, 9,*91 - 108.

Philip, H. L., Vaughan, C. J., Terry, L. J., Sally, D. A.., Diane, H. M., & Belinda, B. M. (1999). Effortful and automatic information processing in boys with ADHD and specific learning disorders. *Journal of Child Psychology and Psychiaty,40,*275-286

Rolls E.(2000). The orbitofrontal cortex and reward. *Cerebral Cortex., 10* :284 - 294.

Scheres, A., Oosterlaan, J.,& Sergeant, J.(2001). Response inhibition in children with DSM-IV subtypes of AD/HD and related disruptive disorders: The role of reward. *Child Neuropsychology, 7,* 172 - 189.

Sergeant, J.(2000). The cognitive-energetic model: An empirical approach to attention-deficit hyperactivity.

*Neuroscience Biobehavioral Review, 24:*7 - 12.

Sonuga-Barke, E.J. (2002). Psychological heterogeneity in AD/HD:a dual pathway model of behaviour and cognition. *Behavioural Brain Research, 130,* 29 - 36

Wilding, J. (2005).Is attention impaired in ADHD? British *Journal of Developmental Psychology, 23*(4):487-505

Xavier, F., Castellanos, E., & Tannock. R. (2002). Neuroscience of attention deficit/Hyperactivity disorder: The search for endophenotypes. *Nature reviews. Neuroscience ,3,* 617-628

Zelazo, P., & M ü ller U. （2002）.Executive function in typical and atypical development. Hand book of childhood cognitive development. *Oxford; Blackwell.*

Zhang, W. (2008). *The unique deficit of working memory of ADHD: the evidence come from the comparison of ADHD and RD in memory updating (in Chinse).* Unpublished Doctoral dissertation. Beijing Normal University.

[张微 .(2008). *注意缺损多动障碍（ADHD）儿童的特异性工作记忆缺陷：来自 ADHD 和阅读障碍的记忆刷新比较研究证据* . 博士学位论文 , 北京师范大学].

Zhang, W. ,Liu, X., Liao, R. ,& Gu, Q.（2007）. An epidemiological investigation of ADHD in six cities (in Chinse). *Chinese journal of clinical psychology. 1,* 23-25.

[张微 , 刘翔平 , 廖冉 , 顾群 .（2007）. 六城市 ADHD 流行病学调查 , *中国临床心理学杂志，1,* 23-25].

Zhu, Z. (2006). *The effect of reward and punishment on affective decision-making on two subtypes of children with attention deficit hyperactivity disorder (ADHD)* . Unpublished Master thesis Shanxi Normal University.

[朱昭红 (2006). 奖惩对两种亚型儿童情感性决策的影响 . 硕士学位论文 , 陕西师范大学].

The influence of "Hot" executive function on the verbal working memory of attention deficit hyperactivity disorder and reading disability children

ZHANG Wei[1]; LIU Xiangping[2]; SONG Hongyan[3]

([1] School of Psychology, Huazhong Normal University , Hubei Province Key Lab for Human Development and Mental Health, Wuhan,
430079,China)

([2] School of Psychology, Beijing Normal University, Beijing, 100875, China)

([3] Center of Psychological counseling, Hankou Branch of Huazhong Normal University, Wuhan, 430000,China)

Abstract The deficit of "cool" executive function (EF) associated with dorsolateral prefrontal cortex (DLPFC) in attention deficit hyperactivity disorder (ADHD) has been substantially confirmed. But whether ADHD children show the deficit of "hot" EF associated with orbital and medial prefrontal cortex (OMPFC) remains unknown, and till now no research made an explicit exploration for interaction models between the two EFs. Commonly different from some studies related to the children's gambling task, in which the "hot" EF impeded "cool" EF, this study aims to explore the facilitation of the "hot" EF to "cold" EF in the entertaining verbal N−back task. Pure cognitive processions were involved in boring N−back task while both "hot" EF and "cold" EF were involved in the entertaining N−back task. Participants were 77 children age between seven and twelve, of whom 60 were classified as having ADHD and /or reading disability (RD). All the disorder participants were recruited at a clinic and normal children were recruited from a elementary school. A four−group mixed design consisting of reading disabilities only (RD, $n =15$), reading disabilities and ADHD (RD+ADHD, $n =24$),ADHD only (ADHD, $n =21$) and a comparison group ($n =17$) was utilized. In the experiment , two adapted N−back working memory paradigms were used to explore verbal working memory ability, one was a traditional N−back task, another was entertaining N−back task. There are the same difficulty and materials between the two tasks.These results indicate that ADHD and RD groups behaved worse than comparison group and no significant differences had been detected between ADHD and RD groups in the boring task. A significant increase in ADHD had been found when comparing entertaining task with boring task. No significant differences had been detected between ADHD and comparison groups. Also, no significant changes related to the task types had been found in RD children. All these findings suggest that ADHD and RD children both show verbal working memory problems, however they have different mechanisms. The "hot" EF facilitates the performance of ADHD in verbal working memory task while not to RD. These results support the Haber model indirectly. According to this model, "hot" EF modulates "cool" EF by a special pathway.

Keywords ADHD; reading disability; "hot" executive; verbal working memory

心理学报 , 2015, 47(6), 774 - 786.

幼儿对威胁性刺激蛇的注意觉察：来自眼动证据 *

王福兴[1]，李文静[1]，颜志强[2]，段朝辉[1]，李　卉[3]

（[1]华中师范大学心理学院，武汉 430079）

（[2]北京大学心理学系，北京 100871）

（[3]天津师范大学心理与行为研究院，天津 300074）

摘　要　前人研究发现相对于中性刺激花，没有经验的婴幼儿会对威胁性刺激蛇产生更快的觉察反应。研究选取 4-6 岁幼儿和成人被试，改进了刺激材料呈现范式和线画的刺激材料，采用 3×3 刺激矩阵呈现的方式，利用眼动仪记录被试的视觉搜索过程，探索威胁性刺激蛇是否被更快注意定向、作为干扰刺激的蛇是否同样能被更快觉察，以及蛇的特殊外形是否在快速觉察中具有重要作用。实验 1 发现，相对于目标物花，成人和幼儿对蛇的首次注视到达时间更短，注视到目标前的注视点个数更少，首次注视的时间更短。实验 2 采用线画的方式去除了刺激材料的色彩和纹理，只保留了蛇蜿蜒的外形，结果发现儿童和成人仍然以更短的注视到达时间、更少注视次数注意到蛇，对蛇的首次注视时间更短。此外，对干扰物分析发现，蛇作为干扰物（花为目标物）仍然表现出更快注意定向。两个实验对比发现，被试对彩色、真实蛇的注视快于线画的蛇。结论认为，即使是对蛇具有较少经验的幼儿，也表现出了对蛇的快速注意偏向；蛇的色彩和纹理会促进蛇的快速觉察；蛇的低水平知觉特征（蜿蜒外形）确实对蛇的注意觉察具有重要作用。

关键词　蛇；威胁性刺激；幼儿；觉察；眼动

分类号　B842

1 引言

恐惧是人类的基本情绪之一，我们对外界事物所产生的恐惧随着年龄的发展而不断变换恐惧对象。婴儿通常会对陌生人恐惧 (Feiring, Lewis, & Starr, 1984)；学龄前的儿童会害怕鬼怪；之后儿童便开始发展出对动物的恐惧，比如：蛇 (LoBue, Rakison, & DeLoache, 2010)。人们对某些事物的恐惧（如：陌生人和鬼怪）可能只在某个发展阶段存在，但是有些恐惧可能会在人的一生中都存在 (Berger, 2010)，比如：蜘蛛 (Purkis & Lipp, 2009)、 蛇 (Isbell, 2006; Soares, Esteves, Lundqvist, & öhman, 2009) 和狮子 (Penkunas & Coss, 2013a)。

* 国家自然科学基金青年项目 (#31300864)、国家自然科学基金面上项目（#31170979）、2012 年大学生创新实验计划项目"恐惧还是偏好？儿童对危险刺激的优先加工"和华中师范大学教学研究项目 (#201115) 资助。

通讯作者：王福兴，E-mail: fxwang@mail.ccnu.edu.cn

蛇在很多文化中都被当作恶魔的象征，出现在各种文学、影视作品中 (Isbell, 2006)。研究发现，当要求人们在一些由动物图片组成的刺激矩阵（3×3 或 2×2）中搜索蛇、蜘蛛等威胁性动物时，对蛇和蜘蛛这类威胁性刺激搜索的反应时间更短 (öhman, Flykt, & Esteves, 2001)。对于人类为什么惧怕蛇，心理学研究却有不同的解释。如果作为一个有生活经验的成人，我们从书籍、电视、电影或别人的经验中习得了蛇是恐怖的、有毒的（尽管不是所有蛇都分泌毒素）、被咬到会致命。所以，一些研究者认为，我们对蛇这类威胁性动物的快速觉察可能来自后天的学习或生活经验。比如：öhman 和 Mineka(2001, 2003) 用猴子作为被试研究发现，在实验室养大的猴子对蛇并不会产生恐惧，但是通过观察自然中长大的猴子对蛇的恐惧反应，实验室饲养的猴子可以很快习得对蛇的恐惧。对儿童和成人的对比发现，由于成人对蛇具有更多与恐惧有关的经验，成人对蛇的觉察速度要快于儿童 (LoBue & DeLoache, 2008; LoBue et al., 2010)。此外，在对蜘蛛恐惧研究中也发现后天的经验和学习在恐惧中扮演着重要作用 (Gerdes, Alpers, & Pauli, 2008; Waters, Lipp, & Randhawa, 2011)。

也有一些研究者认为人类天生恐惧蛇，对蛇这类刺激的注意偏向是由于人类具有将蛇和恐惧联结在一起的先天倾向 (innate predisposition)，所以对这些刺激的反应时间更短，觉察速度更快 (öhman & Mineka, 2001, 2003; Quinlan, 2013)。人类对蛇这种威胁性刺激的视觉搜索优势可能是这种行为对人类有其特殊的生物学或进化意义 (Blanchette, 2006; Isbell, 2006; öhman & Mineka, 2003)。从进化的观点来看，人类如果能更有效、更快速地搜索令人感到恐惧的威胁性刺激（蛇、蜘蛛等），将更有可能成功地回避这种危险而生存下来（相关评述见：Isbell, 2006; LoBue et al., 2010; öhman, 2009)。实验室饲养的猴子研究为这种假设提供了很好的实验支持。因为这些猴子完全是在实验室养大，没有任何与蛇有关的经验，它们仍然会对威胁性刺激产生更快的觉察（参见：öhman & Mineka, 2001; Shibasaki & Kawai, 2009)。 比如：Shibasaki 和

Kawai(2009) 发现实验室饲养的日本恒河猴对威胁性刺激（蛇）图片的搜索反应时要显著快于非威胁性刺激（花）的图片。此外，近期有关生物对威胁性刺激恐惧的神经生物学研究也为这种假设提供了支持 (Keil et al., 2013)。虽然"先天倾向"假设看似合理，但是它并没有为揭示我们对蛇的快速觉察和反应机制是什么。

由于婴、幼儿没有与蛇等威胁性刺激有关的经验，如果婴、幼儿也能表现出类似于成人的快速觉察反应，这就为人类可能天生恐惧蛇（对蛇的觉察反应更快）提供了强有力的支持。LoBue 和 DeLoach(2008) 在实验中要求 3 ~ 5 岁的儿童在中性刺激（花、青蛙和毛毛虫）中搜索威胁性刺激（蛇）或在威胁性刺激中搜索中性刺激，发现儿童对威胁性刺激蛇的反应显著快于花，表现出了与成人一致的反应模式。Waters，Lipp 和 Spence (2008) 在 9 ~ 12 岁的儿童中也发现了相似的结果。随后，DeLoache 和 LoBue(2009) 采用视觉偏好范式对 7~9 和 14~16 个月婴儿的研究发现，当播放蛇和其它非蛇的视频片段时，婴儿能够将恐惧的声音和蛇的视频片段进行匹配。此外，她们对 8-14 个月婴儿也进行了类似的研究，发现婴儿对蛇的注视反应要快于中性刺激花 (LoBue & DeLoache, 2010)。类似地，Rakison 和 Derringer(2008) 采用视觉偏好和习惯化范式研究了 5 个月大婴儿对蜘蛛和花的注视模式，结果发现婴儿对蜘蛛图像的注视时间要长于花。这些结论似乎都支持人类对蛇具有天生的敏感和快速觉察。

到底是什么原因导致我们对蛇这种特殊的爬行动物的快速觉察？ LoBue 等人 (2013) 认为，蛇的低水平知觉特征（low-level perceptual features）可能起到了重要作用。LoBue 和 Deloache(2011) 实验结果发现，儿童与成人在卷曲的电线和卷曲的蛇之间的反应时没有差异；儿童与成人在非卷曲的蛇与花之间的反应时没有差异，这些都显示可能是蛇的形状在起重要作用。研究者提出了"知觉模板"（perceptual template）假设来解释蛇的外形在幼儿对蛇的注意偏向中的作用 (LoBue, 2013; LoBue et al., 2010; Rakison &

Derringer, 2008)。该假说认为，一些恐惧刺激（蛇、蜘蛛）在人脑中存在知觉模板，它可能包含恐惧刺激的一些特有特征，人们一旦看到这些特征信息就会自动激活这个模板，进而导致更快的注意觉察。如果人对于这类进化相关的恐惧刺激具有天生的快速觉察，那么无论成人还是儿童应该都表现出相同的知觉偏向。"知觉模板"假设在"先天倾向"基础上给出了更加合理的解释，而且也让研究对此类问题的验证更加可操作化，因而受到了研究者的关注。

从行为反应结果（按键）来看，成人研究确实发现威胁性刺激觉察要快于非威胁性刺激。但是，也有研究者认为，这种按键反应时差异可能反应的是对威胁性刺激的快速行动而不是快速觉察 (Flykt & Caldara, 2006; LoBue, Matthews, Harvey, & Stark, 2014)。所以，采用眼动来了解威胁性蛇的视觉搜索对于揭示是否被试真的在视觉搜索阶段就表现出更快的注意定向具有重要意义 (LoBue et al., 2014)。此外，以往研究中对于婴儿注意的测量，更多是采用婴儿对两个事物的注视偏好，并且依据录像编码来确定其加工时间 (参见：DeLoache & LoBue, 2009; LoBue & DeLoache, 2010; Rakison & Derringer, 2008)。尽管在年龄较大的 3~5 岁儿童实验中，研究者使用了改进的触屏范式，要求儿童把手放在一个手模 (handprints) 上，以保证收集到可靠的反应时数据 (具体见：LoBue & DeLoache, 2008)。但也仍然存在从看到刺激到做出动作反应的误差，尤其是当儿童的精细动作反应没有完全发展前。行为方法中的反应时测量的是从感觉器官接收到刺激到动作器官做出反应的全部过程时间。但是，本研究更加关心在做出反应前的视觉搜索阶段（从目标刺激呈现到注视到目标刺激），被试是否也能够更快的定位或觉察到刺激，以及对目标刺激的首次加工。婴、幼儿群体的语言尚未发展成熟，不能进行自我报告，基于眼动技术提供的注视数据就为了解婴、幼儿的认知加工提供了有效的手段 (评述见：Bornstein, Mash, & Arterberry, 2011; Feng, 2011; Gredeback, Johnson, & von Hofsten, 2010; Oakes, 2012)。眼动仪可以记录过程性信息或观看过程，可以为直接测量儿童注视或搜索威胁性刺激提供更加客观和准确的数据。

基于以上论述，本研究实验 1 为验证性实验，目的在于验证先前实验中的行为反应时结果是否会表现出相同的视觉搜索差异，即探索儿童、成人对威胁性刺激蛇是否会表现出更短的首次到达时间、更少的注视次数以及更短的首次注视持续时间。实验 2 在实验 1 的基础上采用线画图 (line drawing)，保留了蛇蜿蜒的外形特征，探讨蛇的低水平知觉特征在快速觉察中作用，并且进一步验证"知觉模板"假设。在先前研究中，LoBue 和 DeLoache(2011) 的实验中所用的图片是真实的蛇和真实的电线图片。实验 2 关心的是如果知觉模板假设成立，即蛇特殊的外形导致了其快速加工。实验使用线画图片把蛇的颜色、纹理结构、头部特征等最大化去除后，只保留蛇的外形特征。这就有助于进一步证实这样一个结论：对蛇的快速觉察不是由于其恐惧性颜色、花纹，而是其特殊的蜿蜒的外形 (LoBue & Deloache, 2011; LoBue et al., 2010)。由于婴、幼儿对蛇具有更少的社会经验，研究者推测婴、幼儿所表现出来的对蛇等威胁性刺激的快速注意偏向可能是由于知觉模板在起作用。而蛇特有的蜿蜒的外形就是知觉模板的关键特征之一。如果人类真的是对蛇弯曲的体形具有更快的反应，我们预期儿童和成人在实验 2 中所有对目标物注视上表现出一致的快速注视。由于实验 1 和实验 2 采用了相同的蛇~花配对，所以，研究也对比实验 1 和实验 2 的数据差异，从而了解蛇的色彩、纹理以及特殊的外形在其快速注意偏向中作用。

其次，LoBue 和 DeLoache(2011) 的研究发现作为干扰刺激的干扰程度对目标刺激的加工不产生影响。实验使用了相同的干扰刺激（马、鹿、兔子）来搜索目标物蛇和目标物青蛙。结果仍然发现蛇快于青蛙。我们对此持有不同的看法，因为原来的研究都是把目标物（蛇、蜘蛛）和干扰物（青蛙、花）配对呈现的。但是，LoBue 和 DeLoache(2011) 的研究中把所有干扰物替换为非威胁性刺激后，得出结论具有局限性和误导性。其最近的研究也发现，蛇作为目标物和干扰物的首次注视时间没有区别 (LoBue et al., 2014)。如果蛇这类威胁性动物作为目标物会有更快的反应时或觉察，那么蛇作为干扰物仍然会有更快的觉察。对此，有研究者曾提出了"注意脱离困难"(disengaging difficulty) 假设，即对花的反应变慢可能是由于对蛇的注视导致注意脱离困难 (Fox, Russo, Bowles, & Dutton, 2001)。但是这个假设没有在现有同类刺激研究中得到验证。所以，本研究预期当蛇作为干扰物时，对蛇的反应同样会快于作为干扰物的花。对与目标物的配对干扰物进行分析，有助于验证实验的预期。为了验证当威胁性刺激作为干扰物时仍然会具有更快注视，我们对作为干扰物的蛇和花进行了分析。

最后，两个实验在实验范式和被试选取上做了一些改进。第一，有关威胁性刺激搜索研究发现，3×3 矩阵搜索范式的中间位置的刺激加工速度更快 (Blanchette, 2006)。其他研究发现，3×3 矩阵搜索中中心位置和四周位置存在反应时间差异 (Brosch & Sharma, 2005)。而以往采用搜索范式的研究中，都没有对中心位置效应进行平衡或处理。另外，考虑到眼动仪校准的注视点会出现在屏幕中央，如果目标刺激位置在矩阵中心，对目标物快速注视或搜索可能不是威胁性本身导致，而是其位置有利于加工导致的。所以，研究修改了原有的视觉搜索范式，在 3×3 矩阵搜索中，去除了中心位置刺激。第二，以往研究中大多数都是方便取样，选取被试的父母作为成人对照组（如：DeLoache & LoBue, 2009; LoBue & DeLoache, 2008, 2010）。相关研究发现，父母对蛇的恐惧经验、家族恐惧历史对个体恐惧都具有影响作用 (Fredrikson, Annas, & Wik, 1997; Murray & Foote, 1979)。所以本研究拟选取没有任何血缘关系的大学生被试群体作为对照的成人组，这样可以避免因为儿童与父母之间的生活经验相似性或遗传的影响。

2 实验 1

以往研究发现，儿童对威胁性刺激蛇的觉察反应要快于中性刺激花。实验 1 关注在做出反应前的视觉

搜索阶段，威胁性刺激是否先于非威胁性刺激被优先注视到？实验修改呈现范式，去除了中间位置刺激，采用经典的蛇～花配对作为刺激材料，记录被试眼动数据，验证儿童、成人对威胁性刺激是否有更快的、更短的注视。此外，还分析了当蛇作为干扰物（目标物为花）时，是否也会有更快的视觉搜索。

2.1 被试

24 名 4~6 岁的儿童。由于实验过程中不专心、动作幅度过大（眼动采样率均小于 75%），8 名儿童数据没有进入分析，有效被试 16 人（女 7 人），平均年龄为 5.1 岁（$SD=0.6$）；成人被试选自某师范大学本科生 22 人（女 11 人），平均年龄 19.6 岁（$SD=1.2$）。所有被试视力或者矫正视力正常，无色盲、色弱。

关于蛇的先验知识问卷参照 LoBue 等人的研究自己编制 (LoBue, 2010b; LoBue & DeLoache, 2008)。问卷采取是否式计分（选"是"或"消极"计 1 分，"否"或"积极"记 0 分，最高分 4 分，具体见附录 1）。由于已有研究发现儿童记忆不准确 (Braun, Ellis, & Loftus, 2002; Cole & Loftus, 1987; Loftus & Davies, 1984)，以及幼儿言语和表达能力发展不够完善，所以参照以往研究方式，儿童关于蛇的先验知识问卷由父母填写。实验 1 父母报告的儿童先验知识平均得分 1.4（$SD=1.2$），成人自我报告的平均得分 2.4（$SD=1.0$），成人的经验显著高于儿童，$t(36)=2.70$, $p<0.05$, Cohen's $d=0.9$。此外，实验 1 与实验 2 的成人在先验知识问卷上无显著差异，$t(43)=-1.74$, $p>0.05$, $d=0.5$；实验 1 与实验 2 儿童群体也无显著差异，$t(43)=-0.47$, $p>0.05$, $d=0.1$。先验问卷分析发现性别、城乡对蛇的恐惧程度没有差异 ($p_s>0.05$)，因此在两个实验结果中均没有包含这些变量。

2.2 材料

本研究所有材料图片均来自于互联网，由研究者仿照 LoBue 等人研究 (DeLoache & LoBue, 2009; LoBue, 2010b; LoBue & DeLoache, 2008) 自己制作。图片刺激类型为蛇和花，共 72 张图片，均去除图片背景。在每个实验试次中，显示器会呈现一个 3×3 的图片矩

图1 实验1材料图片示例（左图目标物为蛇，右图目标物为花）

阵（见图1），每个矩阵中心位置不呈现图片，共包括8个刺激物。每个刺激矩阵中包括1个目标物（蛇或花）和7个干扰物（花或蛇）。实验目标物每种类型各8张，不重复出现（随机呈现），干扰图片每种类型（蛇或花）各28张，为了保证实验效果，所有干扰物图片采取伪随机，即在与目标图片进行匹配时重复一次。用于视觉搜索的3×3矩阵共16个，其中蛇和花作为目标物各8个。矩阵中每张图片大小为325×245像素。

采取单盲方式，请20位心理学专业学生（女10人）对72张刺激图片的恐惧性程度进行7点（1，一点都不害怕；7，非常害怕）以及明亮度进行5点（1，非常暗；5，非常亮）评定。结果显示恐惧唤起程度有显著性差异$t(70) = 56.51$，$p<0.001$，$d=13.5$，蛇的恐惧唤起更高；亮度不存在差异$t(70) =-0.70$，$p>0.05$，$d= 0.2$。

2.3 仪器、设计与程序

实验仪器为Tobii T120 Eye-tracker（Tobii Technology, Sweden）。双眼红外追踪，采样率120Hz，眼睛距屏幕距离60cm（17英寸显示器，1024×768分辨率）。单张图片像素大小为200×150像素，单张图片刺激物的水平视角约3.3度，垂直视角约2.5度；矩阵刺激图片的像素为800×600，矩阵图片的对侧视角约14.3度。

实验为2（年龄：成人、儿童）×2（目标类型：蛇、花）的混合实验设计，实验范式修改自LoBue等人用于儿童的搜索范式（DeLoache & Lo Bue, 2009; LoBue,2010b; LoBue & DeLoache, 2008）。为了控制中心位置的影响，去除了3×3矩阵中间位置的刺激。

在正式实验进行前，被试先要完成7个试次的练习（练习材料均不出现在正式实验材料中），最初要求儿童依次观看两个单张的图片，第一个是目标类型，第二个是干扰类型，目的是熟悉实验材料；接下来的两个练习是同时呈现一幅目标图片和一幅干扰图片，要求儿童观看目标图片；最后的3个是呈现3×3的矩阵图片（与正式实验一致），让儿童在干扰图片中寻找目标图片（与其他刺激不同的图片），找到后口头报告，然后主试操作进入下一个试次。正式实验中每个矩阵图片最长呈现3s，在矩阵图片呈现之前先呈现注意吸引图片（卡通米奇），当主试确认被试确实是在看图片刺激时才进入到下一个试次，共16个试次（呈现顺序随机）。在每一个矩阵图片呈现的间隔都会呈现一张米奇老鼠的图片以吸引被试的注意。实验过程中，一名主试陪伴在儿童身边指导其完成练习测试。成人实验程序与儿童相同，但是指导语表述方式有所更改，去掉了儿童化语言表述。实验结束后，每位被试都需要完成一份先验知识问卷（儿童的问卷由父母完成）。

2.4 结果

为了探讨目标刺激对人的注意捕获能力和干扰物对人的注意干扰能力，对刺激图片划分了两类兴趣区（Area of Interest,AOI）：目标物（蛇或花）和干扰物（花或蛇），并且把所有干扰物看作一个整体进行数据导出和分析。采用的眼动指标为：首次注视到达时间（Elapse Time of First Fixation to AOI，该指标计算的是从刺激呈现到第一次注视到目标物的时间，时间越短表明目标被越早注视到）；首次进入兴趣区之前的注视点个数（Fixation Count Before Enter

AOI，是指在被试的首个注视点进入兴趣区之前的注视点个数，次数越少说明目标物被识别越快）；首个注视点的持续时间（First Fixation Duration of AOI，指的是进入目标兴趣区的第一个注视点的注视持续时间，此指标说明了被试对目标物的首次加工时间）（见表1）。本研究剔除了三个标准差之外的极端数据（原始眼动数据），对所有眼动数据采用2（年龄）×2（目标类型）的重复测量方差分析。

2.4.1 目标物注视

在首次到达兴趣区时间上，目标类型主效应显著（$F(1,36)=45.14$，$p<0.001$，$\eta_p^2=0.56$，partial η^2，下同），事后检验（Bonferroni，下同）发现目标刺激蛇（$M=638$）被首次注视到时间显著快于目标刺激花（$M=905$）。年龄主效应显著（$F(1,36)=86.14$，$p<0.001$，$\eta_p^2=0.71$），成人（$M=479$）首次到达目标刺激时间显著快于儿童（$M=1174$）。目标类型与年龄的交互效应显著（$F(1,36)=22.74$，$p<0.001$，$\eta_p^2=0.39$），儿童首次到达目标物蛇的时间显著短于花（$F(1,36)=56.98$，$p<0.001$）。

首次进入兴趣区之前的注视次数，目标类型主效应显著（$F(1,36)=36.42$，$p<0.001$，$\eta_p^2=0.50$），到达蛇（$M=2.5$）之前的注视次数显著少于花（$M=3.4$）。年龄主效应显著（$F(1,36)=86.73$，$p<0.001$，$\eta_p^2=0.71$），成人（$M=2.2$）到达目标物之前的注视次数显著少于儿童（$M=4.0$）。目标类型与年龄的交互效应显著（$F(1,36)=12.84$，$p<0.01$，$\eta_p^2=0.26$），相对于目标物花，儿童以更少注视次数就注视到了目标物蛇（$F(1,36)=39.94$，$p<0.001$），成人也表现出相同的结果，即更少注视次数就锁定了目标物蛇（$F(1,36)=3.57$，$p=0.067$）。

兴趣区首个注视点持续时间，目标类型主效应显著（$F(1,36)=12.01$，$p<0.01$，$\eta_p^2=0.25$），蛇的首个注视点的注视时间（$M=459$）显著短于花（$M=627$）。年龄主效应显著（$F(1,36)=26.15$，$p<0.001$，$\eta_p^2=0.42$），成人（$M=695$）首个注视点持续时间显著长于儿童（$M=334$）。目标类型与年龄的交互效应

显著（$F(1,36)=10.45$，$p<0.01$，$\eta_p^2=0.23$），成人对蛇的首个注视点的持续时间显著短于花（$F(1,36)=26.64$，$p<0.001$）。

2.4.2 干扰物注视

在兴趣区首次注视到达时间上，干扰类型主效应显著（$F(1,36)=22.83$，$p<0.001$，$\eta_p^2=0.39$），干扰刺激蛇（$M=381$）被首次注视到时间显著快于干扰刺激花（$M=515$）。年龄主效应不显著（$F(1,36)=1.33$，$p>0.05$，$\eta_p^2=0.04$）。干扰物与年龄的交互效应显著 $F(1,36)=6.82$，$p<0.05$，$\eta_p^2=0.16$），成人对蛇的首次注视到达时间显著短于花（$F(1,36)=32.43$，$p<0.001$）；在干扰刺激蛇上，成人的首次注视到达时间显著短于儿童 $F(1,36)=7.73$，$p<0.01$（见表1）。

首次进入兴趣区之前的注视点个数上，干扰类型主效应显著（$F(1,36)=27.71$，$p<0.001$，$\eta_p^2=0.44$），在看到蛇（$M=1.1$）之前注视点个数显著少于花（$M=1.6$）。年龄主效应不显著（$F<1$，$p>0.05$，$\eta_p^2<0.01$）。干扰物与年龄的交互效应显著（$F(1,36)=9.07$，$p<0.01$，$\eta_p^2=0.20$），在干扰刺激蛇上，成人到达兴趣区之前的注视点个数显著少于儿童（$F(1,36)=5.85$，$p<0.05$）；相对于花，成人以较少注视次数就注视到了蛇（$F(1,36)=40.66$，$p<0.001$）。

在兴趣区的首个注视点的注视持续时间上，干扰类型（$F(1,36)=2.33$，$p>0.05$，$\eta_p^2=0.06$）和年龄主效应（$F(1,36)=1.45$，$p>0.05$，$\eta_p^2=0.04$）均不显著。干扰类型与年龄的交互效应显著（$F(1,36)=8.73$，$p<0.01$，$\eta_p^2=0.20$）。在干扰刺激花上，成人首个注视点的注视持续时间短于儿童 $F(1,36)=11.34$，$p<0.01$）；成人对蛇的首次注视持续时间长于花 $F(1,36)=11.92$，$p<0.01$。

2.5 讨论

当蛇作为搜索目标时，实验1的眼动结果重复且证实了威胁性刺激蛇的首次注视时间要明显快于中性刺激花，即从刺激呈现到注视点落到蛇上，被试的首次注视到达时间更短。而且，在注视到目标物蛇上所用的注视点个数更少，对蛇的首次观看持续时间也更短。这些结果都与之前行为反应的结

表 1　　儿童和成人对威胁性刺激和非威胁性刺激的眼动数据

实验	兴趣区眼动	年龄	目标物				干扰物			
			蛇		花		花		蛇	
			M	SD	M	SD	M	SD	M	SD
实验 1	首次注视到达时间 (ms)	儿童	917	206	1432	533	503	142	447	161
		成人	435	62	522	74	524	176	334	89
	首次进入前注视点个数	儿童	3.2	1.0	4.7	1.1	1.5	0.3	1.3	0.4
		成人	2.0	0.4	2.4	0.5	1.7	0.5	1.0	0.3
	首个注视点持续时间 (ms)	儿童	329	77	339	52	258	35	248	59
		成人	553	244	836	386	222	31	254	47
实验 2	首次注视到达时间 (ms)	儿童	1126	187	1314	508	485	157	445	79
		成人	725	140	756	141	357	93	317	89
	首次注视前注视点个数	儿童	3.9	0.7	4.7	1.9	1.5	0.4	1.3	0.2
		成人	3.3	0.7	3.6	0.5	1.5	0.4	1.4	0.4
	首个注视点的持续时间 (ms)	儿童	341	88	398	112	284	54	274	58
		成人	350	77	386	88	220	32	205	35
	行为反应（ms）	儿童	3116	605	3233	602	/	/	/	/
		成人	1568	286	1625	276	/	/	/	/

果相一致，即被试对蛇这种威胁性刺激觉察更快，搜索到的时间更短 (LoBue & DeLoache, 2008; LoBue & Deloache, 2011;öhman, Flykt, etal., 2001; öhman & Mineka, 2001)。这个结果也与情绪面孔眼动研究一致，负性面孔的首次注视潜伏期更短 (Reynolds, East wood, Partanen, Frischen,& Smilek,2009)。LoBue 近期对于成人研究也发现，相对于花，首次注视到蛇的潜伏期更短 (Lo Bue etal.,2014)。

儿童表现出了和成人一致的趋势，即无论是儿童还是成人对蛇的觉察都要快于花。在交互作用上，儿童也是对蛇的搜索注视要快于花，这个结果与 LoBue 和 De Loache（2008）LoBue 和 Deloache（2011）采用触屏搜索范式结果是一致的。虽然本实验所选的成人被试不是儿童的父母，成人对目标物（蛇或花）的注视仍要快于儿童，这个结果也与以往研究类似 (Lo Bue & De Loache, 2008; LoBue & Deloache, 2011)。但是，在注视点持续时间上发现成人更长。可能是由于成人具有更成熟的视觉搜索技能以及对蛇具有更多的社会经验。所以，成人的反应要明显快于儿童，并且对蛇加工也更多。有关成人的研究发现，经验可以调节被试对威胁性刺激的视觉搜索 (Gerdes et al, 2008; Peira, Golkar, Larsson, & Wiens, 2010)，通过调查问卷数据也可以发现，儿童缺少与蛇有关的后天经验。

对干扰物的蛇进行分析，眼动数据显示无论是儿童还是成人对蛇的首次注视到达时间要短于花，以较少的注视次数就注意到蛇。这个结果与蛇和花作为目标物进行分析时结果是一致的。该结果证实了预期的假设，即不管作为目标物还是干扰物，威胁性刺激的注意定向都更快。即当蛇作为干扰物出现时，其作为恐惧刺激所具有的快速觉察仍然得到了体现。但是，对干扰物分析的眼动结果并没有支持"注意脱离困难"假设 (Fox et al., 2001)。虽然蛇作为干扰物时的视觉搜索快于花，但是在首次注视持续时间上并没有发现花与蛇的区别。

3　实验 2

在实验 1 基础上，借鉴情绪面孔研究中采用线画或模式化来突出刺激的形态特征方式 (Fox et al., 2000; LoBue & Larson, 2010)，使用线画的蛇和花作为刺激材料，进一步验证蛇低水平的知觉特征（蜿蜒外形）是否会对早期的视觉加工和搜索产生影响，检验"知觉模板"假设。儿童和成人是否会对蛇的形态特征有更快的注意和更短的注视加工？以及当蛇作为干扰物时，这种特征是否仍然起作用？此外，与实验 1 数据进行对比，进一步讨论蛇的外形、颜色、纹理在其快速觉察中的作用。

图 3　实验 2 刺激材料示例（左图目标物为蛇；右图目标物为花）

3.1 被试

22 名 4~6 岁的儿童，有 2 名儿童数据未进入分析（采样率小于 75%），有效被试为 20 人（女 5 人），平均年龄为 5.1 岁（$SD=0.5$）；成人被试选取本科生 23 人（女 10 人），平均年龄 20.0 岁（$SD=1.3$），所有被试视力或者矫正视力正常，无色盲、色弱。实验 2 中父母报告儿童关于蛇的先验知识平均得分 1.58（$SD=0.51$），成人平均得分 2.87（$SD=0.51$），成人经验显著高于儿童，$t(40)=6.34$，$p<0.001$，$d=2.0$。

3.2 材料

本研究作者参照实验 1 中图片，手工画制了所有线画图，将图片类型变量中的真实蛇和花转换成了线画蛇和花（见图 3 及附录 4）。所有实验材料均参照实验 1 中彩色图片绘制。18 位不知实验目的心理学本科生（女 9 人）对这些线画图进行了评定（评定方式和计分同实验 1），结果发现恐惧唤起程度有显著性差异 $t(70)=28.16$，$p<0.001$，$d=6.7$，恐惧刺激蛇的唤起度显著高于花；亮度不存在差异 $t(70)=-1.06$，$p>0.05$，$d=0.3$（数值见附录 3）。

3.3 仪器、设计与程序

仪器同实验 1；实验设计同实验 1。实验基本流程同实验 1。不同的是，在实验 2 中记录了被试的行为反应时。由于儿童被试年龄较小且无法精确操作鼠标，实验要求被试在口头报告发现目标物后，由主试及时点击鼠标（鼠标图标不在屏幕上出现），记录从刺激呈现到点击鼠标时间作为被试的行为反应时间。主试点击鼠标后自动进入下一实验序列。成人采用相同方式记录反应时间。

3.4 实验 2 结果

3.4.1 目标物注视

在兴趣区首次注视到达时间上发现，目标主效应显著（$F(1,41)=4.34$，$p<0.05$，$\eta_p^2=0.10$），目标物蛇（$M=911$）被首次注视到时间短于花（$M=1016$）。年龄主效应显著（$F(1,41)=50.56$，$p<0.001$，$\eta_p^2=0.55$），成人（$M=740$）快于儿童（$M=1220$）。目标类型与年龄的交互效应不显著（$F(1,41)=2.22$，$p>0.05$，$\eta_p^2=0.05$）。

到达兴趣区之前的注视次数，目标主效应显著（$F(1,41)=5.98$，$p<0.05$，$\eta_p^2=0.13$），蛇（$M=3.6$）在首次被注视到之前注视次数显著少于花（$M=4.1$）。年龄主效应显著（$F(1,41)=12.08$，$p<0.01$，$\eta_p^2=0.23$），成人（$M=3.4$）在到达目标刺激前注视次数显著少于儿童（$M=4.3$）。目标类型与年龄的交互效应不显著（$F(1,41)=1.12$，$p>0.05$，$\eta_p^2=0.03$）。

兴趣区首次注视持续时间，目标主效应显著（$F(1,41)=9.94$，$p<0.01$，$\eta_p^2=0.19$），蛇（$M=346$）的首次注视持续时间显著短于花（$M=392$）。年龄主效应不显著（$F<1$，$p>0.05$，$\eta_p^2<0.01$）。目标类型与年龄的交互效应不显著（$F<1$，$p>0.05$，$\eta_p^2=0.01$）。

3.4.2 干扰物注视

兴趣区首次注视到达时间上，干扰类型主效应显著（$F(1,41)=4.63$，$p<0.05$，$\eta_p^2=0.10$），干扰刺激蛇（$M=376$）被首次注视到时间显著短于干扰刺激花（$M=417$）。年龄主效应显著（$F(1, 41)=22.17$，$p<0.001$，$\eta_p^2=0.35$），成人（$M=337$）显著短于儿童（$M=465$）。干扰类型与年龄的交互效应不显著（$F<1$，$p>0.05$，$\eta_p^2<0.001$）。

首次进入兴趣区之前的注视点个数上，干扰类型主效应边缘显著（$F(1, 41)=3.52$，$p=0.068$，$\eta_p^2=0.08$），蛇（$M=1.4$）注视次数少于花（$M=1.5$）。年龄主效应（$F<1$，$p>0.05$，$\eta_p^2<0.01$）以及干扰类型与年龄的交互效应（$F<1$，$p>0.05$，$\eta_p^2<0.01$）均不显著。

首个注视点的注视持续时间上，干扰类型主效应边缘显著（$F(1,41)=3.31$，$p=0.076$，$\eta_p^2=0.08$），蛇（$M=237$）的首个注视点持续时间短于花（$M=249$）。年龄主效应显著（$F(1,41)=30.08$，$p<0.001$，$\eta_p^2=0.42$），成人（$M=212$）的首次加工时间显著短于儿童（$M=279$）。干扰类型与年龄的交互效应不显著（$F<1$，$p>0.05$，$\eta_p^2<0.01$）。

3.4.3 行为反应

目标类型主效应边缘显著（$F(1,41)=3.78$，$p=0.059$，$\eta_p^2=0.08$），目标物蛇（$M=2288$）的反应快于花（$M=2373$）。年龄主效应显著（$F(1,41)=140.31$，$p<0.001$，$\eta_p^2=0.77$），成人（$M=1596$）反应快于儿童（$M=3174$）。目标类型与年龄的交互效应不显著（$F<1$，$p>0.05$，$\eta_p^2=0.01$）。

对目标物蛇的首次注视到达时间和行为反应进行配对样本 t 检验，发现眼睛首次到达时间显著短于行为反应，$t(42) = -12.32$，$p < 0.001$，$d = 1.9$。目标物花的对比也发现，眼睛首次到达时间要显著快于行为反应，$t(42) =-11.73$，$p < 0.001$，$d =1.8$。

3.5 实验1和实验2对蛇的注视比较

为了进一步揭示蛇的外形在快速觉察中的作用，对实验1和实验2的数据进行了对比分析。探讨去掉了颜色和纹理后，实验2中的蛇是否被注视的更早。

由于实验2中增加了被试报告及要求主试点击鼠标记录被试的反应时间，为了确保实验1和实验2数据对比的有效性，计算了儿童和成人对目标物蛇的首次到达时间和首个注视点持续时间均值之和（实验1儿童：917+329=1246 ms；实验1成人：435+553=988 ms；实验2儿童：1126+341=1467 ms；实验2成人：725+350=1075 ms），发现数值远低于实验2中儿童反应时均值3116 ms和成人反应时均值1568 ms。由于研究关注的是眼动的初期反应指标（首次注视到达时间、首次注视前注视点个数、首个注视点持续时间），而不考虑后期指标（比如：平均注视时间、总注视时间、总注视次数）。所以，实验2中记录行为反应时过程并不会对3个眼动指标产生影响。因为儿童和成人的反应时均值都滞后于对目标物蛇首次到达时间和首个注视点持续时间均值之和，说明被试必须先搜索到目标刺激才能做出反应。此外，在最近采用成人被试眼动研究中，发现成人被试从首次注视结束到做出按键反应（反应时）的延迟时间为748 ms(恐惧刺激)和826 ms(非恐惧刺激)(LoBue et al., 2014, p.820)，这也说明是在首次注视结束后做出行为反应，而不影响眼动初期反应指标。数据采用了2（蛇的特征：真实、彩色的蛇，蛇的线画图）×2（年龄：成人，儿童）的被试间分析。其中，真实、彩色的蛇的数据来源于实验1，蛇的线画图数据来自实验2（见表1）。

3.5.1 蛇作为目标物

首次注视到达时间上，特征主效应显著（$F(1,77)=52.49$，$p<0.001$，$\eta_p^2=0.41$），实验1中彩色、真实的蛇（$M=638$）首次注视到达时间显著短于实验2中线画的蛇（$M=911$）。年龄主效应显著（$F(1,77)=164.24$，$p<0.001$，$\eta_p^2=0.68$），成人（$M=583$）快于儿童（$M=1033$）。特征与年龄的交互效应不显著（$F(1,77)=1.37$，$p>0.05$，$\eta_p^2=0.02$）。

首次注视到蛇之前的注视次数，特征主效应显著（$F(1,77)=39.35$，$p<0.001$，$\eta_p^2=0.34$），彩色、真实的蛇（$M=2.5$）注视次数显著少于线画的蛇（$M=3.6$）。年龄主效应显著（$F(1,77)=33.70$，

$p<0.001$，$\eta^2_p=0.30$），成人（$M=2.7$）少于儿童（$M=3.6$）。特征与年龄的交互效应不显著（$F(1, 77)=2.78$，$p>0.05$，$\eta^2_p=0.04$）。

首个注视点持续时间上，特征主效应显著（$F(1, 77)=8.63$，$p<0.01$，$\eta^2_p=0.10$），实验1中彩色、真实的蛇（$M=459$）首次注视时间显著长于实验2中线画的蛇（$M=346$）。年龄主效应显著（$F(1, 77)=12.92$，$p<0.01$，$\eta^2_p=0.14$），成人（$M=449$）长于儿童（$M=335$）。特征与年龄的交互效应显著（$F(1, 77)=10.94$，$p<0.01$，$\eta^2_p=0.12$），成人对真实、彩色的蛇（实验1）首个注视点时间长于线画的蛇（实验2）（$F(1, 77)=21.60$，$p<0.001$）。

3.5.2 蛇作为干扰物

首次注视到达时间上，年龄主效应显著（$F(1,77)=26.19$，$p<0.001$，$\eta^2_p=0.25$），成人（$M=325$）快于儿童（$M=446$）。特征主效应（$F<1$，$p>0.05$，$\eta^2_p<0.01$）、特征与年龄的交互效应（$F<1$，$p>0.05$，$\eta^2_p<0.01$）均不显著。

首次注视到干扰物前注视次数，特征主效应显著（$F(1,77)=8.03$，$p<0.01$，$\eta^2_p=0.09$），彩色、真实的蛇（$M=1.1$）注视次数显著少于线画的蛇（$M=1.3$）。年龄主效应不显著（$F(1,77)=2.59$，$p>0.05$，$\eta^2_p=0.03$）。特征与年龄的交互效应边缘显著（$F(1,77)=3.86$，$p=0.053$，$\eta^2_p=0.05$）。成人首次注视到真实、彩色蛇前注视次数少于线画蛇（$F(1,77)=12.88$，$p<0.01$）。

首个注视点持续时间上，特征主效应不显著（$F(1,77)=1.07$，$p>0.05$，$\eta^2_p=0.01$）。年龄主效应显著（$F(1,77)=7.79$，$p<0.01$，$\eta^2_p=0.09$），成人（$M=229$）短于儿童（$M=262$）。特征与年龄的交互效应显著（$F(1,77)=10.94$，$p<0.01$，$\eta^2_p=0.13$），成人对真实彩色蛇的首次加工时间长于线画蛇（$F(1,77)=11.20$，$p<0.01$）。

3.6 讨论

实验2结果与实验1类似，即相对于非威胁性的花，蛇被首次注视到时间更短，所用注视次数更少，且首次加工的持续时间更短。这个结果证实了预期

假设，并且与采用其他刺激物对比的结论存在一致性（LoBue & Deloache, 2011），为儿童和成人对蛇的快速加工是由于低水平知觉特征提供了直接证据。这种对蛇的轮廓外形效应，在其他研究中也有类似验证。比如：Blanchette (2006) 的研究发现，被试对威胁性刺激的符号表征（蛇的卡通图片）搜索反应时显著得快于中性刺激。

对干扰物的分析仍然发现了与实验1类似结果。即虽然作为干扰物，蛇仍然表现出更短的首次注视时间、更少注视次数和更短的首次注视持续时间。这不仅进一步证明了威胁性刺激蛇确实具有更短的觉察时间，同时也说明当保留蛇蜿蜒的形态特征后，作为干扰物蛇的反应仍然更快。实验2记录了被试觉察目标的行为反应速度，虽然过程不够精细，但是结果仍然重复了已有研究结论，发现蛇的反应要快于花、成人要快于儿童 (LoBue & DeLoache, 2008; LoBue & Deloache, 2011)，也说明了结论的有效性。但是，从统计显著性水平来看，相对于真实、彩色的蛇，线画的蛇导致的反应强度还是降低了，即彩色、真实蛇具有更快的觉察效应。这个结论可以在实验1和实验2对比中得到验证。

对比行为反应和首次注视到达时间可以发现，首次到达时间要明显快于反应时，但是结果趋势具有一致性。说明行为反应的过程需要经过视觉搜索、信息传递、神经中枢决策、肌肉反应等过程，而视觉搜索和定向更加快速。也证明了眼动在威胁性刺激快速觉察研究中可以提供更加直接和有效的反应指标。此外，也有研究者认为，眼动记录视觉搜索（比如：首次注视到达时间）反应了被试对于威胁性刺激觉察的自下而上的加工过程，而行为反应则涉及到自上而下的加工过程(LoBue et al., 2014)。从客观上而言，眼动自下而上的加工过程能够更好地反应人们对蛇的快速觉察。

在实验2中线画蛇相对于真实的彩色蛇去除了许多额外信息（如：颜色、背景、突出的头部特征、纹理等）而保留了低水平知觉特征（蜿蜒的形状）。与实验1数据对比分析发现，当蛇作为目标物进行搜

索时，实验 1 中真实、彩色蛇的被注视到时间明显快于线画的蛇。这个结论证实了我们的推测，说明蛇所具有的颜色、纹理等额外信息还是促进了对蛇的识别。该结论提示我们，蛇作为一种特殊爬行动物，其独有的特征不仅仅包括其弯曲的外形，其颜色和纹理还是突出蛇的特殊性，使其被快速识别和觉察 (Isbell, 2006, 2009)。在 LoBue 和 DeLoache(2011) 研究的一个实验中，研究者控制了刺激呈现的颜色（采用黑白图片）探讨蛇特殊的色彩和外表对变化刺激觉察影响，发现无论成人还是儿童对蛇的觉察仍然快于青蛙。她们的结论认为颜色并不影响对蛇的快速觉察。但是，在改变威胁刺激材料呈现方式后，我们发现色彩仍具有一定促进作用。当蛇作为干扰物时，这种色彩和纹理的效应明显下降，可能是由于当蛇作为干扰物时，被试不是主动去识别蛇。此外，相关研究发现情绪唤醒有助于视觉知觉 (Phelps, Ling, & Carrasco, 2006; Vaish, Grossmann, & Woodward, 2008)。实验 2 的线画刺激相比于真实图片，情绪唤醒度可能较低，可能导致其觉察效应下降。

4 总讨论

本研究使用了眼动仪来采集被试对威胁性刺激的反应数据。根据以往的婴幼儿视觉搜索研究 (DeNicola, Holt, Lambert, & Cashon, 2013; Koster, Crombez, Van Damme, Verschuere, & De Houwer, 2004; Lipp & Waters, 2007; Peltola, Leppänen, Vogel-Farley, Hietanen, & Nelson, 2009)，可以把被试的首次注视到达时间和首次进入兴趣区前注视点个数界定为目标物注意定向（attention orienting），即被试对目标刺激（蛇或花）的注意定向速度，时间越短或注视次数越少，说明被试更早或更快地就锁定了目标刺激；把首个注视点持续时间界定为目标物注意维持（attention holding），即对目标物的注视维持时间越短，说明被试对目标加工更少。实验 1 重复和扩展了 LoBue 等人的实验 (LoBue & DeLoache, 2008, 2010)，发现无论是成人还是儿童对威胁性刺激蛇具有更快的注意定向，更短的注意维持。

实验 2 在去除了蛇的颜色等外部特征后也发现了相同的结论。这些基于眼动的研究结果，也为理解威胁性刺激优先觉察提供了更加客观的实证数据。这说明无论是有经验的成人被试，还是较少经验的 3~5 岁儿童，他们都表现出对蛇更快的觉察。

恐惧研究是一个很广泛的领域，人们不仅对蛇、蜘蛛、蜥蜴这类威胁性动物产生注意偏向或快速反应，同时也对恐惧面孔 (LoBue & Larson, 2010; öhman, Lundqvist, & Esteves, 2001)、威胁性武器枪 (Fox, Griggs, & Mouchlianitis, 2007)、刀和注射器 (Blanchette, 2006; LoBue, 2010b) 等产生注意偏向。研究者把对蛇和蜘蛛这类刺激称为进化相关的威胁（evolutionary relevant threats）或种系相关的（phylogenetical）刺激 (Fox et al., 2007; LoBue et al., 2010)。通过以上的两个实验发现，虽然儿童对蛇具有更少的经验，但是他们仍然表现出了与成人一样的注视模式，对蛇表现出了更快的觉察。这就让我们进一步思考一个重要的问题，即我们为何会对蛇觉察更快，是因为我们恐惧蛇，还是蛇特殊的外形。öhman 等人提出了"先天倾向"假说来解释蛇的快速觉察，他们假设人类认知系统中存在一个恐惧模块（evolved fear module），它是一个相对独立的行为、心理和神经系统，某些特定的刺激会自动化激活的神经系统，这些神经系统会对一些特定的威胁刺激（蛇）有选择性地、自动地激活，并产生防卫性反应。他们认为这个恐惧模块是进化适应的结果，在大部分哺乳动物中都存在 (öhman & Mineka, 2001, 2003; öhman, Soares, Juth, Lindström, & Esteves, 2012)。虽然先天倾向的进化模块可以解释现实情境中的一些恐惧刺激觉察现象，但是却不能很好地解释人类对一些低水平特征的威胁性刺激的快速觉察现象。即这个恐惧模块很难解释我们对一些特定威胁刺激的注意偏向或快速觉察机制。

LoBue 等人在恐惧模块基础上使用了低水平知觉偏向和知觉模板假设来解释婴、幼儿对蛇的觉察 (LoBue, 2013; LoBue et al., 2010)。该假说认为，对蛇具有较少经验的婴、幼儿所表现出来的快速注意偏向可能是由于知觉模板在起作用。知觉模板主要是基于威胁刺激的基本特征或混合的低水平的图示化特征，例如，蛇的知觉模板就是其连续的曲线轮廓以及尾部弯曲为一团；蜘蛛的视觉模板是居于中心的身体及其发散到四周的曲线 (LoBue, 2013; LoBue et al., 2010)。实验 2 中结果支持了蛇弯曲的外形在快速觉察中的作用，与知觉模板假设相符合。因为经过线画处理后，蛇本身的曲线特征更加突出，可以说蛇特殊的轮廓特征导致了其快速注意定向和更短的注意维持。但是，仍然需要深入探讨的一个问题是，这种觉察是否是真的由于其特殊的外形特征，还是人类本身就对特殊的形状反应更快。因为在人类基本视觉搜索中，弯曲的线相较于直线确实更容易引起人们的视觉注意和反应 (Treisman & Gormican, 1988; Wolfe, Yee, & Friedman-Hill, 1992)。对于这个问题解释，LoBue 等人也对于知觉模板假设持有比较谨慎的态度 (LoBue, 2013; LoBue et al., 2010)，后续研究仍然需要对这一问题进行深入探讨。

蛇作为爬行动物，其所具有的特殊外形、头部特征、颜色和纹理是其区别于其他爬行动物重要特征 (Isbell, 2009)。对比实验 1 和实验 2 数据发现，儿童和成人对于实验 1 中真实、彩色蛇的反应明显要快于实验 2 中去掉这些特征点的线画蛇。这说明蛇作为一种特殊的爬行动物，之所以被大部分哺乳动物和人类恐惧和快速觉察，是因为其所具有的颜色、花纹、特殊外形、头部特征作为一个综合体仍然起重要作用。比如，相对于同为爬行动物的蜥蜴，虽然两者在外形一些外部特征上具有相似性，但是研究发现蛇的觉察仍然要快于蜥蜴 (Penkunas & Coss, 2013b)。这也可能是 LoBue 等人前期研究中使用蛇的真实图片作为刺激材料的原因，即现实生活环境中，无论是在动物园还是在野外环境中见到的蛇都是带有色彩的、有纹理的、弯曲的。

研究对于干扰刺激注视情况分析证实了研究的预期，即虽然作为搜索刺激的干扰物，威胁性刺激蛇仍然被更早和更快的注意定向，并且有更短的注意维持时间。这进一步间接证实了恐惧性刺激快速知觉的普遍性。同时，这个结果确实不同于已有研究

认为干扰物属性不影响对目标刺激觉察的结论，但是也不能完全支持"注意脱离困难"假设 (Fox et al., 2001)。比如：Penkunas 和 Coss(2013a) 将蛇和蜥蜴配对，同样被作为威胁性刺激快速觉察的蜥蜴在遇到蛇时反应变慢了，他们认为蛇作为干扰物可能会使得非威胁性刺激或低威胁性刺激觉察变慢。但是，对于可以支持这个"注意脱离困难"的首次注视加工时间来说，在实验 1 和实验 2 的分析中都没有发现蛇和花之间有显著差异。即虽然蛇作为干扰物的时候会被更快觉察，但是并没有发现在首次加工上会存在注意脱离困难。类似地，LoBue 等人 (2014) 采用成人被试得出眼动结果也不支持"注意脱离困难"假设。作为固定时间分析来说，干扰物分析结论仍然需要谨慎，这也提示后续研究探讨干扰物属性的影响作用。

本研究由于幼儿被试限制，在数据采集、实验刺激上仍然存在改进地方，未来研究应该关注以下几个问题：第一，针对幼儿被试无法精确操作键盘和鼠标的情况，未来研究需要利用声音反应盒、特殊按键等设备精确记录幼儿的行为反应时间；在幼儿被试可以接受范围内，增大实验的试次。第二，恐惧性刺激诱发的情绪在其中扮演的作用 (Peira et al., 2010)。如果对蛇的觉察归结为其特殊外形，那么这种快速觉察就归结为基本的模式差别导致的反应差异。那么，蛇作为一类"恐惧动物"，它在儿童的快速觉察中有没有引发恐惧情绪？如果诱发了恐惧情绪，那么蛇的外形是否还起作用？第三，对蛇的快速觉察是否可以概括化到蜘蛛、蟑螂、狮子等其他动物恐惧中 (LoBue, 2010a; Penkunas & Coss, 2013a)。比如：以往很多研究都发现人类对蜘蛛这种特有的生物具有与蛇类似的注意偏向 (Blanchette, 2006; ?hman & Mineka, 2001)。第四，儿童经验对恐惧性刺激刺激觉察的影响。未来研究可以尝试选取更加广泛的年龄段探讨经验对蛇快速觉察的影响。第五，关于蛇的特殊外形导致了它的注意偏向和快速觉察。以往研究对蛇的颜色、生命性、体态特征、纹理等这些因素都没有很好控制，未来研究仍然需要进一步严格控制变量进行深入探讨。

参考文献

Berger, M. (2010). 'It's the sight not the bite': A model and reinterpretation of visually-based developmental fears. *Clinical Psychology Review, 30*(6), 779 - 793.

Blanchette, I. (2006). Snakes, spiders, guns, and syringes: How specific are evolutionary constraints on the detection of threatening stimuli? *The Quarterly Journal of Experimental Psychology, 59*(8), 1484 - 1504.

Bornstein, M. H., Mash, C., & Arterberry, M. E. (2011). Perception of object - context relations: Eye-movement analyses in infants and adults. *Developmental Psychology, 47*(2), 364 - 375.

Braun, K. A., Ellis, R., & Loftus, E. F. (2002). Make my memory: How advertising can change our memories of the past. *Psychology and Marketing, 19*(1), 1 - 23.

Brosch, T., & Sharma, D. (2005). The role of fear-relevant stimuli in visual search: A comparison of phylogenetic and ontogenetic stimuli. *Emotion, 5*(3), 360 - 364.

Cole, C., & Loftus, E. (1987). The memory of children. In S. Ceci, M. Toglia & D. Ross (Eds.), *Children's Eyewitness Memory* (pp. 178 - 208): Springer US.

DeLoache, J. S., & LoBue, V. (2009). The narrow fellow in the grass: Human infants associate snakes and fear. *Developmental Science, 12*(1), 201 - 207.

DeNicola, C. A., Holt, N. A., Lambert, A. J., & Cashon, C. H. (2013). Attention-orienting and attention-holding effects of faces on 4- to 8-month-old infants. *International Journal of Behavioral Development, 37*(2), 143 - 147.

Feiring, C., Lewis, M., & Starr, M. D. (1984). Indirect effects and infants' reaction to strangers. *Developmental Psychology, 20*(3), 485 - 491.

Feng, G. (2011). Eye tracking: A brief guide for developmental researchers. *Journal of Cognition and Development, 12*(1), 1 - 11.

Flykt, A., & Caldara, R. (2006). Tracking fear in snake and spider fearful participants during visual search: A multi-response domain study. *Cognition and Emotion, 20(8)*, 1075 - 1091.

Fox, E., Griggs, L., & Mouchlianitis, E. (2007). The detection of fear-relevant stimuli: *Are guns noticed as quickly as snakes? Emotion, 7*(4), 691 - 696.

Fox, E., Lester, V., Russo, R., Bowles, R. J., Pichler, A., & Dutton, K. (2000). Facial expressions of emotion: Are angry faces detected more efficiently? *Cognition and Emotion, 14*(1), 61 - 92.

Fox, E., Russo, R., Bowles, R., & Dutton, K. (2001). Do threatening stimuli draw or hold visual attention in subclinical anxiety? *Journal of Experimental Psychology: General, 130*(4), 681 - 700.

Fredrikson, M., Annas, P., & Wik, G. (1997). Parental history, aversive exposure and the development of snake and spider phobia in women. *Behaviour Research and Therapy, 35*(1), 23 - 28.

Gerdes, A. B. M., Alpers, G. W., & Pauli, P. (2008). When spiders appear suddenly: Spider-phobic patients are distracted by task-irrelevant spiders. *Behaviour Research and Therapy, 46*(2), 174 - 187.

Gredeback, G., Johnson, S., & von Hofsten, C. (2010). Eye tracking in infancy research. *Developmental Neuropsychology, 35*(1), 1 - 19.

Isbell, L. A. (2006). Snakes as agents of evolutionary change in primate brains. *Journal of Human Evolution, 51*(1), 1 - 35.

Isbell, L. A. (2009). The fruit, *the tree, and the serpent: Why we see so well*. USA: Harvard University Press.

Keil, M. F., Briassoulis, G., Nesterova, M., Miraftab, N., Gokarn, N., Wu, T. J., et al. (2013). Threat bias in mice with inactivating mutations of Prkar1a. *Neuroscience, 241*, 206 - 214.

Koster, E. H. W., Crombez, G., Van Damme, S., Verschuere, B., & De Houwer, J. (2004). Does imminent threat capture and hold attention? *Emotion, 4(3)*, 312 - 317.

Lipp, O. V., & Waters, A. M. (2007). When danger lurks in the background: Attentional capture by animal fear-relevant distractors is specific and selectively enhanced by animal fear. *Emotion, 7(1)*, 192 - 200.

LoBue, V. (2010a). And along came a spider: An attentional bias for the detection of spiders in young children and adults. *Journal of Experimental Child Psychology, 107(1)*, 59 - 66.

LoBue, V. (2010b). What's so scary about needles and knives? Examining the role of experience in threat detection. *Cognition & Emotion, 24(1)*, 180 - 187.

LoBue, V. (2013). What are we so afraid of? How early attention shapes our most common fears. *Child Development Perspectives, 7(1)*, 38 - 42.

LoBue, V., & DeLoache, J. S. (2008). Detecting the snake in the grass: Attention to fear-relevant stimuli by adults and young children. *Psychological Science, 19(3)*, 284 - 289.

LoBue, V., & DeLoache, J. S. (2010). Superior detection of threat-relevant stimuli in infancy. *Developmental Science, 13(1)*, 221 - 228.

LoBue, V., & Deloache, J. S. (2011). What's so special about slithering serpents? Children and adults rapidly detect snakes based on their simple features. *Visual Cognition, 19(1)*, 129 - 143.

LoBue, V., & Larson, C. L. (2010). What makes an angry face look so?···?angry? Examining visual attention to the shape of threat in children and adults. *Visual Cognition, 18(8)*, 1165 - 1178.

LoBue, V., Matthews, K., Harvey, T., & Stark, S. L. (2014). What accounts for the rapid detection of threat? Evidence for an advantage in perceptual and behavioral responding from eye movements. *Emotion, 14(4)*, 816 - 823.

LoBue, V., Rakison, D. H., & DeLoache, J. S. (2010).

Threat perception across the life span: Evidence for multiple converging pathways. *Current Directions in Psychological Science, 19(6)*, 375 - 379.

Loftus, E. F., & Davies, G. M. (1984). Distortions in the memory of children. *Journal of Social Issues, 40(2)*, 51 - 67.

Murray, E. J., & Foote, F. (1979). The origins of fear of snakes. *Behaviour Research and Therapy, 17(5)*, 489 - 493.

Oakes, L. M. (2012). Advances in eye tracking in infancy research. *Infancy, 17(1)*, 1 - 8.

Öhman, A. (2009). Of snakes and faces: An evolutionary perspective on the psychology of fear. *Scandinavian Journal of Psychology, 50(6)*, 543 - 552.

Öhman, A., Flykt, A., & Esteves, F. (2001). Emotion drives attention: Detecting the snake in the grass. Journal of Experimental Psychology: *General, 130(3)*, 466 - 478.

Öhman, A., Lundqvist, D., & Esteves, F. (2001). The face in the crowd revisited: A threat advantage with schematic stimuli. *Journal of Personality and Social Psychology, 80(3)*, 381 - 396.

Öhman, A., & Mineka, S. (2001). Fears, phobias, and preparedness: Toward an evolved module of fear and fear learning. *Psychological Review, 108(3)*, 483 - 522.

Öhman, A., & Mineka, S. (2003). The malicious serpent: Snakes as a prototypical stimulus for an evolved module of fear. *Current Directions in Psychological Science, 12(1)*, 5 - 9.

Öhman, A., Soares, S. C., Juth, P., Lindstr?m, B., & Esteves, F. (2012). Evolutionary derived modulations of attention to two common fear stimuli: Serpents and hostile humans. *Journal of Cognitive Psychology, 24(1)*, 17 - 32.

Peira, N., Golkar, A., Larsson, M., & Wiens, S. (2010). What you fear will appear: Detection of schematic spiders in spider fear. *Experimental Psychology, 57(6)*,

470 - 475.

Peltola, M. J., Lepp?nen, J. M., Vogel-Farley, V. K., Hietanen, J. K., & Nelson, C. A. (2009). Fearful faces but not fearful eyes alone delay attention disengagement in 7-month-old infants. *Emotion, 9*(4), 560 - 565.

Penkunas, M. J., & Coss, R. G. (2013a). A comparison of rural and urban Indian children's visual detection of threatening and nonthreatening animals. *Developmental Science, 16*(3), 463 - 475.

Penkunas, M. J., & Coss, R. G. (2013b). Rapid detection of visually provocative animals by preschool children and adults. *Journal of Experimental Child Psychology, 114*(4), 522 - 536.

Phelps, E. A., Ling, S., & Carrasco, M. (2006). Emotion facilitates perception and potentiates the perceptual benefits of attention. *Psychological Science, 17*(4), 292 - 299.

Purkis, H. M., & Lipp, O. V. (2009). Are snakes and spiders special? Acquisition of negative valence and modified attentional processing by non-fear-relevant animal stimuli. *Cognition & Emotion, 23*(3), 430 - 452.

Quinlan, P. (2013). The visual detection of threat: A cautionary tale. *Psychonomic Bulletin & Review, 20*(6), 1080 - 1101.

Rakison, D. H., & Derringer, J. (2008). Do infants possess an evolved spider-detection mechanism? *Cognition, 107*(1), 381 - 393.

Reynolds, M. G., Eastwood, J. D., Partanen, M., Frischen, A., & Smilek, D. (2009). Monitoring eye movements while searching for affective faces. *Visual Cognition, 17*(3), 318 - 333.

Shibasaki, M., & Kawai, N. (2009). Rapid detection of snakes by Japanese monkeys (Macaca fuscata): An evolutionarily predisposed visual system. *Journal of Comparative Psychology, 123*(2), 131 - 135.

Soares, S. C., Esteves, F., Lundqvist, D., & ?hman, A. (2009). Some animal specific fears are more specific than others: Evidence from attention and emotion measures. *Behaviour Research and Therapy, 47*(12), 1032 - 1042.

Treisman, A., & Gormican, S. (1988). Feature analysis in early vision: Evidence from search asymmetries. *Psychological Review, 95*(1), 15 - 48.

Vaish, A., Grossmann, T., & Woodward, A. (2008). Not all emotions are created equal: The negativity bias in social-emotional development. *Psychological Bulletin, 134*(3), 383 - 403.

Waters, A. M., Lipp, O., & Randhawa, R. (2011). Visual search with animal fear-relevant stimuli: A tale of two procedures. *Motivation and Emotion, 35*(1), 23 - 32.

Waters, A. M., Lipp, O., & Spence, S. H. (2008). Visual search for animal fear-relevant stimuli in children. *Australian Journal of Psychology, 60*(2), 112 - 125.

Wolfe, J. M., Yee, A., & Friedman-Hill, S. R. (1992). Curvature is a basic feature for visual search tasks. *Perception, 21(4)*, 465 - 480.

Children's attention detection to snakes: Evidence from eye movements

WANG Fuxing[1]; LI Wenjing[1];YAN Zhiqiang[2]; DUAN Zhaohui[1]; LI Hui[3]

([1] School of Psychology, Central China Normal University, Wuhan 430079 China)

([2] Department of Psychology, Peking University, Beijing 100871 China)

([3] Academy of Psychology and Behavior, Tianjin Normal University, Tianjin 300074 China)

Abstract　Previous research shows that preschool children detect snakes quickly than non-threating stimuli (e.g. flowers). In this study, we used eye tracking technology to provide direct evidences about the superior detection about threat-relevant stimuli. Two experiments were designed to testify whether the snakes would be fixated faster and quickly by preschool children and adults. In addition, we also used line drawing snakes and flowers as stimuli to control the shape of snakes and to testify the perceptual template hypothesis. In experiment 1, sixteen 4- to 6-year-old preschool children and 22 undergraduates were recruited as participants. A revised 3 × 3 matrices of color photographs of threat-relevant (snakes) and threat-irrelevant (flowers) stimuli were presented to both preschool children and adults. All participants were asked to find the threat target (snake) among seven non-threat distractors (flowers) and vice versa. Sixteen matrices with 8 pictures (1 target and 7 distractors) were presented to the participants. We changed the standard visual search task that did not present stimuli in the middle of the 3 × 3 matrices to control the central location effect and make the procedure appropriate for eye tracking calibration. It's a 2 (age: children, adults)× 2 (target: snake, flower) mixed design, and age was the between subject variable. In experiment 2, we improved the stimuli with line drawings to pop out the continuous curvilinear contour of snakes. The design, presentation method was the same as experiment 1. In two experiments, Tobii T120 Eye tracker was used to record the viewing behaviors of adults and children. The results of experiment 1 indicated that both the preschool children and adults fixated snakes faster and with less fixation counts than flowers, and their first fixation duration was shorter on snakes than flowers. Adults performed faster fixation, much less fixation counts than children. As distractors (flower was target), snakes were also fixated quickly than flowers (snake was target). In experiment 2, the same results were found that both children and adults located line drawing snakes quicker than line drawing flowers. And the first fixation duration to snakes were much shorter than flowers. Adults still fixated faster than children. For the snakes as distractors, we found the same results as experiment 1. Compared to the fixations of the line drawing snakes without color and pattern in experiment 2, real and colorful snakes in experiment 1 were fixated faster. The real and colorful snakes were detected faster and with less fixation counts before they were located as distractors. In conclusion, even preschool children who have little snake experience also show faster attention orienting and shorter attention holding. Based on the eye movements evidences, the continuous curvilinear shape plays an important role in the snake relevant threat detection. The eye fixations of line drawing snakes provide direct evidence to the perceptual template theory. To be a special reptile, the color, patterns can boost the quick detection.

Keywords snakes; threat–relevant stimuli; preschool children; detection; eye movements

心理发展与教育 , 2014, 4, 403 - 410.

装饰图片影响多媒体学习的眼动研究 *

龚少英 [1,2]，段 婷 [1,2]，王福兴 [1,2]，周宗奎 [1,2]，卢春晓 [1,2]

（[1] 青少年网络心理与行为教育部重点实验室 武汉 430079）

（[2] 华中师范大学心理学院 武汉 430079）

摘 要 为探讨装饰图片对多媒体学习效果和认知加工过程的影响，本研究采用眼动仪追踪 30 名低知识经验大学生在有装饰图片和无装饰图片条件下学习多媒体课件的视觉注意过程。结果发现：（1）有装饰图片组的保持和迁移成绩显著低于无装饰图片组；（2）有装饰图片组被试在认知兴趣图的注视次数、文本与认知兴趣图之间的注意转换次数显著少于无装饰图片组；（3）在装饰图片组内，80% 学习者报告被装饰图片吸引，并回忆出与装饰图片有关的先前知识经验。这些结果表明，装饰图片干扰学习者对主要学习内容的记忆与理解；装饰图片可能主要通过干扰学习者对主要学习内容的一致性理解以及激发不恰当的先前知识经验而阻碍学习。

关键词 多媒体学习；装饰图片；眼动

1 问题提出

装饰材料 (seductive details) 是指学习材料中插入的兴趣水平高、与学习主题有关联，但与学习目标无关的内容 (Garner, Gillingham, & White, 1989)。相对主要学习内容而言，装饰材料容易唤起学习者的兴趣（高兴趣水平），但它提供了与当前学习目标无关的信息，是在当前目标学习中不重要的内容（低重要性）(Lehman, Schraw, McCrudden, & Hartley, 2007)。

以往研究在关于装饰材料是否影响学习效果上存在很大争议。多数研究发现装饰材料干扰了学习 (Garner, et al., 1989; Harp & Mayer, 1998; Lehman, et al.,

2007; Peshkam, Mensink, Putnam, & Rapp, 2011; Sanchez, Wiley, 2006；Myayer, Heiser & Lonn, 2001)， 有 装 饰材料组学生在主要内容的回忆成绩和理解成绩上显著低于无装饰材料组。然而，也有一些研究 (Goetz & Sadoski, 1995; Sadoski, Goetz, & Fritz, 1993; Sanchez, Wiley, 2006) 发现装饰材料促进了学习，装饰图片能吸引学生更多地关注主要学习内容，进而加强对主要学习内容的记忆与理解。同时，也有研究发现，装饰材料对学习成绩并无显著影响 (Lusk, 2008)。关于装饰材料影响学习成绩的研究存在争议的主要原因可能是：（1）装饰材料的界定不一致。大部分研究 (Lehman, et al., 2007; Wade, Schraw, Buxton, & Hayes,

* 基金项目: 教育部人文社科项目（13YJA190005），国家自然科学基金青年项目（31300864），中央高校基本科研业务费协同创新重大项目：基于教育信息服务的社交网络建构关键技术研究

通讯作者: 龚少英 gongsy_psy@163.com; 周宗奎 zhouzk@mail.ccnu.edu.cn

1993) 根据装饰材料的兴趣水平和重要性水平两个显著特征界定装饰材料，发现装饰材料干扰了学习。也有研究 (Lusk, 2008) 未按照装饰材料的特征严格界定装饰材料，发现装饰材料对学习无阻碍作用。（2）装饰材料的类型和呈现方式不一致。装饰材料干扰学习的情形主要为：装饰文本插入解释性文本或动画中 (Lehman, et al., 2007)、装饰文本和装饰图片同时插入解释性文本中 (Harp & Mayer, 1998)；而装饰材料促进学习或者不影响学习的主要情形为：装饰文本插入描述性的文本中 (Peshkam, et al., 2011)、装饰声音加入动画教学材料中 (Thalheimer, 2004)。（3）学习者的个体差异。工作记忆容量小或知识经验少的学习者其多媒体学习效果更易受装饰性材料干扰（Sanchez & Wiley, 2006; Magner, Schwonke, Aleven, Popescu & Renkl, 2012）。

多媒体学习的认知理论为解释装饰材料对多媒体学习效果的影响提供了理论基础。首先，根据多媒体学习认知理论的通道容量有限假设，每个通道在同一时间加工的信息数量有限。如果呈现的不同类型材料（如文字和图片）占用同一通道，可能导致容量超限，对学习产生干扰作用（Mayer, 2001）。与此同时如果学习过程中呈现与学习目标无关的内容，或者学习材料的呈现方式不当，将导致外部认知负荷增加，可用于加工学习内容和促进知识整合的相关认知负荷减少，从而对学习产生阻碍作用 (Mayer, 2001)。前述发现装饰材料干扰了文本学习的研究中，研究者将装饰文本或装饰文本加装饰图片插入解释性文本 (Lehman, et al., 2007; Harp & Mayer, 1998)，这些装饰材料干扰多媒体学习的一个可能原因是装饰文本或装饰文本加装饰图片和要学习的目标文本都占用视觉通道，导致通道容量超载，干扰了学习。此外，工作记忆容量小的学习者其学习容易受到装饰性材料干扰（Sanchez, Wiley, 2006），可能也与装饰材料占用了其有限的资源，导致相关认知负荷减小有关。

其次，多媒体学习的认知理论提出有意义的学习包括选择、组织和整合三个主动加工阶段 (Mayer, Heiser, & Lonn, 2001)，为研究者探查装饰材料影响

多媒体学习的过程提供了理论基础。当学习内容以多媒体方式呈现时，学习者需要进行主动加工，包括选择相关的信息、将所选择的信息组织成一致的心理表征，最后将心理表征和已有知识经验整合成一个整体。为了探讨装饰材料具体影响学习的哪个阶段，研究者 (Harp & Mayer, 1998; Lehman, et al., 2007) 提出了相对应的三种假设：（1）注意减少假设 (reduced attention hypothesis)，是指装饰材料吸引学生更多注意，减少对主要学习内容的注意和阅读，从而导致对主要内容的记忆成绩下降；（2）一致性中断假设 (coherence break hypothesis) 是指装饰材料的加入导致学习者对学习材料的一致性理解中断，因而导致理解成绩下降；（3）注意转移假设 (diversion hypothesis)，后又被称为不恰当图式激活假设 (inappropriate schema hypothesis)，是指装饰材料会激活与当前的主要学习内容不太匹配的先前知识，阻碍学习者在新旧知识间建立恰当的联系，影响学习者对学习内容的全面加工，从而干扰记忆与理解成绩。关于装饰材料影响多媒体学习的上述三个假设都得到了研究证据的支持。Harp 和 Mayer(1998) 的研究主要采用间接测量的方法，通过学习成绩来推测装饰材料对学习中的认知加工过程的影响，其结果支持注意转移假设。Mayer 等人 (2008) 的研究通过探讨不同兴趣水平的装饰材料对学习成绩的影响间接探讨装饰材料对多媒体学习认知加工阶段的影响，研究结果支持一致性中断假设。为了更直接地检验三个研究假设，Lehman 等人 (2007) 在 Harp 和 Mayer(1998) 研究的基础上加入了测量每个句子的阅读时间这一客观的指标，结果发现，有装饰材料组对主要内容的阅读时间显著少于无装饰材料组，且对主要内容的回忆成绩和理解成绩均显著低于无装饰材料组，结果支持注意减少假设和一致性中断假设。

前述关于装饰材料影响多媒体的研究主要采用间接测量法探查了装饰材料对多媒体学习的影响，但对于装饰材料如何影响多媒体学习的过程还没有得到明确的结论，本研究欲采用直接的眼动记录法和间接的学习效果测量法，探查装饰材料对多媒体

学习过程及效果的影响 (Hyona, 2010)，揭示装饰材料影响多媒体学习效果的机制。眼动追踪研究为检验多媒体学习的认知理论提供了独特的方法 (Mayer, 2010)，可以为研究者揭示多媒体学习中的认知过程提供直接的证据。如总的眼睛注视时间反映被试对多媒体学习中的各部分的注意（Hyona, 2010），注视次数反映学习材料的认知加工负荷（闫国利等，2013），在文本和图片之间的眼动转换次数反映正在整合和理解所阅读的内容 (Holsanova, Holmberg, & Holmqvist, 2009)，使用被试眼动轨迹作为线索要求被试回忆当时所思所想来探查眼动背后的认知加工过程 (De Koning, Tabbers, Rikers, & Paas, 2010; Scheiter, Gerjets & van Gog, 2010) 等等。本研究欲将学习者在多媒体学习中的眼动轨迹、眼动轨迹回放报告和学习结束后的学习效果测量结合起来，揭示装饰图片对多媒体学习过程和学习效果的影响。本研究采用的装饰材料严格按其操作定义进行界定，满足高兴趣水平和低重要性两个标准。其次，装饰材料只使用单一的装饰图片，而不是混合的装饰材料。这主要是考虑到在多媒体学习中，采用图片作为装饰材料是一种典型的情况。因此，本研究先从单一的装饰图片类型入手进行研究。本研究中的眼动指标包括总的注视时间、注视次数、注视点比率、文本和认知兴趣图之间的眼跳次数等。本研究将从两方面对已有研究进行扩展：一是将直接的眼动追踪和间接的学习结果测量结合起来，既可以揭示在一定条件下装饰材料是否影响多媒体学习的效果，更可以揭示装饰材料是如何通过影响多媒体学习过程，从而影响最终学习效果的；二是使用眼动轨迹回放报告可以为解释眼动过程背后的认知机制提供证据。

根据前人研究和多媒体学习的认知理论，装饰材料虽然可以激发学习者的兴趣，但可能增加低知识经验者的外部认知负荷，影响其多媒体学习加工过程，从而影响多媒体学习的效果，因此，本研究提出如下假设：（1）装饰图片组在主要内容（文本区、认知兴趣图区）上的总注视时间、注视次数和注视比率以及在文本区与认知兴趣图间的眼跳次

数将显著低于无装饰图片组；（2）在装饰图片组内，大部分被试将在学习后报告回忆出与装饰图片有关的先前知识经验。本研究采用眼动轨迹回放报告 (cued retrospective reporting) 的方式来验证此假设；（3）无装饰图片组的学习效果将好于有装饰图片组。

2 方法

2.1 被试

被试为某师范大学学生 30 名，其中男生 13 人，女生 17 人，平均年龄为 20.60 ± 1.67。被试的视力或矫正视力正常，无色盲色弱。根据前测（闪电形成原理的经验测验）结果，选择先前知识经验少（前测少于 10 分，满分 35 分）的被试，随机分配到两种实验条件下：无装饰图片组，有装饰图片组。

2.2 材料

先前知识经验测验是修订 Harp 和 Mayer(1997) 研究中的测验内容，共包括 4 个关于气象学知识熟悉度的主观评定题和 1 个关于闪电形成原理的问答题。其中，主观评定采用 0（完全不符合）到 4（完全符合）点评价，满分 16 分；问答题为："请写下你所知道的或想到的关于闪电形成的原理（即闪电是如何形成的），19 个计分点，答对一个计 1 分"。

学习材料：翻译并修订 Harp 和 Mayer(1997) 研究材料"闪电的形成过程与原理"，制成 PPT 课件的形式，共 5 页。其中，无装饰图片组学习材料仅有文本和认知兴趣图（重要性和兴趣水平都高的图片），有装饰图片组学习材料包括文本、认知兴趣图和装饰图片（兴趣水平高重要性低的图片）。学习时间为 5 分钟。

后测内容依次为：算术题、心理努力评价、认知负荷评价、再认测验、保持测验、迁移测验、眼动轨迹回放报告。保持测验和迁移测验题目是通过翻译并修订 Harp 和 Mayer(1997) 研究中采用的内容。心理努力和认知负荷采用 9 点计分量表进行测试，要求被试报告在学习课件时的努力程度，或者报告学习

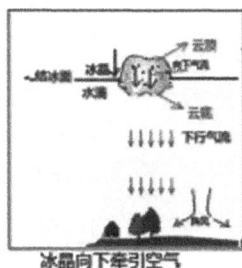

第2页

当云顶端的高度超过结冰面所在高度时，云顶水滴变成微小的冰晶。（在结冰面及其以上的高度，温度低于零度，水能迅速变成冰晶。）当形成的冰晶越来越大，无法漂浮在云顶时，从云顶掉下来。向下掉的冰晶从云层内牵引出部分空气，产生向下运动的气流。

a（无装饰图片）

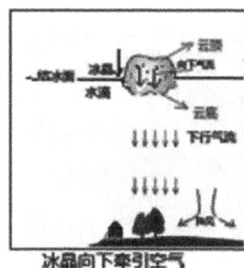

第2页

当云顶端的高度超过结冰面所在高度时，云顶水滴变成微小的冰晶。（在结冰面及其以上的高度，温度低于零度，水能迅速变成冰晶。）当形成的冰晶越来越大，无法漂浮在云顶时，从云顶掉下来。向下掉的冰晶从云层内牵引出部分空气，产生向下运动的气流。

b（有装饰图片）

图1　学习材料示例（图的上部为文本区，左下图为认知兴趣图，右下图为装饰图）

内容的难度，分值越高，表示越努力或认知负荷越大。再认测验是7个多选题，答案完全来自所学课件；保持测验是让学生写出所记住的与闪电有关的所有内容；迁移测验是4个需要应用闪电原理来解决的问题，例如，如何才能降低闪电的强度。

前测和后测的计分：根据以往研究 (Harp & Mayer, 1998; Schmidt-Weigand, Kohnert, & Glowalla, 2010) 的评分标准进行修订后，保持测验内容为学习材料中的重要知识点，共有19个知识点，每回忆出1个知识点计1分。迁移成绩的评分则根据学习者回答的正确数目计分，每答对1点得2分，答错不扣分。再认测验的评分标准按学习者选择了多少正确答案计分，每选对一个选项计1分。所有成绩的评分均由两位经过严格培训并对评分标准非常熟悉的心理学研究生担任，评分者的一致性系数均在0.95以上。

为确保装饰图片符合装饰材料的操作定义，以及所选的学习材料难度适中，在正式实验之前，本研究另招募了20名被试进行预实验，要求学生在多媒体学习后对各部分学习材料（文本、认知兴趣图、装饰图片）的重要性（从1~9表示从一点也不重要至非常重要）和兴趣（从1~9表示从非常枯燥至非常有趣）进行评定。评定结果为：被试对装饰图片的兴趣程度显著高于文本 ($p<0.001$) 和认知兴趣图 ($p<0.05$)，对认知兴趣图的兴趣程度显著高于文本区的兴趣程度 ($p<0.001$)；而在重要性程度上装饰图片显著低于文本 ($p<0.001$) 和认知兴趣图 ($p<0.001$)，而文本区和认知兴趣图在重要性上无显著差异 ($p>0.05$)，表明所选装饰图片满足装饰材料的操作定义。此外，预实验发现学习者学习每页PPT的时间约为60秒，因此在正式实验中将学习每页PPT的时间定为1分钟，以确保所有学生均能阅读完学习内容。

2.3 设计

本实验为单因素实验设计，自变量为装饰图片的类型（无装饰图片、有装饰图片），因变量为心理努力程度、认知负荷、再认测验成绩、保持测验成绩、迁移测验成绩以及眼动指标。其中具体的眼动指标为：兴趣区的总注视时间、注视次数、注视比率、兴趣区间眼跳次数与次序。

2.4 仪器与程序

本实验采用的仪器为 EyeLink 1000（SR Research, Canada) 眼动仪，采样率为250 Hz。19英寸显示器，分辨率为 1280×1024。实验材料的水平视角为28.7度，垂直视角15.3度，被试眼睛与屏幕之间的距离为75cm。

实验开始前，对被试进行9点校准。然后，呈现实验指导语。在确认被试已熟悉实验任务并做好了

表 1　有无装饰图片组在后测成绩上的差异分析

	无装饰图片组 (n=13)		有装饰图片组 (n=15)		t	d
	M	SD	M	SD		
再认成绩	12.46	1.33	11.53	2.42	1.23	0.47
保持成绩	10.73	3.01	6.73	1.51	4.53**	1.68
迁移成绩	8.38	2.48	6.43	1.72	2.44*	0.91
心理努力	7.92	1.12	7.53	1.46	0.8	0.30
难度	4.15	1.41	4.80	1.66	−1.1	0.42

注：* $p<0.05$，** $p<0.01$，*** $p<0.001$，下同。

实验准备后，让被试开始学习 PPT 课件。学习结束后，先进行 5 分钟算术题测验，以排除短时记忆对学习成绩的影响。然后依次进行心理努力与认知负荷评价、再认测验、保持测验、迁移测验，最后要求有装饰图片组的被试根据回放的眼动轨迹报告学习过程中的思维内容。

2.5 结果

2.5.1 后测成绩分析

正式实验中有两名被试由于实验过程中头部移动程度较大，导致眼动追踪中断，剔除了这两名被试的所有数据。有、无装饰图片组的后测成绩差异分析结果（表 1）发现，无装饰图片组的保持成绩和迁移成绩显著高于有装饰图片组，但两组被试的再认成绩无显著差异。

2.5.2 心理努力和认知负荷

有、无装饰图片组被试的心理努力程度不存在显著差异，且两组被试的心理努力平均值大于 7 分，表明两组被试的心理努力均较高；有、无装饰图片组在学习难度上不存在显著差异，两组被试在任务难度的均值在 4-5 分之间，表明学习内容难度中等偏低。

2.5.3 有无装饰图片组在眼动指标的差异分析

对眼动数据进行分析前，先将每一页学习内容划分为三个兴趣区，分别是文本区、认知兴趣图、装饰图片。眼动数据分析结果见表 2。有、无装饰图片组在文本区和认知兴趣图的注视时间以及注视比率上均不存在显著差异。无装饰图片组在认知兴趣图的注视次数显著多于有装饰图片组，但在文本区的注视次数无显著差异。无装饰图片组在从文本区到认知兴趣图的眼跳次数 [$t(26)=3.38$，$p<0.01$，$d=1.27$] 以及从认知兴趣图到文本区的眼跳次数上 [$t(26)=3.46$，$p<0.01$，$d=1.30$] 都显著高于有装饰图片组。

2.5.4 眼动轨迹回溯报告结果分析

根据已有研究 (Harp & Mayer, 1998; Lehman, et al., 2007) 中的注意转移假设（或不恰当图式激活假设）：装饰材料激活了学生先前知识经验中与之有关的内容而阻碍学习。本研究将学习时被试回忆出与装饰图片有关而与闪电形成原理无关内容的现象界定为发生了注意转移，反之则为无注意转移。正式实验中，有装饰图片组的 15 名被试均参与了眼动轨迹回溯报告，其中有 12 人 (80%) 报告被装饰图片吸引，并回忆出与装饰图片有关的先前知识经验，有 3 人 (20%) 报告装饰图片与文字没什么关联，并未进行过多思考。

表 2　有、无装饰图片组在眼动指标上的差异分析结果

兴趣区	眼动指标	无装饰图片组		有装饰图片组		t	d
		M	SD	M	SD		
文本区	总注视时间 (ms)	37295	5250	37045	4950	0.13	0.05
	注视次数	155.83	24.38	145.83	26.40	1.04	0.39
	注视点比率	0.73	0.10	0.72	0.08	0.54	0.11
	文本向认知兴趣图的眼跳次数	5.28	2.24	2.69	1.80	3.38**	1.27
认知兴趣图	总注视时间 (ms)	12349	4332	10525	4322	1.12	0.42
	注视次数	54.85	22.81	38.91	15.26	2.20**	0.82
	注视点比率	0.25	0.10	0.19	0.07	1.83	0.69
	认知兴趣图向文本的眼跳次数	6.00	2.20	3.33	1.88	3.46**	1.30

3 讨论

3.1 装饰图片对多媒体学习效果的影响

本研究发现,有装饰图片组在保持成绩和迁移成绩上显著低于无装饰图片组,但在再认成绩上没有显著差异。两组被试在保持的差异效应量最大,在再认的差异效应量最小。这些结果部分验证了装饰图片对学习有阻碍作用的预期。这与以往的部分研究结果一致 (Harp & Mayer, 1997, 1998; Lehman, et al., 2007; Moreno & Mayer, 2000),支持装饰图片对学习有干扰作用的假设。装饰图片不仅阻碍了学习者对主要学习内容的保持记忆,同时也阻碍了学习者对重要内容的理解和应用。装饰图片对多媒体学习的不同方面干扰作用不同,表明装饰图片很可能对简单的再认任务没有干扰作用,而对需要理解和应用的任务有显著干扰。相对来说,再认任务只需从四个答案中选择在文本中出现的信息,不需要深度加工,而保持任务需要被试在没有线索的情况下回忆出学习过的内容,迁移任务则需要被试将学习过的原理应用于新情境,需要将已学原理整合到已有的图式中,并提取出来加以运用,两者都需要学习者对所学内容进行深度加工。这些结果表明装饰图片可能主要影响学习者对内容的理解、保持和应用,而不是简单的再认学习。

但本研究结果与 Park 等人(2011)按照高兴趣水平和低重要性两个特征定义的装饰材料影响多媒体学习的研究结果不一致。在 Park 等人的研究中,装饰材料为装饰文本加装饰图片,装饰材料只在 11 页多媒体课件中的 4 页出现,被试评定的认知负荷为中等,学习效果的测量任务是保持、迁移、问题解决任务的混合,结果发现在主要学习内容为文本加认知兴趣图条件下有无装饰材料组的学习结果无显著差异。本研究和 Park 等的研究在装饰材料类型、装饰材料的呈现频率(是否在每页出现)和学习结果测量方式不同可能是造成这种差异的一个原因。其次,不管有无装饰材料,Park 等人研究中的被试在解说加认知兴趣图条件下的认知负荷显著高于文本条件

(两种条件下的认知负荷皆为中等水平),而且在解说和认知兴趣图加装饰材料条件下的学习效果显著好于其他三种条件(文本和认知兴趣图加装饰材料、文本和认知兴趣图、解说和认知兴趣图),出现了装饰材料逆转效应。这一方面表明解说和认知兴趣图加装饰材料条件下的积极学习效果伴随着更高水平的认知负荷,意味着认知负荷的适度增加并不必然对学习效果产生干扰作用;另一方面从不同通道呈现主要学习内容和装饰材料可能会减少装饰材料的消极作用,甚至起到促进作用。综合本研究和前人研究,装饰材料对多媒体学习的影响可能受到任务难度、装饰材料类型、装饰材料的呈现次数安排、主要内容和装饰材料的呈现通道以及个体因素等多方面因素的影响。装饰材料因素如何和其他因素相互作用影响多媒体学习的过程和效果还有待进一步深入系统的研究探查。

3.2 装饰图片对多媒体学习过程的影响

本研究采用眼动追踪技术实时记录学习者进行多媒体学习时的认知加工模式,结果发现,有、无装饰图片组在文本区的总注视时间、注视次数和注视比率以及在认知兴趣图的注视时间和注视比率上均不存在显著差异,仅在认知兴趣图的注视次数上,无装饰图片组显著高于有装饰图片组。因此,本研究结果并不完全支持注意减少假设。可能装饰图片本身只需很少的注意努力就能使学习者在短时内获得整体理解 (Shirey & Reynolds, 1988),因而并不会占用过多的阅读时间和注视次数。学习者在有、无装饰图片条件下的心理努力和难度评价都没有显著差异,表明装饰图片的呈现没有给学习者增加显著的认知负荷,也为上述装饰图片可能只需要很少注意这一推论提供了额外支持。这一研究结果与 Mayer 等人 (Harp & Mayer, 1998; Mayer, Griffith, Jurkowitz, & Rothman, 2008) 的结果一致,即装饰材料并不是通过减少对主要内容的注意而干扰学习;但与 Lehman 等人 (2007) 的研究结果(有装饰文本组对主要内容的阅读时间显著少于无装饰文本组)矛盾。主要原因可能是 Lehman 等人 (2007) 的研究

采用的是装饰文本，发现学习有装饰文本的学习者对基本文本的注意和阅读时间减少。而与文字不同的是，图片在获取信息上具备独特的高效性和丰富性（Dansereau & Simpson, 2009; 沈德立 & 陶云, 2001），装饰图片相对于装饰文本而言在短时间内容易理解且可以获得整体信息而不需要花费过多注意来精加工，因而装饰图片并未导致对主要内容的注视时间显著减少。

虽然两组被试在文本区的注视时间、注视次数和注视点比率无显著差异，但和认知兴趣图相比，两组被试都在文本区有更多的注视次数、注视时间和注视点比率。这表明学习者优先注意文本而不是图片，与以往研究发现的文本优先趋势是一致的（Schmidt – Weigand, Kohnert, & Glowalla, 2010）。

眼动数据分析发现无装饰图片组在文本区与认知兴趣图之间的眼跳次数显著高于有装饰图片组。被试在文本与图片之间的眼跳（又称为注意转换）反应了被试试图在语义上连接文本和图形信息形成整合性理解的程度 (Holsanova, et al., 2009)。这一结果表明在无装饰图片的条件下，学习者可能更多地在文本与认知兴趣图上进行组织和深加工，而当遇到装饰图片时，学习者对主要学习内容的整合加工次数显著减少，从而导致保持和理解成绩下降，这一发现支持一致性中断假设。

眼动轨迹回放报告分析结果表明，有80%的学习者（12名被试）在看到装饰图片时想起了与图片有关的内容（例如：有的学习者看到闪电中的自由女神像，当时的心理活动是"自由女神像在哪个国家？建筑师是谁？"）。研究发现，仅有20%的学习者（3名被试）报告"仅仅是瞟了一眼，发现有趣但与学习任务无关，于是没有详细浏览"，即并没有激活与装饰图片有关的先前知识经验。这一方面说明，装饰图片是否对多媒体学习产生影响与学习者的抑制干扰能力有关。装饰图片对学习的影响程度存在个体差异，学习者在学习过程中的自主调节学习影响了装饰图片对学习的作用方向与程度。更重要的是，在外界材料足够吸引学习者的情境下，大部分

学习者被装饰图片所吸引，并回忆起了与装饰图片有关的内容。可见，总的来看，装饰图片可能激发了学习者先前知识经验中相关的内容而干扰学习。眼动轨迹回放报告的结果支持不恰当图式激活假设。综上，本研究结果主要支持一致性中断假设与不恰当图式激活假设，表明装饰图片对多媒体学习的干扰主要发生在组织和整合阶段。对于装饰图片是否影响多媒体学习中的选择过程、什么样的学习者（不）会产生注意转移，以及认知负荷增加对学习效果的影响受到哪些因素的调节等问题还有待未来进一步的研究探查。

本研究将直接的眼动轨迹和间接的学习表现结合起来探查装饰图片对多媒体学习过程和效果的影响，揭示了装饰图片影响多媒体学习过程中的组织和整合阶段，表明装饰图片对多媒体学习的不同阶段产生阻碍作用，为多媒体学习认知理论解释装饰材料效应提供了基于过程和结果的证据，这是本研究的一个特色。另一个特色是本研究将实时的眼动轨迹指标和事后的眼动轨迹回放报告结合起来，初步揭示了在包含有装饰图片的多媒体学习中眼动背后的认知加工过程。但是，本研究也存在一些局限。首先，本研究的结果来自于先前知识经验较少的被试，学习材料的难度中等，对于具有较多先前知识经验的学习者，或者学习不同难度的内容，情况是否也如此，还有待进一步探查。其次，虽然本研究揭示了对于低先前知识经验的学习者来说，在学习内容中插入装饰图片对其学习过程和效果产生了干扰，但未来还有待进一步探查不同类型装饰材料（如文本、声音、图片、动画）或不同类型装饰材料的组合对多媒体学习的影响，深入系统地揭示装饰材料影响多媒体学习的机制。

4 结论

（1）本研究发现在以科学主题为学习内容的多媒体学习情境中，装饰图片阻碍学习者对主要学习内容的记忆与理解；

（2）装饰图片主要通过干扰低知识经验学习者对主要学习内容的一致性理解以及激发学习者不恰当的先前知识经验而阻碍学习者对主要学习内容的记忆与整合性理解。

参考文献

沈德立, & 陶云. (2001). 初中生有无插图课文的眼动过程研究. 心理科学, 24(4), 385－389.

闫国利, 熊建萍, 臧传丽, 余莉莉, 崔磊, 白学军（2013）。阅读研究中的主要眼动指标评述。心理科学进展, 21（4）, 589-605

Dansereau, D. F., & Simpson, D. D. (2009). A picture is worth a thousand words: The case for graphic representations. *Professional Psychology: Research and Practice, 40*(1), 104－110.

De Koning, B. B., Tabbers, H. K., Rikers, R. M. J. P., & Paas, F. (2010). Attention guidance in learning from complex animation: Seeing is understanding? *Learning and Instruction, 20*(2), 111－122.

Garner, R., Gillingham, M. G., & White, C. S. (1989). Effects of "seductive details" on macroprocessing and microprocessing in adults and children. *Cognition and Instruction, 6*, 41－57.

Goetz, E. T., & Sadoski. (1995). The perils of seduction: Distracting details or incomprehensible abstractions? *Reading Research Quarterly, 30*(3), 500－511.

Harp, S. F., & Mayer, R. E. (1997). The role of interest in learning from scientific text and illustrations: On the distinction between emotional and cognitive interest. *Journal of Educational Psychology, 89*, 92－102.

Harp, S. F., & Mayer, R. E. (1998). How seductive details do their damage: A theory of cognitive interest in science learning. *Journal of Educational Psychology, 90*(3), 414－434.

Holsanova, J., Holmqvist, K., & Holmberg, N. (2009). Reading information graphics: The role of spatial contiguity and dual attentional guidance. *Applied Cognitive Psychology, 23*(9), 1215－1226.

Hyona, J. (2010). The use of eye movements in the study of mutimedia learning. *Learning and Instruction, 20,*(2), 172-176

Lehman, S., Schraw, G., McCrudden, M. T., & Hartley, K. (2007). Processing and recall of seductive details in scientific text. *Contemporary Educational Psychology, 32*(4), 569－587.

Lusk, D. L. (2008). The effects of seductive details and segmentation on interest, recall and transfer in a multimedia learning environment. Doctoral *Dissertation. Virginia Polytechnic Institute and State University.*

Magner, U. I. E, Schwonke, R., Aleven, V., Popescu, O., Renkl, A. (2014). Triggering situational interest by decorative illustrations both fosters and hinders learning in computer-based learning enviornments. *Learning and Instruction, 29*, 141－152

Mayer, R. E. (2001). Multimedia learning. New York: *Cambridge University Press.*

Mayer, R. E., Griffith, E., Jurkowitz, I. T. N., & Rothman, D. (2008). Increased interestingness of extraneous details in a multimedia science presentation leads to decreased learning. *Journal of Experimental Psychology: Applied, 14*(4), 329－339.

Mayer, R. E., Heiser, J., & Lonn, S. (2001). Cognitive constraints on multimedia learning: When presenting more material results in less understanding. *Journal of Educational Psychology, 93*(1), 187－198.

Mayer, R. E.(2010). Unique contributions of eye-tracking research to the study of learning with graphics. *Learning and Instruction, 20*, 167-171

Moreno, R., & Mayer, R. E. (2000). A coherence effect in multimedia learning: The case for minimizing irrelevant sounds in the design of multimedia instructional messages. *Journal of Educational Psychology, 92*(1),

117 - 125.

Paas, F., Renkl, A., Sweller, J. (2003). Cognitive load theory and instructional design: Recent developments. *Educational Psychologist, 38*(1), 1–4

Park, B., Moreno, R., Seufert, T., & Brünken, R. (2011). Does cognitive load moderate the seductive details effect? A multimedia study. *Computers in Human Behavior, 27*(1), 5 - 10.

Peshkam, A., Mensink, M. C., Putnam, A. L., & Rapp, D. N. (2011). Warning readers to avoid irrelevant information: When being vague might be valuable. *Contemporary Educational Psychology, 36*(3), 219 - 231.

Sadoski, M., Goetz, E. T., & Fritz, J. B. (1993). Impact of concreteness on comprehensibility, interest, and memory for text: Implications for dual coding theory and text design. *Journal of Educational Psychology, 85*(2), 291 - 304.

Sanchez, C. A., & Wiley, J. (2006). An examination of the seductive details effect in terms of working memory capacity. *Memory & cognition, 34*(2), 344 - 355.

Schmidt - Weigand, F., Kohnert, A., & Glowalla, U. (2010). A closer look at split visual attention in system and self-paced instruction in multimedia learning. *Learning and Instruction, 20*(2), 100 - 110.

Shirey, L. L., & Reynolds, R. E. (1988). Effect of interest on attention and learning. *Journal of Educational Psychology, 80*(2), 159 - 166.

Thalheimer, W. (2004). Bells, whistles, neon, and purple prose: When interesting words, sounds, and visuals hurt learning and performance, Retrieved June 10, 2011, *from http://www.oktopusz.hu/domain2019/ files/ modules/module*2015/28283C28732CAE28682.pdf.

Wade, S. E., Schraw, G., Buxton, W. M., & Hayes, M. T. (1993). Seduction of the strategic reader: Effects of interest on strategies and recall. *Reading Research Quarterly, 28*(2), 93 - 114.

The effects of seductive illustrations on multimedia learning: An eye movement study

GONG Shaoying[1,2]; DUAN Ting[1,2]; WANG Fuxing[1,2]; ZHOU Zongkui [1,2]; LU Chunxiao [1,2]

([1] Key Laboratory of Adolescent Cyberpsychology and Behavior (CCNU), Ministry of Education, Wuhan, 430079)

([2] School of Psychology, Central China Normal University, Wuhan 430079)

Abstract　This study aimed to explore the effects of seductive illustrations on performance and cognitive processes by using eye movement to track thirty low prior experience learners' process of visual attention during multimedia learning with or without seductive illustrations. Findings revealed that relative to learners in the non–seductive illustration condition, learners in the seductive illustration condition (1) recalled significantly fewer main ideas and generated fewer problem–solving solutions, (2) got fewer number of fixations in cognitive interest illustrations and also fewer attention switches between text and cognitive interest illustrations. Moreover, in the seductive illustration condition, 80% learners reported attracted to seductive illustrations, and recalled prior knowledge related to seductive illustrations. The result indicates that seductive illustrations may interfere students' memory and understanding of the main learning content by disrupting the coherent understanding of the materials and by priming inappropriate prior knowledge around seductive illustrations.

Keywords　multimedia learning; seductive illustrations; eye–movement

脑与认知研究所团队成员

周治金，博士，教授，脑与认知研究所所长，主要研究领域为创造性思维的认知神经基础，创新性教学等。

刘思耘，博士，教授，主要研究领域为语言的加工与表征，言语知觉，二语习得过程，具身认知。

赵庆柏，博士，副教授，脑与认知研究所副所长，主要研究领域为认知神经科学，网络语言，基于网络的问题解决。

定险峰，博士，副教授，主要研究领域为从具身认知的角度探讨数量、时间和权力等抽象概念的心理表征和神经机制。

范炤，博士，副教授，主要研究领域为视知觉，注意与工作记忆，网络心理学，虚拟现实在心理学中的应用。

高闯，副教授，研究领域为数理心理学、认知心理学，认知神经科学，实验心理学，网络心理学。试图从学理理论，对心理学经验研究领域进行重构，确立数理心理学理论体系。

程晓荣，博士，副教授，主要研究领域为儿童语言发展，语言与阅读的认知加工机制，视知觉与注意。

成良，博士，副教授，主要研究领域为利用神经电生理方法（TDT、ERP、fMRI 等）研究感觉信息加工及脑高级认知活动的神经机制，网络成瘾对脑高级认知活动的影响机制，以动物为研究对象采用分子生物学和遗传学手段研究学习记忆和创造力的神经机制。

李玉杰，博士，讲师，主要研究领域为注意，老龄化，刻板印象。

心理学报 , 2013, 45(1), 35 – 46.

汉语成语谜语问题解决中思路竞争的眼动研究

黄福荣[1,2] , 周治金[2] , 赵庆柏[1]

([1] 青少年网络心理与行为教育部重点实验室 ; 湖北省人的发展与心理健康重点实验室 ; 华中师范大学心理学院 , 武汉 430079)

([2] 北京市 "学习与认知" 重点实验室 ; 首都师范大学心理系 , 北京 100048)

摘　要　在谜语问题解决过程中 , 可能存在着通过简单联想和新异联想寻找答案这两种思路。两个实验中设置了包括寻常答案与新颖答案在内的若干备择答案 (实验 1 为 4 个 , 实验 2 为 6 个) 供被试选择 , 利用眼动技术记录被试在解题过程中的不同时间段内对新颖答案与寻常答案的平均注视时间 , 考察汉语成语谜语问题解决中两种思路之间冲突的过程。实验 1 操纵了任务要求 , 实验 2 操纵了规则线索的有效性。实验结果表明 : （ 1 ）在成语谜语问题解决中的一段时间内新异联想和简单联想能够同时发生且形成竞争 ; （ 2 ）选择 "新颖且合适答案" 的任务要求 , 提高了成功形成新颖语义联结的概率 , 但是并没有加快新异联想发生、发展的进程 , 也没有改变两种思路相互竞争的局面 ; （ 3 ）有效的规则线索可以抑制简单联想 , 阻止其发生 , 同时可以加快新异联想发生、发展的进程。

关键词　成语谜题 ; 新异联想 ; 简单联想 ; 眼动

分类号　B842

1　引言

在顿悟问题的解决过程中 , 问题解决者往往会经历一个明显的 "僵局" (impasses), 感到思维停滞不前 (Schooler, Ohlsson, & Brooks, 1993)。关于僵局产生的原因 , "心理定势理论" 认为思维习惯会使问题解决者不再致力于寻找新的、可能是更为有效的问题解决途径。"心理成规理论" 认为 , 僵局的产生是由于反复尝试使用某种错误的搜索路径增强了该路径的激活程度 , 相应地降低了搜索其他解题路径的可能性 , 它强调问题解决过程中的 "强迫症" 倾向。"功能固着理论" 认为物体的心理表征与它的常用功能属性相关 , 如果在特定的问题中 , 需要利用它的不寻常的功能属性 , 自动提取的常用功能属性会阻碍不寻常功能属性的激活和利用 , 僵局就产生了 (罗劲 , 2004)。

那么 , 僵局又是怎样被打破的？ "表征转换理论" 认为顿悟问题的成功解决取决于问题表征方式的转换 , 它可以通过限制消除、组块破解等多种方式对无效的、错误的初始表征进行重构 (Knoblich, Ohlsson, Haider, & Rhenius, 1999), 也可能只需要额外注意和编码问题中的其他关键信息 , 对初始表征进行修复就可以打破僵局 (Kaplan & Simon, 1990)。"进程监控理论" 的基本观点是 , 问题解决者会依据将要达到的问题目标状态 , 确定一些看似有效的内在标准 , 并根据这些标准来监控每一个局部行动的有效性。一旦个体意识到这些行动不会成功时 , 就会产生一种内驱力 , 促使个体去解除过去知识经验的限制 , 寻找其它解题途

通讯作者 : 周治金 , E-mail: zhouzhijin64@yahoo.com.cn

径 (Chronicle, MacGregor, & Ormerod, 2004; MacGregor, Ormerod, & Chronicle, 2001)。

上述有关顿悟问题解决的两类理论中，前一类理论着眼于说明采用常规的思维方式，就会形成无效的解题思路，从而使问题解决陷入僵局；后一类理论则着眼于说明新颖（有效）的解题思路是如何形成的。由此可见，在顿悟问题解决过程中，可能存在着无效的常规解题思路与有效的新颖解题思路之间的冲突。有关顿悟神经生理机制的研究为两种思路之间的冲突提供了证据。例如，脑功能成像研究发现，当顿悟发生时，与认知冲突的监控相关的前扣带回被激活了 (Aziz-Zadeh, Kaplan, & Iacoboni, 2009; Kounios et al., 2006; 罗劲，2004)。有关顿悟感产生的 ERP 研究也发现，与没有顿悟感的条件相比，顿悟感出现的条件下，新颖答案诱发出 N320/N380 成分，该成分溯源于前扣带回（买晓琴，罗劲，吴建辉，罗跃嘉，2005; 邱江，罗跃嘉，吴真真，张庆林，2006)，这说明顿悟问题解决过程中包含着两种解题思路之间的冲突。

在两种思路的冲突中，常规的无效思路对新颖的有效思路必然起着干扰和阻碍作用 (öllinger, Jones, & Knoblich, 2008)。虽然表征转换理论和进程监控理论认为，顿悟的产生需要解除过去知识经验的限制，但是没有阐明过去知识经验（即无效解题思路）与新颖有效的解题思路之间冲突的发生、发展过程。毕竟，新颖有效的思路需要战胜常规无效的思路才能出现顿悟。所以，仅描述新颖有效思路形成过程不足以全面反映顿悟产生过程，而深入探究新颖有效思路的发生、发展过程，以及新颖思路战胜常规无效思路的过程，才能更深入地揭示顿悟问题解决的规律。

从表征转换理论可以推论，新颖思路与常规思路之间转换可能是瞬间完成的。罗劲 (2004) 认为，顿悟是问题解决视角的瞬间"新旧交替"过程。有关顿悟问题解决中预热感判断和对关键信息注视时间的变化为此观点提供了实验证据。例如，Metcalfe 考察了被试在问题解决中接近问题解决的预热感 (feeling of warmth, FOW)，研究发现被试对常规问题解决的预热感是逐渐上升的，但是对顿悟问题解决的预热感在问题解决前一直很低，在问题解决瞬间突然上升 (Metcalfe, 1986; Metcalfe & Wiebe, 1987)。Knoblich, Ohlsson 和 Raney (2001) 分析了被试解决"火柴棒算式问题"时的注视时间的变化趋势，发现关键元素的注视时间在问题解决前陡增。这些研究结果倾向于支持顿悟问题解决中新颖有效的解题思路是突然形成，并且在瞬间替代了常规无效的解题思路。但是，这些研究并未直接考察新旧两种思路之间冲突的发生、发展过程。

从进程监控理论可以推论，新颖思路是一个逐渐发生的过程。一些实验研究也确实发现新颖思路不是突然发生的。例如，Yaniv 和 Meyer (1987) 认为 Metcalfe (1986, 1987) 的结论只适用于语义信息不丰富的顿悟问题，在语义信息丰富的顿悟问题解决过程中，与答案相关的语义信息在酝酿期是逐渐积累的。Ellis, Glaholt 和 Reingold (2011) 通过分析被试在解决英语字谜问题过程中的注视时间发现，被试对关键词素和干扰词素的注视时间比例在问题解决初期是没有显著差异的，直到问题解决前几秒钟，干扰词素的注视时间比例逐渐下降。这说明，在语义信息丰富的顿悟问题解决中，新颖有效的解题思路和常规无效的解题思路有可能同时发生，并相互竞争。

由此可见，深入研究顿悟问题解决过程中，新颖有效思路的发生、发展过程及其影响因素，以及新颖思路战胜常规无效思路的过程，是揭示顿悟问题解决认知机制的一条重要途径。谜语问题属于语义丰富的顿悟问题，在谜语问题解决中存在两种解题思路，一种是根据谜面语义所进行的简单联想，另一种是寻找并激活谜面（或谜底）不寻常的语义信息所进行的新异联想（买晓琴，罗劲，吴建辉，罗跃嘉，2005; 邱江，罗跃嘉，吴真真，张庆林，2006; 沈汪兵，刘昌，张小将，陈亚林，2011; 朱新秤，李瑞菊，周治金，2009)。所以，本研究拟采用汉语成语谜语问题为实验材料，探讨实验任务要求与规则线索的有效性对谜语问题解决中两种解题思路发生、发展过程的影响。在实验任务的选择上，虽然答案生成任务最能体现问题解决的过程与特点，但是采用此实验任务难以监测两种思路发生、发展的进程。采用呈现答案并确定顿悟感的判断任务，又难以反映完整的问题解决过程。所以，本研究采用选择答案的任务范式，为每一道谜题设置寻常答案、新颖答案（谜底）、似是而非答案和无关答案等几类备择答案，采用眼动追踪技术记录被试在解题过程中对新颖答案与寻常答案的注视时间，来推测谜语问题解决中新异联想的发生、发展过程，以及新异联想与简单联想之间冲突的过程。

之所以选择采用眼动追踪技术，是因为传统认知实验心理学研究方法，以及 ERP 或 fMRI 等认知神经科学技术，都难以直接考察顿悟问题解决中两种解题思路发生、发展及其相互竞争的过程；而眼动记录仪可以直接"记录"问题解决的进程。近几年来，越来越多的研究者使用这种技术研究问题解决过程 (Bilalic, McLeod, & Gobet, 2008; Kaller, Rahm, Bolkenius, & Unterrainer, 2009; Patsenkoa & Altmanna, 2010; Thomas & Lleras, 2007) 和顿悟问题解决过程 (Ellis, Glaholt, & Reingold, 2011; Grant & Spivey, 2003; Jones, 2003; Knoblich, Ohlsson, & Raney, 2001; Thomas & Llerasb, 2009)。Knoblich 等人 (2001) 除了统计分析顿悟问题解决过程中的总注视时间以外，还把整个问题解决过程分成三个相等的时间阶段，分析不同阶段中注视时间的变化。Ellis, Glaholt 和 Reingold (2011) 把英语字谜问题解决过程按照 100ms 的标准划分成若干相等的时间阶段，分析注视时间的变化趋势，这种数据分析方法能更好地考察创造性问题解决的进程。所以，本研究拟将谜语问题解决过程中的眼动数据进行分时间段分析，考察成语谜语问题解决中新颖有效思路和常规无效思路的发生、发展进程。

2 实验 1

2.1 实验目的

采用眼动仪记录被试解决成语谜题的眼动轨迹，首先对新颖答案和寻常答案的相对注视时间与选择比例进行相关分析，证实对新颖答案和寻常答案的注视能够敏感反映新异联想和简单联想。然后分析解

题过程中的每一时间段内对新颖答案、寻常答案与无关答案的注视时间差异，考察新异联想和简单联想的发生、发展进程，检验两种联想之间是否会发生竞争。本实验还操纵了指导语，尝试检验任务要求是否会改变两种联想的发生、发展进程和竞争过程。

2.2 实验方法

2.2.1 实验设计　实验 1 采用单因素组间设计，自变量为指导语中的任务要求（合适组 vs 新颖组），合适组要求被试选择合适答案，新颖组要求被试选择新颖且合适的答案，并且提示被试不是所有与谜面之间存在语义关联的备择答案都满足新颖性要求。新颖组被试在解题过程中可能会意识到存在两个或两个以上的答案，但是实验中并没有给予如何处理的指示。因变量为选择新颖答案和寻常答案的百分数，以及对各类备择答案兴趣区的注视时间。

2.2.2 被试　44 名来自武汉某大学的本科生参加了本实验，其中男生 20 名，女生 24 名。将被试随机分成两组，每组 22 人。所有被试的视力或矫正视力均正常，实验结束后被试获得一件小礼物。

2.2.3 实验仪器　实验仪器是加拿大 SR 公司生产的 Eyelink 1000 型眼动仪，采样频率设置为 1000 Hz，采用瞳孔 + 角膜模式采集数据。呈现刺激的显示器为 21 inch，刷新频率为 60Hz，分辨率为 1024 × 768。

2.2.4 实验材料　从汉语成语谜语库中 (朱新秤，李瑞菊，周治金，2009) 筛选了 48 道谜题，为每道谜题设置了 4 个备择答案，包括一个新颖答案、一个寻常答案和两个无关答案。新颖答案是成语谜语词典中所提供的谜底，它与谜面之间存在内隐而非直接的语义关联。在预实验中请一组被试根据谜面进行自由联想，生成与谜面之间存在语义关联的答案，然后从被试根据谜面所生成的答案中筛选出生成概率最高的成语作为该谜题的寻常答案。无关答案是从被

表 1　实验 1 中使用的实验材料举例

谜面	寻常答案	新颖答案	无关答案 I	无关答案 II
破晓	旭日东升	毁于一旦	良莠不齐	寿终正寝
越做越快	熟能生巧	积劳成疾	察言观色	言简意赅
善战而多谋	足智多谋	精打细算	旧地重游	孤苦伶仃
交通业的兴起	四通八达	应运而生	洁身自好	精雕细刻

试根据其它谜面生成的答案中选择的，且与谜面不存在语义关联的成语。表 1 列举了实验 1 中使用的 4 道谜题及其备择答案。

2.2.5 实验程序 实验在隔音的眼动实验室内完成。实验开始前，被试的前额和下颚放在托架上以固定头部，眼睛距离显示器大约 75 cm。进行校准和确认后，被试先进行 12 道题的练习，熟悉实验程序之后进入正式实验。正式实验流程是：首先在屏幕中央呈现一个"十"注视点 500 ms，然后呈现谜面 4000 ms，之后谜面不消失，并且在谜面下方呈现 4 个备择答案（4 个备择答案的位置进行了平衡），要求被试在 6000 ms 内按要求选择答案。间隔 2000 ms 后，呈现下一道谜题。

正式实验中呈现备选答案的同时，眼动仪开始记录被试的眼动轨迹等数据，被试按键选择答案之后，眼动仪停止记录。每一道谜题呈现之前，都会进行一次漂移校正以保证眼动记录的准确性。

2.2.6 数据处理 首先在备择答案图片上划分出 5 个兴趣区：谜面、寻常答案、新颖答案、无关答案 I、无关答案 II。通过 SR 公司提供的 Dataviewer 数据分析软件导出各个兴趣区的眼动数据，然后使用 SPSS 11.5 统计软件包进行统计分析。

2.3 结果与分析

2.3.1 选择答案的百分数 表 2 所示的是两组被试选择寻常答案与选择新颖答案的百分数。

表 2　选择寻常答案和新颖答案的百分数

任务要求	选择寻常答案	选择新颖答案
合适组	0.62（0.11）	0.32（0.11）
新颖组	0.38（0.22）	0.48（0.18）

注：括号内数据为标准差。

对寻常答案的选择百分数进行独立样本 T 检验，结果显示，$t(42)=4.75, p < 0.001$，合适组的选择百分数更高。对新颖答案的选择百分数进行独立样本 T 检验，结果显示，$t(42)=3.63, p < 0.005$，新颖组的选择百分数更高。

2.3.2 问题解决过程中的总注视时间

表 3 所示的是两组被试选择寻常答案与选择新颖答案过程中，分别对新颖答案、寻常答案和无关答案兴趣区的总注视时间（该兴趣区内所有注视点的总时间）。其中无关答案区的总注视时间是两个无关答案选项兴趣区的总注视时间的平均值。

为了检验对新颖答案区、寻常答案区的注视与新异联想、简单联想的关系，对新颖答案和寻常答案的选择百分数与三类兴趣区注视时间比例之间分别进行相关分析。兴趣区注视时间比例是指每类答案区的总注视时间占三类答案兴趣区的总注视时间的百分数。结果显示，新颖答案的选择百分数与对新颖答案兴趣区注视时间的比例呈显著正相关，$r = 0.86, p < 0.01$，与对寻常答案区注视时间的比例呈显著负相关，$r = -0.85, p < 0.01$，与对无关答案区注视时间的比例也呈显著负相关，$r = -0.85, p < 0.01$。寻常答案的选择百分数与对新颖答案区注视时间的比例呈显著负相关，$r = -0.83, p < 0.01$，与对寻常答案区注视时间的比例呈显著正相关，$r = 0.91, p < 0.01$，与对无关答案区注视时间的比例呈显著负相关，$r = -0.91, p < 0.01$。

2.3.3 选择新颖答案过程中的不同时间阶段内的注视时间 表 4 所示的是两组被试在选择新颖答案过程中的不同时间阶段内对不同兴趣区的注视时间。

为了进一步考察谜语问题解决的加工进程，本实验参考 Ellis 等人（2011）和沃建中等人（2006）所采用的分时间阶段考察注视轨迹的数据分析方法。具体做法是：取每 500 ms 为一个时间样本，计算被试在每一个时间样本内对三类答案区的注视时间。实验 1 统计分析了 0~3.5 s 的时间范围注视轨迹的数据，作此选择的理由是：被试选择新颖答案的平均时间是 3441(±514) ms，我们对实验数据的预处理发现，在 3.5 s 前后的三个时间阶段内，对各兴趣区注视时间之间的数量关系比较稳定，说明无论完成答案选择的时间是短于平均选择时间，还是长于平均选择时间，答案选择过程中所涉及的思维成分与特点具有一致性。

对于选择新颖答案的解题过程的注视时间进行 2（任务要求：合适组；新颖组）× 3（兴趣区类型：寻常答案区；新颖答案区；无关答案区）× 7(时间段：0~0.5 s; 0.5~1 s; 1~1.5 s; 1.5~2 s; 2~2.5 s; 2.5~3 s; 3~3.5 s) 三因素方差分析，重点分析兴趣区与时间段的交互

表3　问题解决过程中的总注视时间 (ms)

任务要求	选择类型	寻常答案区	新颖答案区	无关答案区
合适组	选择寻常答案	884（0.47）	556（0.30）	442（0.23）
	选择新颖答案	548（0.26）	1098（0.53）	433（0.21）
新颖组	选择寻常答案	1031（0.47）	653（0.29）	524（0.24）
	选择新颖答案	532（0.24）	1226（0.56）	438（0.20）

注：括号内数据为注视时间比例。

表4　选择新颖答案过程中的不同阶段内的注视时间 (ms)

任务要求	兴趣区	0~0.5 s	0.5~1 s	1~1.5 s	1.5~2 s	2~2.5 s	2.5~3 s	3~3.5 s
合适组	新颖答案区	7	62	130	153	146	146	139
	寻常答案区	4	21	95	83	110	76	50
	无关答案区	4	50	89	73	62	54	36
新颖组	新颖答案区	4	53	138	155	170	170	145
	寻常答案区	2	40	98	89	90	74	56
	无关答案区	6	48	75	67	54	56	46

作用，考察在不同时间段内对三类兴趣区注视时间的差异。若对新颖答案区或寻常答案区的注视时间长于无关答案区，说明新异联想或简单联想已经发生了，若对新颖答案区的注视时间也长于寻常答案区，说明新异联想开始占据主导地位。结果显示，任务要求的主效应不显著，$F(1,42) = 0.60$, $p = 0.44$; 兴趣区类型的主效应显著，$F(2,84) = 295.18$, $p < 0.001$; 时间段的主效应显著，$F(6,252) = 207.10$, $p < 0.001$; 任务要求与兴趣区类型的交互作用不显著，$F(2,84) = 1.03$, $p = 0.36$; 任务要求与时间段的交互作用不显著，$F(6,252) = 0.72$, $p = 0.63$; 兴趣区类型与时间段的交互作用显著，$F(12,504) = 14.25$, $p < 0.001$, 从1s开始，对新颖答案区的注视时间长于无关答案区，在2~2.5s和2.5~3s阶段，对寻常答案区的注视时间长于无关答案区，并且从0.5s开始对新颖答案区的注视时间长于寻常答案区；三因素交互作用不显著，$F(12,504) = 1.19$,

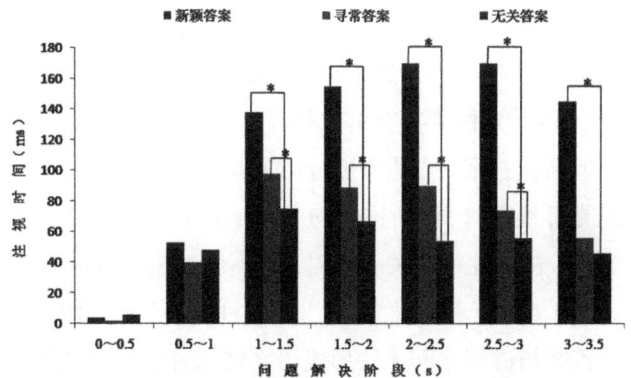

图1　合适组选择新颖答案的注视时间进程

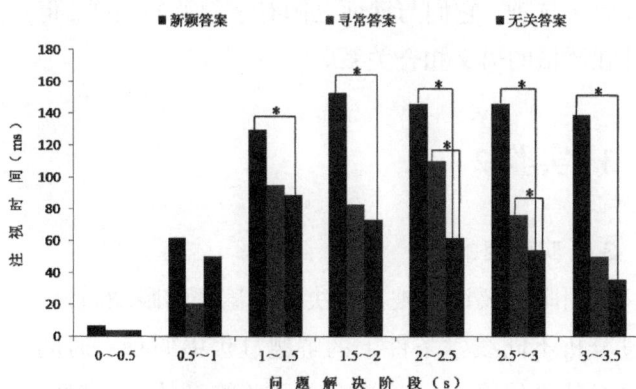

图2　新颖组选择新颖答案的注视时间进程

$p = 0.28$。合适组和新颖组的注视时间进程分别如图1、图2所示。

在三因素方差分析基础之上，选取新颖答案区和寻常答案区的注视时间都显著长于无关答案区的时间阶段，进一步比较新颖答案区与寻常答案区注视时间之差。结果显示，在合适组解题的2~2.5s和2.5~3s时间段内对新颖答案区与寻常答案区的注视时间之差没有显著差异，$t(21)=1.44$, $p =0.17$。在新颖组解题的1~1.5s、1.5~2s、2~2.5s和2.5~3s时间段内，对新颖答案区与寻常答案区的注视时间之差存在显著差异，$F(3,63)=3.18$, $p < 0.05$, 其中1~1.5s阶段的注视时间之差显著小于2.5~3s阶段，其余时间段之间的差异均没有达到显著水平。如图3所示。

2.4 讨论

实验1的结果表明，被试选择新颖答案的百分数

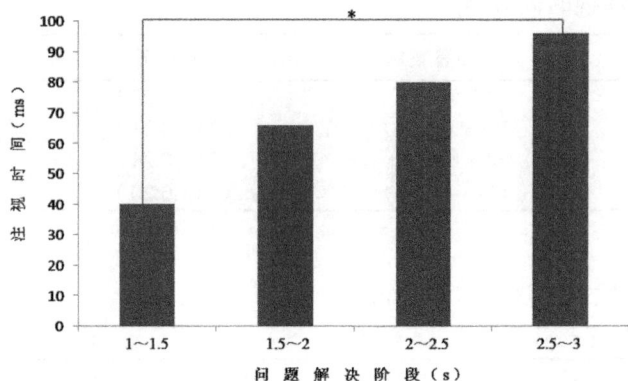

图 3　新颖组的新颖答案与寻常答案区注视时间之差

与其对新颖答案的注视时间呈高度正相关，而与其对寻常答案和无关答案的注视时间呈高度负相关。反过来，被试选择寻常答案的百分数与其对寻常答案的注视时间呈高度正相关，而与其对新颖答案和无关答案的注视时间呈高度负相关。这说明，被试对两类答案的注视（时间）与其对两类答案的选择高度一致。进而言之，被试注视新颖答案时，他所进行的主要是新异联想，被试注视寻常答案时，他所进行的主要是简单联想。

当然，被试注视无关答案时，并不代表他没有进行思考，他可能进行着各种尝试、探索性的思考。当他发现无关答案与谜面之间无法形成语义联结时，就会放弃该答案，转而注视其它答案。这种尝试建立备择答案与谜面之间语义联结的过程，是本实验中较具代表性的思考过程。所以，如果被试对新颖答案或寻常答案的注视时间长于对无关答案的注视时间，就说明新异联想或简单联想发生了，并处于比较稳定的发展过程之中。

在合适组选择新颖答案过程中，从 1s 开始新异联想发生了，在 2~3 s 阶段简单联想也发生了。在新颖组选择新颖答案过程中，从 1s 开始新异联想发生了，在 1~3 s 阶段简单联想也发生了。由此可知，在汉语成语谜语问题解决过程中的某些时间阶段，新异联想和简单联想能够同时发生。而新异联想与简单联想都是针对同一谜题进行的，并且被试往往更习惯进行简单联想，新异联想的顺利进行需要抑制简单联想的思维习惯，所以，两种联想之间可能存在着竞争关系。在新异联想和简单联想都发生的阶段，新颖答案区和寻常答案区的注视时间之差越小表示两种联

想之间的竞争程度越高，随着问题的解决，新异联想与简单联想之间的竞争程度逐渐减弱。在选择新颖答案过程中的思路竞争阶段，对新颖答案的注视时间一直长于寻常答案，表明新异联想占居主导地位。

注视时间的三次交互作用不显著，说明实验任务要求没有改变简单联想和新异联想的发生、发展进程，新颖组和合适组被试的新异联想都从 1 秒后开始占据主导地位，两组被试选择新颖答案的"相对速度"也是一致的，到 3441 ms（平均解题时间）时的累积百分数分别为 50% 和 51%。实验任务要求也没有改变两种解题思路可以同时发生且形成相互竞争的局面，但是影响了两种思路竞争的周期，合适组的思路竞争发生在 2~3 s 阶段，而新颖组的竞争发生在 1~3s 阶段。可能原因在于，新颖组根据指导语中的任务要求知道谜题既有新颖答案，也有寻常答案，能够更加有意识地主动抑制简单联想。

实验 1 中，在选择新颖答案过程中，短于平均选择时间三个标准差的两个时间阶段内（1~1.5 s 和 1.5~2.0 s)，对新颖答案区的注视时间显著长于无关答案区，可能反映了被试对新颖答案的直觉偏好。

为了进一步探索谜语问题解决过程中常规思路与新颖思路发生、发展的特点及其影响因素，实验 2 操纵了规则线索的有效性，重点考察有效和无效规则线索是否影响新颖思路的发生、发展过程，以及常规思路是否同时发生并与新颖思路形成竞争。在实验 1 中，被试有可能采用"排除法"策略来选择答案，为了尽量减少"排除法"策略的使用，提高考察新颖思路发生、发展过程的有效性，实验 2 增加了两个似是而非答案选项，它们与谜面之间有字面上的关联，但不存在严格的语义扣合关系。

3 实验 2

3.1 实验目的

采用眼动仪记录被试解决成语谜题的眼动轨迹，通过分析不同线索条件下的解题过程中的每一时间段内对新颖答案、寻常答案与无关答案的注视时间

表5　实验2中使用的实验材料举例

谜面	总体规则	具体规则	寻常答案	新颖答案	似是而非答案	无关答案
降落伞	正扣	性质特点联想	从天而降	随机应变	落落大方	明察秋毫
五指	正扣	象形联想	长短不一	三长两短	首屈一指	以身作则
泣别	反扣	会意联想	依依不舍	不欢而散	曲终人散	门庭若市
都成眷属	反扣	结果联想	成双成对	无独有偶	独善其身	水泄不通

的差异，考察总体规则线索和具体规则线索如何影响新异联想和简单联想的发生、发展进程，以及两种联想的竞争过程。

3.2 实验方法

3.2.1　实验设计　实验采用2（总体规则线索有效性：有效 vs 无效）× 2（具体规则线索有效性：有效 vs 无效）的两因素组内设计。为了确保被试信任并利用线索解题，两类规则线索有效的数量都占2/3，无效的数量占1/3，实验中为每一道谜题同时提供一条总体规则线索和一条具体规则线索。实验指导语要求被试选择新颖且合适的答案，并且提示被试不是所有与谜面之间存在语义关联的备择答案都符合新颖性要求。因变量为选择新颖答案的百分数，以及选择新颖答案过程中对各类备择答案的注视时间。

3.2.2　被试　22名来自武汉某大学的本科生参加了本实验，其中男生10名，女生12名。所有被试的视力或矫正视力均正常，此前未参加过类似实验，实验结束后获赠一件小礼物。

3.2.3　实验仪器　实验仪器同实验一

3.2.4　实验材料　从汉语成语谜语库中抽取90道谜题，分别属于正扣和反扣两种总体规则，以及会意、结果、象形和性质特点等四种不同的具体规则。为每道谜题设置6个备择答案，其中包括一个新颖答案、一个寻常答案、两个似是而非答案和两个无关答案。新颖答案、寻常答案和无关答案的编制方法同实验1，似是而非答案是在预实验中从被试根据谜面所生成的答案中选取生成概率非常低、与谜面中的某个字存在关联的成语。表5列举了实验2中使用的4道谜题及其备择答案。

3.2.5　实验程序　实验在隔音的眼动实验室内完成。实验开始前，被试的前额和下颚放在托架上以固定头部，眼睛距离显示器大约75 cm。进行校准

和确认后，被试先进行8道题的练习，熟悉实验程序后进入正式实验。正式实验流程是：首先在屏幕中央呈现一个"十"注视点500 ms，然后呈现谜面3000 ms，之后在谜面下方同时呈现两条规则线索，其中具体规则线索在左侧，总体规则线索在右侧，2000 ms之后再在线索下方呈现6个备择答案（6个备择答案的位置进行了平衡），要求被试在15s内选择谜题的新颖答案。间隔2000 ms后，呈现下一道谜题。

正式实验中呈现备择答案的同时，眼动仪开始记录被试的眼动轨迹等数据，被试按键选择答案之后，眼动仪停止记录。每一道谜题呈现之前，都会进行一次漂移校正以保证眼动记录的准确性。

3.2.6　数据处理　将整个注视画面划分为9个兴趣区：谜面、总体规则线索、具体规则线索、新颖答案、寻常答案、似是而非答案I、似是而非答案II、无关答案I、无关答案II。通过SR公司提供的Dataviewer数据分析软件导出各个兴趣区的眼动数据，使用SPSS 11.5统计软件包进行统计分析。

3.3 结果与分析

3.3.1　选择新颖答案的百分数　表6所示的是被试在不同规则线索条件下选择新颖答案的百分数。

对新颖答案的选择百分数进行2（总体规则线索有效性）× 2（具体规则线索有效性）的两因素方差分析，结果显示，总体规则线索有效性的主效应显著，$F_{(1,21)} = 28.33$，$p < 0.001$，有效水平的选择比例更高；具体规则线索有效性的主效应不显著，$F_{(1,21)} = 1.33$，$p = 0.26$；交互作用不显著，$F_{(1,21)} = 0.84$，$p = 0.37$。

3.3.2　选择新颖答案过程中的不同时间段内的注视时间　表7所示的是被试在选择新颖答案过程中的不同时间段内对不同兴趣区的注视时间。

为了考察谜语问题的正确解决过程中新异联想

表 6　选择新颖答案的百分数

总体规则线索	具体规则线索	选择新颖答案的百分数
有效	有效	0.54（0.17）
	无效	0.50（0.14）
无效	有效	0.37（0.17）
	无效	0.36（0.15）

注：括号内数据为标准差。

和简单联想的发生、发展进程，依然采用分时间段数据分析方法，进行 2（总体规则线索有效性）× 2（具体规则线索有效性）× 3（兴趣区类型：新颖答案区，寻常答案区，无关答案区）× 8（时间段：0~1 s，1~2 s，2~3 s，3~4 s，4~5 s，5~6 s，6~7 s，7~8 s）四因素组内设计的方差分析。重点分析总体规则线索有效性、具体规则线索有效性、兴趣区类型、时间段的四次交互

9.16，$p < 0.01$；总体规则线索有效性与时间的交互作用显著，$F(7,147) = 3.68$，$p < 0.005$；兴趣区类型与时间段的交互作用显著，$F(14,249) = 6.73$，$p < 0.001$；总体规则线索有效性、具体规则线索有效性与时间段的交互作用显著，$F(7,147) = 8.12$，$p < 0.001$。具体规则线索有效性、兴趣区类型与时间段的交互作用显著，$F(14,294) = 2.05$，$p < 0.05$。四因素的交互作用显著，$F(14,294) = 3.19$，$p < 0.001$。其他交互作用均不显著。

对于四因素的交互作用，重点分析总体规则线索有效性与具有规则线索有效性形成的四种组合条件下，不同时间段上，对三类答案兴趣区注视时间的差异。简单效应分析发现：(1) 当总体规则线索和具

表 7　选择新颖答案过程中的不同阶段内的注视时间 (ms)

总体规则线索	具体规则线索	兴趣区	0~1s	1~2s	2~3s	3~4s	4~5s	5~6s	6~7s	7~8s
有效	有效	新颖答案区	105	149	196	190	179	189	157	141
		寻常答案区	103	82	97	78	67	62	50	47
		无关答案区	69	75	61	64	57	50	50	42
	无效	新颖答案区	30	128	147	173	170	160	204	189
		寻常答案区	39	83	77	79	80	87	47	32
		无关答案区	77	66	84	64	57	49	44	37
无效	有效	新颖答案区	53	120	180	168	209	183	199	116
		寻常答案区	48	112	60	70	87	68	77	44
		无关答案区	90	85	55	56	54	55	41	37
	无效	新颖答案区	217	143	124	214	165	232	237	151
		寻常答案区	198	70	71	84	92	58	29	27
		无关答案区	38	46	61	81	80	50	59	30

作用，即考察四种规则线索有效性组合条件下，在不同时间段，对新颖答案区、寻常答案区和无关答案区的注视时间的差异。被试选择新颖答案的平均时间是 8300(± 1636) ms，所以实验 2 分析 8s 以内时间范围注视时间数据，理由同实验 1。

方差分析的结果表明，总体规则线索有效性的主效应边缘显著，$F(1,21) = 4.05$，$p=0.057$；具体规则线索有效性的主效应不显著，$F(1,21) = 1.32$，$p = 0.26$；兴趣区类型的主效应显著，$F(2,42) = 129.98$，$p < 0.001$，对新颖答案区的注视时间长于无关答案区 ($p < 0.001$)，也长于寻常答案区 ($p < 0.05$)；对寻常答案区的注视时间长于无关答案区 ($p < 0.001$)。时间段的主效应显著，$F(7,147) = 5.81$，$p < 0.001$；总体规则线索有效性与具体规则线索有效性的交互作用显著，$F(1,21) =$

体规则线索都有效时，各时间段上，对新颖答案区的注视时间显著长于无关答案区；在 0~1 s 和 2~3 s 时间段上，对寻常答案区的注视时间显著长于无关答案区；从 1~2 s 时间阶段开始，对新颖答案区的注视时间就显著长于寻常答案区。(2) 当总体规则线索有效但具体规则线索无效时，在 1~2 s 和 2~3 s 时间阶段对新颖答案区的注视时间长于无关答案区的注视时间，差异边缘显著，3 s 以后对新颖答案区的注视时间显著长于无关答案区；对寻常答案区和无关答案区的注视时间在各时间阶段都没有显著差异；从 2~3 s 时间阶段以后，对新颖答案区的注视时间显著长于寻常答案区。(3) 当总体规则线索无效但具体规则线索有效时，从 2~3 s 时间阶段以后，对新颖答案区的注视时间显著长于无关答案区；各时间阶段上，对

寻常答案区和无关答案区的注视时间都没有显著差异；从 2~3 s 时间阶段以后，对新颖答案区的注视时间显著长于寻常答案区。（4）当总体规则线索和具体规则线索都无效时，各时间阶段上，对新颖答案区的注视时间显著长于无关答案区；在 0~1s 阶段，对寻常答案区的注视时间显著长于无关答案区，1s 之后对寻常答案区和无关答案区的注视时间又没有显著差异；从 3~4s 时间阶段以后，对新颖答案区的注视时间显著长于寻常答案区。如图 4、图 5、图 6 和图 7 所示。

3.4 讨论

实验 2 结果表明，有效的总体规则线索提高了新颖答案选择的比例，促进了新颖语义联结的形成；当总体规则线索无效时，它明显阻碍了新颖语义联结的形成。具体规则线索的有效性对新颖答案的选择并没有显著的作用。可能是因为总体规则线索提供的总体联想方向的真伪难以区分，而具体规则线索提供

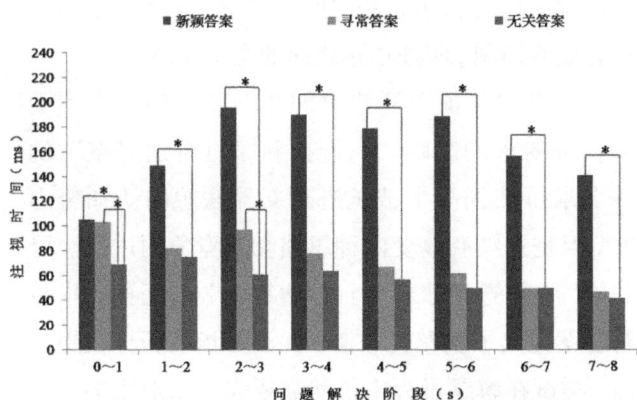

图 6　总体规则无效、具体规则有效

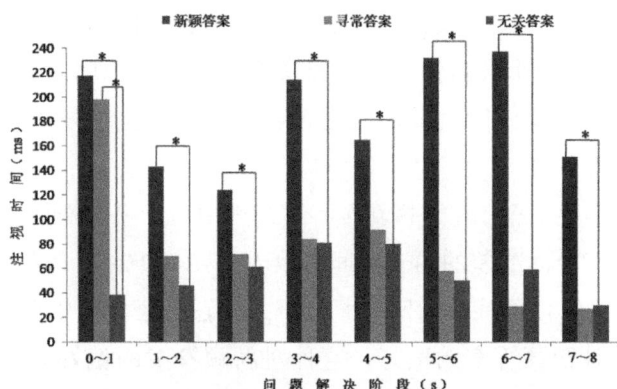

图 7　总体规则无效、具体规则无效

的联想的具体形式的真伪相对容易区分，被试能够批判性地利用具体规则线索，所以只发现了总体规则线索有效性对于新颖答案选择的显著影响。在两条规则线索都无效时，他们选择新颖答案的百分数 (36%) 显著高于随机选择的概率 (16.7%)。这也能够说明，被试并没有完全受线索的约束，他们能够主动调整思路，寻找新颖答案，建立谜面与答案之间的新颖语义联结。

在选择新颖答案过程中，无论两条规则线索是否有效，一般在 3 秒后，被试对新颖答案区的注视时间就显著长于无关答案区，而被试对寻常答案区与无关答案区的注视时间都没有显著差异，这说明新异联想发生了并占据主导地位，而稳定的简单联想基本上没有发生。可能的原因是，在提供了解题线索并要求被试选择新颖且合适答案的条件下，更加激发了被试寻找新颖答案的动机，他们利用两种规则线索努力进行新异联想，同时抑制简单联想。所以，在正确选择新颖答案的解题过程中，没有出现稳定的新异联想与简

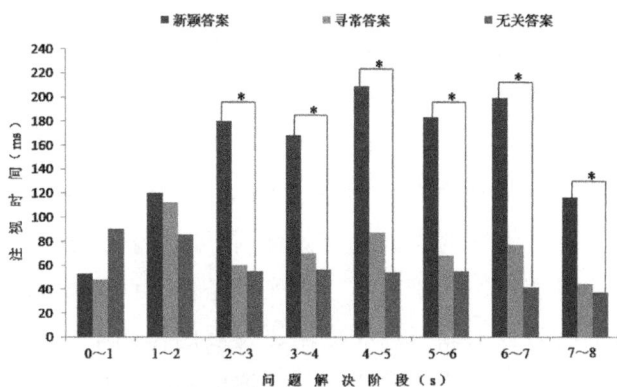

图 4　总规规则有效、具体规则有效

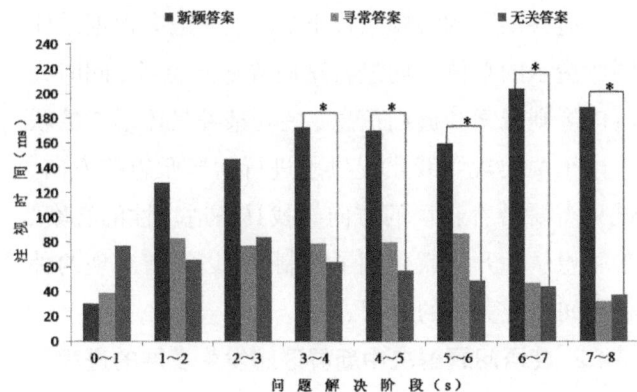

图 5　总体规则有效、具体规则无效

单联想同时发生且相互竞争的过程。

规则线索有效性的作用也反映在新异联想的发生、发展进程上。到8300 ms（平均解题时间）时，在两类规则线索所形成的四种有效性条件下，选择新颖答案的累积百分比分别为54%、47%、46%和48%。在两条规则线索都有效性时，选择新颖答案的"相对速度"快于其它三种实验条件。两类规则线索的有效性对新异联想发生、发展的作用不是独立的，因为正确的新异联想不仅需要确定正向思考或反向思考的总体方向，还需要确定具体的会意、性质特点、结果或象形联想等具体方向，任何一条线索无效，都会对恰当的新异联想产生误导，所以两类规则线索同时有效才能促进新异联想发生、发展的进程。

实验2中，在选择新颖答案过程中短于平均选择时间三个标准差的两个时间阶段内(1~2s 和 2~3s)，对新颖答案区的注视时间显著长于无关答案区，反映了被试对新颖答案的偏好。这种对新颖答案的偏好并不受规则线索有效性的影响，因此可以进一步认为，在这两个时间阶段对于新颖答案更长时间的注视属于直觉加工过程。

4 总讨论

4.1 谜语问题解决中新异联想与简单联想竞争的过程

联结主义认为创造就是把头脑中的观点按照不寻常的、新颖独特且有用的方式加以组合，从而形成一种新颖的联结的过程，或者说通过远距离联想形成信息间新颖联结。但是，由于"思维惰性"，人们往往习惯于进行一些简单的联想。本研究提供了包括寻常答案与新颖答案在内的4个（或6个）备择答案，在尝试发现并建立谜面与各个答案之间可能的语义联结过程中，通过语义激活扩散就可以自动建立谜面和寻常答案之间的寻常语义联结，这是简单联想；而打破常规思维方式，发现并建立谜面和新颖答案之间的新颖语义联结，这是新异联想。在谜语问题解决中，被试需要努力尝试解决问题的新颖思路，发现并建立谜面与谜底之间的新颖语义联结。

显然，通过新异联想所形成的新颖且有效的解题思路战胜常规的解题思路是顿悟产生的关键。表征转换理论、进程监控理论以及后续的研究都没有直接考察两种思路冲突的过程，本研究采用眼动技术发现汉语成语谜语问题解决中的一段时间内，新异联想与简单联想可以同时发生并形成竞争（实验1）。出现这种结果的可能原因是，在选择答案的解题过程中，被试首先需要发现和建立谜面与各个答案之间可能的语义关联，并评估各答案的适切性，在此基础上进一步比较与评估它们的新颖性。在发现并建立谜面与各个答案之间可能的语义联结过程中，虽然简单联想与新异联想都可能满足适切性要求，但是只有新异联想可能满足新颖性要求，而新异联想的发生与发展需要打破简单联想的限制。所以，在选择新颖答案的过程中，简单联想与新异联想可以在一段时间内同时发生，形成了常规思路与新颖思路之间的竞争，直到简单联想得到有效的抑制，两种思路之间的竞争才结束。

新异联想与简单联想之间相互竞争的过程不受实验任务要求的影响。"选择新颖且合适答案"的任务要求虽然提高了被试进行新异联想的认知努力程度，但是并没有改变两种思路相互竞争的过程。可能的原因是，选择"新颖且合适答案"的任务要求只是对答案提出了新颖性限制，并没有指明新异联想的方向，被试在解题过程中依然需要先尝试形成多种可能的语义联结，然后再比较不同联结之间的新颖性。

然而，在提供解题规则线索并要求被试选择"新颖且合适答案"的实验条件下，被试一般会根据总体规则所指示的总体方向进行正向或逆向思考，同时根据具体规则线索和谜面信息，展开某种具体形式的联想。由于两类规则线索提供了进行新异联想的方向，被试只需要沿着新颖的方向寻找具备适切性的答案，简单联想从一开始就受到了抑制，所以没有出现两种思路之间相互竞争的过程。

4.2 谜语问题解决中新异联想发生发展的进程

在提供备择答案的实验条件下，通过语义激活扩散可以自动建立谜面与寻常答案之间的寻常语义联结，所以，简单联想很快就能发生。然而，发现并建立谜面与新颖答案之间新颖语义联结，则需要抑制谜面或谜底关键字（词）常见意义，而激活其非寻常意义，所以新异联想发生、发展的过程应该较晚。然而，本研究中两个眼动实验的数据都表明，在各种实验条件下选择新颖答案过程中，在问题解决的早期阶段都存在对新颖答案的直觉偏好，反映了对新颖语义关联的无意识加工。国外研究中有类似的结果，例如，Ellis, Glaholt 和 Reingold (2011) 在英语字谜问题解决的眼动研究中发现，答案相关的知识在问题解决早期就已经被无意识地激活了。Bowden 和 Jung-Beeman 也发现，顿悟问题的正确答案在被试有意识地提取之前是处于无意识激活状态的 (Bowden & Jung-Beeman, 1998; Jung-Beeman & Bowden, 2000)。

在被试最终选择新颖答案的情况下，对新颖答案的注视时间（各时间阶段）都显著长于对无关答案的注视时间，这说明，本研究中新颖语义联结的形成是一个逐渐积累的过程。Yaniv 和 Meyer (1986) 也认为，在语义信息丰富的顿悟问题的解决过程中，与答案相关的语义信息是逐渐积累的。本研究的结果比较符合进程监控理论的预期。

本研究还发现，选择"新颖且合适答案"的任务要求增强了问题解决者进行新异联想的认知努力，提高了成功建立新颖语义联结的概率，但是并没有加快建立新颖语义联结的"速度"。可能的原因是，选择"新颖且合适答案"的任务要求虽然对谜面与答案之间的语义联结有了"新颖性"限制，但是，实验任务要求并没有说明如何进行新异联想，也没有说明朝哪个方向进行新异联想，所以实验任务要求没有加快新异联想发生、发展的进程。

在要求选择"新颖且合适答案"的条件下，提供两类规则线索，加快了新颖语义联结形成的"速度"。因为总体规则线索能够引导被试进行正向或逆向思考，具体规则线索能够引导被试根据谜面信息进行象形、会意、因果等具体方式的联想。遵循有效规则线索的引导，被试能够更快地进行恰当的新异联想，所以有效的规则线索加快了新异联想的发生、发展进程。

4.3 本研究的创新与不足

实验提供新颖答案、寻常答案、似是而非答案和无关答案等几类备择答案，利用眼动技术记录了被试在解题过程中对于新颖答案和寻常答案的注视时间，实验发现对新颖答案的总注视时间与选择新颖答案比例具有高度的正相关，说明对新颖答案注视时间的长短可以作为新异联想的敏感指标。本研究进一步将谜语问题解决时间分为若干阶段，通过比较不同时间阶段内对新颖答案注视时间、对寻常答案的注视与对无关答案的注视时间，直接考察了谜语问题解决过程中新颖有效的解题思路发生、发展的过程，以及常规的解题思路与新颖的解题思路之间冲突过程。本研究所发现的顿悟问题解决过程具有逐渐积累的性质，对于揭示语义信息丰富的顿悟问题解决过程的特性，具有一定的意义。

虽然采用选择答案的实验任务可以较好地操纵（或引导）被试进行新异联想与简单联想，也能有效地控制顿悟问题解决的时间，但是，毕竟产生答案才能最好地体现顿悟问题解决过程。

本研究对被试解决几十道谜题的眼动轨迹，按不同解题时间阶段分别进行平均统计分析，这有利于从整个解题过程上描述新异联想与简单联想发生、发展的特点，但是，采用这种计算方法，难以说明某道谜题的解决过程中是否存在新异联想瞬间替代简单联想的过程。

5　结论

根据两个实验的结果，可以得出如下三个结论：(1) 在谜语问题解决过程中的一段时间内新异联想和简单联想能够同时发生并形成竞争。(2) 选择"新颖且合适答案"的任务要求，提高了成功形成新颖语义联结的概率，但是并没有加快新异联想发生、发展的进程，也没有改变两种思路相互竞争的局面。(3) 有效的规则线索可以抑制简单联想，阻止其发生，同时可以加快新异联想发生、发展的进程。

参考文献

Aziz-Zadeh, L., Kaplan, J. T., & Iacoboni, M. (2009). "Aha!": The neural correlates of verbal insight solutions. *Human Brain Mapping*, *30*(3), 908–916.

Bilalic, M., McLeod, P., & Gobet, F. (2008). Why good thoughts block better ones: The mechanism of the pernicious Einstellung (set) effect. *Cognition, 108*(3), 652–661.

Bowden, E. M., & Jung-Beeman, M. (1998). Getting the right idea: Semantic activation in the right hemisphere may help solve insight problems. *Psychological Science, 9*(6), 435–440.

Chronicle, E. P., MacGregor, J. N., & Ormerod, T. C. (2004). What makes an insight problem? The roles of heuristics, goal conception, and solution recoding in knowledge-lean problems. *Journal of Experimental Psychology: Learning, Memory, and Cognition, 30*(1), 14–27.

Ellis, J. J., Glaholt, M. G., & Reingold, E. M. (2011). Eye movements reveal solution knowledge prior to insight. *Consciousness and Cognition, 20*(3), 768–776.

Grant, E. R., & Spivey, M. J. (2003). Eye movements and problem solving: Guiding attention guides thought. *Psychological Science, 9*(5), 462–466.

Jones, G. (2003). Testing two cognitive theories of insight. Journal of Experimental Psychology: Learning, *Memory, and Cognition, 29*(5), 1017–1027.

Jung-Beeman, M., & Bowden, E. M. (2000). The right hemisphere maintains solution-related activation for yet-to-be-solved problems. *Memory & Cognition, 28*(7), 1231.

Kaller, C. P., Rahm, B., Bolkenius, K., & Unterrainer, J. M. (2009). Eye movements and visuospatial problem solving: Identifying separable phases of complex cognition. *Psychophysiology, 46*(4), 818–830.

Kaplan, C. A., & Simon, H. A. (1990). In search of insight. *Cognitive Psychology, 22*, 374–419.

Knoblich, G., Ohlsson, S., & Raney, G. (2001). An eye movement study of insight problem solving. *Memory & Cognition, 29*(7), 1000.

Knoblich, G., Ohlsson, S., Haider, H., & Rhenius, D. (1999). Constraint relaxation and chunk decomposition in insight problem solving. Journal of Experimental Psychology: Learning, *Memory, and Cognition, 25*(6), 1534–1555.

Kounios, J., Frymiare, J. L., Bowden, E. M., Fleck, J. I., Subramaniam, K., Parrish, T. B., & Jung-Beeman M. (2006). The prepared mind: Neural activity prior to problem presentation predicts subsequent solution by sudden insight. *Psychological Science, 17*, 882–890.

Luo, J. (2004). Neural correlates of insight. *Acta Psychologica Sinica, 36*(2), 219–234.

[罗劲. (2004). 顿悟的大脑机制. *心理学报*, 36(2), 219–234.]

MacGregor, James N., Ormerod, T. C., & Chronicle, E. P. (2001). Information processing and insight: A process model of performance on the Nine-Dot and related problems. Journal of Experimental Psychology: Learning, *Memory, and Cognition, 27*(1), 176–201.

Mai, X. Q., Luo, J., & Wu, J. H. (2005). "Aha!" effects in a guessing riddle task: An event-related potential study. *Acta Psychologica Sinica, 37*(1), 19–25

[买晓琴, 罗劲, 吴建辉, 罗跃嘉. (2005). 猜谜作业中顿悟的 ERP 效应. *心理学报*, 37(1), 19–25.]

Metcalfe, J. (1986). Feeling of knowing in memory and problem solving. Journal of Experimental Psychology: Learning, *Memory, and Cognition, 12*(2), 288–294.

Metcalfe, J., & Wiebe, D. (1987). Intuition in insight and noninsight problem solving. *Memory & Cognition, 15*(3), 238.

öllingera, M., Jones, G., & Knoblich, G. (2008). Investigating the effect of mental set on insight problem solving. *Experimental Psychology, 55*(4), 269 – 282.

Patsenkoa, E. G., & Altmanna, E. M. (2010). How planful is routine behavior? A selective–attention model of performance in the Tower of Hanoi. *Journal of Experimental Psychology: General, 139*(1), 95 – 116.

Qiu, J., Luo, Y. J., Wu, Z. Z., & Zhang, Q. L (2006). A further study of the ERP effects of 'insight' in a riddle guessing task. *Acta Psychologica Sinica, 38*(4), 507 – 514.

[邱江 , 罗跃嘉 , 吴真真 , 张庆林 . (2006). 再探猜谜作业中 "顿悟" 的 ERP 效应 . *心理学报* , *38*(4), 507 – 514.]

Schooler, J. W., Ohlsson, S., & Brooks, K. (1993). Thoughts beyond words: When language overshadows insight. *Journal of Experimental Psychology: General, 122*(2), 166 – 183.

Shen, W. B., Liu, C., Zhang, X. J., & Chen, Y. L. (2010). The time course and hemispheric effect of "insight" in three–character Chinese riddles task: An ERP study. *Acta Psychologica Sinica, 43*(3), 229 – 240.

[沈汪兵 , 刘昌 , 张小将 , 陈亚林 . (2011). 三字字谜顿悟的时间进程和半球效应 : 一项 ERP 研究 . *心理学报* , *43*(3), 229 – 240.]

Thomas, L. E., & Lleras, A. (2007). Moving eyes and moving thought: On the spatial compatibility between eye movements and cognition. *Psychonomic Bulletin and Review, 14*(4), 663 – 668.

Thomas, L. E., & Llerasb, A. (2009). Covert shifts of attention function as an implicit aid to insight. *Cognition, 111*(2), 168 – 174.

Wo, J. Z., Li, Q., & Tian, H. J. (2006). The eye movements study on children with different ability levels in the figure reasoning problem. *Developmental and Educational Psychology, 3*, 6 – 10.

[沃建中 , 李琪 , 田宏杰 . (2006). 不同推理水平儿童在图形推理任务中的眼动研究 . *心理发展与教育* , *3*, 6 – 10.]

Yaniv, I., & Meyer, D. E. (1987). Activation and metacognition of inaccessible stored information: Potential bases for incubation effects in problem solving. Journal of Experimental Psychology: Learning, *Memory, and Cognition, 13*(2), 187 – 205.

Zhu, X. C., Li, R. J., & Zhou, Z. J. (2009). The role of clues in Chinese idiom riddle solving. *Acta Psychologica Sinica, 41*(5), 397 – 405.

[朱新秤 , 李瑞菊 , 周治金 . (2009). 谜语问题解决中线索的作用 . *心理学报* , *41*(5), 397 – 405.]

An eye movement study of associate competition in Chinese idiom riddles solving

HUANG Furong[1,2]; ZHOU Zhijin[1]; ZHAO Qingbai[1]

([1]Key Laboratory of Adolescent Cyberpsychology and Behavior (CCNU), Ministry of Education;

Key Laboratory of Human Development and Mental Health of Hubei Province;

School of Psychology, Central China Normal University, Wuhan 430079, China)

([2] Beijing Key Lab of Learning and Cognition, Department of Psychology; Capital Normal University, Beijing 100048, China)

Abstract Most Chinese idiom riddles require insightful thoughts to solve. Novel and simple associations can be formed during the process of idiom resolutions. Insightful thought occurs only when novel associations overwhelm simple ones. However, it is unclear how this happens in the mind. According to the Representational Change Theory, the competition happens in a sudden way. But according to the Process Monitoring Theory, it is completed gradually. By using eye–tracking technology, we intended to investigate the time course of insightful problem resolution, using Chinese idiom riddles as experimental materials. In this study, an option selection task was adopted. Chinese idiom riddles were presented, together with four types of options serving as the spare answers to the riddle (novel, ordinary, plausible and absolutely wrong). Participants were asked to make a choice among options. At the same time, the fixation times of the participants spent on different options were recorded. In Experiment 1, the participants were randomly divided into an appropriate group and a novel group. In different groups, the participants were asked to make a choice between an appropriate answer and a novel and appropriate one. In Experiment 2, a 2 (general solution rule: effective vs. ineffective) × 2 (special solution rule: effective vs. ineffective) experimental designs was adopted. Besides, a general solving rule and a specific solving rule were presented at the same time. Results showed that there was a positive correlation between the fixation times of the participants spent on novel or ordinary answers and the percentage of the corresponding selections. The participants were found to have formed novel associations while fixating on novel answers, and simple associations while fixating on ordinary answers. The result further revealed that: (1) Novel associations and simple associations were formed simultaneously and competed to each other for a while before the idiom riddles were solved. (2) The demand of choosing a novel and appropriate answer induced people to make more efforts on novel association formation, and to select more novel answers. However, the task demand did not accelerate the time course of novel association formation, or change the competitive situation. (3) The effective solution rules, which promoted the novel answer selections, not only accelerated the time course of novel association formation, but also inhibited simple association formation, and eliminated the competition between them.

Keywords Chinese idiom riddle; novel association; simple association; eye movement

心理学报，2015, 47(3), 285－299.

语言标签和自我关联对新颖客体类别知觉的影响 *

刘思耘，孟健欣

（青少年网络心理与行为教育部重点实验室；华中师范大学心理学院，武汉 430079）

摘　要　类别知觉 (Categorical perception) 是人类最基本的认知活动之一，探讨语言对类别知觉的影响是心理语言学领域的热门话题之一。在这个研究的 3 个实验中，分别在高、低不同的语言标签表征强度下、高、低不同的客体自我关联程度下观察新颖客体类别知觉的过程。研究发现，语言标签表征程度的增强可促进新颖客体类别知觉的右视野优势效应；客体与自我关联程度的提高会促进左、右视野的类别知觉效应；语言标签的表征程度和客体与自我关联程度同时增强时，语言标签的作用依旧表现出来，但与自我关联的影响产生权衡分配，且其影响力并不足以产生右视野优势效应。

关键词　类别知觉；语言标签；自我关联；右视野优势效应

分类号　B842

1　前言

迄今为止，研究者对语言与思维关系的争论逐步形成两派，一派持语言普遍论 (linguistic universalism) 观点，认为尽管不同语言在表层结构上具有多样性，但其深层内容是一样的，与人类的认知加工并无直接联系，不会造成不同语言不同文化下人们不同的认知行为 (Malt & Wolff, 2010)；另一派则持语言相对论 (linguistic relativism) 观点 (Sapir, 1921; Whorf, 1956)，认为不同语言代表了不同的世界观，语言间的巨大差异会导致人们体验及思维上的差异。持这两种观点的研究者用不同的方法在不同范围内分别找到了大量支持性证据，其中对类别知觉的探讨是最主要的一部分。

类别知觉是指人们对于连续变化的客体刺激感知到的是非连续的、属于不同类别客体的一种现象，如连续变化的可见光波在人类感知的方式下形成了界限分明的七种色彩 (McCullough & Emmorey, 2009)。在类别知觉过程中，类别间客体的差异得到强化，而类别内客体的差异则被弱化，由此人们对类别间客体的感知要优于对类别内客体的感知，即类别知觉效应 (Goldstone & Hendrickson, 2010)。人类对很多客体的感知过程都存在类别知觉效应，如颜色（如 Gilbert, Regier, Kay, & Ivry, 2006)、线条 (Franklin, Catherwood, Alvarez, & Axelsson, 2010)、音符和弦 (Klein & Zatorre, 2011)、语音 (Peng et al., 2010)、情绪面孔（如 Fugate, 2013) 等。

语言作为人类认知系统中特有的结构复杂的心理表征系统，不仅是人际交流的重要媒介和工具，更是整个认知系统操纵各种心理表征并协调各个层次

---

* 中央高校基本科研业务费专项资金（项目号：CCNU14A02015)，教育部人文社科项目 (13YJA190005)。

通讯作者：刘思耘，E-mail: liusy@mail.ccnu.edu.cn

认知加工的工具 (Gentner & Goldin-Meadow, 2003)。探讨语言对类别知觉过程的影响有助于人们更好地了解语言在整个认知加工过程中，特别是在初级认知加工过程中，所起到的重要作用。

1.1 语言与类别知觉的关系

前人在探讨语言与类别知觉关系时，发现了截然不同的两种结果：一种是语言对类别知觉没有影响；而另一种则相反，认为语言对类别知觉有显著的影响。这两种结果分别得到了不同实证研究的支持。

1.1.1 支持语言普遍论，认为语言对类别知觉无影响的研究

早期研究者在幼儿及前语言阶段婴儿群体中就发现了类别知觉效应，认为如果未习得语言的婴幼儿都可以表现出类别知觉效应，那么该效应很有可能是不受语言影响的 (Franklin & Davies, 2004; Franklin, Pilling, & Davies, 2005; Franklin, Clifford, Williamson, & Davies, 2005a, Franklin et al., 2005; Franklin, Drivonikou, Bevis, Davies, Kay, & Regier, 2008; Ozturk, Shayan, Liszkowski, Majid, 2013)。在成人被试中也有研究者发现，语言标签并未对类别知觉起到促进作用 (Holmes & Wolff, 2012)。如 Sauter 等人比较了德语母语者和玛雅人母语者对愤怒和恶心情绪面孔的类别知觉反应，其中玛雅语中并没有两个不同的词汇分别对应恶心和愤怒两种情绪，结果发现这两种被试对上述两种情绪及其它情绪均可做到类别知觉 (Sauter, LeGuen, & Haun, 2011)。

分析这类研究的一个共性，发现大部分由于被试（为婴幼儿）的特殊性，研究中的数据采集也不同于很多其它相同主题的研究。如为了针对婴儿被试而采用习惯化范式 (Franklin & Davies, 2004)，或者采用眼动技术 (Franklin et al., 2005; Ozturk et al., 2013)，分析婴儿对同类或异类刺激注视时长是否有显著差异。进一步分析这两种研究范式所涉及的被试任务及所采集的数据类型，可以发现，无论是对新颖刺激做出新奇的反应，还是注视点停留在差异性更大的刺激上多一些时间，这些反应类型更多体现的是一种初期自动化知觉加工的结果。而早期知觉加工不受到语言

因素的影响并不表示输入刺激在后期认知加工中不受语言因素的影响。当被试在测试中被要求完成的任务类型涉及更为复杂的后期认知加工时，那么很有可能会出现不一样的结果。

1.1.2 支持语言相对论，认为语言对类别知觉有影响的研究

尽管上述研究发现语言与类别知觉是相对独立的，但更多的研究发现语言和类别知觉有着密不可分的关系。具体来说，是语言中的语义类别扩大了客体类别间的差异而缩小了类别内的差距，导致在一定程度上影响了人们对客体的知觉 (Gumperz & Levinson, 1996)。研究者认为在类别知觉过程中，语言的标签作用使得类别知觉效应更加显著（如 Gilbert et al., 2006; Roberson, Damjanovic, & Pilling, 2007; Roberson, Pak, & Hanley, 2008)。

首先，前人的研究发现，如果某种语言用不同的符号标记出两个客体属于不同的类别，则这种语言标签的作用似乎能够促进个体更好地对类别间和类别内客体进行再认和知觉判断（如：Roberson, Davies, & Davidoff, 2000; Roberson, Davidoff, Davies, & Shapiro, 2005; Lupyan, 2012; Lupyan, Rakison, & McClelland, 2007; Peng et al., 2010; Zhou et al., 2010)，甚至在记忆上造成差异 (Roberson & Davidoff, 2000)。这些研究通过对比不同母语者在类别知觉过程中的不同表现，推断出语言标签在类别知觉中起到的重要作用，相关证据来自于多个语言的母语者对相同材料进行类别知觉的不同表现 (Roberson, Davies, & Davidoff, 2000; Roberson et al., 2005; Goldstein, Davidoff, & Roberson, 2009; Winawer et al., 2007; Roberson et al., 2008; 张积家，刘丽红，陈曦，和秀梅，2008; 谢书书，张积家，和秀梅，林娜，肖二平，2008; McCullough & Emmorey, 2009)。

研究者还利用视觉搜索范式发现了客体类别知觉的右视野优势效应，该效应是语言对类别知觉影响的最直接体现。研究者认为由于语言加工的主要区域在左半球，因此如果语言对类别知觉过程产生影响，那么在左-右视野靶子刺激分别呈现的视觉搜索

过程中, 右视野 – 左脑这个通道会由于得到语言的促进作用, 使得被试能更快地对类别间或类别内的客体做出判断, 即产生右视野优势效应 (Gilbert et al., 2006; Gilbert, Regier, Kay, & Ivry, 2008; Paluy, Gilbert, Baldo, Dronkers, & Ivry, 2011)。而右视野优势的产生随着语言在人类婴幼儿时期的发展也有一个逐步变化的过程, 即婴幼儿在习得语言之前对类别知觉呈现出右脑优势效应 (Franklin, Drivonikou, et al., 2008), 而随着语言的逐步习得, 其类别知觉加工的优势才逐步从右脑向语言加工所在区的左脑转移 (Franklin, Drivonikou, et al., 2008)。

语言标签对类别知觉的影响不仅得到行为学研究结果的支持, 还得到大量来自认知神经机制研究的支持。研究者们通过采集脑电 (EEG 或 ERP) 和脑成像 (MRI 或 fMRI) 的数据进一步验证了语言对类别知觉过程的影响。如 ERP 脑电研究发现, 负责语言加工的左半脑在刺激左 / 右视野呈现时能够产生更大波幅的 N2pc 成分, 表明左半脑给予了刺激更多的注意分配, 刺激得到更强劲的加工 (Liu, Li, et al., 2009; 姚树霞, 杨东, 齐森青, 雷燕, Ding, 2012)。还有研究发现, 在语义上有差别 (即类别间) 的颜色会诱发视觉失匹配负波 (vMMN), 而无差异的则不会 (Thierry, Athanasopoulos, Wiggett, Dering, & Kuipers, 2009; Mo, Xu, Kay, & Tan, 2011)。

1.2 自我关联对客体类别知觉的影响

作为高度社会化的人类来说, 其思维的方式不仅仅只受到语言的影响, 还受到其它诸多来自社会和人格特质方面因素的影响, 如客体与自我关联的程度被发现有可能是影响类别知觉的重要影响因素。如 Lupyan 等人 (2007) 让被试在与客体建立高自我关联的情况下, 分别在有或无语言标签的条件下对新颖客体进行分类学习, 发现学习了语言标签的被试表现出了更好的分类能力。而 Holmes 和 Wolff (2012) 在研究中仅仅是要求被试对毫无意义的新颖图片进行分类, 新颖的客体与被试自我之间毫无关联, 被试在学习或未学习新颖客体语言标签的条件下完成视觉搜索任务, 结果并未发现两组被试的类别知觉效应有显著差

异。由此可推断出, 这两个研究结果的差异很可能是由于一个研究操纵了客体与自我的关联, 而另一个则没有。自我在类别知觉过程中有可能起到重要影响作用。但 Lupyan 等人 (2007) 的研究并没有系统地操纵自我关联与新颖客体之间的关系, 原因在于作者在该研究中的主要目的是探讨冗余的语言标签是否会促进新颖客体分类的学习, 而且作者让被试与新颖客体建立高自我关联是促使被试进行类别学习的一种手段, 自我关联并不是作者探讨的重点。因此, 自我关联与新颖客体类别之间的关系本质还不是很清晰。

早期研究 (Klein & Loftus, 1988; Klein, Rozendal, & Cosmides, 2002; Rogers, Kuiper, & Kirker, 1977) 发现, 自我的意识一旦参与到个体的信息加工过程, 其加工的模式和效率都会有所不同, 即出现自我参照效应 (self-reference effect), 如记忆效果会明显得到提升。后期多个研究为自我参照效应提供了支持性证据 (Kesebir & Oishi, 2010; Kim & Johnson, 2012; Sui, He, & Humphreys, 2012; Turk et al., 2011)。研究者认为这是因为与自我关联程度高的加工伴随有独特的动机和情感意义 (袁翠平, 卢光明, 2010), 或伴随出现自我意识 (Ma & Han, 2010), 从而促进个体对这些信息的反应。

研究者们还从神经机制的角度去探讨自我的加工脑区, 发现了相关的主要大脑结构及其分布状况。其中有研究发现, 关于自我的神经机制是广泛分布于左、右脑区的 (Northoff et al., 2006; 程蕾, 陈熙海, 黄希庭, 2011; 杨帅, 黄希庭, 傅于玲, 2012), 个体所表征的关于自我的大部分信息都是全脑表达的 (Powell, Macrae, Cloutier, Metcalfe, & Mitchell, 2010), 这一点与语言加工的脑区分布是不同的。

1.3 问题提出及研究假设

在前人的相关研究中, 越来越多的证据表明语言标签对客体类别知觉的影响是显著的、稳定的, 但同时也有研究对这个结论不断提出了质疑, 并提供了确切的数据进行反驳。如果说人类在习得语言之前其类别知觉加工系统已经形成, 当习得语言后, 语言逐步对这个过程产生了影响, 而且这个影响最终发生在早期无意识阶段 (Mo et al., 2011), 即达到高

度自动化程度，那么在语言标签形成的初期，语言对类别知觉的影响应该有一个从控制性加工逐步转变成自动加工的过程，其影响力也是由小变大的，但对于这个过程似乎前人的研究中还未涉及到。因此，现在还不清楚语言标签的表征在形成的初期是否会对类别知觉产生影响或产生何种程度的影响，这个影响过程是否是一个随着表征程度逐步稳定而趋于显著的过程。尽管前人采用不同的技术手段和实验范式，从多个角度为语言相对论提供了大量支持性证据，但毕竟还是有些证据显示语言在类别知觉过程中的影响力是缺失的，由此推断语言在整个认知加工的过程中应该呈现一种动态的影响模式，而不是固定不变的。

在前人研究的基础之上，当前研究从动态的角度探讨语言标签的不同表征程度是否会对类别知觉有不同的影响；同时，根据前文的分析，自我很可能也是影响类别知觉的因素之一，因此将语言与自我这两个因素放在一起进行观察，可以进一步揭示语言与其它因素综合影响类别知觉过程的机制。这样做的目的是，首先，前人在探讨类别知觉过程中很少将语言与其它影响因素综合起来进行观察，而在现实情况下，任何的认知加工都掺杂了复杂的多个影响因素；其次，如果类别知觉会受到来自高级认知层面的语言的影响，那么很有可能也会受到其它抽象概念的影响，那么在与其它因素产生交互的过程中，语言的影响力又是如何表现的呢？该结果将有助于人们更好地理解语言与类别知觉的关系本质。

鉴于此，当前研究将围绕个体在新颖客体类别学习过程中与客体、语言标签产生的不同关系提出以下3个假设：

第一，语言标签与客体之间不同的表征程度对类别知觉过程有影响。

第二，语言标签表征程度较低情况下，客体与自我关联程度高的对类别知觉的影响更大。

第三，语言标签表征程度较高情况下，语言标签和自我关联两个因素对客体类别知觉均有影响，但语言标签的影响作用更大。由于前人大量研究显示语言

标签对类别知觉是有影响的，而且采用的语言标签都是表征程度高且相对稳定的，如熟悉的颜色的名字，熟悉客体的名字等；而仅仅只有 Lupyan 等人 (2007) 及 Holmes 和 Wolff (2012) 两个研究的结果发现了自我对类别知觉的影响。因此，从前人研究的范围和深度来预测，语言的影响力有可能更大、更稳定，这也是在假设中预测语言标签的影响力更大的主要原因。

2 实验前测：研究材料的测评及筛选

2.1 前测目的

由于整个研究均采用连续变化的新颖客体，但前人在探讨连续变化的客体类别知觉时仅采用了颜色，因此为了确保实验所选的材料刺激在类别学习之后会产生典型的类别知觉效应，所以首先对所制作的材料进行了前测。同时，在这个前测过程中，由于被试所学习的客体未被赋予任何语言标签或其它社会意义，因此对于被试来说，其分类判断的标准应该主要是客体本身所展示的不同的特征，即形状上的差异。也就是说，被试必须要能够根据一系列客体在形状上的差异就可以把它们区分成两种类别，产生明显的类别知觉效应。而在后面正式实验中，这一系列客体被赋予了某种语言标签，具备了某种与人有关的社会意义。在这种情况下，被试进行类别判断的标准除了形状上的差异之外，还有语言标签所带来的差异。

2.2 前测方法

2.2.1 被试

招募大学生被试10名，均为右利手，母语为汉语，且视力或矫正视力正常。每位被试完成实验后均可获得精美礼品一份。

2.2.2 实验材料

原始图片下载自卡内基梅隆大学相关网址 (www.tarrlab.org) 中的新异物体刺激 Greeble Generator。原始图片材料原本是红色，经过灰白处理之后，再使用专门软件 (Abrosoft FantaMorph, www.fantamorph.com) 制作成两个新颖客体之间连续等距渐变的 12 张图片 (Fugate, Gouzoules, & Barrett, 2010)，相邻图片无论是类

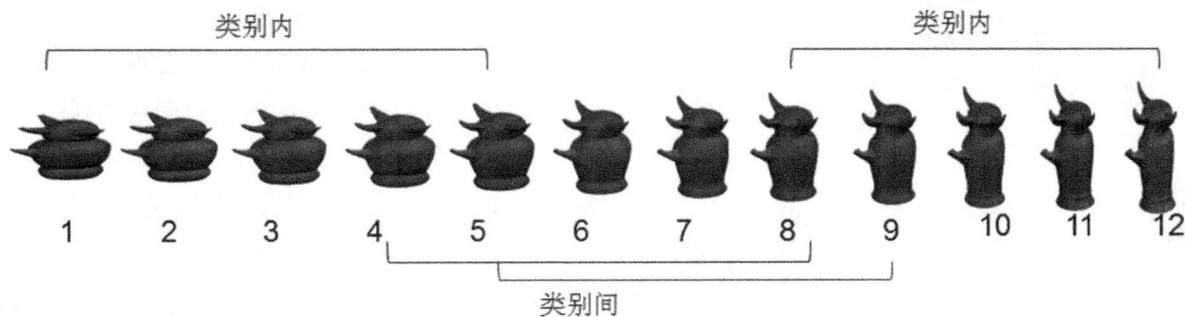

图 1　新颖连续渐变剪影图形

别内还是类别间的差异都是相等的 (如图 1)。这 12 张新颖的图片是使用软件 (AbrosoftFantaMorph, www.fantamorph.com) 在两个原始图片的基础上制作成连续等距渐变的 12 张图片。其中 1 至 6 的新颖图片更多是基于同一个客体，因此相互间为类别内关系；而 7 至 12 则更多基于另一个客体，因此相互间也为类别内关系。但 1~6 中的任意一张图片和 7~12 中的任意一张图片则为类别间的关系。这 12 张图片两两间类别内、类别间关系是由它们的原始图片决定的。

2.2.3　实验设计

采用(客体类别: 类别间 vs. 类别内) 单因素设计，因变量为被试判断刺激所在视野的反应时。

2.2.4　实验程序

采用 E-Prime 2.0 软件对实验设计进行编程，整个实验流程在电脑上进行。被试端坐于实验室中，

片。熟悉过后实验开始。12 张图片将单个随机出现，被试快而准确的按键判断所出现的图片属于哪一类，并按左、右键进行反应。每张图片呈现前有 1 s 的"+"注视点呈现时间，随后图片呈现 3 s。被试在 3 s 内做出反应的话，则程序自动跳转到下一个试次。每次按键后屏幕上将会出现反馈 (300 ms)。如果被试反应错误或未及时反应，则随后原图会再呈现一次 (700 ms)，被试熟悉过刚才看到的图片后才进入下一个试次。每两个试次间的空屏时间为 500 ms。每个实验单元包含 12 个试次，连续渐变的每个客体图片均只出现一次。直到某个实验单元中的正确率达到 100%，屏幕上才显示类别学习阶段结束，被试停止学习进入测试阶段。类别学习阶段的实验流程图见图 2。

在测试阶段，被试完成类别判断的经典视觉搜

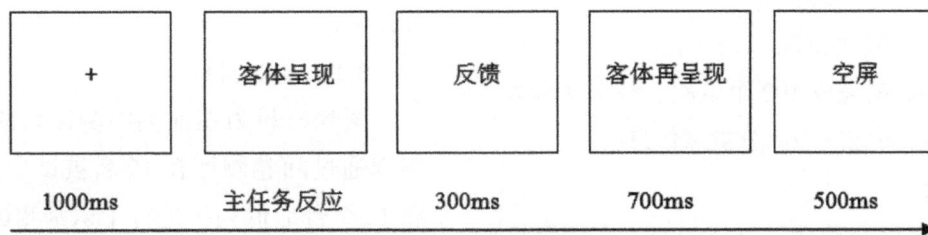

图 2　类别学习阶段实验流程图

与电脑屏幕保持合适距离，屏幕中央位置与被试视线平齐。实验分为两个阶段：类别学习阶段和测试阶段。类别学习阶段的主要目的是为了让被试学习并掌握新颖客体的不同类别。在类别学习阶段之前先给被试同时呈现两个类别客体的典型图片，让被试熟悉图片，并告知被试每类客体都有多张不同图

索任务 (见 Gilbert et al., 2006) (如图 3)。为确保目标刺激投射在大脑的半侧，搜索圈的视角设定为 8.5 至 10 度，偏心视角为 4.25 至 5 度 (蔡厚德，1999)。被试的眼睛与屏幕距离得到控制，以确保靶子刺激分别投射到左右视野。被试需要在快速呈现的搜索圈中判断不同的那一张图片是在左侧 ("1"或"2"的位置)

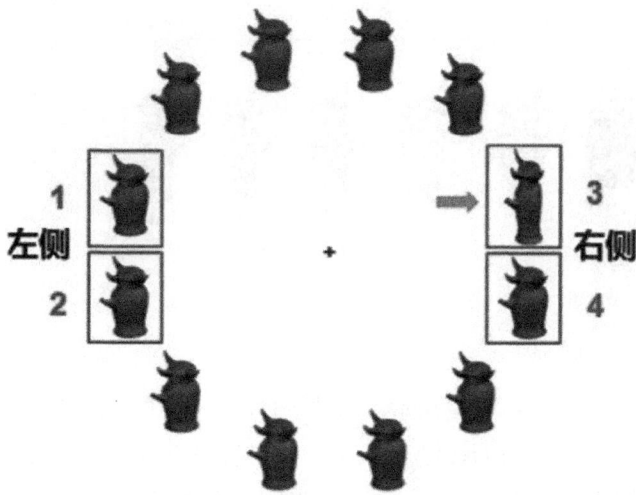

图 3　视觉搜索圈

还是右侧（"3"或"4"的位置），并按键盘上相应的"左"、"右"键反应。

实验流程如下：首先屏幕中央呈现一个红色的"＋"，紧接着快速呈现一个图片圈，要求被试在实验中始终保持对图片圈中央"＋"的注视，不随意转移目光。搜索圈消失后尽量快而准的做出判断，按键反应之后"＋"消失，进入下一个判断任务。测试阶段的实验流程图见图4。每个实验单元的试次设置具体如下：选取4对图片作为制作图片圈的原始图片，类别内和类别间各2种。每个图片均要充当一次目标和分心刺激，因此产生8种搭配情况。此外，由于靶子刺激会出现在4个不同的位置上，因此在原来8种搭配的基础上共产生32种搭配。每种情况再重复6次，那么每个被试总共需要完成192个试次。整个实验被均衡切分成三部分，每完成一部分被试休息一次。

2.3　结果分析

被招募的10名参与前测的被试中有一名因测试阶段正确率低于50%而被剔除，最终有9份有效数

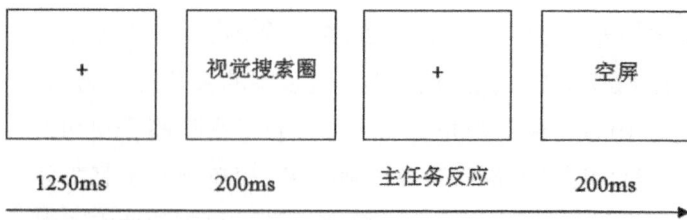

图 4　测试阶段实验流程图

据进入统计分析。运用 SPSS 17.0 对被试的类别变量（类别间 vs. 类别内）进行 t 检验，发现类别的主效应显著（$t(8) = 2.63, p = 0.03$），被试对图片类别间的反应显著快于类别内图片刺激的反应，表明所制作的新颖客体在类别学习后可以产生经典类别知觉效应。

3　实验1：语言标签表征的不同程度对新颖客体类别知觉的影响

3.1　实验目的

探讨语言标签的不同表征程度对新颖客体的类别知觉效应的影响。通过操纵两组不同被试学习新颖客体语言标签的次数，使新颖客体的语言标签在两组被试中的表征程度不一样，从而将被试分成语言标签表征程度高和表征程度低的两个组。实验1的假设是，如果语言标签的表征程度对类别知觉效应有影响，那么两组被试在类别知觉过程中的表现应该是不一样的：语言标签表征程度高的被试会表现出更为显著的类别知觉效应。

3.2　实验方法

3.2.1　被试

招募大学生被试34名（男性9名，女性25名，平均年龄23.4），均为右利手，母语为汉语，且视力或矫正视力正常。每位被试完成实验后均可获得精美礼品一份。

3.2.2　实验材料

图片材料为在前测中测试好的图片。而语言标签则通过问卷测评在10名被试中对12个超低频字和12个普通低频字进行了识别测试。最终根据笔画数、结构和音节等多方面考虑，选取了识别率均为0的两个字"先"和"丑"，并告诉被试这两个字分别念 lí 和 jǐ。

3.2.3　实验设计

采用2（语言标签学习程度：高 vs. 低）×2（客体类别：类别间 vs. 类别内）×2（视野：左 vs. 右）三因素混合设计。其中语言标签学习程度为被试间变量，客体类别和视野为被试内变量。因变量为被试判断

靶子刺激所在视野的反应时。

本实验采用与预实验中相同的视觉搜索研究范式。与预实验不同的是,在类别学习之后、视觉搜索任务之前插入语言标签学习任务。语言标签学习分高、低两种表征程度在被试间进行,其中语言标签低表征程度学习组的被试对每个客体的名称学习 2 遍,而语言标签高表征程度学习组的被试对每个客体的名称学习 7 遍。在实验 1 中,类别学习阶段和测试阶段的视觉搜索任务要求与预实验相同。

3.2.4 实验程序

采用 E-Prime 2.0 软件对实验设计进行编程,整个实验流程在电脑上进行。实验分 3 个阶段:类别学习阶段,标签学习阶段和测试阶段。类别学习和测试阶段过程和前测中一样。

在语言标签学习阶段,首先告知被试两类图片各自的名字("先"和"丑"),要求被试认真学习并记住它们相应的名称。实验开始时,首先屏幕中央出现一个红色的注视点"+"(1000 ms),随后出现一张图片 (1000 ms),同时下方呈现其相应的语言标签名称。要求被试尽量快而准确地判断名称是否与图片相匹配,并按"是"或"否"键进行反应。如被试未在图片出现后 3 s 内做出反应,则程序自动跳转到下一个试次。每次按键后无论对错,屏幕上将立即出现反馈 (300 ms)。每两个试次间的空屏时间为 200 ms。语言标签学习阶段的实验流程图见图 5。每个实验单元 12 个试次,每张图片出现 1 次。被试每学习晚所有图片后休息一次,休息时屏幕中呈现正确匹配的客体和语言标签。语言标签学习程度高的被试组对每个客体的标签学习 7 遍,语言标签学习程度低的被试组对每个客体标签学习 2 遍。

主任务反应

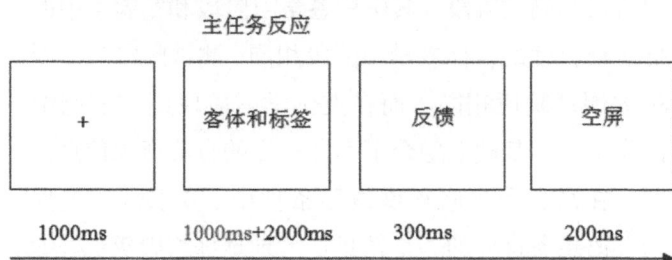

图 5   语言标签学习阶段实验流程图

3.3 结果分析

本实验中类别学习与标签学习阶段均未剔除数据,而测试阶段视觉搜索任务中的正确率和反应时的极端数据处理标准如下:首先,测试阶段正确率方面:为了排除被试猜测判断,根据二项分布定律,删除正确个数为 108 以下的被试数据。猜对与猜错的概率 p 和 q 均为 0.5,n=192,根据公式:$N=u+1.645\alpha$ ($u= np$ =96,α 的平方 =npq) 得出完全凭猜测猜对 108 次以下的概率为 95%,因此将剔除正确率低于 108/192=0.56 的被试数据。同时测试阶段反应时方面:也将反应时低于 20ms 和超过 3000ms 的极端反应时试次进行了剔除。

数据经整理后共有 34 份有效数据进入统计分析,高、低标签学习条件下各 17 名被试。运用 SPSS 17.0 软件首先进行 2(标签学习程度:高 vs. 低)×2(客体类别:类别间 vs. 类别内)×2(视野:左 vs. 右) 的三因素 ANOVA 分析。重复测量的方差分析发现,标签学习程度、客体类别和左、右视野三者之间交互作用

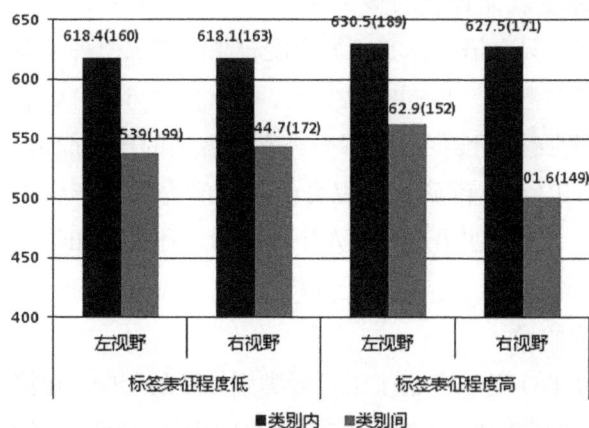

图 6   高、低语言标签学习程度下类别知觉平均反应时 (ms)

显著,$F(1,32)$ =4.79,p = 0.036,η_p^2=0.13。进一步进行简单交互效应分析发现,低标签学习组的客体类别(类别间 vs. 类别内)和视野(左 vs. 右)的交互作用不显著,$F(1,32)$=0.08,p=0.77,η_p^2= 0.01;而高标签学习组的客体类别 (类别间 vs. 类别内) 和视野 (左 vs. 右) 的交互作用则显著,$F(1,16)$=7.43,p=0.015,η_p^2=0.32。在高标签学习条件下,客体类别:(类别间 vs. 类别内) 与视野:(左 vs. 右) 的反应时的交互作用显著,$F(1,32)$=7.87,p

=0.008, η_p^2=0.20。进一步进行简单简单效应分析得出：在左视野中，被试对类别间目标刺激的平均反应时显著快于类别内，$F(1,33)$ =5.05, p=0.031, η_p^2=0.13; 右视野中，被试对类别间目标刺激的平均反应时要比类别内的快，$F(1,33)$ =17.68, p<0.001, η_p^2=0.35。表明标签学习程度较高条件下产生了类别知觉右视野优势效应（见图6）。

此外，在标签学习阶段，高标签学习组被试的第七遍正确率(71.1%)显著高于低标签学习组被试的第二遍正确率(57.8%), t=2.47, p=0.019。客体类别知觉的主效应非常显著，$F(1,30)$=23.50, p<0.001, η_p^2=0.42。在低标签学习条件下 [$F(1,16)$ =5.83, p=0.028, η_p^2=0.27] 高标签学习条件下均分别显著 [$F(1,16)$=34.21, p< 0.001, η_p^2=0.68], 即无论语言标签学习情况，类别间的判断总是优于类别内的判断。

3.4 讨论

实验1的结果证实了实验的假设，即高、低不同语言标签表征程度的被试在类别知觉过程中有不同的表现，只有语言标签表征程度较高的被试出现了右视野优势效应。但同时也发现，增强客体与语言标签的表征强度只能促进右视野类别间客体的辨别能力，对类别内客体辨认和左视野的辨别能力均不产生影响，进一步表明语言通过左半大脑对客体类别知觉产生作用，这在一定程度上支持了 Sapir-Whorf 的语言相对论假设。

实验1单独考察了语言对类别知觉的影响，在接下来的实验中，将观察语言与其它因素（自我关联）交互影响类别知觉的过程。但为了更好地观察这两者的交互作用，首先单独观察自我关联程度对类别知觉的影响。

4 实验2：客体与自我不同关联程度对类别知觉的影响：语言标签表征较低情况下的探讨

4.1 实验目的

探讨在新颖客体标签表征程度低的条件下，被分类客体与自我的关联程度与类别知觉效应及右视野优势效应的关系。前人在探讨语言标签对类别知觉影响的研究中得到不同结论，如当客体涉及自我关联时，其类别知觉会得到促进 (Lupyan et al., 2007); 而当客体与自我无关时，有无语言标签的学习结果均无显著差异 (Holmes & Wolff, 2012)。导致结论不一致的原因有可能是由于客体与自我关联程度在两个实验中并未得到统一处理的缘故所致。因此，在实验2中，语言标签的表征程度被控制在一个相对较低的水平，自我与客体知觉的关联程度得到相对独立地系统地操纵，以期观察到类别知觉效应如何受到自我关联的影响。

4.2 实验方法

4.2.1 被试

招募大学生被试30名（男性10名，女性20名，平均年龄22.6)。被试均为右利手，且母语为汉语。所有被试视力或矫正视力正常。每位被试完成实验后均可获得精美礼品一份。

4.2.2 实验材料

通过谷歌搜索引擎找到代表被试自我及外星人研究专家的人形剪影图片，而语言标签材料（"尣"和"卂"）和外星人图片材料则与实验1相同。

4.2.3 实验设计

采用2(客体与自我的关联程度：高 vs. 低)×2(客体类别：类别间 vs. 类别内)×2(视野：左 vs. 右)三因素混合设计。其中客体与自我关联程度为被试间变量，客体类别和视野为被试内变量。因变量为被试判断刺激所在视野的反应时。

4.2.4 实验程序

实验2分为3个部分：类别学习阶段、标签学习阶段和测试阶段。其中标签学习阶段和实验1中的低表征程度语言标签学习阶段相同，测试阶段与前测中的测试阶段相同。而在类别学习阶段则在前测中的类别学习基础上融合了不同程度的自我意识信息。

在高自我关联程度诱导条件中，要求被试根据指导语想象这样的一个情景："地球即将毁灭，科学家及时寻找到另一个适合人类居住的星球，人类开始

初始状态：　　　　　　　　　当你靠近外星人时：　　　　　　　　当你远离外星人时：

图 7　自我关联学习阶段使用的材料

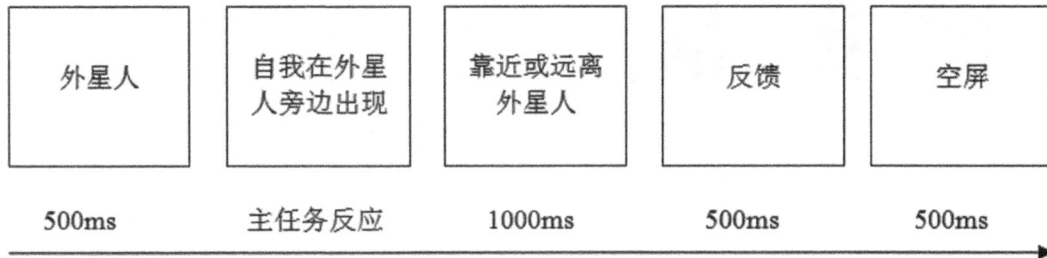

外星人	自我在外星人旁边出现	靠近或远离外星人	反馈	空屏
500ms	主任务反应	1000ms	500ms	500ms

图 8　高、低语言标签学习程度下类别知觉平均反应时 (ms)

逐渐往该星球迁移。而你刚刚抵达星球，得知星球上有两种外星人，其中一种是安全型较友好的，人类可以靠近；另一种是危险型，遇到就会对人类会发起致命攻击，所以见到了必须远离。你只知道两种外星人的样貌，而不清楚哪种是安全的，哪种是危险的。为了在星球上生存下去，被试必须学会辨认出哪种外星人是安全友好可以靠近的，哪一种又是危险需要远离的"（见图 7）。在实验过程中，首先出现一个外星人（即前测中选取的新颖图片），紧接着一个代表人的人形剪影随机出现在外星人左边或右边，该人形剪影代表被试的虚拟自我。要求被试通过在键盘上按"←"键或"→"键来控制自己是远离还是亲近所呈现的外星人。主试事先并不告诉被试哪种外星人是友好的、哪种是致命的，需要被试亲自通过尝试及得到的反馈逐步学习到。

在低自我关联程度诱导条件中，被试读到的指导语基本与高自我关联条件下相同，唯一不同之处在于，不是被试自己而是一位外星人研究专家来到了外星球，这位专家必须对外星人进行安全型和危险型的分类工作。被试在屏幕上控制的人形剪影为他人（外星人研究科学家）而并非自我。被试的任务和具体实验流程与高自我关联条件下的类别学习阶段完全相同（见图 8）。

每个实验单元中的 12 张客体图片均将单个随机出现一次，每一个都判断正确后类别学习阶段才能结束，否则被试将再学习一遍直到某一个实验单元正确

率达到 100% 才进入语言标签学习阶段。

4.3 结果分析

在本实验中，类别学习与标签学习阶段产生的数据均有效，没有数据被剔除。而在测试阶段视觉搜索任务中，正确率和反应时的极端数据处理标准如实验 1。数据经初步整理后发现，参与实验的 30 名被试中有 2 名被试因测试阶段正确率过低，其相应数据被剔除，最后总共有 28 份有效数据进入统计分析。采用 SPSS 17.0 软件对数据进行重复测量方差分析发现，客体类别和客体与自我的关联程度的交互作用显著，$F(1,26)=5.77$, $p=0.024$, $\eta_p^2=0.18$；客体类别的主效应显著，$F(1,26)=21.94$, $p<0.001$, $\eta_p^2=0.46$。对类别与自我关联程度的交互作用做简单效应分析发现，在低自我关联程度下，被试并未产生类别知觉效应，$F(1,26) = 2.60$, $p=0.12$, $\eta_p^2=0.09$；而在高自我关联程度下才产生了显著的类别知觉效应，$F(1,26)=25.11$, $p < 0.001$, $\eta_p^2=0.49$。但在两种关联程度条件下均未产生右视野优势效应（见图 9）。

4.4 讨论

实验 2 结果表明，在控制语言标签表征程度较低的情况下，当客体与自我关联程度较高时，被试表现出了非常显著的类别知觉效应，即客体类别间的辨认优于类别内的辨认；而当客体与自我关联程度较低时却未出现类别知觉效应。研究还发现，相对于低自我关联条件下的被试，高自我关联条件下的被试对类别

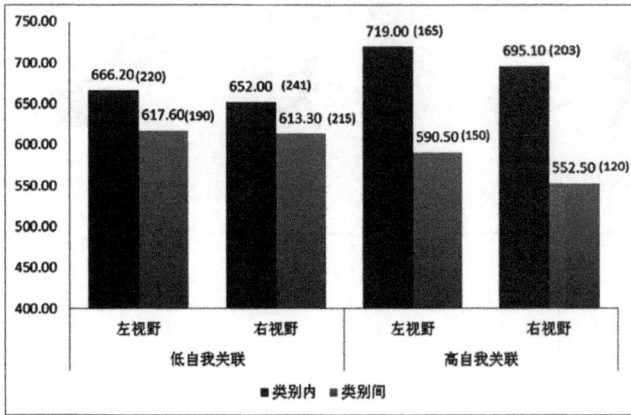

图 9 语言标签表征程度低时，高、低自我关联不同程度条件下类别知觉平均反应时 (ms)

间目标刺激的辨别更快了，且对类别内的目标刺激辨别变慢了。这似乎表明自我关联不仅增加了被试对类别间刺激的心理差异且进一步模糊了被试对类别内刺激的心理差异。这一点与前人对自我意识的探讨结果较为一致，表明由于自我关联导致了自我意识的提升，使得个体的内向注意力增大，从而进一步使得类别间的心理差异放大，类别内的心理差异缩小。同时这一点与语言标签对类别知觉的影响是不同的，语言标签在表征程度高的情况下并没有进一步模糊类别内客体间的差异，也没有进一步放大类别间客体间的差异，表明语言标签和自我关联对类别知觉的影响本质是不同的。

同时前人多个研究表明，大脑关于自我觉知的神经机制是同时分布于大脑的左右半球多个区域的 (Northoff et al., 2006; 程蕾等，2011; 杨帅，黄希庭，傅于玲，2012)，并未出现大脑偏侧化表征的现象，因此实验 2 没有发现左或右视野优势效应与自我的脑机制研究结果是一致的。

5 实验 3：客体与自我不同关联程度对类别知觉的影响：语言标签表征较高情况下的探讨

5.1 实验目的

实验 2 结果表明，当语言标签作用不显著时，高自我关联程度的类别知觉效应是显著的，自我意识对新颖客体的类别知觉有显著影响效果。但实验 2 并未探讨当语言标签也对客体的类别知觉有影响时，自我与客体的关联程度和语言标签的表征程度如何起到一个交互的作用。实验 3 假设在语言标签表征程度较高情况下，当语言标签和自我关联程度同时参与作用时，语言标签和自我关联两个因素对客体类别知觉均有影响，但语言标签的影响作用会更大。

5.2 实验方法

5.2.1 被试

37 名大学生被试（男性 20 名，女性 17 名，平均年龄 21.9)，均为右利手，母语为汉语，且视力或矫正视力正常。每位被试完成实验后获得精美礼品一份。

5.2.2 实验材料

与实验 2 相同。

5.2.3 实验设计

采用 2(标签自我关联程度：高 vs. 低)×2(客体类别：类别间 vs. 类别内)×2(视野：左 vs. 右) 三因素混合设计。其中客体与自我的关联程度为被试间变量，客体类别和视野为被试内变量。因变量为被试判断刺激所在视野的反应时。

5.2.4 实验程序

实验 3 的 3 个部分与实验 2 基本相同，所不同的是语言的标签学习阶段是学习 7 遍的高程度语言标签表征，而不是学习 2 遍的低程度语言标签表征。

5.3 结果分析

在本实验中，类别学习与标签学习阶段所产生的数据均有效，而在测试阶段视觉搜索任务中，正确率和反应时的极端数据处理标准如实验 1。数据经初步整理后发现，在所招募的 37 名被试中有 1 名被试在测试阶段的正确率过低，其相应数据被剔除，最后共有 36 份有效数据 (高、低标签学习条件下各 18 名被试) 进入最终的统计分析。采用 SPSS 17.0 软件对有效数据进行重复测量的方差分析，发现客体类别和视野的交互作用显著，$F(1,34) =4.42$, $p =0.043$, $\eta_p^2=0.12$。对类别与视野的交互作用做进一步简单效应分析，发现左、右视野分别产生了显著的类别知觉效应 [F 左视野 $(1,34)=28.60$, $p<0.001$, $\eta_p^2=0.46$; F 右视野 $(1,34)= 35.62$,

$p<0.001$, $\eta_p^2=0.51$]。其中左视野中类别间的反应时比类别内的快 65ms，右视野中类别间的反应时比类别内的快 104ms，从某种程度上来说右视野的判断更具优势，将左右视野类别内与类别间的平均反应时差值进行 t 检验，结果发现两者差异显著，$t(35) = -2.11$, $p=0.04$。客体类别的主效应显著 [$F(1,34)=50.03$, $p<0.001$, $\eta_p^2=0.59$]，即类别间的辨别比类别内的要快，而左、右视野的主效应并不显著 [$F(1,34)=1.07$, $p=0.31$, $\eta_p^2=0.03$]，这可能是由于在低自我关联中左视野的类别内和类别间的差异不够大所致。另外，自我关联程度这个变量与客体类别 [$F(1,34) =1.84$, $p=0.18$, $\eta_p^2=0.05$] 和左、右视野 [$F(1,34)=0.05$, $p=0.82$, $\eta_p^2=0.002$] 这两个变量均未产生交互作用，其主效应也不显著，$F(1,34)=0.16$, $p=0.69$, $\eta_p^2=0.01$（见图 10）。

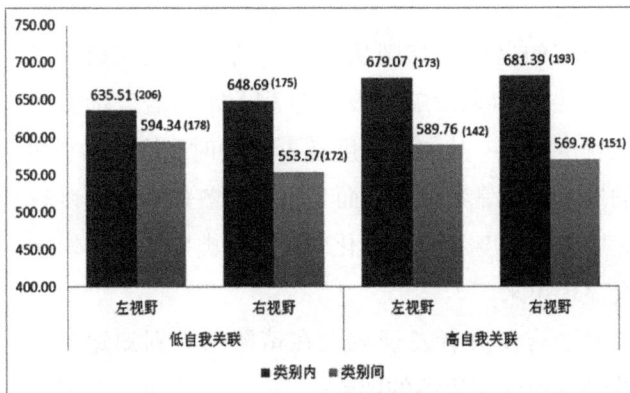

图 10　语言标签表征程度高时，高、低自我关联不同程度条件下类别知觉平均反应时 (ms)

5.4 讨论

在实验 2 中（语言标签表征程度低的情况下），在低自我关联条件下被试并没有表现出类别知觉效应，即类别间的辨认并不占优势；而在高自我关联条件下，客体类别知觉效应才显示出来，即类别间的辨别明显优于类别内的辨认。但在实验 3 中（语言标签表征程度高的情况下），由于自我关联主效应并不显著，且只发现客体类别与左、右视野两因素交互作用显著，表明自我关联无论高低，右视野的类别间辨认要优于左视野的类别内辨认。但由于左、右视野并未表现出主效应，而客体类别的主效应是显著的，因此右视野的类别知觉优势效应并没有表现得很明显。这有

可能是因为语言标签的高表征程度在低自我关联条件下的作用促进了右视野的类别知觉效应，而在高自我关联条件下，客体与自我之间的高关联又促进了左视野的类别知觉效应，缩小了与右视野类别知觉之间的差异。同时，对比实验 2 在低自我关联条件下类别知觉效应不显著，高表征程度的语言标签在实验 3 的低自我关联条件下的作用使得类别间的辨认比类别内的辨认更具优势。因此，综合来看，语言标签和自我关联对类别知觉的影响并不是两者相互简单叠加，而是相互间制约权衡的结果。

但自我关联和语言标签影响力未能叠加的原因并不能通过这个研究来得到确定，这可能有两种原因：一个是两个影响因素在类别知觉过程中有可能会损耗认知资源，如两种力量的权衡过程，从而导致这两个影响力大打折扣；另一个是当两个强有力的因素存在时，只有一个因素起到主要作用。对于这个问题的本质，还有待将来进一步的研究和探讨

6　总讨论

当前研究就类别知觉过程中语言标签表征程度的影响进行了较为深入的探讨。同时，为了进一步揭示语言标签对客体类别知觉影响的本质，引入客体的自我关联变量，并将语言标签和自我关联放在一起观察两者的交互作用，以更好地探讨语言标签和其它变量如何共同影响客体的类别知觉过程。在观察语言标签的影响力时，选择渐变等距的新颖客体作为类别知觉的对象，以最大的限度接近前人对颜色的研究过程。

当前研究设计了 3 个实验。实验 1 首先探讨了语言标签的不同表征程度对新颖客体类别知觉效应的影响，结果发现仅当语言标签表征程度高时，被试才表现出显著的右视野优势效应。该结果虽然总体上支持了语言对类别知觉影响的假设，但也指出一个客体在和一个语言标签建立关联的初期，语言标签对该客体的类别知觉的影响是微乎其微的。只有当该客体的语言标签表征相对稳定之后，才会开始对其类

别知觉产生影响。

实验 2 在语言标签表征程度较低情况下探讨客体与自我之间的不同关联程度如何影响客体类别知觉。该实验结果发现当个体与客体建立自我关联时，其对客体的类别知觉显著受到自我意识的影响，即出现了类别知觉效应，且类别间刺激的心理差异进一步扩大，类别内刺激的心理差异被进一步缩小。而当自我关联较低时，客体则没有表现出类别知觉效应。但实验 2 并未发现右视野优势效应，这一点应该与自我并未出现大脑偏侧化现象、且在大脑左 / 右均有分布有关。

实验 3 探讨在语言标签表征较高条件下，客体与自我的不同关联程度对类别知觉产生的影响。结果显示当语言标签和自我同时作用于客体类别知觉时，两个变量同时对客体类别知觉产生影响。与实验 1 结果不同的是，尽管语言标签表征程度与实验 1 中的一样，但右视野优势效应并没有很明显地表现出来，只发现右视野的类别间辨别力比左视野的类别内辨别力有显著优势，表明语言标签和自我关联在共同对类别知觉起作用。与实验 2 结果不同的是，该实验中自我关联的高低在类别知觉上没有表现出差异，进一步表明语言标签和自我关联对类别知觉起到一种联合作用。

6.1 语言标签对类别知觉过程的影响

对于语言标签与类别知觉的关系，持语言普遍论者认为既然类别知觉可以脱离语言而存在，那么这两者之间应该就没有必然的联系。对这一论点的大部分证据来自于语言仍处于发展期的婴幼儿被试，少量来自成人情绪类别知觉的研究。但是，由于人的高度社会化属性，语言的发展是必然，且语言的加工在后期人的社会化过程中不断得到强化而变得越来越自动化。语言加工的这种逐步自动化属性使得它最终在整个认知加工过程中得到优先处理并进一步影响其它加工过程（如经典的 Stroop 效应），这一点从前人探讨语言与类别知觉关系时分化成两派不同的观点得到了印证。

对于认为语言标签对类别知觉没有影响的研究者来说，他们所招募的被试大多数是语言系统仍处于发展期的婴幼儿，其语言系统的建构并不成熟、完整，其加工还远未达到自动化程度，因此相比其它信息的认知加工并不占据优势，还无法构成对其它认知加工的影响。而对于语言系统发展完善的成人来说，语言符号加工的高度自动化使得语言不可避免参与到其它认知加工的过程中，毕竟语言符号或者概念的建构是基于初级的感知觉加工 (Barsalou, 1999, 2008)，且两者在神经加工机制上享有共同的通道和激活脑区 (Niedenthal, Barsalou, Winkielman, Krauth-Gruber, & Ric, 2005; Niedenthal, Winkielman, Mondillon, & Vermeulen, 2009)，因此认知加工的这种自上而下的加工属性决定了语言符号影响的必然性，这与大量支持语言影响类别知觉过程的研究结果是一致的。因此可以说，前人发现的所谓的相互矛盾的证据实际上是统一的，符合人的语言发展和影响的规律的。

当前的这个研究则从动态的角度在线观察到了语言对类别知觉影响的变化过程，即当语言标签的表征程度较低，其加工速度还不够快时，其对类别知觉的影响是非常有限的；而当语言标签的表征程度较高，其加工速度足够自动化时，其对类别知觉的影响则会显现出来。

至于有研究者发现成人在情绪的类别知觉过程中并未发现语言标签的影响作用，这很有可能是因为涉及情绪信息加工的脑区主要位于边缘系统的杏仁核，而涉及语言符号加工的脑区则在大脑皮层，这与其他研究者所探讨的客体类别知觉过程 (如颜色，线条，音符等) 主要也位于大脑皮层是有区别的。情绪信息的加工既然在脑机制分布上独立于语言符号加工的脑区，那么其与在大脑皮层进行的认知加工过程也很有可能是相对独立的。从进化的角度来看，人类对情绪面孔的识别应该早于语言的加工和表征，且前者对人类生存更具重要意义，其在生理结构和加工上的独立性可以使其可以得到相对完善的保护。因此人类的情绪加工系统相对独立于其它信息的加工 (Sauter et al., 2011) 是符合生物进化要求和规律的，但这个证据不能用来否定语言系统对其它在大脑皮层完成的认知加工的影响。

6.2 自我关联对类别知觉过程的影响

前人研究发现，人类自我意识的能力和程度会影响到人的行为、判断及心理健康 (Wheeler, Morrison, DeMarree, & Petty, 2008)，而探讨自我意识问题的研究者一直对自我意识在认知加工过程中的影响的本质感兴趣 (Klein & Loftus, 1988; Ma & Han, 2010; Symons & Johnson, 1997; Wang, Zhang, & Sui, 2011)。当前研究发现，自我关联对类别知觉有影响，能够产生类别知觉效应，且相比于非自我关联条件进一步扩大了类别间的心理差异、缩小了类别内的心理差异，表明自我关联所导致的自我意识的增强，很有可能是通过注意力内向化导向的方式使得类别知觉过程在心理上的认知消耗减小，从而使类别知觉过程变得更容易。

同时当前研究发现，自我关联与语言标签共同对类别知觉产生影响，且两者的影响作用呈现权衡较量结果，并非简单作用力的叠加。该结果表明：(1) 在探讨语言对其它认知加工影响的时候，有必要考虑除语言以外的其它影响因素的存在。由于语言并不是现实生活中唯一对认知加工过程有显著影响的一个因素，综合考虑其它因素与语言的交互作用是语言认知研究更具生态效应的一个标志。(2) 根据前人研究发现，自我意识对自动化的认知加工是有影响的。那么自我意识在激活过程中，很有可能同时激活了与自我相关的概念表征，而自我概念表征的激活意味着相关语言符号的激活，从而产生类似语言标签的作用。这或许就是造成语言标签和自我关联对客体类别知觉的作用不是简单叠加，而是有重叠的原因。

6.3 对外界因素影响类别知觉过程本质的思考

在人类所生存的客观世界中存在各种各样的客体，这些客体之间的差别是多样化的，有些在外形上有区别，有些在属性或功能上有所不同；有些差别很明显，而有些差别则不易为人所察觉。人类在与这些不同客体的交互作用过程中，为了更高效地处理外界的信息，必然会动用各种资源来帮助自己更有效地对这些客体进行分类。也就是说，在高级认知层面，凡是与分类有关的过程都有可能对客体的类别知觉产生促进作用，而语言标签的建立和自我意识产生的过程应该恰好都具备了这样的属性，所以才导致它们在对客体类别知觉的过程中产生相似的促进作用。

首先，根据前人对语言标签功能的研究，语言标签可以从心理层面对任意两个被划分到不同类别的客体间的差异起到放大的作用，这个作用从本质上来说是一种主观上的分类过程，在两个差异不明显的客体知觉过程中显得尤为重要。人类在前语言阶段的婴儿时期就可以做到类别知觉，表明客体的客观差异在足够明显时，是不必依赖主观上的帮助来完成分类的。但在人类适应外部世界的过程中，有很多时候需要对客观差异不够明显或客观差异过大的一些客体进行类别的划分。针对前者，人类要将看上去相似的两个客体分成两类；而针对后者，则是要将看上去不怎么相似的两个客体分成一类。这时，语言标签的作用在于放大前者的差异、缩小后者的差异，使得前者的微小差异得到注意，而后者的巨大差异得到忽略。语言的这种调节作用，表明类别知觉过程融合了自上而下和自下而上的认知加工过程。

其次，自我意识的过程实际上是将自我及与自我密切相关的因素和他人及任何其它外界因素进行分类的一个过程，这一点从个体对自我和他人具有不同的表征结构和脑区激活模式就可以推断出来 (Burris & Rempel, 2008; Dijksterhuis & Bargh, 2001; Macrae, & Roseveare, 2002; Mitchell, Macrae, & Banaji, 2006; Ruby & Decety, 2001; 杨红升，朱滢，2004; Han et al., 2008; Heatherton et al., 2006; Jenkins & Mitchell, 2011; Kelley et al., 2002; Lombardo et al., 2010; Moran, Heatherton, & Kelley, 2009; Wu, Wang, He, Mao, & Zhang, 2010; Zhu, Zhang, Fan, & Han, 2007)。因此可以说，自我意识的过程在某种程度上也是一种分类的过程，那么这个过程对外界客体的分类过程产生影响就不足为奇了。

而自我意识的产生对客体类别知觉产生促进作用而不是抑制作用，其原因很可能在于当个体对所加工的信息进行自我关联之后，会在随后的信息通达和记忆中会表现出显著优势 (Klein & Loftus, 1988; Ma & Han, 2010; Symons & Johnson, 1997; Wang et al., 2011)。前人 (Vogeley & Fink, 2003) 认为这种优势是由个体看待事物的视角所致。个体在自我关联之后，他们看待事物的角度就变成了第一人称视角 (the first-person perspective)，而这个视角位于个体基于自身为参照所建构的多模态体验空间的中心，是自我意识形成的重要指标。而自我意识又与个体注意力内向化导向紧密相关，自我意识越强，则表明个体的内向注意力越强，个体越不容易受到外界事物的干扰或启动 (Dijksterhuis & van Knippenberg, 2000)，即自我意识的提升或降低会对行为的启动效应 (prime-to-behavior effects) 产生影响，而且对自动化行为的影响也是多方面、多样化的 (Wheeler et al., 2008)。综合这些研究结果，可以推测出，一旦将某个客体或事件自我关联之后，自我意识会随之提升，这种自我意识的提升对同时进行的其它认知行为，特别是已自动化的认知行为，产生促进作用。

7 结论

当前研究通过采用经典的视觉搜索范式，从语言标签和客体的自我关联两个角度探讨了高级概念层面的信息对新颖客体类别知觉的影响。研究结果显示：(1) 语言标签表征程度的增强可促进新颖客体类别知觉的右视野优势效应，且这种优势只体现在对右视野类别间客体的辨认上。(2) 客体与自我关联程度的提高会促进左、右视野的类别知觉效应；且不仅增大了类别间客体的心理差异，还缩小了类别内客体的心理差异。(3) 语言标签的表征程度和客体与自我关联程度同时增强时，语言标签的作用依旧表现出来，但与自我关联的影响产生权衡分配，且其影响力并不足以产生右视野优势效应。

参考文献

Barsalou, L. W. (1999). Perceptual symbol systems. *Behavioral and Brain Sciences, 22*(4), 577–609.

Barsalou, L. W. (2008). Grounded cognition. *Annual Review of Psychology, 59*, 617–645.

Burris, C. T., & Rempel, J. K. (2008). Me, myself, and us: Salient self-threats and relational connections. *Journal of Personality and Social Psychology, 95*(4), 944–961.

Cai, H. D. (1999). Some methodological issues of tachistoscopic visual half-field technique. *Psychological Science, 22*(3), 265–272.

[蔡厚德. (1999). 半视野速示技术的若干方法学问题. *心理科学, 22*(3), 265–272.]

Cheng, L., Chen, X. H., & Huang, X. T. (2011). Role of the right or the left hemisphere? Dispute on neural mechanism of Self-awareness. *Advances in Psychological Science, 19*(9), 1319–1327.

[程蕾, 陈煦海, 黄希庭. (2011). 左脑还是右脑——自我觉知神经机制的述评. *心理科学进展, 19*(9), 1319–1327.]

Dijksterhuis, A., & Bargh, J. A. (2001). The perception-behavior expressway: Automatic effects of social perception on social behavior. *Advances in Experimental Social Psychology, 33*, 1–40.

Dijksterhuis, A., & Van Knippenberg, A. (2000). Behavioral indecision: Effects of self-focus on automatic behavior. *Social Cognition, 18*(1), 55–74.

Franklin, A., & Davies, I. R. L. (2004). New evidence for infant colour categories. British *Journal of Developmental Psychology, 22*(3), 349–377.

Franklin, A., Pilling, M., & Davies, I. R. L. (2005). The nature of infant color categorization: Evidence from eye movements on a target detection task. *Journal of Experimental Child Psychology, 91*(3), 227–248.

Franklin, A., Clifford, A., Williamson, E., & Davies, I. R. L. (2005). Color term knowledge does not affect categorical perception of color in toddlers. *Journal of Experimental Child Psychology, 90*, 114–141.

Franklin, A., Drivonikou, G. V., Bevis, L., Davies, I. R. L., Kay, P., & Regier, T. (2008). Categorical perception of color is lateralized to the right hemisphere in infants, but to the left hemisphere in adults. *Proceedings of the National Academy of Sciences of the United States of America, 105*(9), 3221–3225.

Franklin, A., Drivonikou, G. V., Clifford, A., Kay, P., Regier, T., & Davies, I. R. L. (2008). Lateralization of categorical perception of color changes with color term acquisition. *Proceedings of the National Academy of Sciences of the United States of America, 105*(47), 18221–18225.

Franklin, A., Catherwood, D., Alvarez, J., & Axelsson, E. (2010). Hemispheric asymmetries in categorical perception of orientation in infants and adults. *Neuropsychologia, 48*, 2648–2657.

Fugate, J. M. B. (2013). Categorical perception for emotional faces. *Emotion Review, 5*(1), 84–89.

Gentner, D., & Goldin-Meadow, S. (Eds.). (2003). Language in mind: *Advances in the study of language and thought*. Cambridge, MA: MIT Press.

Gilbert, A. L., Regier, T., Kay, P., & Ivry, R. B. (2006). Whorf hypothesis is supported in the right visual field but not the left. *Proceedings of the National Academy of Sciences of the United States of America, 103*(2), 489–494.

Gilbert, A. L., Regier, T., Kay, P., & Ivry, R. B. (2008). Support for lateralization of the whorf effect beyond the realm of color discrimination. *Brain and Language, 105*(2), 91–98.

Goldstein, J., Davidoff, J., & Roberson, D. (2009). Knowing color terms enhances recognition: Further evidence from English and Himba. *Journal of Experimental Child Psychology, 102*(2), 219–238.

Goldstone, R. L., & Hendrickson, A. T. (2010). Categorical perception. Wiley *Interdisciplinary Reviews: Cognitive Science, 1*, 69–78.

Gumperz, J. J. & Levinson, S. C. (1996). Introduction to part I. In J. J. Gumperz, & S. C. Levinson (Eds.), *Rethinking linguistic relativity* (pp. 21–35).

Cambridge: Cambridge University Press.

Han, S., Mao, L., Gu, X., Zhu, Y., Ge, J., & Ma, Y. (2008). Neural consequences of religious belief on self-referential processing. *Social Neuroscience, 3*, 1–15.

Heatherton, T. F., Wyland, C. L., Macrae, C. N., Demos, K. E., Denny, B. T., & Kelley, W. M. (2006). Medial prefrontal activity differentiates self from close others. *Social Cognitive and Affective Neuroscience, 1*, 18–25.

Holmes, K., & Wolff, P. (2012). Does categorical perception in the left hemisphere depend on language? *Journal of Experimental Psychology: General, 141*(3), 439–443.

Jenkins, A. C., & Mitchell, J. P. (2011). Medial prefrontal cortex subserves diverse forms of self-reflection. *Social Neuroscience, 6*, 211–218.

Kelley, W. M., Macrae, C. N., Wyland, C. L., Caglar, S., Inati, S., & Heatherton, T. F. (2002). Finding the self? An event-related fMRI study. *Journal of Cognitive Neuroscience, 14*, 785–794.

Kesebir, S., & Oishi, S. (2010). A spontaneous self-reference effect in memory. *Psychological Science, 21*, 1525–1531.

Kim, K., & Johnson, M. K. (2012). Extended self: Medial prefrontal activity during transient association of self and objects. *Social Cognitive and Affective Neuroscience, 7*, 199–207.

Klein, S., & Loftus, J. (1988). The nature of self-referent encoding: The contributions of elaborative and organizational processes. *Journal of Personality and Social Psychology, 55*, 5–11.

Klein, S., Rozendal, K., & Cosmides, L. (2002). A social-cognitive neuroscience analysis of the self. *Social Cognition, 20*, 105–135.

Klein, M. E., & Zatorre, R. J. (2011). A role for the right superior temporal sulcus in categorical perception of musical chords. *Neuropsychologia, 49*(5), 878–887.

Lombardo, M. V., Chakrabarti, B., Bullmore, E. T., Sadek, S. A., Pasco, G., Wheelwright, S. J., ··· Baron-Cohen, S. (2010). Atypical neural self-representation in autism. *Brain, 133*, 611-624.

Lupyan, G. (2012). Linguistically modulated perception and cognition: The label-feedback hypothesis. *Frontiers in Psychology, 3*, 54.

Lupyan, G., Rakison, D. H., & McClelland, J. L. (2007). Language is not just for talking: Redundant labels facilitate learning of novel categories. *Psychological Science, 18*(12), 1077-1083.

Ma, Y., & Han, S. (2010). Why respond faster to the self than others? An implicit positive association theory of self advantage during implicit face recognition. Journal of Experimental Psychology: *Human Perception and Performance, 36*(3), 619-633.

Macrae, C. N., & Roseveare, T. A. (2002). I was always on my mind: The self and temporary forgetting. *Psychonomic Bulletin & Review, 9*(3), 611-614.

Malt, B., & Wolff, P. (2010). Words and the mind: *How words capture human experience*. Oxford University Press.

Mitchell, J. P., Macrae, C. N., & Banaji, M. R. (2006). Dissociable medial prefrontal contributions to judgments of similar and dissimilar others. *Neuron, 50*(4), 655-663.

McCullough, S., & Emmorey, K. (2009). Categorical perception of affective and linguistic facial expressions. *Cognition, 110*(2), 208-221.

Mo, L., Xu, G. P., Kay, P., & Tan, L.H. (2011). Electrophysiological evidence for the left-lateralized effect of language on preattentive categorical perception of color. *Proceedings of the National Academy of Sciences of the United States of America, 108*, 14026-14030.

Moran, J. M., Heatherton, T. F., & Kelley, W. M. (2009). Modulation of cortical mid-line structures by implicit and explicit self-relevance evaluation. *Social Neuroscience, 4*, 197-211.

Niedenthal, P. M., Barsalou, L. W., Winkielman, P., Krauth-Gruber, S., & Ric, F. (2005). Embodiment in attitudes, social perception, and emotion. *Personality and Social Psychology Review, 9*(3), 184-211.

Niedenthal, P. M., Winkielman, P., Mondillon, L., & Vermeulen, N. (2009). Embodiment of emotion concepts. *Journal of Personality and Social Psychology, 96*(6), 1120-1136.

Northoff, G., Heinzel, A., de Greck, M., Bermpohl, F., Dobrowolny, H., & Panksepp, J. (2006). Self-referential processing in our brain—A meta-analysis of imaging studies on the self. *NeuroImage, 31*(1), 440-457.

Ozturk, O., Shayan, S., Liszkowski, U., & Majid, A. (2013). Language is not necessary for color categories. *Developmental Science, 16*(1), 111-115.

Paluy, Y., Gilbert, A., Baldo, J., Dronkers, N., & Ivry, R. (2011). Aphasic patients reveal a reversal of hemispheric asymmetries in categorical color perception. *Brain and Language, 116*, 151-156.

Peng, G., Zheng, H. Y., Gong, T., Yang, R-X., Kong, J. P., & Wang, W. S. Y. (2010). The influence of language experience on categorical perception of pitch contours. *Journal of Phonetics, 38*(4), 616-624.

Powell, L. J., Macrae, C. N., Cloutier, J., Metcalfe, J., & Mitchell, J. P. (2010). Dissociable neural substrates for agentic versus conceptual representations of self. *Journal of Cognitive Neuroscience, 22*(10), 2186-2197.

Roberson, D., Damjanovic, L., & Pilling, M. (2007). Categorical perception of facial expressions: Evidence for a "category adjustment" model. *Memory & Cognition, 35*(7), 1814-1829.

Roberson, D., & Davidoff, J. (2000). The categorical perception of colors and facial expressions: the effect of verbal interference. *Memory and Cognition, 28*(6),

977–986.

Roberson, D., Davies, I., & Davidoff, J. (2000). Color categories are not universal: Replications and new evidence from a stone age culture. *Journal of Experimental Psychology: General, 129*(3), 369–398.

Roberson, D., Davidoff, J., Davies, I. R. L., & Shapiro, L. R. (2005). Color categories: Evidence for the cultural relativity hypothesis. *Cognitive Psychology, 50*(4), 378–411.

Roberson, D., Pak, H., & Hanley, J. R. (2008). Categorical perception of colour in the left and right visual field is verbally mediated: Evidence from Korean. *Cognition, 107*(2), 752–762.

Rogers, T. B., Kuiper, N. A., & Kirker, W. S. (1977). Self-reference and the encoding of personal information. *Journal of Personality and Social Psychology, 35*(9), 677–688.

Ruby, P., & Decety, J. (2001). Effect of subjective perspective taking during simulation of action: A PET investigation of agency. *Nature Neuroscience, 4*(5), 546–550.

Sapir .E(1921) Language : An Introduction to Study of Speech. NewYork :Harhcount,Brance & Company

Sauter, D. A., LeGuen, O., & Haun, D. B. (2011). Categorical perception of emotional facial expressions does not require lexical categories. *Emotion, 11*(6), 1479–1483.

Sui, J., He, X., & Humphreys, G. W. (2012). Perceptual effects of social salience: Evidence from self-prioritization effects on perceptual matching. *Journal of Experimental Psychology: Human Perception and Performance, 38*, 1105–1117.

Symons, C. S., & Johnson. B.T. (1997).The self-reference effect in memory: A meta-analysis .*Psychological Bulletin, 121*(3), 371–394

Thierry, G., Athanasopoulos, P., Wiggett, A., Dering, B., & Kuipers, J. R. (2009). Unconscious effects of language-specific terminology on preattentive color perception. *Proceedings of the National Academy of Sciences of the United States of America, 106*(11), 4567–4570.

Turk, D. J., Van Bussel, K., Brebner, J. L., Toma, A. S., Krigolson, O., & Handy, T. C. (2011). When "it" becomes "mine" : Attentional biases triggered by object ownership. *Journal of Cognitive Neuroscience, 23*, 3725–3733.

Vogeley, K., & Fink, G. R. (2003). Neural correlates of the first-person-perspective. *Trends in Cognitive Sciences, 7*, 38–42.

Wang, L. Y., Zhang, M., & Sui, J. (2011). Self-face advantage benefits from a visual self-reference frame. Acta *Psychologica Sinica, 43*, 494–499

Winawer, J., Witthoft, N., Frank, M. C., Wu, L., Wade, A. R., & Boroditsky, L. (2007). Russian blues reveal effects of language on color discrimination. *Proceedings of the National Academy of Sciences of the United States of America, 104*(19), 7780–7785.

Wheeler, S. C., Morrison, K. R., DeMarree, K. G., & Petty, R. E. (2008). Does self-consciousness increase or decrease priming effects? It depends. *Journal of Experimental Social Psychology, 44*, 882–889.

Whorf, B. L. (1956). Language, thought, and reality: *Selected writings of Benjamin Lee Whorf.* New York: MIT Press.

Wu, Y., Wang, C., He, X., Mao, L., & Zhang, L. (2010). Religious beliefs influence neural substrates of self-reflection in Tibetans. *Social Cognitive and Affective Neuroscience, 5*, 324–331.

Xie, S. S., Zhang, J. J., He, X. M., & Xiao, E. P. (2008). Culture's effects on 'black' and 'white' Color cognition of undergraduates from Yi Nation, Bai Nation, Naxi Nation and Han Nation. *Acta Psychologica Sinica, 40*(8), 890–901.

[谢书书，张积家，和秀梅，林娜，肖二平 (2008). 文化差异影响彝、白、纳西和汉族大学生对黑白的认知.

心理学报, *40*(8), 890–901.]

Yang, H. S., & Zhu, Y. (2004). The self and retrieval—induced forgetting. *Acta Psychologica Sinica, 36*(2), 154–159.

[杨红升, 朱滢. (2004). 自我与提取诱发遗忘现象. *心理学报*, *36*(2), 154–159.]

Yang, S., Huang, X.T., & Fu, Y. L.(2012).Medial Prefrontal Cortex: Neural Basis of the Self. Advances in Psychological Science,*20*(6),853–862

[杨帅, 黄希庭, 傅于玲.（2012）内侧前额叶皮质——"自我"的神经基础, *心理科学进展*, *20*(6), 853–862.]

Yao, S. X., Yang, D., Qi, S. Q., Lei, Y., & Ding, C. (2012). Studies on the N2pc component in visual spatial attention. *Advances in Psychological Science, 20*(3), 365–375.

[姚树霞, 杨东, 齐森青, 雷燕, Ding, C. (2012). 视觉空间注意研究中的 N2pc 成分述评. *心理科学进展*, *20*(3), 365–375.]

Yuan, C. P., & Lu, G., M. (2010). The advance of Self-reference processing of fMRI. *Chinese Journal of Medical Imaging Technology*, *26*(12), 2382–2384.

[袁翠平, 卢光明. (2010). 自我参照加工的功能磁共振成像研究进展. *中国医学影像技术*, *26*(12), 2382–2384.]

Zhang, J. J., Liu, L. H., Chen, X., & He, X. M. (2008). The relationship of color cognition and color terms in Naxi people. *Minority Languages of China*, (2), 49–55.

[张积家, 刘丽红, 陈曦, 和秀梅 (2008). 纳西语颜色认知关系研究. *民族语文*, (2), 49–55.]

Zhou, K., Mo, L., Kay, P., Kwok, V. P. Y., Tiffany, N. M. I., & Tan, L. H. (2010). Newly trained lexical categories produce lateralized categorical perception of color. *Proceedings of the National Academy of Science of the United States of America, 107*(22), 9974–9978.

Zhu, Y., Zhang, L., Fan, J., & Han, S. (2007). Neural basis of cultural influence on self-representation. *NeuroImage, 34*, 1310–1316.

The influence of language labels and self-reference on new object categorical perception

LIU Siyun; MENG Jianxin

(Key Laboratory of Adolescent Cyberpsychology and Behavior (CCNU), Ministry of Education;

School of Psychology, Central China Normal University, Wuhan 430079 China)

Abstract　Categorical perception is one of the most basic cognitive processes of human beings. When humans process incoming information, categorization help them further clarify and simplify the sophisticated inputs. The categorical perception effect refers to the phenomenon that individuals will respond faster or more accurately when discriminating two stimuli that cross a category boundary than when discriminating two stimuli from the same category, despite between- and within-category stimuli being equated in distance (Bornstein & Korda, 1984). However, inconsistent evidence has been obtained from previous studies on how language would influence categorical perception. The language label theory suggested that the language label is a cue that helps individuals to categorize information unconsciously and automatically; whereas perceptual feature theory suggested that categorical perception is based on a pure perceptual process, which arises from life experiences that eventually changes the mappings of perceptual neurons. The current study systematically investigates how language labels would interact with self-reference factor to play a mutual role on new object categorical perception. The hypothesis of this study is that language labels are not the only important factors that would influence categorical perception, other social or personality factors may also play a role, but it is not sure whether language labels would play a more important role. In this study, three experiments were conducted. In Experiment 1, a 2 (language label learning times：more vs. less) × 2 (object category：between vs. within) × 2 (visual field：left vs. right) mixed experiment was designed, and it is found that only the participants who have learned the labels for more times showed a significant right visual field advantage effect. In Experiment 2, a 2 (object self-reference connection: tight vs. loose) × 2 (object category：between vs. within) × 2 (visual field：left vs. right) mixed experiment was designed. This time, all the participants learned the language labels for only twice. As a result, it is found that only the participants who had built a tight self-reference connection with the new object revealed a significant categorical perception effect, however, on both left- and right-visual fields. Furthermore, higher level of self-reference made the participants show a better discriminating ability for between-category objects, but not for within-category objects. In Experiment 3, a 2 (object self-reference connection：tight vs. loose) × 2 (object category：between vs. within) × 2 (visual field: left vs. right) mixed experiment was designed. This time all the participants learned the language labels for seven times. As a result, it is found that under higher level representation of language labels, both language label and the self-reference play a role on categorical perception. In summary, this study revealed an important and complicated role of language labels on categorical perception, and a nonetheless very important influence of self-reference as well.

Keywords　categorical perception; language label; self-reference; right visual field advantage effect

心理学报，2015, 47(8), 992 – 1003.

网络使用经验对动作动词加工的影响 *

刘思耘，周宗奎，李　娜

（青少年网络心理与行为教育部重点实验室；华中师范大学心理学院，武汉 430079）

摘　要　具身认知理论认为高级概念认知和低级感知觉认知紧密关联，且两者共享相同的神经系统。本研究设计了 3 个实验：实验 1 观察面部表情动词加工是否受面部情绪表达的影响，结果发现网络使用经验多的被试在促进和抑制面部积极表情条件下对动词的反应没有表现出显著差异。实验 2 探讨网络使用经验对肢体动作词汇加工的影响，结果发现网络使用经验多的被试在动词－名词转换时没有产生显著认知损耗。实验 3 探讨两类被试对正常序列和随机序列图式动词加工是否有差异，结果发现网络使用经验多的个体对随机序列和正常序列的回忆成绩并无显著差异。本研究从动词认知加工角度验证了高级认知与低级感知觉加工间的紧密联系，揭示了网络行为与高级认知加工间的相关关系。

关键词　具身认知；转换消耗效应；图式；网络使用经验

分类号　B842

1　前言

根据 2012 年中国互联网络信息中心（CNNIC）发布的中国互联网网络发展状况，中国 30 岁以下的青少年网民占整体比例的 56.8%，而网瘾者在 2007 年就达到 2.4%，可见上网人数及成瘾人数之多。自从有互联网以来，人们花在网络上的时间越来越多，并越来越依赖于网络以处理各种日常事宜。那么网络科技给人们生活模式带来的改变是否会影响人的高级认知加工呢？如果有影响，会在哪方面造成影响？影响程度如何？本研究将比较不同程度网络使用经验者的语言加工过程，探讨网络使用经验与高级认知加工间关系，揭示网络行为对人们生活影响的实质。

1.1　网络成瘾行为对高级认知加工的影响

当网络成瘾行为极大程度地影响到人们的生活时，研究者开始研究其形成机制及对认知功能的影响，并提出相应的理论模型（牛更枫，孙晓军，周宗奎，魏华，2013）。对网络成瘾行为的研究主要从成瘾者的人格特质的角度及影响网络成瘾环境等因素展开讨论（Weinstein & Lejoyeux, 2010; 贺金波，郭永玉，向远明，2008），对网络成瘾者认知功能变化的探讨则主要采用脑成像和脑电信号采集的方法，但网瘾形成和维持机制还不太明确（Weinstein & Lejoyeux, 2010; 戴珅懿，马庆国，王小毅，2011; 黄敏等，2010）。

研究者对网络成瘾的认知神经机制探讨的结果发现，成瘾者在认知的多个方面具有功能受损的表

* 中央高校基本科研业务费专项资金（项目号：CCNU14A02015), 国家社科基金重大项目 (11&ZD151) 。

通讯作者：周宗奎，E-mail: zhouzk@mail.ccnu.edu.cn

现，具体如下：1）成瘾者的大脑信息加工能力降低（郁洪强等，2009；贺金波，郭永玉，柯善玉，赵仑，2008）。2）成瘾者面孔和表情识别能力降低（高文斌，陈祉妍，2006；赵仑，高文彬，2007）。3）成瘾者的记忆和注意能力受损，但该结论是基于对芬兰15-64岁的2000名被试的问卷调查（N?si & Koivusilta, 2013）。4）成瘾者的语言加工能力受损（金璞，傅先明，钱若兵，牛朝诗，韩晓鹏，2009）。5）成瘾者的思维认知能力受损（郁洪强等，2009）。6）成瘾者的决策能力受损（Pawlikowski & Brand, 2011；梁三才，游旭群，2010；Sun, Chen, Ma, Zhang, Fu, & Zhang, 2009）。

尽管多个研究发现网络成瘾者的高级认知加工能力受到严重损坏，但研究者大多采用认知神经的方法和技术，从脑神经机制的角度来推断网络成瘾者的认知功能变化，而较少从行为学的角度系统探讨网络成瘾者认知加工能力的状况。究其原因，可能是网络成瘾者在行为上已经与正常个体的行为产生了明显分化，而如果在实验室环境下仍只探讨成瘾者群体的行为特异性，其启示意义远不如探究其脑神经机制的改变。但是，对于大多数对网络有一定程度依赖但还远未达到病理性成瘾状态的网络使用者来说，首先从行为层面去界定他们的行为本质及与正常行为的偏差程度是具有重要意义的。本研究将关注这类群体，他们不仅使用网络时间较长，且在一定程度上对网络已形成心理依赖，但其心理健康受损程度并没有达到病理性的严重程度。我们称这类群体为网络使用经验多的群体。

1.2 具身认知为网络行为研究提供的新视角

我们真实的社交空间是具身的，而网络虚拟空间是非具身的（Hanlon, 2001）。当人们在虚拟空间中进行各种交际活动时，很多社交场合所必需的肢体动作、面部表情等都大大削弱和减少，这使得人们不得不更多地依赖于抽象的语言符号或者人工表情符号去弥补交流时线索的缺乏（Gackenbach, 2011）。网络的使用还使得人们的身体活动范围和频率也受

到限制，使一些必要的肢体的动作和运动逐步减少，最终导致具身体验机会的减少（Alessi, 2001; Hanlon, 2001; Kang, 2007; Kim, et al., 2012）。本研究将采用具身认知的视角探讨网络使用经验与高级认知加工间的关系。

1.2.1 具身认知的主要理论及其相关支持性证据

高级认知加工的理论中影响最大的有两种观点：一个是传统的符号加工理论，它认为高级的认知加工只涉及抽象的符号运算，并不涉及低级感知 - 运动系统的参与（Fodor, 1975）；另一种观点认为在概念加工过程中，低级感知觉系统也会得到激活（Damasio, 1989; Snodgrass, 1984），并逐渐形成具身认知理论（embodied cognition theory）。具身认知理论认为认知是基于身体的各种体验形成，个体高级的认知加工过程会受到身体状态的直接影响，大脑对先前的感知 - 运动经验的再激活（re-activation）在认知加工过程中扮演着至关重要的作用。

具身认知理论包括身体观和模拟观，这两个观点并不对立，而是相辅相成，从不同侧面印证身体体验和高级认知加工之间的紧密关系。身体观强调自下而上的加工路径，而模拟观则强调自上而下的加工路径。身体观是指实际的身体状态（如身体的各种感知觉体验、姿势、动作等）会影响高级认知加工（Larsen, Kasimatis, & Frey, 1992; Strack, Martin, & Stepper, 1988）；而模拟观则指高级的认知加工会激活先前经验所涉及的感知觉、运动、内省系统（Barsalou, 1999, 2008; Barsalou, Kyle Simmons, Barbey, & Wilson, 2003; 谢久书，张常青，王瑞明，陆直，2011）。

前人从模拟观的角度提供了大量行为学层面的支持性证据（Marques, 2006; Pecher, Zeelenberg, & Barsalou, 2003; Vermeulen, Corneille, & Niedenthal, 2008）以及神经机制方面的证据（Goldberg, Perfetti, & Schneider, 2006; Kellenbach, Brett & Patterson, 2001; Simmons et al., 2007）。在行为学实验中，研究者主要采用转换消耗效应（switching costs effect）和知觉负荷效应（sensory load effect）（Pecher et al., 2003; Marques, 2006）范式。如有研究者（Pecher et al.,

2003）发现当要求被试对所呈现的词进行属性判断，且先后呈现的特征词均属相同感觉通道时（如头发 – 黑色的 vs. 芒果 – 黄色的）被试的反应比不同通道时（如头发 – 黑色的 vs. 草莓 – 酸酸的）要快。在采用知觉负荷效应范式时，研究者（Vermeulen et al., 2008）发现，当第一次呈现的感觉刺激占用了概念特征判断任务所需的资源时，被试对随即呈现的相同感觉通道的刺激加工会变缓，表明概念加工对低级感觉通道有再次激活的功能（Vermeulen et al., 2008）。在神经机制方面，研究者用 PET 和 fMRI 技术均验证了高级概念加工对与该概念相关的感知觉脑区的再激活（Goldberg et al., 2006; Kellenbach et al., 2001; Simmons et al., 2007）。

不仅客体的概念加工会再激活低级感知觉通道，复杂的社会认知加工也会出现类似现象。许多有关情绪信息的加工研究（Halberstadt, Winkielman, Niedenthal, & Dalle, 2009; Oosterwijk, Rotteveel, Fischer, & Hess, 2009; Price, Peterson, & Harmon-Jones, 2012; Tan, Walter, Scheck, Hrabal, Hoffmann, Kessler, & Traue, 2012; Vermeulen, Niedenthal, & Luminet, 2007; Wiswede, Münte, Kr?mer, & Rüsseler, 2009）和社会知觉研究（Andersen, Reznik, & Manzella, 1996; Vanman, Paul, Ito, & Miller, 1997）均提供了支持性证据。如当由感觉通道属性判断转向情绪信息判断时，反应时会延长（Vermeulen et al., 2007）；情绪概念的加工会激活面部肌肉运动系统（Halberstadt et al., 2009; Price et al., 2012; Tan et al., 2012; Wiswede et al., 2009），个体的身体体验也会受到高级社会认知的影响（Andersen et al., 1996; Vanman et al., 1997）。

研究者从身体观的角度也提供了诸多支持性证据。研究者发现个体的身体体验会反过来影响社会认知加工，其中面部肌肉运动、身体姿态和各种感知觉体验对情绪、社会知觉、态度都会产生显著影响。如研究者通过对比面部肌肉受损和正常被试两大群体发现，面部肌肉运动会影响个体对情绪面孔的识别（Ponari, Conson, D'Amico, Grossi, & Trojano, 2012; Oberman, Winkielman, & Ramachandran,

2007; Niedenthal, 2007; Neal & Chartrand, 2011; Havas, Glenberg, Gutowski, Lucarelli, & Davidson, 2010; Williams, Whiten, & Singh, 2004; Oberman et al., 2007; Stel & van Knippenberg, 2008）；个体身体姿态的具身体验对其情绪的表达（Duclos et al., 1989; Van den Stock, Righart, & de Gelder, 2007; Wallbott, 1998）和社会态度的形成（Briñol & Petty, 2008; Cacioppo, Priester, & Berntson, 1993; Förster & Strack, 1998; Taylor, Lord, & Bond, 2009; Wells & Petty, 1980）也有重要影响；个体的各种感知觉体验如温度感觉、触觉、物理距离知觉等也能影响某些社会知觉（Ackerman, Nocera, & Bargh, 2010; IJzerman & Semin, 2009; Jostmann, Lakens, & Schubert, 2009; Williams & Bargh, 2008）。

1.2.2 与动作动词加工相关的具身研究

语言中动词表征的是人们在实际生活中逐步形成的典型动作，是联接不同客体间关系的重要逻辑枢纽（Tanenhaus, Carlson, & Trueswell, 1989）。研究者从行为学和脑神经机制两方面为动词加工提供了大量实证证据。在行为学方面，研究者发现动词短语的理解和动作系统存在着相互作用，即当被试执行一个动作时，若同时阅读一个和该动作方向一致的动词短语，则被试执行该动作的速度会得到促进（Glenberg & Kaschak, 2002; Lindemann, Stenneken, Van Schie, & Bekkering, 2006）。研究者认为被试在短语理解过程中对所描述的动作进行了模拟，语言理解过程再激活了相应的运动系统，从而对共享相同运动系统的动作产生了促进作用。这个效应表明动词理解是以低级运动系统的再激活为基础的，需要动作经验的参与。

研究者还采用 fMRI（Hauk, Johnsrude, & Pulvermüller, 2004; Filimon, Nelson, Hagler, & Sereno, 2007）、TMS（Buccino, Riggio, Melli, Binkofski, Gallese, & Rizzolatti, 2005; Pulvermüller, Hauk, Nikulin, & Ilmoniemi, 2005）、MEG（Hauk & Pulvermüller, 2004; Hauk, Shtyrov, & Pulvermüller, 2008）等技术手段对动词加工过程进行了深入的探讨，从脑机制层面进一步验证了动词加工过程中个体对相应动作的模拟（De Zubicaray, Postle, McMahon, Meredith, & Ashton,

2010）。

1.2.3 长期具身经验对高级认知加工的影响

研究发现不仅短暂的身体体验变化对高级认知有影响，长期具身经验也具类似影响。其中躯体特异性理论（body-specificity hypothesis）（Casasanto, 2009）就指出主体与外界环境互动经验的不同会形成不同的概念加工方式。如右利手被试倾向于将喜欢的、好的物体放在右侧，将讨厌的、不好的物体放在左侧；而左利手被试则正好相反。这种不同的身体经验造成了他们对"好"、"坏"概念的不同空间表征（Casasanto, 2009; Casasanto & Henetz, 2012）。

概念隐喻理论（Lakoff & Johnson, 1999）也认为身体经验是概念的投射源，概念系统是以低级的感知觉经验为基础形成的。如果作为投射源的身体经验不足，那么身体经验和概念之间的隐喻联接强度就会减弱，最终会影响对相应概念的理解。如早期缺乏与母亲身体接触的个体会表现出较弱的物理温度 - 人际间情感（physical warmth – interpersonal warmth）隐喻联接强度（Fay & Maner, 2012; Ijzerman, Karremans, Thomsen, & Schubert, 2013）。也有研究发现由于女性比男性具有更丰富的面部表情体验，因此女性对情绪的表达与识别都强于男性（LaFrance & Hecht, 2000）。

1.3 问题提出与研究假设

基于以上文献的回顾和分析，我们提出如下实验假设：一、面部动作具身体验的减少将会影响个体对面部动作词汇的加工；二、肢体具身体验的减少将会影响人们对肢体动作词汇的加工；三、肢体具身体验的减少将会影响人们对动作图式中动词系列的加工。

上述的三个假设将从不同的角度和层面探测网络使用经验不同的群体在动词加工过程中的差异。其中前两个假设的提出是基于人们在真实场景的社交过程中会涉及大量面部表情动作和肢体动作，而网络使用经验多的群体会由于将更多的时间用于网络，从而缺失足够的真实社交体验。假设三是在前两个假设的基础上进一步深入探测网络使用经验多的群体是否在一系列具有逻辑关联的动词加工上会有所不同表现。3个实验从一定的广度和深度探讨了网络使用经验与高级认知加工之间的关系。

2 实验1：网络使用经验对面部动作词汇加工的影响

2.1 实验目的

探讨网络使用经验的多少与面部表情动词加工间关系。实验一假设，网络使用经验过多的个体不易受具身状态的影响，即当这类群体的面部表情动作受到抑制或促进时，其相应概念的加工不会发生改变；而网络使用经验少的群体则会有明显的变化。

2.2 实验方法

2.2.1 被试

本研究中被试备选库的准备：抽取出白羽和樊富珉（2005）的《大学生网络依赖测量量表》中的人际健康维度、周治金和杨文娇（2007）的《网络成瘾类型》中网络游戏成瘾维度和网络信息成瘾维度编制成一份新问卷，作为筛选出本研究所需的网络使用经验多和少的被试的测量工具。此问卷共19个项目，每个项目采用5点计分方式，从"极不符合"到"非常符合"分值依次增加。从某高校中随机选取149名大学生进行施测，问卷回收率100%。最终分别选取总分在两端的60名被试作为备用被试，其中低分组30名（男生11名，女生19名）作为网络使用经验少的被试群体；高分组30名（男生15名，女生15名）作为网络使用经验多的被试群体。被试的年龄范围在19岁~25岁之间（M=22.42，SD=2.16）。经独立样本t检验，网络使用经验多的备用被试群体在总分、人际健康维度、网络信息成瘾维度、网络游戏成瘾维度及日平均上网时间上的平均得分均显著高于网络使用经验少的备用被试群体（$t_{总分}(58)$=18.48, $p<0.001$; $t_{健康维度}(58)$=10.98, $p<0.001$; $t_{信息成瘾维度}(58)$=9.10, $p<0.001$; $t_{游戏成瘾维度}(58)$=8.36, $p<0.001$; $t_{上网时间}(58)$=7.11, $p<0.001$）。

实验一被试：从60名备用被试中随机选取低分

组 18 名（男生 6 名，女生 12 名）作为网络使用经验少的被试群体，高分组 20 名（男生 11 名，女生 9 名）作为网络使用经验多的被试群体。被试的年龄范围在 19 岁 ~25 岁之间（$M=22.84$，$SD=2.01$）。经独立样本 t 检验，网络使用经验多的被试群体在总分、人际健康维度、网络信息成瘾维度、网络游戏成瘾维度和日平均上网时间上的平均得分均显著高于网络使用经验少的被试群体（$t_{总分}(36)=17.90$, $p<0.001$; $t_{健康维度}(36)=9.50$, $p<0.001$; $t_{信息成瘾维度}(36)=7.32$, $p<0.001$; $t_{游戏成瘾维度}(36)=6.89$, $p<0.001$; $t_{上网时间}(36)=5.97$, $p<0.001$）。

2.2.2 实验材料

从《现代汉语词典》（2005）中选取 140 个四字成语或三字惯用语（见附录 1）。在 15 名随机招募的被试中进行情绪效价的 7 点量表测评，最后确定符合情绪效价要求的材料共 132 个（消极词 71 个，积极词 61 个）。然后再随机选取 15 名被试对这 132 个词是否和面部表情动作有关进行 7 点量表评定，最终选取相关程度均分在 5.80 以上的 80 个成语和惯用语（消极和积极词各 40 个）作为实验材料，且在与面部表情动作的相关程度上没有显著差异（$F(1,78)=0.03$, $p=0.87$），而在情绪效价上有显著差异（$F(1,78)=1585.10$, $p<0.001$）。

2.2.3 实验设计

采用 2（被试类别：网络使用经验少 vs. 网络使用经验多）× 2（面部操纵状态：促进积极情绪 vs. 抑制积极情绪）× 2（词的情绪效价：积极 vs. 消极）三因素混合设计。其中被试类别为被试间变量，词的情绪效价和面部操纵状态为被试内变量。因变量为被试判断刺激情绪效价的反应时和正确率。

2.2.4 实验程序

采用 E-Prime2.0 软件对实验设计进行编程，整个实验流程在电脑上进行。正式实验中，按照对被试面部肌肉的不同操纵方法分为两个反应时段，一个为促进积极情绪时段，另一个为抑制积极情绪时段。促进积极情绪的实现手段是，要求被试用门牙紧紧咬住一支筷子，且在这个过程中不能让嘴唇碰到

筷子（见图 1 左图）；抑制积极情绪的实现手段是，要求被试用嘴唇紧紧固定住一支筷子，且在这个过程中不能让牙齿碰触到筷子（见图 1 右图）（Niedenthal, 2007; Strack et al., 1988）。要求被试在实验进行中时刻保持正确的咬筷子方式，并明确告知被试有录像设备对他们是否正确操作进行监控。这两种控制条件的先后顺序在被试间做了平衡处理。被试的任务

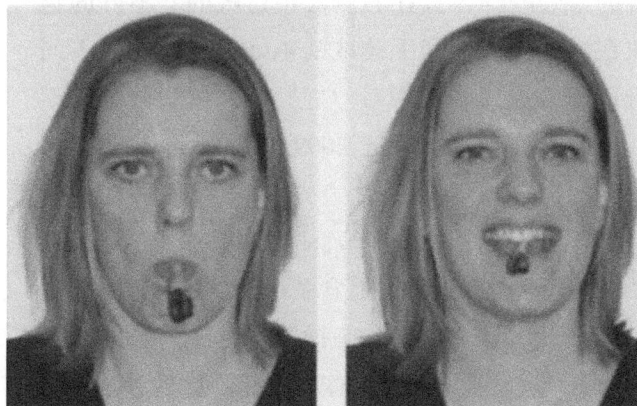

图 1　抑制积极情绪和促进积极情绪动作模拟的示意图（摘选自 Niedenthal, 2007）

是对所呈现的词汇进行积极和消极情绪效价的判断。一半的被试用左手按"积极"键，右手按"消极"键，另一半则相反。

每个反应时段由练习和正式实验组成。正式实验中每个试次的具体流程如下：首先在电脑屏幕正中呈现一个红色的"+"注视点，500ms 后注视点消失，随即在相同位置呈现一个词语，时长为 800ms。被试需在准确的基础上尽量快的做出情绪效价的判断。反应后空屏时长为 1000ms，随即下一个试次开始。每个反应时段有 80 个试次，且两个反应时段中实验材料相同。刺激在每个反应时段内随机呈现，且每个反应时段内被试休息两次，每次 5min。被试在休息时，安排被试完成一份无关问卷，以削弱被试对上一个时段刺激的记忆。整个实验一共 160 个试次，共耗时 15min 左右。

2.3 实验结果

38 名被试中有 2 名被试的数据因正确率低于 80% 而被剔除，最终有 36 份有效数据进入统计分析。其中网络使用经验多的有 19 名，网络使用经

验少的有 17 名。运用 SPSS17.0 对有效数据进行三因素重复测量的方差分析，发现被试类别、面部操纵状态和词的情绪效价三因素交互作用显著（$F_{(1,34)}=4.53$, $p=0.04$, $\eta_p^2=0.12$）。进一步进行简单交互作用分析发现，被试类别和面部操纵状态的交互作用在积极词上显著（$F_{(1,34)}=4.63$, $p=0.039$, $\eta_p^2=0.12$），而在消极词上不显著（$F_{(1,34)}=0.03$, $p=0.88$, $\eta_p^2=0.001$）。进一步进行简单简单效应分析发现，网络使用经验少的被试在促进和抑制积极情绪状态下对积极词的反应显著不同（$F_{(1,34)}=7.59$, $p=0.009$, $\eta_p^2=0.18$）。而网络使用经验多的被试则无显著差异（$F_{(1,34)}=0.05$, $p=0.83$, $\eta_p^2=0.001$）。同时，被试类别的主效应显著（$F_{(1,34)}=5.16$, $p=0.03$, $\eta_p^2=0.13$），即网络使用经验少的被试对词

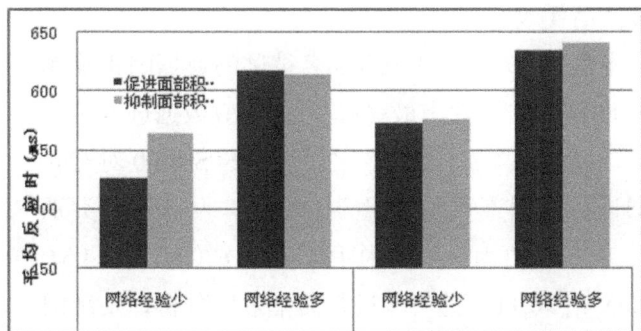

图 2 两类被试群体在不同面部状态下的平均反应时

汇的反应显著快于网络使用经验多的被试。词的情绪效价主效应显著（$F_{(1,32)}=19.04$, $p<0.001$, $\eta_p^2=0.36$），即所有被试对积极词的反应要快于对消极词的反应。

2.4 讨论

实验 1 的结果表明网络使用经验多的被试对面部动作词汇的反应与网络使用经验少的被试有显著不同。前者在积极情绪面部表达受促进或受抑制情况下，对词汇的判断无差别；而后者则有显著不同，实验 1 的假设得到了证实。但两组被试只在积极情绪词汇的判断上表现出差异，在消极情绪的词汇判断中则未表现出不同，表明积极情绪概念的加工似乎比消极情绪概念的加工更易受具身经验减少的影

响，但也有可能是在本实验中被试受到操纵的只是积极情绪的表达，而不是消极情绪的表达。与此同时，网络使用经验多的被试对所有词汇的总体反应都要比网络使用经验少的被试要慢，表明前者整体的概念加工已不如后者。

3 实验 2：肢体具身体验的减少对肢体动作概念加工的影响

3.1 实验目的

网络使用经验有可能不仅影响面部表情词汇的加工，还对肢体动作词汇加工有影响。实验 2 比较网络使用经验不同被试在加工肢体动作动词时是否有所不同。实验 2 假设，网络使用经验多的被试由于在感知觉层面的动作表征减弱，因此会比网络使用经验少的被试在词类转换时产生更少的认知转换消耗。

3.2 实验方法

3.2.1 被试

从被试备用库中随机选取低分组 20 名（男生 5 名，女生 15 名）作为网络使用经验少的被试，高分组 19 名（男生 8 名，女生 11 名）作为网络使用经验多的被试。被试的年龄范围在 19 岁 ~25 岁之间（$M=22.69$，$SD=2.12$）。经独立样本 t 检验，网络使用经验多的被试群体在总分、人际健康维度、网络信息成瘾维度、网络游戏成瘾维度和日均上网时间上的平均得分均显著高于网络使用经验少的被试群体（$t_{总分}(37)=16.22$, $p<0.001$; $t_{健康维度}(37)=9.81$, $p<0.001$; $t_{信息成瘾维度}(37)=6.51$, $p<0.001$; $t_{游戏成瘾维度}(37)=6.85$, $p<0.001$; $t(37)=6.25$, $p<0.001$）。

3.2.2 实验材料

从《现代汉语常用词词频词典》（刘源等，1990）中选取 140 个双字词，并随机招募 15 名被试对所选词汇属于动词还是名词进行 3 点量表评定，最后选取符合要求的词汇 60 个（动词、名词各 30 个）。根据 Pecher 等人（2003）实验中对刺激的配对方法对这 60 个词进行配对，具体方法为：从 60 个词里选择 10 个动词和 10 个名词作为目标词，使其出现在配对

词汇中后面的位置，剩下的 40 个词汇作为启动词汇出现在配对词汇中前面的位置。当目标词为动词（如：抽打）时，从启动词汇中选取动词（如：按摩）与之配对成具有相同加工通道（均为动词）的词对（如：按摩－抽打），或选取名词（如：戒指）与之配对成具有不同通道的词对（如：戒指－抽打）。用同样的方法另外组成目标词为名词的相同通道的词对和不同通道的词对。每个目标词会得到两次配对的机会，最后一共配成 40 个真词－真词对作为实验二的靶刺激。所有词对的词频两两差异不显著（$F(2,57)=0.04$，$p=0.96$），且靶刺激中所有名词和所有动词的平均词频差异不显著（$F(1,58)=0.001$, $p=0.98$）。最后另外准备 80 个真词和 160 个假词随机配成真词－假词、假词－假词、假词－真词各 40 对作为填充刺激，由此组成 160 个词对作为实验二的实验材料。

3.2.3 实验设计

采用 2（被试类别：网络使用经验少 vs. 网络使用经验多）× 2（转换方式：相同通道 vs. 不同通道）两因素混合设计，其中转换方式为被试内变量。因变量分为两类：一类是被试对启动词和目标词均正确反应条件下对目标词的反应时和正确率，用于分析两个被试群体的转换消耗效应的大小；另一类是被试对所有动词的反应时和正确率，用于分析两个被试群体对动词加工是否存在差异。

3.2.4 实验程序

采用 E-Prime2.0 软件对实验设计进行编程，整个实验流程在电脑上进行。实验的单个试次流程如下：首先在电脑屏幕中间呈现一个红色的"+"注视点，500ms 后注视点消失，随即在相同的位置上呈现一个双字词，时长 500ms。要求被试在准确的基础上尽快地按键判断呈现的词是真词还是假词。在指导语中，我们给被试指出"真词是指符合语义、常见的词语，如"挠头、衣服"等；假词则指不符合语义、本来不存在的词语，如"片和、灯度"等。"。一个试次结束后有 1000ms 的空屏时间。整个实验由 160 对词对、320 个反应试次组成。被试对所呈现的刺激进行真词和假词的判断。整个实验分两个时段完成，所有实验材料被均分成两份并以预先配好的词对顺序伪随机逐个呈现。即词对中的两个词分别先后呈现，但词对的呈现顺序则完全随机。两个时段间被试休息 5min。整个实验需要 15min 左右。

3.3 实验结果

首先分析两类被试群体是否均出现了转换消耗效应，如果都出现了，两者是否有显著差异。由于所有刺激均以配对假随机形式呈现，而配对刺激中的第一个刺激被认为是类似启动的一种刺激，在计算转换消耗效应时，被试对其的反应并不被考虑。每个配对中的第二个刺激为靶刺激，是计算转换消耗效应时的关键刺激。比较被试在相同通道条件下对靶刺激的平均反应时与不同通道条件下对靶刺激的平均反应时是否有显著差异，从而确定是否产生了转换消耗效应。

本实验 39 名被试中有 3 名被试的数据因正确率低于 80% 而被剔除，最终有 36 份有效数据进入统计分析，其中两类被试各 18 名。运用 SPSS17.0 对有效数据进行两因素重复测量方差分析，发现被试类别和转换方式两因素交互作用显著，$F(1,34)=4.31$，$p=0.05$，$\eta_p^2=0.11$。进一步进行简单效应分析，发现网络使用经验少的被试在相同通道条件下对目标词的反应比不同通道条件下显著快些（$F(1,34)=18.38$，$p<0.001$，$\eta_p^2=0.35$），表明出现了转换消耗效应；而

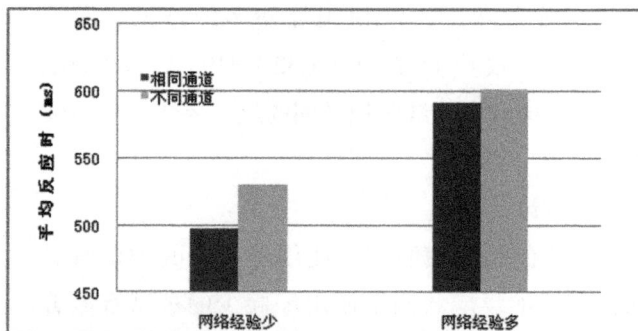

图 3 两类被试群体在两种转换方式下对目标词的平均反应时

网络使用经验多的被试在两种不同通道条件下对目标词的反应没有显著差异（$F(1,34)=1.83$, $p=0.19$, $\eta_p^2=0.05$），表明没有出现转换消耗效应。

然后分析两类被试群体对动词加工是否存在整

体上的差异。所有 39 名被试的数据均为有效数据，且正确率均高于 80%。采用 SPSS17.0 对两类被试群体加工动词的反应时进行独立样本 t 检验，发现被试类别的主效应显著（$t(37)=3.16, p=0.003$），即网络使用经验少的被试对动词的反应要快于网络使用经验多的被试。

3.4 讨论

实验二的结果表明两类被试群体对肢体动作动词的反应有显著不同，网络使用经验多的被试没有表现出应有的转换消耗效应，实验二的假设得到验证，即网络使用经验多的被试在加工动词和名词时，表现出更多的相似性，导致其认知转换消耗减少。根据具身认知理论，由于在肢体动作动词的加工过程中，个体需要激活先前的运动经验，而网络使用经验多的被试其肢体运动具身体验的感知觉表征有可能减弱了，从而导致他们在肢体动作词汇加工过程中对肢体运动系统的再激活变得困难，使得其加工更趋同于名词。但这一结论还有待更多的研究加以证实。

实验二还发现网络使用经验多的被试从总体上来说对动词加工要比网络使用经验少的被试要慢，这表明尽管网络使用经验多的群体还未达到病理性的网络成瘾状态，但由于长期的网络使用经验，其高级动作概念的加工已经产生了异化。

4 实验3：肢体具身体验的减少对动作图式加工的影响

4.1 实验目的

前面两个实验探讨了网络使用不同经验者在加工单个动作概念时的差异。实验三将比较这两类群体在加工一系列有内在逻辑关联的动词时的不同表现。实验三采用图式理论中动作图式概念（action schema）。图式（schema）是从过去经验中抽取出来的重要且稳定的元素，是一种高水平、有组织的概念表征方式（Barsalou, 2000）。动作图式是图式的一种，是个体在观察和不断练习的基础上，

在大脑中形成的一种概括化的动作结构（Galotti, 2009）。动作图式中的动作结构的核心成分是由动词组成的、逻辑上归属于同一个事件的若干动词（短语）。实验三假设，网络使用经验多的被试对以正常顺序呈现和以随机顺序呈现的系列动词的系列回忆成绩没有显著差异，而网络使用经验少的被试则会受打乱顺序的影响。

4.2 实验方法

4.2.1 被试

由于被试备用库中有部分被试流失，因此实验三依据实验一筛选被试的方法补充了一些被试，形成低分组 24 名（男生 6 名，女生 18 名）作为网络使用经验少的被试，高分组 23 名（男生 12 名，女生 11 名）作为网络使用经验多的被试。被试的年龄范围在 19 岁 ~25 岁之间（$M=22.62$, $SD=2.10$）。经独立样本 t 检验，网络使用经验多的被试群体在总分、人际健康维度、网络信息成瘾维度、网络游戏成瘾维度和日平均上网时间上的平均得分均显著高于网络使用经验少的被试群体（$t_{总分}(45)=16.12, p<0.001$; $t_{健康维度}(45)=10.38, p<0.001$; $t_{信息成瘾维度}(45)=7.89, p<0.001$; $t_{游戏成瘾维度}(37)=7.25, p<0.001$; $t_{上网时间}(45)=5.44, p<0.001$）。

4.2.2 实验材料

正式实验前准备 22 个情景事件（见附件 3），让随机招募到的 15 名被试根据这些情景写出最能表示这一情景的 6-8 个符合逻辑的连续动词（即动作图式），每个词不要超过 4 个字。如在描述"去餐馆吃饭"情景中，与之相匹配的动作图式可以是：开门、看菜单、点餐、吃东西、付账、离开。问卷有效回收率为 100%。对被试所列的词语进行频次统计，把每个事件中的动词系列按频次从高到低进行排列，选出事件中排在前五位的动词作为动作图式中的代表性动词，作为本实验的材料。

符合本实验要求的动作图式共准备了 16 个，每个图式中的系列动词有 5 个。16 个图式中的系列动词分为正常序列和随机序列两种，共组成 32 个动词序列。每个被试所接受的随机序列动词系列无

论在项目上还是在序列上都是一样的。将这 32 个动词序列按照拉丁方的方法分配给被试，使每个被试只接受一个动作图式中的一种序列。每个被试共接受 16 组动作序列，8 组正常序列，8 组随机序列。这 16 组正常与随机序列在每个被试内的呈现顺序完全随机。

4.2.3 实验设计

采用 2（被试类别：网络使用经验少 vs. 网络使用经验多）× 2（动词序列类型：正常序列 vs. 随机序列）两因素混合设计，其中网络使用经验为被试间变量。因变量为被试对每个词汇回忆的项目正确率和项目位置正确率。

4.2.4 实验程序

采用 E-Prime2.0 软件对实验设计进行编程，整个实验流程在电脑上进行。正式实验中每个试次的具体流程如下：首先在电脑屏幕中间呈现一个红色的"+"注视点，500ms 后，屏幕中央依次呈现 5 个词汇，每个词汇呈现 1000ms，ISI 为 500ms。词汇呈现完毕后，屏幕上呈现"请按词语呈现的先后顺序依次作答"字样，提醒被试马上在答题纸上由左至右按词汇的呈现顺序尽量准确和快地写出刚才呈现的词语。作答完毕后按空格键进入下一个试次。实验共有 16 个试次，做完 8 个试次后被试休息几分钟。整个实验持续 15min 左右。

4.3 实验结果

本实验的 47 名被试中有 1 名被试由于操作不当数据被剔除，最终有 46 份有效数据进入统计分析，其中两类被试各有 23 名。采用 SPSS17.0 对项目正确率（只要项目被正确回忆出来即视为正确）、项目和位置正确率（项目和位置均被正确回忆出来才视为正确）进行两因素重复测量的方差分析发现，被试类别和动词序列类型的两因素交互作用显著（$F_{项目正确}$（1,44）=5.276, $p_{项目正确}$=0.026, $\eta_p^2{}_{项目正确}$=0.107; $F_{项目和位置正确}$（1,44）=5.57, $p_{项目和位置正确}$=0.02, $\eta_p^2{}_{项目和位置正确}$=0.11）。进一步进行简单效应分析得出，网络使用经验少的被试对随机序列动词的项目正确率显著低于正常序列动词的项目正确率（$F_{项目正确}$（1,44）

=12.98, $p_{项目正确}$=0.001, $\eta_p^2{}_{项目正确}$=0.22; $F_{项目和位置正}$

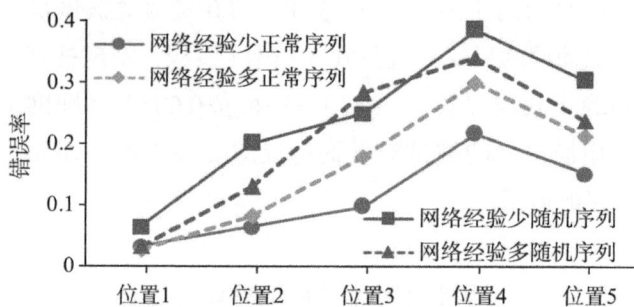

图 4 两类被试群体的项目正确回忆的系列位置曲线

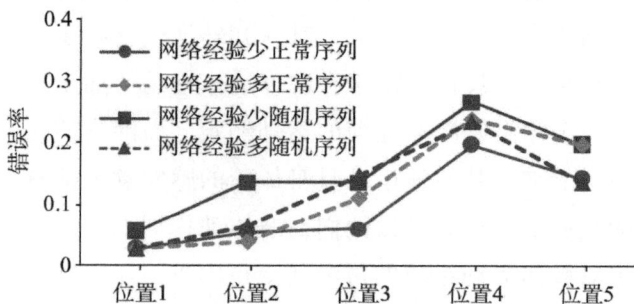

图 5 两类被试群体的项目和位置正确回忆的系列位置曲线

确（1,44）=26.60, $p_{项目和位置正确}$<0.001, $\eta_p^2{}_{项目和位置正确}$=0.37），而网络使用经验多的被试则没有表现出不同（$F_{项目正确}$（1,44）=0.08, $p_{项目正确}$=0.78, $\eta_p^2{}_{项目正确}$=0.002; $F_{项目和位置正确}$（1,44）=3.31, $p_{项目和位置正确}$=0.08, $\eta_p^2{}_{项目和位置正确}$=0.06）。

4.4 讨论

实验三发现网络使用经验多的被试在对动作图式中的系列动词作系列回忆时，无论是对动词本身还是对系列动词中的序列信息，都表现出与网络使用经验少的被试有显著不同。网络使用经验多的被试对以正常顺序和以随机序列呈现的系列动词没有表现出回忆成绩的不同，而网络使用经验少的被试则在这两种条件下表现出了明显差异。根据具身认知理论，网络使用经验多的被试很可能由于长期缺乏足够的具身动作体验，对动作图式中的动作及其顺序表征已经不够敏感，从而在复杂概念的表征与加工上产生异化，但具体原因还有待进一步深入的研究。

5　总讨论

有研究表明网络环境和使用的特殊性及人们对网络本身的迷恋程度对人类尤其是青少年的身心健康造成了严重损伤，甚至改变了相应的大脑神经加工机制。如网络经验越多，被试体验到的孤独感和抑郁感越强，体验到的社会支持度也越低（Kang, 2007）；网络成瘾者在体验离身刺激时，所激活的脑区与精神分裂、抑郁症患者特有的脑激活区一致，且网瘾时间越长，这些变异性的脑区激活程度越高（Kim et al., 2012）。本研究正是在这样一个大的社会环境背景之下，从具身认知理论出发，探讨网络使用经验的不同程度与个体加工高级认知概念之间的关系。

5.1　网络使用经验对单个动作词汇加工的影响

根据具身认知理论，个体在加工动作词汇时，会再激活大脑中的感知运动系统，对动词描述的动作进行模拟。本研究中的实验一和实验二采用表示面部表情或肢体动作的动词作为刺激，来测试具有不同网络使用经验的被试群体是否有不同的行为反应。这两个实验的结果均验证了我们的假设，即网络使用经验多的人无论是对面部表情动作词汇还是对肢体动作词汇，其加工方式都与网络使用经验少的有所不同。分析其原因，我们认为这很可能是由于网络使用经验多的个体由于将更多的时间投入到网络上，且在一定程度上对网络产生了心理依赖，使得他们对面部表情的体验和肢体运动的体验严重减弱，导致低级的感知觉系统的再激活有可能受到阻碍，从而在相应的概念加工过程中产生异化。本研究的结果进一步证实了具身感知觉经验的不同与高级认知活动之间的密切关系，支持了具身认知理论。但网络使用经验多的群体与高级认知加工之间关系的实质还有待进一步研究。

实验二的结果进一步验证了具身理论的观点。前人对失语症病人的研究发现不同部位的大脑皮层区受损会影响名词和动词的加工，暗示名词和动词的神经表征系统是相对独立的 (Shapiro & Caramazza,

2003; 刘涛 等, 2008)。因此当先后加工动词和名词时，由于不同的脑区得到激活，在认知上会产生一定损耗；而如果脑区激活模式趋同的话，则损耗会减少。实验二中发现的网络使用经验多的被试并未产生显著的转换消耗效应，表明他们对动词加工的模式已逐步趋同于名词的模式。而最值得引起注意的是，由于动词加工在语义知识理解、存储及提取的过程中都承载着重要作用（Kemmerer & Gonzalez-Castillo, 2010），因此如果这种差异的减小真的是由于对动词加工的减弱引起的，那么长期沉迷于网络的的行为有可能最终在语义知识体系建构与表征上产生本质性改变。

5.2　网络使用经验对系列动词加工的影响

本研究的实验三发现，网络使用经验多的被试对一系列具有逻辑顺序的动词的加工已产生异化。网络使用经验多的被试在刺激系列的顺序被打乱后，其系列回忆成绩并无显著变化，这表明被试对动作动词逻辑顺序的表征变弱，导致序列信息对其系列回忆成绩产生不了影响。根据身体图式（body schema）（Assaiante, Barlaam, Cignetti, &Vaugoyeau, 2014; Dijkerman & de Haan, 2007）理论，人类自出生以来就通过本体感受和视觉获取的方式，为将来与外界的交流开始建构身体各部件之间及与周围环境间的内部表征（Head & Holmes, 1911），这种表征包含了方位、速度等维度。由于身体图式主要基于身体的移动及与他人的互动所产生的感知觉信息建构起来的，因此实际的身体体验对于身体的图式表征就显得非常重要。由此，我们推断网络使用经验过多者对身体动作图式中系列动词加工产生的异化现象，很可能是由于他们对动作的具身体验减弱所导致，这一结果进一步验证了网络使用经验与高级认知加工之间的关系。

5.3　未来研究展望

虽然本研究针对不同程度网络使用经验的被试设计了三个实证实验，验证了网络使用经验的多少与动作动词加工之间的关系。但由于我们在实验过程中并未操纵被试的网络使用行为，所以从本质上来说本研究是一个相关研究，我们还不清楚最终导致两组被试在动作动词加工上显著差异的是网络使用经验本身，还是网络的长期使用引起了其它认知能力的变化所导致。因此，未来的研究除了进一步操纵和控制个体网络使用的某些变量，以进一步探讨网络使用中的哪些行为是引起高级认知功能变差的原因，还应该找到相应的预防措施，以防止人类在无法避免的网络接触过程中将面临的认知能力的逐步衰退。

6 结论

本研究从具身认知理论的角度，探讨了网络使用经验与汉语动词加工间的关系。研究的结果表明当前的网络行为与高级认知加工有着紧密关联，至少在动词加工方面表现在：1）网络使用经验多的被试其面部肌肉动作的内部表征可能已经弱化；2）网络使用经验多的被试其肢体动作的内部表征可能已经弱化；3）网络使用经验多的被试其身体图式的内部表征可能已经弱化。

参考文献

Ackerman, J. M., Nocera, C. C., & Bargh, J. A. (2010). Incidental haptic sensations influence social judgments and decisions. *Science, 328*(5986), 1712–1715.

Alessi, N. (2001). Disembodiment in cyberspace is not a myth. *CyberPsychology & Behavior, 4*(4), 537–538.

Andersen, S. M., Reznik, I., & Manzella, L. M. (1996). Eliciting facial affect, motivation, and expectancies in transference: Significant–other representations in social relations. *Journal of Personality and Social Psychology, 71*(6), 1108–1129.

Assaiante, C., Barlaam, F., Cignetti, F., & Vaugoyeau, M. (2014). Body schema building during childhood and adolescence: A neurosensory approach. *Clinical Neurophysiology, 44,* 3–12.

Bai, Y., & Fan, F. M. (2005). A study on the internet dependence of college students: The revising and applying of a measurement. *Psychological Development and Education, 21*(4), 99–104.

[白羽, 樊富珉. (2005). 大学生网络依赖测量工具的修订和应用. *心理发展与教育, 21*(4), 99–104.]

Barsalou, L. W. (1999). Perceptual symbol systems. *Behavioral and Brain Sciences, 22*(4), 577–660.

Barsalou, L. W. (2008). Grounded cognition. *Annual Review of Psychology, 59,* 617–645.

Barsalou, L. W., Kyle Simmons, W., Barbey, A. K., & Wilson, C. D. (2003). Grounding conceptual knowledge in modality–specific systems. *Trends in Cognitive Sciences, 7*(2), 84–91.

Briñol, P. & Petty, R. E. (2008). Embodied persuasion: Fundamental processes by which bodily responses can impact attitudes. *Embodiment Grounding: Social, Cognitive, Affective, and Neuroscientific Approaches* (pp. 184–207). Cambridge: Cambridge University Press.

Buccino, G., Riggio, L., Melli, G., Binkofski, F., Gallese, V., & Rizzolatti, G. (2005). Listening to action–related sentences modulates the activity of the motor system: A combined TMS and behavioral study. *Cognitive Brain Research, 24*(3), 355–363.

Cacioppo, J. T., Priester, J. R., & Berntson, G. G. (1993). Rudimentary determinants of attitudes: II. Arm flexion and extension have differential effects on attitudes. *Journal of Personality and Social Psychology, 65*(1), 5–17.

Casasanto, D. (2009). Embodiment of abstract concepts: Good and bad in right–and left–handers. *Journal of Experimental Psychology – General, 138*(3), 351–367.

Casasanto, D. & Henetz, T. (2012). Handedness shapes children's abstract concepts. *Cognitive Science, 36*(2), 359–372.

Dai, S. Y., Ma, Q. G., & Wang, X. Y. (2011). Attentional

bias to addiction-related stimuli in internet addiction patients: An ERP study. *Journal of Psychological science,34* (6), 1302-1307.

[戴珅懿 , 马庆国 , 王小毅 . (2011). 网络游戏成瘾者对成瘾相关线索的注意偏向：一项 ERP 研究 . *心理科学 , 34*(6), 1302-1307.]

Damasio, A. R. (1989). Time-locked multiregional retroactivation: A systems-level proposal for the neural substrates of recall and recognition. *Cognition, 33*(1-2), 25-62.

De Zubicaray, G., Postle, N., McMahon, K., Meredith, M., & Ashton, R. (2010). Mirror neurons, the representation of word meaning, and the foot of the third left frontal convolution. *Brain and Language, 112*(1), 77-84.

Dictionary Department of the Institute of Linguistics of the Chinese Academy of Social Sciences (Ed.). (2005). *Modern Chinese Words Dictionary* (5th Ed.). Beijing, China: Commercial Press.

[中国社会科学院语言研究所辞典编辑室 (编). (2005). *现代汉语词典* (第五版). 北京：商务印书馆 .]

Dijkerman, H. C. & de Haan, E. H. (2007). Somatosensory processes subserving perception and action. *Behavioral Brain Science, 30*(2): 189-201.

Duclos, S. E., Laird, J. D., Schneider, E., Sexter, M., Stern, L., & Van Lighten, O. (1989). Emotion-specific effects of facial expressions and postures on emotional experience. *Journal of Personality and Social Psychology, 57*(1), 100-108.

Fay, A. J. & Maner, J. K. (2012). Warmth, spatial proximity, and social attachment: The embodied perception of a social metaphor. *Journal of Experimental Social Psychology, 48*(6), 1369-1372.

Filimon, F., Nelson, J. D., Hagler, D. J., & Sereno, M. I. (2007). Human cortical representations for reaching: mirror neurons for execution, observation, and imagery. *Neuroimage, 37*(4), 1315-1328.

Fodor, J. A. (1975). *The Language of Thought.*

Cambridge, MA: Harvard University Press.

Förster, J., & Strack, F. (1998). Motor actions in retrieval of valenced information: II. Boundary conditions for motor congruence effects. *Perceptual and Motor Skills, 86*(3 Pt 2), 1423-1426.

Gackenbach, J. (Ed.). (2011). *Psychology and the internet: Intrapersonal, interpresonal, and transpersonal implications* (2nd Ed.). Burlington, MA: Academic Press.

Galotti, K. M. (2009). *Cognitive Psychology: In and Out of the Laboratory.* Toronto: Cengage Learning.

Gao, W. B., & Chen, Z. Y. (2006). A Study on Psychopathology and Psychotherapy of Internet *.Advances in Psychological Science, Addiction, 14*(4), 596-603.

[高文斌 , 陈祉妍 . (2006). 网络成瘾病理心理机制及综合心理干预研究 . *心理科学进展 , 14*(4), 596-603.]

Glenberg, A. M. & Kaschak, M. P. (2002). Grounding language in action. *Psychonomic Bulletin & Review, 9*(3), 558-565.

Goldberg, R. F., Perfetti, C. A., & Schneider, W. (2006). Perceptual knowledge retrieval activates sensory brain regions. *The Journal of Neuroscience, 26*(18), 4917-4921.

Halberstadt, J., Winkielman, P., Niedenthal, P. M., & Dalle, N. (2009). Emotional conception: How embodied emotion concepts guide perception and facial action. *Psychological Science, 20*(10), 1254-1261.

Hanlon, J. (2001). Disembodied intimacies: Identity and relationship on the Internet. *Psychoanalytic Psychology, 18*(3), 566-571.

Hauk, O., Johnsrude, I., & Pulvermüller, F. (2004). Somatotopic representation of action words in human motor and premotor cortex. *Neuron, 41*(2), 301-307.

Hauk, O. & Pulvermüller, F. (2004). Neurophysiological distinction of action words in the fronto-central cortex.

Human Brain Mapping, 21(3), 191–201.

Hauk, O., Shtyrov, Y., & Pulvermüller, F. (2008). The time course of action and action–word comprehension in the human brain as revealed by neurophysiology. *Journal of Physiology–Paris, 102* (1–3), 50–58.

Havas, D. A., Glenberg, A. M., Gutowski, K. A., Lucarelli, M. J., & Davidson, R. J. (2010). Cosmetic use of botulinum toxin–A affects processing of emotional language. *Psychological Science, 21*(7), 895–900.

He, J. B., Guo, Y. Y., Ke, S. Y., & Zhao, L. (2008). Cognition deficit in internet game addicts: an auditory oddball P300 study. *Journal of Psychological science, 31*(2), 380–384

[贺金波 , 郭永玉 , 柯善玉 , 赵仑 . (2008). 网络游戏成瘾者认知功能损害的 ERP 研究 . *心理科学* , *31*(2), 380–384.]

He, J. B., Guo, Y, Y., & Xiang, Y. M. (2008) Forming mechanism of Adolescents' Internet– game Addiction. *Chinese Journal of Clinical Psychology, 16*(1), 46–48.

[贺金波 , 郭永玉 , 向远明 . (2008). 青少年网络游戏成瘾的发生机制 . *中国临床心理学杂志* , *16*(1), 46–48.]

Head, H. & Holmes, G. (1911). Sensory disturbances from cerebral lesion. *Brain, 34*:102–254.

Huang, M., Qian, R. B., Fu, X. M., Wang, C. X., Liu, Y., Han, X. P. ··· Wang, Y. H. (2010). Location of related brain activation in internet game addicts: an fMRI study. *Chin J Neurome, 9*(2), 167–171

[黄敏 , 钱若兵 , 傅先明 , 王昌新 , 刘影 , 韩晓鹏 , ··· 汪业汉 . (2010). 网络游戏成瘾者相关脑区功能定位的 fMRI 研究 . *中华神经医学杂志* , *9*(2), 167–171.]

IJzerman, H., Karremans, J.C., Thomsen, L., & Schubert, T. W. (2013). Caring for sharing. *Social Psychology, 44*(2), 160–166.

IJzerman, H. & Semin, G.R. (2009). The thermometer of social relations: Mapping social proximity on temperature. *Psychological Science, 20*(10), 1214–1220.

Jin, P., Fu, X. M., Qian, R. B., Niu, C. S., & Han, X. P. (2009). Event–related potentials N400 in adolescents with internet addiction disorder. *Chin J Stereotact Funct Neurosurg, 21*(6), 333–335

[金璞 , 傅先明 , 钱若兵 , 牛朝诗 , 韩晓鹏 . (2009). 青少年网络成瘾的事件相关电位 N400 研究 . *立体定向和功能性神经外科杂志* , *21*(6), 333–335.]

Jostmann, N. B., Lakens, D., & Schubert, T. W. (2009). Weight as an embodiment of importance. *Psychological Science, 20*(9), 1169–1174.

Kang, S. (2007). Disembodiment in online social interaction: Impact of online chat on social support and psychosocial well–being. *CyberPsychology & Behavior, 10*(3), 475–477.

Kellenbach, M., Brett, M., & Patterson, K. (2001). Large, colorful, or noisy? attribute– and modality–specific activations during retrieval of perceptual attribute knowledge. *Cognitive, Affective, & Behavioral Neuroscience, 1*(3), 207–221.

Kemmerer, D. & Gonzalez–Castillo, J. (2010). The two–level theory of verb meaning: an approach to integrating the semantics of action with the mirror neuron system. *Brain and Language, 112*(1), 54–76.

Kim, Y. R., Son, J. W., Lee, S. I., Shin, C. J., Kim, S. K., Ju, G., ··· Jo, S. (2012). Abnormal brain activation of adolescent internet addict in a ball–throwing animation task: possible neural correlates of disembodiment revealed by fMRI. *Progress in Neuro–Psychopharmacology and Biological Psychiatry, 39*(1), 88–95.

LaFrance, M., & Hecht, M. A. (2000). Gender and smiling: a meta–analysis. In *Gender and Emotion: Social Psychological Perspectives* (pp. 118–142). New York: Cambridge University Press.

Lakoff, G., & Johnson, M. (1999). *Philosophy in the flesh:*

The Embodied Mind and Its Challenge to Western Thought. New York: Basic Books.

Larsen, R. J., Kasimatis, M., & Frey, K. (1992). Facilitating the furrowed brow: An unobtrusive test of the facial feedback hypothesis applied to unpleasant affect. *Cognition & Emotion*, 6(5), 321–338.

Liang, S. C. & You, Y. Q. (2010) Affective Decision-making in Patients with Internet Addiction, *Chinese Journal of Clinical Psychology*, 18(5), 597–599.

[梁三才, 游旭群. (2010). 网络成瘾者情感决策能力的对照研究. *中国临床心理学杂志*, 18(5), 597–599.]

Lindemann, O., Stenneken, P., Van Schie, H. T., & Bekkering, H. (2006). Semantic activation in action planning. *Journal of Experimental Psychology: Human Perception and Performance*, 32(3), 633–643.

Liu, T., Yang, Y. M., Zhang, H., Zhang, S. S., Liang, D. D., Gu, J. X., & Hu, W. (2001). Neural distinction between Chinese nouns and verbs in the grammatical context: an ERP study. *Acta Psychologica Sinica*, 40(6), 671–680.

[刘涛, 杨亦鸣, 张辉, 张珊珊, 梁丹丹, 顾介鑫, 胡伟. (2008). 语法语境下汉语名动分离的 ERP 研究. *心理学报*, 40(6), 671–680.]

Liu, Y. (Ed.). (1990). *Modern Chinese Frequency Dictionary of common words*. Beijing, China: Aerospace Press.

[刘源 (编). (1990). *现代汉语常用词词频词典*. 北京: 宇航出版社.]

Marques, J. F. (2006). Specialization and semantic organization: evidence for multiple semantics linked to sensory modalities. *Memory & Cognition*, 34(1), 60–67.

Näsi, M. & Koivusilta, L. (2013). Internet and everyday life: the perceived implications of internet use on memory and ability to concentrate. *Cyberpsychology, Behavior, and Social Netwroking*, 16(2), 88–93.

Neal, D. T. & Chartrand, T. L. (2011). Embodied emotion perception amplifying and dampening facial feedback modulates emotion perception accuracy. *Social Psychological and Personality Science*, 2(6), 673–678.

Niedenthal, P. M. (2007). Embodying emotion. *Science*, 316(5827), 1002–1005.

Niu, G., Sun, X., Zhou, Z., & Wei, H. (2013). A review of cognitive neuroscience studies on internet addiction. *Advances in Psychological Science*, 21(6), 1104–1111.

[牛更枫, 孙晓军, 周宗奎, 魏华. (2013). 网络成瘾的认知神经科学研究述评. *心理科学进展*, 21(6), 1104–1111.]

Oberman, L. M., Winkielman, P., & Ramachandran, V. S. (2007). Face to face: Blocking facial mimicry can selectively impair recognition of emotional expressions. *Social Neuroscience*, 2(3–4), 167–178.

Oosterwijk, S., Rotteveel, M., Fischer, A.H., & Hess, U. (2009). Embodied emotion concepts: How generating words about pride and disappointment influences posture. *European Journal of Social Psychology*, 39(3), 457–466.

Pawlikowski, M. & Brand, M. (2011). Excessive internet gaming and decision making: Do excessive World of Warcraft players have problems in decision making under risky conditions? *Psychiatry Research*, 188(3), 428–433.

Pecher, D., Zeelenberg, R., & Barsalou, L.W. (2003). Verifying different-modality properties for concepts produces switching costs. *Psychological Science*, 14(2), 119–124.

Ponari, M., Conson, M., D'Amico, N. P., Grossi, D., & Trojano, L. (2012). Mapping correspondence between facial mimicry and emotion recognition in healthy subjects. *Emotion*, 12(6), 1398–1403.

Price, T. F., Peterson, C. K., & Harmon-Jones, E. (2012). The emotive neuroscience of embodiment. *Motivation and Emotion*, 36(1), 27–37.

Pulvermüller, F., Hauk, O., Nikulin, V. V., & Ilmoniemi, R. J. (2005). Functional links between motor and language systems. *European Journal of Neuroscience*, *21*(3), 793–797.

Shapiro, K. & Caramazza, A. (2003). Grammatical processing of nouns and verbs in left frontal cortex? *Neuropsychologia*, *41*(9), 1189–1198.

Simmons, W. K., Ramjee, V., Beauchamp, M. S., McRae, K., Martin, A., & Barsalou, L. W. (2007). A common neural substrate for perceiving and knowing about color. *Neuropsychologia*, *45*(12), 2802–2810.

Snodgrass, J. G. (1984). Concepts and their surface representations. *Journal of Verbal Learning and Verbal Behavior*, *23*(1), 3–24.

Stel, M., & van Knippenberg, A. (2008). The role of facial mimicry in the recognition of affect. *Psychological Science*, *19*(10), 984–985.

Strack, F., Martin, L. L., & Stepper, S. (1988). Inhibiting and facilitating conditions of the human smile: A nonobtrusive test of the facial feedback hypothesis. *Journal of Personality and Social Psychology*, *54*(5), 768–777.

Sun, D. L., Chen, Z. J., Ma, N., Zhang, X. C., Fu, X. M., & Zhang, D. R. (2009). Decision–making and prepotent response inhibition functions in excessive internet users. *CNS Spectr*, *14*(2), 75–81.

Tan, J. W., Walter, S., Scheck, A., Hrabal, D., Hoffmann, H., Kessler, H., & Traue, H. C. (2012). Repeatability of facial electromyography (EMG) activity over corrugator supercilii and zygomaticus major on differentiating various emotions. *Journal of Ambient Intelligence and Humanized Computing*, *3*(1), 3–10.

Tanenhaus, M. K., Carlson, G., & Trueswell, J. C. (1989). The role of thematic structures in interpretation and parsing. *Language and Cognitive Processes*, *4*(3–4), SI211–SI234.

Taylor, C. A., Lord, C. G., & Bond, C. F. (2009). Embodiment, agency, and attitude change. *Journal of Personality and Social Psychology*, *97*(6), 946–962.

Van den Stock, J., Righart, R., & de Gelder, B. (2007). Body expressions influence recognition of emotions in the face and voice. *Emotion*, *7*(3), 487–494.

Vanman, E. J., Paul, B., Ito, T., & Miller, N. (1997). The modem face of prejudice and structural features that moderate the effect of cooperation on affect. *Journal of Personality and Social Psychology*, *73*(5), 941–959.

Vermeulen, N., Corneille, O., & Niedenthal, P. M. (2008). Sensory load incurs conceptual processing costs. *Cognition*, *109*(2), 287–294.

Vermeulen, N., Niedenthal, P. M., & Luminet, O. (2007). Switching between sensory and affective systems incurs processing costs. *Cognitive Science*, *31*(1), 183–192.

Wallbott, H. G. (1998). Bodily expression of emotion. *European Journal of Social Psychology*, *28*(6), 879–896.

Weinstein, A., & Lejoyeux, M. (2010). Internet addiction or excessive internet use. *The American Journal of Drug and Alcohol Abuse*, *36*(5), 277–283.

Wells, G. L. & Petty, R. E. (1980). The effects of over head movements on persuasion: Compatibility and incompatibility of responses. *Basic and Applied Social Psychology*, *1*(3), 219–230.

Williams, J. H., Whiten, A., & Singh, T. (2004). A systematic review of action imitation in autistic spectrum disorder. *Journal of Autism and Developmental Disorders*, *34*(3), 285–299.

Williams, L. E. & Bargh, J. A. (2008). Experiencing physical warmth promotes interpersonal warmth. *Science*, *322*(5901), 606–607.

Wiswede, D., Münte, T. F., Krämer, U. M., & Rüsseler, J. (2009). Embodied emotion modulates neural signature of performance monitoring. *PLoS One*, *4*(6), e5754.

Xie, J. S., Zhang, C. Q., Wang, R. M., & Lu, Z. (2011).

Paradigms in the study of perceptual symbol systems. *Advances in Psychological Science, 19*(9), 1293–1305.

[谢久书 , 张常青 , 王瑞明 , 陆直 . (2011). 知觉符号理论及其研究范式 . *心理科学进展* , *19*(9), 1293–1305.]

Yu, H. Q., Zhao, X., Li, N., Wang, M. S., & Zhou, P. (2009). Effect of excessive Internet use on the time–frequency characteristic of EEG. *Progress in Natural Science*, 19(10), 1383–1387.

[郁洪强 , 汪曈 , 赵欣 , 李宁 , 刘海婴 , 王明时 . (2009). 网络成瘾患者的 EEG 小波熵与复杂度特征分析 . *中国生物医学工程学报* , *28*(1), 157–160.]

Zhao, L. & Gao, W. B. (2007). Early face processing of internet addiction patients by face specific N170. *Space Medicine & Medical Engineering. 20*(1), 72–74.

[赵仑 , 高文彬 . (2007). 网络成瘾患者早期面孔加工 N170 的研究 . *航天医学与医学工程* , *20*(1), 72–74.]

Zhou, Z. J. & Yang, W. J. (2007). Different types of internet addiction scale for undergraduates. *The Handbook of Psychological Assessment Scales pp.* 503–507.

[周治金 , 杨文娇 . (2007). 大学生网络成瘾类型问卷 . *心理评定量表手册* （第一版）, 503–507.]

The impact of cyber-experience on action verb processing

LIU Siyun; ZHOU Zongkui; LI Na

(Key Laboratory of Adolescent Cyberpsychology and Behavior (CCNU), Ministry of Education;

School of Psychology, Central China Normal University, Wuhan 430079 China)

Abstract Two major different approaches have been held when researchers study higher level cognitive processes. The classic symbolic approach suggested that our higher level cognitive processing belongs to an independent system from that of lower level cognitive processes; whereas embodied cognition theory proposed that our conceptual knowledge is grounded in our sensorimotor systems and shares common neural systems with them. Up till now, a growing number of behavioral and neurological data have provided supporting evidence for embodied cognition theory. In this study, we explored whether different degree of cyber-experience would affect higher level of cognitive processing. Three experiments were designed to investigate the impacts of cyber-experience on the processing of facial expression verbs, body action verbs and action schema verbs. In Experiment 1, two experimental groups of participants were presented with facial expression verbs while their facial positive expressive capability was either facilitated or inhibited. Results showed that participants who had excessive cyber experiences showed no different performance under two different facial muscle controlling conditions, while the participants with less cyber experience recognized positive facial verbs more quickly under facilitated condition than that under inhibited condition. In Experiment 2, the switching costs paradigm was used to explore the impact of cyber-experience on body action verb processing. Results showed that the participants with excessive cyber experience did not show any cost while they switched between verbs and nouns, but the participants with less cyber experience showed significant cognitive cost while switching. In Experiment 3, the serial recall experimental paradigm was applied to explore the impact of cyber-experience on action schema verb processing. Results showed that the excessive cyber-experience participants' recall performance of the action schemas verbs was no different between logic sequence and random sequence conditions, whereas those with less cyber-experience showed significantly worse performance in random sequence condition than that in logic sequence condition. In summary, current findings suggested that excessive cyber behaviors may hurt individuals' higher level of cognitive processing, in that their verb processing may be weakened or delayed as a result of less normal conceptual representations. Our study also provided supportive evidence for the close relationship between the sensorimotor systems and the higher level of conceptual processing.

Keywords embodied cognition; switching cost effect; schema; cyber-experience

Frontiers in Psychology, 2015, 6, 1 - 8.

Are past and future symmetric in mental time line?

Xianfeng Ding[1,2], Ning Feng[1,2], Xiaorong Cheng[1,2], Huashan Liu[1,2], Zhao Fan[1,2,*]

([1] Key Laboratory of Adolescent Cyberpsychology and Behavior, Ministry of Education, Wuhan, China)

([2] School of Psychology, Central China Normal University,Wuhan,China)

Abstract A growing body of evidence has suggested that time, from early to late, or from past to future, was represented in a spatially oriented mental time line. However, little is known about its characteristics. The present study provided the first empirical evidence to explore the symmetry of spatial representations of past and future in the mental time line. Specifically, we compared the Spatial–Temporal Association Response Codes (STARC) effects and distance effects of past and future in four experiments. Results showed that for near past and near future, STARC effects were similar (Experiment 1). For distant past, the STARC effect was significant, but not for distant future (Experiment 2). Furthermore, the distance effect in the past was significantly stronger than in the future (Experiments, 3,4). These findings supported the idea that time points are not evenly distributed in mental time line. Spatial representations of the past and the future are asymmetric, and the spatial representation of past seems stronger than future. The logarithmic pattern of internal spatial representation of past or future is also discussed.

Keywords mental time line; asymmetry; STARC effect; distance effect

1 Introduction

Human beings often represent abstract concepts in concrete visual–spatial images. The spatial representation of number is a typical instance. It was suggested that numbers are represented in a continuous mental number line based on the extensive research on the Spatial–Numerical Association of Response Codes (SNARC) effect (Dehaene et al., 1993; Fischer et al., 2003; Schwarz and Keus, 2004; Hubbard et al., 2005; Nuerk et al., 2005 a,b;). Small numbers are represented at the left side of the line, while large numbers are represented at the right side. Time is also tightly connected with space. Specifically, researchers recently observed a SNARC like effect with time, which was labeled as the Spatial–Temporal Association of Response Codes (STARC) effect (Ishihara

*Correspondence : Zhao Fan, School of Psychology, Central China Normal University,152 Luoyu Road, Hongshan District,Wuhan 430079, China. E–mail: z.fan@mail.ccnu.edu.cn

This research was supported by grants from National Social Science Foundation of China to Xianfeng Ding (13CSH075), National Natural Science Foundation of China to Zhao Fan and Xianfeng Ding (31170979), China Scholarship Council Funding to Xianfeng Ding (201306775018),SRF for ROCS, SEM, and the Humanities and Social Sciences Foundation of the Ministry of Education, China to Xiaorong Cheng (11YJC190006).

et al., 2008;Vallesi et al., 2008). Therefore, time was analogically thought to be represented in a mental time line similar to the mental number line. In other words, time is represented in a continuous spatial line with a left-to-right orientation, where time flows from early to late, or from past to future (Bonato et al., 2012).

The mental time line hypothesis was supported by three categories of spatial-temporal congruency effects. The first type of congruency effect was based on temporal duration or interval. Vallesi et al.(2008) found that left responses were faster when associated with a short duration, while right responses were faster when associated with a long duration. The authors thought this compatible effect was a result of the spatial representation of elapsing time. When a temporal duration has to be estimated, elapsing time may be represented progressively from the left to the right. Then a short duration would be represented relatively to the left, and a long duration relatively to the right (See also Vallesi et al.,2011; Fabbri et al.,2012). Furthermore, duration estimation or judgment can also be influenced by spatial attention. Time duration would be underestimated when attention was directed to the left space, and be overestimated when attention was directed to the right space (Vicario et al., 2007; Frassinetti et al., 2009). A short duration was responded faster when a visual prime was in the left space, whereas a long duration was responded faster when a visual prime was in the right space (Di Bono et al.,2012). These findings suggested that elapsing time was represented in a mental time line from the left to the right.

The second type of congruency effect was based on temporal order. Santiago et al.(2010) found a space-time congruency effect when meaningful event sequences were presented by means of naturalistic movie clips or picture sequences. Order judgments between two events were faster when the left hand was used to respond "before" and the right hand to respond "after" than when

responded with the opposite mapping (see also Fuhrman & Boroditsky, 2010; Boroditsky et al., 2011). However, inherent and logical associations between successive stimuli may be confounded with temporal order in these studies. Some researchers found the STARC effect when using stimuli without logical or internal links et al. (2010) used nine words to explore the congruency effect of order and space in a serial learning paradigm. After an over-learned training phase, these nine words showed a SNARC similar effect for both order-relevant and order-irrelevant tasks. Moreover, a STARC effect based on mere temporal order was also found in working memory paradigms (Ding et al., 2014). These findings indicated that we could represent temporal order information in a spatial line.

The third type of congruency effect was based on abstract time words. Gevers et al. (2003, 2004) found that early months of a year or early days of a week were responded faster with the left key, whereas late months or days were responded faster with the right key. In addition, words referring to the past were responded faster with the left hand; words referring to the future were responded faster with the right hand (Santiago et al., 2007). This effect was found in both visual and auditory modalities (Lakens et al., 2011; Kong & You, 2012). Furthermore, time words can shift attention. Words related to the past can shift attention to the left and words related to the future can shift attention to the right in priming tasks (Weger and Pratt, 2008, 2009; Ouellet et al., 2010a). These findings indicated that the abstract concept of time could also be represented in a spatially oriented mental time line. Past is at the left side of this line, and future is at right side of this line. Time flows from past (left) to future (right) in the mental time line.

Taken together, three categories of evidence strongly supported that time can be represented in a mental time line. Time, from early to late, or from past to future, appears to be represented in a left-to-right spatially continuous line. Previous studies have provided a lot of evidence for the existence of the mental time line, however, little is known about its characteristics. Are time points distributed evenly in this mental time line? Specifically, are past and future symmetric in the mental time line? The past is time we have actually experienced while the future is time that we have never experienced. Could this difference in reality for past and future lead to different spatial representations? A temporal asymmetry of past and future was suggested by evidence in some other paradigms. Representations of past events were associated with more specific details than representations of future events (D'Argembeau and Van der Linden, 2004, 2006; Addis and Schacter, 2008; Wang et al., 2011), and future events were more prototypical than past events (Kane et al., 2012). Thus we hypothesized that the spatial representations of past and future with the same temporal distance from the present are not identical but asymmetric in the mental time line. Examining symmetry would provide the first empirical evidence of characteristics of the mental time line, which is important to the construction of a theory of the spatial representation of time. Moreover, it will enhance the understanding of difference or similarity of past and future from the aspect of spatial linear representations and further provide a more specific spatial frame for past-future related theories.

To investigate this issue, we compared the spatial representations of past and future in the mental time line. According to previous studies, the STARC effect is the most important index of the spatial representation of time. Thus, we explored the symmetry of past and future in the mental time line by comparing the STARC effects of past and future. Another typical index of the spatial representation is the distance effect, in which the distance discrimination of two points located on a spatial line would be faster when the two points are far from each other than when they are near from each other (Moyer and Landauer,

1967; Dehaene et al.,1990; Dehaene, 1997). So we also compared the distance effects between past and future. We hypothesized that if the past and future are symmetric in the mental time line, there should be no differences on STARC effects and distance effects between past and future.

2 Experiment 1

Experiment 1 was designed to examine whether the STARC effects for near past (yesterday) and near future (tomorrow) were different. If the STARC effects were the same, it would support that spatial representations of near past and near future were symmetric in the mental time line.

2.1 Methods

2.1.1 Participants

Thirty six undergraduate students (13 male and 23 female) from Central China Normal University participated in the experiment for course credits. All participants signed a consent form according to the requirements of Institutional Review Board of CCNU. They were 19.6 years old on average (range 18 to 21). All participants were naive to the purpose of the experiment.

2.1.2 Stimuli and apparatus

Sixteen Chinese time words were used, 8 referring to the past time of yesterday (e.g., yesterday morning, yesterday afternoon, yesterday evening, etc.), and the other 8 referring to the future time of tomorrow (e.g., tomorrow morning, tomorrow afternoon, tomorrow evening, etc.). Time for the two groups of words were same except that past time was labeled with yesterday, and future time was labeled with tomorrow.

Participants viewed words on a 17-in. CRT screen (refresh rate 75 Hz and resolution 1280 × 1024 pixels) from a distance of 70 cm. The experiment procedure was programmed in Visual C++.

2.1.3 Experimental design

We used a 2 × 2 × 2 mixed design. A between-subjects factor was Type of Time Words (yesterday vs. tomorrow) and two within-subjects factors were Temporal Position (early vs. late), and Response Congruence (congruence vs. incongruence, congruence means early stimuli responded with the left key and late stimuli responded with the right key; incongruence means early stimuli responded with the right key and late stimuli responded with the left key). Response times (RTs) and accuracy rates were dependent variables.

2.1.4 Procedure

Half of participants took part in the past or yesterday condition and the other half in the future or tomorrow condition. In the past condition, a trial started with a central fixation cross, lasting for 500 ms. Following that cross, a time word of yesterday was presented for 300 ms. Participant were required to judge whether the time of word was earlier or later than yesterday noon. For example, yesterday evening was later than yesterday noon. In one session, the participant pressed the left key (left arrow on the keyboard) if earlier and pressed the right key (right arrow on the keyboard) if later. In the other session participants responded in the opposite way. The order of the two sessions was counterbalanced across participants. The participants were required to respond as fast and accurately as possible using two fingers of the right hand only. After responses to stimuli, a 1000 ms blank separated one trial from another. Each session included 10 trials of practice and 4 blocks of 160 trials in the formal experiment.

In the future condition, the procedure was the same as in the past condition, except that the stimuli were time words of tomorrow and participants were required to judge whether the time was earlier or later than tomorrow noon.

2.1.5 Data analysis

Trials were treated as errors and discarded from the RT analyses if a response was made during the first 100

ms after the stimuli onset (anticipated responses), if the RT was slower than 2000 ms or no response was detected (delayed and null responses), or if the judgment was incorrect. RT outliers of correct trials (out of 3 standard deviations) were also filtered on a per-participant basis and excluded from analyses. A 2 × 2 × 2 repeated measures MANOVA was performed both for accuracy rates and mean RTs of correct trials.

2.2 Results and discussion

The mean error rate in judging the time words was 2.31%. No significant effect was observed in the MANOVA concerning accuracy. The results of RTs indicated that the main effect of response congruence was significant (See Figure.1), $F_{(1,34)}=29.33$, $p<0.001$, partial $\eta^2=0.46$. The main effect of temporal position was significant, $F_{(1,34)}=8.13$, $p<0.01$, partial $\eta^2=0.19$. The main effect of type of time words was not significant, $F_{(1,34)}=0.94$, $p>0.05$. The interaction between response congruence and type of time words was not significant, $F_{(1,34)}=2.99$, $p>0.05$. The interaction between temporal position and response congruence was not significant, $F_{(1,34)}=0.01$, $p>0.05$. The interaction between temporal position and type of time words was not significant, $F_{(1,34)}=3.06$, $p>0.05$. The three-way interaction was not significant either, $F_{(1,34)}=0.32$, $p>0.05$.

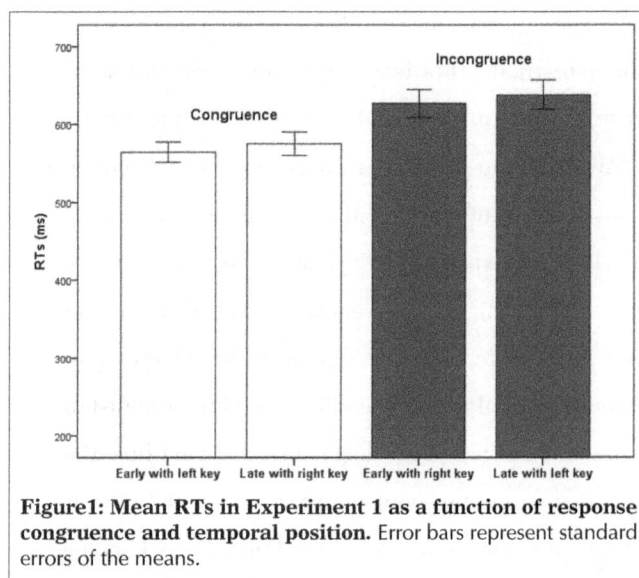

Figure1: Mean RTs in Experiment 1 as a function of response congruence and temporal position. Error bars represent standard errors of the means.

The results of Experiment 1 revealed a typical STARC effect. The RT of congruence (M=569 ms, SD=83) was significantly shorter than the RT of incongruence (M=632 ms, SD=110). In other words, early time of a day was responded faster with the left key and late time was responded faster with the right key. However, there was no significant interaction effect between response congruence and type of time words. The STARC effects were the same for near past (yesterday) and near future (tomorrow). This result is consistent with the idea that the spatial representations of past and future in near space are symmetric in the mental time line.

3 Experiment 2

The STARC effects for near past and near future did not show any difference in Experiment 1, 2 was designed to further examine whether the STARC effects of distant past (last year) and distant future (next year) were different.

3.1 Methods

3.1.1 Participants

Thirty six undergraduate students (16 male and 20 female) from Central China Normal University participated in the experiment for course credits. All participants signed a consent form according to the requirements of Institutional Review Board of CCNU. They were 20.1 years old on average (range 18 to 21). All participants were naive to the purpose of the experiment.

3.1.2 Stimuli and apparatus

Sixteen Chinese time words of festivals were used, 8 referring to past time of last year (e.g., Lantern Festival of last year, Labor Day of last year, National Day of last year, etc.) and the other 8 referring to future time of next year (e.g., Lantern Festival of next year, Labor Day of next year, National Day of next year, etc.). Times of the words were same except that past time was labeled with last year, and future time was labeled with next year.

3.1.3 Experimental design

The design was similar to the design in Experiment 1. The only change was the time words. Distant time words were used: last year vs. next year instead of yesterday vs. tomorrow.

3.1.4 Procedure

The procedure was the same as in Experiment 1. The task was to judge whether the time of word was earlier or later than July of last year or July of next year. For example, National Day of last year was later than last July. Half of participants took part in the past or last year condition and the other half in the future or next year condition.

3.1.5 Data analysis

Data analysis was the same as in Experiment 1.

3.2 Results and discussion

The mean error rate in judging the time words was 3.74%. No significant effect was observed in the MANOVA concerning accuracy. The results of RTs indicated that the only significant main effect was for response congruence, $F_{(1,34)}$=12.59, p<0.001, partial η^2=0.27. The interaction between response congruence and type of time words was significant, $F_{(1,34)}$=6.64, p=0.014, partial η^2=0.16 (See **Figure 2**). No other interaction was significant, ps>0.05. Simple effect analysis revealed that the response congruence effect was significant only for last year, $F_{(1,34)}$=18.76, p<0.001; but not significant for next year, $F_{(1,34)}$=0.47, p=0.49.

The results of Experiment 2 revealed a significant STARC effect as in Experiment 1. However, there was an interaction between response congruence and type of time words. The STARC effect was only found in distant past (last year) condition, but not in distant future (next year) condition. Early times of last year were responded faster with the left key and late times of last year were responded faster with the right key, while this was not true for next year. Thus, it suggested that spatial representations of

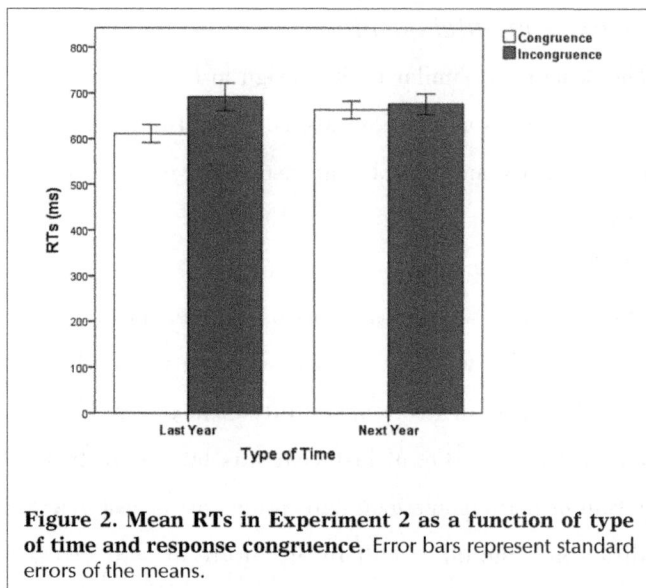

Figure 2. Mean RTs in Experiment 2 as a function of type of time and response congruence. Error bars represent standard errors of the means.

distant past and future were asymmetric in the mental time line, and the spatial representation of distant past was stronger than that of distant future.

4 Experiment 3

The results of Experiment 1, 2 showed that spatial representations of past and future in the mental time line were symmetric in near space, but asymmetric in distant space. If this was the case, a specific distance of past or future (two time points from either near space or distant space with same temporal distance) might be represented asymmetrically in the mental time line. Consequently, the distance effect for past and future might be different. Experiment 3 was designed to further examine whether the distance effects of past and future were the same.

4.1 Methods

4.1.1 Participants

Thirty six undergraduate students (14 male and 22 female) from Central China Normal University participated in the experiment for course credits. All participants signed a consent form according to the requirements of Institutional Review Board of CCNU. They were 21.2 years old on average (range 18 to 22). All participants were naive to the purpose of the experiment.

4.1.2 Stimuli and apparatus

Sixteen Chinese time words were used. Eight words were near distance time words, 4 of them referring to past time of yesterday (e.g., yesterday morning, yesterday evening, etc.) and the other 4 referring to future time of tomorrow (e.g., tomorrow morning, tomorrow evening, etc.). Eight words were far distance time words, 4 of them referring to past time of last year (e.g., Labor Day of last year, National Day of last year, etc.) and the other 4 referring to future time of next year (e.g., Labor Day of next year, National Day of next year, etc.).

4.1.3 Experimental design

We used a $2 \times 2 \times 2$ mixed design. Three independent variables were temporal distance (near vs. far), between-subjects factor; type of time words (past vs. future), within-subjects factor; response congruence (congruence vs. incongruence, congruence means past time with left key and future time with right key; incongruence means past time with right key and future with left key), within-subjects factor. RTs and accuracy rates were dependent variables.

4.1.4 Procedure

Half of participants were in the far distance condition (last year and next year). A trial started with the central fixation cross, lasting for 500 ms. Following that cross, the time word of last year or next year was presented, lasting for 300 ms. Participants were required to judge whether the time of word was earlier or later than present. For example, National Day of last year was earlier than present. In one session, the participant pressed the left key (left arrow in keyboard) if earlier and pressed the right key (right arrow in keyboard) if later. In the other session participants were required to respond in the opposite way. The order of the two sessions was counterbalanced across participants. The participants were required to respond as fast and accurately as possible using two fingers of the right hand only. After responding to stimuli, a 1000 ms blank separated one trial from another. Each session included 10 trials of practice and 4 blocks of 160 trials in the formal experiment. The other half participants were in the near distance condition (yesterday and tomorrow). The procedure was the same as in the far distance condition.

4.1.5 Data analysis

Data analysis was the same as in Experiment 1.

4.2 Results and discussion

The mean error rate in judging the time words was 2.56%. No significant effect was observed in the MANOVA concerning accuracy. The results of RTs indicated that the main effect of response congruence was significant, $F_{(1,34)}=13.09$, $p<0.001$, partial $\eta^2=0.29$. The main effect of temporal distance was significant, $F_{(1,34)}=7.17$, $p=0.011$, partial $\eta^2=0.17$. The main effect of type of time words was not significant, $F_{(1,34)}=1.64$, $p>0.05$. The interaction between temporal distance and type of time words was significant (See **Figure3**), $F_{(1,34)}=5.51$, $p=0.025$, partial $\eta^2=0.14$. No other interactions were significant, ps>0.05 Simple effect analysis revealed that the distance effect was greater for past time, $F_{(1,34)}=9.69$, $p=0.004$; smaller for future time, $F_{(1,34)}=4.74$, $p=0.036$.

The results of Experiment 3 revealed a typical STARC effect as in Experiment 1, 2. Past time words were responded faster with the left key, whereas future time words were responded faster with the right key. Moreover, significant distance effect was observed. The time words of near distance (yesterday and tomorrow) were responded slower than far distance (last year and next year). Most importantly, there was an interaction between temporal distance and type of time words. The distance effect was greater for the past than for the future. These results further suggest that the spatial representations of past and future are asymmetric in the mental time line and that spatial representation of the past seems to be stronger than that of future.

However, some characteristics of time words, such as familiarity, could be confounded with temporal distance

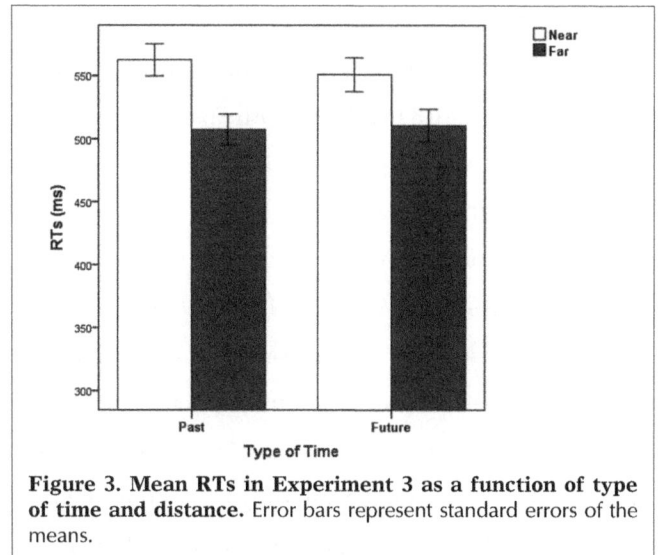

Figure 3. Mean RTs in Experiment 3 as a function of type of time and distance. Error bars represent standard errors of the means.

in Experiment 3. Separately we found that time words of near distance (yesterday or tomorrow) were more familiar than time words of far distance (last year or next year) in Experiment 3 through a questionnaire. Since familiar words were usually responded faster than unfamiliar words, it seemed that this distance effect could not be explained by familiarity of words. Nonetheless, we ran another experiment to balance the familiarity of time words in different distance.

5 Experiment 4

Experiment 4 was designed to balance the familiarity of time words in different distance. Both familiar and unfamiliar time words were chosen in near and far distance condition. Moreover, we changed the response way from left–right direction to an orthogonal up–down direction in the keyboard. If distance effects of past and future were actually different, the way of response would not affect it.

5.1 Methods

5.1.1 Participants

Eighteen undergraduate students (6 male and 12 female) from Central China Normal University participated in the experiment for course credits. All participants signed a consent form according to the requirements of Institutional Review Board of CCNU. They were 19.2 years old on average (range 18 to 20). All participants were naive to the purpose of the experiment.

5.1.2 Stimuli and apparatus

Thirty two Chinese time words were used. Sixteen words were near distance time words, 8 referring to past time of yesterday (e.g., yesterday morning, yesterday evening, etc.) and the other 8 referring to future time of tomorrow (e.g., tomorrow morning, tomorrow evening, etc.). Sixteen words were far distance time words, 8 referring to past time of last year (e.g., Labor Day of last year, National Day of last year, etc.) and the other 8 referring to future time of next year (e.g., Labor Day of next year, National Day of next year, etc.). In a pre-experimental questionnaire investigation, 46 subjects rated the familiarity of 32 time words from 1 (unfamiliar) to 5 (familiar). The results showed that the familiarity of time words in different temporal distances were not significantly different (M_{near}=3.48 vs. M_{far}=3.31), $F_{(1,45)}$=1.81, p=0.19.

5.1.3 Experimental design

We used a $2 \times 2 \times 2$ within-subjects design. Three independent variables were time distance (near vs. far), type of time words (past vs. future), response key (up arrow vs. down arrow). RTs and accuracy rates were dependent variables.

5.1.4 Procedure

The procedure was similar to the previous experiments. Participants were required to judge whether the time of word was earlier or later than present. For example, Yesterday morning or National Day of last year was earlier than present. In one session, the participant pressed up key (up arrow in the keyboard) if earlier and pressed the down key (down arrow in the keyboard) if later. In the other session participants were required to respond in the opposite way. The order of the two sessions was counterbalanced across participants. The participants were required to respond as fast and accurately as possible using the middle finger of the right hand only. Each session included 10 trials of practice and 8 blocks of 320 trials in the formal experiment.

5.1.5 Data analysis

Data analysis was the same as in Experiment 1.

5.2 Results and discussion

The mean error rate in judging the time words was 3.13%. No significant effect was observed in the MANOVA concerning accuracy. The results of RTs indicated that the main effect of time distance was significant, $F_{(1,17)}$=29.04, p<0.001, partial η^2=0.63. The interaction between time distance and type of time words was significant (See **Figure 4**), $F_{(1,17)}$=8.05, p=0.011, partial η^2=0.32. All other main effects and interactions were not significant, ps>0.05. As the interaction between time distance and type of time words was significant, a simple effect analysis revealed that distance effect for past time, $F_{(1,17)}$=24.57, p<0.001; and a smaller effect for future time, $F_{(1,17)}$=4.58, p=0.047.

The results of Experiment 4 were similar as in Experiment 3. After controlling the familiarity of time words, the distance effect was still observed. The time words of near distance (yesterday and tomorrow) were responded slower than words of far distance (last year and next year). More important, the distance effect of past was also greater than that of future. Again, this result suggest that spatial representations of past and future were asymmetric in the mental time line and the spatial representation of the past seemed stronger than that of the future.

6 General discussion

Previous findings supported that representation of time flows from past to future in a continuous spatial line with a left-to-right orientation. The present study provided the first empirical evidence for a fundamental characteristic of the mental time line: Are the spatial representations of past

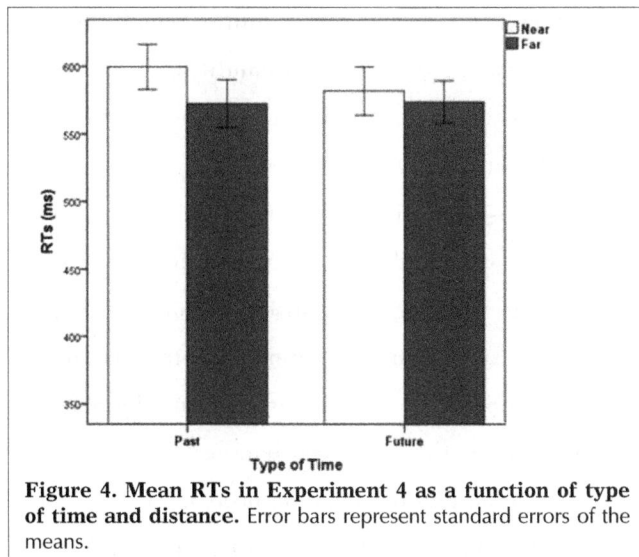

Figure 4. Mean RTs in Experiment 4 as a function of type of time and distance. Error bars represent standard errors of the means.

and future symmetric in the mental time line?

In Experiment 1, we compared STARC effects under near past and near future conditions. As expectedly, a typical STARC effect was observed. Early time was responded faster with the left key, whereas late time was responded faster with the right key. Moreover, STARC effects were the same between yesterday and tomorrow. This result indicated that spatial representations of past and future were symmetric in near past and near future in the mental time line. In Experiment 2, STARC effects were further compared under distant past and distant future condition. However, the STARC effect was only observed in the distant past condition, not in the distant future condition. This result showed that spatial representations of distant future and distant past were asymmetric in the mental time line and the spatial representation of past seemed stronger than that of future, as the STARC effect disappeared in distant future. Therefore, it seemed that past and future in the mental time line were symmetric in near space, but not in distant space.

In Experiment 3, 4, distance effects were compared under past and future conditions. Results showed that there were both a significant STARC effect and a distance effect in Experiment 3. Past time words were responded faster with the left key, whereas future time words were responded faster with the right key. When compared with

the present, time points in the far distance (last year or next year) were responded faster than in the near distance (yesterday and tomorrow). Moreover, the distance effect in the past was greater than in the future. The same result was observed even when the response was changed to an orthogonal direction in Experiment 4. These results about distance effects support the idea that past and future are represented asymmetrically in the mental time line. Again, the spatial representation of the past seemed stronger than that of the past, as the distance effect for the future was smaller than that of the past.

These findings revealed that the mental time line is not evenly distributed and the past and future were asymmetric in the mental time line. According to our results, the STARC effect was significant for the distant past, but not for the distant future. And the distance effect was stronger for the past than for the future. Why is past different from future? A possible reason is that the past is more concrete or clear than the future. The past is time we have actually experienced. It is true and available for us. We can store the past information in our memory and retrieve it. The construction of the representation of past could be based on real events. However, the future is not yet true. It is obscure, abstract and fictional for us. The construction of the representation of future could only be based on fictional events.

This temporal asymmetry was in line with findings in some other paradigms. For example, Vallesi et al. (2008) found a stronger leftward representation of short durations than rightward representation of long durations in their fourth experiment at least numerically, suggesting a similar asymmetrical effect though no further statistics were provided in this literature. In a neuroimaging study, Okuda et al. (2003) found that anteromedial frontal pole and medial temporal areas showed a significant effect of temporal distance from the present. Specifically, the increase in brain activity in the left parahippocampal gyrus (BA 36) from the near future task to the far future task was smaller than that from the near past task to the far past task. It suggested that distance effect was smaller for the future than for the past task. Addis and Schacter (2008) found that representations of past events were associated with more specific details than representations of future events (See also D'Argembeau and Van der Linden, 2004, 2006; Wang et al., 2011). Future events were also found more prototypical than past events (Kane et al., 2012). These findings supported that the representation of past was more concrete or clear than that of future.

Although the future is different from the past, it is similar in that the past and future are represented as a spatial line. STARC effects and distance effects were found for both past and future. These findings were consistent with constructive episodic simulation hypothesis (Schacter and Addis, 2007a,b). Since future is what we have never experienced before, how do we construct representations that we never truly experienced? Schacter and Addis (2007a,b) thought that one of important functions of constructive episodic memory is to allow individuals to simulate or imagine future episodes. We construct future based on past that we have experienced. Therefore, there should be considerable overlap in the psychological and neural processes involved in remembering the past and imagining the future. Neuroimaging evidence from Mental Time Traveling (MTT) supported that underlying neural mechanisms for past and future were similar. Remembering the past and imaging the future may activate same brain areas (Okuda et al., 2003; Schacter et al., 2007; Szpunar et al., 2007).These findings have led to the concept of the prospective brain and an idea that a crucial function of the brain is to use stored information to imagine, simulate and predict possible future events (Schacter et al., 2007).

The present study further indicated that the internal spatial representation of past or future seemed to be unevenly-distributed in the mental time line. Taken the results of Experiment 1,2 together, the STARC effect was significant in tomorrow (near future) condition but not in next year (distant future) condition. Thus, the spatial representation of the near future seemed stronger than that of distant future in the mental time line. This finding was consistent with some studies about a loglinear characteristic of mental time. When participants were asked to judge whether an event of past or future was before or after an imagined "location" on the time line, the reaction time of this "self-projection" decreased logarithmically as the temporal distance between this imagined location and the location of another imagined event from the time line increased (Arzy et al., 2009). In addition, logarithmic curves were also found to fit the relation between temporal distance and memory, as the distribution of the correct recall of events from different points in time was logarithmic (Rubin and Schulkind, 1997; Spreng and Levine, 2006).

This logarithmic pattern suggested that time points (past or future) near the present were relatively sparse, and time points far from the present were relatively dense in the mental time line. In other words, if a set of two time points is near the present, the spatial distance would be larger, as the reaction times decrease sharply with the increase of the temporal distance; if the set of two time points was far from the present, the spatial distance would be smaller, as the reaction times decrease slowly with the increase of the temporal distance. Interestingly, a logarithmic pattern was also found in the mental number line. Humans map numbers into space line in logarithmic scaling (Siegler and Booth, 2004; Dehaene and Cohen, 1995; Dehaene et al., 2008). The mechanisms of processing time and number are similar, in line with the A Theory of Magnitude, i.e. ATOM (Walsh, 2003; Bueti and Walsh, 2009). Nonetheless, we should be cautious with these inferences and further research is needed on this logarithmic pattern of spatial representation in the mental time line.

Finally, it was worth noting that culture may play an

important role on the representations of past and future. For instance, reading and writing habits can change the direction of the mental time line. The direction is from left to right in English or Italian speakers, whereas the direction is from right to left for Arabic or Hebraic speakers (Fuhrman and Boroditsky 2010; Ouellet et al., 2010b; Vallesi et al., 2014). Most importantly, culture may shape the characteristics of the mental time line. Westerners exhibit greater episodic specificity than East Asians (Wang et al., 2011). The spatial representations of mental time for Westerners might be stronger than those of East Asians. Age and gender may also influence the representations of past and future. Older adults generated fewer internal details than younger adults for both past and future events (Addis and Schacter, 2008). Women exhibit greater episodic specificity than men for both past and future events (Wang et al., 2011).

In summary, the present study provided the first empirical evidence for the characteristics of the mental time line. Time points are not evenly distributed in the mental time line. The differences on STARC effect and distance effect supported that the spatial representations of past and future are asymmetric in the mental time line. And the spatial representation of past seemed stronger than that of future. Importantly, future studies should focus more on the characteristics of internal spatial representation of past or future (e.g. logarithmic pattern) and how the culture and some other factors shape the characteristics of the mental time line.

References

Addis,D.R.,and Schacter,D.L.(2008).Effects of detail and temporal distance of past and future events on the engagement of a common neural network. *Hippocampus 18*, 227–237.doi:10.1002/hipo.20405

Arzy,S.,Adi-Japha,E.,and Blanke,O.(2009).The mental time line:an analogue of the mental number line in the mapping of life events. *Conscious.Cogn. 18*, 781–785. doi:10.1016/j.concog.2009.05.007

Bonato,M.,Zorzi,M.,andUmiltà,C.(2012).When time is space:evidence for a mental time line. *Neurosci. Biobehav.Rev. 36*, 2257–2273.doi: 10.1016/ j.neubiorev.2012.08.007

Boroditsky,L.,Fuhrmana,O.,and McCormicka,K.(2011). Do Englisha nd Mandarin speakers think about time differently? *Cognition 118*, 123–129.doi: 10.1016/ j.cognition.2010.09.010

Bueti,D.,and Walsh,V.(2009).The parietal cortex and the representation of time, space,number and other magnitudes. *Philos.Trans.R.Soc.Lond.B Biol.Sci. 364*, 1831–1840.doi:10.1098/rstb.2009.0028

D'Argembeau,A.,and van der Linden,M.(2004). Phenomenal characteristic sassociated with rojecting oneself back into the past and forward into the future: influence of valence and temporal distance. *Conscious. Cogn.13*, 844–858.doi: 10.1016/j.concog.2004.07.007

D'Argembeau,A.,and van der Linden,M.(2006). Individual differences in the phenomenology of mental time travel:the effects of vivid visual imagery and emotion regulation strategies. *Conscious.Cogn. 15*, 342–350.doi: 10.1016/j.concog.2005.09.001

Dehaene, S.(1997). *The Number Sense:How the Mind Creates Mathematics*. New York,NY:Oxford University Press.

Dehaene, S.,Bossini,S.,and Giraux,P.(1993).The mental representation of parity and number magnitude. *J. Exp.Psychol.Gen. 122*, 371–396.doi:10.1037/0096-3445.122.3.371

Dehaene, S.,and Cohen,L.(1995).Towards an anatomical and functional model of number processing. *Math. Cogn. 1*, 83–120.

Dehaene, S.,Dupoux,E.,and Mehler,J.(1990).Is numerical comparison digital? Analogical and symbolic effects in two-digit number comparison. J. Exp.Psychol.Hum.

Percept.Perform.16,626–641.doi:10.1037/0096–1523. 16.3.626

Dehaene, S.,Izard,V.,Spelke,E.,and Pica,P.(2008).Log or linear?Distinct intuitions of the number scale in Western and Amazonian indigene cultures. *Science 320*, 1217–1220.doi:10.1126/science.1156540

Di Bono,M.G.,Casarotti,M.,Priftis,K.,Gava,L.,Umilt à ,C .,and Zorzi,M. (2012). Priming the mental time line. J. Exp.Psychol.Hum.*Percept.Perform. 38*, 838–842. doi:10.1037/a0028346

Ding,X.,Cheng,X.,Fan,Z.,and Liu,H.(2015).Is elapsing time really recoded into spatial linear representation in working memory? *Exp.Psychol. 62*,11–19.doi: 10.1027/1618–3169/a000269

Fabbri,M.,Cancellieri,J.,and Natale,V.(2012).The A Theory Of Magnitude (ATOM)model in temporal perception and reproduction tasks. *Acta Psychol. 139*, 111–123. doi:10.1016/j.actpsy.2011.09.006

Fischer,M.H.,Castel,A.D.,Dodd,M.D.,and Pratt,J.(2003). Perceiving numbers causes spatial shifts of attention. *Nat.Neurosci. 6*, 555–556.doi:10.1038/nn1066

Frassinetti,F.,Magnani,B.,and Oliveri,M.(2009).Prismatic lenses shift time perception. *Psychol.Sci. 20*, 949–954. doi:10.1111/j.1467–9280.2009. 02390.x

Fuhrman,O.,and Boroditsky,L.(2010).Cross–cultural differences in mental representations of time:evidence from an implicit nonlinguistic task. *Cogn.Sci. 34*, 1430–1451. doi:10.1111/j.1551–6709.2010.01105.x

Gevers,W.,Reynvoet,B.,and Fias,W.(2003).The mental representation of ordinal sequences is spatially organized. *Cognition 87*, 87–95.doi:10.1016/S0010–0277(02)00234–2

Gevers,W.,Reynvoet,B.,and Fias,W.(2004).The mental representation of ordinal sequences is spatially organized:evidence from days of the week. *Cortex 40*, 171–172. doi:10.1016/S0010–9452(08)70938–9

Hubbard,E.M.,Piazza,M.,Pinel,P.,and Dehaene,S.(2005).

Interactions between number and space in parietal cortex. *Nat.Rev.Neurosci. 6*, 435–448.doi: 10.1038/ nrn1684

Ishihara,M.,Keller,P.E.,Rossetti,Y.,and Prinz,W.(2008). Horizontal spatial rep representations of time:evidence for the STEARC effect. *Cortex 44*, 454–461.doi: 10.1016/j.cortex.2007.08.010

Kane,J.,Van Boven,L.,and Mcgraw,A.P.(2012).Prototypical prospection:future events are more prototypically represented and simulated than past events. Eur. *J. Soc.Psychol. 42*, 354–362.doi:10.1002/ejsp.1866

Kong,F.,and You,X.(2012).Space–time compatibility effects in the auditory modality. *Exp.Psychol. 59*, 82–87.doi:10.1027/1618–3169/a000129

Lakens,D.,Semin,G.R.,and Garrido,M.V.(2011).The sound of time:cross– modal convergence in the spatial structuring of time. *Conscious.Cogn. 20*, 437–443. doi:10.1016/j.concog.2010.09.020

Moyer,R.S.,and Landauer,T.K.(1967).Time required for judgments of numerical inequality. *Nature 215*, 1519–1520.doi:10.1038/2151519a0

Nuerk,H–C.,Bauer,F.,Krummenacher,J.,Heller,D.,and Willmes,K.(2005a).The power of the mental number line:how the magnitude of unattended numbers affects performance in an Eriksen task. *Psychol.Sci. 47*, 34–50.

Nuerk,H.C.,Wood,G.,and Willmes,K.(2005b).The universal SNARC effect. *Exp. Psychol. 52*, 187–194. doi:10.1027/1618–3169.52.3.187

Okuda, J.,Fujii,T.,Ohtake,H.,Tsukiura,T.,Tanji,K.,Suzuki ,K.,etal.(2003). Thinking of the future and the past:the roles of the frontal pole and the medial temporal lobes. *Neuroimage 19*, 1369–1380.doi:10.1016/S1053–8119(03)00179–4

Ouellet, M.,Santiago,J.,Funes,M.J.,and Lupianez,J.(2010a). Thinking about the future moves attention to the right. J. Exp.Psychol.Hum.*Percept.Perform. 36*, 17–24.

doi:10.1037/a0017176

Ouellet, M.,Santiago,J.,Israeli,Z.,and Gabay,S.(2010b).Is the future the right time? *Exp.Psychol. 57*, 308–314. doi:10.1027/1618–3169/a000036

Previtali,P.,deHevia,M.D.,and Girelli,L.(2010).Placing order in space: the SNARC effect in serial learning. *Exp.BrainRes. 201*, 599–605.doi: 10.1007/s00221–009–2063–3

Rubin, D. C., & Schulkind, M. D. (1997). The distribution of autobiographical memories across the lifespan. *Mem. Cognit. 25*, 859 – 866. doi: 10.3758/BF03211330

Santiago, J., Lupianez, J., Perez, E., & Funes, M. J. (2007). Time (also) flies from left to right. *Psychon. Bull. Rev. 14*, 512 – 516. doi: 10.3758/BF03194099

Santiago, J., Roman, A.,Ouellet, M., Rodriguez, N., & Perez–Azor, P. (2010). In hindsight, life flows from left to right. *Psychol. Res. 74*, 59 – 70. doi: 10.1007/s00426–008–0220–0

Schacter, D. L., & Addis, D. R. (2007a). The cognitive neuroscience of constructive memory: remembering the past and imagining the future. *Philos. Trans. R. Soc. Lond B Biol. Sci. 362*, 773 – 786. doi: 10.1098/rstb.2007.2087

Schacter, D. L., & Addis, D. R. (2007b). The ghosts of past and future. *Nature 445*, 27. doi: 10.1038/445027a

Schacter, D. L., Addis, D. R., & Buckner, R. L. (2007). Remembering the past to imagine the future: the prospective brain. *Nat. Rev. Neurosci. 8*, 657 – 661. doi: 10.1038/nrn2213

Schwarz, W., & Keus, I. M. (2004). Moving the eyes along the mental number line comparing SNARC effects with saccadic and manual response. *Percept. Psychophys. 66*, 651 – 664. doi: 10.3758/BF03194909

Siegler, R. S., & Booth, J. L. (2004). Development of numerical estimation in young children. *Child Dev. 75*, 428 – 444. doi: 10.1111/j.1467–8624.2004.00684.x

Spreng, R. N., & Levine, B. (2006). The temporal distribution of past and future autobiographical events across the lifespan. *Mem. Cognit. 34*, 1644 – 1651. doi: 10.3758/BF03195927

Szpunar, K. K., Watson, J. M., & McDermott, K. B. (2007). Neural substrates of envisioning the future. *Proc. Natl. Acad. Sci. U.S.A. 104*, 642 – 647. doi: 10.1073/pnas.0610082104

Vallesi, A., Binns, M. A., & Shallice, T. (2008). An effect of spatial–temporal association of response codes: understanding the cognitive representations of time. *Cognition 107*, 501 – 527. doi: 10.1016/j.cognition.2007.10.011

Vallesi, A., McIntosh, A. R., & Stuss, D. T. (2011). How time modulates spatial responses. *Cortex 47*, 148 – 156. doi: 10.1016/j.cortex.2009.09.005

Vallesi, A., Weisblatt, Y., Semenza, C., & Shaki, S. (2014). Cultural modulations of space–time compatibility effects. *Psychon. Bull. Rev. 21*, 666 – 669. doi: 10.3758/s13423–013–0540–y

Vicario, C. M., Caltagirone, C., & Oliveri, M. (2007). Optokinetic stimulation affects temporal estimation in healthy humans. *Brain Cogn. 64*, 68 – 73. doi: 10.1016/j.bandc.2006.12.002

Walsh, V. (2003). A theory of magnitude: common cortical metrics of time, space and quantity. *Trends Cogn. Sci. 7*, 483 – 488. doi: 10.1016/j.tics.2003.09.002

Wang, Q., Hou, Y., Tang, H., & Wiprovnick, A. (2011). Travelling backwards and forwards in time: culture and gender in the episodic specificity of past and future events. *Memory 19*, 103 – 109. doi: 10.1080/09658211.2010.537279

Weger,U.W.,& Pratt,J.(2008).Time flies like an arrow: space–time compatibility effects suggest the use of a mental timeline. *Psychon. Bull. Rev. 15*, 426 – 430. doi: 10.3758/PBR.15.2.426

Weger, U. W., & Pratt, J. (2009). Time–words guide spatial attention. *J. Vis. 7*, 447. doi: 10.1167/7.9.447

Psychophysiology, 2012, 49(8), 1133 - 1144.

The role of sustained posterior brain activity in the serial chaining of two cognitive operations: A MEG study

Zhao Fan[1,2], Suresh D. Muthukumaraswamy[3], Krish D. Singh[3], Kimron Shapiro[4]

([1] Key Laboratory of Adolescent Cyberpsychology and Behavior (CCNU), Ministry of Education, Wuhan, China)

([2] School of Psychology, Central China Normal University, Wuhan,China)

([3] CUBRIC, School of Psychology, Cardiff University, Cardiff,UK)

([4] School of Psychology, Bangor University, Bangor, UK)

Abstract A fundamental necessity in human cognition is to link sequential mental operations where appropriate execution of the second task requires input from the first. The present study explores the neural basis of such 'chaining' using a novel psychological refractory period(PRP) task. Participants were required to make speeded responses to two sequential visual tasks that were chained or independent. Magnetoencephalography (MEG) signals were recorded simultaneously to reveal the brain's response to these similar but fundamentally different conditions. RTs to Task 1 and 2 were slower in the Chained condition, and their temporal coupling weakened, relative to the Independent condition. MEG analysis of the accompanying event–related fields (ERFs) revealed an increased sustained posterior component in the Chained condition beginning approximately 350 ms after Task 2 onset and lasting for 450 ms. Beamformer localisation of this ERF effect revealed a left hemisphere source near the junction of the temporal, parietal and occipital lobes. These results extend our understanding of the behavioural and corresponding neural mechanisms required by everyday decision making.

Keywords Chained sequential operations; Dual task; Magnetoencephalography (MEG); PRP

1 Introduction

Experimental investigations of making a simple decision, for example, which of two buttons to press to signify a chosen response in a two-alternative situation, have yielded detailed behavioural and neural insights into the cognitive operations that underlie such a task. But rarely do real world decisions involve such few choices, indeed often they involve antecedent or consequent decisions that form a chain of cognitive operations. The purpose of the present report is to understand the neural basis of such mental processes, which are crucial to many forms of decision making.

This project was supported by a grant from the Human Frontier Science Program to S. Dehaene, K. Shapiro, P. Roelfsema, M. Sigman, and W. Vanduffel and by the National Natural Science Foundation of China (Grant No. 31170979), the Humanities and Social Sciences Foundation of the Ministry of Education, China (Grant No. 11YJC190006), and the selfdetermined research funds of CCNU from the colleges' basic research and operation of MOE, China (Grant No. CCNU11A02019) to Z. Fan. Funding was also provided by the Wales Institute for Cognitive Neuroscience (WICN) to K. Singh and K. Shapiro.

The consequences of performing two temporally adjacent cognitive operations have been studied for many years. Indeed, several lines of behavioural observations, such as the psychological refractory period (PRP; cf. Pashler, 1984; Sigman & Dehaene, 2005, 2008) and the attentional blink (AB; Raymond, Shapiro, & Arnell, 1992), converge to highlight the 'bottleneck' that arises when we attempt to perform multiple sequential operations. The term 'bottleneck' refers to the finding that aspects of the second task are prevented from execution until aspects of the first are completed. Although both PRP and the AB have been studied extensively, both share the commonality that each of the sub-tasks comprising them is independent. By 'independent' we refer to the fact that the decision from the first sub-task has no informational bearing on the second. Whereas the AB and PRP phenomenon do characterise many real-world cognitive tasks, they fall short in yielding insight into the more complex, yet frequently encountered tasks, which require the input from one to direct decision-making on the second. We refer to such tasks as 'chained', an example of which arises when a tennis player is required to judge whether to rush the net or remain at the baseline, then having decided to rush the net now must decide whether to prepare for the returned ball to come high or low.

In 1936, Alan Turing published his seminal essay "On computable numbers" and introduced the Turing machine (1936), which served as a computational blueprint for the subsequent invention of the electronic computer. Turing explicitly aimed to capture the basic operations of the human mind. Yet surprisingly, 70 years after Turing's paper, we still have little knowledge of the brain mechanisms by which humans carry out even the simplest of sequential computations postulated in Turing's framework. The present report is an attempt to understand the neural basis of the cost of performing two sequential chained operations over and above the cost of performing the same two independent operations as assessed by the standard PRP paradigm.

A recent behavioural study (Fan, et al., 2011) reveals a number of important aspects of this additional cost using a novel PRP paradigm. In two experiments, we required participants to give speeded responses to two sequential visual tasks separated by variable SOAs (Stimulus Onset Asynchrony). By requiring information to be passed from Task 1 to Task 2 in one condition and comparing it to another condition where the same two tasks did not require information from the first to perform the second, we explored the behavioural costs of chaining. Our results revealed two costs in the chained condition over and above classical dual-task PRP costs: (1) an altered distribution of response times RTs (particularly for the second task) in terms of an increased mean and variance; and (2) a disrupted temporal coupling between the first and the second tasks, as evidenced by a reduced correlation between the response times of both tasks. Based on these findings, we proposed that serial chaining costs originate from two sources: the first is an early stage of task setting, and the second, a later stage of information buffering and result-passing. Our results are consistent with a recent report (Sackur & Dehaene, 2009), which provides evidence that serial chaining is a slow and effortful process that consumes central processing resources and requires conscious control. Together these two behavioural reports shed light on the behavioural correlates of serial chaining induced by limits on human information processing.

Interest in the neural correlates of dual-task processing has been intense in recent years, thus we will attempt only a brief overview to set the stage for the current investigation. Brisson & Jolicoeur (2007a, 2007b) revealed the attenuation of an ERP component (N2pc) known to index the deployment of spatial attention as the SOA was reduced between the two tasks comprising a standard PRP paradigm. In addition to showing that this reduction was due to central bottleneck constraints as are indexed by the PRP paradigm (see also Lien, Croswaite, & Ruthruff, 2011), Brisson & Jolicoeur. also found the onset latency of another ERP component (sustained posterior contralateral negativity, or SPCN[1]) to be delayed with decreasing Task1-Task2 SOA. The SPCN has been shown to index visual short-term memory (VSTM) capacity limits and probably reflects maintenance of the array in VSTM. As we shall show here, the present investigation finds the SPCN distinguishes between chained and independent tasks.

The SPCN component remains a current topic of investigation, as evidenced by recent reports from Jolicoeur's lab (Robitaille, Grimault, & Jolicoeur, 2009; Robitaille, et al., 2010) who examined the SPCN using different approaches, including for the present purpose magnetoencephalography (MEG). The MEG equivalent, referred to as SPCM, was determined to arise from a cortical network including bilateral parietal loci, likely intraparietal/intraoccipital cortex, and contralateral parietal sources, paralleling findings from neuroimaging (e.g., Xu & Chun, 2006). Moreover, as its equivalent (SPCN) in the electrophysiological domain, it was elicited by VSTM maintenance.

Although we have an understanding of the behavioural costs, we have surprisingly little knowledge of the brain mechanisms underlying the process by which multiple cognitive operations are assembled to implement serial chaining, in spite of the fact that such operations are vital to support the daily requirements of human decision making. Attempts to understand the locus and function of the central bottleneck are still being sought, for example, recent reports have specified brain areas involved in a "unified central bottleneck" (Tombu et al., 2011) and have even shown the bottleneck can be overcome

with practice (Oberauer & Kliegl, 2004). The present report seeks to characterise the cognitive and neural mechanisms of serial chaining by comparing a variant of the PRP task, which requires transferring information from the first to the second task, to a traditional PRP task, which does not require information passing. We achieve this goal by conducting simultaneous behavioural and magnetoencephalographic (MEG) measurements to test the hypothesis that serial chaining requires additional resources and recruits additional brain areas relative to serial independent tasks. We employ the same PRP paradigm as was used successfully in a prior behavioural report (Fan et al., 2011).

2 Methods

2.1 Participants

Twenty-six right-handed participants took part in this experiment, with five excluded due to excessive head / body movement during MEG recording and thus lack of enough valid trials per cell during subsequent data analysis. Twelve females and nine males were included in the data analyses, aged from 19 to 38. All participants had normal or corrected-to-normal vision, with no history of visual disorders and gave their informed consent to participate.

2.2 Stimuli

The participant was seated in a shielded-MEG recording room that was dark except for the display. All stimuli were produced within MATLAB (The Mathworks, Inc.) using the Psychophysics Toolbox (Brainard, 1997; Pelli, 1997), and displayed on a Mitsubishi Diamond Pro 2070 colour monitor running at 60 Hz and located 208 cm from the participant's eyes (10.7 ° × 8.0 ° visual angle).

---

[1] The SPCN appears to be the same component as the contralateral delay activity (CDA) ERP component (Vogel & Machizawa, 2004).

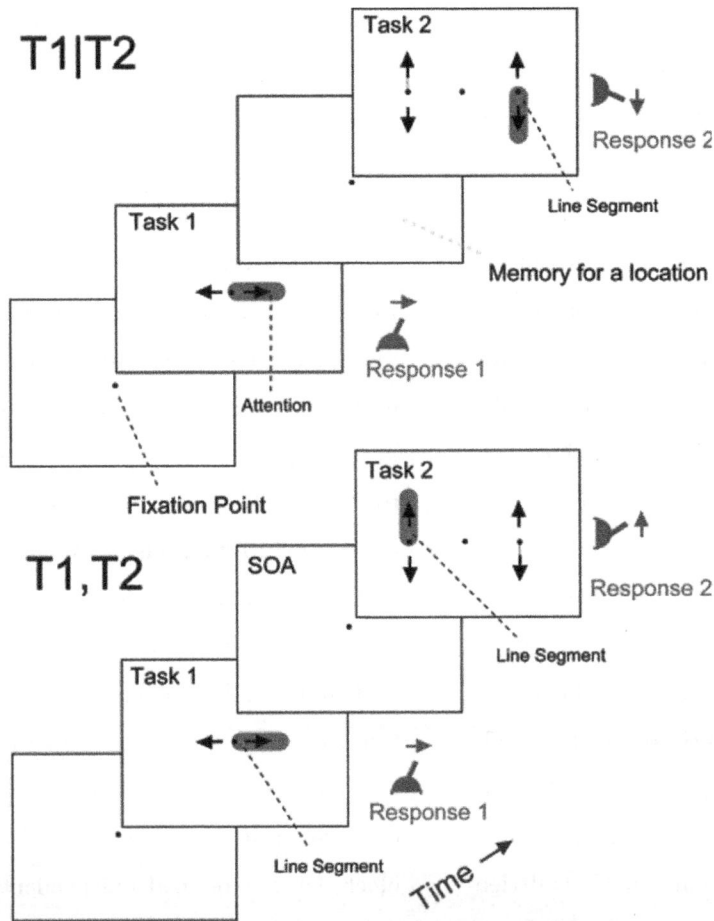

Figure 1. Schematic representation of the tasks used to study serial processing. Upper: in Task 1 of the Chained task (T1|T2), the participant saw two horizontal arrows; one of these arrows was connected by a low contrast line segment to the fixation point. The participant was instructed to push a joystick as quickly as possible with the left hand to the corresponding side of the connected line segment and remember this location. For Task 2 two vertical arrows and one line segment appeared on both sides simultaneously and the participant indicated the direction of the arrow connected to the T2 line segment of the remembered side by pushing a joystick with his/her right hand. The grey region (figure only) indicates the anticipated distribution of attention as well as the location that had to be remembered. Lower, in the independent task (T1,T2), Task 1 is identical. Task 2 was independent from Task 1 because the participant reported either the left or right side T2 line segment discrimination for a block of trials. In this example, the participant must report the left side T2 line segment and it involves a switch of spatial location from the left side to the right side (a trial in the Independent-switch condition).

Participant's responses were recorded by two MEG-compatible fORP fibre-optic joysticks (Current Designs Inc.).

In the first task (Task 1; T1), two horizontal arrows and a single (low-contrast) line segment were presented (see Figure 1). Each arrow contained an equal-sided triangle head (0.91° length for the sides and 0.82 ° length for the base) and a horizontal arrow line (1.1 ° length). The apex of each arrow head was 2.91° away from the central fixation point horizontally. The luminance of the background and that of the arrows were 80 and 1.2 cd/m^2, respectively. The contrast (C) of the (low-contrast) line segment was −15.

This contrast was defined as Weber contrast (C=100*(L-Lb)/Lb), that is, with reference to the mean luminance of the background, Lb. For each trial, Target 1 was to be judged "left" or "right" by virtue of which end of the arrow line was connected to the central fixation point by a horizontal low-contrast line segment (1° length). For T1, participants were instructed to push the left-hand joystick horizontally (left or right) according to the location of the target as quickly and also as accurately as possible after the target appeared.

The stimuli in the second task (Task 2; T2) were composed of four vertically oriented arrows and two vertical line segments/targets (see Figure 1). The size of the arrows and line segments were the same as those in T1, however the orientation and presentation locations were different. There were two targets in T2, each with two possibilities for their locations (either top-left and bottom-right quadrants or top-right and bottom-left quadrants of the visual field). By this arrangement, one target was always on the left and the other on the right side of the display. For T2 participants were instructed to respond to either the left- or right- side target and push the right-hand joystick vertically up or down, depending on whether the selected target was in the upper or lower visual field.

The additional information about which target (left or right) to which a response was required in T2 varied. In the Chained condition (T1|T2; upper display in Figure 1), this extra precuing information was derived from the result of T1. For example, if the single target in T1 was located on the left side of the display, then a discrimination had to be made on the left side of the display in T2, and vice versa. By contrast, in the Independent group, (T1,T2; lower display in Figure 1), this precuing information (left or right), was provided at the beginning of each (Independent) block. For example, in the beginning of an Independent-left block, participants were told through screen instructions to respond to the left side target in T2 during this block, and vice versa for an Independent-right block.

Half of the trials were dual-task trials requiring two tasks to be performed sequentially. The other half were single-task trials requiring the first task only. This allowed us in the MEG analysis to isolate the effects due to T2 from T1.

2.3 Design

A two-factor within-snbjects design was used in this experiment. One factor was whether a trial was a single-task trial (T1 only) or dual-task trial (T1 and T2). The other factor was the relationship (Chained vs. Independent) between the two tasks comprising a dual-task trial. The SOA between T1 and T2 stimuli was kept constant at

100ms. The location(s) of the target(s) in both tasks was counterbalanced across trials. Each combination of trial type (single or dual task) and dual-task relationship (Independent or Chained) was repeated 80 times, resulting in a total of 320 trials in each experimental session.

The experimental session was divided into eight blocks with 40 trials in each block. Half of the trials in each block were single-task trials and the other half were dual-task trials (randomly mixed) and the participant did not know in advance whether a trial was a single- or dual- task trial. Four out of eight blocks were Chained blocks, each containing 20 dual-task trials of the Chained condition (A|B) and 20 single-task trials. The other four blocks were Independent blocks, each containing 20 dual-task trials of the Independent condition (A, B) and 20 single-task trials. The left target of T2 was predefined in two of four Independent blocks, and in the other two blocks the right target of T2 was predefined. The order of Chained block, left pre-defined Independent block, and right pre-defined Independent block was counterbalanced across participants.

Unlike in the Chained condition where T1 and T2 by definition occurred in the same spatial location, in the Independent condition (T1,T2) the two tasks occurred on half the trials in the same location and on the other half in a different location. Thus the Independent condition had two subconditions, Independent-stick (without a location switch-called "Stick") and Independent-switch (with a location switch-called "Switch"), both of which were randomly mixed within each Independent block of trials. Since the Chained condition involved no switch of spatial attention, the Independent-stick condition is a more appropriate comparison to the Chained condition.

2.4 Procedure

Participants were instructed to respond as quickly and also as accurately as possible. Response order in the dual-task trials was also emphasized in that a T1 response should be always followed by a T2 response. A verbal instruction was presented on the display in the beginning of each block to inform participants in which block they were (Chained, Independent Left, or Independent Right). At the beginning of each trial, participants pressed a button on the joystick to initiate the trial whenever they were ready. A black fixation point (radius 0.16 °) was then presented immediately in the centre of the display and remained until the feedback was presented at the end of each trial. Participants were instructed to perform the tasks while maintaining the central fixation point. The T1 stimulus appeared 1000 ms after the onset of the fixation point and lasted for 100 ms. For a dual-task trial, the onset of the T2 stimulus was always 100ms after onset of T1, i.e., immediately after T1 offset. T2 stimuli remained on the display until participants made their response. Feedback was provided 1000ms after the occurrence of the T2 response and the fixation point was replaced by the feedback. Feedback was provided by two coloured dots appearing to the left and right side of the location previously occupied by the fixation point. The dot on the left side indicated the accuracy of T1, and the dot on the right side indicated the accuracy of T2. If the participant's response to T1 was correct, the left dot was green (red if it was incorrect). The same rule applied for the right dot. The left dot was yellow if participants made no response to T1. The feedback dots remained on the display for 1000 ms before the initiation of the next trial. For a single-task trial, the fixation point remained on the display for another 1800ms from the offset of T1 but before the onset of the feedback. The right feedback point was green in default and only became red if participants mistakenly pushed the right-hand joystick either up or down.

Response Grouping[2] between T1 and T2 was

discouraged by presenting a yellow warning point in the location of the fixation point if T2 response (RT2 + SOA) was made in less than 125% of the response to T1, measured from T1 onset. This operational definition of response grouping was adapted from Van Selst and Jolicoeur (1994). One practice block (containing 40 trials with mixed conditions) was given to participants before the formal test. All the aspects of the practice block were exactly the same as the subsequent experimental blocks. To reduce fatigue, self-controlled breaks between two continuous blocks were provided. The entire MEG recording took about 50 min.

2.5 Behavioural analysis

For RT analysis, only trials with correct T1 and T2 responses and correct response order were entered into the analysis. An outlier screen procedure (Van Selst & Jolicoeur, 1994) was used to exclude outliers of RTs in each cell for each participant. By using this approach, less than 4% of trials were labelled as outliers in the RT analysis. Post hoc analyses in each analysis of variance (ANOVA) of each experiment were conducted with the Bonferroni correction for multiple comparisons.

2.6 MEG acquisition and analysis

Whole-head MEG recordings were performed using a 275-channel axial gradiometer system (VSM MedTech Ltd, Port Coquitlam, Canada) with a 600 Hz sample rate and a 0-150Hz band-pass filter. Three out of the 275 channels were turned off due to excessive sensor noise. An additional 29 reference channels were used during the recording for the purpose of noise cancellation, and the primary sensors were analysed as synthetic third-order gradiometers (Vrba & Robinson, 2001). Three fiduciary markers (coils) were attached at fixed distances from anatomical landmarks (tragus, eye centre, which are identifiable in participants' anatomical MRIs) before the MEG data acquisition in order to achieve MRI/MEG co-registration in later source localization. Their location was

---

[2] Response qrouping is the tendency for participarts to wait for the reponding of T2 before initiating their response to T1.

recorded with digital photographs before and after MEG acquisition. MEG data were first visually inspected, and trials with eye movement artefacts or sensor drifts were removed from the analysis.

2.7 ERF data preparation

To generate event-related field (ERF) time series, raw MEG data were first filtered using a 40 Hz low-pass filter. The averaging epoch was then defined from 200 ms before T1 onset (300ms before T2 onset) to 1200ms after T1 onset (1100ms after T2 onset), and data were baseline corrected based on a 200-ms pre-T1 onset interval. Since the CTF MEG system has 275 first-order axial gradiometer sensors that measure the gradient of the magnetic field in the radial direction, that is, orthogonal to the scalp, we also performed an axial to planar transformation using the Fieldtrip toolbox (neuroimaging.ruhosting.nl/fieldtrip/), which rearranged MEG data to a synthetic planar gradient configuration. One advantage of the axial to planar transformation is that the signal amplitude typically is then largest directly above a source.

In this experiment, half the correct T1 trials were on the left side of the display and half on the right side (Note that we did not have enough trials to further split the data into T1-left and T1-right conditions for further analysis). In order to avoid biasing the ERF result by an uneven number of valid T1 trials, we performed a counterbalancing procedure in the analysis for left-/right-T1 responses. For the ERF waveform analysis, from each condition of each individual participant the same number of T1-left and T1-right trials (depending on which condition had the minimum valid trial number) were randomly selected and fed into the ERF analysis, resulting in equal numbers of T1-left and T1-right trials. The total number of trials in the Chained dual condition was twice that of either the Stick or Switch dual conditions. In order to avoid a potential bias due to the uneven number of valid trials (particularly in the contrast of Chained vs. Independent-

stick), the valid trials of the Chained dual condition were randomly sampled (according to the minimum valid trial number of Stick and Switch conditions) to equalize trial numbers of Chained, Stick and Switch conditions for each individual participant. Similar random sampling procedures were performed for the other two contrasts (Stick vs. Switch; Chained single vs. Independent single) to equalize trial numbers for contrasts. Only trials with correct responses were used in ERF and source-localisation analyses. The minimum number of epochs used to produce any ERF waveform for any individual participant in any condition was 24. Otherwise this individual was excluded from analysis. On average, 32.1 epochs (out of 40) were used to produce an individual ERF for Chained dual, Independent-stick dual and Independent-switch dual conditions (SD=4.14), and 69.05 epochs (out of 80) were used to produce an individual single-task ERF for both Chained and Independent blocks (SD=6.65).

2.8 Multisensor ERF analysis

In order to avoid bias in selecting any particular sensor or sensor groups in the ERF analysis, we first performed a whole-head, multisensor ERF analysis with all the sensors included. To achieve this, a cluster-based permutation test was used to perform statistical group contrasts between three pairs of whole, 272-sensor, ERFs, (i.e., Chained dual ERF vs. Independent-stick dual ERF, Chained Single ERF vs. Independent Single ERF and Independent-stick ERF vs. Independent-switch ERF). By comparing activities induced by single-task trials within Chained blocks and that within Independent blocks, we can investigate whether other factors, such as different expectations and different T1 processing levels between Chained and Independent blocks contaminated our comparisons of interest (Chained dual task vs. Independent dual task). 1000 permutations were calculated for each statistical group analysis by using the Fieldtrip toolbox, and each resulting paired-samples Tstatistic was compared with a primary threshold (0.05) to form clusters on the basis of temporal adjacency. A minimum neighbour sensor of 2 was used as an independent criterion in the formation of clusters to prevent all selected samples being included into a cluster if they had less than two neighbour sensors that also were selected. Cluster mass, that is, the sum of Tstatistics within each cluster, was used as the cluster–level statistic to assess the significance of suprathreshold clusters. The Monte Carlo method was used to calculate the distribution of the maximum of cluster mass in order to estimate the significance probability for each cluster. Only clusters with a significance probability (i.e., p value) less than a false alarm rate of .05 (two–tailed) were accepted to correct multiple comparisons (Maris & Oostenveld, 2007).

2.9 SAMerf analysis

For source localization, we used an event–related beamformer (Cheyne, Bakhtazad, & Gaetz, 2006), specifically SAMerf (Robinson, 2004) to explore the temporal sequence of cortical source activity underlying serial chaining. To perform the SAMerf analysis, unaveraged data (filtered 0 to 100Hz) from the whole of all trials (–200ms to 1200ms) were used to establish a global covariance matrix and set of SAM beamformer weights for all conditions (Robinson & Vrba, 1999). For headmodelling, a multiple, local–spheres–forward head model was derived by fitting spheres to each individual's brain surface extracted by BET(Brain Extraction Tool) (Huang, Mosher, & Leahy, 1999). Each individual had a structural image MRI scan (3D FSPGR [fast spoiled gradient echo] , 1 mm isotropic) acquired on a 3–T General Electric HDx scanner. From these beamformer weights, ERFs for individual conditions were passed through the beamformer weights so that SAMerf images of source power could be computed. Source power is defined as mean squared amplitude in a particular response time window (Robinson, 2004). Since we were interested in the stimulus–locked sustained activity (based on the multi-

sensor ERF results, see below), we used a time window covering from 450ms to 850ms after T1 onset to create SAMerf images, referenced to a baseline period of –200 to 0ms.

For group analysis, SAMerf images were spatially normalized onto the MNI (T1) average brain using FMRIB's Linear Affine Registration Tool (FLIRT; Jenkinson & Smith, 2001). This was achieved by first obtaining a set of warping parameters by registering each participant's anatomical MRI with the MNI template brain and thereafter applying these parameters to that individual's SAMerf maps. Nonparametric permutation tests (Singh, Barnes, & Hillebrand, 2003) (corresponding to parametric paired–sample t tests) were conducted in FSL (www.fmrib. ox.ac.uk/fsl/) using Threshold–Free Cluster Enhancement (TFCE) to perform cluster–level thresholding (Smith & Nichols, 2009). A 5mm Gaussian kernel was used to perform variance smoothing. The visualization program mri3dX was used to identify activation foci, to automatically label areas based on the Talairach Daemon database (Lancaster, et al., 2000), and to render group results onto the MNI averaged brain. Three groups of contrasts (i.e., Independent dual vs. Chained dual, Independent single vs. Chained single and Independent–stick vs. Independent–switch) were performed.

3 Results

3.1 Behavioural analysis

Effect of chaining operations on RT & accuracy. Since the comparison between the Chained and Stick conditions is the best way to evaluate the effect of chaining (i.e., neither condition requires a switch of spatial location), we carried out six paired–sample ttests (Chained vs. Independent–stick) for RT1, RT2, T1 Accuracy, T2 Accuracy, standard deviation (*SD*) of RT1, and SD of RT2. RT1 for the Chained

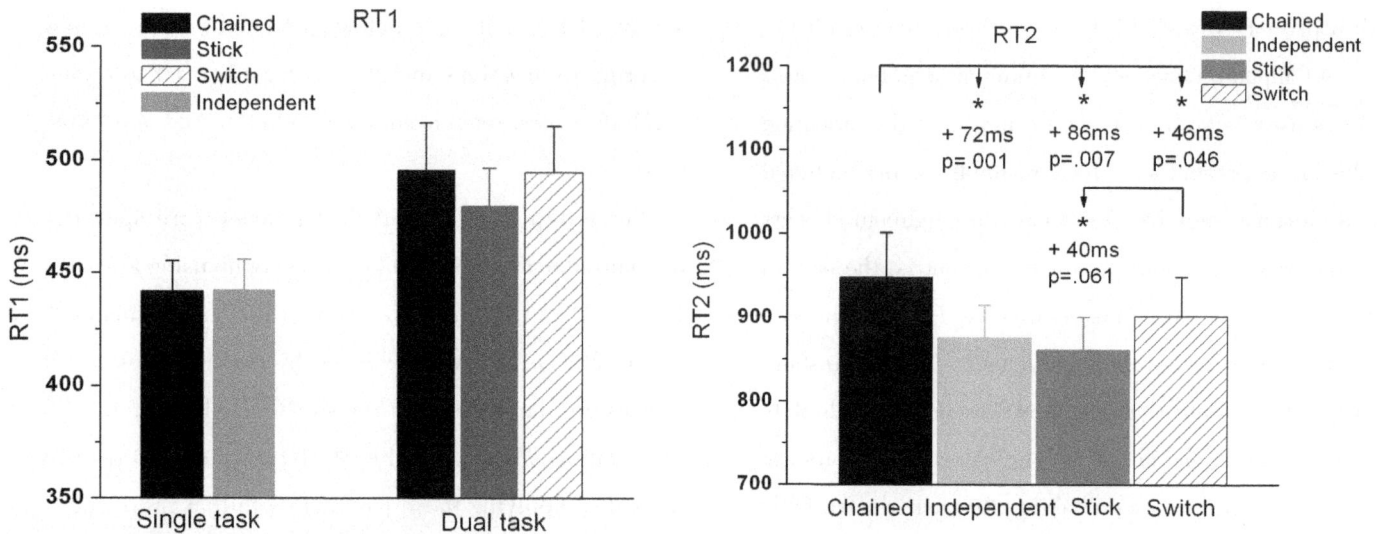

Figure. 2. Grand means of RT1 (left panel) / RT2 (right panel) of Chained and Independent conditions. On the X axis (left panel), the first column group shows RT1 from single task trials (no T2); the second column group shows RT1 from the Dual-task condition. The right panel shows RT2 from the four conditions. Vertical bars represent one standard error.

condition was slower than RT1 for the Independent-stick condition (Difference=16ms, SE=8.5ms, t=1.863, df=20, p=0.038, one-tailed). RT2 for the Chained condition was significantly slower than RT2 for the Independent-stick condition (Difference=86ms, SE=29ms, t=2.988, df=20, p=0.007) (see Figure2). T1 Accuracy for Independent-stick was not significantly different from T1 Accuracy for the Chained condition (p=0.477). However, T2 Accuracy for the Chained condition was significantly worse than T2 Accuracy for the Independent-stick condition (Difference =-6%, SE = 1.8%, t=-3.349, df= 20, p=0.003). SD of RT1 for the Chained condition was significantly larger than SD for RT1 of the Independent-stick condition (Difference= 20ms, SE= 6ms, t=3.253, p=0.005). The SD of RT2 for the Chained was also significantly larger than the SD of RT2 for the Independent-stick condition (Difference=88ms, SE=41ms, t=2.988, p=0.045). Given that RT and accuracy effects are in the same direction we can conclude there was no speed-accuracy trade-off.

Effect of chaining on task correlations. We first computed the within-trial Pearson's correlation between RT1 and RT2 for each participant and then submitted all individual mean correlation coefficients to paired samples

ttests, which revealed that the correlation between RT1 and RT2 was smaller (Difference=-.08, SE=0.039, t=-1.934, df=20, p=0.033, one-tailed) in the Chained condition (Mean=0.36, SD=0.20) relative to that in the Independent-stick condition (Mean=0.44, SD=0.23). The correlation difference between the Independent-stick (Mean=0.44, SD=0.23) and Independent-switch conditions (Mean=0.37, SD=0.25) did not reach significance (p=0.13), neither did the correlation difference between the Chained and the Independent-switch conditions (p= .899).

Effect of switching attention. We also explored the effect of switching attention by comparing the Independent-stick to the Independent-switch condition. Both mean RT1 (p=0.035) and RT2 (p=0.061, marginally) were significantly slower in the Independent-switch than their counterparts in the Independent-stick condition. However, unlike the effect of serial chaining, the SD of RT1 (p=0.130) and RT2 (p=0.527) in the Independent-stick condition were not significantly different from those in the Independent-switch condition.

Effect of T2 on T1: dual versus single tasks. There were only two types of single task trials (e.g. in Chained blocks and in Independent blocks) with no

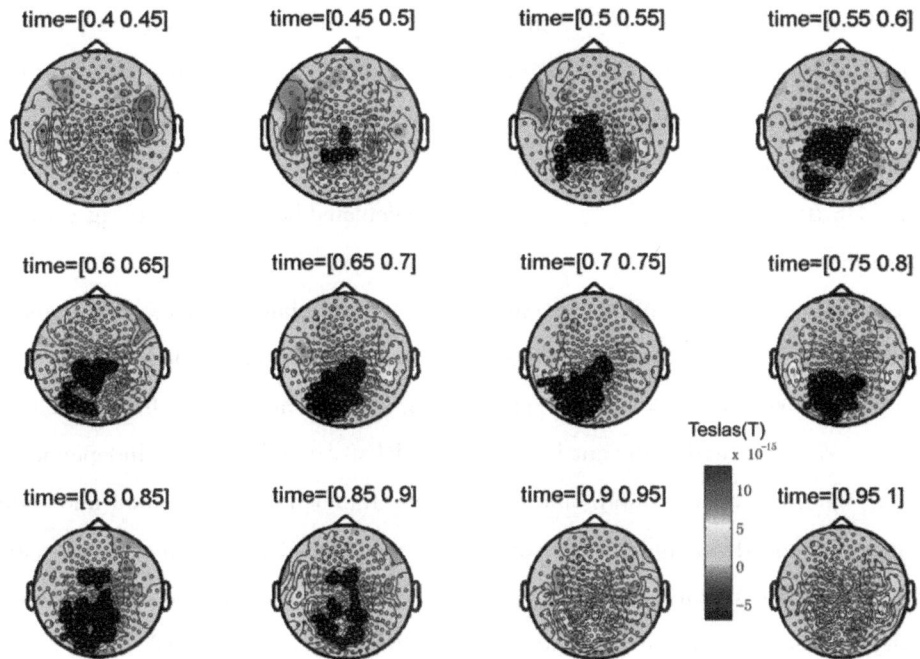

Figure 3. Results of the multi-sensor ERF analysis. Topography of the ERF difference between Chained (dual) and Independent-stick (dual) is overlaid with the topography of the significant clusters (p<.05, two-tailed, corrected by a nonparametric permutation test). Sensors that belong to significant positive clusters, i.e., higher amplitude in Chained (dual) ERF relative to Stick (dual) ERF, are indicated by black circles. Each subplot represents the topography during a time window of 50ms. Data were submitted to ERF analysis after axial to planar transformation.

significant difference between them on both RT ($p=0.928$) and percentage correct ($p=0.926$). Additional analyses were performed to explore how T2 affected T1. RT1 in the single task was significantly faster than RT1 in the dual task, $F (1, 20) =24.408$, $p= .000$; Difference $=-49$ms, $SE=9.8$ms, $p=0.000$. T1 Accuracy in the single task was significantly greater than T1 Accuracy in the dual task, $F(1, 20)=8.626$, $p=0.008$; Difference$=3\%$, $SE=0.8\%$, $p= .008$. Both RT and Accuracy data suggest that T2 produced an extra cost to the central processing stage of T1 when the two tasks were temporally adjacent (100ms SOA).

3.2 Multisensor ERF analysis

Three pairs of whole-head multisensor ERF contrasts were performed with a cluster-based permutation test at the grouplevel to correct for multiple comparisons along a time window from 200 ms before T1 onset (300ms before T2 onset) to 1200ms after T1 onset (1100ms after T2 onset). No cluster was significant before the onset of T1 for each of the three contrasts, which suggests

baseline levels were the same for all conditions. For the contrast of Chained dual vs. Independent-stick dual, only positive (higher amplitude in Chained condition) but not negative (higher amplitude in Stick condition) clusters were significant. For the other two contrasts (i.e., Chained vs. Independent Single and Independent-stick ERF vs. Independent-switch ERF), neither positive nor negative clusters were significant.

Figure 3 shows the group results of multisensor ERF analysis for the contrast of (dual) Chained versus Independent Stick over time. The amplitude difference between these two conditions reaches maximum in the posterior part of the brain and reveals a sustained effect. This sustained posterior activity distinguishing serial chaining from traditional PRP (independent) dual-task processing emerges from central posterior sensors approximately 450ms after T1 onset and spreads gradually to left parietal/temporal/occipital sensors. The sustained activity lasts for several hundred milliseconds and is not fully attenuated until 900ms after T1 onset (800ms after T2 onset), near the end of T2 execution (average RT2 was around 900ms). There are no other significant clusters outside the time window of this activity.

3.3 Sensor-Group ERF analysis

The previously described multi-sensor ERF analysis reveals this sustained activity lasting from 450 – 900 ms after T1 onset, which appears to be linked with serial chaining. Inspection of this analysis suggests that, whereas 116 out of 272 sensors were statistically relevant in the contrast between (dual) Chained versus Independent-stick conditions (Figure 4), not all of the 116 sensors are unambiguously linked to the sustained activity we observed. For example, some sensors above motor areas likely represent the effect of the RT difference between these conditions. For this reason, we used three criteria to restrict our subsequent analyses to sensors that were directly linked with the sustained activity. The three criteria included were: First, based on the result of the multi-sensor ERF analysis, the time window used in this sensor selection was 450–900 ms after T1 onset. Second, only sensors in which the sustained effect lasted continuously longer than half of the maximum effect were selected, resulting in a cut-off value of 185ms since the maximum sustained effect of the input sensors was 370ms at sensor MLP31. Third, any individually selected sensor in the final sensor group had to have at least two neighbouring sensors that were also included. Only 19 of the 116 sensors meeting these criteria were used in the analysis.

By applying such criteria, 19 sensors (see Figure 4) met the above criteria and were used for the sensor-group analysis. These 19 sensors were derived from left and central parietal, left occipital, and left temporal clusters. A standard classification approach to the CTF system was used to define 14 sensor clusters, including central frontal, central central, central parietal, central occipital, left frontal, left central, left parietal, left occipital, left temporal, right frontal, right central, right parietal, right occipital and right temporal.

For each individual participant, 5 ERF curves were calculated based on the averaged ERFs of all sensors in the selected sensor group (N=19). These five synthetic ERFs were calculated for five conditions (i.e., Chained single, Independent single, Chained dual, Independent-stick dual and Independent-switch dual). Three contrasts of paired ERFs(Chained single vs. Independent single, Chained dual vs. Independent-stick dual and Independent-stick dual vs. Independent-switch dual) were also performed using a paired-samples ttest at each sample point of each paired ERFs (i.e., moving ttests)

The results of the sensor-group ERF analysis revealed clear sustained ERF activity from left occipital, left temporal and left parietal sensors, starting from 450ms and lasting until 900ms after T1 onset, congruent with the multi-sensor ERF analysis (see Figure 5). Importantly, this difference is manifest only in the contrast between the Chained and Independent-stick conditions, suggesting this activity is directly linked with serial chaining and not modulated by other factors such as switching the locus of spatial attention. Moving t-tests in the contrast Stick vs. Switch suggest that the field amplitude for the Independent-switch condition was significantly higher in a time window from 900–960 ms after T1 onset, likely reflecting the switching of attention required in the latter condition.

3.4 SAMerf analysis

Similar to the ERF analysis in sensor space, we performed the same three contrasts (Chained dual vs. Stick dual; Stick dual vs. Switch dual, and Chained single vs. Independent single) in source space to localize event-related and phased-locked source activity. The results of the SAMerf approach reveal that only one out of the three contrasts (Chained dual vs. Stick dual, showed a significant

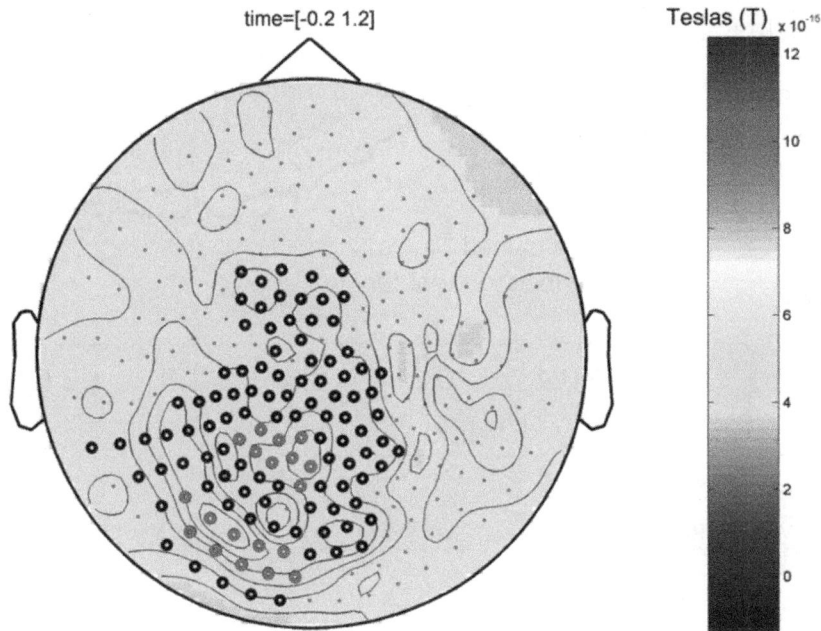

Figure. 4. The selected 19 sensors (labelled in red), out of a cluster of 116 sensors (labelled with black), that were used in the sensor–group ERF analysis. The 116 sensors (out of total 272 sensors) were those belonging to a significant cluster based on multi–sensor ERF analysis of Chained dual vs. Stick dual. Background topography was the ERF of Chained (dual) minus the ERF of Independent–stick (dual). Data were submitted to ERF analysis after axial to planar transformation.

effect, consistent with the ERF analysis. Figure 6 shows the group results of this analysis for the time window of 450ms to 850ms from T1 onset. The results suggest that source power for the Chained condition was significantly larger than that for the Independent–stick condition between 450ms to 850ms in the left superior temporal gyrus (Talairach coordinates at peak value = −45.2 −63.2 19.0). This result was consistent with the ERF analysis, which demonstrated sustained higher ERF amplitude for posterior sensors in the same time window. Importantly, this analysis points to a left lateralized source consistent with the finding in the sensor–group ERF analysis, that is, 18 out of 19 sensors in the selected sensor group were located on the left side.

Based on the source localization group results, we reconstructed the time course of source power activity at the left superior temporal gyrus (Talairach coordinates −45.2 −63.2 19.0), for each individual participant for five conditions. The T1–onset locked timecourse of source power activity was calculated based on the average of the

source power profile for each condition. The top panel of Figure 7 shows the T1–locked activity (normalized onto the baseline period from 200 ms pre–T1 to T1 onset) for each of the different conditions for the localized source shown in Figure 6 (i.e., a virtual sensor). Global covariance matrices and beamformer weights were used to produce the virtual sensor. The moving t–test profiles of the three contrasts of interest are presented in the lower panel. The data suggest that the left superior temporal source indicates a sustained higher level of activity in the contrast Chained dual vs. Independent–stick dual, but not in the other two contrasts (Stick dual vs. Switch dual and Chained single vs. Independent single). As in the ERF analysis, an equal number of T1–left and T1–right epochs were entered into the SAMerf analysis for each condition and for each participant in order to avoid a potential bias from the lateralized stimulus distribution.

3.5 Discussion

The behavioural data from the present experiment reveal the dramatic performance cost of serial chaining

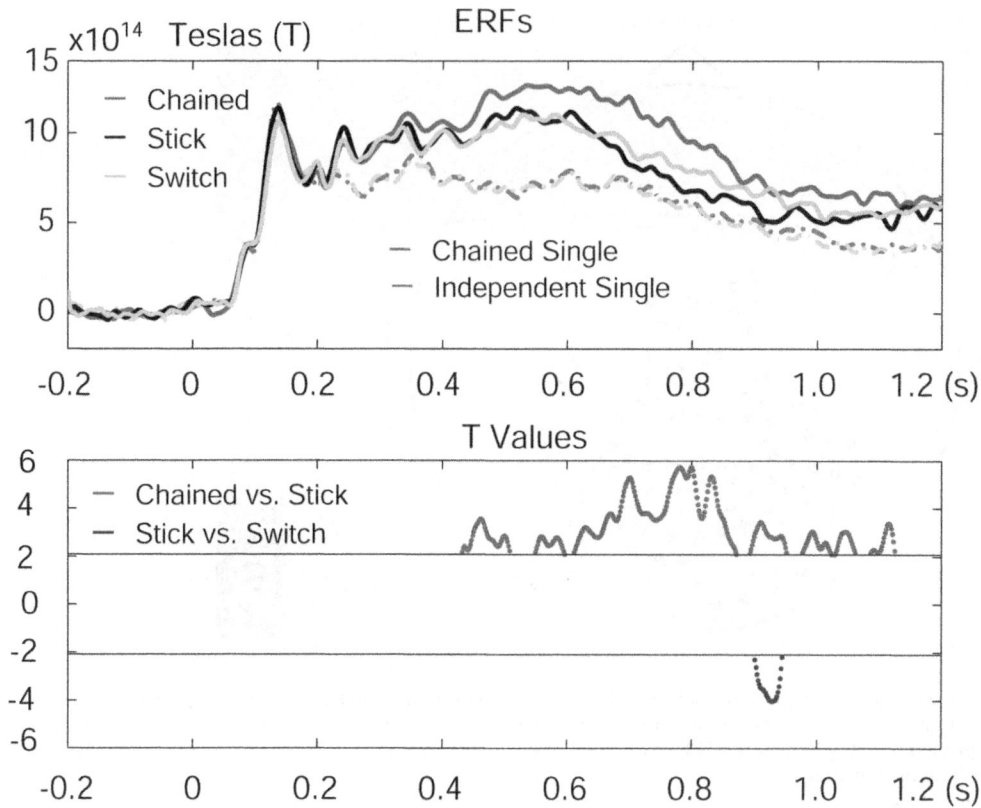

Figure. 5. Sensor–group ERF analysis for 19 selected sensors. The upper panel shows five synthetic ERFs for the Chained single condition (magenta dashed), Independent single condition (blue dashed), Chained dual (red solid), Independent–stick dual (black solid) and Independent–switch dual (green solid). The bottom panel shows the results of a moving t–test between the Chained dual vs. Independent–stick dual (red) and Independent–stick dual vs. Independent–switch dual (blue). There was no significant effect for the contrast of Chained single vs. Independent single. T–values between the two horizontal lines (t(20)= 2.086, e.g. two–tailed alpha value at 0.05), were masked and not shown. A 30ms time window was used to threshold the moving t–test values along the temporal dimension. Waveforms shown are the result of axial to planar transformation, i.e. synthetic planar gradient data.

in two important aspects: First, a slowed RT and corresponding increase in variability; and second, a decreased temporal coupling between the two sequential tasks. Importantly, this behavioural cost is not revealed when the chaining operation is not required, that is in the independent task with and without the requirement to switch the location of attention, even though location switching has an effect on RT. Our behavioural results are consistent with a recent report (Sackur & Dehaene, 2009), revealing that serial chaining consumes central processing resources and replicates our previous behavioural results (Fan, et al., 2011).

Importantly, the present report reveals the neural correlates of the large behavioural cost associated with chaining cognitive operations. Sensor and source space MEG analyses converge to reveal sustained posterior activity occurring from 450ms to 900ms after T1 onset (350ms to 800ms after T2 onset) that is linked to the serial chaining requirement. This activity appears only in the contrast between Chained vs. Independent–stick; not in the other two contrasts (Chained single vs. Independent single; Independent–switch vs. Independent–stick), mirroring the effects found in both the behavioural and subsequent SAMerf (source space) analyses. Localisation in source space using an event–related beamformer approach suggests the source of this activity is distributed along the boundary of left temporal, left parietal and left occipital brain areas. The event–related beamformer (Robinson,

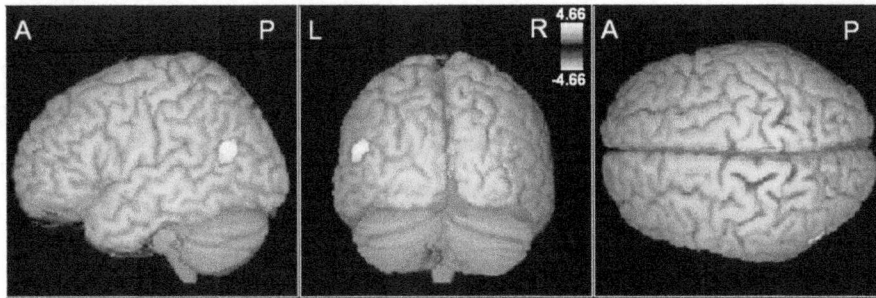

Figure. 6. Group–level SAMerf analysis showing the results of the contrast between Chained dual vs. Independent–stick dual conditions for ERFs during the time window 450ms to 850ms after T1 onset. A TFCE approach of thresholding was used to correct for multiple comparisons. Labelled voxels are significant at the p<0.05 level (two–tailed) after correction. Units are pseudo–*t* values. Yellow/orange areas indicate relatively larger source power in the Chained dual condition.

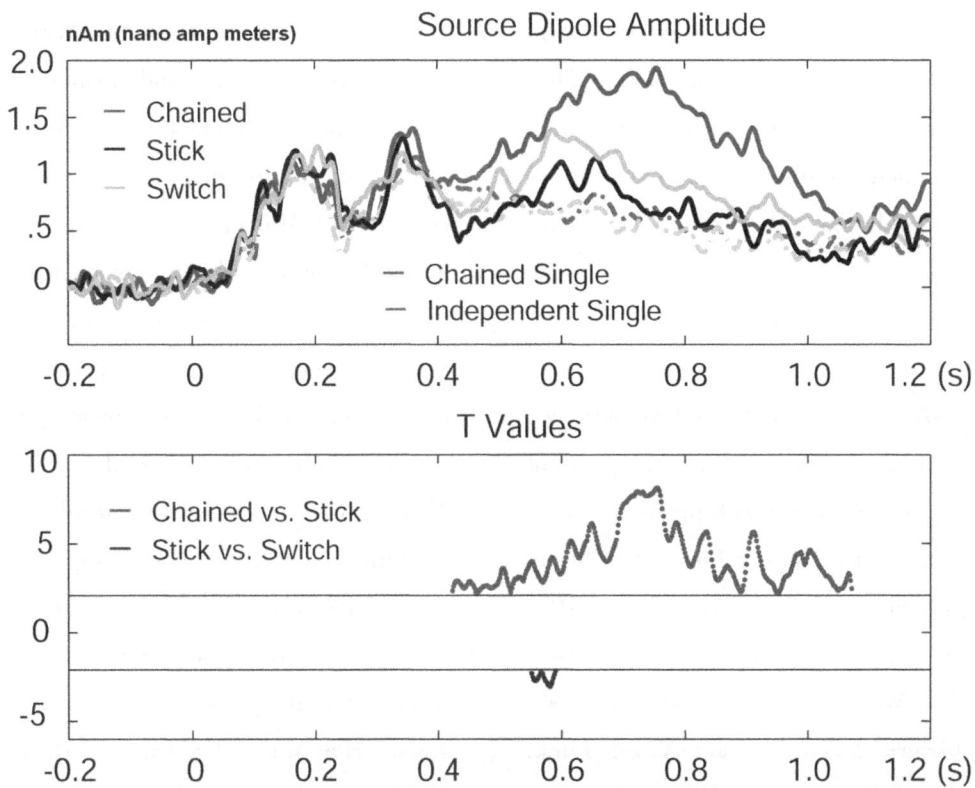

Figure. 7. Group–level time series of source dipole amplitude for different conditions and their contrasts. The upper panel shows the time series of activation for the Chained single condition (magenta dashed), Independent single condition (blue dashed), Chained dual (red solid), Independent–stick dual (black solid) and Independent–switch dual (green solid). Bottom panel: *t* values of a moving *t*– test analysis, i.e., Chained dual vs. Independent–stick dual (red) and Independent–stick dual vs. Independent–switch dual (blue). There was no significant effect for the contrast of Chained single vs. Independent single. A 30ms time window was used to perform extent thresholding on *t*–test values along the temporal dimension. *T*values between two horizontal lines ,*t*(20)= ± 2.086, e.g. two–tailed alpha value at 0.05, were masked and are not shown.

2004) is a powerful, relatively recent method for localising ERFs with very few assumptions required (Cheyne, et al., 2006). One shortcoming of the beamformer approach is that it is insensitive to sources that are highly correlated, although simulations have shown that beamformers can detect correlated sources depending on the degree of correlation and SNR (Gross, et al., 2001). It is therefore theoretically possible that the present source localisation analysis has missed some source activity which is both temporally correlated and spatially distributed that other approaches such as minimum norm–based solutions might detect. However, the consistency both spatially, temporally and experimentally of the event-related beamformer results with the left lateralisation pattern observed in the sensor space analysis suggests this not a major issue with the present analyses.

It could be argued that the sustained activity observed in the Chained, relative to the Independent-stick, condition occurred as a result of the longer display duration for Task 2 stimuli in the former. The longer duration is a consequence of the fact that RT2 for the Chained condition is significantly slower than RT2 for the Independent-stick condition. The temporal characteristics of this activity revealed by the multichannel and SAMerf analyses, however, argue against this account. The larger magnitude in the Chained condition emerges as early as 350ms after T2 onset, well before the stimulus terminates in either condition. Thus, the increased activity in the Chained condition cannot be explained by a low-level, sensory-based account.

The characteristics of the sustained activity we observed, which distinguish chained from independent dual-task processing, are consistent with previous reports from human ERP studies (Brisson & Jolicoeur, 2007a, 2007b; Dell'Acqua, Sessa, Jolicoeur, & Robitaille, 2006; Jolicoeur, Sessa, Dell'Acqua, & Robitaille, 2006a, 2006b; Klaver, Talsa, Wijers, Heinze, & Mulder, 1999; McCollough, Machizawa, & Vogel, 2007; Vogel, Luck, & Shapiro, 1998; Vogel & Machizawa, 2004) and MEG (Robitaille, et al., 2009; Robitaille, et al., 2010) where a temporally and topographically similar component, the SPCN (see also Note 1) or SPCM, respectively, have strongly been linked with maintenance in VSTM (or 'working' memory). The SPCN typically begins at about 300ms after target onset and is sustained for at least several hundred milliseconds into the period during which the stimulus generating this component has to be retained (McCollough, et al., 2007).

Although the sustained activity witnessed in the Chained condition likely relates to the SPCM and its ERP counterpart (i.e., SPCN or CDA), we speculate it may represent a marker for information buffering over and above the reflection of VSTM activity normally ascribed to it. First, we note that the SPCM is typically believed to index increases in working memory (WM) capacity as it reflects the number of stimuli stored in posterior brain areas. As we are comparing the Chained to Independent conditions, where the number of stimuli is not being manipulated, the differences observed may reflect other cognitive machinery. Second, although the WM requirement for the Chained relative to the Independent conditions requires the content to be changed more frequently; even in the latter condition participants must actively maintain the contents of WM in order to correctly make a (T2) response. Yet, behavioural and electrophysiological indices suggest the former exacts an increased cost that we suggest reflects the chaining operation, per se. Third, Mitchell and Cusack (2011) note that the effects of memory load using MEG are predominantly bilateral (see also Grimault, et al., 2009). This suggests that the present result, which did not reveal bilateral activity, may not simply index VSTM load but is sensitive to the more complex operation required by chaining. Finally, we note that using MEG to assess VSTM usually reveals generators in the superior intraparietal sulcus (IPS; Mitchell & Cusack, 2011; Robitaille, et al., 2010). We, on the other hand, localised MEG activity on the boundary of the temporal, occipital, and parietal lobes, further supporting our contention we are not merely revealing differences due to increased memory load.

An interesting observation from the present study is the left lateralization of the serial chaining-related neural activity in both sensor and source spaces. Since these analyses contained exactly the same number of T1-left and T1-right trials, it is unlikely that this left lateralization effect is due to an uneven distribution of T1 stimuli or other artefact. Instead, the leftlateralization of serial chaining is consistent with a well-known phenomenon in clinical neurology known as the left hemispheric lateralization for praxis planning. In the early 20th century, Hugo Liepmann observed that stroke patients with a lefthemispheric lesion are more likely to develop apraxia than those with a right-hemispheric lesion (reviewed in Goldenberg, 2003). Clinical neurological studies (Alexander, Baker, Naeser, Kaplan, & Palumbo, 1992; Geschwind, 1965) suggest that praxis planning, particularly the programming of a motor sequence, is largely dominated by the left hemisphere. In the past decade, this suggestion has been confirmed repeatedly using functional imaging (Bohlhalter, et al., 2009; Buccino, et al., 2004; Fridman, et al., 2006; Hermsdorfer, Terlinden, Muhlau, Goldenberg, & Wohlschlager, 2007; Johnson-Frey, Newman-Norlund, & Grafton, 2005). In particular, a recent fMRI study (Bohlhalter, et al., 2009) revealed a left hemispheric lateralization in posterior parietal cortex (PPC) and premotor cortex (PMC) association areas when participants were planning to perform a pantomime task (such as 'wave good-bye'), regardless of whether the planning occurred for the right or left hand.

We suggest our results provide evidence of a supervisory control system, over and above that revealed by typical dual-task (e.g., PRP) demands, by the existence of sustained activity in a left posterior brain area. This area, likely in combination with prefrontally based executive control centres, subserves planning and execution demands by organising hierarchies of cognitive subunits into complex goal-directed behaviour. This cognitive organisation, though essential to complex human behaviour, comes at a behaviourally measurable cost, as evidenced by slower response times and disrupted response coupling between relevant tasks. The present findings lend new understanding to the brain processes that underlie the moment-by-moment need for humans to chain sequential cognitive operations as they go about their daily behaviour.

References

Alexander, M. P., Baker, E., Naeser, M. A., Kaplan, E., & Palumbo, C. (1992). Neuropsychological and neuroanatomical dimensions of ideomotor apraxia. *Brain*, 115, 87–107. doi: 10.1093/brain/115.1.87

Bohlhalter, S., Hattori, N., Wheaton, L., Fridman, E., Shamim, E. A., Garraux, G., ... Hallett, M. (2009). Gesture subtype-dependent left lateralization of praxis planning: An event-related fMRI Study. *cerebral Cortex*, 19, 1256–1262. doi: 10.1093/cercor/bhn168

Brainard, D. H. (1997). The psychophysics toolbox. *Spatial vision*, 10, 433–436. doi: 10.1163/156856897X00357

Brisson, B., & Jolicoeur, P. (2007a). Cross-modal multitasking processing deficits prior to the central bottleneck revealed by event-related potentials. *Neuropsychologia*, 45(13), 3038–3053. doi: 10.1016/j.neuropsychologia.2007.05.022

Brisson, B., & Jolicoeur, P. (2007b). A psychological refractory period in access to visual short term memory and the deployment of visual-spatial attention: Multitasking processing deficits revealed by event-related potentials. *Psychophysiology, 44*, 323–333. doi: 10.1111/j.1469-8986.2007.00503.x

Buccino, G., Vogt, S., Ritzl, A., Fink, G. R., Zilles, K., Freund, H. J., ... Rizzolatti, G. (2004). Neural circuits underlying imitation learning of hand actions: An event-related fMRI study. *Neuron, 42*, 323–334. doi: 10.1016/S0896-6273(04)00181-3

Cheyne, D., Bakhtazad, L., & Gaetz, W. (2006). Spatiotemporal mapping of cortical activity accompanying voluntary movements using an event-related beamforming approach. *Human. Brain Mapp, 27*(213–219). doi: 10.1002/hbm.20178

Dell'Acqua, R., Sessa, P., Jolicoeur, P., & Robitaille, N.

(2006). Spatial attention freezes during the attentional blink. *Psychophysiology, 43*, 394–400. doi: 10.1111/j.1469–8986.2006.00411.x

Fan, Z., Singh, K. D., Muthukumaraswamy, S. D., Sigman, M., Dehaene, S., & Shapiro, K. (2011). The cost of serially chaining two cognitive operations. *Psychological Research*. doi: 10.1007/s00426–011–0375–y

Fridman, E. A., Immisch, I., Hanakawa, T., Bohlhalter, S., Waldvogel, D., Kansaku, K., ... Hallett, M. (2006). The role of the dorsal stream for gesture production. *NeuroImage, 29*, 417–428. doi: 10.1016/j.neuroimage.2005.07.026

Geschwind, N. (1965). Disconnexion syndromes in animals and man. *I. Brain, 88*, 237–294. doi: 10.1093/brain/88.2.237, II doi: 10.1093/brain/88.3.585

Goldenberg, G. (2003). Apraxia and beyond: life and work of Hugo Liepmann. *Cortex, 39*, 509–524. doi: 10.1016/S0010–9452(08)70261–2

Grimault, S., Robitaille, N., Grova, C., Lina, J. M., Dubarry, A. S., & Jolicoeur, P. (2009). Oscillatory activity in parietal and dorsolateral prefrontal cortex during retention in visual short–term memory: Additive effects of spatial attention and memory load. *Human Brain Mapping, 30*, 3378–3392. doi: 10.1002/hbm.20759

Gross, J., Kujala, J., Hamalainen, M., Timmermann, L., Schnitzler, A., & Salmelin, R. (2001). Dynamic imaging of coherent sources: studying neural interactions in the human brain. *Proceeding of the National Academy of Science, 98*, 694–699. doi: 10.1073/pnas.98.2.694

Hermsdorfer, J., Terlinden, G., Muhlau, M., Goldenberg, G., & Wohlschlager, A. M. (2007). Neural representations of pantomimed and actual tool use: Evidence from an event–related fMRI study. [Supplement 2].*NeuroImage, 36*, T109–T118. doi: 10.1016/j.neuroimage.2007.03.037

Huang, M. X., Mosher, J. C., & Leahy, R. M. (1999). A sensor–weighted overlapping–sphere head model and exhaustive head model comparison for MEG. *Physics in Medicine and Biology. 44,*, 423–440. doi: 10.1088/0031–9155/44/2/010

Jenkinson, M., & Smith, S. (2001). A global optimisation method for robust affine registration of brain images. *Medical Image Analysis, 5*, 143–156. doi: 10.1016/S1361–8415(01)00036–6

Johnson–Frey, S. H., Newman–Norlund, R., & Grafton, S. T. (2005). A distributed left hemisphere network active during planning of everyday tool use skills. *Cerebral Cortex, 15*, 681–695. doi: 10.1093/cercor/bhh169

Jolicoeur, P., Sessa, P., Dell'Acqua, R., & Robitaille, N. (2006a). Attentional control and capture in the attentional blink paradigm: Evidence from human electrophysiology. *European Journal of Cognitive Psychology, 18*, 560–578. doi: 10.1080/09541440500423210

Jolicoeur, P., Sessa, P., Dell'Acqua, R., & Robitaille, N. (2006b). On the control of visual spatial attention: Evidence from human electrophysiology. *Psychological Research, 70*, 414–424. doi: 10.1007/s00426–005–0008–4

Klaver, P., Talsa, D., Wijers, A. A., Heinze, H. J., & Mulder, G. (1999). An eventrelated brain potential correlate of visual short–term memory. *Neuroreport, 10*, 2001–2005. doi: 10.1097/00001756–199907130–00002

Lancaster, J. L., Woldorff, M. G., Parsons, L. M., Liotti, M., Freitas, E. S., Rainey, L., ... Fox, P. T. (2000). Automated Talairach Atlas labels for functional brain mapping. *Human Brain Mapping, 10*, 120–131. doi: 10.1002/1097–0193(200007)10:3<120::AID–HBM30>3.0.CO;2–8

Lien, M.–C., Croswaite, K., & Ruthruff, E. (2011). Controlling spatial attention without central attentional resources: Evidence from event-related potentials. *Visual Cognition, 19*, 37–78. doi: 10.1080/13506285.2010.491643

Maris, E., & Oostenveld, R. (2007). Nonparametric statistical testing of EEG- and MEG-data. *Journal of Neuroscience Methods, 164*(1), 177–190. doi: 10.1016/j.jneumeth.2007.03.024

McCollough, A. W., Machizawa, M. G., & Vogel, E. K. (2007). Electrophysiological measures of maintaining representations in visual working memory. *Cortex, 43*, 77–94. doi: 10.1016/S0010-9452(08)70447-7

Mitchell, D. J., & Cusack, R. (2011). The temporal evolution of electromagnetic markers sensitive to the capacity limits of visual short-term memory. *Frontiers in Human Neuroscience, 5*, 1–20. doi: 10.3389/fnhum.2011.00018

Oberauer, K., & Kliegl, R. (2004). Simultaneous cognitive operations in working memory after dual-task practice. *Journal of Experimental Psychology: Human Perception & Performance, 30*, 689–707. doi: 10.1037/0096-1523.30.4.689

Pashler, H. (1984). Processing stages in overlapping tasks: Evidence for a central bottleneck. *Journal of Experimental Psychology: Human Perception & Performance, 10*, 358–377. doi: 10.1037/0096-1523.10.3.358

Pelli, D. G. (1997). The VideoToolbox software for visual psychophysics: Transforming numbers into movies. *Spatial vision, 10*, 437–442. doi: 10.1163/156856897X00366

Raymond, J. E., Shapiro, K. L., & Arnell, K. M. (1992). Temporary suppression of visual processing in an RSVP task: an attentional blink? *Journal of Experimental Psychology: Human Perception & Performance, 18*, 849–860. doi: 10.1037/0096-1523.18.3.849

Robinson, S. E. (2004). Localization of event-related activity by SAM(erf). *Neurolagy & Clinical Neurophysiology, 109*, 109.

Robinson, S. E., & Vrba, J. (1999). Functional neuroimaging by synthetic aperture magnetometry (SAM). In T. Yoshimoto, M. Kotani, S. Kuriki, H. Karibe & N. Nakasato (Eds.), *Recent Advances in Biomagnetism* (pp. 302–305). Sendai: Tohoku University Press.

Robitaille, N., Grimault, S., & Jolicoeur, P. (2009). Bilateral parietal and contralateral responses during maintenance of unilaterally encoded objects in visual short-term memory: Evidence from magnetoencephalography. *Psychophysiology, 46*, 1090–1099. doi: 10.1111/j.1469-8986.2009.00837.x

Robitaille, N., Marois, R., Todd, J., Grimault, S., Cheyne, D., & Jolicoeur, P. (2010). Distinguishing between lateralized and nonlateralized brain activity associated with visual short-term memory: fMRI, MEG, and EEG evidence from the same observers. *NeuroImage, 53*, 1334–1345. doi: 10.1016/j.neuroimage.2010.07.027

Sackur, J., &Dehaene, S. (2009). The cognitive architecture for chaining of two mental operations. *Cognition, 111*, 187–211. doi: 10.1016/j.cognition.2009.01.010

Sigman, M., & Dehaene, S. (2005). Parsing a cognitive task: A characterization of the mind's bottleneck. *PLoS Biology*, 3, e37. doi: 10.1371/journal.pbio.0030037

Sigman, M., & Dehaene, S. (2008). Brain mechanisms of serial and parallel processing during dual-task performance. *Journal of Neuroscience, 28*, 7585–7598. doi: 10.1523/JNEUROSCI.0948-08.2008

Singh, K. D., Barnes, G. R., & Hillebrand, A. (2003). Group imaging of task-related changes in cortical synchronisation using nonparametric permutation testing. *NeuroImage, 19*, 1589–1601.

Smith, S., & Nichols, T. E. (2009). Threshold-free cluster enhancement: addressing problems of smoothing, threshold dependence and localisation in cluster inference. *NeuroImage, 44*, 83–98. doi: 10.1016/

j.neuroimage.2008.03.061

Tombu, M. N., Asplund, C. L., Dux, P. E., Godwin, D., Martin, J. W., & Marois, R. (2011). A Unified attentional bottleneck in the human brain. *Proceeding of the National Academy of Science, 108*, 13426–13431. doi: 10.1073/pnas.1103583108

Turing, A. M. (1936). On computable numbers, with an application to the Entscheidungs problem. *Proceedings of the London Mathematical Society ,42.* 230–265.

Van Selst, M., & Jolicoeur, P. (1994). A solution to the effect of sample size on outlier elimination. *The Quarterly Journal of Experimental Psychology, 47*, 631–650. doi: 10.1080/14640749408401131

Vogel, E. K., Luck, S. J., & Shapiro, K. L. (1998). Electrophysiological evidence for a postperceptual locus of suppression during the attentional blink. *Journal of Experimental Psychology: Human Perception & Performance, 24*, 1656–1674. doi: 10.1037/0096-1523.24.6.1656

Vogel, E. K., & Machizawa, M. G. (2004). Neural activity predicts individual differences in visual working memory capacity. *Nature, 428*, 748–751. doi: 10.1038/nature02447

Vrba, J., & Robinson, S. E. (2001). Signal processing in magnetoencephalography. *Methods, 25*, 249–271. doi: 10.1006/meth.2001.1238

Xu, Y., & Chun, M. M. (2006). Dissociable neural mechanisms supporting visual short-term memory for objects. *Nature, 440*(7080), 91–95. doi: 10.1038/nature04262

Neuroscience, 2014, 256, 334 - 341.

Neural pathway in the right hemisphere underlies verbal insight problem solving

Qingbai Zhao[1, 2], Zhijin Zhou[1, 2,*], Haibo Xu[3,*], Wenliang Fan[3], Lei Han[4]

([1] Key Laboratory of Adolescent CyberPsychology and Behavior (CCNU), Ministry of Education, Central China Normal University,Wuhan 430079, China)

([2] Key Laboratory of Human Development and Mental Health of Hubei Province, School of Psychology, Central China Normal University, Wuhan 430079, China)

([3] MRI Center of Union Hospital, Tongji Medical College, Huazhong University of Science and Technology, Wuhan 430022, China)

([4] School of Psychology, Shandong Normal University, Jinan 250014, China)

Abstract Verbal insight problem solving means to break mental sets, to select the novel semantic information and to form novel, task-related associations. Although previous studies have identified the brain regions associated with these key processes, the interaction among these regions during insight is still unclear. In the present study, we explored the functional connectivity between the key regions during solving Chinese 'chengyu' riddles by using event-related functional magnetic resonance imaging. Results showed that both insight and noninsight solutions activated the bilateral inferior frontal gyri, middle temporal gyri and hippocampi, and these regions constituted a frontal to temporal to hippocampal neural pathway. Compared with noninsight solution, insight solution had a stronger functional connectivity between the inferior frontal gyrus and middle temporal gyrus in the right hemisphere. Our study reveals the neural pathway of information processing during verbal insight problem solving, and supports the right-hemisphere advantage theory of insight.

Keywords insight; fMRI; functional connectivity; Chinese 'chengyu' riddle; right hemisphere

* Corresponding authors. Address: Key Laboratory of Adolescent CyberPsychology and Behavior (CCNU), Ministry of Education, Central China Normal University, Wuhan 430079, China. Tel: +86-02767865767 (Z. Zhou).

E-mail addresses: zhouzj@mail.ccnu.edu.cn (Z. Zhou), xuhaibo1120@hotmail.com (H. Xu).

This study was supported by the National Natural Science Foundation of China (Grant Nos. 81171386 and 30770623), the Social Science and Humanities Research Youth Foundation Project of the Educational Ministry of China (Grant No. 10YJCXLX065), and the Fundamental Research Funds for Central Universities (Grant Nos. CCNU11C01005 and CCNU13A05043).

1 Introduction

Since Köhler observed (1925) that chimpanzees could resolve problems suddenly rather than by an approach of trial and error, the processing of insight has attracted attention of many researchers. Since unsuitable representations of problem would lead to the failure of effective problem solving in many situations, some cognitive psychologists proposed that the representation change such as constraint relaxation and chunk decomposition should be the crucial process of insight (Kaplan and Simon, 1990, Knoblich et al., 2001, Ormerod et al., 2002). Using some visual-representation-based problems such as nine-dot problem and Chinese chunk decomposition problem, researchers found that there were multiple sources of difficulty of particular insight problems, and that early perceptual processes could crucially affect thinking and problem solving (MacGregor et al., 2001, Kershaw and Ohlsson, 2004, Luo et al., 2006, Wu et al., 2013).

With the development of neuroimaging technique, especially from 1990s onwards, the investigations on the neural correlates of insight flourished (Dietrich and Kanso, 2010). However, since a number of homogenous mental events which can be repeatedly observed are required for the neuroimaging approach, the classic insightful paradigms such as nine-dot problem and six-matchstick problem are no longer suitable (Luo, 2004). Thus, a variety of verbal problems have been applied in the studies of insight, such as riddles, logogriphs and compound remote associates problems (Luo and Niki, 2003, Jung-Beeman et al., 2004, Qiu et al., 2010, Zhao et al., 2013). In these studies, researchers focused on two key components of insight processing, that is to break the mental sets and to form novel, task-related associations among the old nodes of concepts or cognitive skills (Luo and Niki, 2003,

Bowden and Jung-Beeman, 2007).

Studies indicated that the frontal cortex played a key role in breaking mental sets. Therein, the anterior cingulate cortex was highlighted and proposed to monitor the cognitive conflicts resulted from mental sets (Luo et al., 2004, Mai et al., 2004, Qiu et al., 2008, Aziz-Zadeh et al., 2009). After detecting the cognitive conflicts, one should break its mental sets to solve the conflicts in insight. It was found that the lateral prefrontal cortex, including the inferior and middle frontal gyri, was activated in chunk decomposition of Chinese characters (Luo et al., 2006), set-shift problems (Goel and Vartanian, 2005) and insightful riddle solving (Luo and Niki, 2003, Luo et al., 2004, Qiu et al., 2010). Due to its role in establishing and shifting the attentional sets (MacDonald et al., 2000, Luks et al., 2002), the lateral prefrontal cortex was thought to be associated with conflict resolution.

The breaking of mental sets would result in the retrieval of new information pieces. Then, the selection of novel information pieces and forming novel, task-related associations were the keys of insight problem solving (Bink and Marsh, 2000). According to the coarse semantic coding theory, the right-hemisphere engages in coarse semantic coding, weakly and diffusely activating alternative meanings and more distant associates (Faust and Chiarello, 1998, Beeman and Bowden, 2000, Bowden and Jung-Beeman, 2003). Therefore, some researchers highlighted the role of the right hemisphere (especially the right anterior superior temporal gyrus) in making connections across distantly related information during insight (Bowden and Jung-Beeman, 2003, Jung-Beeman et al., 2004). In fact, the right temporal cortex might be mainly in charge of processing novel semantic information (Faust and Mashal, 2007, Mashal et al., 2008, Pobric et al., 2008), while the hippocampus should be the key brain region in forming the novel associations (Luo and Niki, 2003, Zhao et al., 2013), due to its function in path reorientation (Redish, 2001), relational memory (Cohen et al., 1999, Luo and Niki, 2002) and response to novel stimuli (Knight, 1996, Johnson et al., 2008).

Obviously, verbal insight problem solving activates a distributed neural network including the anterior cingulate cortex, lateral prefrontal cortex, right temporal areas and hippocampus. Although previous studies have identified the roles of these brain regions, the information integrations among them are still unclear. The electroencephalograph study showed that good performance in the divergent thinking task was related to increased functional connectivity of central-parietal areas of both hemispheres and greater ipsilateral connections between the cortex regions of the right hemisphere in the beta2 band (Razoumnikova, 2000). And the study using diffusion tensor imaging reported significant positive relationships between individual creativity as measured by the divergent thinking test and fractional anisotropy in the white matter in or adjacent to the bilateral prefrontal cortices, the body of the corpus callosum, the bilateral basal ganglia, the bilateral temporal-parietal junction and the right inferior parietal lobule (Takeuchi et al., 2010). These studies indicated the importance of the information integration of different brain regions in creativity.

In verbal insight problem solving, the forming of novel associations is dependent on the selection of novel semantic information. Since the lateral prefrontal cortex, temporal areas and hippocampus are respectively associated with conflict resolution, semantic processing and relational memory, it is speculated that the functional connectivity between the lateral prefrontal cortex and temporal areas might reflect the information selection process in insight and that between right temporal areas and hippocampus might underlie the forming of novel association. The current work aims to reveal the functional connectivity among the key brain regions which underlies the cognitive processing in insight.

Additionally, according to the coarse semantic coding theory, the right temporal cortex should play a crucial role in verbal insight problem solving, and this is supported by several studies (Bowden and Jung-Beeman, 2003, Jung-Beeman et al., 2004, Zhang et al., 2011, Zhou et al., 2011). However, there are also some studies revealing bilateral activation patterns associated with insight events (Luo and Niki, 2003, Aziz-Zadeh et al., 2009, Zhao et al., 2013). There might be several reasons why the latter studies do not find the right hemisphere advantage. First, verbal insight problems can not be solved by the conventional semantic information processing, and then the process of retrieving the novel semantic information is the key of insight solving. However, some of the latter studies adopted the paradigm of providing triggers to catalyze the insight processes. This would simplify the retrieval of the novel meanings and distant associates, and then weaken the activation of the right temporal cortex. Second, since insight solution comes to mind suddenly, the right temporal cortex should show greater activation at the time just prior to the solution (Jung-Beeman et al., 2004). However, most of the latter studies focused on the activation throughout the solving period, not exactly catching the key period of the activation in right temporal cortex. It is noticed that all these findings, no matter supporting the right hemisphere theory or not, are from the location analysis of brain functions. Thus, as one of the two patterns of brain functional organization (Tononi et al., 1994), the functional integration analysis may provide something new for the discussions on hemisphere difference in insight.

2 Experimental procedures

2.1 Participants

As paid volunteers, 20 undergraduates or graduates (13 women, 7 men), aged 21 – 35 years (mean age, 23.6 years) from Central China Normal University, participated in the experiment, and gave their informed consent according to the requirements of Institutional Review Board of Central China Normal University. All participants were healthy, right-handed, and had normal or corrected to normal vision. Two participants were excluded from analysis due to their experiencing of less than 15% normal associations during the experiment. Another participant was excluded due to the excessive head motion during functional magnetic resonance imaging (fMRI) scanning.

2.2 Stimuli and task

In the present study, we adopted the Chinese 'chengyu' (in Chinese pinyin) riddles to explore the underlying neural mechanism of insight. A Chinese 'chengyu' riddle may be a phrase, or a saying, and its answer is a four-character 'chengyu' which is a type of traditional Chinese idiomatic expressions. As each 'chengyu' only has one meaning, the meanings of its four component characters are constrained by the chunk of the 'chengyu'. This prevents the successful riddle solving because the riddles aim at an unconstrained meaning of the key character rather than the meaning of the 'chengyu' as a whole. To solve the riddle, the chunk of 'chengyu' must be decomposed, and extensive meanings of individual characters must be explored and retrieved. For example, the answer of the riddle 'shan zhan er duo mou' (善战而多谋 , means adept at fighting and planning) is the chengyu 'jing da xi suan' (精打细算 , means being very careful in reckoning). The key character in this riddle is 'da' (打 , one of its meanings is to hit), corresponding to 'zhan' (战 , with the meaning to fight). However, inside the 'chengyu', the 'da' (打) is bound with 'suan' (算). And the meaning of their combination 'da suan' (打算) is to plan or to reckon. Obviously, the successful riddle solving is relied on the successful constraint relaxation to the key character or the successful chunk decomposition. This is theoretically similar with the visual chunk decomposition of Chinese characters (Luo et al., 2006). Once extensive meanings of key character were retrieved,

a number of temporary connections between the riddle and the 'chengyu' would be formed, and then the riddle would be solved by the selection process of task-related connections. Since there is a process of representation change when the participants tried to associate the riddle with the original answer, it is considered as the answer with novel association.

In the current work, a control with normal association was produced in a pretest. A group of subjects were asked to report the four-character 'chengyu' that came to mind first when they saw the riddle in the pretest. Mostly, they could not find the novel answer and gave some different answers. The 'chengyu' with the highest frequency was chosen as the control. Thus, there are two answers, one of which is novel, and the other is normal. For example, the novel answer to the riddle 'shan zhan er duo mou' (善战而多谋 , means adept at fighting and planning) is the 'chengyu' of 'jing da xi suan' (精打细算 , means being very careful in reckoning), while its normal answer is 'zu zhi duo mou' (足智多谋 , means being able and adept at planning).

In order to determine the difference between the answers with novel and normal associations, we had another group of subjects (totally 32) to rate their understanding of the Reasonability (matching with the answers to riddles) and Novelty on a scale of 1 to 5 for each of the 120 riddles. In the end, 84 riddles whose answers (both novel and normal ones) were evaluated as reasonable (mean scores > 3.5) were selected as the test riddles. Results showed that there was a significant difference in novelty [paired t-test, $t(83) = 16.84$, $p < 0.001$] between the answers with novel (mean score = 3.6) and normal association (mean score = 2.6).

To familiarize the participants with the procedure and pace of this task, participants were trained with another set of 10 similar materials before they were put into the scanner. In the formal experiment, 84 test riddles were

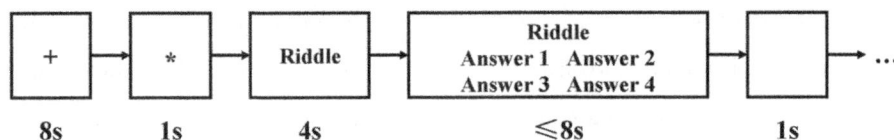

Fig. 1. The flow map of the formal experiment

presented one by one with an event-related design. There was not any repetition of stimuli in the test. The Chinese characters, appearing in both the riddles and answers, had a font size of 28 (Song Ti font). The experimental paradigm was illustrated in Fig.1 The trial began with an 8-s black plus, a sign for rest, and a star sign for 1s, followed by a warning of the presentation of riddle. After the riddle was displayed for 4s, the novel association answer, normal association answer and two answers with no associations were presented. Participants were asked to select a novel and reasonable answer among these options within a limited 8-s period. Then was a 1-s blank followed by the next trial. The spatial positions were balanced among the different answers.

Since it is difficult to solve the riddle on participants' own initiative, the paradigm of providing a trigger to catalyze insight processes was adopted by some studies (Luo and Niki, 2003, Luo et al., 2004, Mai et al., 2004). This paradigm is helpful in investigating the processes of breaking mental sets and forming novel associations, but of little effect in exploring how participants retrieve the extensive information to solve problems. Therefore, the present study introduced the answer selection paradigm described above to investigate the information selection in insight. Participants were asked to select the novel and reasonable answer from four options, in which the novelty was more emphasized. According to the selections of participants, the trials are classified into insight and noninsight solutions, respectively. Since the normal answer points to the conventional thinking, it is easy to understand and should be found first. However, the normal answer is of less novelty, and then participants might actively look for the novel one from the others. If participants ultimately could not find the association between the novel answer and the riddle, they might select the normal one, in which case the problem solving is a simple process in the conventional thinking, and it is considered as noninsight-based solution. Once participants found the association between the novel answer and the riddle, they would select it as asked. In this case, there is a competition between the novel and normal answer, and the selection of novel answer just reflects the breaking of conventional thinking. Additionally, because of including a process of representation change, the selection of novel answer is indeed an insight-based solution.

Indeed, although the answer selection makes participants more active than solution recognition, it is different from the actual problem solving on participants' own. The adoption of the answer selection paradigm is an inevitable compromise.

2.3 fMRI acquirement

During MRI scanning, whole brain T2*-weighted echo planar imaging, based on blood oxygenation level-dependent contrast (EPI-BOLD) fMRI data, was acquired with a Siemens Trio 3.0-T MR-scanner using a standard head coil at the MRI Center of Wuhan Union Hospital. 32 interleaved slices, covering the entire brain, were acquired using a gradient-echo echo-planar pulse sequence. The slice thickness was 3.75 mm and the voxel size was 3 mm × 3 mm (TR = 2s, TE = 30 ms, FA=78°, FOV=192 × 192mm, Matrix size = 64 × 64). Head motion was restricted with plastic braces and foam padding. The whole scanning sequence was divided into two runs, each consisted of 42 trials.

2.4 fMRI data analysis

Preprocessing. The statistical parametric mapping

(SPM5, http://www.fil.ion.ucl.ac. uk/spm/) was used for image preprocessing and voxel-based statistical analysis. Scans were first slice-time corrected, realigned, normalized (using the functional EPI template provided in SPM5), and smoothed (a Gaussian kernel with a full width at the full width at half maximum - FWHM of 8 mm). The resultant images had cubic voxels of $3 \times 3 \times 3$ mm.

Mapping brain activation. Two types of events, the insight solution and the noninsight solution, were defined according to participants' selections to the answers. Since the participants could not solve the riddle by themselves in the initial 4s of riddle presentation (Zhu et al., 2009), the event was defined as the answer selection process for each type which began at the onset of the answers' presentation and ended in the participants' pressing. Then, the time vector of the event was convoluted by the classic haemodynamic response function. Finally, by the general linear model, the activated brain regions associated with the insight solution and the noninsight solution, as well as the differences between the two conditions, were obtained for each participant, and then combined in a random effect analysis to identify differences consistent

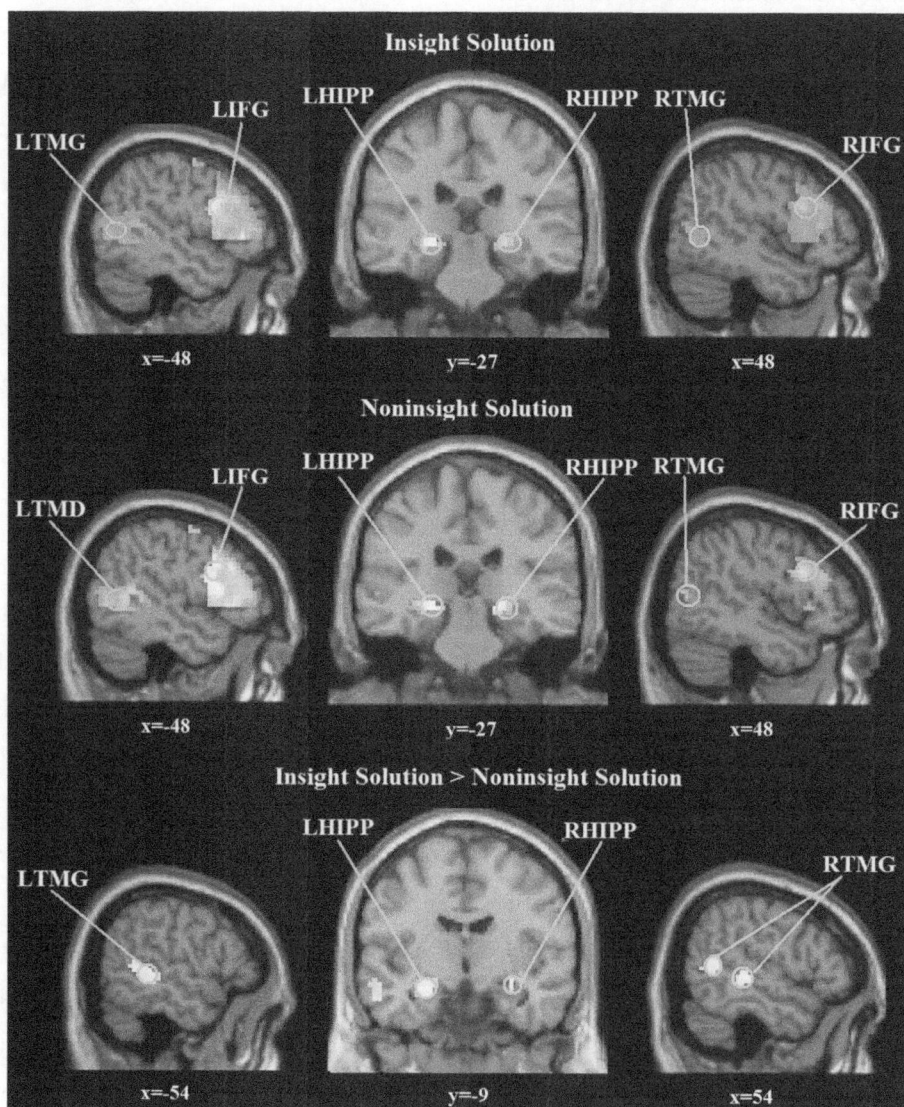

Fig. 2. Brain areas activated for insight solution and noninsight solution as well as the difference between the two conditions in the six regions of interest. For insight vs. rest and noninsight vs. rest, the thresholds were set at $p<0.001$ (False Discovery Rate control for multiple comparisons) and 30 or more contiguous voxels, while for insight vs. noninsight solution, the threshold was set at $p<0.05$ (False Discovery Rate control for multiple comparisons) and 30 or more contiguous voxels.

across all participants. For insight vs. rest and noninsight vs. rest, the thresholds were set at p<0.001 (False Discovery Rate control for multiple comparisons) and 30 or more contiguous voxels, while for insight vs. nonisight solution, the threshold was set at p<0.05 (False Discovery Rate control for multiple comparisons) and 30 or more contiguous voxels.

Functional connectivity. In the current work, we were interested in the functional connectivity among several brain regions including bilateral inferior/middle frontal gyri, middle/superior temporal gyri and hippocampi. Using the Automated Anatomical Labeling (Tzourio-Mazoyer et al., 2002), the activated voxels in these regions were obtained. The region of interest (ROI) was defined as a cube of $9 \times 9 \times 9$ mm centered at the activation peak in each brain region, and its BOLD time series was defined as the mean of all the voxels in the cube. If there were more than one peak in a region, the most central one located in the activation cluster was selected. Then, the time series associated with insight and noninsight solution

Table 1. Brain areas activated for insight solution and noninsight solution as well as the difference between the two conditions in the six regions of interest

Area	BA	Voxels	x	y	z	T	Z
Insight Solution							
Left inferior frontal gyrus	47	1051	−33	30	3	20.66	7.21
	44		−48	9	24	15.59	6.59
Left hippocampus	20	147	−24	−27	−9	18.13	6.93
	20		−33	−18	−9	9.47	5.42
Right hippocampus	27	123	21	−33	−3	14.28	6.40
	20		24	−27	−9	11.19	5.83
Right inferior frontal gyrus	47	731	36	27	3	10.51	5.68
	48		48	15	24	9.56	5.45
Left middle temporal gyrus	21	524	−60	−42	0	10.10	5.58
	22		−54	−45	9	9.94	5.54
Right middle temporal gyrus	37	113	45	−69	9	7.21	4.74
	37		45	−63	−3	6.72	4.57
Noninsight Solution							
Left hippocampus	20	82	−21	−27	−9	13.80	6.32
Left inferior frontal gyrus	48	914	−48	15	12	13.55	6.28
	44		−42	9	27	13.19	6.21
Right hippocampus	27	65	21	−30	−3	11.41	5.87
Right inferior frontal gyrus	47	348	36	27	3	10.82	5.75
	48		51	18	30	8.42	5.13
Left middle temporal gyrus	21	392	−57	−45	9	9.41	5.41
	21		−60	−33	3	7.58	4.87
Right middle temporal gyrus	37	32	48	−72	12	5.82	4.20
	37		42	−63	9	5.45	4.04
Insight Solution > Noninsight Solution							
Left hippocampus	20	48	−27	−9	−18	6.87	4.63
Right middle temporal gyrus	21	205	54	−30	0	6.41	4.45
	21		54	−54	12	5.39	4.01
Left middle temporal gyrus	22	126	−57	−36	3	5.09	3.87
	22		−54	−48	6	5.00	3.82
Left middle temporal gyrus	21	34	−54	−6	−24	4.96	3.81
Right hippocampus	20	6	30	−9	−12	4.74	3.69

BA, Brodmann area. Coordinates (x,y,z) were the MNI (Montreal Neurological Institute) coordinates. For insight vs. rest and noninsight vs. rest, the thresholds were set at $p<0.001$ (False Discovery Rate control for multiple comparisons) and 30 or more contiguous voxels, while for insight vs. nonisight solution, the threshold was set at $p<0.05$ (False Discovery Rate control for multiple comparisons) and 30 or more contiguous voxels. T- and Z-scores of the activations were also shown. Note that the 6 voxels in right hippocampus were included in a larger cluster with 116 contiguous voxels mostly located in right amygdala. At a looser threshold ($p<0.005$ uncorrected), there were 84 contiguous voxels more activated in right hippocampus for insight than noninsight solution.

was respectively segregated from the whole. Since the BOLD signal delayed for six seconds according to the haemodynamic response function by SPM, and the reaction time was about 4s (see in Results), the time series of each condition was defined from 0s to 10s after the presentation of optional answers. For each condition, the functional connectivity was evaluated by computing the temporal partial correlation between all pair-wise combinations of ROIs controlling the effects of the others (Friston et al., 1993). Finally, the correlation coefficients were compared across conditions after the Fisher transformation.

3 Results

On average, in 61.7% of trials participants selected the answers with novel associations (average reaction time was 3.70s with a standard deviation of 0.78s), and in 26.3% of trials they selected the answers with normal associations (average reaction time was 4.06s with a standard deviation of 0.94s). The larger trial percentage of insight solutions might result from the instruction before the experiment that asked participants to select a novel and reasonable answer.

Our fMRI results demonstrated that both of the insight and noninsight solution induced extensive changes of brain activity in the bilateral inferior frontal gyri, middle temporal gyri and hippocampi (see in Fig.2 and Table 1). Compared with noninsight solution, insight solution activated more in bilateral middle temporal gyri and hippocampi. The activation of the bilateral inferior frontal gyri showed no significant differences between two conditions.

As mentioned in Methods, the peaks of ROIs for insight solution were located at (-60, -42, 0) in left middle temporal gyrus (LMTG), (-48, 9, 24) in left inferior frontal gyrus (LIFG), (-24, -27, -9) in left hippocampus (LHIPP), (21, -33, -3) in right hippocampus (RHIPP), (45, -63, -3) in right middle temporal gyrus (RTMD) and (48, 15, 24) in

right inferior frontal gyrus (RIFG). And the peaks of ROIs for noninsight solution were located at (-57, -45, 9) in LMTG, (-42, 9, 27) in LIFG, (-21, -27, -9) in LHIPP, (21, -30, -3) in RHIPP, (42, -63, 9) in RTMD and (51, 18, 30) in RIFG.

The functional connectivity analysis showed that the ROIs were significantly connected with the ipsilateral regions and the homologous regions in the other hemisphere. In each hemisphere, there was a neural pathway from the inferior frontal gyrus to middle temporal gyrus to hippocampus for both insight and noninsight solution. The functional connectivity between the RIF6, and RTMD was stronger for insight than noninsight solution, while that between the LMTG and RHIPP was stronger for noninsight than insight solution (p<0.05, paired t-test).

4 Disscussion

4.1 Information selection in verbal Insight problem solving

As a type of traditional Chinese idiomatic expressions, the 'chengyu' could be regarded as a chunk. Then, the meanings of four component characters enhanced by the 'chengyu' should be their normal meanings, and their other meanings could be considered as novel ones. Although the left hemisphere is dominant in language processing, it is the right temporal cortex that processes the novel meaning of idioms (Faust & Mashal, 2007; Mashal, Faust, Hendler, & Jung-Beeman, 2008; Pobric, Mashal, Faust, & Lavidor, 2008). In the current work, the greater activation in right temporal gyrus for insight than noninsight solution indicated more retrieval of novel semantic information.

However, the novel semantic information is generally weak in retrieval (Giora, 1997). This conflict might be resolved by the lateral prefrontal cortex due to its role in establishing and shifting the attentional sets (MacDonald et al., 2000, Luks et al., 2002). Although there was no significant difference of activation strength in bilateral prefrontal cortices at the given threshold between insight and noninsight solution, the RIFG was stronger functionally connected to the RTMD for insight than noninsight solution. The frontal–temporal neural pathway might reflect the selection of novel semantic information in verbal insight problem solving.

4.2 Forming novel association in verbal insight problem solving

Previous studies indicated the hippocampus was involved in the relational memory and its activation strength was associated with the novelty degree of stimuli (Luo and Niki, 2002, 2005, Johnson et al., 2008). Therefore, it was proposed as the key brain region of forming novel association (Luo and Niki, 2003, Zhao et al., 2013). In the present study, the bilateral hippocampi were involved in both insight and noninsight solution, but more activated for insight solution. These might reflect that the associations between the riddle and the selected answer were formed in both insight and noninsight solution, but the novel associations only existed in insight solution.

In insight 'chengyu' riddle solving, the forming of novel associations was dependent on the retrieval of the novel meaning of the key character. Our results showed that the RHIPP was functionally connected with the RTMD in insight solution, Fig.3 but the strength was not larger than that in noninsight solution. However, the functional connectivity between the RHIPP and LMTG was stronger in noninsight than insight solution. This might reflect that the retrieval of normal semantic information resulted in the normal associations in noninsight solution.

4.3 Right–hemisphere theory in verbal insight problem solving

Since the key of creative problem solving is the selection of novel information (Bink and Marsh, 2000), and the right temporal cortex engages in relatively coarse and novel semantic processing (Beeman and Bowden, 2000, Mashal et al., 2008), it is proposed that there is a right–hemisphere advantage in insight problem solving (Bowden and Jung–Beeman, 2003). In particular, the right superior temporal gyrus is found in several studies and it is considered to facilitate the formation of remote associations (Jung–Beeman et al., 2004, Kounios et al., 2008, Zhang et al., 2011). However, the present study showed the bilateral middle temporal gyri involved in insight solution, even at the time just prior to the solution (Zhao et al., 2013). The paradigm adopted by the present study might be the reason why the advantage of right hemisphere was not found. Although the answer selection paradigm made participants

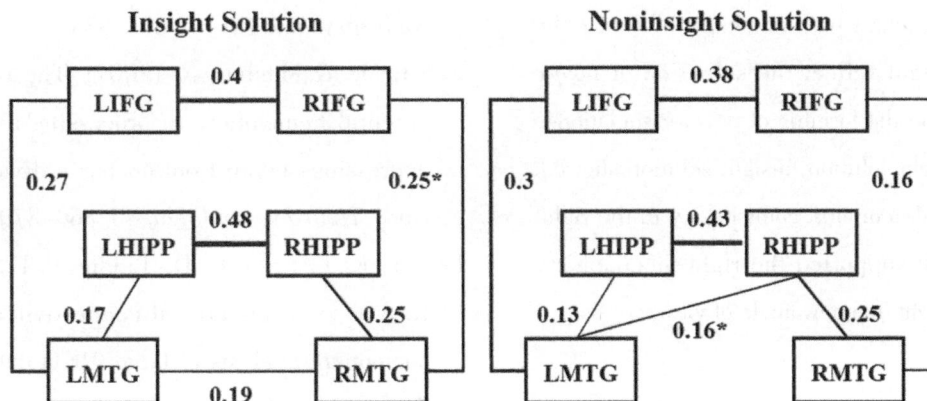

Fig. 3. The functional connectivity among the ROIs for insight and noninsight solution. A line between two regions indicates that the region–to–region correlation is statistically significant, and the thickness of the line reflects the strength of functional connectivity. The star means significant difference between two conditions (p<0.05, paired t–test).

more active than solution recognition, it still simplified the retrieval of the novel meanings and distant associates, and then weakened the activation of the right temporal cortex.

However, the brain activity contains two aspects: the activation of local regions and the interaction among them. Most previous debates about the right-hemisphere advantage in insight were only based on the results of the activation in local regions. In the present study, the functional connectivity among the key brain regions in insight and noninsight solution was evaluated. Results showed that there was a neural pathway from the inferior frontal gyrus to middle temporal gyrus to hippocampus for insight solution in the both hemispheres, but insight solution had greater ipsilateral frontal-temporal connectivity in the right hemisphere than noninsight solution. This implied that verbal insight problem solving needed the participations of both the left and right hemispheres, in which the right hemisphere might be more important compared with common verbal problem solving.

5 Conclusion

Our results demonstrated that verbal insight problem solving activated broad brain regions including the lateral prefrontal cortex, middle temporal gyrus and hippocampus in both hemispheres. There regions constituted a frontal-temporal-hippocampal neural pathway, especially in right hemisphere, which might reflect the selection of novel semantic information and the forming of novel associations. Compared with noninsight solution, insight solution showed greater ipsilateral frontal-temporal connectivity in the right hemisphere. Our result supported the right-hemisphere advantage theory of insight in a new angle of view.

References

Aziz-Zadeh, L., Kaplan, J. T., & Iacoboni, M. (2009). "Aha!": The neural correlates of verbal insight solutions. *Hum Brain Mapp, 30,* 908-916.

Beeman, M. J., & Bowden, E. M. (2000). The right hemisphere maintains solution-related activation for yet-to-be-solved problems. *Memory & cognition, 28,* 1231-1241.

Bink, M. L., & Marsh, R. L. (2000). Cognitive regularities in creative activity. *Review of General Psychology, 4,* 57-78.

Bowden, E. M., & Jung-Beeman, M. (2003). Aha! Insight experience correlates with solution activation in the right hemisphere. *Psychonomic bulletin & review, 10,* 730-737.

Bowden, E. M., & Jung-Beeman, M. (2007). Methods for investigating the neural components of insight. *Methods, 42,* 87-99.

Cohen, N. J., Ryan, J., Hunt, C., Romine, L., Wszalek, T., & Nash, C. (1999). Hippocampal system and declarative (relational) memory: summarizing the data from functional neuroimaging studies. *Hippocampus, 9,* 83-98.

Dietrich, A., & Kanso, R. (2010). A review of EEG, ERP, and neuroimaging studies of creativity and insight. *Psychological bulletin, 136,* 822-848.

Faust, M., & Chiarello, C. (1998). Sentence context and lexical ambiguity resolution by the two hemispheres. *Neuropsychologia, 36,* 827-835.

Faust, M., & Mashal, N. (2007). The role of the right cerebral hemisphere in processing novel metaphoric expressions taken from poetry: a divided visual field study. *Neuropsychologia, 45,* 860-870.

Friston, K. J., Frith, C. D., Liddle, P. F., & Frackowiak, R. S. (1993). Functional connectivity: the principal-component analysis of large (PET) data sets. *J Cereb Blood Flow Metab, 13,* 5-14.

Giora, R. (1997). Understanding figurative and literal language: the graded salience hypothesis. *Cognitive*

Linguistics, 7, 183–206.

Goel, V., & Vartanian, O. (2005). Dissociating the roles of right ventral lateral and dorsal lateral prefrontal cortex in generation and maintenance of hypotheses in set-shift problems. *Cereb Cortex, 15,* 1170–1177.

Huang, F. R., Zhou, Z. J., & Zhao, Q. B. (2013). An eye movement study of associate competition in Chinese idiom riddles solving. *Acta Psychologica Sinica, 45,* 36–46.

Johnson, J. D., Muftuler, L. T., & Rugg, M. D. (2008). Multiple repetitions reveal functionally and anatomically distinct patterns of hippocampal activity during continuous recognition memory. *Hippocampus, 18,* 975–980.

Jung-Beeman, M., Bowden, E. M., Haberman, J., Frymiare, J. L., Arambel-Liu, S., Greenblatt, R., Reber, P. J., & Kounios, J. (2004). Neural activity when people solve verbal problems with insight. *PLoS Biol, 2,* E97.

Köhler, W. (1925). The mentality of apes. London: Routledge & Kegan Paul.

Kaplan, C. A., & Simon, H. A. (1990). In search of insight. *Cognitive Psychology, 22,* 374–419.

Kershaw, T. C., & Ohlsson, S. (2004). Multiple causes of difficulty in insight: the case of the nine-dot problem. *Journal of experimental psychology Learning, memory, and cognition, 30,* 3–13.

Knight, R. (1996). Contribution of human hippocampal region to novelty detection. *Nature, 383,* 256–259.

Knoblich, G., Ohlsson, S., & Raney, G. E. (2001). An eye movement study of insight problem solving. *Memory & cognition, 29,* 1000–1009.

Kounios, J., Fleck, J. I., Green, D. L., Payne, L., Stevenson, J. L., Bowden, E. M., & Jung-Beeman, M. (2008). The origins of insight in resting-state brain activity. *Neuropsychologia, 46,* 281–291.

Luks, T. L., Simpson, G. V., Feiwell, R. J., & Miller, W. L. (2002). Evidence for anterior cingulate cortex involvement in monitoring preparatory attentional set. *Neuroimage, 17,* 792–802.

Luo, J. (2004). Neural correlates of insight. *Acta Psychologica Sinica, 36,* 219–234.

Luo, J., & Niki, K. (2002). Role of medial temporal lobe in extensive retrieval of task-related knowledge. *Hippocampus, 12,* 487–494.

Luo, J., & Niki, K. (2003). Function of hippocampus in "insight" of problem solving. *Hippocampus, 13,* 316–323.

Luo, J., & Niki, K. (2005). Does hippocampus associate discontiguous events? Evidence from event-related fMRI. *Hippocampus, 15,* 141–148.

Luo, J., Niki, K., & Knoblich, G. (2006). Perceptual contributions to problem solving: Chunk decomposition of Chinese characters. *Brain Res Bull, 70,* 430–443.

Luo, J., Niki, K., & Phillips, S. (2004). Neural correlates of the 'Aha! reaction'. *Neuroreport, 15,* 2013–2017.

MacDonald, A. W., Cohen, J. D., Stenger, V. A., & Carter, C. S. (2000). Dissociating the role of the dorsolateral prefrontal and anterior cingulate cortex in cognitive control. *Science, 288,* 1835–1838.

MacGregor, J. N., Ormerod, T. C., & Chronicle, E. P. (2001). Information processing and insight: a process model of performance on the nine-dot and related problems. *Journal of experimental psychology Learning, memory, and cognition, 27,* 176–201.

Mai, X. Q., Luo, J., Wu, J. H., & Luo, Y. J. (2004). "Aha!" effects in a guessing riddle task: an event-related potential study. *Hum Brain Mapp, 22,* 261–270.

Mashal, N., Faust, M., Hendler, T., & Jung-Beeman, M. (2008). Hemispheric differences in processing the literal interpretation of idioms: converging evidence from behavioral and fMRI studies. *Cortex, 44,* 848–860.

Ormerod, T. C., MacGregor, J. N., & Chronicle, E.P. (2002). Dynamics and constraints in insight problem solving. *Journal of experimental psychology Learning, memory, and cognition, 28*, 791–799.

Pobric, G., Mashal, N., Faust, M., & Lavidor, M. (2008). The role of the right cerebral hemisphere in processing novel metaphoric expressions: a transcranial magnetic stimulation study. *Journal of cognitive neuroscience, 20*, 170–181.

Qiu, J., Li, H., Jou, J., Liu, J., Luo, Y., Feng, T., Wu, Z., & Zhang, Q. (2010). Neural correlates of the "Aha" experiences: evidence from an fMRI study of insight problem solving. *Cortex, 46*, 397–403.

Qiu, J., Li, H., Yang, D., Luo, Y., Li, Y., Wu, Z., & Zhang, Q. (2008). The neural basis of insight problem solving: an event–related potential study. *Brain and cognition, 68*, 100–106.

Razoumnikova, O. M. (2000). Functional organization of different brain areas during convergent and divergent thinking: an EEG investigation. *Brain Res Cogn Brain Res, 10*, 11–18.

Redish, A. D. (2001). The hippocampal debate: are we asking the right questions? *Behavioural brain research, 127*, 81–98.

Takeuchi, H., Taki, Y., Sassa, Y., Hashizume, H., Sekiguchi, A., Fukushima, A., & Kawashima, R. (2010). White matter structures associated with creativity: evidence from diffusion tensor imaging. *Neuroimage, 51*, 11–18.

Tononi, G., Sporns, O., & Edelman, G. M. (1994). A measure for brain complexity: relating functional segregation and integration in the nervous system. *Proc Natl Acad Sci USA, 91*, 5033–5037.

Tzourio–Mazoyer, N., Landeau, B., Papathanassiou, D., Crivello, F., Etard, O., Delcroix, N., Mazoyer, B., & Joliot, M. (2002). Automated anatomical labeling of activations in SPM using a macroscopic anatomical parcellation of the MNI MRI single–subject brain. *Neuroimage, 15*, 273–289.

Wu, L., Knoblich, G., & Luo, J. (2013). The role of chunk tightness and chunk familiarity in problem solving: Evidence from ERPs and fMRI. *Hum Brain Mapp, 34*(5), 1173–86.

Zhang, M., Tian, F., Wu, X., Liao, S., & Qiu, J. (2011). The neural correlates of insight in Chinese verbal problems: an event related–potential study. *Brain Res Bull, 84*, 210–214.

Zhao, Q. B., Zhou, Z. J., Xu, H. B., Chen, S., Xu, F., Fan, W. L., & Han, L. (2013). Dynamic neural network of insight：A functional magnetic resonance imaging study on solving Chinese 'chengyu' riddles. *PloS one, 8*, e59351.

Zhou, Z. J., Xu, H. B., Zhao, Q. B., Zhao, L. L., & Liao, M. J. (2011). The processing of novel semantic association in Chinese: Converging evidence from behavior and fMRI studies. *The 4th International Conference on Image and Signal Processing (CISP 2011), Shanghai, 3*, 1588–1592.

Zhu, X. C., Li, R. J., & Zhou, Z. J. (2009). The role of clues in Chinese idiom Riddle solving. *Acta Psychologica Sinica, 41*, 397–405.

人格与社会心理研究所团队成员

郭永玉，博士，教授，人格与社会心理研究所所长，主要研究领域为中西人格理论，社会历史文化背景下的人格研究，社会治理中的心理学问题。

佐斌，博士，教授，华中师范大学社会心理研究中心主任，主要研究领域为文化与社会心理学理论及其应用、青少年儿童对社会与文化的理解、青少年人格发展与心理健康教育等。

李晔，博士，教授，人格与社会心理研究所副所长，主要研究领域为教师心理、师生关系、文化与社会认知；组织公平感与社会公平感等。

贺金波，博士，副教授，主要研究领域为人格的生理机制，网络成瘾的发生机制，网络与青少年心理健康，前注意的神经机制，酒精对前注意的影响等。

王伟，副教授，主要研究领域为社会心理学理论。

黄飞，博士，讲师，主要研究领域为青少年人格发展与评估，人格知觉，人际关系与社会发展，测量等值性等。

温芳芳，博士，讲师，主要研究领域为社会认知刻板印象，社会与群体认同，青少年网络心理与行为，面孔吸引力等。

李静，博士，讲师，主要研究领域为物质主义，收入与幸福，社会阶层心理学等。

中国科学院院刊, 2012, 27(心理学理论体系与方法论专辑), 88 - 97.

人格心理学：人性及其差异的研究 *

郭永玉，李　静，胡小勇

（ 华中师范大学心理学院，青少年网络心理与行为教育部重点实验室，武汉 430079 ）

摘　要　人格心理学是心理学学科体系中注重从整体的视角探究人性本质的一个分支，它以人性及其差异作为其核心，研究对象是作为整体的人。人格心理学不仅在心理学的学科体系内部处于重要地位，而且在关于人的所有生命科学、社会科学和人文学科中也处于基础性的位置。经过 100 多年曲折的发展，人格心理学已经步入了一个新的发展和繁荣时期。近 20 年来西方人格心理学的研究进入了快速发展阶段，在研究范式、研究方法以及研究内容（包括人格结构、人格动力、人格发展）方面都有了较大的进展。与此同时，我国学者在大量介绍西方人格心理学的基础上，开始着手研究中国人的人格问题。当前中国社会文化背景下的人格心理学研究主题主要包括人格与创造力、人格与人员选拔及安置、人格与贪腐行为、人格与暴力犯罪、人格与疾病、以及和谐社会的健全人格建构问题。鉴于人格心理学具有重要的科学价值及其在社会进步中所能做出的重要贡献，建议未来中国人格心理学研究应从研究现代化背景下中国人的人格和研究方法多元化等方面予以加强。

关键词　人格心理学；人性及其差异；现代化人格研究

1 引言

我们中国人说到“人格”，神情往往不由自主地严肃起来，因为这个词往往具有法律和道德的涵义。法律上讲“保护人格尊严”，“不能侮辱人格”，是将人格视为权利义务的主体。日常话语中讲“人格高尚”或“人格低下”，甚至“没有人格”，是将人格视为道德品质，与人品、品格或品德同义。事实上，古汉语中并无“人格”一词，这个词是近代从日文中来的，而日文“人格”一词又是对英文“personality”一词的翻译。这个英文词也可以译为“人性”，是指人（person）的各种特征，并没有道德（以至“道德高尚”与否）的含义。它首先是一个事实性的概念，而不是一个评价性的概念。因此，心理学探讨人格和探讨感知、记忆、思维、情绪、智力等心理现象一样，也是认识人类自身的一种研究活动。只是心理学家研究感知是为了了解感知现象和规律，研究记忆是为了了解识记、保持和遗忘等现象和规律……，人格心理学就是将完整的人作为研究对象，不仅仅是研究人的某一种心理或行为。

人格心理学的研究对象是整体的人。但要研究整体，仍需要对其加以分析。人格心理学家大体从 3 个层面分析一个人：第一，人类本性的层面（ the human nature level ），即一个人首先是人，与所有人相似（ like all others ）；第二，个体差异和群体差异层面（ the level of individual and group differences ），

* 本文得到国家自然科学基金项目（批准号 71171094）资助。

即一个人与部分他人是相似的（like some others），个体之间的差异仅仅是程度的差异，如外向的程度不同而已，并且一个人与其所在群体的其他成员具有相似性，但与其他群体的成员明显不同；第三，个人独特性层面（the individual uniqueness level），即一个人不同于任何人的（like no others）、独特的、不可重复、不可替代的特征。

人格心理学的任务或目的与其他心理学分支一样，都是寻求准确地描述，合理地解释，有效地预测和控制。但人格心理学的独特性在于其是心理学学科体系中特别注重从整体的视角探究人性本质的一门学科，它将人性作为其核心，关注整体的人。正如 Hergenhahn 所说："在把人作为一个整体来研究的心理学中，人格心理学处于独特的地位。绝大多数其他分支的心理学家往往只深入研究人的某一方面。比如，他们只研究儿童发展、老年问题，或知觉、智力、学习、动机、创造性等。只有人格心理学家才试图描绘出人的完整的图画。"人格心理学不仅在心理学的学科体系内部处于重要地位，而且在关于人的所有生命科学、社会科学和人文学科中也处于基础性的位置。它与所有关于人性的学科有关，并整合这些学科关于人性的知识。

人格心理学的研究不仅对于从整体上把握人的心理和行为的独特模式具有深刻的理论意义，而且还具有广泛的实践意义，研究人格有助于提高教育、生产、管理、医疗、资讯、司法、体育和军事等各种活动的效率和绩效。

2 人格心理学的发展脉络

一般认为，现代人格心理学的正式诞生以 Gordon W. Allport（1897-1967）所著《人格：心理学的解释》（1937）和 Henry A. Murray（1893-1988）所著《人格探究》（1938）两书的出版为标志。自这两本书问世后，关于人格心理学的研究才得以蓬勃开展，而且大学心理学系也因此开设了人格心理学课程。人格心理学的历史可以追溯到科学心理学

的创始人 Wundt 等人的工作，大致经历了 4 个阶段，即奠基期、理论体系形成期、基本人格结构确立期、质疑与复兴时期。

2.1 奠基期（1937 年之前）

Wundt、James、Freud 等人的工作，无论是对于整个心理学还是人格心理学都是开创性的，意义深远。其中 Freud 对人格心理学发展的贡献尤其巨大。Freud 创立了精神分析理论，该理论内容十分庞杂，是所有人格理论中内容最丰富、影响最大的人格理论。该理论强调潜意识、性本能等人格动力的重要性，其研究的主要问题集中在焦虑、防御机制、早期经验对日后人格发展的重要性以及个人发展出来的处理内部驱力和外部刺激的自我适应功能。精神分析理论的研究方法是临床的个案研究，其评鉴技术主要采用梦的分析、自由联想和投射测验等。

2.2 理论体系形成期（1937-20 世纪 50 年代）

20 世纪 30—50 年代，人格心理学发展成为独立的分支。其标志性事件是 Gordon W. Allport（1897-1967）所著《人格：心理学的解释》（1937）和 Henry A. Murray（1893-1988）所著《人格探究》（1938）两书的出版。这一时期，人格心理学家试图将那些通过相关、临床和实验手段得到的研究结果整合起来，建构可以解释整体人格的"大理论"。其中较具影响力的包括 Allport、Murray 和 Raymond B. Cattell 等人的理论。

Allport 最主要的贡献在于，试图在理论上提出一种架构，用以解释每个人身上共性和特性的方面。他强调个体人格的结构和组织，对特质词汇研究有开拓性贡献，在研究方法上强调特殊规律研究法与一般规律研究法相结合。他的理论直接促成了人格心理学的建立。Murray 经由精神分析和变态心理学而进入人格研究领域，采用折衷的、多方法的研究取向，对 Freud 的精神分析和实验心理学进行了开创性的结合，他的工作对于拓宽人格心理学的研究领域具有广泛的影响。Murray 最主要的理论是动机理论，提出了大约 20 种需要，并创造了主题统觉测验这种动机研究方法。Cattell 对人格特质进行了深入的研究，是

特质心理学史上的伟大人物之一。他深受 Spearman 因素分析法的影响，把因素分析视为确定人格基本单元的最好方法，据此编制了著名的 16PF 问卷。

这一阶段，精神分析学派也在蓬勃发展中，著名的心理学家有 Adler、Jung、Horney、Fromm、Erikson 等人。

2.3 基本人格结构确立期（1950-20 世纪 70 年代）

这一阶段，人格心理学的理论构建基本完成，面临的主要问题是确定基本的人格结构，因此有关人格测量的一些问题就越来越为研究者所关注。这一时期的人格问卷以明尼苏达多项人格问卷（MMPI）为典型代表。此外，著名的人格问卷还有加利福尼亚心理问卷（CPI）、爱德华个人偏好调查表（EPPS）以及前文提及的 16 人格因素问卷（16PF）等。

这一时期，4 大人格主题（也是人格结构的重要方面）备受关注，获得了大量的实证研究成果，分别是：权威主义（authoritarian）、成就动机（achievement motive）、焦虑（anxiety）和场独立性（field-independence）。这些主题跟美国当时的社会文化环境以及欧美各国对二战历史的反思有关。这段时期人格心理学的研究一方面注重人格结构的理论与实证研究，成果累累，另一方面逐渐远离社会的应用需求，所以也埋下了后一个阶段面临重重危机的种子。

2.4 质疑与复兴时期（20 世纪 70 年代至今）

20 世纪 50 年代中后期，越来越多的自相矛盾的实证研究结果、测量中不断出现的错误以及学科内部统一性的缺乏，使学术界开始质疑人格心理学。随着 Mischel（1968）等人对特质论的批评，对人格心理学的质疑与不满在 70 年代达到顶峰。Mischel 指出，我们几乎不可能依据人格特质去预测一个人的行为。此外，学术界以外的社会与文化变革似乎也不利于人格心理学的发展。70 年代的美国社会发生了巨变，确定人们的基本类型和稳定的人格差异变得不合时宜。种种不利条件的作用，使人格心理学步入低谷时期。

到 80 年代，局面逐步有所改观。自我的研究，认知取向人格心理学的兴起，"大五"的出现，行为遗传学、神经科学和进化心理学的研究，以及各种研究方法的完善，为人格心理学的发展提供了新的动力。由于这些进步，人格心理学在 20 世纪后期步入了一个新的发展和繁荣时期。

3 西方人格心理学的研究进展

万晓霞以美国科技信息研究所出版的《科学引文索引》（SCI）为数据源检索人格心理学文献，对近 10 年 SCI 人格心理学研究文献进行计量分析，结果显示，人格心理学研究近 10 年进入了快速发展阶段，文献量呈逐步上升趋势（如表 1 所示），说明目前人格心理学的研究呈现出一派繁荣景象。

从学科分布来看，人格心理学研究广泛分布在 138 种学科中，其中既包含社会科学领域又涉及到自然科学领域。收录人格心理学论文排名前 10 名的学科为：社会心理学、多学科心理学、精神病学、临床心理学、神经科学、心理学、临床神经学、应用心理学、教育心理学、医学遗传心理学。说明人格心理学具有自然科学和社会科学的交叉学科的特征。

从研究范式来看，精神分析、特质论、行为主义和人本主义是人格研究的传统范式，近 20 年来，这 4 种范式都已扩展了各自的领域，并繁衍出一些新的人格研究范式，包括：社会—认知范式、生物学范式、积极心理学范式，这三者分别源自于行为主义、特质论和人本主义。当代的依恋研究则得益于精神分析的发展。除此以外，进化心理学范式和后现代

表 1　人格心理学 SCI 十年载文量统计

年份	1999	2000	2001	2002	2003	2004	2005	2006	2007	2008	总计
文献量	69	61	89	207	169	185	188	223	268	282	1741
%	3.96	3.50	5.11	11.89	9.70	10.63	10.80	12.82	15.39	16.20	100.00

心理学范式（如叙事心理学）被视为人格心理学的新范式。

从研究内容来看，当代人格心理学在人格结构、人格动力、人格发展等领域都有较大的进展，分别简要介绍如下：

3.1 人格结构

20世纪末，人格领域最令人欢欣鼓舞的进展应该是2个相似的人格分类系统——"大五"结构（"Big Five" Structure）和五因素模型（five-factor model, FFM）的出现。两种模型分别是词汇学取向和理论取向研究成果的结晶，但让人惊叹的是，两种取向的研究殊途同归，最终在人格结构的问题上达成了初步共识。毫无疑问，人格特质的五因素模型（"大五"）是当前人格研究的主导范式，在整个心理学界都是最有影响力的模型之一。正如McCrae等人所言，"五因素模型"就像一颗圣诞树，与综合性、稳定性、遗传性、会聚效度、跨文化普适性和预测效度有关的研究成果正是满缀其间的圣诞礼物。直到近两年，五因素人格模型仍然是活跃于权威期刊《人格与社会心理学杂志》（*Journal of Personality and Social Psychology*）上的重要研究主题，具体内容涉及到一些更为细小的争议较多的问题，例如，大五人格因素的代际差异、年龄差异、大五人格因素之间的相关是由测量误差造成的还是由更高阶因素导致的、使用简式问卷测量大五人格特质的有效性，等等。随着这些研究的开展，人们对于人格结构的认识会越来越深入。

3.2 人格动力

人格动力领域有两大较为突出的理论进展。一是Mischel基于传统特质理论无法准确地预测和解释跨情境的行为变化的主要缺陷，提出了著名的认知—情感人格系统（cognitive-affective personality system, CAPS）理论。该理论试图在人格研究中引入情境因素，强调个体的人格系统与外部环境的动态交互作用，将人格的基本单元（如特质）视为"如果……那么……"的系统，人格便是大量"如果……那么……"的集合，于是出现在什么样的情境中，不同"如果……

那么……"图式便会指引着人们做出不同的行为。这样不仅考虑了人格的稳定结构，而且还兼顾了人格的动力过程。事实上，研究者已经通过实验研究证实了人际知觉中"如果……那么……"图式的有效性，具体来说，在对他人的社会行为和人格倾向进行解释的过程中，人们会考虑人与情境的交互作用，并以"如果……那么……"的方式进行描述。此外，由于认知—情感人格系统理论是一个元理论（meta-theory），即它是不包含具体内容的。近年来，研究者已将此元理论应用于建构特定领域（如戒烟、精神病理学、组织行为）的包含具体内容的模型。

另一重大的理论进展是由Deci和Ryan提出的自我决定论（self-determination theory），使人们对人类动机的普遍性有了新的认识。Deci和Ryan认为，人类具有3种基本的普遍的心理需要，即自主（autonomy）、胜任（competence）和关系（relatedness）的需要。这3种心理需要的满足对于个体的幸福感、心理健康甚至生理健康都是必需的。在自我决定论的理论框架下，研究者开展了大量的实证研究，内容涉及到动机的类型，社会环境对不同类型动机的影响，不同类型动机对一系列结果变量如学习、绩效、认知功能和幸福感的影响，不同的抱负或生活目标与基本心理需要的满足及绩效和幸福感等结果变量的关系，基本心理需要满足的跨文化研究，以及自我决定论在养育、教育、工作和医疗等具体生活领域的应用。

3.3 人格发展

稳定与变化是人格发展的永恒主题。从研究的数量和规模来看，人格特质的发展是当前研究的主要内容，从婴儿气质怎样发展为成人特质，特别是成年期人格特质如何发展，是当今研究得最为广泛的问题。目前研究者进一步关注人格特质稳定性与可变性的深层影响因素，如年龄、生活事件等。除了特质的发展以外，动机和目标的发展及叙事认同的发展也成为人格发展领域新的研究内容。对于人格发展的影响因素，当代的行为遗传学研究、神经科学研究和进化人格心理学研究已为天性的作用提供了越来越多的证据，而教养的作用如家庭环境对人格的影响也积累了丰富的成果。随着文化心理学的兴起，研究者越来越重视社会文化因素对人格发展的影响。将人格置于特定的社会文化背景下进行研究，有助于获得对人格的更为生动、具体、深刻的理解。总之，人格发展的个人与情境交互作用的观点已深得人心。

4 中国社会文化背景下的人格心理学研究

20 世纪 70 年代末，我国大陆地区开始恢复心理学教学和研究，西方人格心理学也得到介绍；90 年代以来，我国心理学者随着反思西方人格心理学的理论和研究方法论问题，开始着手研究中国社会文化背景下的人格问题。

在理论研究层面，体现为学者对中国人的人格结构、动力和影响因素等一些具体问题开展了探索性的研究。人格结构的研究是人格心理学的一个重要范畴，是了解人格的基本特点、类型以及对个体进行有效评估的基础。杨国枢等较早地进行了相关的本土研究。他从中文人格特质形容词入手，得到了 4-5 个独立的人格维度；王登峰将杨国枢收集到的用于描述稳定人格的形容词与从现代汉语词典和刊物中收集到的词汇合并，用因素分析法进行研究，最后确定中国人人格结构的 7 个维度，并编制了中国人人格量表；张建新等人将他们自己编制的《中国人人格测量表（CPAI）》与西方的五因素问卷（NEO-PI）合起来进行联合因素分析，研究结果显示出一个六因素结构。许燕等运用词汇学方法，通过抽取中文动词 2012 个分析，形成 120 个词汇的动词词表，因素分析结果发现，以动词建立的人格结构有控制、施爱、追求成功 3 个因素；控制与施爱属于关系特质，追求成功属于个人特质。此人格动态结构是以行为为基础的模型，称为中国人人格 CLP 模型。所有这些结果都表明，中国人与西方人的人格结构有共同性也有特殊性。此外，还有许燕、张进辅对价值观等人格动力进行了研究；申继亮对人格发展进行了研究等。

在应用研究层面，体现为我国心理学者们立足本国实际，借鉴西方心理学的方法，去解决我国经济社会发展中有关人格心理学的问题。这些问题构成了当前中国人格心理学的研究主题。大致说来，有如下一些主题。

4.1 人格与创造力

创造型人物的新发现、新发明和新成果，对整个社会文明进步有着重要的意义。心理学家关注人格对个体创造活动的影响。王极盛用自评法调查了 28 位学部委员和 127 位一般科学工作者，发现影响创造活动的主要人格因素有：事业心、勤奋、兴趣、责任心、求知欲、进取心、意志等。张景焕对 34 位院士进行访谈，发现创造人才心理特征排在前 4 位的是一般智力强、勤奋努力、内在兴趣和研究技能策略。更有研究者在与国外学者的研究结果对照后提出了两类创造型人格特征的假设：一类称作创造型人格特征的内核，是与创造力关系最为密切且比较稳定的部分，另一类称为创造型人格特征的外壳，它是较多受到文化背景影响的创造型人格特征。

4.2 人格与人员选拔及安置

人格测验对组织中的人员安置和选拔具有重要的意义。从 20 世纪 80 年代初，随着改革开放，外资企业进入中国，为中国带来了先进的管理思想、观念和技术，推动了人格测验在人事管理中的应用，一批学者和专家开始关注和着手人格测验在中国企业人事管理中的应用，并致力于研究开发具有自主知识产权、体现中国特色、适用于中国文化的人格测验。例如，王重鸣和陈民科建立的管理胜任力模型，王登峰对中国党政干部的胜任特征的研究。更有研究者考察了 MBTI 人格测验对陆军指挥院校学员心理选拔的预测性，发现 MBTI-G 人格类型量表对陆军指挥院校学员胜任特征评价有一定预测性，可以作为选拔工作使用；他们还建立了初级军官、航天员和陆军学院学员等军队人员胜任特征模型。这些基于工作特性的人格模型的建构，为人事决策提供更全面和更科学的信息，提高了组织人事决策的效率，可帮助组织更有效地进行人员的聘用、选择、训练、开发等。

4.3 人格与贪腐行为

腐败问题是当今中国的严重社会问题。目前，从宏观层面来说，制度建设在不断完善、监管力度在持续加大；在微观层面，对于腐败主体——个人的腐败心理动因认识不清，对于腐败过程的心理机制了解不够。在西方文化背景下展开的研究表明，存在一种固化和内化的，为达目的不择手段、操纵他人、谋取私利的人格特质，即马基雅弗利主义。中国学者已通过实证方法证实了在中国人身上存在这种人格特质，并且设计了信效度良好的测量工具。在西方文化背景下予以证实的马基雅弗利主义和贪腐行为、经济机会主义是正相关的，在信息不对称的情况下，马基雅弗利主义者倾向于利用手中的优势，使自己的利益最大化。但这一发现，在我国文化背景下迄今为止还没有研究过。在贪腐行为较为严重的当今中国，马基雅弗利主义人格在其中到底起到何种作用？如何起作用？如何抑制？亟待回答。许燕等研究了腐败过程的心理机制——心理绑架，初步建立了心理绑架的现象模型。

4.4 人格与暴力犯罪

通过2010年的《法治蓝皮书》可知，现阶段我国的暴力犯罪现象十分严重，并且发展的趋势也越来越严峻。更令人担忧的是：在越来越多的暴力犯罪中，由于反社会人格而导致严重暴力犯罪的案件数量更是比往年有增无减。反社会人格者由于其易于冲动、不吸取经验教训、不能爱别人和缺乏内化了的社会价值系统或良心的特点，非常容易触犯社会规范和法律。在违法犯罪人群中具有反社会人格的人的数量较多，可达30%以上，远高于一般人群中的比例（1%以下），且屡次犯罪以及罪行特别残酷或情节恶劣的现象非常严重。我国学者通过问卷调查发现了反社会人格者的一些特征，但反社会人格如何导致暴力犯罪，其间的过程是怎样的；是否有其他变量的中介作用或交互作用；如何控制这些因素来减弱或防止它在暴力行为发生过程中起到的作用；如何在暴力犯罪发生之后，更加彻底地去了解暴力犯罪人，有效帮助他们改造，减少累犯的几率等研究刚刚起步，

因此迫切需要开展更深入和更广泛的系统性研究。

4.5 人格与疾病

许多研究表明，一个人的人格特征与疾病之间存在十分密切的关系，人格直接或间接地影响个体的心理和生理健康，具有某些人格特征的人面临患某些特定疾病的风险。国内研究者指出与疾病有关的4组人格因素：易发怒和具有敌意；情绪性压抑；有失望经历；悲观与宿命论的态度。在这4组因素中，敌意倾向与发怒对心脏病的发病有影响，情绪压抑与心脏病和癌症的产生有关。在压抑的情境下不愿表达情感以及对抑郁心情的压抑是癌症产生的最主要原因。

正如 Eysenk 所说："已有足够的证据可以说明在人格与压力以及疾病之间存在着必然的联系，这种联系影响免疫系统的功能……人格与压力因素是癌症产生的重要原因"。然而，该研究领域正处于不断发展和逐步完善的过程中，也面临着许多有待解决和探讨的问题。例如，如何运用心理学知识改进医疗与护理制度，建立合理的保健措施，节省卫生经费和减少社会损失的途径，以及对有关的卫生决策提供建议等。

4.6 和谐社会的健全人格建构问题

当前我们正在致力构建和谐社会，诚如黄希庭所言，构建和谐社会提出了许多亟待解决的人格心理学问题，其中尤其重要的是健全人格的形成问题。健全人格的人能以辨证的态度对待世界、他人、自己、过去、现在和未来、顺境和逆境，是一个自立、自信、自尊、自强、幸福的进取者。黄希庭及其团队开展了一系列研究，对中国人的自我价值感和自立、自信、自强人格等进行探讨；陈建文对健康人格结构、人格功能与人格状态的探讨，为培养和塑造中国人的健康人格提供了理论支持。

5 发展我国人格心理学的建议

中国人格心理学工作刚刚起步，还有许多理论和实证研究要去完成。我国人格心理学应该朝着什么方向发展？我们认为以下两点十分重要。

5.1 现代化背景下的中国人人格研究

英格尔斯在对人的现代化问题作了长达20多年研究之后，如是说："一个国家，只有当它的人民是现代人，它的国民从心理和行为上都转变为现代的人格，它的现代政治、经济和文化管理机构中的工作人员获得了某种与现代化发展相适应的现代性，这样的国家才可真正称之为现代化的国家"。可见，国家现代化首先是人的现代化，而人的现代化，最根本的是人格现代化。

随着改革开放政策的实施和市场经济制度的确立，中国正沿着现代化的道路迅猛发展。作为一场深刻的社会变革，中国的现代化建设一方面带来了市场经济的繁荣发展，另一方面又使得整个社会环境产生了躁动起伏的剧烈变化。与此同时，人的问题也变得越来越突出。虽然传统人格在一定程度上还是具有适应性的，但是我们更应该看到，传统人格确有很多特征是不适应甚至阻碍现代化发展的。因此，在我们这个古老的民族从传统负重之下迈向现代化的今天，研究现代化人格是极具现实意义的课题。心理学家杨国枢等做了一些开创性的工作，初步探讨了现代化人格的内涵、特征及影响因素。但是现代化人格的形成机制是怎样的？该如何塑造？等等，都是摆在我国心理学者面前紧迫的研究任务。同时，前文提及的当前中国人格心理学的一些备受关注的研究主题也是现代化背景下的中国人人格研究的重要范畴。

5.2 研究方法的多元化

人格研究目前使用的方法可大致归纳为实验法、临床法与问卷调查，而这些方法本身都面临一对矛盾，即内部效度与外部效度难以两全的问题。严格控制变量的实验研究保证了内部效度，却很难将复杂的社会文化变量还原为个别实验室变量，导致研究的生态效度低下；相反，临床研究较好地还原了人的生活场景，但由于变量不易控制，研究的内部效度不尽人意，难以精确地刻画出变量间的因果联系。此外，问卷法虽然在一定程度上吸收了前两种方法的优点，但仍存在理论基础与现实情境相脱节、被试回答真实性、量表预测效度、测量目标的含义难以确定等问题，这些问题将直接影响到研究结论的效度。

很显然，上述3种研究方法各有侧重和忽略，各有优势和不足。由于人格现象十分复杂，我们必须多种研究方法（如文献分析法、深度访谈法、问卷调查法、测量法、实验法、叙事研究法、故事谚语分析法，以及遗传学和神经科学的方法等）并用，即采用多元化的研究方法才能对所要研究的人格问题有一个全面而深入的把握。

参考文献

陈建文 . (2008). 人格与社会适应 . 合肥： 安徽教育出版社 .

陈静，苗丹民，罗正学等 . (2007). MBTI 人格测验对陆军指挥院校学员心理选拔的预测性 . *第四军医大学学报*, *28*(16), 1527–1 529

邓晓芒 . (1989). "人格" 辨义 . *江海学刊*, *8*(3), 116–119

郭永玉 .(2005). 人格心理学：人性及其差异的研究 . 北京： 中国社会科学出版社 .

郭永玉，张钊 . (2007). 人格心理学的学科架构初探 . *心理科学进展*, *15*(2), 267–274.

侯玉波，张梦 . (2009). 对中国人自我结构的理论分析 . *心理科学*, *1*, 226–229

黄希庭 . (2007). 构建和谐社会： 呼唤中国化人格与社会心理学研究 . *心理科学进展*, *15*(2), 193–195.

黄希庭 . (2002). *人格心理学* . 杭州：浙江教育出版社, 5.

黄希庭 . (2006). *时间与人格心理学探索* . 北京： 北京师范大学出版社, 321.

蒋奖，许燕 . (2007). 罪犯反社会人格障碍的调查 . *中国特殊教育*, *5*, 80–85

况志华，叶浩生 . (2005). 当代西方心理学的三种新取向及其比较 . *心理学报*, *37*(5), 702–709.

刘邦惠，张庆林，谢光辉 . (1994). *创造型大学生人格特征的研究* . (5), 553–557.

申继亮, 陈勃, 王大华. (1999). 成人期人格的年龄特征: 中美比较研究. *心理科学, 22*(3), 202-205.

汤舒俊. (2011). *厚黑学研究*. 华中师范大学博士学位论文.

王重鸣, 陈民科. (2002). 管理胜任力特征分析: 结构方程模型检验. *心理科学, 25*(5), 513-516.

王登峰. (2002). *心理学研究的中国化: 理论与策略*. 北京: 轻工业出版社, 166-194.

王芳, 刘力, 许燕等. (2011). 社会心理学: 探索人与社会的互动推动社会的和谐与可持续发展. *中国社科院院刊, 26*(6), 640-649.

王极盛, 孙福立. (1984). 科技工作者创造力的研究. *科学学研究, 2*(4), 36-43.

万晓霞. (2009). 近十年 SCI 人格心理学研究文献计量分析. *心理科学进展, 17*(6), 1 291-1 286.

许燕, 王芳. (2008). *社会变迁与大学生价值观的演变. 科学发展: 社会秩序与价值建构——纪念改革开放 30 年论文集》*(上卷). 北京; 北京师范大学出版社, 277-287.

许燕, 王萍萍. (2011). 基于动词分析的人格结构模型探索. 台北. 第七届华人心理学家大会论文集.

杨国枢. (1974). *中国 "人" 的现代化*. 载杨国枢: 中国人的蜕变. 台北: 桂冠图书公司, 389.

杨国枢, 李本华. (1971). 五百五十个中文人格特质形容词之好恶度、 意义度及熟悉度. *国立台湾大学心理学系研究报告, 13*, 36-57.

英格尔斯, 殷陆军译. (1985). *人的现代化*. 成都: 四川人民出版社.

张景焕. (2005). 科学创造人才心理特征及影响因素研究 (博士学位论文). 北京师范大学.

张进辅. (2006). *青少年价值观的特点: 构想与分析*. 北京: 新华出版社.

张兴贵, 郑雪. (2002). 人格心理学研究的新进展与问题. *心理科学, 25*(6), 744-745.

张建新, 周明洁. (2006). 中国人人格结构探索——人格特质六因素假说. *心理科学进展, 14*(4), 574-585.

Chang, L., Connelly, B. S., & Geeza, A. A. (2012). Separating methodfactors and higher order traits of the Big Five: A meta-analytic multitrait multimethod approach. *Journal of Personality and Social Psychology, 102* (2), 408-426.

Credé, M., Harms, P., Niehorster, S. et al. (2012). An evaluation ofthe consequences of using short measures of the BigFive personality traits. *Journal of Personality and Social Psychology, 102*(4), 874-888.

Cheung, F. M., vande, Vijver, F. J. R., Leong, F. T. L. (2011). Toward a new approach to the study of personality in culture. *American Psychologist, 66*(7), 593-603

Deci, E. L., & Ryan, R. M. (2009). Self-determination theory: A consideration of human motivational universals. In P. J. Corr & G. Mathews (Eds.), *The Cambridge handbook of personality psychology* (pp. 441 - 456). New York, NY: Cambridge University Press.

Eysenk, H. J. (1996). Personality and cancer. In C. L. Cooper (Ed.), *Handbook of stress, medicine and health* (193 - 215). New York: CRC Press.

Hergenhahn, B. R. (1980). *An introduction to theories of personality*. Englewood Cliffs, NJ: Prentice-Hall. Hergenhahn, B. R. (1988). *现代人格心理学历史导引* (文一, 郑雪等 编译). 石家庄: 河北人民出版社.

Kammrath, L. K., Mendoza, Denton. R., Mischel, W. (2005). Incor-porating if...then...personality signatures in person perception: Beyond the personsituation dichotomy. *Journalof Personality and Social Psychology, 88*(4), 605-618.

Kluckhohn, C., & Murray, H. A. (1953). Personality formation and its determinants. *In C. Kluckhohn & H. A. Murray (Eds.)*, Personality, its nature, society, and culture (2nd ed., pp. 53 - 67).

Maria, S., Clive, R., & Yves, T. (2007). Machiavellianism and economic opportunism. *The Journal of Applied Social Psychology, 37*(1), 181-190.

McAdams, D. P., & Olson, B. D. (2010). Personality development: Continuityand change over the life course. *Annual Review of Psychology, 61,* 517–542.

McCrae, R. R. (2009). The five–factor model of personality traits: Consensus and controversy. *In P. J. Corr & G. Matthews (Eds.),* The Cambridge handbook of personality psychology (pp.148 – 161). Cambridge: Cambridge University Press.

Shoda, Y., & Mischel, W. (2006). Applying metatheory to achievegeneralisa bility and precision in personality science. *Applied Psychology: An International Review, 55*(3), 439–452.

Mischel, W., & Shoda, Y. (1995). Acognitive affective system theoryof personality: Reconceptualizing situations, dispositions,dynamics, and invariance in personality structure. *Psychological Review, 102,* 246–268.

Smits, I. A. M., Dolan, C. V., Vorst, H. C. M. et al. (2011). Cohort differences in Big Five personality factors over a period of 25years. *Journal of Personality and Social Psychology, 100*(6), 1124–1138.

Soto, C. J., John, O. P., Gosling, S. D. et al. (2011). Age differences inpersonality traits from 10 to 65: Big Five domains andfacets in a large crosssectional sample. *Journal of Personality and Social Psychology, 100*(2), 330–348.

Specht, J., Egloff, B., & Schmukle, S. C. (2011). Stability and change of personality across the life course: The impact of age and major lifeevents on mean level and rank order stability of the Big Five*Journal of Personality and Social Psychology, 101*(4), 862–882.

Xu, Y. (2011). Psychological Kidnap: A New Model of Corruption Process in China. The 9th Biennial Conference of Asian Associationof Social Psychology, Kunming, China.

Yang, K. S., & Bond, M. H. (1990). Exploring implicit personality theorieswith indigenous or imported constructs: The Chinese case. *Journal of Personality and Social Psychology, 58* (6), 1087–1095.

Personality psychology： The study of human nature and its differences

GUO Yongyu, LI Jing, HU Xiaoyong

(School of Psychology, Central China Normal University 430079 Wuhan; Key Laboratory of Adolescent Cyberpsychology and Behavior, Ministry of Education, Wuhan 430079, China)

Abstract　Personality psychology is the discipline that explores the human nature from the integrality perspective in the psychology discipline system. It focuses on the human nature and its differences as its core, and views the man as integrality. Personality psychology is not only important in psychology discipline system but also in science of life, science of social and humanities about human beings. After one hundred years of tortuous development, personality psychology eventually comes into a new developing and prosperous period. In the last decades, the western personality psychology has jumped into a rapid development period. There are considerable advances in research paradigm, research methods and research contents, including personality structure, personality dynamics and personality development. Meanwhile, domestic scholars began to engage in researching the issues of the Chinese personality on the basis of importing and introducing western personality psychology. Currently, the research themes comprise personality and creativity, personality and personnel selection and resettlement, personality and corruption behavior, personality and violent crime, personality and disease, and the shaping of healthy personality in harmonious society with the Chinese cultural background. Considering the important scientific value of personality psychology and its contributions to the progress of society, we propose that the research of Chinese Personality Psychology should be enhanced in many aspects such as the study of Chinese personality in modernization context and the adopting of diversification of research methods in the future.

Keywords　personality psychology; human nature and its differences; modernization personality research

Psychopharmacology, 2013, 225, 353 - 360.

Effects of alcohol on auditory pre-attentive processing of four sound features: Evidence from mismatch negativity

Jinbo He[1,2], Bingbing Li[2], Yongyu Guo[2], Risto Näätänen[3,4,5], Satu Pakarinen[6], Yuejia Luo[1]

([1] State Key Laboratory of Cognitive Neuroscience and Learning, Beijing Normal University, Beijing 100875, China)

([2] Key Laboratory of Adolescent Cyberpsychology and Behavior of Ministry of Education, School of Psychology, Central China Normal University, Wuhan, China)

([3] Department of Psychology, University of Tartu, Tartu, Estonia)

([4] Center of Functionally Integrative Neuroscience (CFIN), University of Aarhus, Aarhus, Denmark)

([5] Institute of Behavioral Sciences, University of Helsinki, Helsinki, Finland)

([6] Finnish Institute of Occupational Health, Topeliuksenkatu 41 a, 00250 Helsinki, Finland)

Abstracts *Rational* Studies have shown that alcohol could impair automatic pre-attentive change detection. However, several earlier studies which investigated alcohol-induced effects on single auditory feature independently were different from each other on the results. Meanwhile, only few auditory features have been investigated yet. Therefore, it is meaningful to investigate effects of alcohol on multiple auditory features in one experiment. Objectives This study investigates the effects of alcohol on automatic pre-attentive change detection of four kinds of auditory features (frequency, intensity, location, and duration) in one experiment. *Methods* This study, using multi-feature oddball paradigm, compares and analyzes mismatch negativity (MMN) elicited by four kinds of auditory features (frequency, intensity, location, and duration), of 12 participants, under alcohol (0.65 g/kg) and non-alcohol condition. *Results* Compared to non-alcohol condition, amplitudes of all the four MMN types significantly declined under alcohol condition, and their amplitude decline ratios decreased as deviant magnitude became larger. Latencies of frequency and intensity MMN were delayed while latencies of location and duration MMN were not delayed significantly. *Conclusion* Alcohol impaired automatic pre-attentive change detection of all the four auditory features (frequency, intensity, location, and duration). However, the alcoholinduced impairment magnitude on automatic pre-attentive detection of the four auditory features was different from each other. According to analysis of amplitude, frequency seems to be affected most among the four auditory features. According to analysis of latency, only frequency and intensity were affected.

Keywords alcohol; event-related potential; mismatch negativity; multi-feature paradigm; pre-attentive change detection

This study was supported by open fund "Effects of Alcohol on Pre-attentive Processing (200926)" of National Key Laboratory of Cognitive Neuroscience and Learning of China, 973 program (2011CB711000), and NSFC (30930031).

1 Introduction

The damage in attention function caused by alcohol may cause a variety of dangerous accidents, such as car accidents (Brewer and Sandow 1980; Näätänen and Summala 1976). Many studies have already reported the negative effects of alcohol on active attention, but there are only a few studies on pre-attentive processing in audition. Individuals not only cope with environmental stimuli by engaging their attentive processes but also through their pre-attentive processing. There processes determine whether an object is worthy of attention (Wei and Yan, 2008). Furthermore, pre-attentive processes are of great evolutionary value because of their capability to cope with large amounts of information automatically with no attentive resources.

Mismatch negativity (MMN, for review, see Näätänen et al. 2007) is a change-specific event-related potential (ERP) component elicited by any discriminable change in auditory stimulation. It is a marker for pre-attentive deviance detection (Grimm et al. 2006; Näätänen et al. 1978,2004). Therefore, MNN provides an objective index to study the effects of alcohol on pre-attentive processing, which have already been studied by several researchers (for review, see Ahveninen et al. 2000; Jääskeläinen et al. 1996b).

Jääskeläinen et al. (1995a) used the traditional oddball paradigm and found that 0.50 g/kg (alcohol/body weight) of alcohol significantly decreases the amplitude of MMN elicited by a change in frequency and significantly delays its peak latency. Thereafter, some other researchers also studied the effects of alcohol on frequency MMN (e.g., 0.55 g/kg, Jääskeläinen et al. 1995b); 0.80 g/kg (Kähkönen et al. 2005); and 0.54 mL/kg (Kenemans et al. 2010), using a similar paradigm, and their results were were similar to those of Jäskeläinen et al. (1995a). In addition,

Jääskeläinen et al. (1995b) found that frequency MMN elicited by smaller frequency changes is significantly impaired by a dose of 0.55g/kg but not 0.35g/kg alcohol, whereas MMN elicited by a larger frequency change is not impaired by both doses. Aside from studies on the effects of alcohol on frequency MMN, Jääskeläinen et al. (1996a) also examined the effects of alcohol on MMN elicited by duration change. At an interstimulus interval (ISI) of 800ms, they found that the amplitude and peak latency of duration MMN in subjects given a dose of 0.55 or 0.85 g/kg alcohol do not change significantly. However, at an ISI of 2,400 ms, the frontal MMN amplitude decreases significantly only under the larger dose, whereas the peak latency does not significantly change under either dose. These results suggest the following: (a) both frequency and duration MMN are impaired by alcohol, (b) the effects of alcohol on the amplitude of MMN differ from those on the peak latency, (c) the effects of alcohol on MMNs of different sound features differ from each other, and (d) alcohol impairment on MMNs is more obvious when the intake exceeds a certain dose. However, further studies should still be conducted to gain a better understanding regarding the effects of alcohol on auditory pre-attentive processing. First, earlier studies only tested the effects of alcohol on frequency and duration MMN. The effects of alcohol on MMNs elicited by other sound features should also be studied. Second, all earlier studies used the traditional single-feature oddball paradigm with pure sinusoidal tones. As a result, the external validity of these studies is affected because different sound features always appear at the same time and elicit mixed effects in real-life situations. Therefore, we believe that using the multi-feature MMN paradigm would allow the comparison of the effects of alcohol on the pre-attentive processing of different sound features under the same dose. In addition, it would also improve external validity by examining the effects of alcohol on MMNs elicited by different sound features in one experiment.

Accordingly, the present study examined the effects of alcohol on the MMNs of four different sound features (frequency, intensity, location, and duration) by applying the multi-feature MMN paradigm promoted by Näätänen et al. (2004) and Pakarinen et al. (2007) to test whether alcohol affects the pre-attentive processing of different sound features differently. In addition, three deviant tones were used to test the effect of deviance magnitude on the effects of alcohol on MMN. In this study, we hypothesized the following: (a) alcohol impairs all MMNs elicited by the four sound features, (b) the effects of alcohol on MMN decrease as the magnitude of deviation of stimuli becomes larger, and (c) the effects of alcohol on MMNs elicited by the four sound features may differ from each other in terms of effect magnitude.

2 Materials and methods

2.1 Subjects

Twelve participants (right-handed, aged 19 to 26 years, one female) with normal hearing took part in this experiment. All of them were older than 18 and thus legitimate to consume alcohol in China. All participants were healthy normal social drinkers (3 to 12 standard alcohol per month during the past year) who had no chronic alcoholism or family history of alcoholism and other mental illnesses. They were asked to abstain from caffeine, alcohol, nicotine, and other psychoactive substances for 24 h prior to the experiment. They signed an informed consent form and committed to obey the above request. All participants were paid for taking part in this study. This study was approved by the institutional ethical committee of National Key Laboratory of Cognitive Neuroscience and Learning of China.

2.2 Treatment

All participants attended two experimental sessions (alcohol and placebo conditions) separated by 2 weeks. The order of alcohol and placebo sessions was counterbalanced across all participants. A single-blind procedure was employed in both sessions. Participants in the alcohol condition received a dose of 0.65 g/kg alcohol (53% v/v white wine) and were provided with little food (e.g., peanuts). They were given 10min to finish the drink. The breath alcohol concentration (BrAC) level was tested every 5 min after they finished drinking until the level was steady. The mean time interval between the finish of drinking and the beginning of the experiment was 18.75 minutes. The BrAC level was also tested immediately after the experiment finished. The average BrAC level was 0.25 ($\pm$ 0.05) mg/L before and 0.22 ($\pm$ 0.06) mg/L after the experiment. Participants in the placebo condition received a dose of 0.02 g/kg alcohol (53% v/v white wine mixed with distilled water) and were provided with little food. The same BrAC level test procedure as in the alcohol condition was performed after drink intake was finished. The time interval between drink intake and the experiment was 15min. The BrAC level was zero both before and after the experiment in the placebo condition. The mean BrAC level in the alcohol condition was significantly higher than that in the placebo condition ($t_{(11)} = 15.70$, $p<0.01$). In addition to BrAC level measurement, the participants were asked to report their intoxication on a five-point Likert-type scale right before and after the experiment in both conditions. The subjective report of intoxication was 2.67 ($\pm$ 0.52) before and 2.50 ($\pm$ 0.37) after the experiment in the alcohol condition, whereas it was 1.00 ($\pm$ 0.00) before and 1.08 ($\pm$ 0.28) after the experiment in the placebo condition. The subjective report of intoxication in the alcohol condition was significantly larger than that in the placebo condition ($t_{(11)} = 6.37$, $p < 0.01$).

2.3 Stimulus design and task

The stimuli and experimental procedure used were similar to Pakarinen et al. (2007). The standard tones were harmonic tones of 75 (± 5) ms composed of three sinusoidal partials (523, 1046, and 1569 Hz), with the second and third partials at 3 and 6 dB lower in intensity, respectively. They were binaurally presented via headphones at an intensity of 70 dB. The deviant tones, with the magnitude of the deviation varying across the three levels, differed from the standards in terms of frequency, intensity, duration, or perceived sound-source location. The frequency deviants differed from the standards by 3/8, 10/8, and 21/8 semitones in the Western musical scale (fundamental frequencies 512, 487, and 450 Hz). The intensity deviants were softer than the standards by steps of 5 dB (65, 60, and 55 dB). The location deviants were tones perceived 10°, 40°, or 90° to the left or right of the participant. The duration deviants were shorter than the standards by steps of 16 ms (59, 43, and 27 ms).

The tones were presented in 5.5 min sequences (6 sequences in total, 628 tones per sequence, each of the 12 deviants was presented 156 times, and all sequences beginning with 4 successive standards), with the presentation order of the sequences randomly varying across the subjects. The stimulus-onset asynchrony was 500 ms. During the period of stimuli presentation (33 min), the subjects were asked to watch a silent video film and ignore the auditory stimuli.

2.4 Data acquisition

The electroencephalogram (EEG) was recorded (0 to 40 Hz, sampling rate of 500 Hz) by the NeuroScan system using a 64-channel Ag/AgCl electrode cap. An electrode was placed on the tip of the nose to serve as a reference channel. Both bipolar horizontal and vertical electrooculograms were recorded between electrodes placed at 1 cm from the canthi of the eyes. The EEG was filtered offline (pass band of 1 Hz to 30 Hz). Eyemovement artifacts were removed using the correlation method. Epochs of 600 ms (including a 100-ms pre-stimulus period served as a baseline for the amplitude measurement) were separately averaged for tones of different

types and levels. Epochs of EEG elicited by the first eight tones of each sequence and exceeding ± 75 μV were omitted from averaging.

The response to the standard tones was subtracted from the response to each type and level of deviant tones to derive MMNs. The most negative peak occurring at 100 to 250 ms after stimulus onset of the Fz channel was selected as the MMN peak amplitude and peak latency. We noticed two peaks during the MMN timewindow of intensity MMN. With reference to an earlier study (Jacobsen and Schröger 2001) and our observation of the wave maps of intensity deviant tones, the first peak was caused by the difference of N1 between intensity deviant tones and standard tones, which reflected the refreshing of neural cells. The second peak was the true MMN caused by the changes of intensity deviant tones compared with the standard tones, which reflected the automatic processing of stimulus change based on sensory memory. Therefore, we selected the amplitude of the second peak as the amplitude of intensity MMN.

2.5 Data analysis

The distribution of dependent measures of our experiment were not significantly different from normal distribution according to Kolmogorov-Smirnova test on the dependent measures($P > 0.05$). To examine the effects of alcohol on the MMNs of different types and levels of deviant tones, a two-way repeated measures analysis of variance (ANOVA) was conducted separately for MMNs of different types, with the factors being condition (placebo and alcohol) and magnitude of deviation (small, medium, and large magnitudes).

To compare the different effects of alcohol on MMNs among the four types, a two-way repeated ANOVA was conducted separately for the MMN amplitude decline ratio, which is calculated as (amplitude$_{placebo\ condition}$ −amplitude$_{alcohol\ condition}$)/amplitudeplacebo condition, and the peak latency delay ratio, calculated as (peak latency$_{alcohol\ condition}$ − peak latency$_{placebo\ condition}$)/ peak latency$_{placebo\ condition}$ for

Table 1. MMN amplitudes and latencies in placebo and alcohol condition (mean ± SD)

Feature	Deviant level	Amplitude (μv)		Latency (ms)	
		Placebo	Alcohol	Placebo	Alcohol
Frequency	Small	−0.96(.32)	−0.57(.39) *	183(19)	202(12) **
	medium	−1.62(.33)	−1.28(.34) *	162(15)	184(14) *
	large	−2.39(.26)	−1.87(.24)	149(17)	160(26)
Intensity	Small	−0.80 (.29)	−0.44(.30) **	184(33)	198(35) **
	medium	−1.35(.33)	−1.07(.34) *	178(22)	188(25)
	large	−1.68(.28)	−1.36(.24)	173(35)	186(36)
Location	Small	−1.02(.54)	−0.76(.45) *	202(26)	200(37)
	medium	−1.45 (.30)	−1.17(.29) *	150(22)	152(15)
	large	−1.52(.36)	−1.31(.29)	152(19)	148(25)
Duration	Small	−0.91(.33)	−0.65(.30) **	184(16)	183(21)
	medium	−2.01(.35)	−1.57(.32) **	174(22)	173(25)
	large	−2.58(.24)	−2.49(.27)	154(17)	156(20)

* $P<0.05$, ** $P<0.01$

MMNs of different types and levels. The factors used were the type (frequency, intensity, location, and duration) and magnitude (small, medium, and large magnitudes) of deviation. Greenhouse–Geisser correction was used in ANOVA when appropriate, and Bonferroni correction tests were carried out as post hoc analysis.

3 Results

3.1 Amplitude

The results of two-way repeated ANOVA showed no significant interaction effects between the condition (placebo and alcohol) and magnitude of deviation (small, medium, and large magnitudes) for frequency, intensity, and location MMNs. The main effect of condition was significant for MMNs of all these three sound features (frequency, $F_{1,11}=8.027$, $p<0.05$; intensity, $F_{1,11}=9.745$, $P<0.05$; location, $F_{1,11}=12.388$, $p<0.01$). The amplitude of MMNs for each magnitude level was larger in the placebo condition than in the alcohol condition for all the three types of MMNs. Meanwhile, the main effect of magnitude of deviation was also significant for MMNs of all these

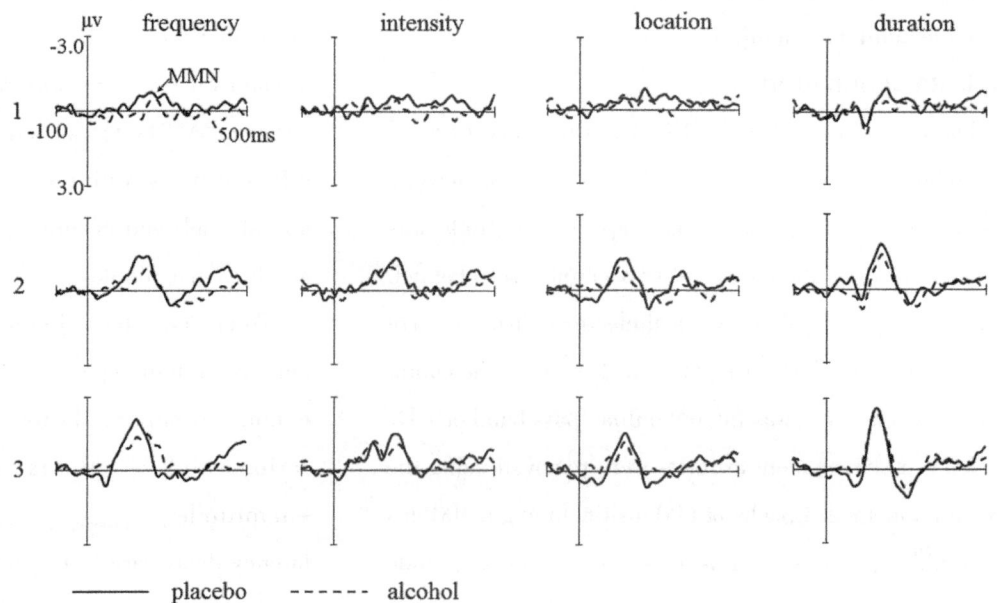

Figure. 1. Grand average MMN for electrode site Fz of different types and levels. From left to right are frequency, intensity, location, and duration MMMs. *Rows 1, 2, and 3* show MMNs of small, medium, and large magnitudes of deviation, respectively.

three types (frequency, $F_{1,11}=26.066$, $p<0.01$; intensity, $F_{1,11}=34.180$, $p<0.01$; location, $F_{1,11}=18.704$, $p<0.01$). The amplitude of MMNs became larger as the magnitude of deviation became larger for all the three types of MMNs. However, the interaction effect between condition and magnitude of deviation was significant for duration MMN ($F_{1,11}=3.586$, $p<0.05$). Analysis of the simple effect of condition showed that duration MMNs elicited by deviant stimuli of small and medium magnitude levels were significantly larger in the placebo condition than in the alcohol condition (small magnitude, $F_{1,11}=14.19$, $p<0.01$; medium magnitude, $F_{1,11}=10.11$, $p<0.01$). However, the amplitude of duration MMN elicited by large magnitude level was not significantly changed in the alcohol condition compared with the placebo condition ($p>0.1$; See Fig1). The absolute value of amplitude of MMNs was illustrated in Table 1(significant level in it was based on paired t test).

3.2 Peak latency

The results of two-way repeated ANOVA showed no significant interaction effects between condition (placebo and alcohol) and magnitude of deviation (small, medium, and large magnitudes) for all the four types of MMNs ($p>0.1$). The main effect of condition was significant for both frequency and intensity MMNs (frequency, $F_{1,11}=20.203$, $p<0.01$; intensity, $F_{1,11}=8.488$, $p<0.05$). The peak latency of MMNs was delayed in the alcohol condition compared with the placebo condition for both frequency and intensity MMNs. Meanwhile, the main effect of magnitude of deviation was also significant for MMNs of both sound features (frequency, $F_{1,11}=20.437$, $p<0.01$; intensity, $F_{1,11}=16.462$, $p<0.01$). The peak latency of MMNs of both types became shorter as the magnitude of deviation became larger. The main effect of condition was not significant for both location and duration MMNs ($p>0.1$), whereas the main effect of the magnitude of deviation was significant (location, $F_{1,11}=15.903$, $p<0.01$; duration, $F_{1,11}=20.396$, $p<0.01$). The peak latency of MMNs of both

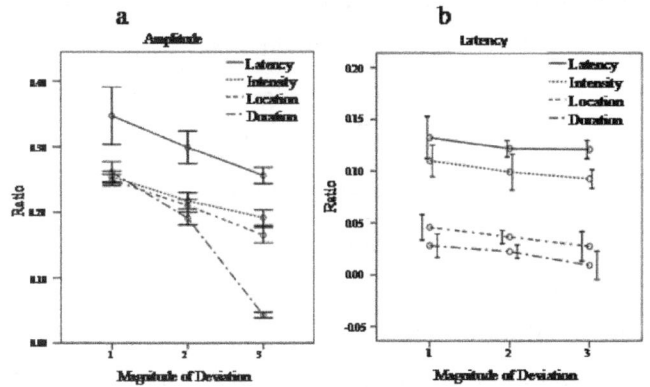

Figure. 2. (a) Amplitude decline ratios of MMNs of different types and levels. (b) Peak latency delay ratios of MMNs of different types and levels. *Numbers 1, 2,* and *3* in the x–axis each represent small, medium, and large magnitudes of deviation, respectively.

types also became shorter as the magnitude of deviation became larger（See Fig1）. The absolute value of latency of MMNs was illustrated in table 1(significant level in it was based on paired t test).

3.3 Amplitude decline ratio

The interaction effect of type (frequency, intensity, location, and duration) and magnitude of deviation (small, medium, and large magnitudes) was significant ($F_{6,66}=7.49$, $p<0.01$). Based on simple effect analysis, the comparison of the different types showed that the amplitude decline ratio was significantly different from one another for MMNs of the four types under medium and large magnitudes of deviation (medium, $F_{3,33}=4.89$, $p<0.01$; large, $F_{3,33}=19.96$, $p<0.01$). The amplitude decline ratio of frequency MMN was larger than all the other three types of MMNs under both medium and large magnitudes of deviation. No differences were observed among the amplitude decline ratios of intensity, location, and duration MMNs under medium magnitude of deviation. However, the amplitude decline ratios of intensity and location MMNs were larger than duration MMNs under large magnitude of deviation, whereas no differences were observed between the amplitude decline ratios of MMNs elicited by both sound features.

The comparison within types showed that the effects of magnitude of deviation were significant for intensity, location, and duration MMNs (intensity, $F_{2,22}=7.26$, $p<0.01$; location, $F_{2,22}=4.79$, $p<0.05$; duration, $F_{2,22}=19.07$, $p<0.01$), whereas the effects of magnitude of deviation were approaching significance ($p=0.068$) for the amplitude decline ratios of frequency MMN. The amplitude decline ratio decreased as the magnitude of deviation became larger (See Fig. 2).

3.4 Peak latency delay ratio

The results of two-way repeated ANOVA showed that neither the interaction effects between type (frequency, intensity, location, and duration) and magnitude of deviation (small, medium, and large magnitudes) nor the main effects of magnitude of deviation were significant for peak latency delay ratio ($p>0.1$). However, the main effect of type was significant ($F_{3,33}=4.73$, $p<0.01$). Post hoc comparison based on Bonferroni correction showed that the peak latency decay ratios of frequency and intensity MMNs were larger than those of location and duration MMNs ($p<0.05$). Neither the differences between the peak latency delay ratios of frequency and intensity MMNs nor the differences between the peak latency delay ratios of location and duration MMNs were significant ($p>0.1$；See Fig. 2）.

4 Discussion

The results from the analysis of amplitude and peak latency of MMNs indicated that alcohol elicited different effects on the amplitude and peak latency of MMNs. The main effect of condition (alcohol and placebo) was significant for MMNs of all the four sound features. This result suggested that the amplitude of MMNs was larger in placebo condition than in the alcohol condition for MMNs of all types and levels, except for the duration MMNs elicited by tones of large magnitude of deviant levels. Moreover, the ability of pre-attentive change to detect changes in sound features (frequency, intensity, location, and duration) was significantly impaired by a

dose of 0.65 g/kg alcohol. This result is consistent with our first hypothesis. In addition, the effects of alcohol on frequency and duration MMNs in this study are similar to those reported in earlier studies (Jääskeläinen et al. 1996a, 1995a,b; Kähkönen et al. 2005; Kenemans et al. 2010). This study confirmed the results of earlier studies. In addition, it also extended the results by proving that alcohol not only impairs the pre-attentive detection of frequency and duration auditory changes but also impairs that of intensity and location auditory changes.

However, the impairment of peak latency was not as steady as that of amplitude. The main effect of condition (alcohol and placebo) was significant for frequency and intensity MMNs but not for location and duration MMNs. Earlier studies also found that alcohol elicits different effects on amplitude and peak latency. Jääskeläinen et al (1996a) found that the amplitude of duration MMN was significantly decreased after alcohol drinking, whereas the peak latency of duration MMN did not significantly change. These results indicated that the amplitude of MMN was more sensitive to alcohol than peak latency.

Amplitude decline ratio decreased as the magnitude of deviation became larger. This result suggested that alcohol impairment of pre-attentive change detection decreased as the magnitude of deviation became larger. Alcohol did not affect simple change detection but only the more complicated and subtle sensory perceptive processing of stimuli. Jääskeläinen et al (1995b) also found that the frequency MMN of more widely deviant stimuli did not change significantly, whereas the MMN of less deviant stimuli decreased significantly after alcohol ingestion. They attributed this result to the higher threshold for the pre-attentive detection of acoustic deviations after alcohol ingestion, which meant that the pre-attentive change detection of subtle changes was impaired. Earlier studies (Jääskeläinen et al. 1996a, b) also found that N1 was not impaired by alcohol, whereas MMNs were significantly impaired. Kenemans et al. (2010) found that the amplitude of visual SFD80 (spatial frequency−dependent difference at 80 ms; Kenemans et al. 2000) did not change after alcohol ingestion, whereas visual mismatch negativity (for reviews, see Czigler 2007; Pazo−Alvarez et al. 2003) decreased significantly. N1 and SFD80 are ERP components that reflect sensory processing, whereas MMN reflects memory−dependent processing.

The main effect of type (frequency, intensity, location, and duration) for the amplitude decline ratio was significant under medium and large magnitudes of deviation level. The decline ratios of frequency MMNs were significantly larger than all the other three MMN types under both levels. The decline ratios of intensity and location MMNs were larger than duration MMNs under both small and medium magnitudes, whereas no significant differences were found between them under both levels. This is somehow unclear. These results indicated that alcohol impaired the frequency MMN the most, intensity and location MMN less, and duration MMN the least of all the four MMN types. The amplitude decline ratio of duration MMNs elicited by tones of large magnitude of deviation level was significantly less than that of the small and medium levels and that of the other MMN types under the same deviant level. No differences were found between the amplitude of duration MMNs elicited by the large magnitude of deviation level in the placebo and alcohol conditions (Fig.1), which indicated that alcohol elicited no effect on the pre−attentive change detection of temporal features of tones, when the magnitude of deviation was excessively large compared with the standard tones that pre−attentive processing was not deep enough.

The effects of feature and magnitude of deviation level on peak latency delay ratio were different from those on amplitude decline ratio. No significant interaction effects were found between the type (frequency, intensity, location, and duration) and magnitude (small, medium, and large magnitudes) of deviation. The main effect of magnitude of deviance level was not significant for any of the four MMN types. Different from the amplitude decline ratio, the peak latency delay ratio did not decrease as the magnitude of deviance level became larger. This result implied that peak latency was not as sensitive to the effects of alcohol as amplitude. In addition, the differences in peak latency delay ratio between the different sound features were also different from those in amplitude decline ratio. Similar peak latency delay ratios were found between frequency and intensity MMNs as well as between location and duration MMNs under the same magnitude of deviation level. However, the peak latency delay ratios of frequency and intensity MMNs were larger than those of location and duration MMNs. Further studies should be conducted to determine why the effects of alcohol on amplitude and peak latency were different. However, earlier studies reported that peak latency is not as steady as amplitude to be used as an index of the effects of alcohol on MMN. Therefore, further studies are necessary to verify the results based on peak latency.

The results showed that effects of alcohol on frequency and intensity MMNs as well as on location and duration MMNs were similar. Moreover, the effects of alcohol on the former two MMN types were different from those of the latter two MMN types. This result indicated that the effects of alcohol on the pre−attentive change detection of spectrum auditory information (frequency and intensity) were different from those of temporal information (duration). This result may be attributed to the fact that pre−attentive change detection of spectrum information is different from that of temporal information. Whether acoustic features are processed independently or pre−attentively integrated has been under debate (e.g. Giard et al. 1995; Schairer et al. 2001; Winkler et al. 1996). Recent studies have supported the independent view. Grimm et al. (2006) found a right hemisphere preponderance for frequency MMN but not for duration MMN. Molholm et al. (2005) found that anatomically distinct networks of auditory cortices are activated by different acoustic features (frequency and duration) using functional magnetic resonance imaging technology. Changes in duration activated both the left and right frontal cortices, whereas changes in frequency only activated the right frontal cortex. Frequency and intensity features reflected spectrum information of tones, whereas location (the different initial time between two ears in this study) and duration features reflected temporal information of tones. Therefore, the results of this study not only supported the view that the pre-attentive processing of acoustic features occurred independently, but they also suggested that pre-attentive change detection of spectrum information was more sensitive to alcohol than temporal information.

This study used the multi-feature MMN paradigm to examine the effects of alcohol on MMNs of four sound features (frequency, intensity, location, and duration). In conclusion, this study showed that: (a) The MMNs of all the four sound features were impaired by alcohol, and the impairment decreased as the magnitude of deviation level became larger. In addition, the amplitude of MMN was more sensitive to alcohol than peak latency; (b) The trends of the effects of alcohol on the MMNs of different sound features differed from each other. In terms of the amplitude decline ratio, frequency MMN was impaired the most, intensity and location MMN were less impaired, and duration MMN was impaired the least. Meanwhile, the peak latency delay ratios of frequency and intensity MMNs were larger than those of location and duration MMNs, whereas no differences were observed between the latter two MMNs.

References

Ahveninen, J., Escera, C., Polo, M. D., Grau, C., & Jääskeläinen, I. P. (2000). Acute and chronic effects of alcohol on preattentive auditory processing as reflected by mismatch negativity. *Audiol Neuro-Otol, 5*, 303 - 311.

Brewer, N., & Sandow, B. (1980). Alcohol effects on driver performance under conditions of divided attention. *Ergonomics, 23*, 185 - 190.

Czigler, I. (2007). Visual mismatch negativity: Violation of nonattended environmental regularities. *Journal of Psychophysiology, 21*, 224 - 230.

Giard, M., Lavikahen, J., Reinikainen, K., Perrin, F., Bertrand, O., Pernier, J., & N??t?nen, R. (1995). Separate representation of stimulus frequency, intensity, and duration in auditory sensory memory: An event-related potential and dipole-model analysis. *Journal of Cognitive Neuroscience, 7*, 133 - 143.

Grimm, S., Roeber, U., Trujillo-Barreto, N. J., & Schroger, E. (2006). Mechanisms for detecting auditory temporal and spectral deviations operate over similar time windows but are divided differently between the two hemispheres. *Neuroimage, 32*, 275 - 282.

Jääskeläinen, I., Pekkonen, E., Hirvonen, J., Sillanaukee, P., & Näätänen, R. (1996a). Mismatch negativity subcomponents and ethyl alcohol. *Biological psychology, 43*, 13 - 25.

Jääskeläinen, I. P., Lehtokoski, A., Alho, K., Kujala, T., Pekkonen, E., Sinclair, J. D., Näätänen, R., & Sillanaukee, P. (1995a). Low dose of ethanol suppresses mismatch negativity of auditory event-related potentials. *Alcoholism: Clinical and Experimental Research ,19*, 607 - 610.

Jääskeläinen, I. P., Näätänen, R., & Sillanaukee, P. (1996b). Effect of acute ethanol on auditory and visual event-related potentials: A review and reinterpretation. *Biological psychiatry, 40*, 284 - 291.

Jääskeläinen, I. P., Pekkonen, E., Alho, K., Sinclair, J. D., Sillanaukee, P., & Näätänen, R. (1995b). Dose-related effect of alcohol on mismatch negativity and reaction time performance. *Alcohol, 12*, 491–495.

Jacobsen, T., & Schr?ger, E. (2001) Is there pre–attentive memory–based comparison of pitch?. *Psychophysiology, 38*, 723–727.

Kähkönen, S., Rossi, E. M., & Yamashita, H. (2005). Alcohol impairs auditory processing of frequency changes and novel sounds: a combined MEG and EEG study. *Psychopharmacology, 177*, 366–372.

Kenemans, J., Baas, J., Mangun, G., Lijffijt, M., & Verbaten, M. (2000). On the processing of spatial frequencies as revealed by evoked–potential source modeling. *Clinical Neurophysiology, 111*, 1113–1123.

Kenemans, J. L., Hebly, W., vanden, Heuvel, E. H. M., Grent, T., & Jong, T. (2010). Moderate alcohol disrupts a mechanism for detection of rare events in human visual cortex. *J Psychopharmacol, 24*, 839–845.

Molholm, S., Martinez, A., Ritter, W., Javitt, D. C.,& Foxe, J. J. (2005). The neural circuitry of pre–attentive auditory change–detection: an fMRI study of pitch and duration mismatch negativity generators. *Cerebral Cortex, 15*, 545–551.

Näätänen, R., Gaillard, A. W. K., & M?ntysalo, S. (1978). Early selective–attention effect on evoked potential reinterpreted. *Acta psychologica, 42*, 313–329

Näätänen, R., Paavilainen, P., Rinne, T., & Alho, K. (2007). The mismatch negativity (MMN) in basic research of central auditory processing: a review. *Clinical Neurophysiology, 118*, 2544–2590.

Näätänen, R., Pakarinen, S., Rinne, T., & Takegata, R. (2004). The mismatch negativity (MMN): towards the optimal paradigm. *Clinical Neurophysiology, 115*, 140–144.

Näätänen, R., & Summala, H. (1976). Road–user behavior and traffic accidents. North–Holland Pub. Co.(Amsterdam and New York)

Pakarinen, S., Takegata, R., Rinne, T., Huotilainen, M., & Näätänen, R. (2007). Measurement of extensive auditory discrimination profiles using the mismatch negativity (MMN) of the auditory event–related potential (ERP). *Clinical Neurophysiology, 118*, 177–185.

Pazo–Alvarez, P., Cadaveira, F., & Amenedo, E. (2003). MMN in the visual modality: a review. *Biological psychology, 63*, 199–236.

Schairer, K. S., Gould, H. J., & Pousson, M. A. (2001). Source generators of mismatch negativity to multiple deviant stimulus types. *Brain topography, 14*, 117–130.

Wei, J. H., & Yan, K. L. (2008). *Fundametals of Cognitive Neuroscience*. People's Education Press, Beijing.

Winkler, I., Karmos, G., & Näätänen, R. (1996). Adaptive modeling of the unattended acoustic environment reflected in the mismatch negativity event–related potential. *Brain Research, 742*, 239–252.

Personality and Social Psychology Bulletin, 2010, 36(5), 583 - 597.

Cultural differences in the representativeness heuristic: Expecting a correspondence in magnitude between cause and effect

Roy R. Spina[1], Lijun Ji[1], Tieyuan Guo[1], Zhiyong Zhang[2], Ye Li[3], Leandre Fabrigar[1]

([1] Queen's University, Kingston, Ontario, Canada)

([2] Beijing University, Beijing, China)

([3] Huazhong Normal University, Wuhan, China)

Abstract Based on previous research on cultural differences in analytic and holistic reasoning, it was hypothesized in these studies that when explaining events, North Americans would be more likely than East Asians to expect causes to correspond in magnitude with those events (i.e., big events stem from big causes and small events stem from small causes). In a series of studies, Canadian and Chinese participants judged the likelihood that high- or low-magnitude events were caused by high- or low-magnitude causes. Overall, Canadians expected events and their causes to correspond in magnitude to a greater degree than did Chinese. Also, Canadians primed to reason holistically expected less cause - effect magnitude correspondence than did those primed to reason analytically.

Keywords heuristics; representativeness; holism; attribution; culture and cognition

1 Introduction

In the 1960s, mathematician and meteorologist Edward Lorenz created a computer simulation of hydrodynamic flow (Lorenz, 1963). Allegedly, while using the program to model weather patterns, Lorenz entered the value .506 rather than the actual value .506127. He was surprised to find that the outcomes stemming from the two ostensibly similar initial values varied substantially. Upon publishing the findings, one meteorologist remarked that if the theory were correct, one flap of a seagull's wings could change global weather patterns forever. Over time, through some

The authors received the following financial support for the research and/or authorship of this article: research supported by grants from Social Science and Humanities Research Council of Canada (SSHRCC accounts 410-2003-1043 and 410-2009-0904) and the Principal's Development Fund from Queen's University to the second author.

Corresponding Authors:

(English) Roy R. Spina, Division of Psychology, Birmingham City University, Birmingham, UK, B42 2SU.Email: roy.spina@bcu.ac.uk

(Chinese) Zhiyong Zhang, Department of Psychology, Beijing University, Beijing, China, 100 871; or Ye Li, Department of Psychology, Huazhong Normal University, Wuhan, Hubei Province, China, 430 079. Email: zzhang@pku.edu.cn or liyehlong@yahoo.com.cn

form of cultural metamorphosis, the ungainly seagull transformed into a delicate butterfly and the term *butterfly effect* was born. Regardless of whether this amusing anecdote is true, Lorenz found that seemingly negligible variations in initial conditions led to dramatic divergences in outcome.

The butterfly effect has surprised many of us because it violates our expectations of cause‐effect magnitude correspondence. Why do we expect big causes to lead to big effects and small causes to lead to small effects? Is it simply because we seldom observe violations of this association and thus have developed a heuristic that serves us well most of the time? The present research indicates that the explanation is far more interesting and complex.

1.1 The representativeness heuristic

categorization. Kahneman and Tversky (1973) demonstrated that when categorizing targets into groups, people relied mostly on the degree of similarity between the target and a prototypical member of each group. For example, when judging a fictional student's academic discipline based on his personality, people relied on the personalities of typical students from each discipline. Furthermore, the authors demonstrated that this strong tendency to rely on the degree of similarity caused errors in judgment because people ignored the percentages of students enrolled in each discipline, the base rates. Kahneman and Tversky (1972) referred to this type of judgment as the *representativeness heuristic.*

*causal judgment*s. Most of the research conducted on the representativeness heuristic has focused on judgments people make in categorization contexts (Gilovich & Savitsky, 2002). However, the representativeness heuristic is not limited to categorization. People also use this heuristic when making causal judgments, such as when searching for causes that are similar to an effect. Relatively little research has been conducted on the representativeness heuristic in the domain of causal

judgments. Nonetheless, across a variety of domains including medicine, pseudoscientific systems, and psychoanalysis, people employ the representativeness heuristic when making causal judgments by relying on similarities between causes and effects (Gilovich & Savitsky, 2002).

Einhorn and Hogarth (1986) divided such similarities into two categories: physical resemblance and congruity of strength. Beliefs that causes physically resemble effects are prevalent in traditional medicine, for example, when cures that resemble diseases are sought (Gilovich & Savitsky, 2002). The second category, the tendency to search for causes that resemble effects in congruity of strength, is especially interesting because strength has been identified as one of the fundamental meaningful dimensions people use when judging entities (Osgood, 1957, referred to as potency and included adjective pairs such as *large−small and strong−wea*k). Nisbett and Ross (1980) also proposed that people may seek causes that correspond in magnitude with events they are trying to explain. However, few studies have empirically tested people's expectations of a correspondence in magnitude between causes and effects.

1.2 Empirical research on expectations of cause−effect magnitude correspondence

In one of the first studies that indirectly provided supporting evidence that people expect cause‐effect magnitude correspondence, Shultz and Ravinsky (1977) demonstrated that French Canadian schoolchildren typically chose causes that were similar to effects. For example, in one scenario, they attributed a loud noise to a heavy rather than a delicate lever. However, the other scenarios used would fall under Einhorn and Hogarth's (1986) physical resemblance category, such as attributing physical retaliation to physical rather than verbal aggression.

McCauley and Jacques (1979) also provided indirect supporting evidence of an expectation of cause‐effect magnitude correspondence. In their study, American participants read about a successful or unsuccessful assassination. Participants then estimated the probabilities that the assassinator was acting alone or acting as a member of a group. The authors found that participants attributed the more consequential effect, the successful assassination, to the group and the less consequential effect, the unsuccessful assassination, to the individual. However, this study does not provide strong evidence supporting an expectation of cause – effect magnitude correspondence because participants could be attributing the more consequential effect to a conjunction of causes, a group, more than a single cause, a lone gunman.

McClure, Lalljee, and Jaspars (1991) examined whether people explained extreme and moderate effects by using a conjunction of causes or a single cause. For example, British participants read an extreme crime involving multiple murders and mutilation of bodies, and a moderate crime involving hitting people with a bottle at a football match. Most participants explained the extreme crime by generating single-cause explanations that tended to correspond in magnitude with the effects.

A recent set of studies by Ebel-Lam, Fabrigar, MacDonald, and Jones (2008) provided more direct support for an expectation of cause – effect magnitude correspondence. Canadian participants read a scenario describing either a high- or moderate-magnitude effect. For example, participants read either that a plane crashed killing everybody onboard (high magnitude) or that with difficulty the pilot successfully landed the plane (moderate magnitude). Participants estimated the likelihood that a number of high- and moderate-magnitude causes had led to the effect. Participants attributed high-magnitude effects to high-magnitude causes and low-magnitude effects to low-magnitude causes.

Either directly or indirectly, the aforementioned studies provided consistent evidence that people (at least,

Americans, British, and Canadians) typically expect a correspondence in magnitude between an effect and its cause. Why is this so and is it true across cultures? Research from the cultural psychology literature may provide some clues.

1.3 Cultural differences in analytic and holistic reasoning

Contrary to a Washington newspaper article mocking Walter Reed's suggestion that yellow fever with all of its devastating effects was caused by a tiny mosquito (Nisbett & Ross, 1980), Asian folk wisdom states, "One tiny insect may be enough to destroy a nation." Research indicates that compared with North Americans (including Americans and Canadians), East Asians (including Chinese, Japanese, and Koreans) differ in their reasoning. In particular, East Asians tend to reason holistically, whereas North Americans tend to reason analytically (Nisbett, Peng, Choi, & Norenzayan, 2001).

Across a variety of domains, East Asians attend to situational factors or contexts more than North Americans do, and North Americans attend to focal people or objects more than East Asians do (Miller, 1984; Morris & Peng, 1994). For example, one set of studies found that when describing an underwater scene, the first element Americans typically mentioned was a focal element, such as a large fish in the center of the picture. In contrast, the first element Japanese typically mentioned was a contextual element, such as seaweed (Masuda & Nisbett, 2001). In addition, Japanese focused on the entire scene as a unit, and their ability to recall a focal object from the scene was impaired when the background was altered. Alternatively, Americans focused on the focal objects independent of the background, and changing the background had little or no effect on their ability to recall a focal object.

Focusing relatively different amounts of attention on focal objects and contexts has implications for other aspects of cognition, such as the ability to detect covariation among stimuli. People who focus on focal objects more than contexts should be less likely to detect covariation between elements in a scene compared with people who attend to both focal objects and contexts. Ji, Peng, and Nisbett (2000) asked participants to estimate the degree of covariation between images on a computer screen and this is exactly what they found. That is, compared with American estimates, Chinese estimates were better calibrated with actual levels of covariation.

If Easterners attend more to contextual elements and notice a greater degree of covariation between elements than do North Americans, then they should explain effects differently. This pattern has been found when people make attributions. When people observe a person's behavior in a social situation, a number of elements are present other than the person. That is, other people are often involved and there is a surrounding context or situation in which the person is acting. If an observer focuses relatively more attention on the situation, then he or she should be more likely to attribute causality to that situation. Alternatively,if an observer focuses relatively more attention on the focal person, then he or she should attribute causality to that person. Consistent with this reasoning, past research has shown that Asians tend to make more situational attributions whereas Americans tend to make more dispositional attributions (e.g., Miller, 1984, with Indian participants; Morris & Peng, 1994, with Chinese participants).

Choi, Dalal, Kim-Prieto, and Park (2003) further explicated the cultural differences in reasoning by providing evidence that Easterners may have more complex causal theories and therefore consider more causal factors in their attributions than do Westerners, who may have relatively simple causal theories. For example, in Choi et al., American and Korean participants read that a graduate student killed his or her advisor, along with 97 pieces of information related to the student or advisor. When asked to select items that were pertinent to establishing a motive for the murder, Koreans considered a greater number of items as relevant than did Americans. Additionally, Choi et al. developed and included a 10-item measure of holistic tendency. The measure included statements such as, "Any phenomenon has numerous numbers of causes, although some of the causes are not known." They found that Koreans were more holistic than were Americans, and within each culture, the higher a participant's holistic tendency, the greater the number of items considered relevant to establishing a motive.

1.4 The present research

To summarize the cross-cultural literature, East Asians reason holistically whereas North Americans reason analytically. It is important to note that holism does not appear to be a simple uniform construct determined by a single cognitive mechanism. For example, researchers have focused on at least four factors under the umbrella term holism: causality (Choi et al., 2003), attitude toward contradiction (e.g., Peng & Nisbett, 1999), perception of change (Ji, Nisbett, & Su, 2001), and locus of attention (e.g., Masuda & Nisbett, 2001; see also Nisbett et al., 2001). The present research focused on the causality factor. With respect to causality, East Asians reason holistically by focusing on many causes, whereas North Americans reason analytically by focusing on relatively fewer causes (Choi et al., 2003).

If an observer tends to focus on one or a few causes only when explaining an effect, then the observer should expect a greater correspondence in magnitude between cause and effect. For example, imagine judging how likely it is that each of two buildings sold for a high price. The two buildings look very similar except that one is larger than the other. If all else were equal, one would expect the larger building to sell for more because size is one factor that determines the value of a building. However, there are a number of other factors that also determine the value of a building, such as location. If an observer tends to focus on only one cause, such as the size of the building in this case, when trying to understand a high-magnitude effect, the high selling price, then that person should expect a high-magnitude cause, the large building, to be far more likely than a low-magnitude cause, the small building.

Alternatively, if an observer tends to focus on many factors when explaining an effect, then the observer should expect a lesser correspondence in magnitude between cause and effect. Considering the previous example, if an observer tends to focus on numerous factors, including size, location of the building, and other factors, when trying to understand a high-magnitude effect, such as a high selling price, then a high-magnitude cause, the large building, is not as necessary. Although this multiple-cause observer may also reason that the large building would likely sell for more than the small building, this effect would tend to be less extreme when compared with the observer with a single-cause focus. The lesser correspondence between the magnitudes of cause and effect makes sense because one of the other factors, such as location, could be working against the larger building and in favor of the smaller building when it comes to selling price.

Therefore, based on previous research on cultural differences in analytic and holistic reasoning, specifically regarding differences in perceptions of causal complexity, we hypothesized that expectations of a correspondence in magnitude between effects and their causes would be stronger among North Americans than among East Asians. Specifically, we hypothesized that when explaining an effect, North Americans would tend to look for causes that correspond in magnitude with the effect. In contrast, East Asians would be less likely than North Americans to expect such a correspondence in magnitude between effects and their causes. Lastly, we predicted that the cultural differences in the tendency to look for causes that correspond in magnitude with effects would be explained by differences in analytic and holistic reasoning. We

conducted a series of studies to test these hypotheses.

2 Study 1

The purpose of Study 1 was to test the hypothesis that Canadians would expect a greater correspondence in magnitude between effects and their causes than would Chinese. Participants read hypothetical scenarios with consequences of high or low magnitude and indicated how likely the effect was due to causes that were high or low in magnitude. We used two different versions of scenarios with two different samples, one in Study 1a and the other in Study 1b.

2.1 Study 1a

2.1.1 Method

Participants. Fifty-nine European-Canadians (45 women) were recruited from Queen's University, and 60 Chinese (30 women) were recruited from Beijing University. In all of the studies in this article, Canadian participants were Caucasians of European descent and Chinese were Chinese nationals, mostly of Han descent. Canadian participants received course credit or $5 for their participation, and Chinese participants received a small gift.

Materials and procedure. Participants read a questionnaire describing a disease outbreak that either killed some people (high-magnitude effect) or hospitalized them (low-magnitude effect). The effect was followed by two potential causes: a highly infectious strain of bacteria (high-magnitude cause) or a standard strain of bacteria (low-magnitude cause; see the appendix). Participants rated the likelihood that each of the two causes had led to the effect on a 9-point scale (1=*not likely at all*, 9 =*extremely likely*). In summary, Study 1a had a 2 (culture: Canadians vs. Chinese) × 2 (effect magnitude: high vs. low) × 2 (cause magnitude: high vs. low) design. The cause magnitude factor varied within participants, and the other factors varied between participants.

Cause and effect magnitude. To operationalize the magnitude of effects, we selected single events, such as a disease outbreak, and varied the severity of the consequences associated with that event, for example, few or many deaths. We then selected causes that people would intuitively associate with that event, such as a virus. To operationalize the magnitude of these causes, we chose causes such that if all else were equal, a more extreme version, such as a treatment–resistant strain of the virus, would be associated with the more extreme effect to a greater extent than would a less extreme version, such as a standard strain of the virus. This same procedure for generating study materials was followed in all of the studies reported in this article.

The study materials were generated by two Canadian and two Chinese researchers to ensure they were familiar and realistic to both cultures. Pilot testing indicated that Canadians and Chinese perceived the magnitudes and independent likelihoods of effects and causes to be equivalent. This same procedure for generating and pilot testing study materials was followed in all of the studies reported in this article.

Translation. All materials were first developed in English and then translated into Chinese. The Chinese and English versions were then compared by two Chinese researchers who have lived in North America for at least 4 years. Additionally, a back–translation procedure was used to check consistency of meaning, and finally the translations were checked by at least three Chinese researchers in China to ensure they were free of error and that they sounded natural. The same procedure for translating study materials was followed in all of the studies reported in this article.

2.1.2 Results and discussion

Preliminary analyses indicated no significant gender effects, $Fs > 1$, as was true for all studies reported in this article. Thus, gender will not be discussed further.

A2 (culture) × 2 (effect magnitude) × 2 (cause magnitude) mixed–model ANOVA revealed a significant main effect of effect magnitude, $F(1, 115)=13.55$, $p<.001$, such that participants in the high–magnitude effect condition gave higher likelihood ratings for the causes ($M=5.89$, $SD=0.91$) in comparison with those in the low–magnitude effect condition ($M=5.27$, $SD=0.92$). This pattern indicated a stronger reaction to the high–magnitude effect compared with the low–magnitude effect. The Effect Magnitude × Cause Magnitude interaction was significant, $F(1, 115)=160.18$, $p<0.001$, $\eta_p^2=0.58$. Over–

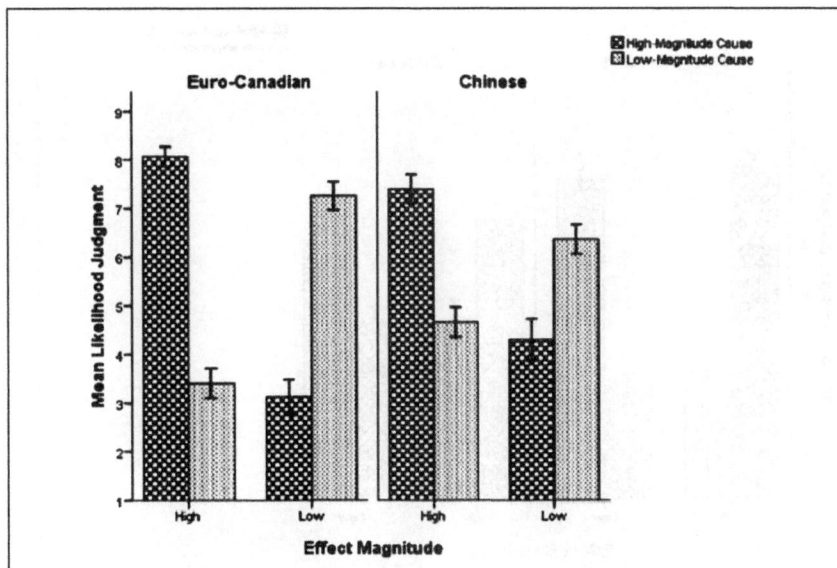

Figure. 1. Canadian and Chinese likelihood estimates (+SE) of high– and low–magnitude causes leading to high– and low–magnitude effects (Study 1a)

all, participants tended to associate the high-magnitude effect with the high-magnitude cause (M=7.73, SD=1.40) more than with the low-magnitude cause (M=4.05, SD=1.77), and the low-magnitude effect with the low-magnitude cause (M=6.82, SD=1.67) more than with the high-magnitude cause (M=3.72, SD=2.21). More importantly, the Culture × Effect Magnitude × Cause Magnitude interaction was significant, $F(1, 115)$ =13.80, p<.001, η_p^2=0.11. As hypothesized, Canadians tended to exhibit a stronger cause-effect magnitude correspondence than Chinese (see Figure 1).

Follow-up, independent-sample t tests on the interaction indicated that to account for the high-magnitude event, Canadians rated the high-magnitude cause as more likely than did Chinese and rated the low-magnitude cause as less likely than did Chinese. For the low-magnitude event, Canadians rated the high-magnitude cause as less likely than did Chinese and rated the low-magnitude cause as more likely than did Chinese, ts>2.10, ps<0.04.

2.2 Study 1b
2.2.1 Method

Participants. Eighty-four European Canadians (57 women) were recruited from Queen's University, and 60 Chinese (33 women) were recruited from Beijing University. Canadian participants received course credit or $5 for their participation, and Chinese participants received a small gift.

Materials and procedure. Study 1b followed a similar design and procedure to Study 1a but had a different scenario. The scenario described either a long negotiation delay (high-magnitude effect) or a brief negotiation delay (low-magnitude effect). The causes were a major disagreement (high-magnitude cause) or a minor one (low-magnitude cause; see the appendix). In summary, Study 1b followed a 2(culture: Canadians vs. Chinese) × 2 (effect magnitude: high vs. low) × 2 (cause magnitude: high vs. low) design, with the cause magnitude varying within participants and the other factors varying between participants.

2.2.2 Results and discussion

The results were similar to those obtained in Study 1a. A 2 (culture) × 2 (effect magnitude) × 2 (cause magnitude) mixed-model ANOVA revealed a significant main effect of effect magnitude, $F(1,140)$=4.12,p=0.04,such that the likelihood ratings for causes were higher in the

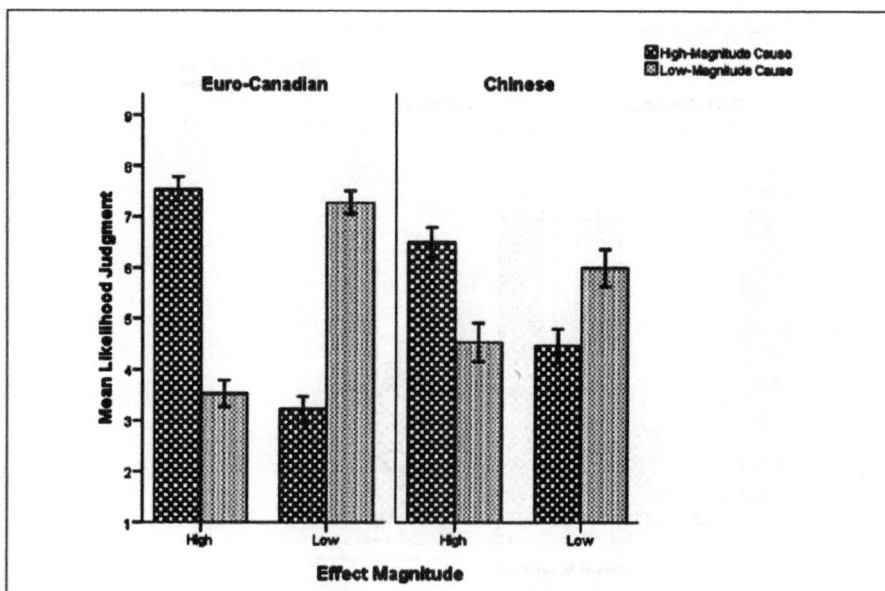

Figure. 2. Canadian and Chinese likelihood estimates (+SE) of high- and low-magnitude causes leading to high- and low-magnitude effects (Study 1b)

high-magnitude effect condition(M=5.53, SD=0.81)than in the low-magnitude effect condition (M=5.25, SD=0.82). The Effect Magnitude × Cause Magnitude interaction was significant, F(1, 140)=133.36, p<0.001, η_p^2=0.49. Overall, participants tended to associate high-magnitude effects with high-magnitude causes (M=7.10, SD=1.63) more than with low-magnitude causes (M=3.96, SD=1.87), and low-magnitude effects with low-magnitude causes (M=6.75, SD=1.81) more than with high-magnitude causes (M=3.74, SD=1.77). More importantly, the Culture × Effect Magnitude × Cause Magnitude interaction was significant, F(1, 140) =20.68, p<.001, η_p^2=0.13. As hypothesized, Canadians tended to exhibit a stronger cause - effect magnitude correspondence than Chinese (see Figure 2).

Follow-up, independent-sample t tests on the interaction indicated that to account for the high-magnitude event, Canadians rated the high-magnitude cause as more likely than did Chinese and rated the low-magnitude cause as less likely than did Chinese. For the low-magnitude event, Canadians rated the high-magnitude cause as less likely than did Chinese and rated the low-magnitude cause as more likely than did Chinese, ts > 2.28, ps < .03

Thus, for both the disease (Study 1a) and the negotiation (Study 1b) scenarios, Canadians associated high-magnitude effects with high-magnitude causes more than with low-magnitude causes and low-magnitude effects with low-magnitude causes more than with high-magnitude causes. And in both scenarios, Chinese exhibited this pattern to a significantly lesser degree.

3 Study 2

The purpose of Study 2 was to replicate the results from Study 1 using a different format. Study 1 manipulated the magnitudes of causes and effects in detailed scenarios.

In Study 2, we used simple scenarios that were described using pictures of common effects. Again, participants read hypothetical scenarios with consequences of high or low magnitude and indicated how likely the effect was due to causes that were high or low in magnitude.

3.1 Method

Participants. Seventy-eight European Canadians (46women) were recruited from Queen's University, and 60 Chinese (27 women) were recruited from Beijing University. Canadian participants received course credit or $5 for their participation, and Chinese participants received a small gift.

Materials and procedure. Participants were randomly assigned to a high- or low-effect magnitude condition. In each condition, they were presented first with a picture of two basketball players, one tall (high-magnitude cause) and one short (low-magnitude cause), and indicated the likelihood that each of the two players had scored the most points (high-magnitude effect) or the least points (low-magnitude effect) in a game. Next, participants were presented with a picture of two tornadoes that had traveled through a city, one wide (high-magnitude cause) and one narrow (low-magnitude cause), and indicated the likelihood that each tornado had caused extensive damage (high-magnitude effect) or no damage (low-magnitude effect). The likelihood judgments were made on a 9-point scale (1=*not likely at all*, 9=*extremely likely*). In summary, Study 2 had a 2 (culture: Canadians vs. Chinese) × 2 (effect magnitude: high vs. low) × 2 (cause magnitude: high vs. low) design. The cause magnitude factor varied within participants, and the other factors varied between participants.

3.2 Results and discussion

Likelihood estimate computation. Each participant completed both scenarios (basketball game and tornadoes). For each scenario, participants gave two likelihood estimates, one for a high-magnitude cause and one for a low-magnitude cause. The pattern of the results was the same for each of the two scenarios. Therefore, we combined the scenarios by averaging the two likelihood estimates for the high-magnitude causes and by averaging the two likelihood estimates for the low-magnitude causes. These two averages were then treated as repeated measures variables.

Test of cause-effect magnitude correspondence. A2(culture) × 2 (effect magnitude) × 2 (cause magnitude) mixed-model ANOVA revealed a significant main effect of effect magnitude, $F(1, 134)=23.72$, $p<.001$, such that participants in the high-magnitude effect condition gave higher likelihood ratings for the causes ($M=5.72$, $SD=0.67$) in comparison with those in the low-magnitude effect condition ($M=5.22$, $SD=0.45$). The Effect Magnitude × Cause Magnitude interaction was significant, $F(1, 134)=50.20$, $p<.001$, $\eta_p^2 =0.27$. Over-all, participants tended to associate high-magnitude effects with high-magnitude causes ($M=6.80$, $SD=1.28$) more than with low-magnitude causes ($M=4.63$, $SD=1.30$), and low-magnitude effects with low-magnitude causes ($M=5.91$, $SD=1.71$) more than high-magnitude causes ($M=4.54$, $SD=1.73$). Replicating the results from Study 1, the Culture × Effect Magnitude × Cause Magnitude interaction was significant, $F(1, 134)=7.48$, $p=0.007$, $\eta_p^2=0.05$. As hypothesized, Canadians exhibited a stronger cause - effect magnitude correspondence than Chinese (see Figure 3).

Follow-up, independent-sample t tests on the interaction indicated that to account for the high-magnitude event, Canadians rated the high-magnitude cause as more likely than did Chinese and rated the low-magnitude cause as less likely than did Chinese. For the low-magnitude event, Canadians rated the high-magnitude cause as less likely than did Chinese and rated the low-magnitude cause as more likely than did Chinese, $ts>1.96$, $ps<.05$.

Thus, in two detailed scenarios represented by words

in Study 1 and two simple scenarios represented by pictures in Study 2, Canadians associated high-magnitude effects with high-magnitude causes more than with low-magnitude causes and low-magnitude effects with low-magnitude causes more than with high-magnitude causes. Furthermore, for all four scenarios, Chinese exhibited this pattern to a significantly lesser degree.

4 Study 3

Studies 1 and 2 found that compared with Chinese, Canadians expect effects and their causes to correspond in magnitude to a greater degree. We argue that the pattern of results is caused by cultural differences in holistic reasoning. However, a simple alternative explanation exists: The results from Studies 1 and 2 could be explained by a stronger preference to choose midpoints on scales by Chinese than by Canadians. In both studies, participants rated their likelihood judgments on a Likert-type scale. Chen, Lee, and Stevenson (1995) found evidence that East Asians typically prefer points closer to the midpoints of such scales, even though the degree of midpoint-response bias in their study was weak, especially comparing Chinese and Canadians. Furthermore, other studies have found no such tendency (e.g., Ji, Schwarz, & Nisbett, 2000).

Nonetheless, we designed Study 3 to rule out the possibility that Chinese were engaging in such a moderacy-response bias. Instead of rating the likelihoods of causes on a scale, participants chose the cause they perceived to have most likely led to the effect. If the same pattern of results as in Studies 1 and 2 emerged regarding participants'choices, then these results would provide strong evidence against the alter-native explanation.

4.1 Method

Participants. Sixty-three European Canadians (48 women) and 63 Chinese nationals living in Canada (44 women) were recruited from Queen's University.

Participants received course credit or $5 for their participation. At the time of the study, the Chinese nationals had lived in Canada for an average of 28.83 months (SD=13.54).

Materials and procedure. Participants were randomly assigned to either the high or the low effect magnitude condition. Within each condition, participants read three scenarios.One scenario was taken from Study 1 (disease outbreak described in words) and two scenarios were taken from Study 2 (basketball game and tornadoes depicted in pictures). In the high–magnitude effect condition, all three scenarios described high–magnitude effects and likewise in the low–magnitude effect condition, all three scenarios described low–magnitude effects. For each scenario, participants chose the more likely cause of the effect between the high– and the low–magnitude causes. Additionally, participants indicated their confidence in each choice on an 8–point scale (1=*not at all confident,8 =extremely confident*). In summary, Study 3 had a 2 (culture: Canadians vs. Chinese) × 2 (effect magnitude: high vs. low) × 2 (cause magnitude: high vs. low) design. The cause magnitude factor varied within participants, and the other factors varied between participants.

4.2 Results and discussion

Test of cause–effect magnitude correspondence. Each participant chose three causes, one for each scenario. Each choice was analyzed separately by conducting a 2 (culture) × 2 (effect magnitude) × 2 (cause magnitude) log–linear analysis. For all three scenarios, the analyses revealed significant interactions between effect magnitude and cause magnitude, all G^2s >25.72, all df=1, all ps<.001, indicating that for each scenario, participants were more likely to associate high–magnitude effects with high–magnitude causes than with low–magnitude causes, and low–magnitude effects with low–magnitude causes than with high–magnitude causes. More importantly, for all three scenarios, the analyses revealed significant Culture × Effect Magnitude × Cause Magnitude interactions, all G^2s >13.58, all df=1, all ps<.001. For all three scenarios, Canadians were more likely to exhibit the magnitude–matching pattern than were Chinese (see Table 1 for frequency counts). Thus, we successfully replicated the results from Studies 1 and 2 using a method that did not rely on a Likert–type scale, suggesting that a midpoint preference by Chinese was unlikely to account for the results obtained in Studies 1 and 2.

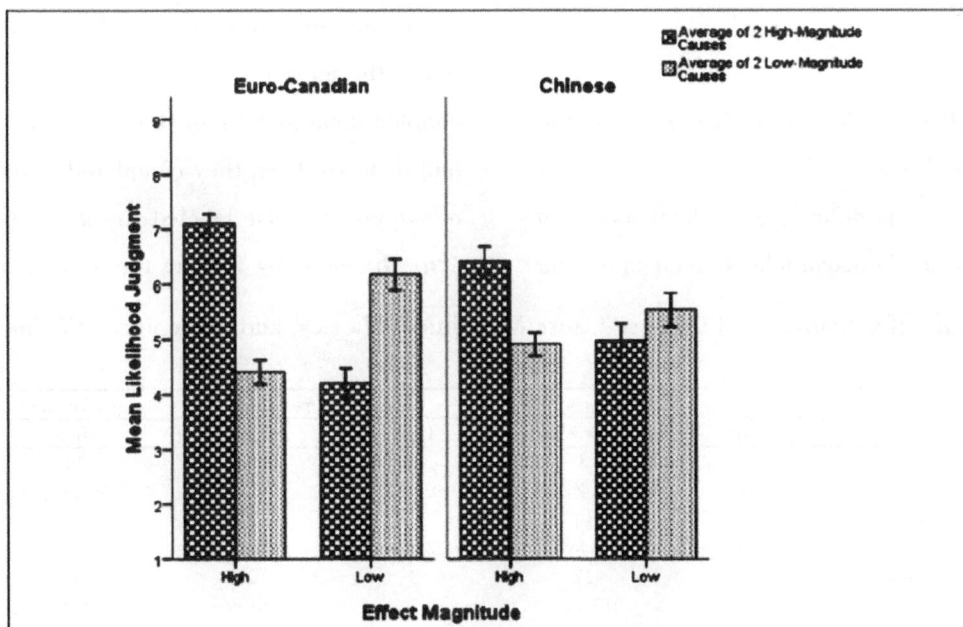

Figure 3. Canadian and Chinese likelihood estimates (+ SE) of high–and low–magnitude causes leading to high– and low–magnitude effects (Study 2)

Confidence ratings. A 2 (culture) × 2 (effect magnitude) × 3 (scenario) mixed-model ANOVA indicated no significant interactions, all $ps > .79$. The effect magnitude main effect was significant, $F(1, 122) = 5.94$, $p = 0.02$, $\eta_p^2 = 0.05$, indicating that participants were more confident in their choices in the low-magnitude condition ($M = 6.25$, $SD = 0.99$) than in the high-magnitude condition ($M = 5.80$, $SD = 1.15$). In addition, the culture main effect was significant, $F(1, 122) = 11.20$, $p < .001$, $\eta_p^2 = 0.08$, such that Chinese were more confident of their choices overall ($M = 6.33$, $SD = 0.82$) than were Canadians ($M = 5.71$, $SD = 1.24$). Therefore, on this scale, Chinese were not more likely than Canadians to prefer responses near the midpoint of the scale, providing further evidence that a midpoint preference by Chinese was unlikely to account for the results obtained in Studies 1 and 2.

In summary, Study 3 replicated the results of Studies 1 and 2 using a different method and ruled out the potential alternative explanation that Chinese were engaging in a moderacy-response bias. In fact, Chinese responded significantly further from the midpoint than did Canadians on confidence ratings of their likelihood judgments.

5 Study 4

The purpose of Study 4 was to directly test the hypothesis that analytic and holistic reasoning was responsible for the cultural differences in the tendency to expect a correspondence in magnitude between cause and effect. We developed an exercise to prime either analytic or holistic reasoning in Canadian participants. Participants completed the exercise, then read a scenario describing a high- or low-magnitude effect, and rated the likelihood that high- or low-magnitude causes had led to that effect. We used two different versions of scenarios with two separate samples, one in Study 4a and the other in Study 4b.

5.1 Study 4a

5.1.1 Method

Participants. Sixty-seven European Canadians (49 women) were recruited from Queen's University. Participants received course credit or $5 for their participation.

Analytic versus holistic prime. We primed analytic and holistic thinking by focusing participants' attention on either a simple or a complex causal field, respectively. For the prime, participants completed an exercise ostensibly unrelated to the rest of the study. All participants read:

Getting into a competitive university such as Queen's University is a major achievement. The majority of high school students do not make it into any university at all, and a large number of applicants to Queen's are turned away every year.

In the analytic prime condition, participants then listed the most significant event in their life that had enabled them to get into Queen's and described how it had done so. Last, they completed a diagram consisting of two ellipses, one labeled *Event* and the other *Getting into Queen's,* by writing the significant event in the

Table 1. Frequencies of Canadian and Chinese Cause Magnitude Choices (and Percentages Within Each Culture): Study 3

Effect	Culture	Disease outbreak causes		Basketball game causes		Tornado damage causes	
		High	Low	High	Low	High	Low
High	Canadian	29	3	30	2	28	4
		(90.6%)	(9.4%)	(9.38%)	(6.2%)	(87.5%)	(12.5%)
	Chinese	18	14	20	12	19	13
		(56.2%)	(43.8%)	(62.5%)	(37.5%)	(59.4%)	(40.6%)
Low	Canadian	4	27	3	28	4	27
		(12.9%)	(87.1%)	(9.7%)	(90.3%)	(12.9%)	(87.1%)
	Chinese	14	17	13	18	13	18
		(45.2%)	(54.8%)	(41.9%)	(58.1%)	(41.9%)	(58.1%)

event ellipse and by drawing an arrow between it and the getting into Queen's ellipse. The exercise was designed to focus each participant's attention on a single cause that had led to a major event in his or her life. The holistic prime was nearly identical to the analytic prime except that participants listed the three most significant events. Holistic reasoning involves not only focusing on numerous causes but also focusing on the interactions between such causes. Therefore, participants also described how the three events had influenced each other.The diagram consisted of four ellipses, three on the periphery labeled *Event* and one in the center labeled *Getting into Queen's*. After writing the three events in the event ellipses, participants drew arrows from each one to the Queen's ellipse. Lastly, they drew arrows connecting the three events to describe how these events had influenced or interacted with each other. The holistic prime was designed to focus participants' attention on a larger causal field and on the connectedness of causes within that field.

Magnitude manipulations. After the prime, participants read the disease scenario from Study 1 and rated the likelihood that each of the two causes had led to the effect on a 9-point scale (1 = *not likely at all*, 9 = *extremely likely*). In summary, Study 4a had a 2 (prime: analytic vs. holistic) × 2 (effect magnitude: high vs. low) × 2 (cause magnitude: high vs. low) design. The causal magnitude factor varied within participants, and the other factors varied between participants.

5.1.2 Results and discussion

Test of cause - effect magnitude correspondence. A 2 (prime: analytic vs. holistic) × 2 (effect magnitude) × 2 (cause magnitude) mixed-model ANOVA revealed a significant main effect of effect magnitude, $F(1, 63)=12.47$, $p<.001$, such that participants in the high-magnitude effect condition gave higher likelihood ratings for the causes ($M=5.97$, $SD=0.94$) in comparison with those in the low-magnitude effect condition ($M=5.35$, $SD=0.92$). The Effect Magnitude × Cause Magnitude interaction was significant, $F(1, 63)=78.92$, $p<.001$, $\eta_p^2=0.56$. Overall, participants tended to associate high-magnitude effects with high-magnitude causes ($M=7.17$, $SD=1.56$) more than with low-magnitude causes ($M=4.51$, $SD=1.98$), and low-magnitude effects with low-magnitude causes ($M=7.00$, $SD=1.30$) more than with high-magnitude causes ($M=4.44$, $SD=1.76$). More importantly, the Prime × Effect Magnitude × Cause Magnitude interaction was significant, $F(1, 63)=12.02$, $p<.001$, $\eta_p^2=0.16$. As hypothesized, Canadians primed to reason analytically exhibited a

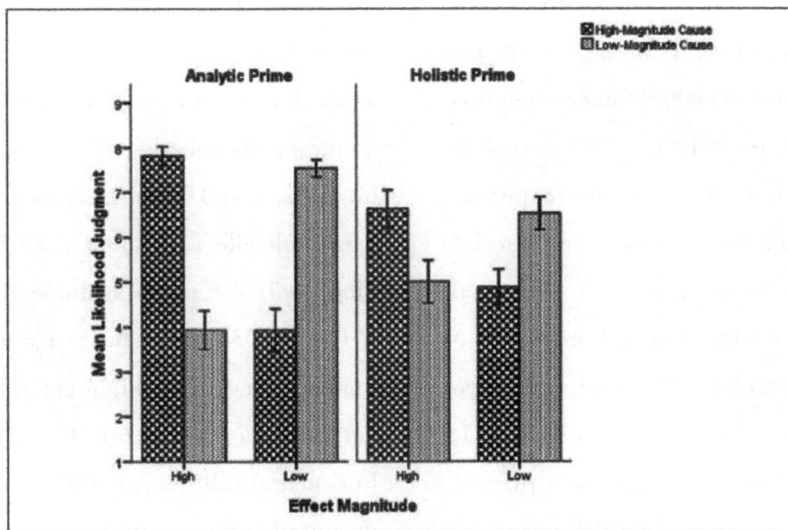

Figure 4. Analytically primed and holistically primed Canadian likelihood estimates (+SE) of high- and low-magnitude causes leading to high- and low-magnitude effects (Study 4a)

stronger cause – effect magnitude correspondence than Canadians primed to reason holistically (see Figure 4).

Follow-up, independent-sample t tests on the interaction indicated that to account for the high-magnitude event, analytically primed Canadians rated the high-magnitude cause as more likely than did holistically primed Canadians and rated the low-magnitude cause as less likely than did holistically primed Canadians. For the low-magnitude event, analytically primed Canadians rated the high-magnitude cause as less likely than did holistically primed Canadians and rated the low-magnitude cause as more likely than did holistically primed Canadians, $ts > 1.96$, $ps < .05$.

Comparison with nonprimed participants from Study 1a. Participants in Study 1a completed the identical disease scenario but without any prime. Therefore, we compared participants from Study 4a with those from Study 1a to determine more specifically what effect the analytic and holistic primes had on participants. First, we compared analytically primed Canadians with those who received no prime. The Prime × Effect Magnitude × Cause Magnitude interaction was not significant, $F(1, 86) = 1.36$, $p = 0.25$, revealing that Canadians primed to reason analytically did not differ from those who received no prime. Next, we compared holistically primed Canadians with those who received no prime. The Prime × Effect Magnitude × Cause Magnitude interaction was significant, $F(1, 91) = 21.39$, $p < .001$, $\eta_p^2 = 0.19$, revealing that Canadians primed to reason holistically expected less cause – effect magnitude correspondence than did Canadians who received no prime. Lastly, we compared holistically primed Canadians with Chinese from Study 1a who received no prime. The Culture × Effect Magnitude × Cause Magnitude interaction was not significant, $F(1, 91) = 1.47$, $p = 0.23$, revealing that Canadians primed to reason holistically did not differ from Chinese who received no prime.

5.2 Study 4b
5.2.1 Method

Participants. One hundred twenty-one European Canadians (75 women) were recruited from Queen's University. Participants received course credit or $5 for their participation.

Materials and procedure. The procedure was nearly identical to Study 4a except that participants read a money scenario after the prime. We also included a control condition in which participants did not receive any prime. The money scenario described a Canadian individual who had either accumulated greater than average savings (high-magnitude effect) or lesser than average savings (low-magnitude effect). The scenario was followed by two potential causes, one of high magnitude (the individual had a higher than average income) and one of low magnitude (the individual had a lower than average income). In summary, Study 4b had a 3 (prime: none, analytic, or holistic) × 2 (effect magnitude: high vs. low) × 2 (cause magnitude: high vs. low) design. The causal magnitude factor varied within participants, and the other factors varied between participants.

5.2.2 Results and discussion

A 3 (prime) × 2 (effect magnitude) × 2 (cause magnitude) mixed-model ANOVA revealed a significant main effect of effect magnitude, $F(1, 115) = 8.72$, $p = 0.004$, such that participants in the high-magnitude effect condition gave higher likelihood ratings for the causes ($M = 5.52$, $SD = 0.97$) in comparison with those in the low-magnitude effect condition ($M = 4.77$, $SD = 0.98$). The Effect Magnitude × Cause Magnitude interaction was significant, $F(1, 115) = 239.66$, $p < .001$, $\eta_p^2 = .68$. Overall, participants tended to associate high-magnitude effects with high-magnitude causes ($M = 6.97$, $SD = 1.53$) more than with low-magnitude causes ($M = 3.43$, $SD = 2.01$), and low-magnitude effects with low-magnitude causes ($M = 6.57$, $SD = 1.51$) more than with high-magnitude causes

(M=3.07, SD=1.64). More importantly, the Prime × Effect Magnitude × Cause Magnitude interaction was significant, $F(2, 115)$=6.81, p=0.002, η_p^2=0.11. We conducted follow-up 2 (prime) × 2 (effect magnitude) × 2 (cause magnitude) mixed-model ANOVA to determine more specifically the effects of the primes. As hypothesized, Canadians primed to reason analytically exhibited a stronger cause - effect magnitude correspondence than Canadians primed to reason holistically, $F(1, 61)$=10.16, p=0.002, η_p^2=0.14. Canadians primed to reason analytically did not differ in the tendency to make this association compared with Canadians who received no prime, $F(1, 87) = .22$, p=0.64. Lastly, Canadians primed to reason holistically tended to exhibit a weaker cause - effect magnitude correspondence than did Canadians who received no prime, $F(1, 82)= 10.26$, p=0.002, η_p^2=0.11 (see Figure 5).

Analytically primed Canadians did not differ from those receiving no prime, and thus we collapsed the two groups together, referred to them as analytic-Canadians, and compared them with holistically primed Canadians. Follow-up, independent-sample t tests on the interaction indicated that to account for the high-magnitude event, analytic-Canadians rated the high-magnitude cause as more likely than did holistically primed Canadians and rated the low-magnitude cause as less likely than did holistically primed Canadians, $ts > 2.10$, $ps < .04$. For the low-magnitude event, analytic-Canadians rated the high-magnitude cause as less likely (M=2.78, SD=1.16) than did holistically primed Canadians (M=3.50, SD=1.28), t=1.96, p<.05, and rated the low-magnitude cause as marginally more likely than did holistically primed Canadians, t=1.91, p=0.07.

In summary, for both the disease (Study 4a) and the negotiation (Study 4b) scenarios, analytically primed Canadians associated high-magnitude effects with high-magnitude causes more than with low-magnitude causes, and low-magnitude effects with low-magnitude causes more than with high-magnitude causes. And in both scenarios, holistically primed Canadians exhibited this pattern to a significantly lesser degree. Additionally, comparing participants from Study 4a with participants from Study 1a who completed the same materials without any prime revealed that the analytic prime had no effect. Instead, the holistic prime was the one that caused Canadians to expect less cause - effect magnitude correspondence. This pattern of results was replicated in Study 4b by comparing the participants who received either the analytic prime or the holistic prime with participants in a control condition who received no prime. Lastly, comparing holistically primed Canadians in Study 4a with Chinese in Study 1a who received no prime revealed no cultural differences in the tendency to expect a cause - effect magnitude correspondence.

6 General discussion

Across three studies, Canadians expected a greater correspondence in magnitude between effects and their causes than did Chinese. In a fourth study, both analytically primed Canadians and Canadians who received no prime were more likely to expect the correspondence in magnitude between cause and effect than were holistically primed Canadians, whereas the holistically primed Canadians showed similar responses to Chinese participants. The results not only demonstrated cultural differences in the extent to which people expect a correspondence in magnitude between cause and effect, but they also demonstrated the underlying factor responsible for such cultural differences, namely, the causal complexity factor of analytic or holistic reasoning.

6.1 Alternative explanations

One potential alternative explanation for the results is that Chinese may be less familiar with the scenarios than are Canadians. If people are reasoning about the cause of an effect in a context where causes and effects tend to correspond in magnitude in the real world, then the tendency to predict such an association would be prudent. Therefore, if Canadians are more familiar than Chinese are with a scenario, and thus the underlying causes of the effect in that scenario, then Canadians should expect a greater degree of correspondence between cause and effect.

This potential alternative explanation has at least two problems. First, when choosing and designing our scenarios, we were very careful to generate scenarios that were familiar and understandable to both cultural groups. Indeed, a number of scenarios that were initially proposed by either the Canadian or Chinese researchers were rejected because they did not meet these criteria. Second, even if a cause‐effect magnitude correspondence existed for all of the scenarios we used, and all of our efforts at ensuring equivalent levels of familiarity with the scenarios failed, the familiarity explanation would not fit with the pattern of results in Studies 4a and 4b. It would be difficult to explain, based on familiarity with the scenarios, why priming Canadians to think holistically would cause them to respond similarly to Chinese, namely, by expecting a

lesser degree of cause‐effect magnitude correspondence.

6.2 Theoretical contributions

Culture and attributions. Gilovich and Savitsky (2002, p. 618) defined the representativeness heuristic as the tendency to process information "on the basis of one overarching rule: 'Like goes with like.' " Most of the research on this heuristic has been conducted in the context of categorization, whereas little research has explored the representativeness heuristic in the context of attributions. Although some have speculated on the tendency to expect cause‐effect magnitude correspondence when making attributions (Einhorn & Hogarth, 1986; Nisbett & Ross, 1980), little research has systematically investigated this speculation. Our studies contribute to the attribution literature by providing strong empirical evidence in support of this proposal.

In addition, we demonstrated that the degree to which people expect cause‐effect magnitude correspondence differs across cultures. Throughout our studies, both cultures associated causes and effects on the basis of magnitude similarity, and this pattern of results indicates that both Canadians and Chinese exhibited judgments consistent with the representativeness heuristic. However, the fact that Canadians expected a greater degree of

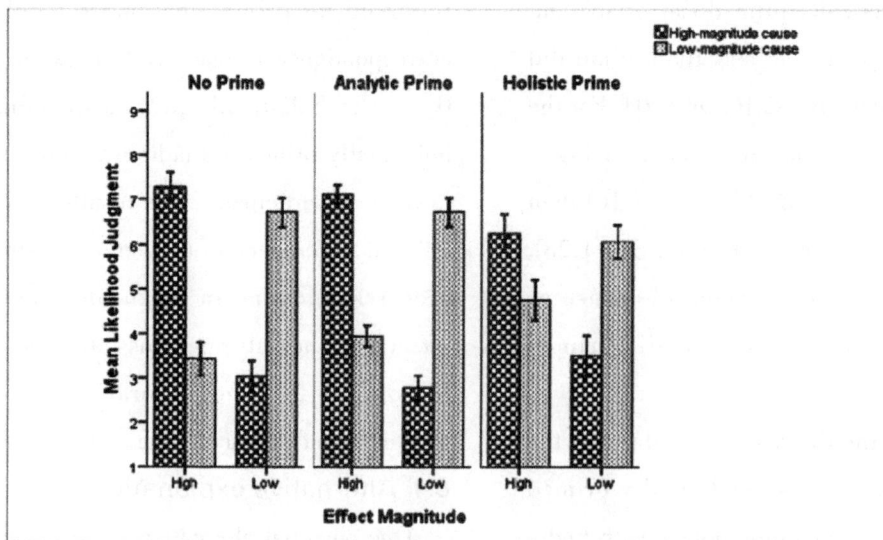

Figure 5. No prime, analytically primed, and holistically primed Canadian likelihood estimates (+ SE) of high‐ and low‐magnitude causes leading to high‐ and low‐magnitude effects (Study 4b)

correspondence in magnitude when making causal judgments indicates that they exhibit a more extreme form of the representativeness heuristic in this context, compared with Chinese. Our study is the first we are aware of to demonstrate cultural differences in the degree to which people employ the representativeness heuristic in the context of attributions.

In addition, in examining the literature on the representativeness heuristic, we are aware of no studies that have demonstrated any causal mechanisms. Our finding that the degree to which people expect cause‐effect magnitude correspondence is in part determined by whether they reason analytically or holistically enhances our understanding of the cognitive underpinnings of the representativeness heuristic in the context of attributions. Furthermore, the fact that this tendency differs across cultures signals the need to investigate whether cultural differences would emerge for other heuristics.

An important caveat here is that we demonstrated cultural differences in the representativeness heuristic in the context of causal judgments based on a similarity in magnitude. We have provided no evidence regarding whether the same pattern of results would emerge with respect to the representativeness heuristic in other contexts, such as attributing causes based on their physical resemblance to effects.

Cultural universals. Notably, within the domain of our scenarios and designs, participants from both cultures expected a cause‐effect magnitude correspondence. What differed across cultures was the degree to which people expected this correspondence. Norenzayan and Heine (2005) outlined a taxonomy of cultural universals, defined as core mental attributes shared by people everywhere. The taxonomy consists of four levels of cultural universality: accessibility, functional, existential, and nonuniversality. Accessibility universals, the most stringent level of universality, are psychological processes that are available to all people, used for the same function, and accessed to the same degree. Functional universals, the second most stringent level of universality, are cognitively available to all people, used for the same function, but accessed to different degrees. Existential universals, the third most stringent level of universality, are cognitively available to all people, but they may be used in markedly different ways and are accessed to different degrees. Nonuniversals are those processes that are not cognitively available to all people.

Although we only sampled people from two cultures, our pattern of results demonstrating that participants from both cultures expected cause‐effect magnitude correspondence, but to different degrees, would potentially qualify as a functional universal. Our studies provide the first evidence we are aware of regarding the universality and cultural variability of the representativeness heuristic. What appears on the surface to be a simple heuristic, the tendency to associate high-magnitude effects with high-magnitude causes and low-magnitude effects with low-magnitude causes, is applied to different degrees depending on how our host culture has shaped our minds to process information.

6.3 Practical implications

The pattern of results could also have important practical implications. One potential application is in the domain of behavioral decision making. According to the U.S. Federal Reserve, per capita personal debt in the United States has increased by at least a factor of 10 between 1945 and 2005, reaching unprecedented levels (Massey, 2008). Canadian debt levels are following a similar trend according to the Bank of Canada. The causes of this trend are surely complex and we do not intend to oversimplify them or claim that our research has solved this problem. However, our results could potentially contribute to better understanding and ameliorating the situation. For example, North Americans may be more likely to believe that to accumulate wealth or to resolve their debt problems, they need to focus on cutting back on major costs and purchases. In doing so, they may pay little attention to the financial impact of minor routine expenses. Consistent with this possibility, financial experts are advising people to cut back on minor expenses, such as the purchase of a daily cup of coffee at a trendy café, because such purchases add up over time. Thus, for people who find themselves in debt, realizing the importance of reducing minor routine expenses could enable them to more quickly and effectively ameliorate their financial hardship. In addition, the results from Study 4, demonstrating that the tendency to expect a magnitude correspondence is reduced when primed to think holistically, provides hope that people can reduce the degree to which they overlook minor expenses in such situations. We are investigating some of these implications in our ongoing research.

Meanwhile, this research could have mirror-image implications for East Asians. For example, East Asians may underestimate the importance of major health-related behaviors because they believe that other minor health-related behaviors will compensate. For example, when told by a medical expert that ceasing a pernicious behavior, such as smoking, is essential for dealing with a major illness, such as lung cancer, Easterners may be more likely to believe that their other health-related behaviors, such as eating well, will compensate for the detrimental effects of smoking.

The present research also has practical implications for problem solving and negotiation. The human world is becoming increasingly interconnected and many nations are becoming more ethnically diverse. People from different cultures are realizing the growing need to work together to solve problems, such as those currently undermining political and financial stability, as well as those plaguing our natural environment. When investigating the cause of a high-magnitude effect, Westerners' tendency to expect high-magnitude causes may lead them to overlook relatively lower magnitude causes that might have played a key role. On the other hand, Chinese might over-emphasize factors that played a minor role at the expense of those that played a major role. These different tendencies to emphasize high- versus low-magnitude causes could lead to disagreement or international conflict. In a world in which cross-cultural interaction is crucial, further understanding cultural differences in reasoning about cause and effect relationships could prove important both for finding solutions to the problems and for negotiating diplomatic resolutions to the inevitable disagreements that will arise between nations.

People associate big causes with big effects and small causes with small effects. However, the degree to which we make this association is at least partially determined by the reasoning processes we are imbued with from the culture in which our minds developed. Some physicists believe that our universe, created by the Big Bang, will end in a big crunch. Perhaps a more holistic interpretation of this scenario would result in the universe that began from the Big Bang ending in a small crunch.

Appendix

Scenarios Used in Studies 1a, 3, and 4a

Effects

- 21 people at a major downtown company became ill. They were stricken with symptoms of nausea and vomiting. Within 3 days, 11 of these individuals had experienced a rapid but horrific death and the other 10 were still in hospital. (*High magnitude*)

- 7 people at a major downtown company became ill. They were stricken with symptoms of nausea and vomiting. Within 3 days, 3 of these individuals had recovered and the other 4 were still experiencing minor symptoms. (*Low magnitude*)

Causes

● An employee came into contact with a highly infectious type of super-bacteria while on a business trip. (*High magnitude*)

● An employee came into contact with a standard type of bacteria while on a business trip. (*Low magnitude*)

Scenarios Used in Study 1b

Effects

● Two parties are negotiating a new contract. This negotiation took over 31 weeks to reach an agreement, which is much longer than usual compared with other similar negotiations. (*High magnitude*)

● Two parties are negotiating a new contract. This negotiation took only 2 weeks to reach an agreement, which is typical compared with other similar negotiations. (*Low magnitude*)

Causes

● Two negotiating parties could not agree on one major point, which amounted to 41% of the total contract value. (*High magnitude*)

● Two negotiating parties could not agree on one minor point, which amounted to 10% of the total contract value. (*Low magnitude*)

References

Chen, C., Lee, S. Y., & Stevenson, H. W. (1995). Response style and cross-cultural comparisons of rating scales among East Asian and North American students, *Psychological Science, 6*, 170–175.

Choi, I., Dalal, R., Kim-Prieto, C., & Park, H. (2003). Culture and judgment of causal relevance. *Journal of Personality and Social Psychology, 84*, 46–59.

Ebel-Lam, A. P., Fabrigar, L. R., MacDonald, T. K., & Jones, S. (2008). *Balancing causes and consequences:The proportionality principle in explanations for complex social events*. Unpublished manuscript, Queen's University, Kingston, Ontario, Canada.

Einhorn, H. J., & Hogarth, R. M. (1986). Judging probable cause. *Psychological Bulletin, 99*, 3–19.

Gilovich, T., & Savitsky, K. (2002). Like goes with like: The role of representativeness in erroneous and pseudo-scientific beliefs. In T. Gilovich, D. Griffin, & D. Kahnemann (Eds.), *Heuristics and biases: The psychology of intuitive judgment* (pp. 617–624). Cambridge, UK: Cambridge University Press.

Ji, L. J., Nisbett, R. E., & Su, Y. (2001). Culture, change, and prediction. *Psychological Science, 12*, 450–456.

Ji, L. J., Peng, K., & Nisbett, R. E. (2000). Culture, control, and perception of relationships in the environment. *Journal of Personality and Social Psychology, 78*, 943–955.

Ji, L. J., Schwarz, N., & Nisbett, R. E. (2000). Culture, autobiographical memory, and behavioral frequency reports: Measurement issues in cross-cultural studies. *Personality and Social Psychology Bulletin, 26*, 585–593.

Kahneman,D.,&Tversky,A.(1972).Subjective probability:A judgment of representativeness. *Cognitive Psychology, 3*, 430–454.

Kahneman, D., & Tversky, A. (1973). On the psychology of prediction. *Psychological Review, 80*, 237–251.

Lorenz, E. N. (1963). Deterministic nonperiodic flow. *Journal of the Atmospheric Sciences, 20*, 130–141.

Massey, D. S. (2008). Globalization and inequality: Explaining American exceptionalism. *European Sociological Review*, 1–15.

Masuda, T., & Nisbett, R. E. (2001). Attending holistically versus analytically: Comparing the context sensitivity of Japanese and Americans. *Journal of Personality and Social Psychology, 81*, 922–934.

McCauley, C., & Jacques, S. (1979). The popularity of conspir-acy theories of presidential assassination: A Bayesian analysis. Journal of Personality and Social Psychology, 37, 637–644.

McClure, J., Lalljee, M., & Jaspars, J. (1991). Explanations of extreme and moderate events. *Journal of Research in Personality, 25*, 146–166.

Miller, J. G. (1984). Culture and the development of everyday social explanation. *Journal of Personality and Social Psychology, 46*, 961–978.

Morris, M. W., & Peng, K. (1994). Culture and cause: American and Chinese attributions for social and physical events. *Journal of Personality and Social Psychology, 67*, 949–971.

Nisbett, R. E., Peng, K., Choi, I., & Norenzayan, A. (2001). Culture and systems of thought: Holistic versus analytic cognition. *Psychological Review, 108*, 291–310.

Nisbett, R. E., & Ross, L. (1980). *Human inference: Strategies and shortcomings of social judgment.* Upper Saddle River, NJ: Prentice Hall.

Norenzayan, A., & Heine, S. J. (2005). Psychological universals: What are they and how can we know? *Psychological Bulletin, 131*, 763–784.

Osgood, C. E. (1957). *Measurement of meaning.* Chicago: University of Illinois Press.

Peng, K., & Nisbett, R. A. (1999). Culture, dialectics, and reasoning about contradiction, *American Psychologist, 54*, 741–754.

Shultz, T. R., & Ravinsky, F. B. (1977). Similarity as a principle of causal inference. *Child Development, 48*, 1552–1558.

Journal of Experimental Education, 2012, 81(1), 105 – 122.

Goal contents and goal contexts: Experiments with chinese students

Ze Wang [1], Xiaoyong Hu [2], Yongyu Guo [2]

([1] University of Missouri)

([2] Central China Normal University, China)

Abstract Using samples of Chinese middle school students, the 2 experimental studies presented here examined the effects of goal content and goal context on test performance, free-choice engagement, and test anxiety within the framework of self-determination theory. Students' learning goals were induced as intrinsic or extrinsic with the learning contexts of either autonomy-supportive or controlling. Results suggested that as the more recent extensions of self-determination theory, goal content and goal context effects existed among our samples of Chinese middle school students. However, there was some inconsistency between the authors' findings and previous findings in Western culture.

Keywords Chinese students; goal content; goal context; self-determination theory

1 Introduction

At educational settings students engage in learning activities for various reasons. Self-determination theory (SDT) examines different orientations of engaging in a task (i.e., what type of motivation; Deci & Ryan, 1985; Ryan & Deci, 2000). SDT researchers focus on how social and cultural factors facilitate or undermine people's sense of volition and initiative. Basic psychological needs such as autonomy, competence, and relatedness are considered as requirements for the healthy development and functioning of human beings. The satisfaction of these needs serves as the foundation in understanding what social and cultural factors affect human motivation and engagement in activities (Deci & Ryan, 2008). Early SDT research differentiated between intrinsic motivation and extrinsic motivation; later research refined extrinsic motivation to reflect different degrees of self-regulation and internalization of an action, and the key differentiation within SDT shifted to a focus on autonomous-versus-controlled motivation (Deci & Ryan, 2008). More recently, goal contents (intrinsic vs. extrinsic), goal framing, and goal contexts have been studied and experimented within the framework of SDT as newer extensions (T. Kasser & Ryan, 1993, 1996; Vansteenkiste, Simons, Lens, Sheldon, & Deci, 2004; Vansteenkiste, Simons, Soenens, & Lens,

Address correspondence to Xiao Yong Hu, School of Psychology, Central China Normal University, 152 Luoyu Road, Wuhan, Hubei 430079, China. E-mail: huxiaoyong8123@yahoo.com.cn

2004). Goal contents reflect the outcomes that people are pursuing, whether for autonomous or controlled reasons (T. Kasser & Ryan, 1996).

The original SDT and its earlier extensions have been studied across cultures (e.g., Levesque, Zuehlke, Stanek, & Ryan, 2004) and contextual settings such as physical education (e.g., Haerens, Kirk, Cardon, De Bourdeaudhuij, & Vansteenkiste, 2010; Standage, Duda, & Ntoumanis, 2005), religious activities (e.g., Neyrinck, Vansteenkiste, Lens, Duriez, & Hutsebaut, 2006), health and medicine (e.g., Halvari & Halvari, 2006) politics (e.g., Losier & Koestner, 1999) and organizations (e.g., Deci, Connell, & Ryan, 1989). However, to date, the more recent extensions of experimental SDT research have mostly been studied in developed countries and in individualistic cultures (e.g., Vansteenkiste, Simons, Soenens, et al., 2004). As the largest population in the world, the Chinese differ in many ways from populations in developed countries in regards to cultural values, educational systems, and interpersonal relationship norms. Therefore, it is necessary to test the applicability of the recent extensions of SDT to the Chinese population. In this paper, we designed studies to examine goal content and goal context effects among Chinese middle school students in mainland China. We believe that testing those effects in a different culture would contribute to the advancement and universality of SDT.

1.1 Goal contents

People's goals vary. Research by Kasser and Ryan (1996) showed that people's long-term goals can be categorized as intrinsic or extrinsic. Intrinsic goals reflect people's inherent growth tendencies and yield an inward-oriented focus (e.g., self-development, health and physical fitness, community contribution, and affiliation), whereas extrinsic goals reflect people's desire to impress others by acquiring outward signs of worth (e.g., financial success, power, status, and physical attractiveness) and are characterized by an outward-oriented frame-of-

reference for viewing the world (T. Kasser & Ryan, 1996; Vansteenkiste, Lens, & Deci, 2006; Williams, Hedberg, Cox, & Deci, 2000). As a subtheory of SDT, goal contents theory distinguishes those two types of goals and examines their effect on behaviors and well-being (T. Kasser, 2002; T. Kasser & Ryan, 1996; Vansteenkiste, Lens, & Deci, 2006). Extrinsic and intrinsic goals are thought to relate differently to basic need satisfaction and therefore produce different psychological outcomes (Grouzet et al., 2005; T. Kasser & Ryan, 1996; V. G. Kasser & Ryan, 1999; Kim, Kasser, & Lee, 2003; Ryan et al., 1999; Schmuck, Kasser, & Ryan, 2000). Along this line, correlational studies have found that the relative importance of extrinsic goals relates negatively to adjustment outcomes and the relative importance of intrinsic goals relates positively to adjustment outcomes (Duriez, Vansteenkiste, Soenens, & De Witte, 2007; T. Kasser & Ryan, 2001; Williams et al., 2000). This basic pattern is called goal content effect and has been found in different cultures, including the United States (T. Kasser & Ryan, 1996), Germany (Schmuck et al., 2000), Spain (Romero, Gómez-Fraguela,& Villar, 2011), Russia (Ryan et al., 1999), and South Korea (Kim et al., 2003).

While long-term goal contents may reflect what people value, they should not be interpreted as goal motives, i.e., either as autonomous motivation that is volitional or as controlled motivation that involves the experience of being pressured or coerced. Vansteenkiste, Lens, and Deci (2006) used an example to demonstrate the difference between those two concepts:"… students could have an after-school job to earn money (extrinsic goal content) because they feel pressured by their parents (controlled motive) or because they value going to college and will need the money (autonomous motive)." Past research has shown that goal content and goal motives have unique effects on well-being and psychological adjustment (Sheldon, Ryan, Deci, & Kasser, 2004).

Nonetheless, short-term goals of students' learning activities may be manipulated and experimental manipulations represent the newer direction of goal contents theory research. Vansteenkiste and colleagues conducted such manipulations in order to induce different goal contents. Learning activities were framed by indicating their instrumentality for attaining future goals (intrinsic vs. extrinsic). It was found that goal contents could be manipulated to some extent and that manipulated goal contents affected behaviors and learning outcomes in much the same way that non-manipulated life goal contents affected well-being and adjustment outcomes (Vansteenkiste, Simons, Lens, Soenens, & Matos, 2005; Vansteenkiste, Simons, Soenens et al., 2004). Although in general, people may establish relatively stable intrinsic or extrinsic goals, goal content manipulations/influences are not rare in daily life. For instance, teachers may orient students' attention towards external signs of success, such as being well-known and admired (i.e., status), or being wealthy and rich (i.e., financial success); alternatively, they may encourage their students to develop their talents and skills (i.e., self-development), or to help people in need (i.e., community contribution). Placing different emphasis on these goal contents may result in different short or long-term outcomes. Findings from past research suggest that intrinsic goals are more conducive to individual functioning and adjustment than extrinsic goals (see Gollwitzer & Moskowitz, 1996; Ryan, Sheldon, Kasser, & Deci, 1996; Vansteenkiste, Timmermans, Lens, Soenens, & Van den Broeck, 2008) and that extrinsic goal framing (manipulation) undermined learning and persistence, regardless of the initial value placed on extrinsic goals (Vansteenkiste, Duriez, Simons, & Soenens, 2006).

1.2 Goal contexts

SDT is an organismic dialectical approach. It emphasizes the interactions between the active organism (the self) and the social context. SDT researchers have explored how social contexts can promote adjustment outcomes (e.g., Assor, Kaplan, & Roth, 2002; Deci, Eghrari, Patrick, & Leone, 1994; Deci & Ryan, 2000; Guay, Ratelle, & Chanal, 2008). Autonomy-supportive contexts tend to facilitate autonomous motivation, and controlling contexts tend to facilitate controlled motivation (Black & Deci, 2000; Sheldon & Krieger, 2007; Williams & Deci, 1996). SDT holds that, in contrast to controlling contexts, an autonomy-supportive environment is associated with more desirable effects (Assor, Roth, & Deci, 2004; Chirkov & Ryan, 2001; Grolnick, Kurowski, Dunlap, & Hevey, 2000; Grolnick & Pomerantz, 2009; Soenens & Vansteenkiste, 2005).

A goal can be introduced and communicated in either type of context. If short-term goal contents are manipulated in different learning contexts, we could test goal content effect, goal context effect, as well as any interaction between them. The effects of manipulating goal contents within autonomy-supportive or controlling learning contexts have been studied by Vansteenkiste and colleagues in Western culture (e.g., Vansteenkiste, Simons, Lens et al., 2004; Vansteenkiste, Simons et al., 2005). They found that when intrinsic goals were pursued in an autonomy-supportive context the outcomes were most conducive to psychological well-being and that even in a controlling context, intrinsic goals were associated with more positive outcomes than extrinsic goals. To explain these findings, Vansteenkiste and colleagues argued that the autonomy-supportive context allowed people to experience the congruence of pursuing an intrinsic goal that was closely aligned with their basic psychological needs, whereas controlling contexts tended to thwart basic need satisfaction.

1.3 Self-determination theory in Chinese culture

China has its own unique history and culture. The pursuit of fortune, political power, and high social status has been traditionally valued and set as the ultimate goal by Chinese scholars and intellectuals. In ancient times, particularly since the Tang Dynasty (circa 700 AD), scholarship attained by vast reading was deemed as the fairest and most efficient way to climb the social ladder. Those excelling in scholarship (and tested via official examinations) were respected by all, and often assigned government positions by the emperor. Once becoming a government official, as a result of the raised status, the scholar would have an official residence and a state-provided salary, along with other benefits (Su, 2002). Because of this, the primary goal of most ancient Chinese scholars was to obtain an official position. Even as a youngster, Chinese were taught that scholarship was superior to all the other types of work and was the surest route to a better personal life (Chen & Uttal, 1988).

In today's China, scholarship is not tied to securing a government job, but is still very highly valued as a way to reach success. Academic achievement and good performance on exams remain the primary path of upward mobility, partly because of urban-rural economic and developmental differences and migration control (Chen & Uttal, 1988). However, the emphasis on achievement as well as high standards imposed by Chinese parents may result in students' relying on extrinsic motivations such as grades to maintain their interest in school.

China's basic education is examination-oriented (Chen & Uttal, 1988). In terms of policies and educational practices in China, nine years of education (Grade 1 through Grade 9; elementary and middle school education) is mandated by law; however, students have to compete very hard to enter top high schools through "Zhongkao," the Senior Secondary Education Entrance Examination. Typical middle school classrooms in China have approximately 60 or more students. Teaching and learning is often done in groups but assessment is almost always based on individual performance. Teaching is usually teacher-centered with students following instructions and passively receiving information. The teaching styles are usually controlling, restrictive, and authoritarian (Chiu, 1986). High-stakes testing, high educational expectations from parents, traditional values, and teaching practices that make comparisons transparent (e.g., test scores or rankings public to all teachers and/or students) contribute to a competitive school environment in China, even in middle schools.

Cross-cultural research has suggested that Chinese students are different from their Western counterparts in terms of psychological constructs such as perceptions of competence, task orientation, anxiety about academics, attributions for failure and success, and human values (Chiu, 1986; Hardré et al., 2006; Hong, 2001; Schwartz & Bilsky, 1990; Stevenson, Chen, & Lee, 1993). It also has been shown that autonomy is less supported in Asian societies (Iyengar & Lepper, 1999; Jang, Reeve, Ryan, & Kim, 2009; Kitayama, Markus, & Kurokawa, 2000; Olsen et al., 2002; Quoss & Wen, 1995; Schwartz, 1992; Triandis, 2001). In Chinese culture, the support of autonomy is not a common socialization practice because of the prevailing Confucian values (e.g., filial piety, humaneness, and ritual). Instead, Chinese culture emphasizes conformity and family interdependence (Bao & Lam, 2008; Chao & Tseng, 2002), and maintaining social harmony and family support is often seen as a lifelong obligation (Tseng, 2004). Thus, some cross-cultural perspectives suggest that the pursuit of autonomy hampers the development of satisfying rela tionships, and such conflicts might be especially problematic for the adjustment outcomes of individuals in collectivistic societies (Iyengar & Lepper, 1999).

Nevertheless, several studies based on SDT have indicated that the East-West differences are not so dramatic in terms of the relationships between autonomy and academic or psychological outcomes (Pu, 2006; Vansteenkiste, Lens, Soenens, & Luyckx, 2006; Vansteenkiste, Zhou, Lens, & Soenens, 2005). For example, Vansteenkiste, Zhou, Lens, and Soenens (2005) found that autonomy was positively related to adaptive learning attitudes, academic success, and personal well-being among Chinese learners. They also argued that within the SDT framework autonomy should be defined in terms of individual, phenomenological experience, rather than in terms of interpersonal, culturally bounded values. Based on this argument, SDT researchers in cross-cultural studies should focus on relationships between, rather than the amounts of, psychological constructs and outcomes. So far, there has been no experimental research examining the goal content and goal context effects among students in China, and the research here filled this gap in the literature.

1.4 Present research

Before this research, we conducted a survey study to identify the goal contents pursued by Chinese scholars. One hundred and three college students responded to the survey, and 75% of those students expressed learning goals as (a) being financially successful, (b) being rich, or (c) having many expensive possessions. We also surveyed 204 high school students and found that 81% of them had similar goals. From these surveys, we concluded that in general the learning goal contents of contemporary Chinese students are mostly extrinsic.

The present research included two experimental studies. In Study 1, we examined differences in learning outcomes when extrinsic or intrinsic goals were induced to Chinese middle school students. Because those two types of goals were thought to relate differently to basic need satisfaction and therefore produce different psychological outcomes, we hypothesized that Chinese students with extrinsic goals would report less positive and more negative learning outcomes than those with intrinsic goals. In Study 2, we examined differences in learning outcomes when extrinsic or intrinsic goals were induced to Chinese middle school students in an autonomy-supportive or controlling

learning context. We hypothesized that the goal content variable would interact with the context variable such that the autonomy-supportive learning context would offset the negative effects of extrinsic goals and enhance positive effects of intrinsic goals.

Dependent variables in Studies 1 and 2 were test performance, free-choice engagement, and test anxiety. Following the positive psychological tradition, we chose test performance and free choice engagement as dependent variables. Those two variables have been used in similar studies of different populations (e.g., Vansteenkiste, Simons, Lens, et al., 2004). More important, they are the key outcome variables in SDT. In addition, test anxiety was measured as an outcome variable because it, too, is a key factor in undermining student performance (Hembree, 1988). Moreover, it is common across a broad spectrum of educational settings (Griffin & Griffin, 1998).

2 Study 1

2.1 Method

2.1.1 Pilot study

Before Study 1, we conducted a pilot study to check the manipulation of goal contents with 133 middle school students from the same city as in Studies 1 and 2. This pilot study was carried out in a similar procedure as used in Pilot Study 1 of Vansteenkiste, Simons, Lens, Sheldon, and Deci (2004). After random assignment, 66 participants were in the intrinsic goal condition and 68 were in the extrinsic goal condition. They received the same instruction sheet in each condition as in Study 1 and read the same text as in Study 1 on improving creativity. All participants answered questions concerning the importance of intrinsic goals (three items on a 5-point Likert scale) and the importance of extrinsic goals (another three items on a 5-point Likert scale) of improving creativity. On the importance of intrinsic goals, the mean was 10.80 (SD=2.58) and 8.94 (SD=2.38) for those in the intrinsic goal condition and the extrinsic goal condition, respectively. The effect size (Cohen's d) was 0.75 (close to large effect). On the importance of extrinsic goals, the mean was 7.42 (SD=2.30) and 8.37 (SD=2.35) for those in the intrinsic goal condition and the extrinsic goal condition, respectively. The effect size (Cohen's d) was 0.41 (close to medium effect). The differences between students in the two goal conditions suggested effective goal content manipulation.

2.1.2 Participants

Participants were 188 middle school students from a city in central China. This city is a traditional agricultural economic zone where people's average social and economic status is at the middle level within China. In addition, this city has a long history of attaching high importance to education. As in most other Chinese middle schools, teaching in sampled schools is structured and teacher-centered. Students spend the majority of their day and evening time on in-class learning. Of those 188 participants in this study, 88 (47%) were female, and 100 (53%) were male. The mean age of the participants was 14.88 years (SD=0.71).

2.1.3 Procedure

The experiment took place during regular classes and students learned text materials about improving creativity as a class activity. We chose the topic of improving creativity because there was a nationwide initiative of quality education when the present study was conducted and cultivating creativity was considered part of this initiative. The teachers were contacted and agreed to participate without knowing anything about what was being examined until the study was completed. The teachers distributed written instructions (in Chinese) to students explaining their task.

There were two types of instruction sheets put in rotation in the same stack for the different manipulations. Those instruction sheets were handed out to students one by one, following the seating in each class to ensure randomization of the experimental manipulations. There were 95 participants in the intrinsic goal condition and 93 in the extrinsic goal condition. All instruction sheets were of similar length and looked similar with different content to ensure fidelity of the experiment. After receiving the instruction sheets, students were told to read them without any discussion. Students then engaged in the target activity of reading a text about improving creativity.

Next, each student was asked to write his or her name on the instruction sheet and then to turn it in at the end of the sessions (along with other materials subsequently explained). The instruction for the intrinsic goal condition read: "Learning how to improve creativity is very important. Reading the text about it will help you to know better about yourself and your potentials and hence contribute to your personal development." The instruction for the extrinsic goal condition included: "Learning how to improve creativity is very important. Reading the text about it will help you to make more money and to buy things you want through applying your knowledge."

After reading the text, students answered questions about their comprehension of it and completed the test anxiety inventory. Subsequently, students were told that there were additional exercise materials about improving creativity that they could practice if they chose to. Last, 1 week later, these exercise materials were collected and graded by their teachers.

2.1.4 Measures

Test performance Students' test performance was measured by eight questions following their reading the text on improving creativity. The first question was a multiple-choice item worth 10 points, and the next seven questions were short essay questions with the first five worth 10 points each and the last two 20 points each. The total possible score was 100 points (we used this scale because it is the most commonly used and familiar scale for classroom testing at schools in mainland China; we also would like to point out that there is little standardized testing in mainland China). Two teachers graded all eight questions. They were given the highest possible score for each question and sample answers for three or four scores were provided for each essay question. However, specific coding criteria were not provided for every possible score. Test items focused on conceptual rather than rote learning and those questions were similar to typical Chinese reading comprehension questions asked during a Chinese test of middle school students. An example essay question was "What does the author try to tell us by using the example of Alfred Nobel in the sixth paragraph?" The total possible score of this question was 10 points. A two-point sample answer was "The author tries to tell us that Alfred Nobel had a notebook." A six-point sample answer was "The author tries to tell us that Alfred Nobel had the habit of writing down spontaneous ideas in a notebook." A 10-point sample answer was "By using the example of Alfred Nobel, the author tries to tell us that it is important to keep track of spontaneous ideas by writing them down and that we can cultivate creativity this way." The two teachers grading test items were blind to students' conditions and did not know about the study design or purpose. The correlation between the two teachers' ratings was .92. Their ratings were averaged to form a performance score for each student.

Free-choice engagement Students were offered additional exercise materials about improving creativity. There were seven problems in the exercise materials. Four of them asked students to describe or to summarize an invention or discovery, and the other three problems asked them to think creatively to provide an answer. An example exercise problem asked students to think creatively about possible uses of a building brick. Students were asked to record time spent on each problem and were given 1 week to choose to work on those exercise materials. Two teachers

rated students' responses to each question on a 0–2 scale with possible scores of 0, 0.5, 1.0, 1.5, and 2.0. A score of 0 was given if the student spent very little time (less than 10 min) or did not write any meaningful sentences. A score of 0.5 was given if the student spent some time (30 min to 2 hr) and provided answers in the most usual way (e.g., writing about how building bricks could be used to build apartments). A score of 1.0 was given if the student spent some time (30 min to 2 hr) and provided more answers involving some creative thinking (e.g., writing about how uses of building bricks could be categorized based on the purposes of the buildings/projects). A score of 1.5 was given if the student spent more than 2 hr and provided unusual but sensible answers involving some creative thinking (e.g., writing about how different attributes of building bricks such as weight, shape, and price should be considered when choosing building bricks). A score of 2.0 was given if the student spent more than 2 hr and provided answers that reflected "thinking outside of the box" (e.g., anthropomorphizing a building brick and having it go on travels to find the meaning of its life). The free-choice engagement score was the total score on the seven exercises. Thecorrelation between the two teachers' ratings was high (r=0.91). The final score was the average of teachers' ratings.

Test anxiety We used the Test Anxiety Inventory to measure participants' test anxiety. The inventory was developed by Spielberger (1980) and was later translated into Chinese by Ye and Rocklin (1988). The Chinese version that was used in this study has been shown to have good reliability (Cronbach's α =0.88) and to be applicable to Chinese students (Ling & Fan, 2008). In this study, students responded to 20 items on a 4-point scale about how much they experienced specific symptoms of anxiety during the test following the reading. The total score was used to measure test anxiety. Cronbach's α of this measure was 0.87 in this study.

2.2 Results

Previous experimental goal content studies did not examine gender differences, either because only one gender group participated (e.g., Vansteenkiste, Simons, Lens, et al., 2004) or because of reasons unexplained (e.g., Sebire, Standage, & Vansteenkiste, 2009). We included gender as an independent variable to see whether it was a significant predictor. Adjusted variable means and associated standard errors of scores for intrinsic and extrinsic goal conditions were calculated for the two gender groups separately (see Table 1). For the free-choice engagement measure, it turned out that the majority of the participants (90%) responded to only one exercise problem. One reason may be that each of these exercises required a longer time to finish than those for the test performance measure (the requirement of each exercise was to write an essay of approximately 200 Chinese characters), and it was unlikely for students to complete all the questions within 1 week (group-learning consumed most of the time in class and there was little free time after class).

To test whether there were significant differences between the intrinsic and extrinsic goal conditions and between gender groups on test performance, free-choice engagement and test anxiety, we conducted a multivariate analysis of variance using a 2 (goal condition) × 2 (gender) design with the three outcome variables as dependent variables. By Wilks' criterion, the combined dependent variables were significantly different by goal condition, $F(3, 182)=17.96$, $p<0.001$, $\eta_p^2 =0.228$ (close to large effect), but not by gender, $F(3, 182)=0.76$, $p=0.54$, or by the interaction between goal condition and gender, $F(3, 182)=2.20$, $p=0.09$.

The main effects of goal condition on each of the dependent variables suggested that students framed in the intrinsic goal condition had better test performance, $F(1, 184)=15.56$, $p<0.001$, $\eta_p^2=0.08$, higher free-choice engagement, $F(1, 184)=14.64$, $p<0.001$, $\eta_p^2=0.07$, and less test anxiety, $F(1, 184)=19.79$, $p<0.001$, $\eta_p^2=0.09$. All those effects were medium or close to medium (Cohen, 1988).

2.3 Discussion

Study 1 provided initial support of goal content effects among Chinese middle school students on three learning-related outcomes: test performance, free-choice engagement, and test anxiety. According to SDT, learning environments that emphasize intrinsic (or extrinsic) goal contents may have similar functional effects on learning and achievement as individuals' relatively stable pursuit of intrinsic (or extrinsic) goals. Vansteenkiste and colleagues have demonstrated effects of manipulated goal contents on deep processing of learning materials, academic achievement, and persistence among students and adults in Western culture (e.g., Vansteenkiste, Simons, Lens, et al., 2004). This study extended the research and has found that goal content (through experimental manipulations) matters among Chinese middle school students as well.

3 Study 2

3.1 Method

3.1.1 Participants

A total of 395 middle school students from the same

Table 1. Adjusted Means and Standard Errors of Scores for Intrinsic and Extrinsic Goal Conditions ($N = 188$)

Goal	Test performance		Free-choice engagement		Test anxiety	
	M	SE	M	SE	M	SE
Intrinsic goal						
Female	45.43	2.25	0.75	0.07	32.89	1.28
Male	46.69	2.23	0.59	0.07	34.73	1.26
Extrinsic goal						
Female	37.81	2.41	0.32	0.07	38.54	1.37
Male	36.50	2.14	0.51	0.07	40.48	1.21

city in China as in Study 1 participated in this study as a regular class activity. One hundred and ninety participants (48%) were female, and 205 (52%) were male. The mean age of the participants was 14.46 years ($SD = 1.11$).

3.1.2 Procedure

Participants' teachers were first contacted and agreed to participate without knowing what the study was about until the study was completed. The teachers distributed written instructions (in Chinese) to their students prepared by the researchers.

There were four types of instruction sheets that were randomly distributed within each class (cell sizes ranged from 96 to 100). The four types of instruction sheets represent the four manipulation conditions: intrinsic goal in an autonomy-supportive learning context, extrinsic goal in an autonomy-supportive learning context, intrinsic goal in a controlling learning context, and extrinsic goal in a controlling learning context. The students and their teachers were not aware that there were different sets of instructions, and all instruction sheets were of similar length and looked similar with different content to ensure fidelity of the experiment. After reading their instructions, students were then engaged in a target activity of reading a text about improving creativity. Each student was asked to write his or her name on the instruction sheet and to turn it in at the end of the sessions (along with other materials explained below). As in Study 1, instructions for participants in the intrinsic goal conditions stated that "Learning how to improve creativity is very important. Reading the text about it will help you to know better about yourself and your potentials and hence contribute to your personal development." The instruction for the extrinsic goal conditions included "Learning how to improve creativity is very important. Reading the text about it will help you to make more money and to buy things you want through applying your knowledge." The learning context was also manipulated. Specifically, in the

autonomy supportive conditions, the instructions included "If you are interested, you may want to learn more about it. The following text provides information on this topic. You can decide to learn more about creativity enhancing strategies." In the controlling conditions, the instructions included "You must learn more about it. You do not have a choice. If you do not finish this required learning task, you will hardly graduate." The experiment had four conditions through manipulation of two factors: goal content and goal context.

After reading the text, students answered the same reading comprehension questions and completed the same Test Anxiety Inventory as in Study 1. They were also told that there were additional exercises about improving creativity that they could practice if they chose. Those exercises were the same as in Study 1. One week later, exercise books were collected and graded by their teachers in the same way as in Study 1.

3.1.3 Measures

Test performance, free-choice engagement, and test anxiety were measured the same way as in Study 1. In this study, the interrater reliability was .94 and .90 for test performance and free-choice engagement, respectively. Cronbach's α for the test anxiety measure in this study was 0.83.

3.2 Results

The adjusted means and standard errors of the three outcome variables for each condition are presented in Table 2. We conducted a multivariate analysis of variance to study how goal content and goal context may affect students' test performance, free-choice engagement, and test anxiety.

Because in this study, male and female students did not differ significantly on any of the three outcomes, and the gender variable or its interaction with the other independent variables was not statistically significant in the multivariate analysis of variance, gender was dropped

Table 1. Adjusted Means and Standard Errors of Scores for Intrinsic and Extrinsic Goal Conditions, by Learning Context ($N = 395$)

Goal condition	Test performance		Free–choice engagement		Test anxiety	
	M	SE	M	SE	M	SE
Autonomy–supportive context						
Intrinsic goal	44.36[ad]	1.23	0.67[be]	0.06	34.68[cf]	0.88
Extrinsic goal	35.84[ag]	1.25	0.28[bh]	0.06	40.44[c]	0.89
Controlling context						
Intrinsic goal	25.80[d]	1.23	0.48[e]	0.06	41.42[f]	0.87
Extrinsic goal	24.14[g]	1.23	0.51[h]	0.06	42.45	0.87

Note. Means with same–letter superscripts are statistically significant at the .05 level.

from the analysis. Therefore, the final multivariate analysis of variance was a 2 (goal content) × 2 (goal context) design with test performance, free–choice engagement, and test anxiety as the dependent variables.

With the use of Wilks' criterion, the combined dependent variables were significantly different by goal content, $F(3, 389) = 14.04$, $p<0.001$, $\eta_p^2 = 0.10$ (medium effect), goal context, $F(3, 389)=58.66$, $p<0.001$, $\eta_p^2=0.31$ (large effect), as well as the interaction between goal content and goal context, $F(3, 389) = 09.21$, $p<0.001$, $\eta_p^2=0.07$ (close to medium effect).

The main effects of goal content indicated that students in the intrinsic goal condition had better test performance, $F(1, 391)=017.00$, $p<0.001$, $\eta_p^2=0.04$ (small effect), higher free–choice engagement, $F(1, 391) =8.33$, $p=0.004$, $\eta_p^2=0.02$ (small effect), and less text anxiety, $F(1,391) = 14.91$, $p<0.001$, $\eta_p^2=0.04$ (small effect) than those in the extrinsic goal condition, averaging across the two types of goal contexts. The main effects of goal context indicated that students in the autonomy–supporting learning context had better test performance, $F(1, 391) =150.31$, $p<0.001$, $\eta_p^2=0.28$ (large effect), and less test anxiety, $F(1, 391)=24.79$, $p<0.001$, $\eta_p^2=0.06$ (small to medium effect), but did not differ in free–choice engagement than those in the controlling learning context, averaging across the two types of goal contents. The interaction between goal content and goal context was statistically significant for all three outcome variables: test performance, $F(1, 391) =7.72$, $p=0.006$, $\eta_p^2=0.02$ (small effect), free–choice engagement, $F(1, 391)=11.28$, $p=0.001$, $\eta_p^2=0.03$ (small effect), and test anxiety, $F(1, 391) = 7.24$, $p=0.007$, $\eta_p^2=0.02$ (small effect).

Simple effects of goal content suggested that goal induction made a difference in the autonomy–supportive learning context, but not in the controlling learning context. In an autonomy–supportive learning context, students in the intrinsic goal condition had better test performance, higher free– choice engagement, and less test anxiety (see Table 2). Simple effects of goal context revealed that the autonomy–supportive learning context produced better test performance, higher free–choice engagement, and less test anxiety than the controlling learning context in the intrinsic goal conditions. In the extrinsic goal conditions, students in the autonomy–supportive learning context had better test performance than those in the controlling learning context. The goal context did not make a difference on test anxiety for those in the extrinsic goal conditions. In addition, the autonomy–supportive learning context was associated with lower free–choice engagement than the controlling learning context in the extrinsic goal conditions.

3.3 Discussion

Study 2 was designed to extend the findings of Study 1 and to examine whether goal content manipulations would work differently in an autonomy–supportive learning context versus a controlling learning context. In the autonomy-supportive learning context, similar goal content effects were observed as in Study 1. However, in the controlling learning context, goal content effects were not statistically significant on the three outcome variables. It is worth nothing that the most positive outcomes were obtained when the task was associated with intrinsic goal induction and was introduced in an autonomy-supportive context. This suggests that intrinsic goals were more fully engaged and accepted by individuals when they were encountered in an autonomy supportive climate. This finding is consistent with findings in Western culture (Vansteenkiste, Simons, Lens, et al., 2004).

In the extrinsic goal conditions, the autonomy-supportive learning context resulted in better test performance than the controlling context; however the learning context did not make a statistically significant difference on test anxiety. It is surprising that the autonomy-supportive/extrinsic-goal combination resulted in lower free-choice engagement than the controlling/ extrinsic-goal condition. This finding of negative effect of autonomy-supportive context in the extrinsic-goal condition did not appear in experimental research among Western samples and was possibly a cross-culture difference. This inconsistency may be the result of various components of psychological control, such as love withdrawal (Ho, 1986), shaming procedures, and threats of abandonment (Wu et al., 2002), which are more frequent in Eastern societies, and the fact that those components are better accepted as a means of regulating Chinese adolescents' behaviors (Chao, 1994; Chao & Tseng, 2002; Olsen et al., 2002). As a consequence, once the application of these controlling strategies disappears, the behaviors may likely withdraw. It is also worth noting that although in the extrinsic goal conditions, free-choice engagement was greater in the controlling context than in the autonomy-supportive context, the greatest free-choice engagement was related to the combination of an intrinsic goal induced in an autonomy-supportive context.

Conclusions and limitations. This was the first experimental research of goal content and goal context effects with samples of Chinese middle school students. The present research included two studies. In Study 1, we aimed to examine the effects of manipulating goal contents on learning outcomes and hypothesized that Chinese students with extrinsic goals would report less positive and more negative learning outcomes than those with intrinsic goals. In Study 2, we aimed to examine the effects of goal content manipulations in two learning contexts and hypothesized that the goal content variable would interact with the context variable such that the autonomy-supportive learning context would offset the negative effects of extrinsic goals and enhance positive effects of intrinsic goals.

The two studies provided evidence for our primary hypotheses and study results are generally in line with SDT and goal contents theory (Deci & Ryan, 2008; Vansteenkiste, Lens, & Deci, 2006). Framing a learning activity in terms of an intrinsic or extrinsic goal attainment resulted in different short-term outcomes. Intrinsic goals were more conducive than extrinsic goals on test performance, free-choice engagement, and test anxiety. Future research is needed to investigate whether similar results would be found in different samples, such as those in different regions of China and whether interventions focused on intrinsic goal contents would generate long-term positive outcomes. We have also found that goal contexts and goal contents interacted with each other such that the negative effect of extrinsic goals could be offset in an autonomy-supportive environment for academic outcomes such as test performance and that the positive effect of intrinsic goals was enhanced in an autonomy-supportive learning environment.

From Study 2, the most positive outcomes were produced in the intrinsic goal/autonomy-supportive condition. However, the controlling context seemed to facilitate free-choice engage mentwhen an extrinsic goal was induced. This may be a cross-cultural difference and further research is needed to examine whether it would be replicated. The controlling context in this study likely made the extrinsic goal contents more salient by stating a consequence for not completing the reading task ("you will hardly graduate"). Under this condition, students might have felt pressure to do whatever they could to alleviate the possibility of the negative consequence. The controlling context may have had a carryover effect on free-choice engagement. The controlling context might also have been unclear, confusing, or unbelievable to students, resulting in ineffective manipulation (unlike in Study 1, there was no pilot study in Study 2 to check goal context manipulation). In the current two studies, the manipulated goal contents might also have served as short-term standards for success that students internalized. In Study 2, when students were offered choice after the controlling context was removed, they probably felt a sense of autonomy and the higher free-choice engagement might be a rebound.

Chinese culture values help to ensure that children will work diligently, and Chinese philosophy has traditionally emphasized malleability and the importance of the environment in the shaping and expression of human potential (Chen & Uttal, 1988). From this perspective, Chinese students are more likely to be cultivated to carry the incremental view of intelligence (Dweck, 1999) and more likely to hold approach-orientated goals (Elliot & Harackiewicz, 1996; Elliot & McGregor, 2001). However, goal orientations are different from goal contents. Goal contents reflect the outcomes that people pursue (Deci & Ryan, 2008), whereas goal orientations focus on the purpose—why an individual engages in certain behaviors (Kaplan, Middleton, Urdan, & Midgley, 2002). While goal contents could be manipulated as demonstrated in this and past research,

studies on achievement goal theory tend to treat goal orientations as attribute-like characteristics.

As a subtheory of SDT, goal contents theory grows out of the distinctions between intrinsic and extrinsic goals and their effect on motivation and wellness. However, the distinctions between intrinsic and extrinsic goals should not be confused with the distinctions between intrinsic and extrinsic motivation, or with the distinctions between autonomous and controlled motivation. Past research has suggested that what goals people pursue (extrinsic goal contents vs. intrinsic goal contents) and why people pursue them (autonomous vs. controlled motivation) have independent contributions to psychological well-being (Sheldon et al., 2004).

There are several limitations of the present research. First, although we used different reading instructions to induce intrinsic or extrinsic goals, students were different in their long-term learning goals. Although our pilot survey study suggested that the majority of Chinese students(in middle schools and in college) had extrinsic life goals, goal induction/framing may work differently for people with intrinsic life goals. Despite studies supporting SDT and showing that promoting extrinsic goals undermines learning regardless of whether the individuals are extrinsically or intrinsically oriented, there is a hypothesis that suggests induced goal contents would yield better learning outcomes when they are consistent with individuals' more stable life goal orientations (Hidi & Harackiewicz, 2000; Sagiv & Schwartz, 2000).

Second, the manipulations of experimental conditions were limited. In both studies, we used altered wording on a set of instructions. The goal content manipulations seemed effective from the pilot study of Study 1. However, we did not check the effectiveness of manipulating goal contexts. In Study 2, the controlling context was framed by including in the instruction "You must learn more about it. You do not have a choice. If you do not finish this required learning task, you will hardly graduate." While using the phrases "must" and "have to" might have resulted in reduced sense of autonomy, the possible consequence ("you will hardly graduate" might have been confusing or unbelievable to students. In addition, although significant results were observed, we do not believe that our manipulations would have long-term effects. For example, the controlling or autonomy-supportive context in Study 2 would likely wane and the general learning context in the classroom would resume.

Third, as a result of limited resources, our samples were from only one city in central China and the participants were all in junior high schools in the year before graduation. Chinese students are usually under great academic pressure during this last year of junior high school. The results may be not generalizable to the larger population of Chinese students.

Fourth, we did not explore the mechanism of goal content and goal context effects. According to SDT, basic psychological needs may be used to explain goal context effects (Deci & Ryan, 2000). Past mediational analyses have found that autonomous motivation mediated goal content and social context effects on learning-related outcomes (Vansteenkiste, Lens, & Deci, 2006; Vansteenkiste, Simons, Lens, et al., 2004). Sheldon, Ryan, Deci, and Kasser (2004) claimed that the negative effects of extrinsic goals could be attributable to personality traits such as high insecurity, low self-esteem, or low cooperation. Vansteenkiste, Neyrionck, Niemiec, Soenens, Witte, and Brock (2007) suspected that traits such as neuroticism could also explain goal content effects.

In sum, we have examined goal content and goal context effects among samples of Chinese middle school students, using experimental studies. Our findings indicate that goal content matters, particularly in autonomy-supportive learning contexts, and that goal context matters, par- ticularly with intrinsic goal contents. These findings may provide some implications for creating optimal learning environments in Chinese middle schools.

References

Assor, A., Kaplan, H., & Roth, G. (2002). Choice is good, but relevance is excellent: Autonomy-enhancing and suppressing teacher behaviours predicting students' engagement in schoolwork. *British Journal of Educational Psychology, 72*, 261 - 278. doi: 10.1348/000709902158883

Assor, A., Roth, G., & Deci, E. L. (2004). The emotional costs of parents' conditional regard: A self-determination theory analysis. *Journal of Personality, 72*(1), 47 - 88. doi: 10.1111/j.0022-3506.2004.00256.x

Bao, X.-H., & Lam, S.-F. (2008). Who makes the choice? Rethinking the role of autonomy and relatedness in Chinese children's motivation. *Child Development, 79*, 269 - 283. doi: 10.1111/j.1467-8624.2007.01125.x

Black, A. E., & Deci, E. L. (2000). The effects of instructors' autonomy support and students' autonomous motiva- tion on learning organic chemistry: A self-determination theory perspective. *Science Education, 84*, 740 - 756. doi: 10.1002/1098-237x(200011)84:6<740::aid-sce4>3.0.co;2 - 3

Chao, R. K. (1994). Beyond parental control and authoritarian parenting style: Understanding Chinese parenting through the cultural notion of training. *Child Development, 65*, 1111 - 1119. doi: 10.1111/1467-8624.ep7252822

Chao, R. K., & Tseng, V. (2002). Asian and American parenting. In M. H. Bornstein (Ed.), *Handbook of parenting* (2nd ed., Vol. 4, pp. 59 – 93). Mahwah, NJ: Erlbaum.

Chen, C., & Uttal, D. H. (1988). Cultural values, parents' beliefs, and children's achievement in the United States and China. *Human Development, 31*, 351 – 358. doi: 10.1159/000276334

Chirkov, V. I., & Ryan, R. M. (2001). Parent and teacher autonomy-support in Russian and U.S. adolescents: Com- mon effects on welling-being and academic motivation. *Journal of Cross-Cultural Psychology, 32*, 618 – 635. doi: 10.1177/0022022101032005006

Chiu, L.-H. (1986). Locus of control in intellectual situations in American and Chinese school children. *International Journal of Psychology, 21*, 167 – 176. doi: 10.1080/00207598608247582

Cohen, J. (1988). *Statistical power analysis for the bahavioral sciences* (2nd ed.). Mahwah, NJ: Erlbaum.

Deci, E. L., Connell, J. P., & Ryan, R. M. (1989). Self-determination in a work organization. *Journal of Applied Psychology, 74*, 580 – 590. doi: 10.1037/0021-9010.74.4.580

Deci, E. L., Eghrari, H., Patrick, B. C., & Leone, D. R. (1994). Facilitating internalization: The self-determination theory perspective. *Journal of Personality, 62*(1), 119 – 142. doi: 10.1111/1467-6494.ep9406221281

Deci, E. L., & Ryan, R. M. (1985). *Intrinsic motivation and self-determination in human behavior*. New York, NY: Plenum.

Deci, E. L., & Ryan, R. M. (2000). The 'what' and 'why' of goal pursuits: Human needs and the self-determination of behavior. *Psychological Inquiry, 11*, 227 – 268. doi: 10.1207/S15327965PLI1104 01

Deci, E. L., & Ryan, R. M. (2008). Facilitating optimal motivation and psychological well-being across life's domains. *Canadian Psychology, 49*(1), 14 – 23. doi: 10.1037/0708-5591.49.1.14

Duriez, B., Vansteenkiste, M., Soenens, B., & De Witte, H. (2007). The social costs of extrinsic relative to intrinsic goal pursuits: Their relation with social dominance and racial and ethnic prejudice. *Journal of Personality, 75*, 757 – 782. doi: 10.1111/j.1467–6494.2007.00456.x

Dweck, C. S. (1999). *Self-theories: Their role in motivation, personality, and development.* Philadelphia, PA: The Psychology Press.

Elliot, A. J., & Harackiewicz, J. M. (1996). Approach and avoidance achievement goals and intrinsic motivation: A mediational analysis. *Journal of Personality and Social Psychology, 70*, 461 – 475. doi: 10.1037/0022-3514.70.3.461

Elliot, A. J., & McGregor, H. A. (2001). A 2 × 2 achievement goal framework. *Journal of Personality and Social Psychology, 80*, 501 – 519. doi: 10.1037/0022-3514.80.3.501

Gollwitzer, P. M., & Moskowitz, G. B. (1996). Goal effects on action and cognition. In E. T. Higgins & A. W. Kruglanski (Eds.), *Social psychology: Handbook of basic principles* (pp. 361 – 399). New York, NY: Guilford Press.

Griffin, M. M., & Griffin, B. W. (1998). An investigation of the effects of reciprocal peer tutoring on achievement, self-efficacy, and test anxiety. *Contemporary Educational Psychology, 23*, 298 – 311. doi: 10.1006/ceps.1998.0971 Grolnick, W. S., Kurowski, C. O., Dunlap, K. G., & Hevey, C. (2000). Parental resources and the transition to junior high. *Journal of Research on Adolescence, 10*, 465 – 488. doi: 10.1207/SJRA1004 05.

Grolnick, W. S., & Pomerantz, E. M. (2009). Issues and challenges in studying parental control: Toward a new concep- tualization. *Child Development Perspectives, 3*, 165 – 170. doi: 10.1177/1477878509104321

Grouzet, F. M. E., Kasser, T., Ahuvia, A., Dols, J. M. F., Kim, Y., Lau, S., . . . (2005). The structure of goal contents across 15 cultures. *Journal of Personality and Social Psychology, 89*, 800 – 816. doi: 10.1037/0022–3514.89.5.800

Guay, F., Ratelle, C. F., & Chanal, J. (2008). Optimal learning in optimal contexts: The role of self-determination in education. *Canadian Psychology, 94*(1), 233 – 240. doi: 10.1037/a0012758

Haerens, L., Kirk, D., Cardon, G., De Bourdeaudhuij, I., & Vansteenkiste, M. (2010). Motivational profiles for secondary school physical education and its relationship to the adoption of a physically active lifestyle among university students. *European Physical Education Review, 16*, 117 – 139. doi: 10.1177/1356336 × 10381304

Halvari, A., & Halvari, H. (2006). Motivational predictors of change in oral health: An experimental test of self-determination theory. *Motivation and Emotion, 30*, 294 – 305. doi: 10.1007/s11031–006–9035–8

Hardre ′ , P. L., Chen, C. H., Huang, S. H., Chiang, C. T., Jen, F. L., & Warden, L. (2006). Factors affecting high school students' academic motivation in Taiwan. *Asia Pacific Journal of Education, 26*, 189 – 207. doi: 10.1080/02188790600937326

Hembree, R. (1988). Correlates, causes, effects, and treatment of test anxiety. *Review of Educational Research, 58*(1), 47 – 77.

Hidi, S., & Harackiewicz, J. M. (2000). Motivating the academically unmotivated: A critical issue for the 21st century. *Review of Educational Research, 70*, 151 – 179. doi: 10.3102/00346543070002151

Ho, D. Y. F. (1986). Chinese patterns of socialization: A critical review *The psychology of the Chinese people* (pp. 1 – 37).New York, NY: Oxford University Press.

Hong, Y. – Y. (2001). Chinese students' and teachers' inferences of effort and ability. In F. Salili, C.–Y. Chiu & Y. Y. Hong (Eds.), *Student motivation: The culture and context of learning* (pp. 105 – 120). New York, NY:

Plenum.

Iyengar, S. S., & Lepper, M. R. (1999). Rethinking the value of choice: A cultural perspective on intrinsic motivation. *Journal of Personality and Social Psychology, 76*, 349 – 366. doi: 10.1037/0022–3514.76.3.349

Jang, H., Reeve, J., Ryan, R. M., & Kim, A. (2009). Can self–determination theory explain what underlies the productive, satisfying learning experiences of collectivistically oriented Korean students? *Journal of Educational Psychology, 101*, 644 – 661. doi: 10.1037/a0014241

Kaplan, A., Middleton, M. J., Urdan, T., & Midgley, C. (2002). Achievement goals and goal structures. In C. Midgley (Ed.), *Goals, goal structures, and patterns of adaptive learning* (pp. 21 – 53). Mahwah, NJ: Erlbaum.

Kasser, T. (2002). Sketches for a self–determination theory of values. In E. L. Deci & R. M. Ryan (Eds.), *Handbook of self–determination research* (pp. 123 – 140). Rochester, NY: The University of Rochester Press.

Kasser, T., & Ryan, R. M. (1993). A dark side of the American dream: Correlates of financial success as a cen– tral life aspiration. *Journal of Personality and Social Psychology, 65*, 410 – 422. doi: 10.1037/0022–3514.65.2. 410

Kasser, T., & Ryan, R. M. (1996). Further examining the American dream: Differential correlates of intrinsic and extrinsic coals. *Personality and Social Psychology Bulletin, 22*, 280 – 287. doi: 10.1177/0146167296223006

Kasser, T., & Ryan, R. M. (2001). Be careful what you wish for: Optimal functioning and the relative attainment of intrinsic and extrinsic goals. In P. Schmuck & K. M. Sheldon (Eds.), *Life goals and well–being: Towards a positive psychology of human striving* (pp. 116 – 131). Ashland, OH: Hogrefe and Huber.

Kasser, V. G., & Ryan, R. M. (1999). The relation of

psychological needs for autonomy and relatedness to vitality, well–being, and mortality in a nursing home. *Journal of Applied Social Psychology, 29*, 935‑954. doi: 10.1111/j.1559‑ 1816.1999.tb00133.x

Kim, Y., Kasser, T., & Lee, H. (2003). Self–concept, aspirations, and well–being in South Korea and the United States. *The Journal of Social Psychology, 143*, 277‑290. doi: 10.1080/00224540309598445

Kitayama, S., Markus, H. R., & Kurokawa, M. (2000). Culture, emotion, and well–being: Good feelings in Japan and the United States. *Cognition and Emotion, 14*(1), 93‑124. doi: 10.1080/026999300379003

Levesque, C., Zuehlke, A. N., Stanek, L. R., & Ryan, R. M. (2004). Autonomy and competence in German and American university students: A comparative study based on self–determination theory. *Journal of Educational Psychology, 96*(1), 68‑84. doi: 10.1037/0022‑0663.96.1.68

Ling, X., & Fan, X. L. (2008). Analysis of application of test anxiety inventory in Changsha middle school students. *Chinese Journal of Clinical Psychology, 2*, 146‑150.

Losier, G. F., & Koestner, R. (1999). Intrinsic versus identified regulation in distinct political campaigns: The consequences of following politics for pleasure versus personal meaningfulness. *Personality and Social Psychology Bulletin, 25*, 287‑298. doi: 10.1177/0146167299025003002

Neyrinck, B., Vansteenkiste, M., Lens, W., Duriez, B., & Hutsebaut, D. (2006). Cognitive, affective and behavioral correlates of internalization of regulations for religious activities. *Motivation and Emotion, 30*, 321‑332. doi: 10.1007/s11031‑006‑9048‑3

Olsen, S. F., Yang, C., Hart, C. H., Robinson, C. C., Wu, P., Nelson, D. A., ... Wo, J. (2002). Maternal psychological control and preschool children's behavioral outcomes in China, Russia, and the United States. In B. K. Barber (Ed.), *Intrusive parenting: How psychological control affects children and adolescents* (pp. 235‑262). Washington, DC: American Psychological Association.

Pu, Z. (2006). *The experimental research on internalization of extrinsic learning motivation of students at junior high school*. Changchun, China: Unpublished doctoral dissertation, Northeast Normal University.

Quoss, B., & Wen, Z. (1995). Parenting styles and children's satisfaction with parenting in China and the United States. *Journal of Comparative Family Studies, 26*, 265‑280.

Romero, E., Go′mez‑Fraguela, J. A., & Villar, P. (2011). Life aspirations, personality traits and subjective well–being in a Spanish sample. *European Journal of Personality*. Retrieved from http://onlinelibrary.wiley.com/doi/10.1002/per. 815/pdfdoi:10.1002/per.815

Ryan, R. M., Chirkov, V. I., Little, T. D., Sheldon, K. M., Timoshina, E., & Deci, E. L. (1999). The American dream in Russia: Extrinsic aspirations and well–being in two cultures. *Personality and Social Psychology Bulletin, 25*, 1509‑1524. doi: 10.1177/01461672992510007

Ryan, R. M., & Deci, E. L. (2000). Intrinsic and extrinsic motivations: Classic definitions and new directions. *Contem‑ porary Educational Psychology, 25*(1), 54‑67. doi: 10.1006/ceps.1999.1020

Ryan, R. M., Sheldon, K. M., Kasser, T., & Deci, E. L. (1996). All goals are not created equal: An organismic perspective on the nature of goals and their regulation. In P. M. Gollwitzer & J. A. Bargh (Eds.), *The psychology of action: Linking cognition and motivation to behavior* (pp. 7‑26). New York, NY: Guilford Press.

Sagiv, L., & Schwartz, S. H. (2000). Value priorities and subjective well–being: Direct relations and congruity ef‑ fects. *European Journal of Social Psychology, 30*, 177‑198. doi: 10.1002/(sici)1099‑0992(200003/04)30:2<177::aid‑ ejsp982>3.0.co;2‑z

Schmuck, P., Kasser, T., & Ryan, R. M. (2000). Intrinsic and extrinsic goals: Their structure and relationship to well-being in German and U.S. college students. *Social Indicators Research, 50*, 225 – 241. doi: 10.1023/ a:100708400- 5278

Schwartz, S. H. (1992). Universals in the content and structure of values: Theoretical advances and empirical tests in 20 countries. In M. P. Zanna (Ed.), *Advances in experimental social psychology* (Vol. 25, pp. 1 – 65). New York, NY: Academic Press.

Schwartz, S. H., & Bilsky, W. (1990). Toward a theory of the universal content and structure of values: Extensions and cross-cultural replications. *Journal of Personality and Social Psychology, 58*, 878 – 891. doi: 10.1037/0022-3514.58.5. 878

Sebire, S. J., Standage, M., & Vansteenkiste, M. (2009). Examining intrinsic versus extrinsic exercise goals: Cognitive, affective, and behavioral outcomes. *Journal of Sport and Exercise Psychology, 31*, 189 – 210.

Sheldon, K. M., & Krieger, L. S. (2007). Understanding the negative effects of legal education on law students: A longitudinal test of self-determination theory. *Personality and Social Psychology Bulletin, 33*, 883 – 897. doi: 10.1177/0146167207301014

Sheldon, K. M., Ryan, R. M., Deci, E. L., & Kasser, T. (2004). The independent effects of goal contents and motives on well-being: It's both what you pursue and why you pursue it. *Personality and Social Psychology Bulletin, 30*, 475 – 486. doi: 10.1177/0146167203261883

Soenens, B., & Vansteenkiste, M. (2005). Antecedents and outcomes of self-determination in 3 life domains: The role of parents' and teachers' autonomy support. *Journal of Youth and Adolescence, 34*, 589 – 604. doi: 10.1007/s10964-005- 8948-y

Spielberger, C. D. (1980). *Test anxiety inventory: Preliminary professional manual*. Palo Alto, CA: Consulting Psychol- ogists Press.

Standage, M., Duda, J. L., & Ntoumanis, N. (2005). A test of self determination theory in school physical education.*British Journal of Educational Psychology, 75*, 411 – 433. doi: 10.1348/000709904 × 22359

Stevenson, H. W., Chen, C., & Lee, S. (1993). Motivation and achievement of gifted children in East Asia and the United States. *Journal for the Education of the Gifted, 16*, 223 – 250.

Su, X. (2002). *Education in China: Reforms and innovations*. Beijing, China: China Intercontinental Press.

Triandis, H. C. (2001). Individualism-collectivism and personality. *Journal of Personality, 69*, 907 – 924. doi: 10.1111/1467-6494.696169

Tseng, V. (2004). Family interdependence and academic adjustment in college: Youth from immigrant and U.S.-born families. *Child Development, 75*, 966 – 983. doi: 10.1111/j.1467-8624.2004.00717.x

Vansteenkiste, M., Duriez, B., Simons, J., & Soenens, B. (2006). Materialistic values and well-being among business students: Further evidence of their detrimental effect. *Journal of Applied Social Psychology, 36*, 2892 – 2908. doi: 10.1111/j.0021-9029.2006.00134.x

Vansteenkiste, M., Lens, W., & Deci, E. L. (2006). Intrinsic versus extrinsic goal contents in self-determination theory: Another look at the quality of academic motivation. *Educational Psychologist, 41*(1), 19 – 31. doi: 10.1207/s15326985ep4101 4

Vansteenkiste, M., Lens, W., Soenens, B., & Luyckx, K. (2006). Autonomy and relatedness among Chinese sojourners and applicants: Conflictual or independent predictors of well-being and adjustment? *Motivation and Emotion, 30*, 273 – 282. doi: 10.1007/s11031-006-9041-x.

Vansteenkiste, M., Neyrinck, B., Niemiec, C. P., Soenens, B., De Witte, H., & Van Den Broeck, A. (2007). On the relations among work value orientations, psychological need satisfaction and job outcomes: A self-determination theory approach. *Journal of Occupational and Organizational Psychology, 80*, 251 – 277. doi: 10.1348/096317906 × 111024

Vansteenkiste, M., Simons, J., Lens, W., Sheldon, K. M., & Deci, E. L. (2004). Motivating learning, performance, and persistence: The synergistic effects of intrinsic goal contents and autonomy-supportive contexts. *Journal of Personality and Social Psychology, 87*, 246 – 260. doi: 10.1037/0022-3514.87.2.246

Vansteenkiste, M., Simons, J., Lens, W., Soenens, B., & Matos, L. (2005). Examining the motivational impact of intrinsic versus extrinsic goal framing and autonomy-supportive versus internally controlling communication style on early adolescents' academic achievement. *Child Development, 76*, 483 – 501. doi: 10.1111/j.1467-8624.2005.00858.x

Vansteenkiste, M., Simons, J., Soenens, B., & Lens, W. (2004). How to become a persevering exerciser? Providing a clear, future intrinsic goal in an autonomy-supportive way. *Journal of Sport and Exercise Psychology, 26*, 232 – 249.

Vansteenkiste, M., Timmermans, T., Lens, W., Soenens, B., & Van den Broeck, A. (2008). Does extrinsic goal framing enhance extrinsic goal-oriented individuals' learning and performance? An experimental test of the match per-spective versus self-determination theory. *Journal of Educational Psychology, 100*, 387 – 397. doi: 10.1037/0022-0663.100.2.387

Vansteenkiste, M., Zhou, M., Lens, W., & Soenens, B. (2005). Experiences of autonomy and control among Chinese learn-ers: Vitalizing or immobilizing? *Journal of Educational Psychology, 97*, 468 – 483. doi: 10.1037/0022-0663.97.3.468.

Williams, G. C., & Deci, E. L. (1996). Internalization of biopsychosocial values by medical students: A test of self-determination theory. *Journal of Personality and Social Psychology, 70*, 767-779. doi: 10.1037/0022-3514.70.4.767

Williams, G. C., Hedberg, V. A., Cox, E. M., & Deci, E. L. (2000). Extrinsic life goals and health-risk behaviors in adolescents. *Journal of Applied Social Psychology, 30*, 1756 – 1771. doi: 10.1111/j.1559-1816.2000.tb02466.x.

Wu, P., Robinson, C. C., Yang, C., Hart, C. H., Olsen, S. F., Porter, C. L., . . . Wu, X. (2002). Similarities and differences in mothers' parenting of preschoolers in China and the United States. *International Journal of Behavioral Development, 26*, 481 – 491. doi: 10.1080/01650250143000436.

Ye, R., & Rocklin, T. (1988). A cross-cultural study of test anxiety. *Psychological Science, 3*, 25 – 29.

Evolutionary Psychology, 2014, 12(4), 719 - 735.

Red is romantic, but only for feminine females: Sexual dimorphism moderates red effect on sexual attraction

Fangfang Wen, Bin Zuo, Yang Wu, Shan Sun, Ke Liu

(School of Psychology, Central China Normal University, Wuhan, Hubei Province, PR China)

Abstract Previous researchers have documented that the color red enhances one's sexual attraction to the opposite sex. The current study further examined the moderating role of sexual dimorphism in red effects. The results indicated that red enhanced men's sexual attraction to women with more feminine facial characteristics but had no effect on ratings of perceived general attractiveness. Red clothing also had a marginally significant effect on men's sexual attractiveness. In addition, regardless of sexual dimorphism cues, male participants rated women with red as warmer and more competent. The underlying mechanisms of the red effect, the limitations of the current study, and suggestions for future directions are discussed.

Keywords red effect; masculine; feminine; sexual attraction; perceived attractiveness

1 Introduction

Color is an omnipresent and inseparable property of objects that constitute our perceptual world. Academic research has abounded on the physics, physiology, and aesthetics of color, but not until recently did psychologists formulate theories on its psychological meaning and effects (Elliot and Maier, 2014). Color-in-context theory, advanced by Elliot and colleagues, argues that color carries psychologically relevant meanings that are rooted either in biological or social learning processes (though they are not necessarily mutually exclusive). Viewing a color in a specific context may automatically evoke several psychological processes, ranging from affect and cognition to behavior intentions (Elliot and Maier, 2012).

Recent studies have documented a unique effect of red on interpersonal attraction (Elliot and Niesta, 2008). Most researchers have concentrated on how and to what extent red enhances men's attraction to women. For example, several studies from America and Europe have consistently shown that women paired with a red background or clad in red are perceived as more sexually attractive by men (Elliot and Niesta, 2008; Guéguen, 2012a; Pazda, Elliot, and Greitemeyer, 2012, 2014; Re, Whitehead, Xiao,

This research is supported by Humanities and Social Science Foundation of Ministry of Education of China for Young Scholars (grant No.13YJC190023), Key Projects of Philosophy and Social Sciences Research, Ministry of Education of China (grant No.11JZD006), and Sub-project (No.CCNU11C01005-24) under Major Cultivation Project for Scientific Research (No.CCNU11C010015), Central Government's Fundamental Research Funds for Universities.

and Perrett, 2011; Roberts, Owen, and Havlicek, 2010; Schwarz and Singer, 2013; Stephen and McKeegan, 2010). However, perceived general attractiveness and other traits were affected less or not at all (Elliot et al., 2010; Elliot, Tracy, Pazda, and Beall, 2013; Schwarz and Singer, 2013). Moreover, research has demonstrated that the effect of red could also be extended to behaviors. For example, when dating a girl dressed in red, men tend to ask questions that are more intimate and show more sexual interest during conversation (Niesta-Kayser, Elliot, and Feltman, 2010). They also offer more help (Gu é guen, 2012a) and are more likely to send a contact solicitation in an online dating context (Gu é guen and Jacob, 2013). Men in bars send more solicitations to women wearing red lipsticks (Gu é guen, 2012b), and male customers to a restaurant tend to tip waitresses wearing red lipstick or red clothes more generously (Gu é guen and Jacob, 2012, 2014). On the other hand, a few studies also investigated the red effect on male attractiveness as perceived by women. Through a series of experiments, Elliot et al. (2010) indicated that women find men paired with a red background or dressed in red more sexually attractive and desirable, and further established that the perception of status mediated the red effect on male attractiveness.

Both social conditioning and evolutionary biology can explain the red effect on sexual attraction. In terms of social conditioning, the red effect may be construed as resulting from the repeated pairing of red and some particular concepts, information, and experiences. Over time, these pairings will form strong implicit associations such that the perception of red alone can activate culturally-conditioned psychological reactions (Elliot and Maier, 2012, 2014). Because red has been associated with romance, passion, lust, and fertility across nearly every long-standing civilization (Hutchings, 2004; Kaya and Epps, 2004; Lee, 2006; Neto, 2002), it could be argued that the effect of red in enhancing sexual attraction is

only a reflection of the color's past associative history. Alternatively, biological evolution can also account for the salience of red in sexuality, as red serves to signal sexual preparedness related to the reproductive process (Deschner, Heistermann, Hodges, and Boesch, 2004; Lynn, McCord, and Halliwell, 2007). During the estrus phase, when heightened sexual receptivity is conducive to conception, the genitals of female chimpanzees swell and redden as a sexual cue to males (Buss, 2008). Therefore, it could also be argued that red may suggest higher sexual opportunity for males. Pazda et al. (2012) found parallel mechanisms in humans, showing that sexual receptivity indeed mediates the red effect on women's sexual attractiveness. With regard to males, red usually indicates higher status in many vertebrates (e.g., Setchell and Dixson, 2001; for a short review, see Elliot and Maier, 2012, p. 94), and females evolve to prefer males who are high in status in order to solve the adaptive problem of securing resources during pregnancy (Buss, 2008). Elliot et al. (2010) verified the mediating role of status in red's effect on male sexual attractiveness. Thus, the effect of red in boosting sexual attractiveness may stem from its concomitance with physiological and psychological processes that are closely related to reproduction. It is crucial to note that although the two sources of red's effects have distinct mechanisms, they are not mutually exclusive and may operate conjointly (Elliot and Maier, 2014). The biological influence might drive or reinforce the cultural association of red, and the cultural norm may generalize or extend its inherent meaning.

Consistent with a biological evolutionary perspective, recently Schwarz and Singer (2013) found a moderator to the red effect. Their study showed that relative to menopausal women, only the younger women were perceived as more sexually attractive when paired with a red rather than white background. If red signals a possibility of conception and fertility and post-menopausal women are unable to conceive, then the color red would lose its status as a signal of reproductive value. Therefore, red would cease to enhance men's sexual attraction. This possibility suggests that sociocultural perspectives alone do not explain the effect of red. If social learning theory is correct, and it is the cultural meaning and the resulting mental association of red that enhances sexual attraction, then red should have an effect on women's attractiveness regardless of whether or not they are menopausal. Additionally, there might be other boundary conditions for the red effect. It could be inferred from above that the factors influencing the perceived reproductive value or future fertility of a target can moderate or change the intensity or directions of the red effect. Consistent with this reasoning, since facial cues of sexual dimorphism (masculine and feminine) are another factor reflecting one's reproductive value in addition to age, they may also moderate the red effect. Therefore, not all individuals would be equally affected by red, and its effect might differ in individuals with different facial traits of masculinity and femininity.

Previous work has already established that the facial cues of sexual dimorphism may be linked with female reproductive value and male status. For female faces, the femininity cues may reflect the level of hormonal secretion, especially estrogen (Law-Smith et al., 2005), implying fertility and an advantage in reproduction (Gesquiere, Wango, Alberts, and Altmann, 2007; Morrison, Clark, Tiddeman, and Penton-Voak, 2010). As for male faces, facial masculinity is positively related to testosterone level (Kasperk et al., 1997; Penton-Voak and Chen, 2004; Pound, Penton-Voak, and Surridge, 2009; Roney, Hanson, Durante, and Maestripieri, 2006), and the level of testosterone is usually associated with status and dominance (e.g., Josephs, Sellers, Newman, and Mehta, 2006). Hence, it came as no surprise that males with masculine facial cues tend to have higher dominance

(Boothroyd, Jones, Burt, and Perrett, 2007; Perrett et al., 1998), which is often a signal to status attainment in groups (Anderson and Kilduff, 2009). Taken together, whereas femininity in females often signals higher reproductive value, masculinity in males is associated with dominance and status.

Given that the majority of recent studies on red effect have been conducted on American or European participants (for an exception, see Elliot et al., 2010, Experiment 4), a replication in a different culture is especially important. In China, red has traditionally been rich in meanings. Red was a symbol for status and stood for official titles in ancient official belief systems, and in folk belief systems red is still related to jubilation, auspiciousness, and fortune, such that traditional festivals like the Spring Festival have always been noted for red decorations (Chen, 2007). Critically, red in China is also closely entwined with images of females and sex. "Red face," a Chinese word, is a byword for young girl, and it could refer to an intimate female friend of a man (yet usually without a sexual relationship) (Shang, 2008). Traditional Chinese weddings are excessive in their use of red, and colloquially "fallen red" refers specifically to the act of "deflowering a virgin" and the blood stain that may be left (Eberhard, 2013). The evolutionary view that the meaning of red may at least partially stem from human biology would be given more credence if we found evidence in China that red can also enhance women's sexual attractiveness, but not general attractiveness, and that this effect is moderated by facial sexual dimorphism cues.

It is currently unknown whether all young targets are perceived as more sexually attractive when presented along with red, regardless of their masculine or feminine cues. It is the intention of this study to investigate this issue. We hypothesize based on evolutionary theory that women with feminine traits are more susceptible to the red effect. Conversely, masculine traits signal relatively lower reproductive value and therefore may lower or even eliminate the red effect. As for men, in line with Elliot et al. (2010), we hypothesize that red only elevates the sexual attractiveness of masculine men. Conversely, if a sociocultural mechanism is the only viable explanation and the red effect stems entirely from the association between red and sex, fertility, or romance, the effect of red should be uniform despite the varying facial cues of sexual dimorphism.

2 Methods

2.1 Participants

A total of 299 students (149 men and 150 women) from a university in Wuhan participated voluntarily for a gift. We deleted the data from 16 participants because they did not follow the instructions, and their responses on the questionnaire were uninterpretable. This resulted in a final sample of 283 participants (139 males and 144 females). The mean age of participants was 20.95 years ($SD = 1.95$), and the range was from 18 to 26 years. All participants were heterosexual and without histories of mental disorder. The participants had normal or corrected to normal vision, and had normal color vision.

2.2 Design

The present study adopted a $2 \times 3 \times 2 \times 2$ mixed design, in which three between–subject variables were gender of participants (male, female), clothing color (red, blue, white), and cues of sexual dimorphism (masculine, feminine). The within–subject variable was the gender of the target (male, female). The dependent variables included 1) sexual attractiveness, 2) general attractiveness, and 3) warmth and competence. The sexual attractiveness was only rated on a target of opposite sex with participants, so the analysis of sexual attractiveness involves only a $2 \times 3 \times 2$ between–subject design.

2.3 Materials

Unlike previous research on red effects, we generated our stimuli using computer– synthesized (rather than natural) photos in order to manipulate the sexual dimorphism cues (DeBruine, Jones, Crawford, Welling, and Little, 2010; Wen and Zuo, 2012). First, we created an averaged facial archetype of both sexes using computer graphic techniques. The face images were obtained from a large database of a university containing facial photographs of male and female graduates posed with the same background, uniform luminance, and neutral facial expressions. A total of 321 images comprising 144 men and 177 women were available, from which 32 images were selected for each gender. Images with eyeglasses, moustaches, or jewelry were excluded. All 64 images were used to generate two average images with FantaMorph 4.0 software. We marked 179 key points in each face that, as a whole, delineated the shape and contour of the face and its delicate features. Two photos were averaged along the values registered by the 179 key points, and the averaged images were further averaged in an identical manner. The processes were repeated several times until we obtained a single averaged image for each sex (Wen and Zuo, 2012), which is shown in Figure 1.

With the archetype images created, we then chose another photo for each sex from the same database to create

Figure. 1. Averaged facial archetype of both sexes used in stimuli construction

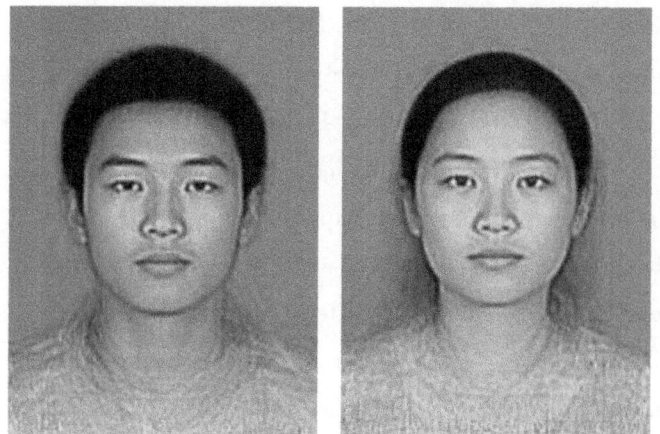

the stimuli. The two photos were rated on attractiveness by 87 undergraduates (41 males and 46 females, M_{age} =20.77, SD=1.63) with a 1 (very unattractive) to 7 (very attractive) Likert scale to ensure that the selected photos have a medium level of attractiveness. The results indicated that the facial attractiveness rating of the male photo was 3.94 (SD=0.89), and the rating of the female photo was 4.03 (SD=0.86). The two photos were then masculinized and feminized using the sexual dimorphism techniques developed by Perrett et al. (1998). In short, we exaggerated or diminished the feature differences between the two photos and the archetypes to create the masculinized and feminized version of faces for both sexes. The operation was performed in DeBruine et al.'s website www.faceresearch. org. Because the people in the original photos were dressed in a white T-shirt, we created a red and a blue version with Adobe Photoshop CS2, using the color selected from materials described in the study by Meier, D'Agostino, Elliot, Maier, and Wilkowski (2012). The parameters for the colors red and blue were LCh(51.1/57.7/27.8) and LCh(51.6/57.6/278.3), respectively (see Figure 2). The resultant 12 photos constituted the material used in the experiment.

2.4 Measures of dependent variables

All measures were rated on a 1 (very un–···) to 7 (very ···) Likert scale.

Sexual attractiveness. We used three items similar to those described in Elliot and Niesta's (2008) study to measure the sexual attractiveness of the target: "How sexually attractive do you think the person is?", "Would you want to have an intimate relationship (euphemism for sexual intercourse in Chinese) with this person?", and "How much do you find this person sexually desirable?" Since the Cronbach's alpha is only .66, we conducted an exploratory factor analysis to test if there is a single factor underlying the three items. The principal axis method was used to extract the common factors. Based on the scree plot, only one factor is the most apposite, accounting for 59.84% of the variance. Moreover, because there was only one factor, no rotation was needed. Based on the results of the exploratory factor analysis, we generated an index of sexual attractiveness using factor scores with the regression method (DiStefano, Zhu, and Mîndrilă , 2009; Grice, 2001). The index is a standardized score with a mean of zero.

General attractiveness. We also used three items to

Figure. 2. Facial stimuli with different sex, color of clothes and sexual dimorphism cues

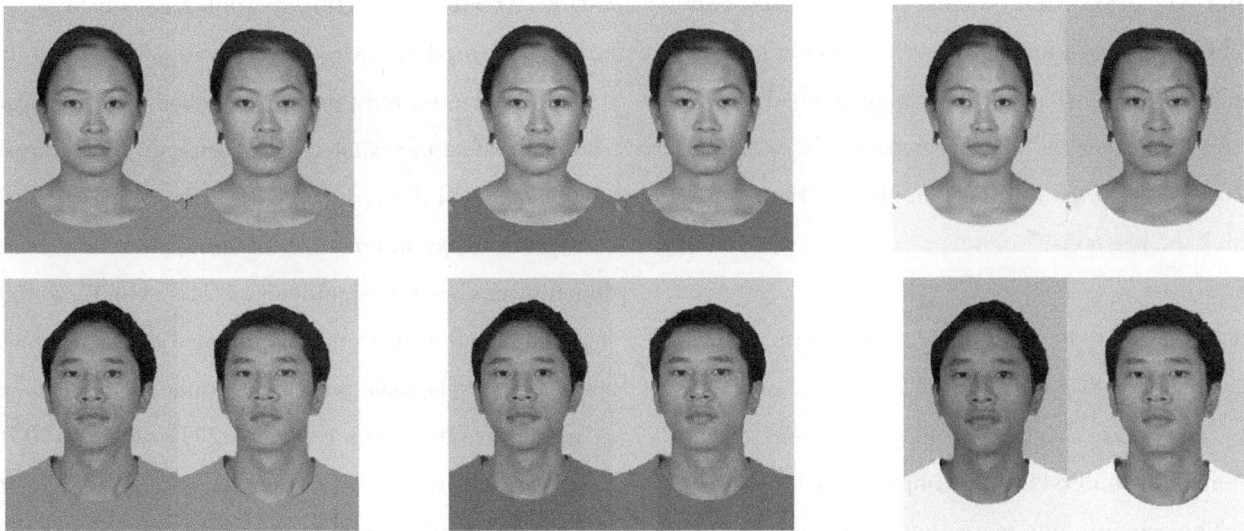

Note. The upper half is female faces and the bottom half is male faces; from left to right is faces with red, blue and white clothes. In each pair, the left image is feminine and the right is masculine.

assess perceived general attractiveness: "How attractive do you find this person?", "How good-looking do you think the person is?", and "If you meet this person face to face, how attractive do you think he/she is?" The Cronbach's alpha is .78 based on the opposite sex rating and .81 in the same sex rating. Due to the medium nature of the reliability, we also generated factor scores in each condition instead of simply summing the three items.

We conducted an exploratory factor analysis using the principal axis method to extract common factors for the opposite sex ratings. The scree plot suggested a single factor as most suitable, and it accounted for 69.17% of the total variances. Because there was only one factor, no rotation was needed. Based on the results, we generated a standardized index of perceived general attractiveness using a regression method. The index has a mean of zero.

The case is similar in same sex ratings. Only one factor emerged, which accounted for 73.07% of total variances. Hence, we generated a similar standardized index.

Warmth and competence. In addition to attractiveness ratings, the studies by Elliot et al. (2008) and Schwarz et al. (2013) both measured other traits of the targets such as sympathy, intelligence, agreeableness, kindness, etc. Based on the stereotype content model (Fiske, Cuddy, Glick, and Xu, 2002), the current study adopted two traits: competence and warmth. These traits could loosely encapsulate the dimensions of the aforementioned traits. The item for competence is "How competent do you think the person is?", and the item for warmth is "How warm do you think the person is?"

2.5 Procedure

Upon arriving at the laboratory, participants were randomly assigned to one of the 12 conditions formed by the $2 \times 3 \times 2$ design and were told that the goal of the experiment is to understand the impression formation process. The participants were then handed a folder containing a facial photo and a questionnaire. The participants were instructed to look at the photo for 5 seconds before filling out the questionnaire, which included demographic information and the measures of dependent variables. After completing the questionnaire, the participants were thanked, debriefed, given a gift, and dismissed. The Committee on Research Ethics approved the entire procedure and materials.

3 Results

3.1 Sexual attractiveness

We conducted a 2 (participant gender: male vs. female) $\times$ 3 (color of cloth: red, blue, and white) $\times$ 2 (sexual dimorphism cues: masculine vs. feminine) three-way between-subject ANOVA on the standardized index of sexual attractiveness. The results indicated a significant main effect of participant gender, $F(1,271) = 7.80$, $p = 0.006$, $\eta_p^2 = 0.03$. Specifically, male participants ($M = 0.14$, $SD = 0.81$) found the opposite sex target more sexually attractive than did female participants ($M = -0.13$, $SD = 0.81$). The main effect of clothes color was marginally significant, $F(2,271) = 2.80$, $p = 0.06$, $\eta_p^2 = 0.02$. Whereas participants found the target wearing red ($M = 0.15$, $SD = 0.90$) slightly more sexually attractive than the one wearing white ($M = -0.13$, $SD = 0.76$) ($p = 0.06$), the differences between the targets wearing blue ($M = -0.02$, $SD = 0.82$) vs. white ($p = 0.75$), and blue vs. red ($p = 0.37$) were both not significant. The main effect of sexual dimorphism cues did not reach significance, $F(1,271) = 1.42$, $p = 0.23$, $\eta_p^2 < 0.01$.

The two-way interaction between gender and sexual dimorphism cues was significant, $F(1,271) = 6.32$, $p = 0.01$, $\eta_p^2 = 0.02$, and further analysis revealed that compared to masculine targets, male participants rated feminine targets as more sexually attractive, $F(1,271) = 6.74$, $p = 0.01$, $\eta_p^2 = 0.02$, but ratings from female participants did not differ, $F(1,271) = 0.89$, $p = 0.35$, $\eta_p^2 < 0.01$. All the remaining two-way interactions were insignificant ($ps > 0.17$).

Figure. 3. Sexual attractiveness as a unction of sexual dimorphism, color of clothes, and participants gender

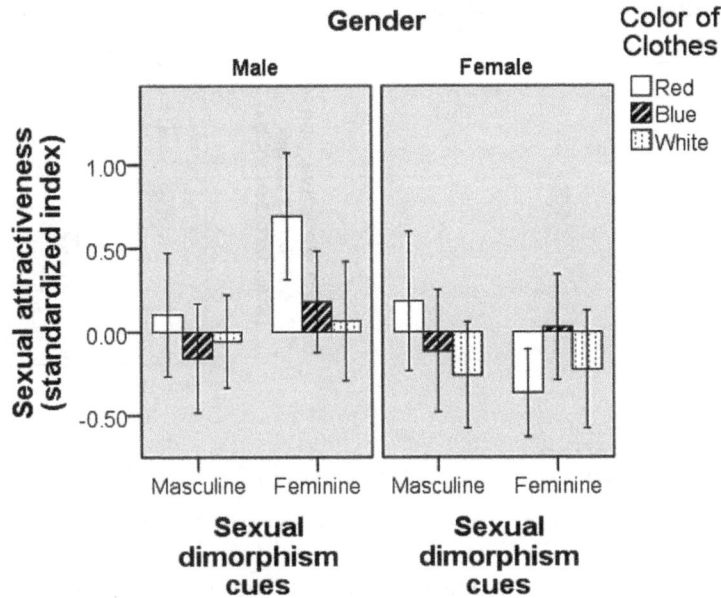

Note. Error bars represent 95% CI.

However, in accordance with our hypothesis we did find a significant three-way interaction among gender, color, and sexual dimorphism cues, $F(2,271) =3.10$, $p=0.047$, $\eta_p^2=0.02$. To decompose the three-way interaction, we first tested the two-way interactions (or simple interaction) between color and sexual dimorphism separately under male and female conditions. For male participants, the two-way interaction was not significant, $F(2,271)=1.00$, $p=0.37$, $\eta_p^2<0.01$, but it indicated a trend toward significance for female participants, $F(2,271) =2.58$, $p=0.08$, $\eta_p^2=0.02$. Female participants rated masculine targets with red clothing more sexually desirable than feminine red targets, $F(1,271)=5.43$, $p=0.02$, $\eta_p^2=0.02$, and showed no such preference for masculinity in other color conditions ($ps>0.5$). To directly test the red effect, we planned a contrast between target with red and the other two colors, and found a trend of red to elevate the sexual attractiveness ratings of masculine targets, as rated by women, though it only reach marginal significance, $F(1,271)=3.25$, $p=0.07$, $\eta_p^2=0.01$.

Alternately, decomposing the three-way interaction from the perspective of sexual dimorphism proved to be clearer. The two-way interaction between gender and color of clothes was not significant under the masculine condition, $F(2,271)=0.45$, $p=0.64$, $\eta_p^2<0.01$, yet it was significant under the feminine condition, $F(2,271)=4.50$, $p=0.01$, $\eta_p^2=0.03$.

We examined the second-order simple effect associated with feminine conditions using Sidak corrections. The difference between photos with different colors was significant when rated by men, $F(2,271)=3.96$, $p=0.02$, $\eta_p^2=0.03$. Consistent with our hypothesis, we planned a contrast between photos with red and the mean of the other two colors, and found that the red target received higher ratings on sexual attractiveness from male participants than targets with the two other colors, $F(1,271)=7.81$, $p=0.006$, $\eta_p^2=0.03$. However, the difference was completely nonsignificant when rated by women, $F(2,271)=1.56$, $p=0.21$, $\eta_p^2=0.01$ (see Figure 3).

Together, the results suggest that the red effect exists only for feminine women rated by male participants, but not for masculine women. Furthermore, we found a discernible trend for red to enhance the sexual attractiveness of masculine men in comparison with feminine men, as

Figure. 4. Perceived general attractiveness as a function of sexual dimorphism, color of clothes, and participants´ gender

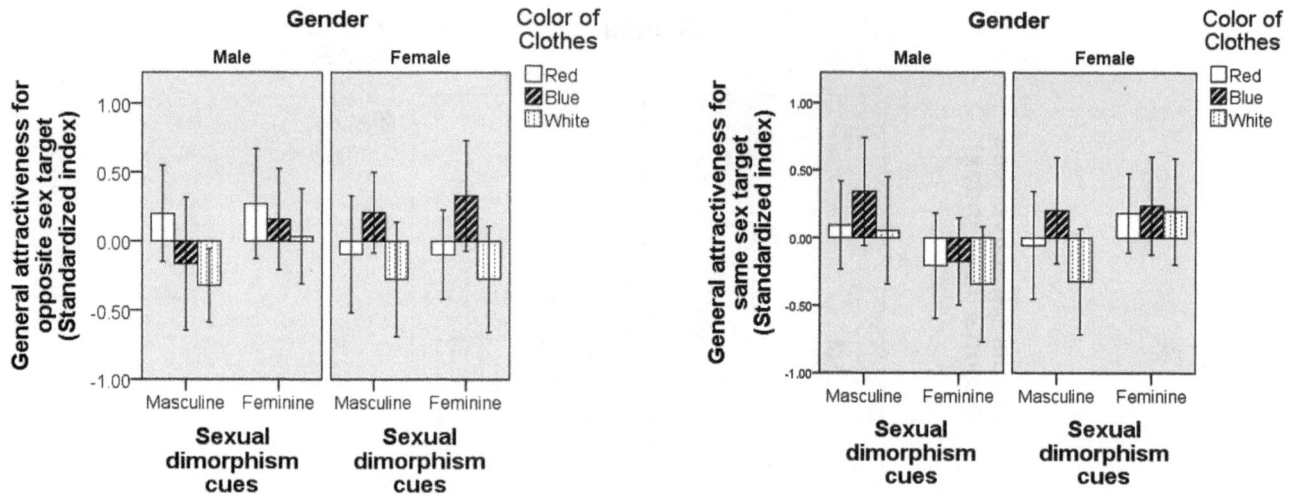

Note. Gender of the target: the left is opposite sex, the right is same sex. Error bars represent 95% CI.

rated by female participants, but the trend did not reach significance.

3.2 General attractiveness

We conducted a 2 (participant gender: male vs. female) × 3 (color of cloth: red, blue, and white) × 2 (sexual dimorphism cues: masculine vs. feminine) × 2 (target gender) four-way mixed design ANOVA on the standardized index of general attractiveness. The target gender is the within-subject variable, and the three other variables are all between-subject. The analysis showed a significant main effect of color of clothes, $F(2,270) = 4.64$, $p=0.01$, $\eta_p^2=0.03$. Post-hoc analysis with Sidak correction found that, overall, participants thought targets wearing blue clothes generally were more attractive than ones wearing white (MBlue−MWhite=0.30, SE =0.10, p= 0.008), and no significant differences were found between red and blue, or white and red (ps>0.13).

Additionally, the three-way interaction among target gender, participant gender, and sexual dimorphism cues was significant, $F(1,270) =11.00$, $p=0.001$, $\eta_p^2=0.04$. Decomposing the three-way interaction based on the target gender, we found that the two-way interaction between participant gender and sexual dimorphism was significant only for same-sex targets, $F(1,270) =10.14$, $p=0.002$, η_p^2

=0.04, but not for opposite sex targets, $F(1,270) =0.92$, p =0.34, $\eta_p^2<0.01$. Further analyzing the second-order simple effects under the same-sex target condition, we found that male participants rated feminine same-sex targets (M=−0.24, SE=0.11) as less generally attractive than masculine same-sex targets (M =0.16, SE =0.10), $F(1,270) =7.25$, $p=0.008$, $\eta_p^2=0.03$; meanwhile, female participants perceived feminine same-sex targets (M= 0.21, SE=0.10) as more generally attractive than masculine ones (M =−0.06, SE=0.11), though the difference was only marginally significant, $F(1,270)=3.25$, $p=0.07$, $\eta_p^2=0.01$, as shown in Figure 4.

All the other main effects, two-way interactions, three-way interactions and four-way interaction were insignificant ($ps > 0.1$).

3.3 Warmth

We conducted a similar 2 (participant gender: male vs. female) × 3 (color of cloth: red, blue, and white) × 2 (sexual dimorphism cues: masculine vs. feminine) × 2 (target gender) four-way mixed design ANOVA on warmth ratings. The target gender is a within-subject variable, and the rest are between-subject variables. As shown in Figure 5, the results revealed a significant main effect of target gender, $F(1,270)=20.73$, $p<0.001$, $\eta_p^2=0.07$; namely, participants rated the opposite sex (M=3.29, SE=0.08) as less warm than same-sex targets (M=3.71, SE =0.07). The main effect of participant gender was also significant, $F(1,270)$=6.23, p =0.01, η_p^2=0.02. Compared with female participants (M =3.35, SE=0.08), male participants tended to give higher warmth ratings to all the targets (M =3.65, SE =0.09). However, the two-way interaction between participant gender and target gender was not significant, $F(1,270)$=2.13, p=0.15, η_p^2=0.01. Furthermore, we found a marginally significant two-way interaction between color of clothes and participant gender, $F(2,270)$=2.70, p=0.07, η_p^2=0.02. Further analysis of simple effects showed that male participants found targets wearing red warmer than did female participants, $F(1,270)$=7.48, p= 0.007, η_p^2=0.03, whereas for targets wearing white, male participants gave only slightly warmer ratings than did female participants, $F(1,270)$ =3.69, p=0.06, η_p^2=0.01; for targets wearing blue, there was no significant gender difference, $F(1,270)$ =0.15, p=0.70, η_p^2<0.01. Alternately, the differences across color of clothes under both male and female conditions were all insignificant, ps >0.11. Planned contrasts between red clothing and the two other colors reached marginal significance in male participants, $F(1,270)$ =2.87, p=0.09, η_p^2=0.01, but not in females

participants (p =0.55), suggesting a slight tendency for men to rate red targets warmer than other targets. All the remaining main effects, two-way interactions, three-way interactions, and four-way interaction were insignificant, ps >0.14.

3.4 Competence

A similar four-way mixed-design ANOVA was conducted on competence ratings. As seen in Figure 6, the results showed a significant two-way interaction between participant gender and target gender, $F(1,270)$ =14.46, p<0.001, η_p^2=0.05. Specifically, rating opposite sex targets, male participants tended to give higher competence ratings (M=4.66, SE=0.09) than did female participants (M =4.25, SE=0.09), $F(1,270)$ =9.81, p=0.002, η_p^2=0.04, yet when rating same-sex targets, male participants tended to give lower competence rating (M=4.41, SE=0.09) than female participants (M=4.65, SE=0.09), though it only reached marginal significance, $F(1,270)$ =3.27, p=0.07, η_p^2=0.01. In addition, the two-way interaction between color of clothes and participant gender was significant, $F(2,270)$=6.09, p=0.003, η_p^2=0.04. Simple effects analysis showed that the difference across color of clothing was significant in male participants, $F(2,270)$=4.29, p= 0.02, η_p^2=0.03, but not in female participants, $F(2,270)$= 2.00,

Figure. 5. Warmth rating as a function of sexual dimorphism, color of clothes and participants gender

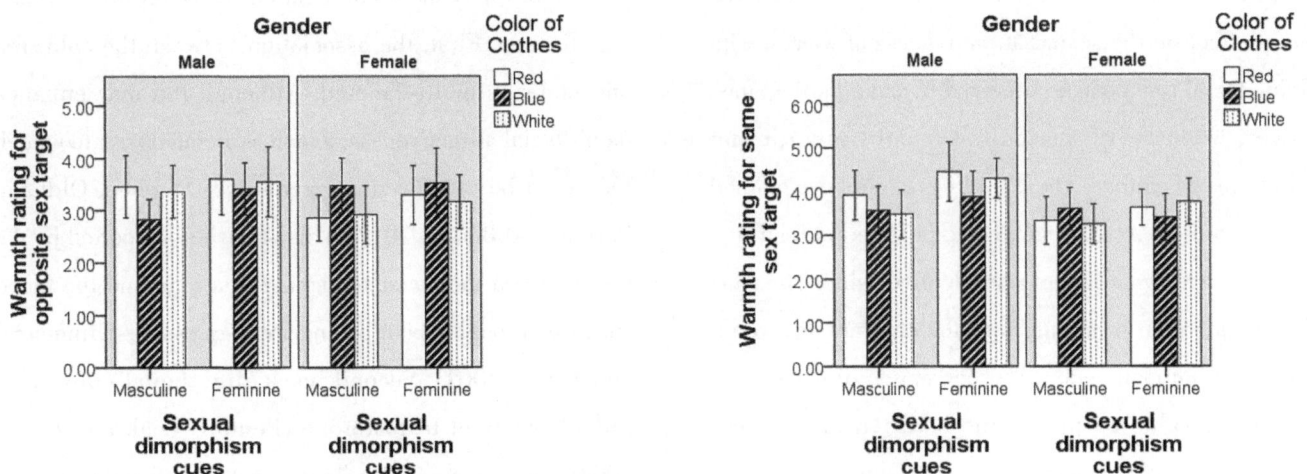

Note. Gender of the target: the left is opposite sex, the right is same sex. Error bars represent 95% CI.

p =0.14, η_p^2 =0.02. Planned contrasts between the color red and the two other colors in the male condition showed that male participants found targets in red more competent than other targets, $F(1,270)$ =6.16, p =0.01, η_p^2=0.02.

All other main effects, two-way interactions, three-way interactions and four-way interaction were not significant, ps >0.18.

relatively lower reproductive value and conflict with the information about fertility presumably imparted by the color red (Law-Smith et al., 2005; Morrison et al., 2010), thereby offsetting or canceling the red effect.

Men with masculine facial cues were perceived by female participants as more sexually attractive when wearing red clothes, but this comparison only reached marginal significance. This tendency is consonant with our

Figure. 6. Competence rating as a function of sexual dimorphism, color of clothes, and participants gender

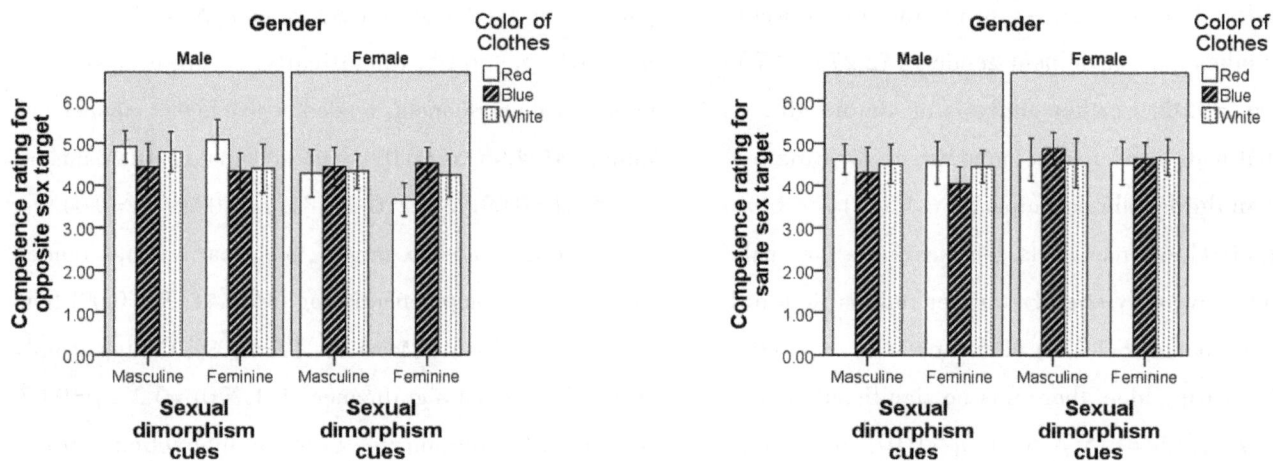

Note. Gender of the target: the left is opposite sex, the right is same sex. Error bars represent 95% CI.

4 Discussion

The present experiment provides strong support for our hypothesized moderation of sexual dimorphism cues on the red effect and also provides a replication of the red effect in a Chinese context. The color red was shown to have a boosting effect on the sexual attractiveness of women with feminine facial traits when perceived by male participants. However, women with masculine facial traits did not benefit from red clothes. These findings could be explained from an evolutionary biological perspective, as feminine traits in women signal higher reproductive value. The most critical information in mating with the opposite sex would be gleaned from many cues. The perception of heightened reproductive value by presenting both red clothes and feminine traits enhances the sexual attractiveness of the female. In contrast, masculine traits in women may signal

hypothesis. Because masculinity and red both signal status in males, and females are likely to favor males with status and resources, in the current study we indeed showed that women tend to rate men with a red-and-masculinity combination as more sexually desirable. Meanwhile, two factors may have contributed to its failure to reach significance. First, the association between the color red and males is multi-faceted. Although red may enhance men's facial attractiveness, a man associated with too much red could be perceived as aggressive (Stephen, Oldham, Perrett, and Barton, 2012). This is further supported by the link between the facial reddening of a man and the blood suffusion brought about by anger or aggression (Drummond and Quah, 2001). Second, masculine facial cues often reflect levels of testosterone (Penton-Voak and Chen, 2004; Roney et al., 2006). Testosterone is related to various personality traits that may have negative connotations, such as dominance or aggressiveness (Boothroyd et al., 2007; Hughes, Dispenza, and Gallup Jr., 2004; Perrett et al., 1998), and the perception of these traits may be more evident when targets are wearing neutral facial expressions (Hareli, Shomrat, and Hess, 2009; Tracy and Beall, 2011), as in the current study. Thus, the masculine cues in the current experiment may have triggered contradictory perceptions of the male target, which, as a result, could have offset the red effect of enhancing the sexual attractiveness of men. Given the dearth of studies on how red affects male sexual attractiveness, future research is needed to clarify the red effect across genders.

The present study also measured the perceived general attraction of the targets. The results demonstrated that male participants did not rate the opposite sex target paired with red as higher on the three measures, relative to the other colors. This result is generally consistent with previous studies (Elliot et al., 2010, 2013; Schwarz and Singer, 2013), yet it should be noted that Elliot and Niesta (2008) did document an effect of red on general attractiveness, though it was smaller in effect size. This slight disparity in results warrants further theoretical and empirical consideration.

Averaging over three colors, we found that whereas female participants did not favor either feminine or masculine women, the male participants perceived feminine men as less generally attractive. This could lend support to the gender dichotomization tendency that is characteristic of a male gender role (Bosson, and Michniewicz, 2013).

As for warmth and competence ratings, we found that, averaging over the sexual dimorphism cues, male participants tended to rate female targets wearing red as warmer and more competent than the targets wearing white or blue. This result on warmth and competence ratings slightly deviated from ratings on other related trait words in Elliot et al. (2010). Considering that it was not

moderated by sexual dimorphism cues, other mechanisms may have been responsible, such as halo effect. The close link between red and general positivity in Chinese culture, along with the status of red as being especially tied with females in Chinese culture, may also exert influences (Eberhard, 2013; Shang, 2008).

This study also has some limitations. First, as in Schwarz and Singer (2013), we used solely self-report ratings that are well documented to be susceptible to social desirability biases, especially on delicate issues such as direct judgment of others' attractiveness and of sexual interest toward a stranger. It could be more accurate and precise to employ biological measures to index one's sexual attraction to the targets.

In same sex ratings, we measured only perception of general attraction, warmth, and competence, dropping sexual attraction. It is counterintuitive to measure heterosexuals' sexual attraction to same-sex targets. Nevertheless, in future studies, same or modified items also measuring the perceived sexual attractiveness of same-sex participants could aid in the discriminant validity of the current measures.

Furthermore, a study to examine the effects of red in different mating or task contexts would be fruitful. Research has already suggested that the red effect may have completely different meanings in working or achievement contexts (Elliot and Maier, 2012, 2014; Elliot, Maier, Moller, Friedman, and Meinhardt, 2007). Thus, the moderation of sexual dimorphism on the red effect may be further moderated by context (i.e., perhaps the feminine female paired with red would be perceived as less competent in working conditions).

So far, studies about the red effect have focused on the perceivers, though it is likely that the color of clothing may also exert influences over the wearers. It is plausible that the color of clothing may change the sexual awareness of its wearer, and it could be the case that wearing red clothes is a reflection of the sexual intention of the wearer from the beginning. The combination of data from the red-wearer and the perceiver would better capture the significance of red and would be an interesting avenue for future studies.

Another limitation of this study is that we did not directly test the alternative explanations. For example, one might argue that the red stimuli simply activated mental representation of traditional wedding ceremonies in China or the "red-light" district in the West due to cultural schema or stereotypes. Female newlyweds or female sex workers are supposed to have more feminine traits. Consistent with this argument, the reason that the red effect is specific to feminine women could be easily sought in social conditioning rather than evolution. Though such reasoning has certain inherent difficulties (e.g., Why did red not invoke an image of the red flag, but activated images of the "red-light" district?), we were not able to test these issues directly in this study. The cultural mechanism cannot be excluded from the red effect on sexual attraction, and it may operate at a different level and exert influence through social learning and acculturation processes, such as through the perceived sexual receptiveness of the individual wearing red clothes (Pazda et al., 2012). Moreover, Roberts et al. (2010) found that in certain settings, black dressing elevated perceived sexual attractiveness as well, which undermines the uniqueness of red and its purported evolutionary underpinnings. In light of these considerations, further research is needed to clarify the contributions of both cultural and biological mechanisms.

The simple perceptual mechanism that reduces the color effect to color contrast or luminance, as in Schwarz and Singer (2013), can be ruled out because the color configuration used in the current study is identical to previous research, and this interpretation could not explain the moderation role of sexual dimorphism cues. However, we did not use a spectrophotometer to check the color parameters of printed materials; this should be corrected in future replications.

Collectively, previous research has shown that red stimuli could enhance an individual's sexual attraction. The current investigation adds to this literature by showing that this effect is moderated by the facial sexual dimorphism cues of the targets. Furthermore, it is the first study to systematically examine the moderating roles of facial cues in explaining the red effect. This finding adds to the growing literature on the red effect and its boundary conditions, and the implications may stimulate future research.

References

Anderson, C., and Kilduff, G. (2009). Why do dominant personalities attain influence in face-to-face groups? The competence-signaling effects of trait dominance. *Journal of Personality and Social Psychology, 96*, 491 - 503.

Boothroyd, L. G., Jones, B. C., Burt, D. M., and Perrett, D. I. (2007). Partner characteristics associated with masculinity, health and maturity in male faces. *Personality and Individual Differences, 43*, 1161 - 1173.

Bosson, J. K., and Michniewicz, K. S. (2013). Gender dichotomization at the level of ingroup identity: What it is, and why men use it more than women. *Journal of Personality and Social Psychology, 105*, 425 - 442.

Buss, D. M. (2008). *Evolutionary psychology: The new science of mind* (3rd ed.). Boston: Allyn and Bacon.

Chen, J. (2007). *Worship and taboos about the colour RED: A study of RED in Chinese belief systems.* Unpublished master's thesis, Sichuan University, Chengdu, China.

DeBruine, L. M., Jones, B. C., Crawford, J. R., Welling, L. L., and Little, A. C. (2010). The health of a nation predicts their mate preferences: Cross-cultural variation in women's preferences for masculinized male faces. *Proceedings of the Royal Society B: Biological Sciences, 277*, 2405 - 2410.

Deschner, T., Heistermann, M., Hodges, K., and Boesch, C. (2004). Female sexual swelling size, timing of ovulation, and male behavior in wild West African chimpanzees. *Hormones and Behavior, 46*, 204 - 215.

DiStefano, C., Zhu, M., and M?ndril?, D. (2009). Understanding and using factor scores: Considerations for the applied researcher. *Practical Assessment, Research and Evaluation, 14*, 1 - 20.

Drummond, P. D., and Quah, S. H. (2001). The effect of expressing anger on cardiovascular reactivity and facial blood flow in Chinese and Caucasians. *Psychophysiology, 38*, 190 - 196.

Eberhard, W. (2013). *Dictionary of Chinese symbols: Hidden symbols in Chinese life and thought*. London: Routledge.

Elliot, A. J., and Maier, M. A. (2012). Color-in-context theory. In M. P. Zanna, P. Devine, J. M. Olson, and A. Plant (Eds.), *Advances in experimental social psychology* (Vol. 45, pp. 61 - 125). Salt Lake City: Academic Press.

Elliot, A. J., and Maier, M. A. (2014). Color psychology: Effects of perceiving color on psychological functioning in humans. *Annual Review of Psychology, 65*, 95 - 120.

Elliot, A. J., Maier, M. A., Moller, A. C., Friedman, R., and Meinhardt, J. (2007). Color and psychological functioning: The effect of red on performance attainment. *Journal of Experimental Psychology: General, 136*, 154 - 168.

Elliot, A. J., and Niesta, D. (2008). Romantic red: Red enhances men's attraction to women. *Journal of Personality and Social Psychology, 95*, 1150 - 1164.

Elliot, A. J., Niesta Kayser, D., Greitemeyer, T., Lichtenfeld, S., Gramzow, R. H., Maier, M. A., and Liu, H. (2010). Red, rank, and romance in women viewing men. *Journal of Experimental Psychology: General, 139*, 399 - 417.

Elliot, A. J., Tracy, J. L., Pazda, A. D., and Beall, A. T. (2013). Red enhances women's attractiveness to men: First evidence suggesting universality. *Journal of Experimental Social Psychology, 49*, 165 - 168.

Fiske, S. T., Cuddy, A. J. C., Glick, P., and Xu, J. (2002). A model of (often mixed) stereotype content: Competence and warmth, respectively, follow from perceived status and competition. *Journal of Personality and Social Psychology, 82*, 878 - 902

Gesquiere, L. R., Wango, E., Alberts, S. C., and Altmann, J. (2007). Mechanisms of sexual selection: Sexual swellings and estrogen concentrations as fertility indicators and cues for male consort decisions in wild baboons. *Hormones and Behaviors, 51*, 114 - 125.

Grice, J. W. (2001). Computing and evaluating factor scores. *Psychological Methods, 6*, 430 - 450.

Guéguen, N. (2012a). Color and women hitchhikers' attractiveness: Gentlemen drivers prefer red. *Color Research and Application, 37*, 76 - 78.

Guéguen, N. (2012b). Does red lipstick really attract men? An evaluation in a bar. *International Journal of Psychological Studies, 4*, 202 - 209.

Guéguen, N., and Jacob, C. (2012). Lipstick and tipping behavior: When red lipstick enhance waitresses tips. *International Journal of Hospital Management, 31*, 1333 - 1335.

Guéguen, N., and Jacob, C. (2013). Color and cyber-attractiveness: Red enhances men's attraction to women's internet personal ads. *Color Research and Application, 38*, 309 - 312.

Guéguen, N., and Jacob, C. (2014). Clothing color and tipping: Gentlemen patrons give more tips to waitresses with red cloths. *Journal of Hospitality and Tourism Research, 38*, 275 - 280.

Hareli, S., Shomrat, N., and Hess, U. (2009). Emotional

versus neutral expressions and perceptions of social dominance and submissiveness. *Emotion, 9*, 378 – 384.

Hughes, S. M., Dispenza, F., and Gallup, G. G., Jr. (2004). Ratings of voice attractiveness predict sexual behavior and body configuration. *Evolution and Human Behavior, 25*, 295 – 304.

Hutchings, J. (2004). Color in folklore and tradition–The principles. *Color Research and Application, 29*, 57 – 66.

Josephs, R. A., Sellers, J. G., Newman, M. L., and Mehta, P. H. (2006). The mismatch effect: When testosterone and status are at odds. *Journal of Personality and Social Psychology, 90*, 990 – 1013.

Kasperk, C., Helmboldt, A., Borcsok, I., Heuthe, S., Cloos, O., Niethard, F., and Ziegler, R. (1997). Skeletal site-dependent expression of the androgen receptor in human osteoblastic cell populations. *Calcified Tissue, 61*, 464 – 473

Kaya, N., and Epps, H. H. (2004). Relationship between color and emotion: A study of college students. *College Student Journal, 38*, 396 – 405.

Law-Smith, M. J., Perrett, D. I., Jones, B. C., Cornwell, R. E., Moore, F. R., Feinberg, D. R., . . . Hillier, S. G. (2005). Facial appearance is a cue to oestrogen levels in women. *Proceedings of the Royal Society Series B: Biological Sciences, 273*, 135 – 140.

Lee, Y. (2006). *Man as the prayer*: The origin and nature of human kind. New York: Trafford.

Lynn, B. M., McCord, J. L., and Halliwell, J. R. (2007). Effects of menstrual cycle and sex on postexercise hemodynamics. *American Journal of Physiology: Regulatory, Integrative and Comparative Physiology, 292*, 1260 – 1270.

Meier, B. P., D'Agostino, P. R., Elliot, A. J., Maier, M. A., and Wilkowski, B. M. (2012). Color in context: Psychological context moderates the influence of red on approach– and avoidance–motivated behavior. PLOS

ONE, 7, e40333.

Morrison, E. R., Clark, A. P., Tiddeman, B. P., and Penton–Voak, I. S. (2010). Manipulating shape cues in dynamic human faces: Sexual dimorphism is preferred in female but not male faces. *Ethology, 116*, 1234 – 1243.

Neto, F. (2002). Colors associated with styles of love. *Perceptual and Motor Skills, 94*, 1303 – 1310.

Niesta–Kayser, D., Elliot, A. J., and Feltman, R. (2010). Red and romantic behavior in men viewing women. *European Journal of Social Psychology, 40*, 901 – 908.

Pazda, A. D., Elliot, A. J., and Greitemeyer, T. (2012). Sexy red: Perceived sexual receptivity mediates the red–attraction relation in men viewing woman. *Journal of Experimental Social Psychology, 48*, 787 – 790.

Pazda, A. D., Elliot, A. J., and Greitemeyer, T. (2014). Perceived sexual receptivity and fashionableness: Separate paths linking red and black to perceived attractiveness. *Color Research and Application, 39*, 208 – 212.

Penton–Voak, I. S., and Chen, J. Y. (2004). High salivary testosterone is linked to masculine male facial appearance in humans. *Evolution & Human Behavior, 25*, 229 – 241.

Perrett, D. I., Lee, K. J., Penton–Voak, I., Rowland, S., Yoshikawa, D. M., Burt, D. M., . . . Akamatsu, S. (1998). *Effects of sexual dimorphism on facial attractiveness. Nature, 394*, 884 – 887.

Pound, N., Penton–Voak, I. S., and Surridge, A. K. (2009). Testosterone responses to competition in men are related to facial masculinity. *Proceedings of the Royal Society B: Biological Sciences, 276*, 153 – 159.

Re, D. E., Whitehead, R. D., Xiao, D., and Perrett, D. I. (2011). Oxygenated–blood colour change thresholds for perceived facial redness, health, and attractiveness. PLOS ONE, 6, e17859.

Roberts, S. C., Owen, R. C., and Havlicek, J. (2010).

Distinguishing between perceiver and wearer effects in clothing color-associated attributions. *Evolutionary Psychology, 8*, 350 - 364.

Roney, J. R., Hanson, K. N., Durante, K. M., and Maestripieri, D. (2006). Reading men's faces: Women's mate attractiveness judgments track men's testosterone and interest in infants. *Proceedings of the Royal Society of London Series B: Biological Sciences, 273*, 2169 - 2175.

Schwarz, S., and Singer, M. (2013). Romantic red revisited: Red enhances men's attraction to young, but not menopausal women. *Journal of Experimental Social Psychology, 48*, 161 - 164.

Setchell, J. M., and Dixson, A. F. (2001). Circannual changes in the secondary sexual adornments of semifree-ranging male and female mandrills (Mandrillus sphinx). *American Journal of Primatology, 53*, 109 - 121.

Shang, X. (2008). Comparison of cultural connotations of "Red" in Chinese and English. *Journal of Linyi Normal University, 30*, 139 - 141.

Stephen, I. D., and McKeegan, A. M. (2010). Lip colour affects perceived sex typicality and attractiveness of human faces. *Perception, 39*, 1104 - 1110.

Stephen, I. D., Oldham, F. H., Perrett, D. I., and Barton, R. A. (2012). Redness enhances perceived aggression, dominance and attractiveness in men's faces. *Evolutionary Psychology, 10*, 562 - 572.

Tracy, J. L., and Beall, A. T. (2011). Happy guys finish last: The impact of emotion expressions on sexual attraction. *Emotion, 11*, 1379 - 1387.

Wen, F., and Zuo, B. (2012). The effects of transformed gender facial features on face preference of college students: Based on the test of computer graphics and eye movement tracks. *Acta Psychologica Sinica, 44*, 14 - 29.

Journal of Environmental Psychology, 2014, 40, 9 – 13.

Gaze direction and brightness can affect self-reported emotion

Xiaobin Zhang [1], Qiong Li [1], Bin Zuo [2,*]

([1] School of Psychology, Northwest Normal University, Lanzhou, China)

([2] School of Psychology, Central China Normal University, Wuhan, China)

Abstract Previous studies revealed that emotion (pleased or depressed) could bias perception in a metaphorically consistent manner (e.g., happy=white (up), depressed=dark (down)). The present study extended this view by investigating whether these metaphors can also affect the emotion of an observer in a metaphorically consistent manner. In Experiment 1, after gazing at a black screen, participants became more depressed and less pleased temporarily. Conversely, after gazing at a white screen, participants became more pleased and less depressed temporarily. Results from Experiment 2 revealed that after gazing at the top of the screen, participants felt more pleased and less depressed temporarily but felt the reverse when gazing at the bottom of the screen. These results suggest that metaphors can, at least temporarily, affect the emotion of an observer along a pleased–depressed dimension.

Keywords metaphor; emotion; vertical position; brightness; pleasede–depressed dimension

1 Introduction

According to metaphor representation theory (Lakoff & Johnson,1999), people conceptualize non–perceptual states in perceptual terms. For example, a growing body of studies has suggested that vertical position and brightness are metaphors of affect: "up" and "brightness" are associated with good, and "down" and "black" have negative connotations (Crawford, Margolies, Drake, & Murphy, 2006; Meier & Robinson, 2004; Meier, Robinson, & Clore, 2004). In addition to affect, emotions of pleased and happy are considered to be light in color or high in vertical position as compared to emotion of depressed states. Specifically, people typically depict "pleased" and "depressed" with its corresponding metaphors as up (light) and down (darkness), respectively (Meier & Robinson, 2006), so we contend that 'brightness'/'spatial position' may only affect emotion on pleaseded–pressed dimension not on other emotions (such as proud, scared). Our prediction is based upon the metaphoric mapping of 'brightness' / 'spatial position' and emotion on happye–depressed dimension. Common, everyday expressions highlight these metaphors: "I feel up today (when one feels happy)" "I' m at a low point (when onefeelsdepressed)," and "A smile brightens one's face (when one feels happy)." These metaphoric mappings seem to be a

---

* Corresponding author. E-mail address: zhangxiaobin624@163.com (B. Zuo).

universal phenomenon, as the expression 'bright smile' is manifested in many languages, including English, German, Italian, Korean, Chinese, and Russian. In addition, some studies have corroborated the association between emotion (on pleasede depressed) and its metaphors, such as spatial position and bright ness (up/bright−down/darkness) (Meier & Robinson, 2006; Song, Vonasch, Meier, & Bargh, 2012). specifically, Meier and Robinson (2006) argue that neuroticism and depressed symptoms can affect participants' vertical selective attention. Subjects who score higher on measures of neuroticism or depressed symptoms are faster at detecting lower (versus higher) spatial attention targets. Song et al. (2012) confirmed the metaphoric mapping of brightness and smiling whereby people judge smiling faces as perceptually brighter than frowning faces.

In fact, prior to psychologists using concepts such as metaphor and embodied cognition to explain the association between emotion and its metaphors, researchers had outlined the association between emotion and certain perceptual inputs (Teasdale,1993;Thayer, 2003).Some earlier studies have found that emotion can affect perception in a way that is congruent with its metaphors (i.e., pleased=up; depressed=down) (Fisher, 1964; Wapner, Werner, & Krus, 1957). One study revealed that participants who felt happy (i.e., they just received an A on a midterm exam) exhibited an upward bias when horizontally bisecting a lu− minous square. However, participants who felt sadness (i.e., just received an F on a midterm exam) exhibited a downward bias on this task (Wapner et al., 1957).

Several prior studies suggest that metaphors might be bidirectional. For instance, abstract metaphorical expressions are often grounded in physical experiences, and concepts within abstract domains influence related physical experiences (Landau, Meier, & Keefer, 2010; Williams, Huang, & Bargh, 2009). Based on the asso

ciation between emotion and its metaphors, prior studies have explored how emotion on the pleused−depressed dimension affects perceptual judgment (vertical position and brightness judgments), and confirmed that emotion on the pleased−depressed dimension were related with the metaphors of up position/ brightness and down position/darkness respectively (Fisher, 1964; Meier & Robinson, 2006; Song et al., 2012; Wapner et al., 1957), such as depressed symptoms can affect participants' vertical se− lective attention (Meier & Robinson, 2006) and smiling faces were judged as perceptually brighter than frowning faces (Song et al.,2012), but there was no study directly testing whether and how these physical metaphors (vertical position or brightness) affect emotion on pleased−depressed dimension. However, the idea that environmental input might affect emotion is not entirely novel. Several studies have revealed that color can temporarily influence emotional judgments (Adams & Osgood, 1973; Fetterman, Robinson, & Meier, 2012; Hevner, 1935; Keith& James, 1975; Valdez & Mehrabian, 1994). Within the field of clinical psychol ogy, several studies have shown that phototherapy (daily exposures to bright light) is effective in combating certain depressed disor− ders, such as seasonal affective disorder (Terman, Terman, & Ross,1998), antepartum depression (Oren et al., 2002), and general depression (Kripke, 1998).

Although several previous studies have shown that environmental input and bodily sensations can influence emotion, we are not aware of a study that has directly explored whether perception relevant to two important metaphors (vertical position or bright ness) of emotion can temporarily influence the emotional state (on the pleased−depressed dimension) of an observer in a way that is congruent with these metaphors. In the current study, we explored whether vertical position and brightness could affect the emotion of an observer on the pleased−depressed dimension.

2 Experiment 1

We conducted Experiment 1 to determine whether brightness could affect the emotion of aperceiver in a metaphorically consistent manner. We expected that prime condition can temporarily affect the emotion of a perceiver along the pleased–depressed dimensions.

2.1 Participants and design

The participants were 53 Chinese undergraduates (10 men) who were offered 10 RMB as compensation. Five participants (three men, two women) were excluded from the analysis, because four of them did not follow instructions when completing the emotion questionnaire and one subject reported not paying much attention to the screen. All participants (age range: 18e25, mean age: 21.5 years) were randomly assigned to the two prime conditions; each prime condition included 24 participants. We used a 2 (prime condition: white or black) independent groups design.

2.2 Stimulus materials and procedure

Participants arrived at the laboratory, individually, and were greeted by a male experimenter. Each participant was told that s/he must complete two unrelated tasks. During the first task, participants were seated facing a computer screen (Dell Computer, 19- inch display monitor, resolution: 1280 x 1024) and were told that the purpose of the task was to explore the ability in focusing one's attention on a screen. In fact, the purpose of the first task was to prime color metaphors (brightness: black or white). On each trial, a fixation cross was first presented at the center of the screen for 1000 ms. Next, the whole screen became white or black depending on the prime condition; the color was displayed for one minute. During this period, participants were required to look at the screen continuously, but are allowed to blink. After 1 min, participants rested for 5 s. Participants completed this process seven times. At the end of the experiment, participants reported the extent to which they focused on the screen on a 9−point scale (1 ="not at all" to 9= "extremely") by pressing a digit key on the keyboard.

Participants were then asked to complete an emotion (pleasededepressed dimension) questionnaire. We chose the words "happy" and "pleased" to represent the pleased dimension, and the words "depressed" and "repressed" to represent the depressed dimen sion; these words were chose according to the BFS (Abele−Brehm & Brehm, 1986), which was a scale of mood. The questionnaire contained four items: "pleased" and "happy" represented the pleased dimension, and "depressed" and "repressed" represented the depressed dimension. Participants were asked to rate how they felt presently and rated the intensity of each mood (four emotion words) separately on 9−point scale (form 1 to 9, 1 ="not at all" to 9 = "extremely"). On completion of the experiment, participants were debriefed as to their awareness of the study hypotheses, and none of the participants were able to identify the purpose of the study.

2.3 Results and discussion

Mean scores on the pleased and depressed dimensions served as the dependent measure. The data were submitted to a MANOVA. The analysis revealed, within the depressed dimension, scores on the black condition (M=3.71, SE=0.35) were significantly higher than scores on the white dimension (M=2.25, SE=0.35), [$F(1,46)$ η^2=8.64, p<0.01, η^2=0.16; see Fig. 1]. Within the pleased dimension, scores on the white dimension (M=4.98, SE=0.32) were higher than scores on the black dimension, but the difference was not significant (M=4.46, SE=0.32), [$F(1, 46)$ =1.29, p>0.05, η^2=0.03; see Fig. 1]. The correlation between items "pleased" and"happy" was significant (r=0.80, p<0.01) and the correlation between items "depressed" and "repressed" was also significant (r=0.79, p<0.01).

We also analyzed the difference of the scores on two dimensions in different prime conditions. The analysis demonstrated that in the white prime condition, scores on the pleased dimension ($M=4.98$, $SE=0.33$) were significantly higher than scores on the depressed dimension ($M=2.25$, $SE=0.34$), [$t(23) =4.45$, $p< 0.01$, $d=1.70$]. The prior study also showed that the scores on the positive emotion items of PANAS were higher than scores on the negative emotion items of PANAS (Watson, Clark, &Tellegen, 1988). Conversely, within the black prime condition, there were no significant differ ences between scores on the pleased ($M=4.46$, $SE=0.32$) and depressed dimensions ($M=3.71$, $SE=0.36$), [$t(23) =1.54$, $p =0.14$, $d=1.17$]. These results indicated that compared to the white prime condition, under the black prime condition

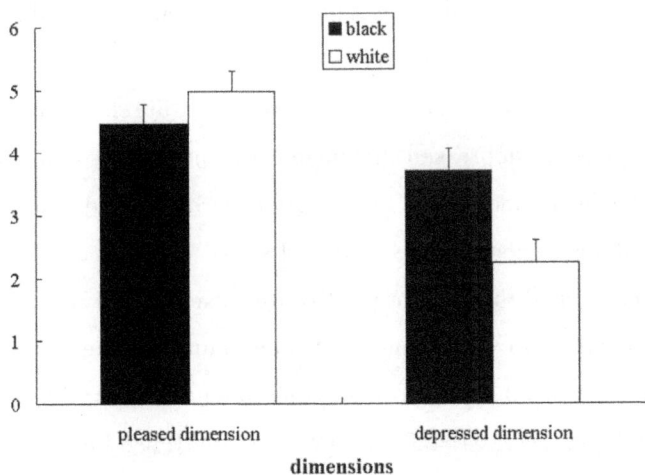

Fig. 1. Mean scores on emotion as a function of brightness and dimensions (pleased–depressed) (study 1).

the difference of scores between depressed dimension and pleased dimension decreased, that was our experimental manipulation was effective.

Results from Experiment 1 revealed that brightness affects emotion felt by a perceiver in a metaphorically consistent manner at least partly.

3 Experiment 2

In addition to brightness, vertical position is also thought to provide a metaphor of emotion along the pleased–depressed dimension. Weconducted Experiment 2 to determine if gaze directed to certain spatial positions could affect emotions along the pleaseded–presseddimension. We expected that participants would feel more pleased and less depressed temporarily if primed by stimuli presented at the top of the computer screen, but would feel the opposite if primed by stimuli presented at the bottom of the computer screen.

3.1 Participants and design

The participants were 73 Chinese undergraduates (44 women)who were offered 10 RMB as compensation. Five participants (3men, 2 women) were excluded from the analysis because they didnot follow instructions when completing the emotion questionnaire. All participants (agerange from18 to 24,mean age 21.1 years)were randomly assigned to the two priming conditions. All par ticipants got (greater than) the correct rate of 80% in the task ofdiscriminating letters from digits. Each priming condition consistedof 34 participants. We implemented a 2 (prime condition: up or down) × 2 (participant gender: male or female) (because the number of male participants was too small and gender was not the important factor we interested in present study, so wedid not makegender as a factor in Experiment 1) between subject design.

3.2 Stimulus materials and procedure

Participants arrived at the laboratory, individually, and were greeted by a male experimenter. Each participant was told that s/hemust complete two unrelated tasks. The purpose of the first task was to prime metaphors (vertical position: top or bottom), but participants were told that the purpose of the first task was to explore the ability to discriminate letters from digits. In the first task, participants were seated facing a computer screen (DellComputer,19-inch display monitor, resolution= 1280×1024), and the height of participants' eyes was the same as the height of the center of the screen (we made it by adjusting the height of chair according to the height of participants). The procedure was as follows: during each trial, a fixation cross was presented at the center of the screen for 300 ms. Following this central cue, a letter (from A to Z but excluding "O" due to difficulty in discriminating between "O" and the digit "0") or digit (from 1 to 25) appeared either at the top or bottom of the screen, depending on the priming condition. Each letter and digit appeared 4 times, so the whole experiment contained 200 trials. Participants were required to report, by means of a key press, whether each target was a letter or digit as quickly and accurately as possible. If the response was inaccurate, the word "incorrect" appeared in red for 1000 ms at the top or bottom of screen, depending on the prime condition. A 500 ms blank screen separated accurate trials.

Next, participants were asked to complete one emotion questionnaire, which contained eight items: two pleased dimension words, "pleased" and "happy," and two depressed dimension words, "depressed" and "repressed." In addition to these four items, we added another four words, which were chosen from the PANAS and were unrelated to the pleased-depressed dimension (Watson et al., 1988). The positive items were "enthusiastic" and "proud," and the negative items were "scared" and "hostile." Participants were asked to rate how they felt presently and rated the intensity of each mood (eight emotion words) separately on 9-point scale (form 1 to 9, 1="not at all" to 9="extremely"). Scores on these items were our dependent measure. On completion of the experiment, participants were debriefed as to their awareness of the study hypotheses, and none of the participants were able to identify the purpose of the study.

3.3 Results and discussion

Mean scores on the pleased and depressed dimensions served as the dependent measure. The data were submitted to a MANOVA. The analysis demonstrated that for the pleased dimension, scores in the top prime condition (M=5.83, SE=0.30) were significantly higher than scores in the bottom prime condition (M=4.99, SE=0.34), [$F(1, 64)$=3.45, p=0.06, η^2=0.05; see Fig. 2]. The main effect of gender [$F(1, 64)$=1.39, p=0.24, η^2=0.02] and the interaction between gender and prime were not significant [$F(1,64)$=0.45, p=0.51, η^2< 0.01]. Conversely, for the depressed dimension, scores in the bottom prime condition (M=3.86, SE=0.38) were significantly higher than scores in the top prime condition (M=2.86, SE=0.34) [$F(1, 64)$=3.89, p=0.05, η^2=0.06; see Fig. 2]. The main effect of gender [$F(1, 64)$=1.85, p=0.18, η^2=0.03] was not significant and the interaction between gender and prime was significant [$F(1, 64)$=4.12, p=0.05, η^2=0.06]. The correlation between items "pleased" and "happy" was significant (r=0.79, p<0.01) and the correlation between items "depressed" and "repressed" was significant (r=0.80, p< 0.01).

Scores for the other four emotion words unrelated to the pleased-depressed dimension (two positive words: enthusiastic and proud; two negative words: scared and hostile) were also analyzed with a MANOVA. The analysis demonstrated that for the positive dimension, there were no significant differences between scores in the top (M=5.29, SE=0.31) and bottom prime conditions (M=5.24, SE=0.35), [$F(1, 64)$=0.01, p=0.91, η^2< 0.01; see Fig. 2]. The main effect of gender [$F(1, 64)$ =7.34, p< 0.01, η^2= 0.10] was significant, and the interaction between gender and prime conditions was not significant [$F(1, 64)$=0.01, p=0.91, η^2< 0.01]. For the negative dimension, there were also no significant differences between scores in the top (M=1.76, SE=0.23) and bottom prime conditions (M=1.89, SE=0.26), [$F(1, 64)$=0.14, p=0.71, η^2< 0.01; see Fig. 2]. The main effect of gender [$F(1, 64)$=0.67, p=0.42, η^2= 0.01] and the interaction between gender and prime were not significant [$F(1, 64)$=0.04, p=0.85, η^2< 0.01]. The correlation between items "enthusiastic" and "proud" was significant (r=0.56, p<0.01) and the correlation between items "scared" and "hostile" was signiifcant (*r*=0.48, *p*<0.01).

Results from Experiment 2 revealed that gaze position (upward vs. downward) also affects emotions experienced along the pleasede depressed dimension. When gazing

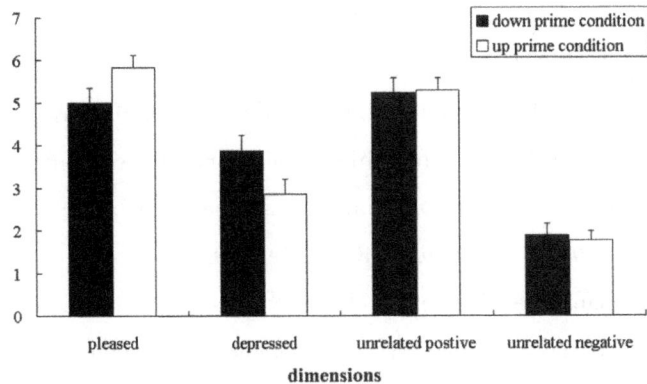

Fig. 2. Mean scores on emotion as a function of gazing vertical positions and dimensions (study 2).

at the top of the screen, participants felt more pleased and less depressed temporarily but felt more depressed and less pleased after gazing at the bottom of the screen temporarily. Most importantly, the gaze position manipulation did not affect ratings on emotional states unrelated to the pleased−depressed dimension, indicating that the prime conditions only affected emotions relevant to their metaphors.

4 General discussion

The present study was the first (to our knowledge) to show that metaphors of emotion (vertical gaze position and brightness) can affect emotional experience along the pleased−depressed dimension in a metaphorically consistent manner. Specifically, gazing at the top of the screen, or looking at a white screen, was associated with participants feeling more pleased and less depressed temporarily, while focusing attention at the bottom of a screen, or viewing a black screen, was associated with feeling more depressed and less pleased temporarily. Prior studies had suggested that emotions do affect

perception (brightness and vertical position judgments) in a metaphorically consistent manner (Meier & Robinson, 2006; Song et al., 2012). Our study has added to this literature by showing that these metaphors (vertical gaze position and brightness) also affect emotional experience along the pleased−depressed dimension. These findings are in line with previous studies suggesting that metaphors of emotion along the pleased−depressed dimension might be bidirectional (Landau et al., 2010; Williams et al.,2009). One prior study found that physical coldness could significantly increase feelings of a specific emotion: loneliness (Bargh & Shalev, 2012). Another recent study found that people use the color red as a perceptual metaphor to understand anger (Fetterman et al., 2012). Our results add to research showing that metaphors (vertical gaze position and brightness) of emotion can infiuence, at least temporarily, emotional experience along the pleased−depressed dimension of an observer.

People typically depict "pleased" and "depressed" with its corresponding metaphors as up (light) and down (darkness), respectively (Meier & Robinson, 2006). Common, everyday expressions highlight these metaphors: "I feel up today," and "A smile brightens one's face." These metaphoric mappings seem to be a universal phenomenon. We chose the pleased−depressed dimension over other positive/negative emotion, because some studies have corroborated the association between emotion (on pleasede-depressed) and its metaphors, such as spatial position and brightness (Meier & Robinson, 2006; Song et al., 2012). There were no study proved the association between the metaphors (such as spatial position and brightness) and general positive and negative emotion. The study 2 also showed that vertical gaze position did not affect ratings on emotional states unrelated to the pleased−depressed dimension, that is positive emotion items (enthusiastic and proud) and negative emotion items (scared and hostile), which were chose from PANAS.

Researchers focusing on metaphor and embodiment have increasingly suggested that the study in this area should attempt to examine mechanisms, moderators, or the implications of embodiment/metaphor in addition to illustrating effects. Our study pushed the envelope by examining the impact of metaphor and embodiment on emotional states. Emotions are important psychological phenomena, and our emotions can affect several cognitive processes and behaviors. Thus, our results have important implications for real life scenarios. Certain environments, such as brightness and the direction of our gaze,can affect our emotions along the pleased–depressed dimension. Thus, we must be mindful of how these factors can inifuence our emotions (i.e., avoid looking at black objects for long periods of time, paint our walls in lighter colors, avoid looking at objects low in vertical position etc.). Furthermore, there was very little cross–cultural work in this area, and our study firstly explored the metaphor of emotion based on the non–western participant samples. Our findings also open up some interesting avenues for future research. Additional studies should explore whether these metaphors of emotion actually inifuence behaviors. For instance, these metaphors could affect our behavior by diverting gaze away from black objects/scenes to avoid aggressive behaviors and focusing on white/light colored objects/scenes to help promote prosocial behaviors. Furthermore, it will be interesting to determine whether the infiuence of these metaphors have any sort of lasting effects on emotional experience and behavior.

References

Abele–Brehm, A., & Brehm, W. (1986). Zur Konzeptualisierung und Messung von Befindlichkeit: Die Entwicklung der "Be?ndlichkeitsskalen" (BFS) [The conceptualization and measurement of mood: The development of the "Mood Survey"]. *Diagnostica,* 32, 209e228.

Adams, F. M., & Osgood, C. E. (1973). A cross–cultural study of the affective meanings of color. *Journal of Cross–Cultural Psychology, 4*(2), 135e156.

Bargh, J. A., & Shalev, I. (2012). The substitutability of physical and social warmth in daily life. *Emotion, 12*(1), 154e162.

Crawford, L. E., Margolies, S. M., Drake, J. T., & Murphy, M. E. (2006). Affect biases memory of location: Evidence for the spatial representation of affect. *Cognition & Emotion, 20*(8), 1153e1169.

Fetterman, A. K., Robinson, M. D., & Meier, B. P. (2012). Anger as "seeing red": Re– action time evidence foran implicit perceptualassociation. *Cognition & Emotion, 1*(1), 1e14.

Fisher, S. (1964). Depressive affect and perception of upedown. *Journal of Psychiatric Research, 2*(1), 25e30.

Hevner, K. (1935). Experimental studies of the affective value of colors and lines. *Journal of Applied Psychology, 19*(4), 385e398.

Keith, W. J., & James, F. S. (1975). Effects of four psychological primary colors on anxiety state. *Perceptual and Motor Skills, 41*(1), 207e210.

Kripke, D. F. (1998). Light treatment for nonseasonal depression: Speed, ef?cacy, and combined treatment1. *Journal of Affective Disorders, 49*(2), 109e117.

Lakoff, G., & Johnson, M. (1999). *Philosophy in the ?esh: The embodied mind and its challenges to western thought*. New York, NY: Basic Books.

Landau, M. J., Meier, B. P., & Keefer, L. A. (2010). A metaphor–enriched social cognition. *Psychological Bulletin, 136*(6), 1045e1067.

Meier, B. P., & Robinson, M. D. (2004). Why the sunny side is up: Associations be– tween affect and vertical position. *Psychological Science, 15*(4), 243e247.

Meier, B. P., & Robinson, M. D. (2006). Does "feeling down" mean seeing down? Depressive symptoms and vertical selective attention. *Journal of Research in Personality, 40*(4), 451e461.

Meier, B. P., Robinson, M. D., & Clore, G. L. (2004). Why good guys wear white automatic inferences about stimulus valence based on brightness. *Psychological Science, 15*(2), 82–86.

Oren, D. A., Wisner, K. L., Spinelli, M., Epperson, C. N., Peindl, K. S., Terman, J. S., et al. (2002). An open trial of morning light therapy for treatment of antepartum depression. *American Journal of Psychiatry, 159*(4), 666–669.

Song, H., Vonasch, A. J., Meier, B. P., & Bargh, J. A. (2012). Brighten up: Smiles facilitate perceptual judgment of facial lightness. *Journal of Experimental Social Psychology, 48*(1), 450–452.

Teasdale, J. D. (1993). Emotion and two kinds of meaning: Cognitive therapy and applied cognitive science. *Behaviour Research and Therapy, 31*(4), 339–354.

Terman, M., Terman, J. S., & Ross, D. C. (1998). A controlled trial of timed bright light and negative air ionization fortreatment of winter depression. *Archives of General Psychiatry, 55*(10), 875–882.

Thayer, R. E. (2003). *Calm energy: How people regulα te mood with food and exercise.* New York, N Y: Oxford University Press.

Valdez P.&MehrabianA.(1994). Effects of co10r on emotions. *journal of Experimental Psycholy: General, 123*(4), 394–409.

Wapner, S., Werner, H., & Krus, D. M. (1957). The etfect of success and failure on space 1ocalization. *journal of Personality, 25*(6), 752–756.

Watson, D., Clark, L. A., & Tellegen, A.(1988). Development and validation of brief measures of positive and negative affect: The PANA5 sca1es. *Journal of Personality and Social Psychology, 54,* 1063–1070.

Williams, L. E., Huang, J. Y., & Bargh, J. A. (2009). The scaffo1ded mind: Higher menta1 processes are grounded in ear1y experience of the physica1 wor1d. *European journal of Social Psychology, 39*(7), 1257–1267.

发展心理研究所团队成员

周宗奎，博士，教授，心理学院院长，青少年网络心理与行为教育部重点实验室主任，主要研究领域为人格与社会性发展、青少年网络心理与行为研究、儿童青少年心理健康教育等。

谷传华，博士，教授，主要研究领域为创造性发展与培养，社会性与人格发展。

范翠英，博士，教授，发展心理研究所所长，主要研究领域为儿童青少年心理发展与心理健康教育、青少年网络欺负、网络道德心理研究等。

莫书亮，博士，副教授，主要研究领域为儿童和青少年社会认知发展，异常儿童社会认知发展及干预。

孙晓军，博士，副教授，发展心理研究所副所长，主要研究领域为人格与社会性发展、青少年网络交往行为、留守儿童的心理问题研究等。

田媛，博士，讲师，主要研究领域为网络心理与行为、媒体学习评测、青少年公正价值观等问题。

刘勤学，博士，讲师，主要研究领域为儿童青少年问题行为与健康发展、网络心理与行为研究、亲密关系与家庭治疗。

孔繁昌，博士，讲师，主要研究领域为社会性与人格发展、自我与健康、网络使用与青少年发展。

李董平，博士，讲师，主要研究领域为处境不利儿童青少年的心理发展，青少年问题行为研究。

心理学报 , 2005, 37(6), 776－783.

童年中期同伴关系与孤独感的中介变量检验 *

周宗奎[1,2]，孙晓军[1]，赵冬梅[1]，Hsueh Yeh[3]

([1]华中师范大学心理学院 , 武汉 430079)

([2]首都师范大学学习与认知实验室 , 北京 100089)

([3] Co llege of Education, University of Memphis)

摘　要　以 571 名小学三、四、五、六年级的儿童为被试 , 考察了儿童社会喜好、友谊质量、社交自我知觉与孤独感的关系 , 检验了社交自我知觉在同伴关系变量与孤独感间的中介作用。结果表明 , 社会喜好、友谊质量、社交自我知觉和孤独感间相关显著 , 并且存在显著的性别差异 ; 社交自我知觉在同伴关系变量与孤独感间存在中介的作用 ; 独立的中介效应检验中 , 社会喜好、友谊质量均通过社交自我知觉的中介作用与孤独感发生联系 , 同时 , 也存在直接的联系 ; 综合模型中 , 社会喜好只通过社交自我知觉的中介作用与孤独感产生联系 , 不存在直接效应 , 而友谊质量与孤独感既存在中介的联系 , 同时也存在直接联系。

关键词　社会喜好；友谊质量；社交自我知觉；孤独感；中介作用。

分类号　**B844**

1　问题提出

孤独是一种消极的、弥漫性的心理状态 , 儿童长期处于此状态会导致适应不良。关于孤独感的研究中 , 有两个主要的理论 , 即社会需要理论和认知加工观。社会需要理论认为 , 人生来就有与人保持交往的需要 , 除非人际交往需要得到满足 , 否则就会产生孤独感。而认知加工理论认为 , 孤独感的产生不是因为人类固有的社会交往需要得不到满足 , 而是当一个人对觉知到的人际关系不满意时 , 孤独感才会产生。换句话说 , 当一个人意识到他想要或希望的人际交往关系与实际现状之间有差距时 , 孤独感才会产生。

在众多对儿童孤独感的研究中 , 研究者普遍发现 , 社会技能、问题行为、人格特征、家庭环境等因素都与儿童的孤独感有关 , 而同伴关系则一直是研究者在研究孤独感时重点考察的一个因素。同伴关系作为同龄人之间或心理发展水平相当的个体间在交往过程中建立和发展起来的一种人际关系 , 可以分为四个水平 : 个体特征水平、人际交互水平、双向关系水平和群体水平。本研究选取的社交自我知觉、友谊质量和同伴接纳就分别处于同伴交往经验的个体水平、双向关系水平和群体水平。

Bush 和 Ladd 的研究表明 , 被拒绝儿童经历了

* 本研究得到国家自然科学基金资助 (项目号 : 30270473), 并获得北京市重点实验室——首都师范大学《学习与认知实验室》经费资助。

通讯作者 : 孙晓军 , E－ mail: sxj_ccnu@hot mai. l com

同伴较多的消极对待，更可能表现出较高的孤独感；Asher 等人的研究则发现，3~6 年级不受欢迎的儿童（被忽视型和被拒绝型儿童）报告了显著高于受欢迎儿童的孤独感；俞国良等人的研究也发现，儿童的同伴接受性与孤独感有显著的负相关。大量的研究都证实了同伴接纳性与孤独感间的密切联系。儿童的社交地位越不利，同伴接纳性越低，其体验到的孤独感就越强。

同时，作为同伴关系另一水平的指标，友谊关系与孤独感间也存在着较强的联系。Hodges 等人的研究发现，友谊质量对儿童的孤独感具有显著的预测作用；受欺负儿童所受到的伤害以及体验到的孤独感会因为拥有一个支持性的朋友而得以减轻；Demir 的研究也发现，对于与异性或同性的友谊不满的青少年报告的孤独感高于感到满意的青少年，并且亲密朋友的数量越多孤独感水平就越低。

而国内研究中，同时从同伴接纳性和友谊关系这两个维度来探讨同伴关系与孤独感的研究较少。实际上这两个指标的具体内涵是有区别的，同伴接纳性考察的是群体水平上的同伴关系，而友谊关系重点强调的是同伴间双向水平的同伴关系。有研究也指出了同伴接纳与友谊在儿童青少年的发展中具有不同的功能。因此，有必要综合这两个维度探讨它们对孤独感的影响，这是本研究所关注的一个方面。

另一方面，Asher 指出，孤独是个体对自己社交状况的一种主观体验。人本主义心理学者也认为当一个人的社会关系网络的数量和质量低于他的期望时，孤独感就产生了。而同伴关系作为一种客观的社交地位，在预测主观孤独感时可能存在偏差。例如，有研究发现，一些受欢迎儿童报告了极高水平的孤独感，而一些被拒绝型儿童却报告了极低水平的孤独感。这种现象一方面印证了孤独是一种主观体验，另一方面说明，只从客观的同伴关系角度来考察孤独感，不能充分解释同一社交地位群体内部的个体差异。

基于上述原因，一些研究者加入了认知因素来研究儿童的孤独感。社交自我知觉作为个体对自身社交地位的主观评价，引起了研究者的广泛兴趣，将客观的社交地位与主观的社交自我知觉结合起来，考察二者对孤独感的影响成为了孤独感研究新的取向。如周宗奎等人在考察二者对孤独感的研究中发现，比起客观社交地位，主观的社交自我知觉对于孤独感具有更强的预测力。在本研究中，我们就试图综合客观的同伴关系和主观的社交自我知觉来考察它们对于孤独感的影响。同时，Hymel 等人曾从社会认知的角度解释孤独感与社会地位的关系，他们认为，儿童的孤独感与儿童在同伴中的实际社交地位之间是以社会认知过程为中介的，个人的人际关系知觉水平是重要的中介变量之一。因此，本研究中，我们假设客观的同伴关系除了与孤独感具有直接联系外，还可以通过主观社交自我知觉的中介作用与孤独感产生间接联系，即主观的社交自我知觉在同伴关系和孤独感间存在部分中介作用。

综上所述，本研究中，我们将综合同伴接纳性和友谊质量这两个指标考察同伴关系对孤独感的影响，在此基础上，检验儿童社交自我知觉在此过程中的部分中介作用。

2 研究方法

2.1 被试的选定

武汉市一所小学的三、四、六年级各两个班，五年级三个班，共 9 个班 580 人。回收有效问卷 571 份，回收率为 98.45%。其中，男生 310 人，女生 261 人；三年级 123 人，四年级 122 人，五年级 200 人，六年级 126 人，各年级男女生分布具体情况如表 1 所示。三、四、五、六年级学生的平均年龄分别为 9.16、10.14、11.11 和 12.12。

2.2 研究工具

2.2.1 同伴提名 给儿童提供一份班级名单表，

表 1 各年级男女生人数分布表

被试	三年级	四年级	五年级	六年级
男生	69	70	100	71
女生	54	52	100	55
总计	123	122	200	126

注：表内各数据均表示被试的人数

要求他们选出自己在班内最喜欢的 3 个同学和最不喜欢的 3 个同学。然后，将每个学生所获得的最喜欢和最不喜欢的提名数除以班级的总人数，分别得到积极提名和消极提名的比例，二者之差表示社会喜好 (sp)，即受欢迎程度，作为被试同伴接纳性的指标，这种记分方法得到了研究者的认可，具有较高的效度。

2.2.2 友谊质量问卷 采用《友谊质量问卷》(Parker & Asher, 1993) 的简表，共 18 个项目，原量表有 40 个项目。包括肯定与关心、帮助与指导、陪伴与娱乐、亲密袒露与交流、冲突解决策略、冲突与背叛这六个友谊维度。Cronbach's α 为 0.76。

2.2.3 儿童自我知觉 (PCSC) 量表 PCSC 量表 (The Perceived Competence Scale for Children) 是 Harter(1982) 编制的儿童自我知觉问卷，原问卷包含的四个维度分别是：社交自我知觉、认知自我知觉、运动技能自我知觉和一般自我知觉，本研究只选用社交自我知觉这一维度，Cronbach's α 为 0.78。该量表给被试同时呈现两个描述性的句子 (如：一些孩子觉得交朋友很困难 vs.另外一些孩子觉得交朋友很容易)，首先要求被试确定他 / 她更符合哪一句的描述，然后再确定他 / 她是有点符合该描述还是完全符合，分别记为 1~4 分。计算该维度所有项目总分的平均分，得到儿童的社交自我知觉分。

2.2.4 儿童孤独量表 采用 Asher 等人 1984 年编制的专用于 3~6 年级学生的儿童孤独量表 (Chilren's Loneliness Scale)，该量表包括 16 个孤独项目 (10 条指向孤独，6 条指向非孤独) 和 8 个关于个人爱好的插入项目 (为使被试在回答时放松一些)，因子分析表明插入项目与负荷于单一因子上的 16 个孤独条目无关，16 个孤独项目的 Cronbach's alpha 为 0.92。计算 16 个项目的平均分 (反向记分的题目先要进行转换)，得到儿童的孤独感得分，得分越高，表示孤独感越强，这一记分方法在国内外都被广泛采用。

2.3 数据收集与分析

由经过培训后的心理学专业研究生主持，采用团体施测的方式进行。施测时以班级为单位，由主试讲明要求，解释指导语，必要时给予个人指导以确保被试正确理解问卷。所有数据于 2003 年 6 月收集完毕，全部数据由 Filemaker 4.0 录入，利用 SPSS11.5 和 LISREL 8.30 软件进行统计处理，主要采用多元方差分析和结构方程模型等统计方法。

3 研究结果与分析

3.1 描述性统计分析结果

本研究中，所考察的各变量间的相关如表 2 所示，由表 2 可知，除了友谊质量的冲突、背叛维度与社会喜好和社交自我知觉相关不显著以外，各变量间均存在极其显著的相关，这就满足了我们进行中介效应检验的前提条件；同时，各问卷的 Cronbach's α 均大于推荐值 (0.70)，表明信度较高。

3.2 性别、班级的差异检验

本研究中，由于各因变量 (社会喜好、友谊质量

表 2 变量相关矩阵 (N=571)

	M	SD	1 社会喜好	2 肯定与关心	3 帮助与指导	4 陪伴与娱乐	5 亲密袒露	6 冲突解决	7 冲突背叛	8 社交知觉	9 孤独
1	0.003	0.098									
2	2.520	0.952	0.124**								
3	2.551	1.052	0.137**	0.459***				(0.76)			
4	3.148	0.902	0.219***	0.436***	0.450***						
5	2.615	1.080	0.129**	0.461***	0.545***	0.417***					
6	2.991	0.991	0.134**	0.371***	0.429***	0.383***	0.391***				
7	4.164	0.991	0.036	0.129**	0.180**	0.150**	0.146**	0.216***			
8	2.760	0.428	0.133**	0.185***	0.107*	0.179***	0.113**	0.146**	-0.036	(0.78)	
9	1.876	0.769	-0.278**	-0.360***	-0.288***	-0.296***	-0.284***	-0.300***	-0.134**	-0.510***	(0.92)

注: *p<0.05 **p<0.01; ***p<0.001; ( )内数据为各量表的内部一致性系数其中, (0.176)是友谊质量整个量表的内部一致性系数。

各维度、社交自我知觉、孤独感）之间存在显著相关，所以不应通过多次方差分析来对班级的差异进行检验，这一过程应通过 MANOVA 进行多元方差分析来完成。多元方差分析的检验统计量通常用 Wilks 的 Λ，得到的是精确的 F 值。班级差异的多元方差分析结果表明，班级的主效应不显著（$\Lambda=0.965$, $F(8, 550) = 2.243$, $p= 0.058$, ns）。

T 检验的结果表明，除了友谊质量的冲突与解决维度和社交自我知觉以外，在所有其它因变量上均存在显著的性别差异。在社会喜好（$M_男=-0.011$, $M_女=0.02$, $t(569) =-3.387$, $p<0.001$）、友谊质量的肯定与关心（$M_男=2.417$, $M_女=2.641$, $t(569)=-2.818$, $p<0.01$）、帮助与指导（$M_男=2.370$, $M_女=2.766$, $t(569) =-4.56$, $p<0.001$）、陪伴与娱乐（$M_男=3.053$, $M_女=3.261$, $t(569) =-2.759$, $p<0.01$）、亲密与袒露（$M_男=2.361$, $M_女=2.917$, $t(569)= -6.332$, $p<0.001$）、冲突与背叛（$M_男=4.051$, $M_女=4.298$, $t(569)=-3.256$, $p<0.001$）维度上，男生的得分均显著低于女生，而在孤独感的得分上，则是男生显著高于女生（$M_男=1.959$, $M_女=1.775$, $t(562)= 2.836$, $p<0.01$），表现出更高的孤独感体验。

3.3 中介效应的检验

如果自变量 X 通过影响变量 M 来影响 Y，则我们就称 M 为中介变量。本研究中，我们假设社交自我知觉是同伴关系与孤独感间的中介变量（如图1所示），图中 a、b、c、c' 均表示相应的标准回归系数。

研究者对中介效应进行检验时较多地运用了3种方法：Sobel 检验（$Z =ab / \sqrt{b^2 S^2 a + a^2_b}$），Goodman I 检验（$Z=ab/ \sqrt{b^2 S^2_b + a^2_{sb} + S^2_a S^2_b}$）和 Goodman II 检验（$Z=ab/ \sqrt{b^2 S^2_b + a^2_b{}^2 - S^2_a S^2_b}$）。本研究中，样本容量较大（$N=571$），因此，3种方法的检验功效差别不大。

通过结构方程模型分析，求得对于社会喜好而言，a、b 值为 0.36 和 -0.76，对应的标准误分别为 0.06 和 0.09，c' 为 -0.10（$\chi^2=105.50$, $df=33$; RMSEA = 0.062; SRMR = 0.043; GFI= 0.96; IFI = 0.94;CFI= 0.94; NNFI=0.92）；相应的，对于友谊质量，a、b 值分别为 0.47 和 -0.51，标准误分别为 0.17 和 0.22，c' 为 -0.13（$\chi^2=220.24$, $df=75$; RMSEA =0.051; SRMR = 0.054; GFI=0.96; IFI = 0.94; CFI =0.96; NNFI=0.93）。将这些结果分别代入上述三个公式，各方法的检验结果见表3。

由表3可知，各种检验的的结果都表明了中介变量的显著作用。即社会喜好或友谊质量都可以通过社交自我知觉的中介作用与孤独感发生间接联系，其中，对于社会喜好来说，中介效应为 -0.273（a × b，下

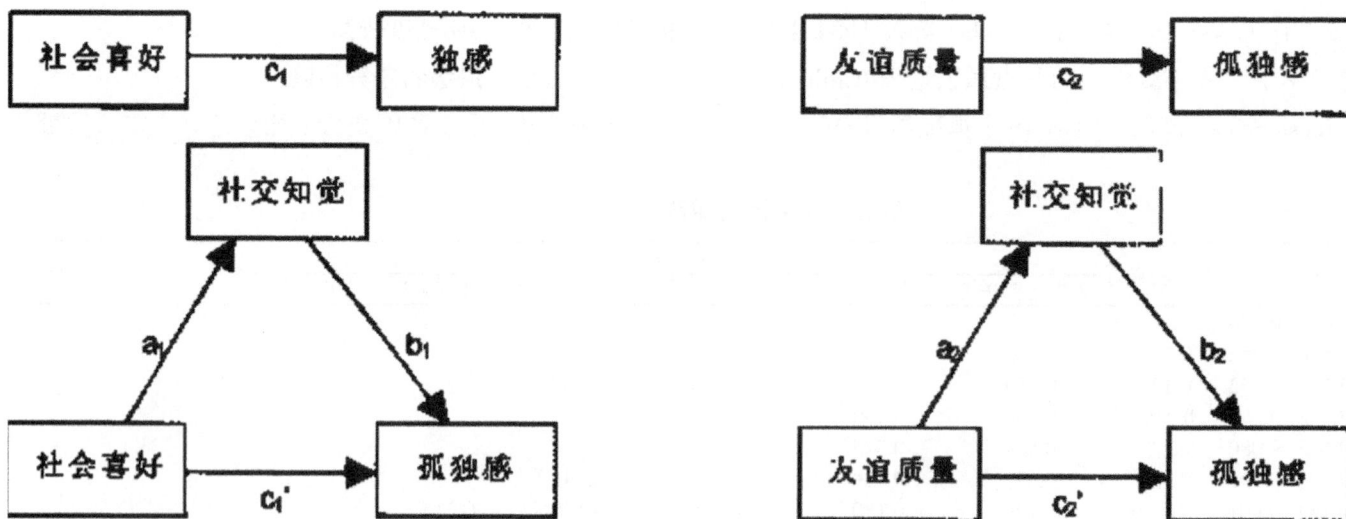

图1 社交自我知觉在同伴关系与孤独感间的中介作用模式图

表 3 社交自我知觉在同伴关系和孤独感之间的中介作用检验

中介作用的路径	社会喜好－社交自我知觉－孤独感	友谊质量－社交自我知觉－孤独感
$a(S_a)$	0.36（0.06）	0.47（0.07）
$b(S_b)$	−0.76（0.09）	−0.51（0.12）
Sobe 脸验（Z）	4.891***	3.591***
Goodman I 检验（Z）	4.868***	3.563***
Goodman II 检验（Z）	4.914***	3.620'**

注：*p< 0.05; **p< 0.01; *** p<0.001。

同），总效应为 − 0.373(a ×b +c' ，下同)，中介效应与总效应的比值为 0.732；而对于友谊质量，其中介效应为 −0.240，总效应为 − 0.37，中介效应与总效应的比值为 0.649，由于中介效应的相对作用均较大（分别为 0.732 和 0.649），因此，这一中介作用的发现更具有重要意义。

上述 3 种检验证实了社交自我知觉在同伴关系与孤独感间的中介作用的存在，同时，结构方程模型的结果表明，无论对于社会喜好或友谊质量，c' 均达到显著性水平，因此，我们可以进一步判断，社交自我知觉在同伴关系与孤独感间存在部分中介的作用。

综合上述分析的结果，我们构建了如图 2 所示的模型，结构方程模型分析表明，该模型能较好地拟合数据（χ 2= 230.34, df=99; RMSEA=0.048; SRMR= 0.045; GFI=0.95; IFI=0.94; CFI=0.94; NNFI =0.93)。

结果表明，友谊质量和孤独感间存在直接联系，直接效应值为 −0.17，同时，友谊质量也可以通过社交自我知觉的中介作用与孤独感产生联系，其中介效应值为 − 0.238(即从友谊质量到孤独感的间接作用路径上的路径系数乘积)，总效应值为 −0.408，中介效应与

总效应的比值为 0.583; 而社会喜好和孤独感间的联系主要是通过社交自我知觉的中介作用来实现的，其中介效应的值为 −0.182, 直接效应不显著 (− 0.07)，因此，是一种完全中介的作用，这一结果与前面证明的部分中介作用是不一致的。

4 分析与讨论

4.1 社会喜好、友谊质量、社交自我知觉和孤独感的性别、班级差异

本研究表明，社会喜好、友谊质量、社交自我知觉和孤独感不存在班级差异，但社会喜好、友谊质量（除冲突与解决维度外）和孤独感的性别差异显著。无论是社会喜好或友谊质量，男生的得分均显著低于女生，这一结果与以往的研究结果是一致的。男生往往表现出较多的问题行为，这种问题行为反应在自身的社交行为和策略上就很容易导致其较差的同伴关系，而女生由于社会环境、文化等因素的影响，往往表现得更为收敛，更愿意建立亲密的同伴关系网络，因此，她们的同伴关系自然就比男生要好。而对于孤独感的性别差异，本研究发现，男生的孤独感要显著高于女生，这一结果与我们前期的一项研究的结果是一致的，也与以往的部分研究结果一致。中国的传统文化要求男生更为独立、自强，不鼓励他们表露自己的情绪；而社会化的过程使女生富于乐群性，她们具有更强的亲和动机，更重视建立亲密的同伴关系，也得到更多的同伴支持，因此，女生体验到的孤独感就比较低。但也有研究表明，儿童的孤独感并不存在性别差

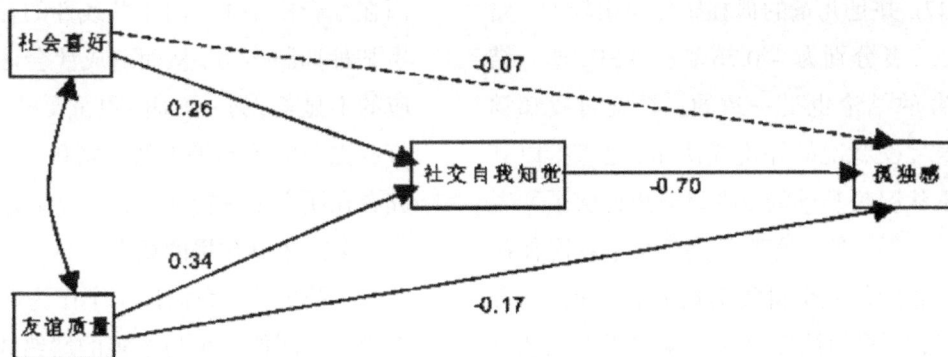

图 2 同伴关系、社交自我知觉与孤独感的关系

异。这可能是由于不同研究中分析的变量不同造成的，我们不能简单地通过这种比较来推断结论，儿童的孤独感是否存在性别差异？为什么存在这些差异？这些都是我们今后的研究中有待解决的问题。

4.2 中介效应的检验

从相关分析的结果来看，本研究中考察的各变量间（同伴接纳、友谊质量、社交自我知觉、孤独感）具有极其显著的相关，其中，同伴接纳、积极友谊质量（友谊质量的前五个维度）和社交自我知觉间存在正相关，表明儿童同伴接纳性越高，其积极友谊质量和社交自我知觉相应也较高，反之，儿童同伴接纳性越低，则其积极友谊质量和社交自我知觉也较低，儿童主观的社交自我知觉是以客观的同伴关系（同伴接纳、友谊质量）为基础的；同时，同伴接纳、积极友谊质量、社交自我知觉三变量与孤独感均存在显著的负相关，表明它们呈互为消长的关系，即同伴接纳性越高、积极友谊质量越好、社交自我知觉越强，则其体验到的孤独感就越低，反之，其体验到的孤独感就越强，这一结论与大多数研究的结果是一致的。

独立的中介效应检验结果表明，社交自我知觉在同伴关系（同伴接纳、友谊质量）与孤独感间起着中介作用，并且是部分中介的作用。由于中介效应的相对作用均较大（分别为 0.732 和 0.649），因此，这一中介作用的发现就更具有实际意义。无论是同伴接纳性或友谊质量，通过社交自我知觉的中介作用对孤独感产生影响的模式是一致的，即较好的同伴关系（同伴接纳性较强或友谊质量较好）背景下，儿童往往具有较高的社交自我知觉水平（路径系数为正数，二者分别为 0.36 和 0.47），并且儿童的孤独感体验也较低（路径系数为负数，二者分别为 -0.76 和 -0.51），这一结果与大多数研究的结论也是一致的。社交自我知觉作为个体对自身社交状况的评价或认知，是直接以其客观的同伴关系状况为基础的，本研究也证实了两者间显著的正相关关系，某一客观的同伴关系状况往往会导致个体产生相应的社交自我知觉水平；而另一方面，研究者发现，主观的社交自我知觉对于孤独感具有较强的预测力，儿童的社交自我知觉水平越高，其

孤独感就越低，二者间存在显著的负相关。因此，我们可以推断，儿童的同伴关系能正向预测其社交自我知觉水平，而社交自我知觉的水平又能负向预测其体验到的孤独感强度，本研究的结果证实了这一推断。

图 2 所构建的模型表明，社会喜好和友谊质量共同对儿童的孤独感产生影响。值得注意的是，从模型中各条路径系数的值来看，我们可以发现，当综合社会喜好和友谊质量来考察同伴关系对孤独感的影响时，同伴接纳性较低的儿童体验到的孤独感会因为较高的友谊质量而降低；相似地，友谊质量较低的儿童体验到的孤独感也会由于其较高的同伴接纳性而降低。Shaffer 指出，拥有一个或多个亲密的朋友可以为儿童提供一个情感上的安全网络，这种安全感可以帮助儿童更积极地迎接新的挑战，而且可以帮助儿童承受所面临的压力（如父母离异、同伴拒绝等）；一些研究则发现，受欺负儿童所受到的伤害以及体验到的孤独感会因为拥有一个支持性的朋友而得以减轻。因此，如果儿童同伴接纳性较低并且友谊质量也较差，那么从整个模型来看，他体验到的孤独感就会更加强烈，对于这类儿童，心理学工作者也应给予更多的关注。

值得注意的是，在综合社会喜好和友谊质量来共同考察社交自我知觉的中介作用时，我们发现了一些不一致的结果。独立的中介效应检验中，社交自我知觉在社会喜好和孤独感间起部分中介的作用，而在综合模型中，社会喜好对孤独感的直接效应并不显著，社交自我知觉起到的是完全中介的作用。造成这一结果的主要原因可能是友谊质量与社会喜好间存在较强的共线性，分别考察时并没有考虑到它们的共线性，而在综合模型中，由于共线性的存在，可能相互间产生某种抑制作用，从而造成社会喜好对孤独感直接效应的不显著；另一方面，更重要的是，心理学研究中一直都强调整体不等于部分之和，这个结果也提醒了研究者在研究中应特别注意研究结论的适用范围。

4.3 中介作用的意义

本研究中，我们证实了作为认知因素的社交自我知觉，在同伴关系与主观的情绪体验（孤独感）间确实存在中介作用。实际上，研究者一直很关注认知成

分在同伴关系和情绪体验间的中介作用，也做过一定的研究，结果基本都证实了这一中介作用的存在。例如，Boivin 等人的研究表明，消极的同伴关系对儿童抑郁的影响依赖于其体验到的孤独感及其对社交环境的认知；有研究者也提出了相似的结论，认为同伴拒绝与情绪体验（孤独、抑郁）间的联系是复杂的，有赖于儿童对他的社交情境的知觉；Valas 和 Sletta 则认为，儿童对待自己的社交情境的知觉有助于解释社会行为和同伴拒绝对其内在情绪体验的作用机制等等。这些研究中，共同之处就是认为认知成分在儿童的同伴关系和情绪体验间存在中介作用，本研究的结果与这些研究的结论也是一致的。

认知成分在同伴关系和情绪体验间的中介作用的发现具有一定的实际意义。一方面，它提示我们，同伴关系对情绪体验的影响是复杂的，并不是简单的一一对应的关系，会受到其它因素（例如认知因素）的影响；另一方面，更为重要的是，这种中介作用的发现将有助于对儿童消极情绪体验的干预，可以从认知层面入手，通过改善儿童的社交认知从而有效地预防其消极的情绪体验。总之，这一中介作用的发现，无论是对于该领域理论的丰富，或是对于实践工作的开展，都具有重要的意义。

需要强调的一点是，本研究中所考察的四个变量（社会喜好、友谊质量、社交自我知觉、孤独感）间可能存在更为复杂的关系，它们之间可能存在与我们的假设完全相反的关系，具体来说，有可能孤独感导致儿童社交自我知觉较低，进一步造成其同伴关系较差等等。本研究中，我们的模型是依据已有理论的支持来构建的，前面我们已经论述过。作为相关研究来说，本研究并不能从严格意义上确定变量间的因果关系，只是提供了这四个变量间可能存在的某种关系，为研究者提供一种选择，严格的因果关系必须通过实验研究才能获取。

5 结论

本研究的结果表明，社交自我知觉在同伴关系和孤独感间存在中介的作用，当单独考察社会喜好或友谊质量与孤独感间的关系时，社交自我自觉起部分中介的作用；而综合考察同伴关系和孤独感间的联系时，友谊质量主要通过社交自我知觉的中介作用与孤独感产生联系，同时也与孤独感间存在直接的联系；而社会喜好与孤独感间的联系则是通过社交自我知觉的中介作用来实现的，不存在直接效应。

参考文献

Asher S R, Hyme lS, Renshaw P D. (1984). Loneliness in children. *Child Development, 55,* 1456-1464

Asher S R, Wheeler V A.(1985). Children' s loneliness: A comparison of rejected and neglected peer status . *Journal of Consulting and Clinical Psychology, 53,* 500-505

BoivinM, Hymel S, BukowskiW M.(1995). The roles of social withdrawal, peer rejection, and victimization by peers in predicting loneliness and depressed mood in childhood. *Development and Psychopathology, 7,* 765-785

Bush E S, Ladd G W.(2001). Peer rejection as an antecedent of Yong Children' s school Adjustment: An Examination of Mediating Processes .*Developmental Psychology, 37,* 550-560

Gauze C, Bukowski W M, Aquan –Assee J, et al .(1996). Interactions between family environment and friendship and associations with well–being during early adolescence. *Child Development, 67,* 2201- 2216

Hodge EV E, BoivinM, Vitaro F, et al .(1999). The power of friendship: Protection against an esca lating cycle of peer victimization. *Developmental Psychology, 35,* 94-104

Hymel S, VaillancourtT, McDougall P,et al.(2004). Peer acceptance and rejection in childhood. In:Smith P K, HartCH. (Ed.) *Blackwell Handbook of Childhood Social Development ,*65–284

Lau S, DennisW K, Patrick C, et al.(1999). Facets of Lone liness and Depression Among Chinese Children and Adolescents . *The Journal of Social Psychology, 139,* 713–725

Margalit M.(1994). Loneliness among children with special need. *New York: Springer-Verlag*, 15 ~ 17

Parker JG, A sherS R.(1993). Friendship and friendship quality in middle childhood: Links with peer group acceptance and feelings of loneliness and social dissatisfaction. *Developmental Psychology, 29*(4), 611–621

Peplau L A, Perl manD.(1984). Loneliness research: A survey of empirical findings. In:L A Peplau, S E Goldston (Eds) .Preventing the harmful consequences of severe and persistent loneliness. Rockville, MD: National Institute of Mental Health, 13–47

Rotenberg K J, Hymel S.(1999). *Loneliness in Childhood and Adolescence*. UK: Cambridge University Press,

Rotenberg K J, Hymel S.(1999). *Loneliness in Childhood and Adolescence*. Cambridge:Cambridge University Press,

RubinK H, BukowskiW, Parker JG. (1998). Peer Interactions , Relationships, and Groups. In W Damon, E Nancy. Handbook of Child Psychology. Vol. 3:Social , Emotional, and Personality Development . Sth ed. New York: W iley, 619–700

Schwatz D, Dodge K A, PettiG S, et al .(2000). Friendship as a moderating factor in the pathway between early harsh home environment and later victimization in the peer group. *Developmental Psychology, 36*, 646–662

ShafferD R. (2002). Developmental Psychology: childhood and adolescence(6th ed.). *Thomson learning*, 575–617

Uruk A C, Demir A.(2003). The ro le of peers and families in predicting the loneliness level of adolescents . *The Journal of Psychology, 137* (2), 179–194

ValasH, Sletta O. (1996). Social behavior, peer relations , loneliness and self-perceptions in middle school children: A mediational model .Paper presented at the Ⅺ Ⅴ thbiennial meeting of the International Society for the Study of Behavioral Development, Quebec City, Canada, WarmanD M, Cohen R.(2000) Stability of Aggressive Behaviors and Children's Peer Relationships. *Social Development, 26*, 277–290

Wang Z Y, Wang J S, Chen H C. (2000). The influence of training on rejected and neglected children (in Chinese) . Psychological Development and Education, 16 (1), 6–11.

[王争艳，王京生，陈会昌.(2000). 促进被拒绝和被忽视幼儿的同伴交往的三种训练方法 .*心理发展与教育，16* (1), 6–11.]

Wen Z L, Chang L, HauK T, et al .(2004). Testing and application of the mediating effects (in Chinese) . *Acta Psychologica Sinica, 36* (5), 614–620.

[温忠麟，张雷，侯杰泰等 . (2004). 中介效应检验程序及其应用 .*心理学报，36* (5), 614–620.]

Xin Z Q, Chi L P. (2003). The relationship between family functioning and children' s loneliness: the role of mediator (in Chinese) .*Acta Psychologica Sinica, 35* (2), 216–221.

[辛自强，池丽萍 . (2003). 家庭功能与儿童孤独感的关系：中介的作用 .*心理学报，35* (2), 216–221.]

Yu G L, Xin Z Q, Luo X L. (2003). The characteristics of loneliness, peer acceptance and their relation to family functioning (in Chinese) *Acta Psycho logica Sinica, 32* (1), 59–64.

[俞国良，辛自强，罗晓路 . (2000). 学习不良儿童孤独感、同伴接受性的特点及其与家庭功能的关系 .*心理学报，32*(1), 59–64.]

Zhou Z K. *Children' s Social Skill (in Chinese)*. (2002). Wuhan: Central China Normal University Press.

[周宗奎 . 儿童的社会技能 . 武汉：华中师范大学出版社 ,2002.]

Zhou Z K, Fan C Y.(2001). Primary school children' s social anxiety and loneliness (in Chinese) . *Psycho* logical Science, 24 (4), 442–442.

[周宗奎, 范翠英. (2001). 小学儿童社交焦虑与孤独感研究. 心理科学, 24 (4), 442–442.]

Zhou Z K, Zhao D M, Chen J, et al .(2003). Lone liness as a fuction of sociometric status and self – perceived social competence in middle childhood (in Chinese) . Psychological Development and Education, 19 (4), 70–74.

[周宗奎, 赵冬梅, 陈晶等 .(2003). 童年中期儿童社交地位、 社交自我知觉与孤独感的关系研究 . 心理发展与教育, 19(4), 70–74.]

Zou H.(1993). Children's loneliness and peer relationships . Psychological Development and Education, 9 (2), 16–18

[邹泓 .(1993). 儿童的孤独感与同伴关系 . 心理发展与教育, 9 (2), 16–18[

Zou H. (1998). Developmental function and influencing factors of peer relationship (in Chinese) . Psychological Development and Education, 14 (2), 40–46 .

[邹泓 . 同伴关系的发展功能及其影响因素 . 心理发展与教育, 14 (2), 40–46]

The test of the mediator variable between peerrelationship and loneliness in middle childhood

Zhou Zongkui[1,2], Sun Xiaojun[1], Zhao Dongmei[1], Hsueh Yeh[3]

([1] School of Psychology, Central China Normal University, Wuhan 430079)

([2] Learning & Cognition Lab, Capital Normal University, Beijing 100089)

([3] College of Education, University of Memphis)

Abstract 571 elementary school children from the third grade to the fifth are investigated in June 2003. The relationships among social preference, friendship quality, self-perceived social competence and loneliness are examined, the mediator effect of self-perceived social competence between peer relationship and loneliness is also tested. The results indicated that, the relationships among social preference, friendship quality, self-perceived social competence and loneliness are significant, and the gender differences is significant;self-perceived social competence has mediator effec tbetween peer relationship and loneliness;in separate analyses, there are not only indirect relation between loneliness and socia lpreference or friendship quality through the mediator effect of self-perceived social competence, but also direct relation;in integrate analysis, there are direct and indirect relations between loneliness and friendship, there are only indirect relation between loneliness and social preference.

Keywords social preference; friendship quality; self-perceived social competence; loneliness; mediator effect

Addictive Behaviors, 2015, 42, 1‐8.

Multi-family group therapy for adolescent Internet addiction: Exploring the underlying mechanisms

Qinxue Liu[1,2], Xiaoyi Fang[3,4],[*] , Ni Yan[5], Zongkui Zhou[1,2] , Xiaojiao Yuan[6], Jing Lan[3] , Chaoying Liu[3]

([1] Key Laboratory of Adolescent Cyberpsychology and Behavior (CCNU), Ministry of Education, Wuhan 430079, China)

([2] School of Psychology, Central China Normal University, Wuhan 430079, China)

([3] Institute of Developmental Psychology, Beijing Normal University, Beijing 100875, China)

([4] Academy of Psychology and Behavior, Tianjin Normal University, Tianjin 300387, China)

([5] Faculty of Psychology, Southwest University, Beibei 400700, China)

([6] School of Sociology and Psychology, Southwest University for Nationalities, Chengdu 610041, China)

Abstract Objective: Internet addiction is one of the most common problems among adolescents and effective treatment is needed. This research aims to test the effectiveness and underlying mechanism of multi‐family group therapy (MFGT) to reduce Internet addiction among adolescents. Method: A total of 92 participants consisting of 46 adolescents with Internet addiction, aged 12–18 years, and 46 their parents, aged 35–46 years, were assigned to the experimental group (six‐session MFGT intervention) or a waiting‐list control. Structured questionnaires were administered at pre‐intervention (T1), post‐intervention (T2) and a three‐month follow‐up (T3). Results: There was a significant difference in the decline both in the average score and proportion of adolescents with Internet addiction in MFGT group at post‐intervention(M_{T1}=3.40, M_{T2}=2.42, p<0.001; 100 versus 4.8%, p<0.001) maintained for three months(M_{T3}=2.06, p<0.001; 100 versus 11.2%, p<0.001). Reports from both adolescents and parents were significantly better than those in the control group. Further explorations of the underlying mechanisms of effectiveness based on the changed values of measured variables showed that the improvement in adolescent Internet use was partially explained by the satisfaction of their psychological needs and improved parent‐adolescent communication and closeness. Conclusions: The six‐session multi‐family group therapy was effective in reducing Internet addiction behaviors among adolescents and could be implemented as part of routine primary care clinic services in similar populations. As family support system is critical in maintaining the intervention effect, fostering positive parent‐adolescent interaction and addressing adolescents' psychological needs should be included in preventive programs for Internet addiction in the future.

Keywords multi‐family group therapy; Internet addiction; family relationships; need satisfaction; effectiveness mechanism

* Corresponding author at: Institute of Developmental Psychology, Beijing Normal University, Beijing 100875, China. Tel.: +86 10 5880 8232; fax: +86 10 5880 8232.   E‐mail address: fangxy@bnu.edu.cn (X.‐Y. Fang).

This study was funded by Project of Social Sciences for Young Scholars from Ministry of Education in China(ProjectNo.12YJC190023), China National Science Foundation(Project No. 31170990), the Program for Changjiang Scholars, and the Fundamental Research Funds of Central China Normal University (CCNU13A05046).

1 Introduction

With the rapid development of the Internet, Internet addiction has become a widespread and problematic phenomenon. Internet addiction, also known as Pathological Internet Use, Problematic Internet Use and Compulsive Internet Use, is characterized by excessive and compulsive Internet use and a preoccupation with and loss of control over this use that interferes with individuals' daily functioning (Caplan, 2002; Davis, 2001; Van den Eijinden, Meerkerk, Vermulst, Spijkerman, & Engles, 2008; Young & Abreu, 2011). Currently, it is one of the most common behavioral problems for adolescents, who are more exposed to Internet use and consequently more vulnerable than adults (Lortie & Guitton, 2013), with a prevalence rate higher than 8% in some countries (Cho, Kim, Kim, Lee, & Kim, 2008; Kuss, Grifths, & Binder, 2013; Van den Eijnden, Spijkerman, Vermulst, van Rooij, & Engels, 2010). In China, approximately 10% of adolesecents (approximately 20 million teenagers) reported a tendency towards or current diagnosis of Internet addiction (China Internet Network Information Block, 2008; Center, 2013). Internet addiction may cause psychological distress, personality development problems, social problems and poor school performance (Brezing, Derevensky, & Potenza, 2010; Young, Pistner, O'Mara, & Buchanan, 2000). In addiction, high comorbidity with effective disorders, impulse control disorders and substance abuse disorders have been reported (Petersen, Weymann, Schelb, Thiel, & Thomasius, 2009; Weinstein & Lejoyeux, 2010). There is significant research around the diagnosis, epidemiology, predicting factors and negative outcomes of Internet addiction, but little is known about treating it, which is an imperative for adolescents, families, schools and society, especially in China (King, Delfabbro, Griffiths, & Gradisar, 2011; Winkler, Drsin,

Rief, Shen, & Glombiewski, 2013).

Petersen et al. (2009) conducted a survey at the request of the German health department and argued that clinical recommendations are not possible due to the lack of studies and that further research is urgently needed. In a systematic review of Internet addiction treatment, only eight studies were included. Half of them were psychological approaches, and two utilized cognitive-behavioral therapy (King et al., 2011). Peukert, Sieslack, Barth, & Batra (2010) also indicate cognitive-behavioral and pharmacological approaches as potentially effective treatments in their review. They suggest that interventions with family members could be useful. Winkler et al. (2013) further examine the efficacy of different treatments for Internet addiction (13 studies included) in their meta-analysis, and their results show that CBT did not perform significantly better than other psychological treatments, even though it appears to be the predominant approach for treating Internet addiction. They also suggest that both individual counseling and group therapy have their shortcomings and that further research around different approaches and modalities is needed. However, there is no study that examines which factors contribute to the efficacy of treatment or what predictors cause the behavior change to happen, which is very important to evaluate and improve interventions (Liu, Fang, & Zhou, 2011).

Family plays a central role in the socializing process for adolescents, and parents provide emotional connection, behavioral constraints and modeling (Gray & Steinberg, 1999; Lau, Quadrel, & Hartman, 1990). Family-based intervention is the most thoroughly studied treatment modality for adolescent substance dependence and addiction, and there is a large body of research to support its efficacy (for a review, see Liddle, 2004). Previous research also proved that a good relationship and communication with parents are protective factors for adolescents from Internet addiction (Kim, Jeong,

& Zhong,2010; Van den Eijinden et al., 2010). Family members involved in interventions facilitate the process of recovery and help the addict maintain a lasting effect of intervention after sessions (Liddle, 2004; Zhong et al., 2011).Grounded in family system theory and integrated in family and group therapy, multi-family group therapy (MFGT) was proposed as a promising new approach to treat Internet addiction behaviors, but no empirical study was conducted (Liu et al., 2011). The effectiveness of MFGT has been empirically demonstrated among adolescents with psychological disorders (Chien & Chan, 2013; McDonell & Dyck, 2004), children at risk for special educational services (Kratochwill, McDonald, Levin, Scalia, & Coover, 2009) and in addiction related areas (Conner et al., 1998; Zubrick et al., 2005). In this field, Zhong et al. (2011) found that family-based intervention is more effective in reducing Internet addiction than group therapy that involved only the adolescents. The multi-family group offers both adults and adolescents the advantages of a peer group, which helps them to get support and learn from peer confrontation. Transferential reactions occur not only within one family but also across family lines, facilitating the group to serve as both as an arena for cross transferences-based on each person's introject and as a reality tester (Leichter, & Schulman, 1974). Connection within family members is also helpful for high treatment attendance (Nieter, Thornberry Jr., & Brestan-Knight, 2013). Moreover, family-oriented intervention might be particularly effective in Chinese culture, where the cohesion between family members is highly emphasized. Therefore, the present study aims to explore both the effectiveness of MFGT on Internet addiction and the underlying mechanisms of the effectiveness.

One mechanism through which MFGT may effectively reduce Internet addiction is improving parent-adolescent communication and closeness. Compared with non-addicts, adolescents with Internet addiction have poorer communication with their parents (Park, Kim, & Cho, 2008) and are more likely to receive rejection and negative feedback from their parents (Van den Eijnden et al., 2010). Poor parent-adolescent communication and low perceived parent-adolescent closeness, in turn, predicted adolescents' Internet addiction (Liu, Fang, Deng, & Zhang, 2012; Liu, Fang, Zhou, Zhang, & Deng, 2013). According to the Circumplex Model of Marital and Family Systems proposed by Olson (Olson, 2000; Olson, Sprenkle, & Russell, 1974;), family communication is critical in facilitating intimacy among family members and strengthening the family's adaptability to change. MFGT emphasizes improving family cohesion and motivation to change within the family; it not only focuses on the parent-adolescent interaction but also values the style and strength of attachment between family members (Dickerson & Crase, 2005). Therefore, it could be a well-suited approach to treat Internet addiction among adolescents.

The second mechanism through which MFGT may take effect in treating adolescents' Internet addiction is by fulfilling their psychological needs through strengthening their communication and relationship with their parents. Psychological need is considered one of the most important driving forces that promotes behavioral change. Fulfillment of psychological needs through Internet use has been proposed as an internal motive in adolescents' Internet addiction (Morris & Ogan, 1996; Suler, 1999). Adolescents' unfulfilled needs for competence and relatedness in life and perceived need satisfaction online are the major precursors of their excessive Internet use (Cai, Cui, & Li, 2007; Shen,Liu, & Wang, 2013; Wan, Zhang, Liu, Deng, & Fang, 2010). Compared with non-addicts, Internet addicts perceived higher need satisfaction online and lower need satisfaction in real life (Deng, Fang, Wan, Zhang, & Xia, 2012). Therefore, if parent-adolescent communication practices and relationships are improved, adolescents' psychological needs for relatedness or competence might be more easily fulfilled through their daily life interactions with their parents, which, in turn, could be helpful to reduce their reliance on the Internet for fulfilling their needs. As fundamental as these two underlying mechanisms appear to be in affecting adolescents' Internet addiction, they have nevertheless been barely examined explicitly in prior Internet addiction intervention studies. In this study, we include these two underlying mechanisms as major intervening variables to examine whether the effectiveness of family group intervention for adolescent Internet addiction depends on them.

Based on a quasi-experimental design, the present study examines the effectiveness of the MFGT for adolescent Internet addiction among 46 pairs of adolescents and their parents. The study aims to examine three hypotheses: First, the intervention group shows a reduction in Internet addiction both at the end of the intervention and at three-month follow-up compared with the control group. Second, adolescents in the intervention group show improved communication and relationship with their parents and psychological needs satisfaction in real life. Third, the effectiveness of the intervention is partially explained by the improved parent-adolescent relationship and communication, and adolescents' psychological need satisfaction in real-life.

2 Method

2.1 Participants

Participants were recruited through advertisements on school websites in Baotou City of Inner Mongolia in China. Related information about the research and a simplified scale of Internet addiction, which is used for clinical diagnosis, were included in the advertisement. Families who were interested and matched the diagnosis were welcome to sign up and have a face-to-face interview one by one. Among the 55 families who signed

up for the intervention study, 46 families were selected based on the Adolescent Pathological Internet Use Scale (APIUS; detailed information about APIUS is provided in the measurement section) and inclusion criteria. The body screening scale, SCL-90 and simplified addiction screening scale were used to exclude participants who possessed physical disabilities, mental disorders or other addictive behaviors. Only one boy was excluded for depression.Twenty-onefamilies were assigned to the intervention group because their schedules matched with the intervention arrangement and the other 25 families were included in the control group because they could not set up a continued intervention schedule. Families in the control group were added to the waiting-list for the intervention study after the informed consent from the parents and the adolescents.

The intervention group had a dominant proportion of male (n=17) over female (n=4) with an average age of 15 years old (SD=1.73). The female-to-male ratio of parents in the intervention group was 16: 5 with an average age of 40.9 years old (SD=2.85). Nine of them held a degree of college or above (42.9%); six of them obtained a high school degree (28.6%) and six of them did not obtain a high school degree (28.6%). The average monthly income ranges from 2,000 to 10,000 Yuan with an average of 4,685, which indicated middle-income families in the city. All parents were first-time married.

The demographic compositions in the control group resembled those in the intervention group. Adolescents in the control group had a male-to-female ratio of 21:4 and the average age was 15.7 years old (SD=1.2). Families in the control group did not did not show significantly differences from those in the intervention group.

2.2 Procedures

First, the Manual of Adolescents Internet Addiction Family Group Therapy was developed with precision based on the theoretic framework of family group therapy, previous intervention practices, and empirical studies. Before the intervention was launched, a pilot study was implemented among six adolescents and their families to assess the operability of the intervention design, potential problems in administering the intervention, and smooth transitions between activity themes in each intervention session. With preliminary results, interview feedback from the pilot studyand consultation with experts on the intervention team, the *Manual of Adolescents Internet Addiction Family Group Therapy* was modified and finalized. Then, recruited parents provided informed consent for their adolescent children's and their own participation.

All participants were asked to complete assessments both before (T1) and after the intervention (T2), and at a three-month follow-up (T3) as well. Participants were assured of the confidentiality of their responses. Procedures were approved by the Institutional Review Board of the Institute of developmental Psychology, Beijing Normal University. The details of the procedures are presented in Figure 1.

2.3 Intervention

After the intervention began, the 21 families in the intervention condition were randomly divided into three intervention groups with seven families in each group. Two therapists were assigned to each group randomly and all therapists had the same clinical background under family and group therapy training. The intervention was given every three days, with each session lasting 2 h. Six sessions were administered for each grouped families.

The intervention was tailored to strengthen parent-adolescent communication and relationships and shift adolescents' fulfillment of psychological needs from the Internet to interactions and building relationships with family members. Specific topics and activities were designed for each intervention session and connected with each other across six sessions, each of which included five parts in 2h : a warm-up exercise, feedback on homework from the last session (except the first session), a main structured activity, a brief summary and the family

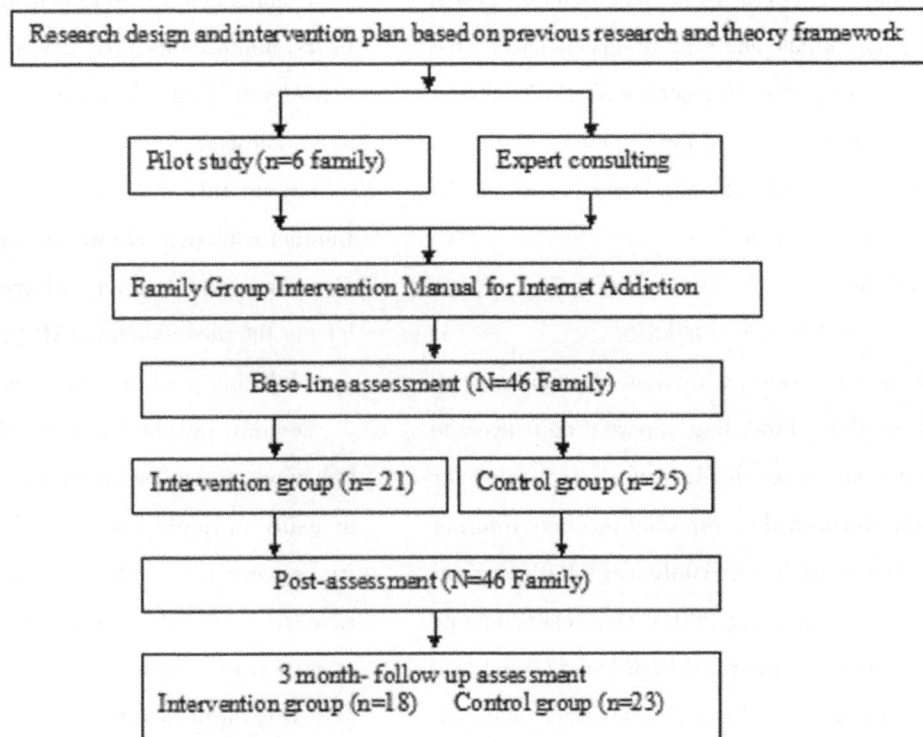

Fig. 1 The intervention process and participants flow diagram.

Table 1. Comparisons of measured variables between the Intervention group and control group at T1, T2 andT3

		T1	T2	T3		
		M(SD)	M(SD)	M(SD)	F[1]	F[2]
Average APUIS	Intervention	3.40(.27)	2.46(.61)	2.06(.73)	38.31***	65.98***
	Control	3.38 (.20)	3.59(.31)	3.27(.26)	7.62**	2.50
	t	−0.54	−7.79***	−6.72***		
Internet use time	Intervention	26.38 (9.6)	11.43(5.75)	7.08(3.98)	40.16***	56.65***
	Control	27.08(11.1)	27.52(11.40)	22.29(6.0)	2.21	3.03
	t	−0.21	−4.73***	−9.39***		
Parental reports	Intervention	3.37(.48)	3.13(.66)	2.70(.44)	6.05**	21.10***
	control	3.36(.52)	3.45(.72)	3.20(.57)	1.43	1.97
	t	0.01	−4.12***	−5.26***		
Addiction rate[a]	Intervention	100%	4.8%	11.1%		
	Control	100%	96%	87 %		

Note. F[1] indicates F statistics from comparison among three assessments; F[2] indicates the F statistics from the linear test; [a]Five Parcitipants were missing at T3.
**p<0.01
***p<0.01.

assignment. During the sessions, the following topics were focused on: understanding a family with Internet addict (session 1), parent–adolescent communication skills training (session 2), parent–adolescent communication practices on Internet addiction (sessions 2 and 3), parent–adolescent relationship building skills training (session 4), associations between psychological needs and Internet use and how to satisfy the unfulfilled need in the family relationships (session 5) and setting up appropriate and healthy expectations for the family system (session 6). One additional session at the three-month follow up was designed to target potential relapse, discuss new issues and generate solutions to maintain the effectiveness of the intervention.

2.4 Measurement

2.4.1 Adolescent internet addiction.

Two indicators of Internet addiction were reported by adolescents themselves. First, they reported their average number of hours spent on the Internet per week over the past month. Second, they reported on their Internet addiction behaviors with the Adolescent Pathological Internet Use Scale (Lei & Yang, 2007). This scale contains 38 items with each item being rated from 1 to 5 (1=not true at all; 5=true all of the time). Six subscales were included: salience, social comfort, mood alteration, tolerance, compulsive Internet use and negative outcomes. The average scores across the 38 items were used as indicators of Internet addiction with higher scores indicating more serious Internet addiction. This scale has high internal consistency (α =0.95 for the whole scale; α between .81 and .91 for all subscales) and high test–retest reliability (r=0.86). Based on the APIUS average scores, adolescents with scores below 3 were considered to be normal Internet users, adolescents with scores between 3 and 3.15 were considered to have a tendency towards Internet addiction, and adolescents with scores higher than 3.15 were defined as having Internet addiction (Lei & Yang, 2007). The Internet addiction rate was calculated based on the number of adolescents defined as having a tendency towards or having Internet addiction (APIUS > 3) and is considered as one of the intervention effectiveness indicators in the study.

Second, parents' reports of children's Internet use behaviors in the last month were used as a supplementary measure of adolescents' Internet use behaviors. Three items were reported: Internet use frequency (1="very rare"to 5 ="very frequently"), parents' observation about adolescents' Internet use appropriateness (1="very appropriate"to 5 ="not appropriate at all"), and parents'satisfaction toward children's Internet use behaviors (1="very satisfied" to 5 ="not satisfied at all"). The average score across the three items was used to present the parents' evaluation of children's Internet use behaviors. A higher score indicates more Internet use and lower behavior control.

2.4.2 Parent–adolescent relationship.

Adolescents reported their relationships with the parent who participated in the intervention on nine items from the Closeness to Parents on a scale (Buchnan, Maccoby, & Dornbush, 1991) from 1 (not at all true) to 5 (very much true). Sample questions included "How openly do you talk with your mother (father)?" or"How close do you feel to your mother (father)?" A Chinese adaption of the scale has been used in previous study (Liu et al., 2013) and it has a high internal consistency in the study (α = .91). Average scores across the nine items were used to represent the adolescent's relationships with the mother or the father during the intervention.

2.4.3 Parent–adolescent communication.

The Parent–Child Communication Scale (Barnes & Olson, 1985) was used to assess the adolescents' perception of their communication with the parent in the intervention. This scale contains 20 items on a scale from 1 (never) to 5 (always). It is composed of two dimensions that measure the degree of openness and the extent of problems in family communications. The responses were identified separately for fathers and mothers. The average score across both dimensions was used to represent the average level of parent–child communication in this study. A Chinese adaption of the scale has been used in previous study (Liu et al., 2012) and the α for the scale is 0.82 in the study.

2.4.4 Adolescent Psychological Needs.

Adolescents rated their psychological needs using a scale modified from the College Students' Psychological Needs and Fulfillment *Scale* (Wan et at., 2010). The scale is composed of three subscales: the degree of psychological needs; needs satisfaction from real life, and needs satisfaction from the Internet. Each subscale includes 35 items that tap into eight dimensions of the targeted subscale:need for autonomy, need for entertainment, need for interaction, need for achievement, need for impact, need for acknowledgement, need for expression and need for information. The rating for each item in the degree of psychological needs subscale ranges from 1 (not strong at all) to 5 (extremely strong). Ratings in the other two need satisfaction subscales range from 1 (very low) to 5 (very high). Unsatisfied psychological needs was calculated by subtracting the subscale scores on need satisfaction in real life from the subscale scores on the degree of psychological needs. The advantage of the Internet in satisfying needs was calculated by subtracting the subscale scores on need satisfaction in real life from the subscale scores on need satisfaction from the Internet. These two scores were two major process variables of interest in the study. The α for the scale is 0.97 in the study.

2.4.5 Analysis plan

The data analysis proceeded in three steps. First, T tests and Repeated–Measures ANOVA analyses were conducted to test the effectiveness of the intervention based on the comparison among adolescents' Internet addiction measures in the intervention and control group at T1, T2 and T3. Second, comparisons on all intervening variables between the two groups at T1, T2 and T3 were conducted, and changed values of the Intervening variables from T1 to T2 and T3 were created. Third, hierarchical multiple regression analyses were conducted to examine whether the change in adolescents' Internet addition behaviors is explained by the change in the measured intervening variables.

3 Results

3.1 Effectiveness of the Intervention

Before examining the effectiveness of the intervention,

Table 2. Comparisons on intervening process variables

		T1 M(SD)	T2 M(SD)	T3 M(SD)	F^1	F^2
P–A relationship	Intervention	2.99(0.45)	3.72(0.81)	3.79(0.64)	14.86***	29.94***
	Control	2.96(0.44)	2.99(0.49)	3.05(0.48)	1.45	2.84
	t	0.21	2.89**	2.96**		
P–A communication	Intervention	2.96(0.71)	3.71(0.49)	3.83(0.62)	13.78***	30.57***
	Control	2.94(0.44)	2.98(0.47)	3.03(0.42)	0.08	
	t	0.19	4.08***	3.97***		
Unsatisfied needs	Intervention	0.67(0.58)	0.19(0.56)	−0.01 (0.45)	11.56***	19.85***
	Control	0.65(0.37)	0.47(0.25)	0.48(0.29)	3.48	4.65
	t	0.18	−2.57*	−3.87**		
Advantage of Internet in satisfying need	Intervention	0.59(0.57)	−0.24(0.59)	−0.81(0.58)	23.68***	39.15***
	Control	0.58(0.44)	0.46(0.25)	0.47(0.29)	3.41	2.61
	t	0.09	−4.83***	−7.78***		

Note. P–A stands for parent–adolescent; F^1 indicates F statistics from comparison among three assessments ; F^2 indicates the F statistics from the linear test.
*$p<0.05$,**$p<0.01$,***$p<0.001$;

the Internet addiction measures at T1 were compared between the intervention and control groups and no significant difference was detected, indicating that the two groups were at the same or similar level of Internet addiction at the baseline of the study (see F values at T1 in Table 1). The effectiveness of the intervention was manifested in three aspects: First, adolescents in the intervention group significantly reduced the time they spent on the Internet by the end of the intervention, spending about half of the time as adolescents in the control group did (see F values at T2 in Table 1). Second, comparisons of APIUS scores demonstrated that the intervention group experienced a decrease in their average APIUS scores from T1 to T2. Third, parents in the intervention group reported more satisfaction with adolescents' Internet use behaviors at the end of the intervention compared with both their satisfaction at the baseline and parents' satisfaction in the control group. In sum, based on reports from both adolescents and their parents, the intervention was effective in terms of reducing adolescents' Internet addiction behaviors by the end of the intervention.

Results from the repeated measures ANOVA (see Table 1) showed that the differences in APIUS scores across the three measurements were significant($F (2, 78) =$ 38.31, $p < .001$) and also displayed a linear decrease over time from Time 1 to Time 3 ($F(1,39)= 65.98$, $p < .001$), which indicates that the intervention effects remained at T3. Time spent on the Internet also displayed a significant decrease from T1 to T3 ($F (2, 78) = 40.16$, $p <.001$) with a linear decrease from T1 to T3 as well ($F (1, 39) =56.65$, $p < .001$). In the control group, significant differences were found among the three APIUS assessments and a further paired sample t test showed that APIUS scores at T2 were significant higher than at T1($t=-4.15$, $p<0.001$), but there was no linear tendency of APIUS scores from T1 to T3 ($F (1, 39)=2.50$, $p=0.13$). Parents' reports on adolescent Internet addiction at the three–month follow–up also substantiated the effect of the intervention. At 3 months after the intervention, parents still reported a significant reduction in their perception of their children's Internet addiction, revealing a significant linear decrease from T1 to T3.

As an important indicator of intervention effectiveness in the whole intervention group, the Internet addiction rate yielded from the APIUS average scores was also compared across the intervention and control groups over time. As displayed in Table 1, all adolescents in both groups were either addicted to the Internet or had the tendency towards Internet addiction before the intervention. By the end of the intervention, only 1 out of 21 adolescents (4.8%) in the intervention group was still addicted to the Internet, compared with as many as 24 out of 25 (96%) in the control group. At the three-month follow-up, two adolescents in the intervention group showed a relapse. However, the intervention effects remained as only 11.1% of adolescents in the intervention group remained addicted after the intervention ended, compared with 87% of their counterparts in the control group.

3.2 Examining mediating effects of intervening variables

Because we measured the perception of parent-adolescent interaction and psychological need that is deeply inside of adolescents, we used the reports from the adolescents themselves to examine the change process of adolescent Internet behaviors. To examine the second hypothesis (that the intervention effects are mediated through major process variables), we first examined the change in parent-adolescent relationships, parent-adolescent communication and adolescents' psychological needs. We measured these attributes at their baseline, at the end of the intervention, and at a three-month follow-up for both the intervention and control groups.

As displayed in Table 2, the measured variables did not differ significantly between the intervention and control groups at T1, but did differ significantly at T2 and T3. Adolescents in the intervention group demonstrated improvement in their relationship and communication with their parents, an increase in their fulfillment of needs, a decrease in Internet use' advantage in fulfilling their needs in a linear function.

With the demonstration of improvement on all process variables in the intervention group, we continued to examine whether the change in those process variables accounted for change in adolescents' Internet addiction behaviors. Before examining those mediating effects, change values (ΔX) were created for adolescent reported Internet addiction behaviors, parent-adolescent communication and relationship, and psychological needs by subtracting the post-intervention assessment values from the baseline values ($\Delta X = X_{T2}-X_{T1}$) .

To examine the contribution of those process variables in the explanation of intervention effectiveness on adolescents' Internet addiction, a hierarchical multiple regression analysis was performed. Process variables were entered in three steps: In step 1, $\Delta APIUS$ was the dependent variable and Δ parent-adolescent relationship was the independent variable. In step 2, Δ parent-adolescent communication was entered into the step 1 equation. In step 3, two indicators of $\Delta Adolescents'$ psychological needs were added in the equation in step 2. The results of step 1 (Table 3) indicated that 31% of the variance in the change in adolescents' Internet addiction (R^2=0.31) was accounted for by improvements in the parent-adolescent relationship. In step 2, after entering changes in mother-adolescent communication, an additional 34% of the variance in the change in adolescents' Internet addiction was explained(ΔR^2=0.34) and this change in R^2 was significant. In step 3, the addition of psychological needs did not explain additional variance in the dependent variable.

To examine the mediating effects of process variables from T1 to T3, change values (ΔX) were created by following the same procedure: subtracting the three-month follow-up assessment values from the baseline values ($\Delta X = X_{T3}-X_{T1}$). A hierarchical multiple regression analysis was again performed (see the right half of Table 3). First, the results of the hierarchical analysis revealed that changes in parent-adolescent communication and relationships explaining significant variance in adolescents' Internet addiction from T1 to T3. Moreover, when predicting longer term effects in reducing Internet addiction, changes in adolescents' psychological needs also added significant prediction of variance in the Internet addiction change

Table 3. Regressions for changed values of measured variables at T2 and T3

	$\triangle X = X_{T2} - X_{T1}$						$\triangle X = X_{T3} - X_{T1}$					
	β	t	R^2	F	$\triangle R^2$	$\triangle F$	β	t	R^2	F	$\triangle R^2$	$\triangle F$
Step 1												
ΔP–A relationship	−.56	−2.93**	.31	8.55**	.31	8.55**	−.61	−2.62*	.36	6.76*	.36	6.76*
Step 2												
Δ P–A relationship	−.03	−.13	.65	16.99***	.34	17.87**	−.03	−.11	.55	6.79*	.19	4.74*
Δ P–A communication	−.78	−4.22**					−.73	−2.1⁹*				
Step 3												
Δ P–A relationship	.03	.17	.70	7.07**	.05	.82	.00	.01	.83	10.01**	.27	6.49*
Δ P–A communication	−.51	−1.98*					−.53	−1.98*				
Δ Unsatisfied Needs	.02	.13					.34	1.00				
ΔAdvantage of Internet in satisfying needs	.39	1.33					.88	3.44**				

Note. P–A stands for parent–adolescent;

*$p<0.05$, **$p<0.01$, ***$p<0.001$.

($\Delta R^2 = 0.27$). This was in large part due to changes in adolescents' satisfaction of their psychological needs through the Internet ($\beta = 0.88$, $p < .01$).

4 Discussion

The current study represents a practical clinical trial of treating adolescents' Internet addiction using the multi–family group approach.Based on prior family group intervention practices in treating adolescent psychopathology (Chien & Chan, 2013; McDonell & Dyck, 2004; Zhong et al., 2011), the current study is the first to our knowledge to apply the approach of MFGT in treating adolescents' Internet addiction. The adolescents' Internet addiction rate dropped from 100% at the baseline assessment to 4.8% at the end of the intervention and remained at 11.1% at the three-month follow–up assessment. Time spent on Internet in the intervention group also significantly declined throughout the intervention until the three-month follow-up. Analyses of the value changes in measured variables indicated which factors were associated with the decrease of adolescent Internet addiction. Improved parent-adolescent communication and need satisfaction in real life were associated with decrease in Internet addiction. If adolescents in the intervention group perceived an improvement in their communication and relationships with parents, learned alternative ways to fulfill their needs and felt less reliance on the Internet,this might promote their motivation to sustainably change their behavior. The results are consistent with prior evidence suggesting the role of feeling supported and trusted in improving the effectiveness of family group intervention (Dickerson & Crase, 2005).

These findings corroborate the idea that intervention programs for adolescents need to get parents actively involved and include them as part of the solution. The family system approach shifts the emphasis on individual family members to the entire family as a unit and the dynamic interactions between family members (Dickerson & Crase, 2005). Further, multi–family group is very helpful in the lasting the effective of intervention. In the multi–family group, each family represents a subsystem with a shared history and current life situation which makes for an enriching and complex process. Then, the multi–group serves both the family system and individual as an arena for cross transferences and as a reality tester (Leichter & Schulman, 1974). Moreover, the participation of other family members in the intervention can create a more supportive environment in which the participants' behavioral changes are valued, encouraged,and maintained even after the intervention ends (McDonell & Dyck, 2004).

The present study takes a further step in unraveling the underlying mechanism through which MFGT took effects. Parent–adolescent interactions and relationship quality and adolescents' increasing satisfaction of psychological needs in real life partially accounted for the effectiveness of the intervention in reducing adolescents' Internet addiction, further supporting prior findings (Liu, et al, 2012, 2013; Olson et al., 1974). Consistent with previous studies, perceived positive interaction with parents protected adolescents from Internet addiction (Liu et al, 2013; Van den Eijnden et al., 2010). Positive communication facilitated emotional connections between family members and helped them to understand and clarify their needs (White, 2000). The alternative ways and skills of need satisfaction that adolescents and parents learned from the sessions would also expand the degree of communication and interaction within the family. All of these changes could be helpful to keeping children away from Internet addiction. However, the decrease in advantage of Internet in satisfying need did not predict change in Internet use behaviors until the three–month follow–up assessment. This delayed effect has been reported before in interventions targeted to treating adolescents' affective disorders (Goldberg-Arnold, Fristad, & Gavazzi, 1999). It suggests that improvement in adolescents' satisfaction of psychological needs through real life interactions may take time to occur. The positive and effective interaction skills and patters between adolescents and parents need to be practiced before adolescents gradually feel comfortable and natural enough to get used to them. It also may take time for adolescents to change and adapt their need satisfaction habits, which may be a potential advantage of MFGT as it may have lasting intervention effects. Once a benign interaction and relationship pattern is established, the family obtains a built-in force to sustain

the intervention effects (McDonell & Dyck, 2004).

Some factors limit the conclusions that can be drawn from these data. First, the study did not take other competing intervention models into consideration. It has been recommended in intervention studies to compare across multiple intervention paradigms to better evaluate the effectiveness of certain intervention approach. For example, a family therapy approach or group intervention approach could be incorporated in a comparison group to further evaluate the effectiveness of the MFGT paradigm. Second, the gender ratio of parents was not balanced, as most of the participating parents were mothers. This might be an obstacle in clarifying the unique role of the mother–adolescent relationship or father–adolescent relationship in preventing Internet addiction among adolescents, as the impact of their relationship with their father and mother differed with adolescent genders on Internet addiction (Liu et al., 2013). It would be clearer if further clinical studies endeavored to recruit parents of both genders. Third, based on a quasi–experimental design, participants were not randomly assigned to an intervention or control group. However, the baseline comparison between the groups did not show a significant difference and indicate that the results of the study are reliable. Fourth, the data were mainly from questionnaires, and social desirability might influence adolescents' reports of their Internet use behavior. Future studies should assess Internet addiction through psychological interviews. Additionally, the present study only measured general Internet use and addiction, limiting the generalization of results to specific Internet addiction. It would be helpful to improving intervention if further studies paid attention to the treatment of subtypes of Internet addiction.

Given the different characteristics of adolescents and their families, multiple dimensions of multi-family group therapy might be more complex and require exploration in future studies.The existence of equifinality in adolescents' Internet addiction behaviors also indicates that multiple factors, including personal, interpersonal and environmental factors, could lead to the emergence of Internet addiction among adolescents. For example, individual differences among children in basic need satisfaction in daily real life were associated with the way in which they engage with the Internet (Shen et al., 2013). Therefore, multi-family group interventions that tailor to different etiologies of Internet addiction might improve its effectiveness.

In the addiction field, prevention is more important than intervention. Future studies might explore the possibility and feasibility of an integrated preventive intervention framework including school, community and family, based on evidence from other related adolescent health-related domain (Dodge & Godwin, 2013; Lovato et al., 2013). In consideration of the particularity of Internet addiction behaviors, namely that Internet serves as a necessary part of life and Internet use cannot be completely cut off but only be guided to a rational and controllable level, it is very important to discern how to guide youth use of the Internet appropriately and optimize the benefits of the Internet for adolescents and children.

References

Barnes, H. L., & Olson, D. H. (1985). Parent - adolescent communication and circumplexmodel. *Child Development, 56*, 438 - 447.

Block, J. J. (2008). Issues for DSM-V: Internet addiction. *The American Journal of Psychiatry,165*, 306 - 307.

Brezing, C., Derevensky, J. L., & Potenza, M. N. (2010). Non-substance-addictive behaviorsin youth: Pathological gambling and problematic internet use. Child and *AdolescentPsychiatric Clinics of North America, 19*(3), 625 - 637.

Buchnan, C. M., Maccoby, E. E., & Dornbush, S. M. (1991). Caught between parents:Adolescents' experience in divorced homes. *Child Development, 62*, 1008 - 1029.

Cai, Y. Y., Cui, L. J., & Li, X. (2007). A research on the psychological needs of teenagers'online game behaviors. *Psychological Science, 30*(1), 169 - 172 (in Chinese).

Caplan,S. E. (2002). Problematic internet use and psychosocial wellbeing: Development ofa theory-based cognitive-behavioral measurement instrument. *Computers in HumanBehavior, 18*, 553 - 575.

Chien, W. T., & Chan, S. W. C. (2013). The effectiveness of mutual support group interven-tion for Chinese families of people with schizophrenia: A randomized controlledtrial with 24-month follow-up. *International Journal of Nursing Studies, 50*,1326 - 1340.

China Internet Network Information Center (2013r). Statistical report on the develop-ment of internet in China. (2013, Jan).

Cho, S. C., Kim, J. W., Kim, B. N., Lee, J. H., & Kim, E. H. (2008). Biogenetic temperamentand character profiles and attention deficit hyperactivity disorder symptoms inKorean adolescents with problematic internet use. *Cyberpsychology & Behavior,11*(6), 735 - 737.

Conner, K. R., Shea, R. R., McDermott, M. P., Grolling, R., Tocco, R. V., & Baciewicz, G. (1998).The role of multifamily therapy in promoting retention in treatment of alcohol andcocaine dependence. *The American Journal on Addictions, 7*(1), 61 - 73.

Davis, R. A. (2001). A cognitive-behavioral model of pathological internet use (PIU).*Computers in Human Behavior, 17*(2), 187 - 195.

Deng, L. Y., Fang, X. Y., Wan, J. J., Zhang, J. T., & Xia, C. C. (2012). The relationship ofpsychological needs and need gratification with internet addiction among collegestudents. *Journal of Psychological Science, 35*(1), 123 - 128 (in Chinese).

Dickerson, A. D., & Crase, S. J. (2005). Parent - adolescent relationships: The influence ofmulti-family therapy group on communication and closeness. *The American Journalof Family Therapy, 33*(1), 45 - 59.

Dodge, K. A., & Godwin, J. (2013). The conduct problems prevention research group.Social-information-processing patterns mediate the impact of preventive interven-tion on adolescent antisocial behavior. *Psychological Science, 24*(4), 456 - 465.

Gray, M. R., & Steinberg, L. (1999). Unpacking authoritative parenting: Reassessing a mul-tidimensional construct. *Journal of Marriage and Family, 61*, 574 - 587.

Goldberg-Arnold, J. S., Fristad, M. A., & Gavazzi, S. M. (1999). Family psychoeducation: Giv-ing caregivers what they want and need. *Family Relations, 48*, 411 - 417.

Kim, D. H., Jeong, E. J., & Zhong, H. (2010). Preventive role of parents in adolescent prob-lematic Internet game use in Korea. *Korean Journal of Sociology, 44*(6), 111 - 133.

King, D. L., Delfabbro, P. H., Griffiths, M. D., & Gradisar, M. (2011). Assessing clinical trials ofinternet addiction treatment: A systematic review and CONSORT evaluation. *ClinicalPsychology Review, 31*, 1110 - 1116.

Kratochwill, T. R., McDonald, L., Levin, J. R., Scalia, P. A., & Coover, G. (2009). Families andschools together: An experimental study of multi-family support groups for childrenat risk. *Journal of School Psychology, 47*(4), 245 - 265.

Kuss, D. J., Griffiths, M. D., & Binder, J. F. (2013). Internet addiction in students: Prevalenceand risk factors. *Computers in Human Behavior, 29*, 959 - 966.

Lau, R. R., Quadrel, M. J., & Hartman, K. A. (1990). Development and change of young adults'preventive health beliefs and behavior: Influence from parents and peers. *Journal ofHealth and Social Behavior, 31*, 240 - 259.

Lei, L., & Yang, Y. (2007). The development and validation of adolescent pathologicalinternet use scale. *Acta Psychologica Sinica, 39*(4), 688 - 696 (in Chinese).

Leichter, E., & Schulman, G. (1974). Multi-family group therapy: A multidimensionalapproach. *Family Process, 13*, 95 - 110.

Liddle, H. A. (2004). Family-based therapies for adolescent alcohol and drug use: Researchcontributions and future research needs. *Addiction, 99*(Suppl. 2), 76 - 92.

Liu, Q. X., Fang, X. Y., Zhou, Z. K., Zhang, J. T., & Deng, L. Y. (2013). Perceived parent - ad-olescent relationship, perceived parental online behaviors and pathological internetuse among adolescents: Gender-specific differences. *PloS ONE*, 8(9), e75642.

Liu, Q. X., Fang, X. Y., Deng, L. Y., & Zhang, J. T. (2012). Parent - adolescent communication,parental internet use and internet-specific norms and pathological internet useamong Chinese adolescents. *Computers in Human Behavior, 28*, 1269 - 1275.

Liu, Q. X., Fang, X. Y., & Zhou, N. (2011). A review of the research on internet addictionamong Chinese adolescents. *Journal of South China Normal University (Social Scienceedition), 3*, 65 - 70.

Lortie, C. L., & Guitton, M. J. (2013). Internet addition assessment tools: Dimensional struc-ture and methodological status. *Addiction, 108*, 1207 - 1216.

Lovato, C., Watts, A., Brown, K. S., Lee, D., Sabiston, C., Nykiforuk, C., et al. (2013). Schooland community predictors of smoking: A longitudinal study of Canadian high school.*American Journal of Public Health, 103*(2), 362 - 368.

McDonell, M. G., & Dyck, D. G. (2004). Multiple-family group treatment as an effective in-tervention for children with psychological disorders. *Clinical Psychology Review, 24*,685 - 706.

Morris, M., & Ogan, C. (1996). The internet as mass medium. *Journal of Communication,46*(1), 29 - 50.

Nieter, L., Thornberry, T., Jr., & Brestan-Knight, E. (2013). The effectiveness of group par-ent - child interaction therapy with community families. *Journal of Child and FamilyStudies, 22*, 490 - 501.

Olson, D. H. (2000). Circumplex model of marital and family systems. *Journal of FamilyTherapy, 22*(2), 144 - 167.

Olson, D. H., Sprenkle, D. H., & Russell, C. S. (1974). Circumplex model of marital and familysystems: I. Cohesion and adaptability dimensions, family types, and clinicalapplications. *Family Process, 18*, 3 - 28.

Park, S. K., Kim, J. Y., & Cho, C. B. (2008). Prevalence of internet addiction and correlationswith family factors among South Korean adolescents. *Adolescence, 172*(43), 895 - 907.

Petersen, K. U., Weymann, N., Schelb, Y., Thiel, R., & Thomasius, R. (2009). Pathologicalinternet use — Epidemiology, diagnostics, co-occurring disorders and treatment.Fortschritte Der Neurologie *Psychiatrie, 77*, 263 - 271.

Peukert, P., Sieslack, S., Barth, G., & Batra, A. (2010). Internet- and computer game addic-tion: Phenomenology, comorbidity, etiology, diagnostics and therapeutic implica-tions for the addictives and their relatives. *Psychiatrische Praxis, 37*, 219 - 224.

Shen, C. X., Liu, R. D., & Wang, D. (2013). Why are children attracted to the internet? Therole of need satisfaction perceived online and perceived in daily real life. *Computersin Human Behavior, 29*, 185 - 192.

Suler, R. (1999). To get what you need: Healthy and pathological internet use.*Cyberpsychology & Behavior, 5*(2), 385 - 393.

Van den Eijinden, R., Meerkerk, G. J., Vermulst, A. A., Spijkerman, R., & Engles, R. (2008).Online communication, compulsive internet use, and psychosocial well-beingamong adolescents: A longitudinal study. *Developmental Psychology, 44*(3), 655 - 665.

Van den Eijnden, R. J. J. M., Spijkerman, R., Vermulst, A. A., van Rooij, T. J., & Engels, R. C. M. E.(2010). Compulsive internet use among adolescents: Bidirectional parent - childrelationships. *Journal of Abnormal Child Psychology, 38*(1), 77 - 89.

Wan, J. J., Zhang, J. T., Liu, Q. X., Deng, L. Y., & Fang, X. Y. (2010). Development of college stu-dents' psychological need internet gratification questionnaire. *Studies of Psychologyand Behavior, 8*(2), 118 - 125 (in Chinese).

Weinstein, A., & Lejoyeux, M. (2010). Internet addiction or excessive internet use.*The American Journal of Drug and Alcohol Abuse, 36*, 277 - 283.

White, F. A. (2000). Relationship of family socialization processes to adolescent moralthought. *The Journal of Social Psychology, 140*(1), 75 - 91.

Winkler, A., D?rsin, B., Rief,W., Shen, Y., & Glombiewski,J.A.(2013). Treatment of internetaddiction: A meta-analysis. *Clinicial Psychological Review, 33*, 317 - 329.

Young, K. S., Pistner, M., O'Mara, J., & Buchanan, J. (2000). Cyber-disorders: The mentalhealth concern for the new millennium. *Cyberpsychology & Behavior, 3*(5), 475 - 479.

Young, K. S., & Abreu, C. (2011). Internet addiction. A handbook and guide to evaluation andtreatment. Hoboken, NJ: John Wiley & Sons.

Zhong, X., Zu, S., Sha, S., Tao, R., Zhao, C., & Yang, F. (2011). The effect of a family-basedintervention model on internet-addicted Chinese adolescents. *Social Behavior andPersonality, 39*(8), 1021 - 1034.

Zubrick, S. R., Ward, K. A., Silburn, S. R., Lawrence, D., Williams, A. A., Blair, E., et al. (2005). Prevention of child behavior problems through universal implementation of agroup behavioral family intervention. *Prevention Science, 6*(4), 287 - 303.

Cognitive Development, 2014, 29, 50 - 61.

Sentential complements and false belief understanding in Chinese mandarin-speaking preschoolers: A training study

Shuliang Mo[1,2,3,*] , Yanjie Su[3] , Mark A. Sabbagh[4], Jiaming Xiu[3]

([1] School of Psychology, Central China Normal University, Wuhan, 430079, China)

([2] Key Laboratory of Adolescent Cyberpsychology and Behavior (CCNU) of Ministry of Education, Wuhan, China)

([3] Department of Psychology, Peking University, Beijing 100871, China)

([4] Department of Psychology, Queen's University, Kingston, Ontario, Canada)

Abstract　We conducted a training study to better understand how Chinese Mandarin-speaking preschoolers' facility with sentential complement grammatical constructions affects performance on false belief tasks. Eighty-four Mandarin-speaking Chinese 34 year-olds who were initially unsuccessful on false belief tasks were randomlyassigned to four training conditions. Two involved training on sentential complement structures, one involved training on understanding of false representations, and one was a control conditionthat involved no specific training. Participants who received trainingon sentential complements with communication verbs performed significantly better on false belief posttests than thos ein the control group. Children in the false representation training group did not show improvement in the sentential complement tests. The findings suggest facility with sentential complement grammatical structures can promote false belief reasoning. However,explicit false belief understanding can emerge even when children have little competence with sentential complement constructions.

Keywords　False belief understanding; Sentential complement; False representation; Training, language

1 Introduction

False belief understanding is a milestone in representational theory of mind development(Wellman, Cross, & Watson, 2001). Children's performance on false-belief tasks is associated with language ability (Astington & Jenkins, 1999; Hughes & Dunn, 1998). A recentmeta-analysis (Milligan,Astington,& Dack, 2007) concluded that children's early-developing language skills predict subsequent development of theory of mind during the

*　Corresponding author at: School of Psychology, Central China Normal University, Wuhan 430079, China.

Tel.:+86 27 67868680; fax: +86 27 67868561.

E-mail addresses: booklight@mail.ccnu.edu.cn (S. Mo), yjsu@pku.edu.cn (Y. Su).

This study was supported by grants from the Educational Science Planning of China, the State General Fund Projects (No. BBA090068) to the first author and partially from National Natural Science Foundation of China (No. 30770728).

preschool years. The goal of this study is to examine Chinese Mandarin-speaking preschoolers to investigate the role of one aspect of language development-facility with the sentential complement grammatical construction-in the development of false belief understanding.

The hypothesis that facility with sentential complement grammatical constructions is critical for false belief understanding was first advanced in a series of influential papers by de Villiers and colleagues (de Villiers & Pyers, 2002; de Villiers & de Villiers, 2000). Noting that mental state verbs (e.g.,think) take sentential complements meaning that the clause that complements the mental verb is itself a complete sentence (e.g., Mary thinks that the chocolate is in the cupboard). De Villiers and colleagues argued that the syntax provides a formal mechanism for contrasting the truth value of two clauses within a single sentence. With reference to the above example, Mary may truly think that the chocolate is in the cupboard, but the content of Mary's belief may or may not be true depending on the preceding events. de Villiers and de Villiers (2000) suggested that this class of syntactic con struction provides a format for thinking about epistemic mental states, and that until facility with this construction is acquired, an understanding of belief may be seriously limited.

There is some evidence that facility with sentential complement constructions is important for the development of preschoolers' false belief reasoning. In one study, de Villiers and Pyers (2002) aske dchildren to report the propositional content of a story character's mistake, lie, or false belief (e.g., He thought he found his ring, but it was really a bottlecap. What did he think?). Ability to report the propositional contents and produce sentential complements themselves showed rapid gains between the ages of 3 and 4, about the time false belief understanding emerges. Indeed, false belief understanding could be uniquely predicted by the earlier onset of sentential complements (de Villiers & Pyers,2002).

A second source of evidence in support of the association between sentential complement facility and theory of mind development comes from studies of deaf children born into hearing families. Schick, de Villiers, de Villiers, and Hoffmeister (2007) reported that deaf children from hearing families showed a significant delay on theory of mind tasks. Importantly, vocabulary and sentential complement comprehension were independent predictors of success on both typical verbal and low-verbaltheory of mind tasks.

In addition to these correlational studies, several studies have shown that training to increase children's facility with sentential complements leads to improvements in false belief performance. For example, Hale and Tager-Flusberg (2003) found that children trained on sentential complements not only acquired the linguistic knowledge fostered by the training, but also significantly increased their scores on a range of theory of mind tasks, including false belief tests. In a typical training session,an experimenter acted out a story with characters (e.g., Big Bird, Grover and a boy). In the story, the boy does one thing, but says that he does another. For instance, in one story, the boy is shown kissing Big Bird, but the boy says, "I kissed Grover." The experimenter then asks the child: "What did the boy say?" Correct responses were responded to with, "That's right. The boy said, 'I kissed Grover,' but he really kissed Big Bird." If the child made an incorrect response, the examiner acted out again and said, "But remember, the boy says, 'I kissed Grover,' but he really kissed Big Bird." The results showed that those who participated in the sentential complement training also improved in false belief performance relativeto a control group.

Hale and Tager-Flusberg's (2003) training protocol was specifically designed to determine whether it was the grammatical features of sentential complement structures, rather than other kinds of more semantic content, that affected children's theory of mind development. Thus, these researchers used "communication"verbs (i.e., said that). However, it is not entirely clear whether the aim of leaving out semantic content was fully achieved. Their training involved a kind of deception in which participants were told that although a story character intentionally said one thing, something else was in fact true. The deceptive aspect of the story was not emphasized in the training protocol, but its presence makes it unclear as to whether the training promoted children's understanding of deception (which is relevant to theory of mind), sentential complement constructions, or both.

With this in mind, Lohmann and Tomasello (2003) developed a training protocol that avoided deceptive communication and obtained the same results, thereby suggesting that the deceptive content of the training was not the sole factor promoting false belief understanding in Hale and Tager-Flusberg's study. Yet, Lohmann and Tomasello's protocol had interpretive difficulties of its own. Specifically, their training involved the use of terms that could, arguably, be interpreted as having mental content (e.g., discussion of what story characters "feel" or "know"). Thus, it remains unclear whether the association between facilitating sentential complement understanding and false belief reasoning is specific to the cognitive-structural "template" that sentential complements provide, or the semantic content of the training.

Other concerns about the empirical relation between sentential complement facility and false belief have been raised by researchers working in languages other than English. For instance, Perner, Sprung, Zauner, and Haider (2003) noted that in German, sentences involving the mental verb want obligatorily take sentential complements; yet, German children (like others) appear to understand the entailments of want well before they understand the entailments of think and believe. Also, in studies with Cantonese-speaking preschoolers that are more analogous to those conducted by de Villiers and colleagues, two groups have shown that once relevant factors are controlled (e.g., general language ability, age, prior theory of mind development), understanding of sentential complements does not make a unique contribution to false belief reasoning (Cheung et al., 2004; Tardif, So, & Kaciroti, 2007,study 2). Taken together, these findings suggested that the relation between sentential complement and false belief understanding may not extend across languages or cultures, which calls into questionits validity as a strong explanatory theory of acquisition of false belief understanding.

With these concerns in mind, the goal of the present study was to provide some clear evidence regarding whether facility with sentential complements is necessary for false belief understanding. We hoped to achieve this goal by conducting a training study with Mandarin speaking Chinese preschoolers. There are important differences between English and Mandarin with respect to the use of mental terms and sentential complements. Relative to English speakers, the use of mentalterms with sentential complements is rare among Mandarin-speaking parents and children (Tardif & Wellman, 2000; Snedeker & Li, 2000). Yet Mandarin-speaking parents and children use sentential complement constructions for communication verbs (e.g., say, in Mandarin) more commonly and earlier in development than their English-speaking counterparts (Tardif & Wellman, 2000).

This comparison raises two interesting possibilities. The first is that perhaps Mandarin-speaking preschoolers trained on sentential complements with the more common communication verbs will show more efficient acquisition of the construction relative to those trained on sentential complements with mental verbs. The second possibility is that training with sentential complements involving mental verbs may be less effective in facilitating children's acquisition of the structure because theuse of mental verbs with sentential complements is non-canonical and potentially confusing. If so,this may allow for a fairly compelling test of the role that sentential complement understanding plays in false belief reasoning. Namely, if sentential complement understanding per se contributes to false belief reasoning, children who receive sentential complement training with communication verbs may perform better on false belief tasks than do children who receive sentential complement training with mental verbs. Investigating this hypothesis is the main goal of the present study.

A second goal is to gain evidence regarding the mechanisms by which sentential complement training may affect false belief understanding. Some have suggested that tests of false belief understanding and tests of sentential complement understanding both rely on an understanding of misrepresentationmore generally–that is, they both require one to consider that the propositional contents of a representation (either a communicative utterance or a belief) may not be consistent with the true state of affairs it is meant to represent (Ruffman, Slade, Rowlandson, Rumsey, & Garnham, 2003). Ifso, sentential complement training may be one way to increase children's understanding of misrepresentation, but perhaps not the only way (Ruffman et al., 2003). In the present study, we included for comparison a condition that trained children on concepts of misrepresentation without the use of sentential complements. Specifically, we borrowed a procedure first used by Wellman, Hollander, and Schult (1996), who investigated preschoolers' understanding of thought bubbles. They found that vast majority of 3-4-year-olds understand that thought bubbles can be used as graphical depictions of"what a person is thinking", and thus can reveal the contents of characters' thought in a variety of situations. More important, thought bubbles can be used to graphically depict a person's misrepresentation of some states of affairs (that is, the propositional content of the thought bubblecan be discrepant with some true state

of affairs). If facility with sentential complements per se is critical to false belief understanding,the misrepresentation training should have little effect on false belief reasoning independent of whatever effects it might also have on sentential complement facility. Alternatively, if sentential complement training has its effects through the more general mechanism of developing a broader understanding of misrepresentation, we might expect a relatively robust effect of misrepresentation training,irrespective of whether it also affects facility with sentential complements.

2 Methods

2.1 Participants

A total of 120 kindergarten children participated (mean age = 46.3 months, SD = 4.9months,range = 40 - 55 months). They came from kindergartens affiliated with two universities in Beijing,China. Most came from working- and middle-class families. All spoke Mandarin. According to parents'or teachers' reports, no children had linguistic or psychological abnormalities. Eighty-four children from this original sample failed at least one of two questions in false belief pre-test (see below) and thus were considered eligible for the training study. Post-test data from three participants were missing due to children's illness (n = 2) or the family relocating (n = 1). Thus, a full set of data was acquired for 81 children (33 boys). Each child received a gift for participation.

2.2 Design

We employed a between-subject design in which participants were randomly assigned to one of the four training groups: (1) sentential complement-communication verb (SC-COMM), (2) sentential complement-mental verb (SC-MENTAL), (3) false representation (FR), and (4) control.

2.3 Pre and posttest measures

2.3.1 *False belief pretest*

A standard unexpected contents false belief task was

Table 1 Mean ages and pretest performance across groups.

	Groups			
	SC–COMM(n = 20)	SC–MENTAL(n = 20)	FR(n = 20)	Control (n = 21)
Age (months)	45.35 (4.40)	47.80 (5.22)	45.45 (5.79)	46.70(4.20)
FB prediction (0–1)	0.20 (0.41)	0.15 (0.36)	0.15 (0.36)	0.14(0.35)
FB ignorance (0–1)	0.50 (0.51)	0.60 (0.50)	0.60 (0.50)	0.52(0.51)
MCwMV(0–1)	0.50 (0.51)	0.45 (0.51)	0.45 (0.51)	0.61(0.49)
MCwCV(0–1)	0.52 (0.50)	0.55 (0.51)	0.50 (0.51)	0.47 (0.50)

Note: Sentential complement–communication verb = SC–COMM, sentential complement–mental verb = SC–MENTAL, false rep–resentation= FR.

used to assess false belief understanding(Perner, Leekam, & Wimmer, 1987). A child was shown a candy box and asked what he or she thought was inside. After children responded saying that they believed the box contained candy, the box was opened to reveal a ball–point pen. After closing the box, the experimenter asked a false belief question:When another child sees this closed box for the first time, what would he say is inside the box? This was followed by an ignorance question: Does he know what is in the box before it is opened? Children received a point for each correct answer.

2.3.2 Sentential complement pretests

Two tasks of memory for sentential complements were administered using the procedure from previous studies (de Villiers & Pyers, 2002; Lohmann & Tomasello, 2003). Children were told a story involving sentential complement with a mental verb (e.g., think) or communication verb (e.g., say) accompanying with line drawings relevant to the sentence. One story involving a mental verb was as follows, "A little rabbit was playing at home, and a wolf was knocking at the door; however, therabbit though it was his/her Mom knocking at the door". The test question on memory for sentential complement for the mental verb think was, "Who did the little rabbit *think* was knocking at the door?"

A story to assess understanding of sentential complements with communication verbs was as follows: Xiaohong was asked to go buy milk by her mother, but she bought a tin of Cola instead of milk. When she went home, her mother asked her, "Xiaohong, what did you buy?" Xiaohong said, "I bought milk". The test question on

memory for sentential complement for the communication verbsay was, "What did Xiaohong say that she bought?"

Children received 1 point for correctly answering the content of complement or for using the entire complement structure.

Children performed better than expected on the tests for sentential complements. Specifically,although four children in each training group did not pass either of the two questions, some children in all groups (4 children in SC–COMM, 4 in SC–MENTAL, 3 in the FR group and 6 in control) answered both correctly. Thus there were children in our final group who performed poorly on false belief tasks although they already showed facility with sentential complements. This pattern would be unexpected if sentential complement understanding were itself sufficient for false belief understanding. Nonetheless, average performance on the sentential complement pretest was equivalent across the different training groups (see Table 1), which allowed us to assess the effects of different kinds training on false belief understanding at the group level.

2.3.3 False belief posttests

Twostandard location–change tasks (Wimmer& Perner, 1983) and two standard unexpected content tasks (Hogrefe, Wimmer, & Perner, 1986; Perner et al., 1987) were administered to children following training.

In the location–change tasks, children were told stories. In one, a boy Xiaogang put his cake into a cupboard and then went out to play. Then his mother transferred the cake from the cupboard to the refrigerator. Children were asked two control questions (Where did Xiaogang put the cake before he left? Where is the cake now?). If a child did not answer correctly the story was repeated up to three times. The false belief questions included a standard behavior prediction question (When Xiaogang comes home, where will he first look for the cake?) and a justification question (Why will he look for it there?). A justification was judged as appropriate if it included references to location (e.g., He put it there before he went out), knowledge (e.g., he does not know it was moved), or belief (e.g., He thinks it still is there). All other justifications were coded as inappropriate (e.g., I do not know; the cake is in the cupboard; he wants to find the ball; he cannot find the ball).

A second story involved a girl who put a ball in a box and in her absence, her classmate moved the ball to a drawer. The two unexpected contents tasks were similar to the false belief pretest except that the false belief prediction question was, "When another child comes in, before having opened it, what does he/she think is inside the box?"

In the false belief tasks, each correct answer to the behavior prediction question was awarded 1 point. Across the four trials, total scores for the prediction question ranged from 0 to 4.

2.3.4 Sentential complement posttests

The sentential complement posttests were identical in structure to the pretests. Children were orally presented ten brief stories in random order, each accompanied by line drawing pictures. Each story involved a character who made a mistake, told a lie, or held a false belief. Of the ten stories, five involved mental state verbs (e.g., think) and five involved communication verbs (e.g., say). The test questions assessed children's memory for a sentential complement presented with each of the verbs. Based on previous work (de Villiers & Pyers, 2002), our Mandarin version was revised slightly for Chinese children (Cheung, Chen, & Yeung, 2009). Two of the stories were as follows:

Story 1: He thought he found his ring, but it was really a bottle cap.

Test question: What did he think he found?

Story 2: Mom asked Limei to buy some milk. But Limei bought some orange juice. Mom asked Limei, "Did you buy milk?" Limei said, "Yes, I bought milk."

Test question: What did Limei say she bought?

As with the pretest, children's responses were scored as correct if they included the content of the complement or use of the entire complement structure (de Villiers & Pyers, 2002; Perner et al.,2003). Children received 1point for each correct answer, resulting in a maximum score of 5 each for sentential complements involving either mental or communication verbs.

2.4 Training protocols[1]

Children interacted with one of three adult experimenters alone in a quiet room in their preschools. The two training sessions took place within two weeks of one another, with each session lasting about 25 min (range: 20–30 min), with a 7–day interval between sessions. Posttests were administered about 4 days after the second training session.

2.4.1 Sentential complement training

The training procedure was modeled on that used by Lohmann and Tomasello (2003) and was revised slightly to make it culturally appropriate for Chinese children. In addition, in order to stimulate children's interest in talking about some topics and to improve the ecological validity of the study, the sentential complement training was conducted in an elaborated, conversational manner.

Each of the two sessions consisted of four trials and lasted about 25 min. In the first session, trials involved the experimenter presenting an object (e.g., apple) which was wrapped in a piece of thin paper and letting children touch it. Children were then asked about characteristics

---

[1] Mandarin materials are available on request from the anthors

of the objects with questions that included a sentential complement whose verb depended on condition (mental vs. communication). The experimenter repeated or corrected children's responses to the questions. For example, if the question was, "What do you think/say this is?" the feedback was, "Okay, you think it is an apple. Yes, I also think it is an apple."

In the second session, trials involved the experimenter telling children two short stories accom panied by relevant pictures. One story was as follows. "Xiaoming wants to get a toy airplane for his birthday gift. On his birthday, his mother bought him a toy airplane. Xiaoming thought/said that his mother bought him a toy airplane. He thought/said that his mother loved him very much. After each story, the experimenter asked the child questions regarding the parts of the story that involved sentential complements (e.g., "Can you tell me, what did Xiaoming think/say his mother bought him?" The experimenter repeated or corrected children's answers. For example, the experimenter said, "Right, Xiaoming thought/said that his mother bought him a toy airplane."

2.4.2 False representation training

A training paradigm based on the "thought bubble" experimental paradigms discussed earlier was adapted for use with Mandarin-speaking children (Flavell, Everett, Croft, & Flavell, 1981; Wellman et al., 1996).

To begin, an experimenter presented children with three black and white line drawings, two of which depicted a familiar object (e.g., horse and turtle); the third was a "thought bubble" picture, i.e., a set of bubbles appearing above the right side of the pictured child's head. The experimenter first presented children with the thought bubble picture and said, "Look at the large bubble that contains a horse. It shows what is in the character's mind. Have you ever seen this kind of thought bubble in cartoons?" Children typically demonstrated understanding of the relation between thought bubbles and mental states

following this description. Children were then shown scenarios in which the contents of thought bubbles did not match some true state of affairs.

The experimenter presented the first picture (e.g., of a horse) and asked children to name the object. Then, presenting the "thought bubble" picture again, the experimenter said to the child, "When this boy/girl saw this picture, does he/she know there is a horse in this picture?" After children answered correctly, the experimenter continued, "Well now this boy/girl turned away from us." (The thought bubble picture was turned away.) The experimenter then showed the second picture, saying, "Look at this picture (showing the second object, e.g., a car), replacing the first picture with this one and asking, 'What is the object in this picture now?' " The experimenter then continued, "Well, the boy/girl turned and does not see the object in this picture." Children were then asked, "Now, what is the object in this picture in his/her mind?"

If children answered correctly, the experimenter asked them to explain why. If not, the experimenter explained that the boy/girl had not seen the picture being replaced and then asked the question again. If children did not answer correctly (i.e., by naming the object in the second picture), the experimenter asked, "Is he/she right (or wrong)? Why?" If children answered incorrectly, the experimenter asked, "If you do not see the picture, do you know what it is in the picture (the experimenter turned the picture away)?"

The false representation training included two sessions, each consisting of four trials, with each trial lasting about 5-7 min.

2.4.3 Control group

The control training condition was based on the procedure used by Lohmann and Tomasello (2003) and the same objects as in the sentential complement training condition. However, the experimenter did not ask

Table 2 Performance on posttest tasks across groups (mean and SD).

	Groups			
	SC–COMM(n = 20)	SC–MENTAL(n = 20)	FR(n = 20)	Control (n = 21)
FB prediction(0–4)	2.45 (1.50)	2.20(1.47)	2.45(1.38)	1.28 (1.27)
MCwMV (0–5)	3.25(1.21)	3.55 (1.14)	2.45(1.43)	2.48 (1.28)
MCwCV (0–5)	4.25 (0.96)	3.80 (0.95)	3.50 (1.27)	3.61 (1.07)

Table 3 Percentage of participants giving appropriate or inappropriate reasons in cake story task.

	Groups			
	SC–COMM(n = 20)(%)	SC–MENTAL(n = 20)(%)	FR(n = 20)(%)	Control (n = 21)(%)
Appropriate justification				
Location or knowledge	40	35	20	9.5
Mental state	10	15	20	9.5
Inappropriate justification	50	50	60	81

children questions involving sentential complement with communication or mental verbs,and only asked simple questions with clear answers to keep their attention. Children's responses were given neutral feedback, although an effort was made to keep children engaged. Finally, an effortwas made to ensure that the control training episodes were roughly as long as in the other training conditions.

3 Results

3.1 Preliminary analyses

Children's mean ages and average performance on the various pretests in each group are presentedin Table 1. A one–way analysis of variance (ANOVA) and Kruskal–Wallis tests showed no significant age differences or any of the pretests across the four groups. Nor were there significant gender differences. These preliminary analyses demonstrated that the four groups were indeed equivalent before training.

3.2 Training effect on false belief posttests

3.2.1 Prediction measures

Performance on the prediction questions in the two false belief tasks (contents and location) was significantly correlated, $r(81) = 45$, $p<.001$. Thus, scores on the two tasks were summed to obtain a total score for each child, ranging from 0 to 4 (see Table 2).

A one–way ANOVA, with training condition as independent variable and the aggregated false belief post–test scores as dependent variable, revealed a significant training condition effect, $F(3, 77)=3.14$, $p=0.03$, partial $\eta^2=0.10$. Post hoc tests revealed that the SC–COMM group and the FR group performed significantly better than the control group (Tukey HSD, $ps<.05$). The SC–MENTAL group did not sig nificantly outperform the control group (Tukey HSD, $p>.1$). No other significant differences were found.

Did improvement reach a level of above–chance responding? Tests comparing posttest performance in each group to chance (i.e.$\mu = 2.00$) showed that the control group performed significantly belowc hance, $t=2.57$, $df=20$, $p=0.018$. Performance in the focal training groups did not exceed chance.

Tables 3 and 4 showed the percentage of children by group who gave appropriate explanations for their responses in the two location–change tasks. In the "cake story" task, both sentential complement training groups outperformed the control group, $x^2(1, N=41)=4.36$, $p=0.037$, for both comparisons. In the "ball story" task, however, only the SC–COMM group gave more appropriate reasons than thecontrol group, $x^2(1, N=41)=5.53$, $p=0.019$.

3.3 Training effect on sentential complement posttests

The scores of the two sentential complement tests (mental and communication verbs) were significantly

Table 4 Percentage of participants giving appropriate or inappropriate reasons in ball story task.

	Groups			
	SC–COMM(n = 20)(%)	SC–MENTAL(n = 20)(%)	FR(n = 20)(%)	Control (n = 21)
Appropriate justification				
Location or knowledge	40	40	35	19.1
Mental state	20	5	10	4.7
Inappropriate justification	40	55	55	76.2

correlated, $r(81) = 0.25$, $p=0.02$ (see Table 2) and therefore combined. Thus, composite scores ranged from 0 to 10. A one-way ANOVA with condition as independent variable showed a significant main effect of condition, $F(3, 77)= 3.99$, $p=0.01$, partial $\eta^2=0.11$. Post hoc comparison revealed that the SC-COMM group and SC-MENTAL group outperformed the FR group and control group (Tukey HSD,the SC-COMM group > control group, $p= .016$; SC-MENTAL group > the control group, $p=0.03$; SC-COMMgroup>FR group, $p=0.009$; SC-MENTAL group >FR group, $p=0.018$). No other significant differences were found (see Fig1). Thus, sentential complement training promoted facility with complement syntax.

3.4 Reduced sample analyses

As noted, several children in each group passed the sentential complement pretests. This made it difficult to test the hypothesis that the sentential complement training involving communication verbs would be more effective than complement training involving mental verbs at promoting both general sentential complement understanding and false belief reasoning. To address this difficulty,we performed some exploratory analyses to determine whether the above patterns would remain when children who performed perfectly on sentential complement pretests were removed from the analysis. A one-way ANOVA with false belief prediction score across all four false belief tasks as dependent variable demonstrated a significant effect of group, $F(3, 60)=3.09$, $p=0.03$. Tukey HSD testshowed that the SC-COMM group ($n=16$, $M= 2.25$, SD $=1.57$) outperformed the control group ($n=15$,$M= 0.93$, SD=1.03, $p=0.05$), and FR group ($n=17$, $M =2.29$, SD=1.49) performed better than the controlgroup ($p<.05$).

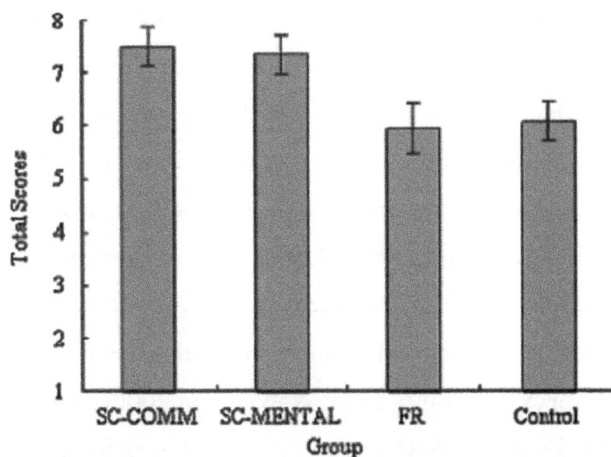

Fig. 1 Total posttest score on sentential complements by group. Note: Error bars represent standard error.

There was no significant difference between the SC-MENTAL group ($n=16$, $M=2.12$,SD=1.58) and the control group ($p>.10$). A similar one-way ANOVA with sentential complementposttest scores as dependent variable showed a significant effect of condition, $F(3, 60) = 3.56$, $p=0.02$. Post hoc tests showed significant differences between the SC-COMM group ($M=7.37$, SD=1.70) and FRgroup ($M=5.76$, SD=2.27) ($p=01$), between the SC-MENTAL group ($M = 7.37$, SD=1.78) and FR group($p=0.01$), between the two sentential complement training groups and the control group ($M=6.00$,SD=1.46, ps<.05). There was no significant difference between FR and control groups.

4 Discussion

This study explored the extent to which sentential complement and false representation comprehension training affected false belief understanding among Mandarin-speaking children. First, we found that training with sentential complement with communication verbs and false representations significantly improved performance on false belief posttests. Second, only the sentential complement training groups performed better than the control group on the sentential complement posttests.

Children who were trained with sentential complements involving communication verbs showed greater improvement on false belief posttests than did children trained on sentential complements with mental verbs. This asymmetry was predicted for Mandarin-speaking children because of the differences in the kinds of verbs that typically take sentential complements in Mandarin. Mandarin-speaking children hear and use sentential complements with the communication verb "*Shuo*1" [/say/] earlier and more commonly than they do with mental verbs (Snedeker & Li, 2000; Tardif & Wellman, 2000). Thus, sentential complementtraining with a communication verb may have provided a more straightforward opportunity for children to capitalize on whatever structural benefits acquiring the sentential complement construction has for reasoning about false beliefs.

These findings have implications for understanding of the mechanisms by which sentential comple ment training might promote false belief understanding. The literature to date left open the possibility that the sentential complement training protocols used by past researchers unintentionally provided informationrelevant to mental state reasoning (such as using deceptive objects or talking about story characters' preferences and feelings). Here, we designed our training to avoid this possibility and exam-ined a group of children who showed a greater benefit of sentential complement training involving verbs that were explicitly non-mental. These findings clarify that sentential complement training can confer benefits on false belief reasoning even when there is no (or minimal) mental state conten twithin the training.

Thus, these findings are generally in line with the view that mastery of sentential complements may provide an important template for reasoning about others' mental states (de Villiers& Pyers, 2002). This view, sometimes called the "syntactic enrichment" view, is that as children acquire the syntacticstructures that allow for embedding one thought in another (e.g., embedded propositions),they gain a format that can facilitate reasoning that requires the explicit separation of two different perspectives on a single situation (de Villiers & Pyers, 2002; Harris, de Rosnay, & Pons, 2005). Our findings here are, at least in part, consistent with this view,given that our training protocols that involvedcommunication verbs were carefully designed toavoid false-belief relevant content but they still promoted false belief understanding.

The enhanced effectiveness of sentential complement training with communicative verbs relative to mental verbs was predicted specifically for Mandarin-speaking preschoolers given the relative frequency with which sentential complements are used for the two verbs in everyday language. An open question is whether, if investigated more directly, a similar advantage might be seen for children who speak other languages. de Villiers and her colleagues argued that an early emerging facility with sentential complements involving communication verbs might provide children with a sort of syntactic bootstrap to developing a more sophisticated (i.e., representational) understanding of mental verbs that are used in sentential complement formats. Evidence from Lohmann and Tomasello (2003) suggests that training with both communication and mental verbs is equally effective and thus that the advantage for communication verb training is specific to Mandarin for the reasons mentioned. Nonetheless, given that our training was designed with the explicit goal of separating the contents of communication and mental verb training, it may be worthwhile to apply our paradigm to other languages to see whether the advantage for communication verbs may be more general, for the reasons described by de Villiers (2003).

A final note about our training protocols concerns the particular mental verbs we used in the training and false belief task protocols. Like some other languages, Mandarin includes several verbs for "think" and "believe". One such verb is/*yi3−wei2*/which is usually translated as "think falsely". Lee,Olson, and Torrance (1999) found that Mandarin−speaking children performed significantly better in the false belief tests when this false belief verb was used, compared with more neutral verbs (see Cheung et al., 2009, for a similar finding in Cantonese). In the present study, we used the neutral mental verbs/*ren4−wei2*/and/*jue2−de*/during the sentential complement training and false belief tasks,respectively. Our choice to use these rarer, semantically neutral verbs may have weakened the extent to which children capitalized on the training protocols. Yet, this caveat does not negate the main finding, which is that sentential complement training involving communication verbs can benefi tfalse belief reasoning, even when that training has no explicit or implicit mental content.

We also found that children in the false representation group showed evidence of improved false belief performance relative to the control group, although the false representation training group did not show any measurable gains in their facility at processing sentential complements. At some level,this finding was expected. As previous work had shown (Wellman et al., 1996), training paradigms that use thought bubbles to emphasize the representational nature of mental states can improve even 3−year−olds' performance on standard false belief tasks. Similar findings have been shown with individuals with autism, a neurodevelopmental disorder in which false belief reasoning is particularly affected (Wellman et al., 2002). Our results extend these findings and clarify that the benefits children receive from thought bubble training likely do not have their effects through promoting facility with sentential complement constructions. In doing so,

our findings are consistent with those of Hale and Tager−Flusberg (2003), who showed that training on false belief only (i.e., without exposure to sentential complement training) improved false belief performance but not sentential complement understanding. We see these findings as having two important implications. First, they show that the ability to reason about false representation is not sufficient for processing sentential complement syntax (Ruffman et al., 2003). Second, they show that explicit false belief understanding can emerge in the absence of sentential complement understanding (Hale & Tager−Flusberg, 2003; Lohmann &Tomasello, 2003).

It should be noted that the false representation training protocol was not fully non−linguistic; we did use a neutral Mandarin epistemic mental verb that we felt was appropriate to the situation (i.e.,/*zhi1dao*4/). Thus, we do not wish to argue against the possibility that naturalistic exposure to mental verbs plays some role in children's false belief development across cultures. Our argument, rather, is that making explicit the representational nature of mental states was effective in promoting false belief understanding, although it apparently did not affect children's facility with sentential complements.

A limitation of the present study is that we have no assessments of children's cognitive and lin-guisticcapacities beyond those investigated. Children's general linguistic abilities arerelated to both sentential complement understanding and false belief performance in Chinese− and English−speaking preschoolers (Tardif et al., 2007). By knowing more about how the training protocols that we used interact with general linguistic abilities, we mightgain greater insight into the specific mechanisms by which these interventions affected false belief development. Also, we did not include measures of representational understanding other than false belief understanding, such as measures of false photograph or false sign performance (Sabbagh, Moses, & Shiverick,

2006). Doing so might have allowed us to determine whether the training protocols that promoted false belief understanding had their effects on representational understanding more broadly or more specifically on mental state understanding.

In conclusion, the present study supports the view that training protocols aimed at improving children's facility with sentential complements can promote false belief understanding, even when the training protocols are structured so as not to include explicit or implicit mental state content. Indeed,there was some evidence that with Mandarin-speaking children, sentential complement training that involved communication verbs was more effective than training with mental state verbs. Yet the findings also showed that false belief performance can be improved by a training protocol that involved no training with sentential complements. Together, these findings suggest that sentential complement understanding may be one path that contributes to false belief understanding, the importance of which may depend on particular sociocultural and linguistic factors.

References

Astington, J. W., & Jenkins, J. M. (1999). A longitudinal study of the relation between language and theory-of-mind development.*Developmental Psychology, 35*, 1311‐1320.

Cheung, H., Chen, H.-C., Creed, N., Ng, L., Wang, S. P., & Mo, L. (2004). Relative roles of general and complementation languagein theory-of-mind development: Evidence from Cantonese and English. *Child Development, 75*, 1155‐1170.

Cheung, H., Chen, H.-C., & Yeung, W. (2009). Relations between mental verb and false belief understanding in Cantonese-speakingchildren. *Journal of Experimental Child Psychology, 104*, 141‐155.

de Villiers, J. G. (2003). Getting complements on your mental state (verbs). In J. Van Kampen, & S. Baauw (Eds.), *Proceedings of2003 GALA conference* (pp. 13‐26). Utrecht: LOT.

de Villiers, J. G., & de Villiers, P. A. (2000). Linguistic determinism and the understanding of false beliefs. In P. Mitchell, & K. J.Riggs (Eds.), *Children´s reasoning the mind* (pp. 191‐228). Hove, England: Psychology Press.

de Villiers, J. G., & Pyers, J. E. (2002). Complements to cognition: A longitudinal study of the relationship between complexsyntax and false−belief−understanding. *Cognitive Development, 17*, 1037‐1060.

Flavell, J. H., Everett, B. A., Croft, K., & Flavell, E. R. (1981). Young children's knowledge about visual perception: Further evidencefor the level 1‐level 2 distinction. *Developmental Psychology, 17*, 99‐103.

Hale, C. M., & Tager-Flusberg, H. (2003). The influence of language on theory of mind: A training study. *Developmental Science,6*, 346‐359.

Harris, P. L., de Rosnay, M., & Pons, F. (2005). Language and children's understanding of mental states. *Current Directions inPsychological Science, 14*, 69‐73.

Hogrefe, G.-J, Wimmer, H., & Perner, J. (1986). Ignorance versus false belief: A developmental lag in attribution of epistemicstates. *Child Development, 57*, 567‐582.

Hughes, C., & Dunn, J. (1998). Understanding mind and emotion: Longitudinal associations with mental−state talk betweenyoung friends. Developmental *Psychology, 34*, 1026‐1037.

Lee, K., Olson, D. R., & Torrance, N. (1999). Chinese children's understanding of false beliefs: The role of language. *Journal of ChildLanguage, 26*, 1‐21.

Lohmann, H., & Tomasello, M. (2003). The role of language in the development of false belief understanding: *A training study.ChildDevelopment, 74*, 1130‐1144.

Milligan, K., Astington, J. W., & Dack, L. A. (2007). Language and theory of mind:Meta-analysis of the relation between languageability and false-belief understanding. *Child Development, 78,* 622－646.

Perner, J., Leekam, S., & Wimmer, H. (1987). Three-year olds' difficulty with falsebelief: The case for a conceptual deficit. *BritishJournalof Developmental Psychology, 5,* 125－137.

Perner, J., Sprung, M., Zauner, P., & Haider,H. (2003). Want That is understood well before Say that Think that, and false belief:A test of de Villiers's linguistic determinism on German－Speaking children. *Child Development, 74,* 179－188.

Ruffman, T., Slade, L., Rowlandson, K., Rumsey, C., & Garnham, A. (2003). How language relates to belief, desire, and emotionunderstanding. *Cognitive Development, 18,* 139－158.

Sabbagh, M. A., Moses, L. J., & Shiverick, S. (2006). Executive functioning and preschoolers' understanding of false beliefs, falsephotographs, and false signs. *Child Development, 77,* 1034－1049.

Schick, B., de Villiers, P., de Villiers, J., & Hoffmeister, R. (2007). Language and theory of mind: A study of deaf children. *ChildDevelopment, 78,* 376－396.

Snedeker, J., & Li, P. (2000). Can the situations in whichwords occur account for cross-linguistic variation in vocabulary com-position? In J. Tai, & Y. Chang (Eds.), *Proceedings of the seventh international symposium on Chinese languages and linguistics.*

Tardif, T., So, C. W.-C., & Kaciroti, N. (2007). Language and false belief: Evidence for general, not specific, effects in Cantonese-speaking preschoolers. *Developmental Psychology, 43,* 318－340.

Tardif, T., & Wellman, H. M. (2000). Acquisition of mental state language in Mandarin- and Cantonese-speaking children.*Developmental Psychology, 36,* 25－43.

Wellman, H. M., Baron-Cohen, S.,Caswell, R., Gomez, J. C., Swettenham, J., Toye, E., et al. (2002). Thought-bubbles help childrenwith Autism acquire an alternative to a theory of mind. *Autism, 6*(4), 343－363.

Wellman, H. M., Cross, D., & Watson, J. (2001). Meta-analysis of theory-of-mind development: The truth about false belief. *ChildDevelopment, 72,* 655－684.

Wellman, H. M., Hollander, M., & Schult, C. A. (1996). Young children's understanding of thought bubbles and thoughts. *ChildDevelopment, 67,* 768－788.

Wimmer, H., & Perner, J. (1983). Beliefs about beliefs: Representation and constraining function of wrong beliefs in youngchildren's understanding of deception. *Cognition, 13,* 103－128.

School Psychology International, 2015, 36(3), 227 - 252.

Psychological development and educational problems of left-behind children in rural China

Xiaojun Sun[1,*], Yuan Tian[1,*], Yongxin Zhang[1], Xiaochun Xie[1], Melissa A. Heath[2], Zongkui Zhou[1]

([1] Central China Normal University, China)

([2] Brigham Young University, USA)

Abstract With China's rapidly developing economy and increasing urbanization, many adults from rural areas migrate to urban areas for better paid jobs. A side effect of this migration is that parents frequently leave their children behind (left-behind children). This research investigated left-behind children's and non-left-behind children's psychological, behav-ioral, and educational functioning. Survey participants included 1,708 adolescents (54.8%female; mean age=15.03 ± 1.93 years) from rural areas in Central China. Additionally,32 left-behind children and 32 head teachers were interviewed. Data indicated that in comparison to non-left-behind children, left-behind children were at a disadvantage in regard to emotional adjustment (i.e. lower life satisfaction, lower self-esteem, and higher depression), but fared better in educational adjustment (greater school engagement). Mitigating factors which positively influenced outcomes of certain subgroups of left-behind children included the presence of one parent, increased parental contact,and shorter length of time since parental migration. Information gathered from interviews with LBC also indicated adverse effects of parent absence on children's develop-ment. Teachers identified education measures and support offered to left-behind children and reported difficulties in communicating with parents. Based on this study's findings, and considering the perspective of educators, implications for school-based interventions are explored.

Keywords Behavior problems; education; left-behind children; migration; psychosocial development; relationships; rural China; school engagement; teacher perspective

1 Introduction

Since the end of the 1970s,China's economy has rapidly expanded due to policy reform and opening up to international trade markets (Jaggi, Rundle, Rosen, & Takahashi,1996; Morrison, 2014). This economic growth has also promoted the process of urbanization, creating a significant influx of rural surplus laborers who pour into cities looking for better jobs. These individuals leave their

* Xiaojun Sun and Yuan Tian are joint first authors.

Corresponding author: Zongkui Zhou, School of Psychology, Central China Normal University, Wuhan, Hubei 430079, China. Email: zhouzk@mail.ccnu.edu.cn

This work was supported by the National Social Science Fund Project of China (Grant No. 11&ZD151) and the National Natural Science Foundation of China (Grant No. 31400887).

rural homes and become migrant workers. This massive rural–to–urban migration constitutes the largest migration in human history (Zhang, 2004). Unfortunately, because of long work hours in the city,destitute living conditions, insufficient income, and the restrictions associated with the binary system between China's rural and urban areas, many migrant workers are forced to leave their children behind in their rural hometown (Luo, Wang, & Gao, 2009).

Parents leaving children behind while one or both of the parents migrate for work is also very common in other countries, such as Japan (Carandang, Sison, &Carandang, 2007; Edillon, 2008; Melgar & Borromeo, 2002), the Philippines(Reyes, 2008), and Mexico (Bryant, 2005; Reyes, 2008; Tarroja & Fernando,2013; Yeoh & Lam, 2007). In China, many children are left in their rural hometown by their parents (one or both) who are hunting for more lucrative work in urban areas. These children are defined by Chinese scholars as left–behind children (LBC) (Luo et al., 2009; Zhao & Shen, 2010).

In recent years, China's migrating population has increased dramatically. Simultaneously, the number of LBC has also increased dramatically. Based on the results of China's Sixth National Population Census, nationwide, there were approximately 69.7 million LBC. The vast majority of China's LBC, 61 million, live in rural areas. In fact, rural LBC account for 87.5% of China's total number of LBCand 21.9% of China's total number of children (Duan, Lu, Guo, & Wang, 2013).

Beyond sharing some common vulnerabilities with unattended and disadvantaged children living in other countries around the world, China's rural LBC also face additional challenging situations. China's dual rural–urban system makes it especially difficult for children to migrate with their parents. Because of the serious imbalance of regional economic development, rural LBC are commonly faced with difficulties associated with low socioeconomic status and the shortage of educational resources.

However, similar to other countries' LBC, the large number of China's LBC continues to be considered a fairly small proportion of the massive population base. This topic has not been adequately researched, nor have these children's needs been adequately addressed (Liu & Wang, 2010; Luo et al., 2009).

By placing this topic in a world–wide context, understanding the developmentand relevancy of China's LBC may contribute to advancing research in the broad overarching field of understanding disadvantaged children's needs. Ultimately this knowledge will assist in identifying and providing appropriate support to address LBC's needs and perhaps the needs of children in other contexts who have been'left behind'. The present study aims to describe China's rural LBC in terms of basic characteristics of children's psychological development and how LBC's present situations relate to and affect their education and social–emotional well–being.

Additionally, research is needed to address serious concerns associated with LBC. For example, due to the lack of parental supervision and care, elevated rates of suicide and sexual abuse are reported for China's rural LBC (Zhou,Wang, & Hong, 2010). The media's criticism of these situations not only exposes LBC's frequent exposure to malignant events, but also to the family's and community's failure to provide adequate supervision of LBC. The media's spotlight on these issues brings the topic of LBC to the visible forefront, fueling criticism aimed directly at Chinese society (Shen, 2009).

In addition to concerns related to failed supervision of LBC, psychological research is keenly interested in the effect of long–term parent–child separation and the ensuing maladjustment of children's social development. Previous studies in other countries and cultures have indicated that paternal absence is correlated with children's maladjustment (Amato & Keith, 1991; Carandang et al., 2007;Cronk, Slutske, Madden, Bucholz, & Heath, 2004).

For example, one research study conducted in China found that left-behind girls were more likely to be unhappy, to contemplate suicide, and to consider leaving home (Gao, Li, Kim,Congdon, Lau, & Griffiths, 2010). Researchers in other countries, such as Indonesia and Mexico, who explored the impact of transnational migration, also found impeded social and psychological development among LBC and a higher incidence of mental disorders (Aguilera-Guzman, de Snyder, Romero, & Medina-Mora, 2004; Hugo, 2002).

Therefore, regardless of whether parents' migration occurs within-country or transnationally, research indicates that separation from parents appears to negatively impact LBC's development and adjustment (Wen & Lin, 2012; Zhao, Liu, &Zhang, 2013). Hence, from an applied frame of reference, it is of practical significance to explore LBC's development from a psychological and educational perspective, focusing on the relationship between the child's individual development and their environment. Investigating these aspects of child development will assist educators in better understanding LBC's needs, helping guide intervention efforts and helping identify and implement teaching strategies aimed at strengthening both healthy psychological development and academic achievement.

In China, the family migrant statuses include non-migrant family, one-parent migrant family, and both-parent migrant family. Another aspect to consider is *who* offers supervision and care in the absence of one or both parents. Caregivers of LBC are divided into five groups: (a) parent who did not migrate—typically the mother; (b) grandparents; (c) kith and kin (extended family); (d) older sibling; and(e) no caregiver in situations where LBC care for themselves (typically adolescent LBC). Other factors which are commonly considered by researchers include the length of time since parental migration (Wei & Chen, 2010), frequency of parental contact (Chen & Xie, 2007), and economic support from parents (Wang, Hu, Shen,2011).

2 Research questions

The following questions will be explored in this study: (a) when taking into account the migrant statuses and demographic characteristics of China's rural LBC, what are the differences in how parental migration impacts these groups?; (b) what aspects of child development are affected by parental migration?; (c) how do LBC describe their living situation and challenges?; and (d) what is the role of school-based education in regard to LBC's development?

Previous research indicates that key areas of children's development include emotional adaption; behavioral development; participation in school activities(school engagement; Fredricks, Blumenfeld, & Paris, 2004); and social relationships (Attili, Vermigli, & Roazzi, 2010; Lin, Fan, Li, & Pan, 2010; McLanahan &Sandefur, 1994; Xiang, 2007; Zhou, Sun, Liu, & Zhou, 2005). In order to examine aspects of rural LBC's emotional adjustment, this study investigated the following emotions: Loneliness, depression, and happiness. Additionally, self-esteem was examined to determine whether being left behind influences children's feelings of self-worth.

Children's problem behaviors were also investigated because difficulties with behavioral adjustment negatively impact both physical and mental health (Fang, Zhen, & Lin, 2001; Qu & Zou, 2009). Previous research indicates that limited parental monitoring is closely related to an increase in children's problem behaviors (Beck, Boyle, & Boekeloo, 2003; Diclemente, Wingood, & Crosby, 2001;Unnever, Cullen, & Travis, 2003), so this study investigated whether LBC whose parent(s) migrated exhibit more problem behaviors than non-LBC. In addition,parent-child relationships and peer relationships—identified as important social relationships—are the main source of adolescents' daily interaction and have a profound influence on adolescent psychological adaptation (Attili et al., 2010;Zhao et al., 2013; Zou, 1998).

In a recent nationwide study of Chinese children and adolescents the above indicators were used to describe the basic characteristics of psychological development (Dong & Lin, 2011). These indicators are well represented in the existing research base. The present study, using the method of psychological measurement, examined the similarities and differences between LBC and children living in non-migrant families. More specifically similarities and differences were investigated in the following areas: Psychological aspects, behavioral aspects, educational outcomes, and social relations.

In order to further understand perceptions of LBC's personal lived experiences and education status, 32 LBC and 32 head teachers were interviewed. The quantitative and qualitative data gathered from this research were considered and used as a basis for making recommendations to assist educators in more adequately meeting LBC's social-emotional, behavioral, and educational needs.

3　Method

3.1　Participants

A total of 1,708 adolescents (54.8% female, mean age=15.03 years; SD=1.93) from two junior high schools (contributing 42.3% of total participants) and two high schools (contributing 57.7% of total participants) were asked to complete an anonymous questionnaire. Participating schools were located in rural areas of Central China.

Of the adolescents who participated, 600 were non-left-behind children (non-LBC) and 1,108 (64.9%) were LBC. Of the LBC participants, 547 (49.4%) were from families of both-parent migration and 561 (50.6%) were from families with one-parent migration.

To supplement information gathered from the questionnaire data, 32 LBC and 32 head teachers who worked with this study's participants were interviewed. These 32 students were selected according to their gender, migrant status, and school performance: 16 boys and 16 girls; 16 from one-parent migration family and 16from both-parents migration family; and 11 with good school performance, 11 with moderately good school performance, and ten with poor school performance.

The 32 head teachers were selected from participating classrooms; their mean age was 38.3 years; and their mean number of years teaching was 17.2 years. However, six teachers reported having less than six years of teaching experience. Minimally, all interviewed teachers had an undergraduate college degree.

3.2　Measures

Loneliness was assessed with Children's Loneliness Scale (Asher & Wheeler, 1985;Bagner, Storch, & Roberti, 2004). The scale consists of 24 items: 16 scored items and eight additional items related to participants' interest. These eight items were intended to help relax the participants when answering questions and were not included in the scored items. Ten of the 16 items were reverse scored, and were transformed before the data were analysed. Participants recorded their responses on a five-point Likert scale, ranging from 1 (always true) to 5 (not at all true). The scale's eight items are internally consistent (Cronbach's α =0.90; Asher & Wheeler,1985; Bagner et al., 2004).

Children's depression was assessed with the Chinese version of the Children's Depression Inventory (CDI; Kovacs, 1992; Yu & Li, 2000). Children were asked to endorse one of three descriptions that best applied to him or her during the last 2 weeks (e.g. 'I feel like crying every day'; 'I feel like crying many days'; 'I feel like crying once in a while'). In this study, the scale contained 26 items, and participants' responses were scored on a three-point

Likert scale ranging from 0 (option which described the least often occurring) to 2 (option which described the most often occurring). Based on previous data describing the CDI, the 26 items are internally consistent (Cronbach's α =0.91).

In addition to a single question ('Do you feel your life is happy in general?'), life satisfaction was measured with the revised Student's Life Satisfaction Scale (SLSS;Huebner, 1991; Huebner, Suldo, & Valois, 2003). Items were rated on a six-point Likert scale ranged from 0 (low) to 5 (high). This scale's internal consistency is considered adequate (Cronbach's α =0.76; Kovacs, 1992; Yu & Li, 2000).

Self-esteem was measured on a ten-item scale, the Self-Esteem Scale (SES;Rosenberg,1965,1979;Schmitt& Allik,2005).However,the eighth item was deleted because of its low validity for Chinese participants (Tian, 2006). The participants' responses were reported on a four-point Likert scale, ranging from 1 (strongly disagree) to 4 (strongly agree). This scale's internal consistency is considered adequate(Cronbach's α =0.84; Baumeister, Campbell, Krueger, & Vohs, 2003; Rosenberg,1965, 1979; Rosenberg, Schooler, Schoenbach, & Rosenberg, 1995).

Problem behaviors were captured by asking the participants to rate each statement as to how often they participated in the described behavior. Questions included asking the student whether or not they participated in socially unacceptable behaviors (e.g. smoking, bingeing on alcohol, or other socially unacceptable behaviors). On this measure, response options range from 1 (not at all) to 4 (very often). This measure included 16 items and has an acceptable reliability(Cronbach's α =0.76; Fang, Li, & Dong, 1996).

School engagement was assessed with a scale of five items. To each of the five items, participants responded with either a 1 (Yes) or 2 (No). Adolescents reported whether or not they followed school rules, followed classroom rules, enjoyed homework, enjoyed going to school, and answered questions in class. Internal reliability of this measure is considered low (Cronbach's α =0.46; Wen & Lin, 2012).

Peer relationships were measured by peer nomination of peer rejection and acceptance. Students were asked to nominate three of their classmates that they liked most and three of their classmates that they liked least. Then, the total number of 'like most' and 'like least' obtained for each child was divided by the total number of students in the class. Proportion of positive nomination minus proportion of negative nomination indicates social preference and the higher the score, the greater the child's popularity. This method of assessing peer relationships is commonly used in Chinese research studies (Zhao et al., 2013).

Parent-child cohesion was measured by the father-child cohesion (Cronbach's α =0.65) and mother-child cohesion (Cronbach's α =0.65) subscales of the Family Adaptation and Cohesion Evaluation Scale (Joh, Kim, Park, & Kim, 2013; Olson,Sprenkle, & Russell, 1979). This portion of the test consisted of two parts (father scale and mother scale) with similar items. Responses were on a five-point Likert scale and ranged from 1 (mostly disagree) to 5 (mostly agree). Examples of the statements included, 'My father/mother are supportive to each other in difficulties';'My father/mother feel intimate to each other'.

Demographic variables included gender (male or female) and education level(junior high school or high school). Additionally adolescents were described as either an only child or one of several children. The participant's birth order was also designated. Finally, to assess each LBC's specific family situation, children were asked to respond to four questions. These questions included parental migration status (one-parent migration, both-parent migration, and no-parent migration); length of time since parent's migration; relationship to primary caregiver;and frequency of parental contact.

3.3 Procedure

Data were collected during spring 2013. Data were collected from two age groups:(a) middle school students from the 7th and 8th grades and (b) high school students from the 10th and 11th grades. More specifically, four classrooms from each of the identified grades in each rural school were randomly selected. Altogether, 32 classrooms were selected. Cluster surveys were used and participants answered anonymously. All questionnaires were administered in school classrooms and were collected on site. Prior to participants agreeing to participate (assent), information about the purposes of the survey and confidentiality of responses were explained. Students were informed that there were no correct or incorrect answers. Extra support was provided for students who had difficulty completing the surveys.

To gather additional information, 32 LBC, one student from each of the 32 classrooms, participated in an interview. The interview consisted of questions regarding the participating adolescent's perception of current life and the changes since their parent's or parents' migration. Additionally, 32 teachers, one from each participating class, were asked to participate in an interview about their education methods when instructing LBC, and the difficulties teachers and LBC faced in regard to schooling.

Table 1. Descriptive statistics reported in means and percentages

	Both–parent migration (N=547)	One–parent migration (N=561)	No–parent migration (N=600)	F/χ2
Demographic information				
Age	15.07 ± 1.94	15.05 ± 1.95	14.96 ± 1.92	0.57
Gender				
Males	55.20%	56.70%	52.70%	
Females	44.80%	43.30%	47.30%	
Only child				23.76***
Yes	23.10%	28.50%	36.20%	
No	76.90%	71.50%	63.80%	
Education level				
Junior high school	44.10%	43.30%	39.70%	
High schoo	55.90%	56.70%	60.30%	
Left–behind characteristics				
Length of parental migration[a]				21.54***
Short–term	23.50%	36.30%	—	
Medium–term	36.60%	29.60%	—	
Long–term	39.90%	34.10%	—	
Frequency of contact[b]				8.37*
Rarely	8.60%	12.50%	—	
Sometimes	36.80%	41.00%	—	
Often	54.50%	46.50%	—	
Primary caregiver				935.42***
Non–migrating parent[c]	—	91.10%	—	
Grandparents	68.70%	—	—	
Others[d]	31.30%	8.90%	—	

Note: *$p<0.05$; **$p<0.01$; ***$p<0.001$.

[a] Short–term means the length of parental migration is less than 1 year, medium–term means the length of parental migration is between 1 to 5 years. Long–term means the length of parental migration is more than 5 years.

[b] Frequency of contact is measured by subjective assessment of their communication with their parents.

[c] Most of non–migrating parents are mothers (more than 90.0%).

[d] Others include four categories: cared for by relatives, cared for by LBC's sibling, cared for by themselves, and cared for by others.

4 Results

4.1 Demographics

Comparing LBC's demographics with non-LBC's demographics, there were no significant differences on the distribution of age, gender, and education level. Table 1 displays the sample's descriptive statistics on selected variables. In this sample (N=1,708), 61.79% of participants reported that at least one parent migrated to an urban area for a job. More than 60% of children in this sample reported being from a one-only-child family. This phenomenon was especially common for LBC's families.

In regard to migrant status, there were significant differences between children of both-parent migration and children of one-parent migration on the following variables: Length of time since parental migration, frequency of contact, and identity of primary caregiver. Children of both-parent migration experienced a longer length of time since parental migration than children of one-parent migration. Among LBC, adolescents generally reported having contact with their migrating parent(s). Most LBC reported they 'often' had contact with their parents. In particular, children of both-parent migration reported often having contact with their parents. The majority of both-parent migration children were raised by their grandparents. Most children of one-parent migration were raised by the remaining parent who did not migrate.

4.2 Psychological development of LBC and their counterparts

There were significant correlations between the majority of variables. In terms of adolescents' loneliness, depression, life satisfaction, self-esteem, social preference, father-child relationship, mother-child relationship, problem behaviors, and school engagement, multivariate analysis of variance (MANOVA) was performed to compare group differences based on (a) migrant status (both-parent migration, one-parent migration, no-parent migration); (b) education level (junior high school, high school); and (c) gender. Findings indicated significant group differences based on migrant status, education level and gender, but no interaction among the variables was found (see Table 2).

Further ANOVA revealed that, based on LBC's migrant status, significant group differences were noted in participants' reports of loneliness, depression, life satisfaction, self-esteem, and school engagement. Based on LBC's education level (junior high versus senior high), differences were noted in social preference, father-child relationship, mother-child relationship, and problem behavior. Based on LBC's gender, significant differences were found in life satisfaction, self-esteem, social preference, father-child relationship, problem behavior, and school engagement (see Table 3).

Results of post-hoc tests indicated that children with both parents who migrated had lower scores on life satisfaction (p<0.001) and self-esteem (p=0.008) than non-LBC. Similarly, children with one parent who migrated had lower scores on life satisfaction than non-LBC (p=0.009). Children with both parents who migrated reported higher levels of depression than non-LBC (p=0.006). However, LBC with one parent who migrated (p=0.001) or both parents who migrated (p=0.003) engaged in more school activities than non-LBC.

Post hoc tests indicated age differences in children's perceptions. Overall, students in high school reported higher scores on social preference (p<0.001), father-child cohesion (p=0.006), mother-child cohesion (p=0.018), and problem behaviors (p=0.019) than students in junior high school.

Table 2. MANOVA of all children on migrant status, education level, and gender (N=1,708)

Independent variable	Wilk's Λ	F	η_p^2
Migrant status	0.97	2.71***	0.014
Education level	0.97	5.30***	0.027
Gender	0.92	16.91***	0.082

Note. *p<0.05 **p<0.01 ***p<0.001.

Table 3. ANOVA of variables of all children on migrant status, education level, and gender(N=1,708)

Independent variable	Dependent variable	F	η_p^2
Migrant status	Loneliness	4.03*	0.005
	Depression	3.95*	0.005
	Life Satisfaction	8.72***	0.010
	Self−esteem	3.51*	0.004
	School Engagement	5.32**	0.006
Education level	Social Preference	13.62***	0.008
	Father−child Relationship	7.52**	0.004
	Mother−child Relationship	5.59*	0.003
	Problem Behavior	5.49*	0.003
Gender	Life Satisfaction	5.81**	0.016
	Self−esteem	7.61**	0.004
	Social Preference	5.17*	0.003
	Father−child Relationship	8.24**	0.005
	Problem Behavior	89.45***	0.050
	School Engagement	9.35**	0.005

Note. *p<0.05, **p<0.01, ***p<0.001.

Post hoc tests revealed gender differences. Boys reported higher self−esteem(p=0.006), stronger father−child cohesion (p=0.004), and more problem behaviors(p<0.001). Girls reported higher levels of life satisfaction (p=0.016), social pref−erence (p0.006), and school engagement (p=0.002).

Comparisons among LBC subgroups. In order to more thoroughly investigate the effects of migration status on LBC, multivariate analysis of covariance (MANCVOA) was employed simultaneously to assess group differences in all dependent variables,considering caregiving type, length of parental migration, and frequency of parental contact while controlling two key demographic characteristics, school and gender. No interaction effect was found between independent variables and covariates.

In regard to LBC's caregiving type, more than 85% of LBC were cared for either by their grandparents or their non−migrating parent. The remaining 15% of LBC children were categorized into four caregiving types, with the subgroups' sizes ranging from 4.1% to 9.5% of the total LBC group. Thus in this study, we merged these four subgroups into a new category ('other') and this subset was not included in the remaining analyses. Further, after

testing for differences between the non−analysed category ('other') and the analysed category (those chidren brought up by a single parent or grandparents) on all dependent variables, no differences were found.

In addition, there were minimal missing data (<5%) on the two variables−length of parental migration and frequency of parental contact. When comparing the two groups, those children with missing data and those children without missing, data were homogeneous on all dependent variables. Therefore,those LBC with missing data were included in the data analyses. Important to consider when interpreting Table 4, the actual number of LBC included in this particular data analyses was 861.

The findings suggest that the main effects by caregiving type, length of parental migration, and frequency of parental contact were significant, as well the interaction between the length of parental migration and the caregiving type (Table 4). Further analyses revealed significant differences between one−parent caregiving and grandparent−caregiving when considering LBC's self−esteem, $F(1, 851)=4.00, p=0.046$, $\eta_p^2=0.005$; and mother−child cohesion, $F(1, 851)=15.58, p<0.001$, $\eta_p^2=0.018$.

Table 4. MANCOVA of differences among left behind children (LBC) based on differences in migrant Status.

Independent variable	Wilk's Λ	F	η^2_p
Caregiving type	0.97	2.45**	0.026
Length of parental migration	0.96	1.84*	0.019
Frequency of parental contact	0.87	6.64***	0.066
Duration × Caregiving type	0.96	1.87**	0.020
Covariates			
Education level	0.97	3.36***	0.035
Gender	0.90	9.93**	0.096

Note. *$p<0.05$; **$p<0.01$; ***$p<0.001$.
$N=861$

Continuing with the results listed in Table 4, when considering the length of parental migration, significant group differences were noted on life satisfaction, $F(2, 851)=5.19$, $p=0.006$, $\eta^2_p=0.012$; and social preference, $F(2, 851)=3.32$, $p=0.037$, $\eta^2_p=0.008$.

When considering the amount of parental contact, there were significant group differences reported on the following variables: Loneliness, $F(2, 851)=12.69$, $p<0.001$, $\eta^2_p=0.029$; life satisfaction, $F(2, 851)=28.45$, $p<0.001$, $\eta^2_p=0.063$; self-esteem, $F(2, 851)=14.35$, $p<0.001$, $\eta^2_p=0.033$; depression, $F(2, 851)=30.67$, $p<0.001$, $\eta^2_p=0.067$; father-child cohesion, $F(2, 851)=21.87$, $p<0.001$, $\eta^2_p=0.049$; mother-child cohesion, $F(2, 851)=16.41$, $p<0.001$, $\eta^2_p=0.037$; school engagement, $F(2, 851)=4.00$, $p=0.019$, $\eta^2_p=0.009$. Additionally, an interaction between the length of parental migration and caregiving type was found on school engagement, $F(2, 851)=4.12$, $p=0.016$, $\eta^2_p=0.010$.

Post-hoc test comparisons revealed that LBC who were brought up by the nonmigrating parent reported higher scores on self-esteem ($p=0.046$) and mother-child relationship ($p<0.001$) than those who were brought up by grandparents. Long-term LBC reported lower life satisfaction scores than medium-term LBC($p=0.001$). However, Owever, on social preference, medium-term LBC reported lower scores than short-term LBC ($p=0.016$).

In regard to the contact frequency, results showed that LBC who often had contact with their migrant parent(s)

felt less lonely and depressed than LBC who sometimes ($p<0.001$) and rarely ($p<0.001$) had contact with migrant parent(s). LBC who sometimes had contact with their migrant parent(s) were less depressed than LBC who rarely had contact ($p<0.001$).

On life satisfaction, self-esteem, father-child cohesion and mother-child cohesion, LBC who often had contact with their migrated parents fared better than those LBC who sometimes ($p<0.001$) and rarely ($p<0.001$) had contact. On father-child cohesion ($p=0.029$) and mother-child cohesion ($p=0.049$), LBCwho sometimes had contact with their parents fared better than LBC who rarely had contact. In comparison to LBC who rarely had contact with their migrant parents, school engagement was significantly better for children who often had contact with their migrant parent ($p=0.005$) and for LBC who sometimes had contact ($p=0.046$). Simple effect analysis for the interactive effects revealed that long-term LBC performed better on school engagement than those medium-term LBC ($p=0.035$) and short-term LBC ($p=0.004$), particularly if they were brought up by their grandparents.

4.3 LBC's perceptions of life

Based on interviews, on the whole, LBC reported holding negative perceptions of life changes after parental migration. Most of them (69%) perceived that their current life was different from life prior to their parents' departure. The main reported change was the harmful effect on the parent-children relationship. For example, care from and communication with migrated parent(s) decreased. Estrangement from migrated parent(s) was either initiated or exacerbated following their departure. On the contrary, after the parents migrated a few LBC reported increased communication with parent(s).

During their interviews, several LBC indicated that they become tougher and more sensible and thoughtful after their parent(s) migrated. However, whether LBC perceived parental migration as positive or negative, the vast majority reported prominent feelings of missing parents and being lonely. LBC commonly reported not wanting their parents to migrate for work and also reported feeling greater happiness when living together with their parents.

In regard to perceptions of their current situation, 56% of LBC indicated that learning was the most important aspect of their current life, followed by family relationships (28%) and friendships with peers (16%). However these aspects of life were both the source of their support and also the source of their dissatisfaction. When asked if LBC preferred to stay at school or at home, half preferred to stay at home and half preferred to stay at school. Each environment had particular benefits and drawbacks. At home, they felt freer, they felt emotionally supported, and they did not feel academic pressure. At school, with their fellow classmates and teachers, LBC reported feeling less lonely. Most LBC reported being dissatisfied with their current life because of interpersonal relationship problems, parent absence, and academic difficulties.

In the interviews almost all of LBC reported having at least one friend. LBC stated that friends made them feel happy. More specifically, friends helped each other with daily life issues and with learning activities; friends shared feelings; friends accompanied and encouraged each other when experiencing difficulties.

From LBC's point of view, teachers were very kind and supportive. Almost all LBC realized the important role teachers played in regard to students' academic guidance. Almost one-third of LBC said that their teachers not only took care of students' daily academic needs, but also taught them how to be 'human'.

4.4 Teachers′ perceptions of LBC′s education

According to information gathered from 32 teachers' interviews, schools have engaged in action to help LBC. Almost half of teachers (41%) reported that each term their schools archive records for LBC. Nearly one-third of teachers reported feeling an obligation to take responsibility for helping LBC in class and, beyond school, making more home visits for LBC. A few teachers (16%) reported that their schools provided a living subsidy for LBC. Additionally, one-fourth of teachers indicated that specific campus activities were developed for LBC, such as themed activities, safety and security education, and psychological counseling. Teachers also indicated that schools provided phone cards for LBC to promote LBC-parent communication.

In regard to how teachers educate LBC, almost half of teachers (44%) reported keeping close contact with students' parents (including grandparents). Teachers indicated that they 'communicate with left-behind students more' and take more responsibility for 'LBC's extracurricular life'. About one-third of teachers indicated that they paid extra attention to LBC. However, in opposition to the majority of teachers' input, one teacher emphasized that all students should be treated equally without discrimination or favoritism.

For most teachers, the greatest difficulty in the process of educating LBC is communicating with LBC's parents. Another major concern for teachers includes expressing concerns about LBC's challenging personal characteristics, such as being intractable and having a mistrustful attitude towards the teacher. Teachers expressed feeling powerless to address LBC's challenging attitudes and behaviors. Moreover, a few teachers pointed out the difficulty posed by the insufficient integration of and cooperation between the LBC's school education and family life.

Thus, based on these educational challenges, teachers also offered some advice. Nearly one-in-four teachers suggested that children should not be left behind, or minimally one parent should stay at home to take care of the children. A similar number of teachers reported the need for teachers to offer LBC more supportive care and encouraging praise. Teachers also stressed the importance of increasing communication between parents and teachers, as well as between parents and children.

Additionally, approximately one-in-five teachers reflected on the healthy development of LBC and the necessary ingredients to support this health development. Teachers cited the need for a combination of LBC's family support, adequate school education, and support from community/government welfare.

5 Discussion

In rural China, parents frequently migrate to urban areas in search of employment. This move often necessitates leaving their children behind. This study conducted in two junior high schools and two senior high schools found significant developmental differences between rural left-behind children (LBC) and non-LBC. LBC experienced less life satisfaction than non-LBC. LBC with both-parent migration report being more depressed and having a lower self-esteem than non-LBC. These findings indicated that rural parents' migration had an overall negative impact on LBC's mental health, particularly when both parents migrated. Findings based on this study's quantitative data (surveys) and qualitative data (interviews) are consist ent with previous studies (Jia & Tian, 2010; Sun, Zhou, Wang, & Fan, 2010; Zhou et al., 2005).

However, in contradiction with the previous findings of Wen and Lin's (2012) study, in the current study LBC reported a higher level of school engagement and adaptation. LBC reported enjoying school activities, doing their homework, and following school rules. This finding was somewhat unexpected. Possibly in this study LBC were more disadvantaged and underwent more stressful life events than non-LBC (Liu & Wang, 2010), and to counter these challenges, LBC may have had more intense motivation to do well in school, to earn good marks, and to become more involved in school activities. Additionally,

these findings may be related to the participating schools' quality of care provided by classroom teachers and the quality of peer relationships unique to the participating schools. These supportive qualities may not be similar across China's rural schools and communities. Therefore, data collected in different schools and communities may not align with the findings of this study.

5.1 Age and gender

The role of two variables—education level (age) and gender—was noteworthy, but an interaction between these two variables and migrant status was not found. However, in comparison to younger LBC (junior high school students), older LBC (high school students) reported better peer relationships and parent-child relationships. On the other hand, problem behaviors of high school students were significantly greater in comparison to younger LBC. This may be related to older LBC's enhanced self-awareness and increased academic frustration (Greene, Krcmar, Walters, Rubin, & Hale, 2000; Siegel & Scovill, 2000). Additionally, older students' school boarding life further weakened parental monitoring and also gave adolescents certain freedoms to independently navigate the cost of living. Moreover, with increasing age and awareness, adolescents naturally become more fully engaged and involved in the complexities of society. This independence and increased freedom might also underlie older LBC's escalating behavior problems (Shaffer & Kipp, 2010).

Gender differences were found in LBC's problem behavior and social relations. In comparison to girls, boys exhibited more problem behaviors, lower levels of school engagement, and worse peer relationships. On the other hand, boys reported better relationships with their fathers and higher levels if self-esteem. Overall, these gender differences were consistent with previous studies (Fan, Fang, Liu, & Liu,2009; Fan, Su, Gill, & Birmaher, 2010; Sun et al., 2010; Wen & Lin, 2012).

This study's findings were also related to parenting

style and social expectations in Chinese culture (Shaffer & Kipp, 2010). Therefore, educators should take into consideration the different developmental stages of youth in educational practice and the type of challenges which are specific to the gender of LBC.

5.2　Migrant status and caregivers

This study also found that different migrant conditions might cause different effects in LBC. Overall, LBC who were cared for by one parent fared better than those who were cared for by grandparent(s). LBC brought up by one parent had a more intimate relationship with their mother and reported higher levels of self-esteem than children brought up by grandparent(s). As previously indicated, more than 90% of LBC who were cared for by the remaining non-migrating parent were brought up by their mother. Additionally, previous research conducted by Zhao, Shen, and Liu (2008) found that mother's support had a positive influence on LBC's self-esteem.

LBC who were cared for by their mothers reported receiving more of their mother's support. The mother's care of LBC was positively related to the individual child's self-esteem and also with a stronger mother-child relationship. Therefore, in the process of raising a child the important role of parent (especially mother) is not easily replicated by a substitute caregiver, such as a grandparent.

5.3　Duration of separation and parental contact

Medium-term LBC had the highest score on life satisfaction and peer acceptance (social preference). Possibly, in comparison to short-term and long-term LBC, medium-term LBC may better cope with and adapt to separation from their parents. In addition, there was an interesting interaction between length of parental migration and caregiving type. For adolescents who were brought up by a grandparent, LBC who had a longer separation from the migrated parent actually reported better school engagement.

Generally speaking, the long-term LBC with both-parent migration (and who were brought up by grandparents) struggled with the most disadvantages and often appeared to show the weakest adaptation to healthy development. However, in this study, LBC who experienced long-term separation from their parents and who were brought up by a grandparent presented the highest school engagement. This might reflect a positive coping style of these LBC who seemingly demonstrate greater resilience in a school environment.

Among the left-behind characteristics investigated in this study, the frequency of parental contact had the broadest impact on LBC's adaption. Parental contact was beneficial to LBC's mental health. Children who had the most frequent contact with their parents suffered less with loneliness and depression and reported the highest life satisfaction and self-esteem. In regard to social relationships, fatherchild cohesion and mother-child cohesion were strongest in the group of LBC who had the most frequent contact with their parents. However, based on frequency of contact with LBC's parents, in regard to peer relationships, there was no significant difference between groups.

Sufficient contact with parents may also improve LBC's school engagement. Adolescents who rarely had contact with their migrated parents performed the worst on school engagement. These results were consistent with previous research (Su, Li, Lin, Xu, & Zhu, 2013). LBC's communication with parents and children's level of disclosure with parents appears to facilitate LBC receiving social support, which enhances healthy emotional adaptation (Chaudoir & Fisher, 2010). Su et al. (2013) also found that high levels of parental communication is one of the strongest protective factors of satisfaction with school and life in general, ultimately related to children's overall happiness and well-being. Thus, frequent contact with parents helps promote LBC's positive mental health and social adaptation.

In the children's interviews, based on the LBC's

perception of their own lives most children reported becoming alienated from their parents following their parents' migration for work. This alienation appears to have the greatest effect on children impacted by parental migration, and subsequently a direct cause of children's loneliness in their parents' absence. According to the 'ecological model of rural LBC's psychological development,' suggested by Zhao and Shen (2010), the proximal factor (such as single parent's migration or both parents' migration) and the distal cause/outcome (such as parent–children relationship) in left–behind environment were differentiated. The distal factors may, through certain proximal factors, contribute to LBC's development. As a result, in the condition where parents migrate for work and leave their children behind, interventions must focus on improving the LBC–parent relationship, helping enhance parent–child communication and building family cohesion.

5.4 School engagement and relationships

Ultimately, educational intervention efforts must focus on encouraging and promoting LBC to actively adapt to their challenging situation. Primarily, LBC's dissatisfaction in life was associated with academic pressure, problems in family relationships, and difficulties in peer relationships. Therefore, school educators and mental health workers should consider these three aspects as they identify school–based strategies to improve LBC's life satisfaction and academic progress. Interestingly, this research study's data indicate that half of LBC like staying at home, while the other half prefer staying at school where they can associate with their peers.

Home and school are the two most important places where children and adolescents develop relationships and personal skills related to life satisfaction. Indeed, peers play an important role in children's socialization. Even when bullied, children's loneliness can be alleviated by associating with a supportive friend (Sun et al., 2010).

Adolescents who had no 'best friend' reported feeling greater loneliness than those who had a 'best friend' (Zou, 1998). Research indicates that the quality of adolescents' friendship is highly correlated with increased self–esteem and sense of self–value, lower levels of depression and anxiety, and social adjustment (Berndt & Keefe, 1995; Hartup, 1996). Friends help care for LBC on a daily basis and accompany them during school and outside of school. Additionally, friends are available to assist LBC with school assignments. Many LBC reported being happy when they were together with friends. This study's findings were in line with previous research which found that good peer relationships help alleviate children's loneliness (Sun et al., 2010; Zhou, Zhao, Chen, Jiang, & Hundley, 2003). The compensation of peer relationship is very important, especially because these relationships act as a protective factor for LBC's adaptation to the school environment.

5.5 School support of LBC

All LBC expressed having a positive attitude towards their teachers. They considered their teachers to play an important role in guiding and improving student learning, providing a daily lift, and helping students develop good character. Above all, it is of critical importance for adolescents to develop emotional intimacy and warmth that ensures good communication between peers and with caring adults. Peers and teachers provide companionship for LBC and serve as guides in day–to–day school life. Based on this study's data, school is an important place for LBC's social development.

According to the teacher interviews, teachers' frontline role in education allowed them to understand the difficulties in educating LBC. The greatest challenge was in bridging the gap between the school's role in education and the family's role in education. For example, teachers reported having difficulty communicating with LBC's parents and perceived that LBC did not have sufficient parent support. Thus, teachers stressed the importance of increasing the communication between parents, teachers, and children. In this way, schools' and parents' combined efforts would better support LBC's educational and social-emotional needs, creating a school environment that nurtured LBC's learning and healthy development.

5.6 Limitations of current research

Considering that Cronbach's α was low for the measurement of School Engagement (Cronbach's α =0.46), we should explain whether it affects our confidence in interpreting data from this measurement. First of all, Cronbach's α is a special case of intra-class correlation coefficient (ICC), comparable to a two-way random model ICC (McGraw & Wong, 1996; Weir, 2005). Thus, when the correlation of items' correlation was low, would be weakened. School Engagement was a variable that measured the degree of students' school activity engagement (e.g. 'Do you like being obedient to school discipline?'). These events often had high incidence rate, so the distribution of these data was skewed, which weakened the item correlation. Second, although these items all belong to one category, each presented item may occur independently from the other items, therefore this traditional estimate of internal consistency may not have been an appropriate measure to judge this measure's reliability and whether these items measured what they purported to measure (Kim, Conger, Elder, & Lorenz, 2003). Previous researchers have also addressed similar concerns with using Cronbach's α (Kim et al., 2003; Straus & Kantor, 2005). Third, Cronbach's α typically increases with an increased number of items. In our research, there were only five items in this scale, thus limiting the scale's α.

Last but not least, Cronbach's α has been criticized as underestimating true reliability (Peterson & Kim, 2013). In fact, Turner and Wheaton (1995) declared that a measure may be valid even if α is zero. Thus we have reason to believe that the school engagement measure's low reliability is not an indication of the measure's validity.

However, we urge caution in drawing conclusions from this aspect of our data.

5.7 Implications for improving prevention and intervention efforts

In line with previous research, our findings support the need for assistance and intervention systems for LBC. Educators and parents must play an important role in providing interventions to address LBC's educational and social-emotional needs. The following seven interventions are recommended and further discussed in the following sections: (a) improve school leadership and management; (b) improve education on the school and classroom level; (c) strengthen school mental health education services; (d) build stronger coalitions between family and school; (e) provide a supportive and caring foundation; (f) support current and future education and (g) increase outside support of LBC. In order to bring about positive change, these recommended interventions should be integrated and complementary.

Improve school leadership and management. In view of the high proportion of left-behind students in rural schools, school leadership must consider and provide accommodations for LBC and strengthen the school's management to ensure the daily care and personal safety of LBC. In addition, according to the needs of some migrated family with good income, the school might consider charging a suitable fee to improve the school's boarding dietary standard and accommodations for children. Moreover, if it is possible, arrange life coaching teachers and teaching tutors to assist the management and education of LBC. In the long run, school administrators should pay close attention to the quality of education, try to improve teacher quality, and provide training opportunities (such as psychological counseling training). Strong academic courses, caring teachers, and a supportive school environment help LBC enjoy school and adapt to their challenging family circumstance.

Improve education on the school and classroom level. First, optimize the educational content. Increase the content such as agriculture, science, technology and other practical technology to help children grasp survival skills prior to graduating from junior high school. Second, adjust the course structure. Strengthen the LBC's survival, safety, and legal education, and improve their self-respect, self-reliance, discipline consciousness and legal knowledge. Third, enrich the extracurricular life of students, especially for the LBC. Entertainment activities increase peer interactions, and offer opportunities to strengthen shape healthy personality. Fourth, mobilize teachers, youth-league members, young pioneers and other students to help LBC by pairing LBC with supportive students to ease stress and encourage psychological adjustment.

Strengthen school mental health education services. As mental health and psychological well-being of LBC are often neglected (Luo, Wang, & Gao, 2009), it is necessary to publicly promote psychoeducational activities and to regularly provide psychological consultation activities. These efforts will be helpful in preventing potential psychological crises in disadvantaged LBC. It is important to establish a psychological counseling room and to encourage students to accept psychological counseling services which will help provide additional social support and enhance LBC's emotional adjustment. For children with serious psychological problems, psychological teachers should conduct tracking observations and offer supportive services to address challenging issues, track progress, and then summarize the effect of the relevant intervention measures.

Build stronger coalitions between family and school. Family supports (both from parents and the primary caregiver) are very important for LBC. Programs should be well designed to strengthen family ties before and following migration. We suggest offering supportive training for parents. Parent-teacher and parent-child communication is critical. Parents should strengthen the exchange and communication with their children and keep in touch with their children's temporary caregiver. Likewise, in a coordinated effort parents and caregivers must regularly communicate with LBC's teachers.

Provide a supportive and caring foundation. Remind parents, caregivers, and teachers to frequently offer LBC more supportive sentiment. Demonstrate concern for LBC's well-being and provide character education to further strengthen social relationships and social skills.

Support current and future education. Even though migrating parents are not present to give learning guidance during their absence, parents must always encourage their children's studying and build children's confidence in their ability to learn and study. Parents should avoid extreme reactions to their children's educational progress—negligence and limited expectations for the children's academic achievement or, on the other hand, exerting too much pressure for children to achieve unrealistic academic goals. Head teachers may take more responsibility for LBC's school involvement by providing special activities, establishing routine home visits, and holding teacher-parent meetings when parents are home during the holidays.

Increase outside support of LBC. Finally, it is worth noting that interventions organized and funded outside of schools are also very important in supporting LBC. Concrete actions must be taken with support from government organizations, community and private enterprises, universities, international communities, and so forth. For example, to compensate for the lack of national fiscal capital investment in rural schools, local governments may try to change the investment system of funding rural education, possibly running a school with subsidized funding from the community and government. One option, public welfare organizations can organize well-educated caring volunteers to set up a Chinese

Pioneer Union, providing strong academic courses and offering supervised activities for LBC, even extending beyond the typical school hours. Additional options include offering parent hotlines or a website with family counseling services. In order to improve resource utilization, these services can be set up in targeted local areas or expanded nationwide and offered to individuals from all walks of life. Additionally, with available internet and technology, LBC and migrated parent(s) should be able to communicate more frequently, countering LBC's feelings of loneliness.

6 Conclusion

Parental migration in China's rural areas appears to be a common situation for many adolescents (Duan & Zhou, 2005; Zhou et al., 2005). Additionally, the negative impact of parent migration on LBC's psychological development and social adaptation is extensive. Understanding the common challenges facing LBC, educators and parents must strategically identify and implement interventions to address these needs, fostering better outcomes for LBC.

References

Aguilera-Guzman, R. M., de Snyder, V. N., Romero, M., & Medina-Mora, M. E. (2004). Paternal absence and international migration: Stressors and compensators associated with the mental health of Mexican teenagers of rural origin. *Adolescence, 39*(156), 711 – 723.

Amato, P., & Keith, B. (1991). Parental divorce and the well-being of children: A metaanalysis. *Psychological Bulletin, 110*(1), 26 – 46. doi: 10.1037/0033-2909.110.1.26.

Asher, S. R., & Wheeler, V. A. (1985). Children's loneliness: A comparison of rejected and neglected peer status. *Journal of Consulting and Clinical Psychology, 53*, 500 – 505. doi: 10.1037/0022-006X.53.4.500.

Attili, G., Vermigli, P., & Roazzi, A. (2010). Children's social competence, peer status, and the quality of mother–child and father–child relationships. *European Psychologist, 15*(1), 23 – 33. doi: 10.1027/1016-9040/a000002.

Bagner, D. M., Storch, E. A., & Roberti, J. W. (2004). A factor analytic study of the Loneliness and Social Dissatisfaction Scale in a sample of African American and Hispanic American children. *Child Psychiatry and Human Development, 34*, 237 – 250. doi: 10.1023/B:CHUD.0000014999.16111.2f.

Baumeister, R. F., Campbell, J. D., Krueger, J. I., & Vohs, K. D. (2003). Does high selfesteem cause better performance, interpersonal success, happiness, or healthier lifestyles? *Psychological Science in the Public Interest, 4*, 1 – 44. doi: 10.1111/1529-1006.01431.

Beck, K. H., Boyle, J. R., & Boekeloo, B. O. (2003). Parental monitoring and adolescent alcohol risk in a clinic population. *American Journal of Health Behavior, 27*(2), 108 – 115. doi: 10.5993/AJHB.27.2.2.

Berndt, T. J., & Keefe, K. (1995). Friends' influence on adolescents' adjustment to school. *Child Development, 66*(5), 1312 – 1329. doi: 10.1111/j.1467-8624.1995.tb00937.x.

Bryant, J. (2005). *Children of international migrants in Indonesia, Thailand and the Philippines: A review of evidence and polices (Innocenti Working Paper No. 2005 – 05)*. Florence: UNICEF Innocenti Research Centre.

Carandang, M. L. A., Sison, B. A., & Carandang, C. (2007). *Nawala ang ilaw ng tahanan: Case studies of families left behind by OFW Mothers*. Pasig City: Anvil Publishing.

Chaudoir, S. R., & Fisher, J. D. (2010). The disclosure processes model: Understanding disclosure decision making and postdisclosure outcomes among people living with a concealable stigmatized identity. *Psychological Bulletin, 136*(2), 236 – 256. doi: 10.1037/ a0018193.

Chen, X., & Xie, Y. (2007). An investigation of rural left-behind children's problematic behaviors and the family factors. *Journal of Inner Mongolia Normal University (Philosophy and Social Science), 36*(6), 29 – 33.

Cronk, N. J., Slutske, W. S., Madden, P. A. F., Bucholz, K. K., & Heath, A. C. (2004). Risk for separation anxiety disorder among girls: Paternal absence, socioeconomic disadvantage, and genetic vulnerability. *Journal of Abnormal Psychology, 112*(2), 237 – 247. doi:10.1037/0021-843X.113.2.237.

Dawei, Y. (2000). Preliminary use of the children s depression inventory in China. *Chinese Mental Health Journal, 14*(4), 225 – 227.

DiClemente, R. J., Wingood, G. M., Crosby, R., Sionean, C., Cobb, B. K., & Harrington, K., et al. (2001). Parental monitoring: Association with adolescents' risk behaviors. *Pediatrics, 107*(6), 1363 – 1368. doi: 10.1542/peds.107.6.1363.

Dong, Q., & Lin, C. D. (2011). *The key index and assessment for mental development of 6 – 15 years old children and adolescents in China*. Beijing: Science Press.

Duan, C., Lu, L., Guo, J., & Wang, Z. P. (2013). Survival and development of left-behind children in rural China: Based on the analysis of sixth census data. *Population Journal, 35*(3), 37 – 49.

Duan, C. R., & Zhou, F. L. (2005). A study on children left behind. *Population Journal, 29*(1), 29 – 36.

Edillon, R. G. (2008). *The effects of parent's migration on the rights of children left behind*. Quezon City: Asia Pacific Policy Center, UNICEF.

Fan, F., Su, L., Gill, M. K., & Birmaher, B. (2010). Emotional and behavioral problems of Chinese left-behind children: A preliminary study. *Social Psychiatry and Psychiatric Epidemiology, 45*(6), 655 – 664. doi: 10.1007/s00127-009-0107-4.

Fan, X. H., Fang, X. Y., Liu, Q. X., & Liu, Y. (2009). A social adaptation comparison of migrant children, rear children, and ordinary children. *Journal of Beijing Normal University (Social Sciences), 5*, 33 – 40.

Fang, X. Y., Li, X. M., & Dong, Q. (1996). The research of adolescent smoking and its relevant factors. *Chinese Mental Health Journal, 10*(2), 77 – 80.

Fang, X. Y., Zhen, Y., & Lin, D. H. (2001). Relationships between family factors and smoking behavior of junior middle school students. *Acta Psychologica Sinica, 33*(3), 244 – 250.

Fredricks, J. A., Blumenfeld, P. C., & Paris, A. H. (2004, Spring). School engagement: Potential of the concept, state of the evidence. *Review of Educational Research, 74*(1), 59 – 109. doi: 10.3102/00346543074001059.

Gao, Y., Li, L. P., Kim, J. H., Congdon, N., Lau, J., & Griffiths, S. (2010). The impact of parental migration on health status and health behaviors among left behind adolescent school children in China. *BMC Public Health, 10*, 56 – 65. doi: 10.1186/1471-2458-10-56.

Greene, K., Krcmar, M., Walters, L. H., Rubin, D. L., & Hale, J. L. (2000). Targeting adolescent risk-taking behaviors: The contributions of egocentrism and sensationseeking. *Journal of Adolescence, 23*(4), 439 – 461. doi: 10.1006/jado.2000.0330.

Hartup, W. W. (1996). The company they keep: Friendships and their developmental significance. *Child Development, 67*(1), 1 – 13. doi: 10.1111/j.1467-8624.1996.tb01714.x.

Huebner, E. S. (1991). Initial development of the student's life satisfaction scale. *School Psychology International, 12*, 231 – 240. doi: 10.1177/0143034391123010.

Huebner, E. S., Suldo, S. M., & Valois, R. F. (2003, March 12 – 13). Psychometric properties of two brief measures of children's life satisfaction: The Students' Life Satisfaction Scale (SLSS) and the Brief Multidimensional Students' Life Satisfaction Scale (BMSLSS) Presented at the Indicators of Positive Development Conference, Washington, DC.Retrieved from http://www.childtrends.org/wp-content/uploads/2013/05/Child_Trends- 2003_03_12_PD_PDConfHSVP.pdf.

Hugo, G. (2002). Effects of international migration on the family in Indonesia. *Asian and Pacific Migration Journal, 11*(1), 13 – 46.

Jaggi, G., Rundle, M., Rosen, D., & Takahashi, Y. (1996). *China´s economic reforms: Chronology and statistics.* Working Paper 96 – 5. Washington, DC: The Peterson Institute for International Economics Retrieved from http://www.iie.com/publications/wp/96-5.pdf.

Jia, Z., & Tian, W. (2010). Loneliness of left-behind children: A cross-sectional survey in a sample of rural China. Child: Care, Health and Development, *36*(6), 812 – 817. doi: 10.1111/j.1365-2214. 2010.01110.x.

Joh, J. Y., Kim, S., Park, J. L., & Kim, Y. P. (2013). Relationship between family adaptability, cohesion and adolescent problem behaviors: Curvilinearity of circumplex model. *Korean Journal of Family Medicine, 34*(3), 169 – 177. doi: 10.4082/kjfm.2013.34.3.169.

Kim, K. J., Conger, R. D., Elder, G. H., & Lorenz, F. O. (2003). Reciprocal influences between stressful life events and adolescent internalizing and externalizing problem. *Child Development, 74*, 127 – 143. doi: 10.1111/1467-8624.00525.

Kovacs, M. (1992). Children's depression inventory manual. North Tonawanda, NY: MultiHealth Systems.Lin, D., Fan, X., Li, X., & Pan, J. (2010). The relationship between environmental and individual factors and adolescent drinking behavior. *Psychological Development and Education, 26*(3), 288 – 293.

Liu, B. Q., & Wang, W. (2010). Study on living stress events and psychological health of children remaining in rural areas. *China Journal of Health Psychology, 18*(2), 210 – 212.

Luo, J., Wang, W., & Gao, W. B. (2009). Review of the studies on rural left-behind children in China. *Advances in Psychological Science, 17*(5), 990 – 995.

McGraw, K. O., & Wong, S. P. (1996). Forming inferences about some intraclass correlation coefficients. *Psychological Methods, 1*(1), 30 – 46. doi: 10.1037//1082-989X.1.1.30.

McLanahan, S. S., & Sandefur, G. (1994). *Growing up with a single parent: What hurts, what helps.* Cambridge, MA: Harvard University Press.

Melgar, G. A., & Borromeo, R. (2002). The plight of children of OFWs. In E. An onuevo, & A. An onuevo (Eds), *Coming home: Women, migration and reintegration* (pp. 106 – 114). Quezon City: Balikbayani Foundation and Atikha Overseas Workers & Communities Initiatives.

Morrison, W. M. (2014, October 9). China's economic rise: History, trends, challenges, and implications of the United States. *CRS report prepared for the members and committees of Congress.* Washington, DC: Congressional Research Service. Retrieved from http://fas.org/sgp/crs/row/RL33534.pdf.

Olson, D. H., Sprenkle, D. H., & Russell, C. S. (1979). Circumplex model of marital and family systems: Cohesion and adaptability dimensions, family types, and clinical applications. *Family Process, 18*(1), 3 – 28. doi: 10.1111/j.1545-5300.1979.00003.x.

Peterson, R. A., & Kim, Y. (2013). On the relationship between coefficient alpha and composite reliability. *Journal of Applied Psychology, 98*(1), 194 – 198. doi: 10.1037/a0030767.

Qu, Z. Y., & Zou, H. (2009). Juvenile delinquency: The role of self-control, family environment and parental monitoring. *Psychological Science, 32*(2), 360‑363.

Reyes, M. M. (2008). *Migration and Filipino children left behind: A literature review.* Quezon City: Miriam College/UNICEF Retrieved from http://www.unicef.org/philippines/ Synthesis_StudyJuly12008.pdf

Rosenberg, M. (1965). *Society and the adolescent self-image.* Princeton, NJ: Princeton University Press.

Rosenberg, M. (1979). *Conceiving the self.* New York, NY: Basic Books.

Rosenberg, M., Schooler, C., Schoenbach, C., & Rosenberg, F. (1995). Global self‑esteem and specific self‑esteem: Different concepts, different outcomes. *American Sociological Review, 60,* 141‑156. doi: doi.org/10.2307/2096350.

Schmitt, D. P., & Allik, J. (2005). Simultaneous administration of the Rosenberg SelfEsteem Scale in 53 nations: Exploring the universal and culture-specific features of global self‑esteem. *Journal of Personality and Social Psychology, 89*(4), 623‑642. doi:10.1037/0022‑3514.89.4.623.

Shaffer, D. R., & Kipp, K. (2010). *Developmental psychology: Childhood and adolescence* (8th ed.). Belmont, WA: Wadsworth.

Shen, J. L. (2009). *Looking in depth of the mental world of disadvantaged children.* Beijing: Beijng Normal University Press.

Siegel, A. W., & Scovill, L. C. (2000). Problem behavior: The double symptom of adolescence. *Development and Psychopathology, 12,* 763‑793. doi: 10.1017/S0954579400004119.

Straus, M. A., & Kantor, G. K. (2005). Definition and measurement of neglectful behavior: Some principles and guidelines. *Child Abuse and Neglect, 29,* 19‑29. doi: 10.1016/j.chiabu.2004.08.005.

Su, S., Li, X., Lin, D., Xu, X., & Zhu, M. (2013). Psychological adjustment among leftbehind children in rural China: The role of parental migration and

parent‑child communication. Child: Care, *Health and Development, 39*(2), 162‑170. doi:10.1111/j. 1365‑2214.2012.01400.x.

Sun, X. J., Zhou, Z. K., Wang, Y., & Fan, C. Y. (2010). Loneliness of children left in rural areas and its relation to peer relationship. *Psychological Science, 33*(2), 337‑340.

Tarroja, M. C. H., & Fernando, K. C. (2013). Providing psychological services for children of overseas Filipino workers (OFWs): A challenge for school psychologists in the Philippines. *School Psychology International, 34*(2), 202‑212. doi: 10.1177/0143034312453399.

Tian, L. M. (2006). Shortcoming and merits of Chinese version of Rosenberg (1965) selfesteem scale. *Psychological Exploration, 26*(2), 88‑91.

Turner, J. R., & Wheaton, B. (1995). Checklist measurement of stressful life events. In L. Gordon (Ed.), *Measuring stress: A guide for health and social scientists* (pp. 29‑58). New York, NY: Oxford University Press.

Unnever, J. D., Cullen, F. T., & Travis, P. C., (2003). Parental management, ADHD, and delinquent involvement: Reassessing Gottfredson and Hirschi's general theory. *Justice Quarterly, 20*(3), 471‑500. doi: 10.1080/07418820300095591.

Wang, X. L., Hu, X. Y., Shen, J. L. (2011). Affection of left children's friendship quality on loneliness and depression. *Chinese Journal of Clinical Psychology, 19*(2), 252‑254.

Wei, X., & Chen, X. (2010). Prosocial tendencies and its relationship to personalities and family functioning among the left‑home and family functioning among the left‑home kids in junior middle school. *Psychological Development and Education, 4,* 402‑408.

Weir, J. P. (2005). Quantifying test‑retest reliability using the intraclass correlation coefficient and the SEM. *Journal of Strength and Conditioning Research,*

19(1), 231 - 240.

Wen, M., & Lin, D. (2012). Child development in rural China: Children left behind by their migrant parents and children of nonmigrant families. *Child Development, 83*(1), 120 - 136. doi: 10.1111/j.1467-8624.2011.01698.x.

Xiang, B. (2007). How far are the left-behind left behind? A preliminary study in rural China. *Population Space and Place, 13*, 179 - 191. doi: 10.1002/psp.437.

Yeoh, B. S. A., & Lam, T. (2007). The costs of (im)mobility: Children left behind and children who migrate with a parent. In *Perspectives on gender and migration* (pp. 120 - 149). In the United Nations Economic and Social Commission for Asia and the Pacific Regional Seminar on Strengthening the Capacity of National Machineries. Bangkok, Thailand: United Nations Economic and Social Commission for Asia and the Pacific (UNESCAP).

Yu, D. W., & Li, X. (2000). Preliminary use of the children's depression inventory in China. *Chinese Mental Health Journal, 14*(4), 225 - 227.

Zhang, W. W. (2004). *Transforming China: Economic reform and its political implications.* New York, NY: St. Martin's Press.

Zhao, J. X., Liu, X., & Zhang, W. X. (2013). Peer rejection, peer acceptance and psychological adjustment of left-behind children: The roles of parental cohesion and children's cultural beliefs about adversity. *Acta Psychologica Sinica, 45*(7), 797 - 810. doi: 10.3724/SP.J.1041.2013.00797.

Zhao, J. X., & Shen, J. L. (2010). The effect of pre-study or post-study emotional arousal on implicit and explicit memory. *Chinese Journal of Special Education, 7*, 65 - 70, 76.

Zhao, J. X., Shen, J. L., & Liu, X. (2008). Left-at-home-adolescents' perception of social support networks and their relevance to individual self-esteem and social initiative: Variable-centered and person-centered perspectives. *Psychological Science, 31*(4), 827 - 831.

Zhou, C., Wang, D. B., & Hong, Q. (2010). Influencing factors and related measures of sex abuse cases among left-behind children. *Medicine and Society, 23*(1), 54 - 56.

Zhou, Z. K., Sun, X. J., Liu, Y., & Zhou, D. M. (2005). Rural left-behind children's psychological development and education problems. *Academic Journal of Beijing Normal University, 1*, 71 - 78.

Zhou, Z. K., Zhao, D. M., Chen, J., Jiang, J. C., & Hundley, R. (2003). Loneliness as a function of sociometric status and self-perceived social competence in middle childhood. *Psychological Development and Education, 19*(4), 70 - 74.

Zou, H. (1998). Developmental function and influence factors of peer relationship. *Psychological Development and Education, 12*(2), 40 - 46.

School Psychology International, 2013, 34(6), 630 - 647.

Cyberbullying and its risk factors among Chinese high school students

Zongkui Zhou [1], Hanying Tang [1], Yuan Tian [1], Hua Wei [1], Fengjuan Zhang [1], Chelsey M. Morrison [2]

([1] Key Laboratory of Adolescent Cyberpsychology and Behavior (CCNU), Ministry of Education, Wuhan, China; School of Psychology, Central China Normal University, Wuhan, China)

([2] Wheaton College, USA)

Abstract Cyberbullying has become a common occurrence among adolescents worldwide; however, it has yet to receive adequate scholarly attention in China, especially in the mainland. The present study investigated the epidemiological characteristics and risk factors of cyberbullying, utilizing a sample of 1,438 high school students from central China. Findings revealed that cyberbullying among high school students in the heartland of central China is relatively common with 34.84% (N=501) of participants reported having bullied someone and 56.88% (N=818) reported having been bullied by online. Significant gender differences were found, suggesting that boys are more likely to be involved in cyberbullying both as perpetrators and victims. Students with lower academic achievement were more likely to be perpetrators online than were students with better academic achievement. Students who spend more time on online, have access to the internet in their bedrooms, have themselves experienced traditional bullying as victims, and are frequently involved in instant-messaging and other forms of online entertainment are more likely to experience cyberbullying. Increased parent and teacher supervision reduced students' involvement in cyberbullying. Implications for intervention are explored.

Keywords Chinese high school students, cyberbullying, intervention, risk factors

1 Introduction

With the popularization of the internet and development of information and communication technology (ICT), online communication has become a common mode of communication. With this change, social phenomena existing off-line have begun to appear online. Cyberbullying is an example of this trend, which has received increasing attention because of its potentially serious consequences and increasing prevalence. However, relatively little is known about cyberbullying (O'Keeffe,

Corresponding author: Zongkui Zhou, School of Psychology, Central China Normal University, Key Laboratory of Adolescent CyberPsychology and Behavior (CCNU), Ministry of Education, Wuhan, Hubei, People's Republic of China. Email: zhouzk@mail.ccnu.edu.cn

Clarke–Pearson, & Council on Communication and Media, 2011) especially in non–Western settings. Although there remains confusion about a standard definition for cyberbullying, most researchers agree that cyberbullying is an intentional, repeated, and aggressive act or behavior carried out by a group or individual instrumentally employing information and communication technology (ICT) (von Marees & Petermann, 2012). Studies show that the prevalence of cyberbullying is high in China; for example, research suggests that 34.9% of Chinese Taiwanese adolescents have been cyberbullied, and 20.4% had cyberbullied others (Huang & Chou, 2010; see Chen & Cheng [2013] for traditional bullying prevalence in a Taiwanese sample). Adolescents who have been bullied online are more likely to experience psychological problems (e.g. anxiety, depression) and engage in problem behaviors (e.g. skipping school, drug and alcohol use) (Beran & Li, 2005; Fosse & Holen, 2006; Juvonen & Gross, 2008; Mitchell, Ybarra, & Finkelhor, 2007; Wolak, Mitchell, & Finkelhor, 2006; Ybarra, Diener–West, & Leaf, 2007; Ybarra & Mitchell, 2007). In extreme cases, cybervictimization has been linked with suicide (Hinduja & Patchin, 2010). These factors have caused cyberbullying to attract attention as an important public health problem.

The majority of previous studies have focused on prevalence and types of cyberbullying. Studies that have systematically examined the risk factors for cyberbullying are rare and recent (e.g. Huang & Chou, 2010; Mishna, Khoury–Kassabri, Gadalla, & Daciuk, 2012; von Marees & Petermann, 2012). To date, most studies investigating risk factors for cyberbullying have focused on demographic variables(Li, 2006; Ortega, Elipe, Mora–Merchán, Calmaestra, & Vega, 2009; Wade & Beran, 2011; Wang, Iannotti, & Nansel, 2009), internet usage (Aricak et al., 2008; Erdur–Baker, 2010; Mesch, 2009; Navarro, Serna, Martínez, & Ruiz–Oliva, 2012; Smith et al., 2008; Topccu, Erdur–Baker, & Capa–Aydin, 2008; Twyman,

Saylor, Taylor, & Comeaux, 2010), and experiences of traditional bullying (Hinduja & Patchin, 2010; Kowalski, Morgan, & Limber, 2012; Li, 2008; Tokunaga, 2010; Wang, Iannotti, & Luk, 2012). There have been many contradictory findings, suggesting that more information is needed to clarify the relationship between these factors and cyberbullying. For example, some studies determined that boys are more likely to be involved in cyberbullying as perpetrators, whereas girls are more likely to be cybervictims (Li, 2006; Ortega et al., 2009; Wade & Beran, 2011; Wang, Iannotti, & Nansal, 2009). However, other studies show that boys are more likely to be involved in cyberbullying as *both* bullies and victims (Aricak et al., 2008; Nansel et al., 2001). In addition, Li (2007) found that students with lower academic status were likely to become cyberbullies, whereas Ma (2001) argued that these students were likely to become victims. It is essential to have a thorough understanding of the risk factors to inform intervention and prevention strategies.

To date, very little is known about cyberbullying among adolescents from the Mainland of China, as the majority of existing studies were conducted in Chinese Taipei (the capital of Taiwan, as it is known in the West) (e.g. Hokoda, Lu, & Angeles, 2006; Huang & Chou, 2010; Wei, Jonson–Reid, & Tsao, 2007). The China Internet Network Information Center (CINIC, 2012) reported that the total number of Chinese *netizens* (or 'internet citizens') reached 537.6 million in June 2012, with adolescents (10– to 19–years–old) accounting for 25.4%. The popularity of internet use among Chinese adolescents makes cyberbullying a social phenomenon deserving of attention.

Culture is a strong predictor for both cyberbullying and cybervictimization (Li, 2007, 2008). Huang and Chou (2010) found that Chinese Taiwanese students usually took no action after being victimized online because of a cultural imperative to avoid conflict so as to maintain group harmony; further, that study found no relationship between cyberbullying and academic achievement. Cultural differences between the West and general Chinese culture may account for some of the differences observed in the empirical literature (e.g. frequency, attitude, setting, motive). Therefore, conclusions about cyberbullying from studies utilizing samples influenced by Western cultures may not be generalizable to Chinese culture. These findings are a reminder that it is essential to examine cultural factors related to the development, presentation, and intervention of cyberbullying.

Due to the limited number of studies on cyberbullying in mainland China, the present study aimed to clarify epidemiological characteristics and risk factors among Chinese mainland high school students with an emphasis on cultural differences between China and Western countries. We also sought to explore effective prevention and intervention measures of cyberbullying.

2 Methods

2.1 Participants

Participants were 1,438 students (42.56% female, 57.44% male, mean age=15.91, SD=1.02) from central China. The grade level of participants ranged from grad 10 to grade 12 (43.91%, 10th grade; 39.81%, 11th grade; 16.28%, 12th grade).

2.2 Measures

Students anonymously completed a survey which included questions on: Demographics; internet usage; Cyberbullying Inventory (CBI); traditional bullying scale; motivation for cyberbullying; and parents' and teachers' supervision of internet usage. With regard to internet usage, participants were asked how long they had been internet users; how many hours they spend online per week; what device they use to access the internet; and where they access the internet. They were also asked how often they engaged in each of the following internet activities: Instant messaging; visiting social networking sites; emailing; phone messaging; online entertainment (e.g. listening to streamed music, watching video); searching for information; playing online games; and online shopping. Answers were provided on a five-point scale ranging from 1=Never to 5=Always (where 1 to 3 was coded as low frequency; 4 to 5, high frequency). Finally, students were asked to indicate whether or not they had a mobile-phone and whether or not it could access the internet.

Students' experiences of cyberbullying were assessed by a Chinese-language questionnaire based on the Cyberbullying Inventory (CBI) (Erdur & Kavsut, 2007) which included two forms: *Cyberbullying* (CB) and *cybervictimization* (CV). Both forms consisted of 18 items that described different forms of CB or CV. Two items were added to the cybervictimization form so that both forms would have an equal number of items. Items described experiences such as sending/ receiving hurtful emails and making/receiving threats in a chat room. In this study, internal consistency of the CB and CV scales was 0.88 and 0.90, respectively.

The traditional bullying and victimization scale was developed by Li, Zhang, and Yu (2012). This scale consists of six items including, 'people make-up cruel nicknames about me', 'people scold me', or 'people tease and mock me', 'students hit, kick, punch or threaten me', and 'students spread rumors about me and try to make others not like me'. The inverse of each of these questions was asked to assess bullying behavior (i.e. 'I/we make-up cruel nicknames for other people', etc.). Each item was assessed using a five-point likert scale. In this study, internal consistency of traditional bullying and victimization was 0.80 and 0.69, respectively.

Eight major motives for cyberbullying were investigated, which had been identified from prior studies. Motives included 'for fun', 'to vent', 'because I dislike someone', 'for revenge', 'out of boredom', 'because it looks cool', 'to attract someone's attention', and 'for other benefits'. Only participants who had already endorsed the statement that they had bullied others online were asked to answer this portion of the questionnaire. Participants could choose more than one motive based on their own experiences. Victims' reactions to cyberbullying were also collected using multiplechoice questions with several options, such as 'ignore/don't care', 'talk about the experience with someone for help', and 'seek revenge on people who hurt me'.

In order to examine the relationship between academic achievement and cyberbullying, we collected information on participants' academic achievement. Due to the difficulty involved in obtaining official academic records for every participant, academic achievement data was obtained only through self-report (a method that has been used by other researchers of Chinese cyberbullying; e.g. Huang & Chou, 2010). Students reported their academic achievement on a three-point scale (1='above average', 2='average', 3='below average').

To assess parental supervision and restriction of internet usage, three questions were asked: 'Do your parents supervise your online activities?', 'Do your parents control your online activities?', and 'Do your parents control your use of your mobile-phone at home?'. The same three questions were asked in reference to teachers' supervision. Additionally, two additional multiple-choice questions were asked to identify parents' methods for controlling children's online activities (e.g. install software to prevent access to some websites, install software to monitor online activities, check history of visited sites, control/limit time online).

2.3 Procedure

Data were collected during Summer 2012 in classroom-settings by trained graduate students and teachers. Information about the purposes of the survey and confidentiality of responses was clearly explained and assent was obtained. Students were informed that there was no right or wrong answer. Extra support was provided for students who had difficulty completing the surveys.

3 Results

3.1 Prevalence rate of cyberbullying

In this sample, 34.84% (N=501) of participants reported that they had bullied someone online and 56.88% (N=818) reported that they had been bullied by someone online within the last semester (one semester continues for about five months in China); in addition, 26.84% (N=386) reported being *both* cyberbullies and cybervictims. A total of 37.34% (N=537) of participants reported that someone in their class had cyberbullied, and 40.33% (N=580) reported that someone in their class had been cybervictimized.

Our study mirrored previous studies that found varying prevalence rates among different types of cyberbullying. Specifically, 'to kick out someone from a chat room' (17.94%), 'to insult someone in a chat room' (12.80%), and 'to spread private information discussed by instant messaging tools (e.g. QQ/MSN)' (7.79%) were the three most frequently reported cyberbullying behaviors in our study. 'Someone stole my passwords of my ICM so that I cannot use them anymore' (24.13%), 'someone kicked me out from a chat room' (22.81%), and 'someone stole my network account and accessed my personal information' (15.09%) were the three most frequently reported forms of cybervictimization.

3.2 Gender differences

Our study found that boys (39.59%) were more likely to report that they were involved in cyberbullying as perpetrators than girls (28.43%), χ^2 (1, N=1438)=19.28, p<0.0001; boys (59.81%) were also more likely to be cybervictims than girls (52.94%), χ^2 (1, N=1438)=6.76, p<0.01. Regarding the relationship between gender and

Table 1. Comparisons of the gender differences on different forms of cyberbullying

Behavior	As a perpetrator			As a victim		
	Boys N=826	Girls N=612	χ^2 df=1	Boys N=826	Girls N=612	χ^2 df=1
Send hurtful e-mails	3.1%	1.0%	7.59**	12.7%	9.6%	3.28
Hiding the name via sending SMS	2.2%	0.8%	4.14	6.8%	5.1%	1.82
Take embarrassing photos	9.1%	4.2%	12.57***	14.6%	8.0%	14.88***
Threaten in chat room	3.5%	1.0%	9.48**	6.3%	4.2%	2.87
Spread rumors	4.2%	1.8%	6.76**	13.3%	9.6%	4.58*
Spread information on SMS	8.0%	7.5%	0.11	10.5%	10.3%	0.02
Kick out from chat room	18.9%	16.7%	1.18	25.5%	19.1%	8.25**
Humiliate by using fake photos	6.7%	1.6%	20.56***	6.9%	3.1%	10.12**
Send hurtful SMS	5.2%	2.6%	6.00*	6.4%	4.7%	1.84
Insult in chat room	17.1%	7.0%	31.78***	16.5%	6.7%	31.06***
Get information without permission	2.4%	0.7%	6.69**	13.6%	12.4%	0.40
Block access by stealing password	2.4%	0.3%	10.24**	11.4%	6.9%	8.38**
Reach messages by stealing passwords	1.5%	0.3%	4.62*	10.0%	6.9%	4.50*
Block from using SMS	2.3%	0.3%	9.51**	26.0%	21.6%	3.82
Use other's username without permission	1.6%	0.2%	7.25**	18.8%	10.1%	20.45***
Violate privacy via webcam	1.1%	0.3%	2.70	3.4%	2.1%	2.03
Humiliate by create Web pages	1.1%	0.2%	4.37	2.7%	0.5%	9.72*
Harm someone known from Internet	5.3%	2.0%	10.64**	7.7%	6.7%	0.57

Note: *** $p<0.001$, ** $p<0.01$, * $p<0.05$

types of cyberbullying or cybervictimization, more boys than girls reported cyberbullying someone using each type of bullying behavior measured with the exception of five types (e.g. 'To spread the information without permission', 'To kick out someone from a chat room', and 'To harm someone you know from the internet') (See Table 1). However, male-cybervictims reported experiencing only about half of all 18 types; these results suggest that the gender differences are weaker among cybervictims. In addition, we also found that' to kick someone from a chat room' was the most frequent bullying behavior perpetrated by both boys and girls online, and 'someone stole my passwords of my ICM so that I cannot use them anymore' was the most frequent type of victimization experienced by both boys and girls online.

3.3 Academic achievement differences

The relationship between academic achievement

difference and participants' experience of cyberbullying was examined using a one-way ANOVA analysis. As presented in Table 2, there were significant differences found between the three academic achievement groups ($F=3.89$, $p<0.05$). Multiple comparisons indicate that compared to students who perform better academically, the students with lower academic achievement were more likely to be online perpetrators. However, there were no differences found between the three academic-achievement groups and experiences of cybervictimization. Therefore, lower academic achievement maybe a potential risk factor for those engaging in cyberbullying as perpetrators.

3.4 Internet usage and cyberbullying

Descriptive data of internet usage was collected. The mean age of when the participants originally gained access to the internet was 5.60 years (SD=2.58). Students reported

Table 2. Cyberbullying experience scores of the three academic-achievement groups

Cyberbullying experience	Above average N=756		Average N=383		Below average N=253		
	Mean	SD	Mean	SD	Mean	SD	F
Cybervictims	1.12	0.23	1.16	0.35	1.14	0.25	2.66
Cyberbullies	1.03	0.12	1.07	0.20	1.07	0.23	3.89*

Note: * $p<0.05$

spending time accessing the online world between one and two days each week (1.87 days; SD=1.77), with 2.13 (SD=2.06) hours per day, on average. This means that on average, students spend about 4.82 hours per week on the internet. Correlation analysis showed that both cyberbullying and cybervictimization were significantly related to total time spent online each week (r=0.22. p<0.01; r=0.19, p<0.01). Time spent online appears to be an important factor for predicting cyberbullying.

For technology use, 71.42% of participants (N=1027) reported having a mobile-phone, and among them 76.73% (N=788) reported that they could access the internet by phone. More than four-in-five of the youth (84.77%) reported having access to the internet by computer, 58.97% used a mobilephone, and 16.41% accessed via tablet/pad computer. Additionally, 83.80% of participants reported having access to the internet at home; 17.45% reported accessing the internet at a commercial setting (e.g. 'coffee-shop'), while 13.70% reported access at school and 9.39% at the home of friends. (As participants could choose more than one option, the totals reported here do not equal 100%).

When asked how often they were involved in any of several major online activities using a five-point scale (from1=never to 5=always), results revealed that participants often spent their time on searching for information (M=3.84), communicating via instant messages (M=3.79), and online entertainment (M=3.61). They were less frequently involved in social networking sites (M=3.13), phone messaging (M=2.93), and online game playing (M=2.55). They were rarely involved in sending emails (M=1.95) or shopping online (M=1.74). Further analyses were run in order to gain a deeper understanding of the relationship between the three online activities most frequently endorsed by participants and their experience of cyberbullying. Students were classified into two groups: Students who endorsed 'often' and 'always' on these items

were classified as Group 1 (often involved in), and all others were classified as Group 2(rarely involved in). The results of t-test showed that students who extensively used instant messaging were more likely to be experienced both as cyberbullies and cybervictims (t=3.46, p<0.001; t=3.73, p<0.0001) and that students who were often involved in online entertainment were more likely to experience cybervictimization (t=2.21, p<0.05).

3.5 Parents and cyberbullying

With regard to parental supervision, 45.83% of the participants reported having their internet usage monitored by parents. The extent of parental restriction was measured on a five-point scale, showing that only 8.41% of parents fail to restrict their child's internet usage, and 73.50% of parents restrict children's internet usage at least a moderate level. T-tests examined the differences in parental restriction between perpetrators and non-perpetrators, as well as victims and non-victims. Non-perpetrators reported that their parents were more restrictive of their internet usage than were perpetrators' parents (t=2.65, p<0.05). No significant differences in parental restriction were found between victims and non-victims (t=0.71, p>0.05).

With regard to specific restriction strategies, 71.77% of youth reported that their parents implemented rules to limit the amount of time they are allowed to spend online, 7.79% reported their parents had installed filtering software to block specific web sites, 6.61% reported that their parents check their browsing history, and 2.36% reported that their parents had installed monitoring software to record online activities.

Participants were asked about the location where they accessed the internet at home: 544 participants (37.83%) reported having access to the internet in their own bedroom and 42.91% of youth reported only having internet access in common areas of their house. Chi square analyses showed that both cyberbullies x^2 (1, N=544)=21.44;

$p<0.0001$) and cybervictims (x^2 (1, N=544)=36.03, $p<0.0001$) were more likely to have access to the internet in a private space at home than were non-cyberbullies and noncybervictims.

3.6 Educators and cyberbullying

This survey also showed that 70.45% of participants reported being monitored by teachers when using the internet at school—suggesting that Chinese teachers have a cautious attitude about students' online activities. The results of t-tests indicate that both perpetrators and victims reported less restriction by teachers while engaging in online activities at school than non-perpetrators and non-victims (t=−2.31, $p<0.05$; t-=2.06, $p<0.05$).

There are limited opportunities for high school students in China to access the internet using a computer while at school; however, as 58.97% of the participants reported having internet access using their mobile-phones, it can be speculated that students use their phones to access the internet while at school. Therefore, we further examined the relationship between teachers' restriction on students' phone usage and students' experiences of cyberbullying. Results indicated that perpetrators reported their mobile-phone use as being less restricted by teachers than non-perpetrators (t=−2.03, $p<0.05$). However, there was no significant difference in teachers' restrictions of phone use between victims and non-victims.

3.7 Traditional bullying and cyberbullying

A significant positive correlation was found between traditional bullying and cyberbullying (see Table 3). To further examine the relationship between traditional and cyberbullying forms, we analysed the overlap between traditional and cyberbullying using the method utilized by Kowalski, Morgan, and Limber (2012). Participants were classified into four groups: Victims only; bullies only; bully/victims; and neither bullies nor victims. Separate categories were created for cyberbullying and traditional bullying. Table 4 presents the overlap between involvement in traditional bullying and involvement in cyberbullying. Participants who were cyberbullies/victims were likely to be involved in traditional bullying as both perpetrators (53.11%) and as victims (68.13%). Similarly, individuals who were bully/victims in traditional bullying were likely to be both bullies (50.49%) and victims (76.83%) online. Individuals who were cybervictims were

Table 3. Correlation matrix between traditional and cyberbullying

	Traditional victimization	Traditional perpetration	Cyber victimization	Cyber perpetration
Traditional victimization	1			
Traditional perpetration	0.48**	1		
Cyber-victimization	0.35**	0.35**	1	
Cyber-perpetration	0.29**	0.44**	0.58**	1
Mean	1.22	1.13	1.13	1.06
SD	0.36	0.31	0.27	0.17

Note: ** $p<0.01$

Table 4. Overlap between traditional and cyberbullying

Cyberbullying status	Traditional victims	Traditional bullies	All participants
Victim	233(53.94%)	146(33.80%)	432(30.04%)
Bully	47(40.87%)	28(24.35%)	115(8.00%)
Bully/Victim	263(68.13%)	205(53.11%)	386(26.84%)
Neither	160(32.32%)	95(19.19%)	495(34.42%)
Traditional status	Cyber victims	Cyber bullies	All participants
Victim	179(61.72%)	101(34.83%)	290(20.17%)
Bully	34(56.67%)	24(40.00%)	60(4.17%)
Bully/Victim	315(76.83%)	207(50.49%)	410(28.51%)
Neither	258(41.41%)	150(24.08%)	623(43.32%)

also likely to be victims of traditional bullying (53.94%). Both bullies (56.67%) and victims (61.72%) of traditional bullying were themselves likely to be victims online. This provides supporting evidence for the predictive effects of traditional bullying on cyberbullying.

3.8 Motives for cyberbullying

We also explored motives for cyberbullying. Based on the previous literature, several types of motives were investigated (Kowalski, Limber, & Agatston, 2012). A frequency distribution showed the following motives as most commonly reported by cyberbullies (N=501): 'I dislike someone' (29.14%); 'for fun' (23.95%); 'out of boredom' (19.16%); 'to vent' (15.77%); 'to get revenge' (7.58%);'to conform/fit in' (5.59%), 'to attract his/her attention' (3.19%); 'it looks cool' (1.60%); and 'to get some other benefit' (1.60%).

3.9 Students´ reactions to cybervictimization

Students may react to cybervictimization in different ways. This survey showed that the most frequent reaction of participants who had experienced cybervictimization was to 'ignore/not react' (45.84%) followed by 'talking about the experience with someone for help' (35.57%). Other outcomes less frequently cited were 'delete the materials which may hurt me' (32.27%), 'change my online account' (24.69%), and'seek revenge on people who hurt me online' (11.86%). Participants who reported'talking about the experience with someone for help' most frequently wanted to talk with classmates/friends (65.64%), followed by parents (28.87%) and siblings (27.84%). Only 2.75% reported they would talk with their teachers for help.

4 Discussion and implications

There are few studies examining the prevalence of cyberbullying in China, especially in the mainland. Li (2008) utilized a study of 197 Chinese students and found that 33% were cybervictims and 7% were cyberbullies.

Another study from Chinese Taipei showed that 34.9% of 545 participants had been cyberbullied, 20.4% had cyberbullied others, and 63.4% reported having witnessed or being aware of cyberbullying (Huang & Chou, 2010). As the majority of the students who participated in these studies were from middle schools, the prevalence rate of cyberbullying among Chinese high school students remains unknown. The results of the present study provide preliminary data indicating that high school students in mainland China are frequently involved in cyberbullying. Studies from Western countries have examined the prevalence rate of cyberbullying among high school students with varying results (von Marees & Petermann, 2012). Based on his review of 14 studies conducted on cyberbullying in Australia, the USA, the UK, and Canada, Kraft (2006) summarized that reported levels of cybervictimization varied between 10% and 42%, and that rates of cyberbullying varied from 6% to 33% (with 11.5% recently confirmed in Australia by Sakellariou, Carroll, & Houghton, 2012).

Many researchers have argued that these discrepancies may be due to differing definitions of cyberbullying utilized by assessment instruments, the age-range of participants, and the timeframe of participants' response (e.g. Kowalski, Limber, & Agatston, 2012). Some researchers have posited that involvement in cyberbullying as perpetrators or victims increases from the age 10- to 16-years-old (von Marees & Petermann, 2012). In the present study, the mean age of participants is 15.93 (SD=1.02), which is very close to 16-years-old. In accordance with the age trend mentioned above, our participants may be more likely to be involved in cyberbullying. Nevertheless, the prevalence rates of cyberbullying revealed in our research are higher than the upper limits summarized by Kraft (2006). Therefore, this suggests that cyberbullying among high school students in China is relatively common and should attract the attention of parents, educators, and public society.

Consistent with the findings of previous studies from Western countries, the most common venues for victimization by cyberbully perpetrators were chat rooms (18.0%) and instant messaging (8.5%) (Hinduja & Patchin, 2008; Kowalski & Limber, 2007; Patchin & Hinduja, 2006). Within these venues, 23.9% of the respondents reported having their ICM passwords stolen and 22.8% reported having been kicked out from chat rooms. This study also found that instant messaging (e.g. QQ/MSN) was one of the most common activities that students participated in online. Among those who were bullied online (N=818), 41.08% claimed that they were aware of the bully's identity. More than one-half (55.06%) of these students reported that the perpetrators were their classmates. This finding may provide support for existing studies that have investigated the anonymity of cyberbullying (Huang & Chou, 2010; Juvonen & Gross, 2008). The proportion of classmates being perpetrators should attract educators' attention in framing prevention and intervention activities.

4.1 Risk factors for cyberbullying among high school students

With regard to gender differences, we found that boys were significantly more likely to be cyberbullies or cybervictims. Although these results are consistent with findings from the traditional bullying literature, they are not aligned with results from early research on cyberbullying. Previous studies have found that girls were more likely to be cybervictims (Smith et al., 2008; Wang et al., 2009) or have found no significant gender differences between cybervictims (Hinduja & Patchin, 2008; Slonje & Smith, 2008; Williams & Guerra, 2007). Our results are supported by the work of Olweus (2003). In a sample of Chinese adolescents from Taipei, Huang and Chou's (2010) findings mirror our own, with male students reporting greater levels of both perpetration and victimization experiences than females. This suggests that these findings

may be due to cultural differences. In traditional Chinese culture, girls are raised to be gentle, polite and kind, while boys are encouraged to be active, brave, and independent. Boys are told that it is not brave to be aggressive towards or even bully girls. This may lead to fewer girls being involved in bullying, whether offline or online.

Our results confirmed that the frequency of internet access may be another risk factor for cyberbullying. Our results are consistent with finings from previous that youth who use the internet more frequently and spend more time online per day may be more likely to become involved in cyberbullying as perpetrators or victims (Mishna et al., 2012; Navarro et al., 2012; Wolak, Mitchell, & Finkelhor, 2007). Chi square analyses showed that having internet access by phone or accessing the internet at a commercial location increased student's risk of involvement in cyberbullying. Descriptive results showed that almost three-quarters of the youth had a phone, and nearly one-fifth of the youth had access to the internet commercially. This implies that youth are more likely to become involved in cyberbullying in unsupervised spaces. As more youth have access to the internet using mobile-phones, additional attention is warranted regarding youths' phone usage. Our results indicate that some types of online activities increase the odds of involvement in cyberbullying. Specifically, the more often students are involved in instant messaging, online entertainment, and information searches, the more likely they are to become involved in cyberbullying. Instant messaging may be an activity in which Chinese high school students are the most likely to experience cyberbullying (Huang & Chou, 2010). In the USA, socialnetwork sites and chat rooms have served as fertile ground for cyberbullying (Mesch, 2009). Although social networking sites are not yet as popular in China as in Western countries, nearly half of Chinese youth (42.98%) report frequently visiting social networking sites. This number is expected to increase which may expose youth to additional risks.

Our study revealed a relationship between traditional bullying and cyberbullying which mirrors the findings of previous studies (Hinduja & Patchin, 2010; Kowaski, Morgan, & Limber, 2012; Li, 2007). Students who report involvement in traditional bullying were found to be at a greater risk for involvement with cyberbullying. We speculate three possible reasons for this result. First, previous studies have suggested traditional bullying and victimization are related to personal traits. Specifically, traditional bullies are typically emotionally impulsive, irritable, and lacking in self-control; in contrast, traditional victims are typically introverted, sensitive, and easily restrained (Zhang, Gu, & Ju, 2001). We speculate that students with these personal traits are also likely to experience bullying online. Second, because bullies and victims often know each other in real life, the social interaction that usually occurs at school may be extended online, along with their status of bully or victim (Kowaski, Morgan, & Limber, 2012). Third, because of the power imbalance between perpetrators and victims, victims may be unable to fight back or transform his/her role in traditional bullying. However, the anonymity of the internet makes this transformation easier and may give traditional victims the courage to counterattack. Therefore, the victims of traditional bullying may become cyberbullies (40.85% in our study); consequently, traditional bullies transition into cybervictims (61.80% in our study). This finding sheds light on the similarities between traditional and cyberbullying while highlighting one of the key differences. Therefore, we contend that there should be a different approach to prevention and intervention strategies for traditional bullying and cyberbullying.

The present study revealed that a relatively high proportion of Chinese parents (73.50%) place at least a moderate level of restriction on their children's internet usage. This suggests that Chinese parents are broadly aware of the risk involved in their children's online activities. Consistent with previous studies (Mesch, 2009; Navarro et al., 2012), our study found that parental restriction reduces the risk of children's involvement in cyberbullying. However, parental restriction was not found to be significantly related to children's experience of cybervictimization. This suggests that although parental restriction of children's internet usage may effectively reduce the risk of children perpetrating online, it may not protect children from being bullied online. This may be due to the unique feature of cyberbullying—that is, unlike traditional bullying, cyberbullying can occur anytime, anywhere (Li, 2008; Tokunaga, 2010). Thus, even though someone may rarely access the internet, he/she may still be vulnerable to online victimization.

The most important agenda for high school students in China is to prepare for their college entrance examinations. In order to achieve a high score in this critical watershed, they must spend almost all of their spare time studying, leaving little time for leisure activities. It is typical for Chinese high school students to spend more than ten hours daily in study. Going online while at school is virtually impossible; our survey revealed that 70.45% of participants reported that their online activities were closely monitored by teachers at school. It is logical to conclude that teacher-supervision and restrictions on students' online behaviors can reduce cyberbullying. While restricting students' computer use at school may be relatively easy, restricting their access to the internet using portable devices may is more complicated. We conclude that teachers' interventions for cyberbullying should include targeted measures to guide students' usage of mobile devices.

4.2 Implications for students

As adolescents are the direct participants of cyberbullying, prevention, and intervention efforts should help them better understand cyberbullying. Our survey indicated that more than two-thirds (69.14%) participants believe that the harm brought to others through cybervictimization is only 'moderate' or 'minimal'. Further, the two most common motives for cyberbullying were 'for fun' and 'out of boredom', indicating that adolescents know little about the seriousness of their online behaviors. Therefore, it is not surprising that the prevalence of perpetrating reported by the participants was relatively high (34.84%). We speculate that if adolescents better understood the potential consequences of their online behaviors, the prevalence of perpetration would decrease.

4.3 Implications for parents

Our results indicate that while three-quarters of parents limit students' time online, relatively few use technological strategies such as installing online-filters or monitoring-software. Though this strategy did have some positive effects preventing cyberbullying among Chinese high school students, it was rudimentary and limited by comparison to Western countries. For example, Mesch (2009) investigated 935 US youth and found that 56% reported that their parents had installed filtering software and had rules on the type of information they were allowed to share over the internet. In addition to direct restriction on internet usage, other studies describe group interventions which involve educating students and creating rules about what personal information is appropriate to share (Navarro et al., 2012). In our study, 83.80% of participants reported having internet access at home. Further, youth who had internet access in the private space of their bedrooms were more likely to be both cyberbullies and cybervictims. In addition to controlling children's time online, Chinese parents may more effective if they intervened by placing the computer in relatively public spaces.

Chinese parents need to better understand cyberbullying; as they gain understanding they can help their children become more aware of the possible negative consequences of online activities. Our study indicated that one-in-three cyberbullied students were willing to talk about their experiences with parents—suggesting that parents could be important supports for children who experience cybervictimization; parents should know how to recognize the signs of cyberbullying and hat to do when they suspect cyberbullying.

4.4 Implications for educators

Supervision and restriction on students' online behaviors in school can effectively reduce cyberbullying (Cassidy, Brown, & Jackson, 2012); however, as mobile devices with internet access become increasingly available, supervision and restriction on students' online access may not be sufficient. The internet is an important part of modern everyday lives; even without computers, youth can go online by using portable devices. The most effective intervention for cyberbullying may be proper guidance. Specifically, schools should re-evaluate their methods of supervision and equip educators to better understand cyberbullying. Although only 4.90% of the cybervictims reported they would talk to their teachers about their experiences, this does not excuse educators from the role they should play to intervene in cyberbullying.

References

Aricak, T., Siyahhan, S., Uzunhasanoglu, A., Saribeyoglu, S., Ciplak, S., Yilmaz, N., & Memmedov, C. (2008). Cyberbullying among Turkish adolescents. *Cyberpsychology and Behavior, 11*(3), 253 - 261. doi:10.1089/cpb.2007.0016.

Beran, T., & Li, Q. (2005). Cyber-harassment: A study of a new method for an old behaviour. *Journal of Educational Computing Research, 32*(3), 265 - 277.

Cassidy, W., Brown, K., & Jackson, M. (2012). 'Under the radar': Educators and cyberbullying in schools. *School Psychology International, 33*(5), 520 - 532.

doi:10.1177/0143034312445245.

Chen, M. L., & Cheng, Y (2013). Prevalence of school bullying among secondary students in Taiwan: Measurements with and without a specific definition of bullying. *School Psychology International*. doi:10.1177/0143034313479694 (in press).

China Internet Network Information Center (CINIC) (2012). 30th China Internet Development Statistics Report. Retrieved August 25, 2012, from http://www.cnnic.cn/ hlwfzyj/hlwxzbg/hlwtjbg/201207/P020120723477451202474.pdf.

Erdur-Baker, O ., & Kavsut, F. (2007). Cyber bullying: a new face of peer bullying. *Eurasian Journal of Educational Research, 27*: 31 - 42.

Erdur-Baker, O. (2010). Cyberbullying and its correlation to traditional bullying, gender and frequent and risky usage of internetmediated communication tools. *New Media Society, 12*(1), 109 - 125. doi:10.1177/1461444809341260.

Fosse, G. K., & Holen, A. (2006). Childhood maltreatment in adult female psychiatric outpatients with eating disorders. *Eating Behaviours, 7*(4), 404 - 409. doi:10.1016/ j.eatbeh.2005.12.006.

Hinduja, S., & Patchin, J. W. (2008). Cyberbullying: An exploratory analysis of factors related to offending and victimization. *Deviant Behavior, 29*(2), 129 - 156. doi:10.1080/01639620701457816.

Hinduja, S., & Patchin, J. W. (2010). Bullying, cyberbullying, and suicide. *Archives of Suicide Research, 14*(3), 206 - 221. doi:10.1080/13811118.20 1.494133.

Hokoda, A., Lu, H.-H. A., & Angeles, M. (2006). School bullying in Taiwanese adolescents. *Journal of Emotional Abuse, 6*(4), 70-90. doi:10.1300/ J135v06n04_04.

Huang, Y. Y., & Chou, C. (2010). An analysis of multiple factors of cyberbullying among junior high school students in Taiwan. *Computers in Human Behavior, 26*(6), 1581 - 1590. doi:10.1016/j.chb.2010.06.005.

Juvonen, J., & Gross, E. F. (2008). Extending the school grounds? Bullying experiences in cyberspace. J*ournal of School Health, 78*(9), 496 - 505. doi:10.1111/ j.1746-1561.2008.00335.x.

Kowalski, R. M., & Limber, S. P. (2007). Electronic bullying among middle school students. *Journal of Adolescent Health, 41*(6), 22 - 30. doi:10.1016/ j.jadohealth.2007.08.017.

Kowalski, R. M., Limber, S. P., & Agatston, P. W. (2012). Cyberbullying: *Bullying in the digital age*. Chicester: Wiley-Blackwell.

Kowalski, R. M., Morgan, C. A., & Limber, S. P. (2012). Traditional bullying as a potential warning sign of cyberbullying. *School Psychology International, 33*(5), 505 - 519. doi:10.1177/0143034312445244.

Kraft, E. (2006). Cyberbullying: A worldwide trend of misusing technology to harass others. *WIT Transactions on Information and Communication Technologies, 36*, 155 - 166. doi:10.2495/IS060161.

Li, H. L., Zhang, W. X., & Yu, F. J. (2012). The relationship between victimization and depression of adolescents. *Psychological Development and Education, 1*, 77 - 82. doi:CNKI:SUN:XL FZ.0.2012-01-011.

Li, Q. (2006). Cyberbullying in schools: A research of gender differences. *School Psychology International, 27*(2), 157 - 170. doi:10.1177/0143034306064547.

Li, Q. (2007). New bottle but old wine: A research of cyberbullying in schools. *Computers in Human Behavior, 23*(4), 1777 - 1791. doi:10.1016/ j.chb.2005.10.005.

Li, Q. (2008). A cross-cultural comparison of adolescents' experience related to cyberbullying. *Educational Research, 50*(3), 223 - 234. doi:10.1080/00131880802309333.

Ma, X. (2001). Bullying and being bullied: To what extent are bullies also victims? *American Educational Research Journal, 38*(2), 351‒370. doi:10.3102/00028312038002351.

Mesch, G. S. (2009). Parental mediation, online activities, and cyberbullying. *CyberPsychology and Behavior, 12*(4), 387‒393. doi:10.1089/cpb.2009.0068.

Mishna, F., Khoury‒Kassabri, M., Gadalla, T., & Daciuk, J. (2012). Risk factors for involvement in cyber bullying: Victims, bullies and bully‒victim. *Children and Youth Services Review, 34*(1), 63‒70. doi:10.1016/j.childyouth.2011.08.032.

Mitchell, K. J., Ybarra, M., & Finkelhor, D. (2007). The relative importance of online victimization in understanding depression, delinquency, and substance use. *Child Maltreatment, 12*(4), 314‒324. doi:10.1177/1077559507305996.

Nansel, T., Overpeck, M., Pilla, R., Ruan, W., Simons‒Morton, B., & Scheidt, P. (2001). Bullying behaviors among US youth: Prevalence and association with psychosocial adjustment. *The Journal of the American Medical Association, 285*(16), 2094‒2100. doi:10.1001/jama.285.16.2094.

Navarro, R., Serna, C., Martnez, V., & Ruiz‒Oliva, R. (2012). The role of Internet use and parental mediation on cyberbullying victimization among Spanish children from rural public schools. *European Journal of Psychology of Education,* 1‒21. doi:10.1007/s10212‒012‒0137‒2.

O'Keeffe, G. S., Clarke‒Pearson, K., & Council on Communication and Media. (2011). Clinical report—The impact of social media on children, adolescents, and families. *Pediatrics, 127*(4), 800‒805. doi:10.1542/peds.2011‒0054.

Olweus, D. (2003). A proEle of bullying at school. *Educational Leadership, 60*(6), 12‒17.

Ortega, R., Elipe, P., Mora‒Merchan, J. A., Calmaestra, J., & Vega, E. (2009). The emotional impact on victims of traditional bullying and cyberbullying. *Zeitschrift fu ⊕r Psychologie/Journal of Psychology, 217*(4), 197‒204. doi:10.1027/0044‒3409.217.4.197.

Patchin, J. W., & Hinduja, S. (2006). Bullies move beyond the school yard: *A preliminary look at cyberbullying. Youth Violence and Juvenile Justice, 4*(2), 148‒169, 376‒385. doi:10.1177/1541204006286288.

Sakellariou, T., Carroll, A., & Houghton, S. (2012). Rates of cyber victimization and bullying among male Australian primary and high school students. *School Psychology International, 33,* 533‒549. doi:10.1177/0143034311430374.

Slonje,R.,&Smith,P.K.(2008).Cyberbullying:Another maintypeofbullying?*Scandinavian Journal of Psychology, 49*(2), 147‒154. doi:10.1111/j.1467‒9450.2007.00611.x.

Smith, P. K., Mahdavi, J., Carvalho, M., Fisher, S., Russell, S., & Tippett, N. (2008). Cyberbullying: Its nature and impact in secondary school pupils. *Journal of Child Psychology and Psychiatry, 49*(4), 376‒385. doi:10.1111/j.1469‒7610.2007.01846.x.

Tokunaga, R. S. (2010). Following you home from school: A critical review and synthesis of research on cyberbullying victimization. *Computers in Human Behavior, 26*(3), 277‒287. doi:10.1016/j.chb.2009.11.014.

Topcu, C., Erdur‒Baker, O., & Capa‒Aydin, Y. (2008). Examination of cyberbullying experiences among Turkish students from different school types. *CyberPsychology and Behavior, 11*(6), 643‒648. doi:10.1089/cpb.2007.0161.

Twyman, K., Saylor, C., Taylor, L. A., & Comeaux, C. (2010). Comparing children and adolescents engaged in cyberbullying to matched peers. Cyberpsychology, *Behavior, and Social Networking, 13*(2), 195‒199. doi:10.1089/cyber.2009.0137.

Von Marees, & Petermann. (2012). Cyberbullying: An increasing challenge for schools. *School Psychology International, 33*(5), 467‐476. doi:10.1177/0143034312445241.

Wade, A., & Beran, T. (2011). Cyberbullying: The new era of bullying. *Canadian Journal of School Psychology, 26*(1), 44‐61. doi:10.1177/0829573510396318.

Wang, J., Iannotti, R. J., & Luk, J. W. (2012). Patterns of adolescent bullying behaviors: Physical, verbal, exclusion, rumor, and cyber. *Journal of School Psychology, 50*(4), 521‐534. doi:10.1016/j.jsp.2012.03.004.

Wang, J., Iannotti, R. J., & Nansel, T. R. (2009). School bullying among adolescents in the United States: Physical, verbal, relational, and cyber. *Journal of Adolescent Health, 45*(4), 368‐375. doi:10.1016/j.jadohealth.2009.03.021.

Wei, H.‐S., Jonson‐Reid, M., & Tsao, H. L. (2007). Bullying and victimization among Taiwanese 7th graders: A multi‐method assessment. *School Psychology International, 28*(4), 479‐500. doi:10.1177/0143034307084137.

Williams, K. R., & Guerra, N. G. (2007). Prevalence and predictors of Internet bullying. *The Journal of Adolescent Health, 41*, S14‐S21. doi:10.1016/j.jadohealth.2007.08.018.

Wolak, J., Mitchell, K., & Finkelhor, D. (2006). Online victimization of youth: 5 years later. *National Center for Missing and Exploited Children, 7*, 06‐025.

Wolak, J., Mitchell, K. J., & Finkelhor, D. (2007). Does online harassment constitute bullying? An exploration of online harassment by known peers and online‐only contacts. *Journal of Adolescent Health, 41*(6), S51‐S58. doi:10.1016/j.jadohealth.2007.08.019.

Ybarra, M. L., Diener‐West, M., & Leaf, P. J. (2007). Examining the overlap in Internet harassment and school bullying: Implications for school intervention. *Journal of Adolescent Health, 41*(6), 42‐50. doi:10.1016/j.jadohealth.2007.09.004.

Ybarra, M. L., & Mitchell, K. J. (2007). Prevalence and frequency of Internet harassment instigation: Implications for adolescent health. *Journal of Adolescent Health, 41*(2), 189‐195. doi:10.1016/j.jadohealth.2007.03.005.

Zhang, W. X., Gu, C. H., & Ju, Y. C. (2001). Review of the relationship between personality and bullying in children. *Advances in Psychological Science, 9*(3), 215‐220.

Journal of School Psychology, 2014, 52(5), 511 - 526.

Friendship quality, social preference, proximity prestige, and self-perceived social competence: Interactive influences on children's loneliness

Fengjuan Zhang[1,2,3], Zhiqi You[1,2,4,5], Cuiying Fan[1,2], Chuang Gao[1,2], Robert Cohen[6], Yeh Hsueh[7], Zongkui Zhou[1,2,*]

([1] Key Laboratory of Adolescent Cyberpsychology and Behavior (CCNU), Ministry of Education, Wuhan 430079, China)

([2] School of Psychology, Central China Normal University, Wuhan 430079, China)

([3] Institute of Complexity Science and Big Data Technology, Guangxi University, Nanning 530004, China)

([4] Department of Sociology, Huazhong Agricultural University, Wuhan 430070, China)

([5] Research Center For Rural Social Construction and Management, Wuhan 430070, China)

([6] Department of Psychology, University of Memphis, Memphis, TN 38152, USA)

([7] College of Education, Health and Human Sciences, University of Memphis, Memphis, TN 38152, USA)

Abstract The purpose of this study was to test an integrative model in which peer relations at different levels of social complexity (friendship quality, social preference, and proximity prestige) are associated with children's loneliness, with children's self-perceived social competence acting as a mediator of these associations. A middle childhood sample of 509 Chinese children (233 girls and 276 boys; 3rd to 6th grade) completed a battery of sociometric and self-report questionnaires. Bootstrap analysis showed that self-perceived social competence mediated the relations between each peer variable and loneliness. In the integrative model tested with SEM, the mediating effect of self-perceived social competence in the relation between friendship quality and loneliness and between social preference and loneliness remained significant. However, self-perceived social competence no longer mediated the association between proximity prestige and loneliness, when considering the simultaneous influences of the three peer variables (friendship quality, social preference, and proximity prestige). The whole model accounted for 56% of the variance in loneliness. These findings suggest that self-perceived social competence played an important role in children's loneliness, that the quality and the quantity of direct peer relations (friendship quality, social preference, and part of proximity prestige) were associated with loneliness, and that indirect friends had a relatively lower but significant influence on children's loneliness. The results are discussed in terms of their implications for preventing children's loneliness.

Keywords friendship quality; social preference; proximity prestige; self-perceived social competence; loneliness

---

*Corresponding author at: Key Laboratory of Adolescent Cyberpsychology and Behavior(CCNU), Ministry of Education, Wuhan 430079, China. School of Psychology, Central China Normal University, Wuhan 430079, China. Tel./fax: +86 2767868632. E-mail address: zhouzk@mail.ccnu. edu.cn (Z. Zhou).

1 Introduction

As a social species, human beings innately have the need for social connection. An absolute or relative lack of social connection can result in the painful emotion of loneliness (Peplau & Perlman, 1982; Weiss, 1973, p. 15). Loneliness is a common emotional experience among children. As many as 80% of children report having experienced loneliness at school (Berguno, Leroux, McAinsh, & Shaikh, 2004). One study of third-grade children found that 23% had a moderate level of loneliness, a feeling that steadily increased during the next two years (Jobe-Shields, Cohen, & Parra, 2011). Substantial evidences show that the feeling of loneliness in childhood not only is associated with children's current life quality but also predicts future maladjustment (Masi, Chen, Hawkley, & Cacioppo, 2011; Rotenberg, 1999; van Dulmen & Goossens, 2013). Lonely children are more likely to experience low self-esteem (Sletta, Valås, Skaalvik, & Sbstad, 1996), increased levels of social anxiety and social avoidance (Vanhalst, Goossens, Luyckx, Scholte, & Engels, 2012), poorer academic performance (Benner, 2011), higher risk of dropping out of school and of delinquency, and more mental and physical health problems (Harris, Qualter, & Robinson, 2013; Heinrich & Gullone, 2006). Loneliness in childhood is a valuable predictor of depressive symptoms in adolescence (Qualter, Brown, Munn, & Rotenberg, 2010).

Two theories have systematically illuminated the nature of loneliness and the factors that influence it: the social needs theory and the cognitive processes approach (Terrell-Deutsch, 1999). The social needs theory claims that loneliness is a response to unmet needs for social connection or to unsatisfactory interpersonal relationships. Consistent with this theory, the majority of research has focused on peer relations to understand children's loneliness, and during the past three decades, peer relations have been found to be a critical factor in children's loneliness (Asher & Paquette, 2003). In contrast to the social needs theory, cognitive processes theory suggests that loneliness is not a result of unmet inherent social needs but of dissatisfaction with one's perceived social relationships. In other words, it is the cognitive awareness of a deficiency in either the quality or the quantity of one's social relationships that leads to the discomfort of loneliness (Peplau & Perlman, 1982; Terrell-Deutsch, 1999). However, there is little research examining the effects of individuals' internal cognitive representations of social relations (e.g., perceived social competence) in the link between peer relations and children's feelings of loneliness.

1.1 Peer relations and loneliness

Experiences with peers constitute an important developmental context for children. Children's peer experiences can be divided into several levels of analysis—individual characteristics, social interactions, dyadic relationships, and group membership and composition. The latter three levels of peer system reflect social participation at different interwoven orders of social complexity (Hinde, 1987; Rubin, Bukowski, & Parker, 2006). Previous research has extensively explored the relations between loneliness and multiple types of peer relations, including peer relations at the dyadic and group levels (Margalit, 2010). Acceptance by the peer group and friendship have been found to be important factors for understanding children's experiences of feeling lonely (e.g., Asher & Paquette, 2003; Margalit, 2010).

The preponderance of research on children's loneliness has focused on the possible influence of acceptance versus rejection by peers (Asher & Paquette, 2003). The group's acceptance of a child refers to the degree to which the child is liked or disliked by group members, and group acceptance is an indicator of the child's social status in the group (Ladd, 1999). Peer acceptance is typically assessed using peer nominations, a sociometric method in which children identify (or "nominate") group members they like most and like least. Acceptance is measured by the number of "like most" nominations a child receives, standardized according to class size, and rejection is the standardized number of "like least" nominations received. These two scores can be combined to create a social preference score (the difference between the acceptance and rejection scores), reflecting the child's relative standing in terms of acceptance by the peer group (Hymel, Vaillancourt, McDougall, & Renshaw, 2002). Previous studies have shown that children who have higher levels of peer acceptance or are more socially preferred are less likely to report feeling lonely (e.g., Mouratidis & Sideridis, 2009; Shin, 2007; Yu, Zhang, & Yan, 2005; Zhou, Sun, Zhao, & Hsueh, 2005).

Friendship lies at the dyadic level of peer experience. A "friend" is defined as a person you know well and like, usually not a member of your family (Hornby, 2010). Friendship is the relationship between friends. The definition of friend suggests that friendships may vary in their degree of mutual knowledge and affection, characteristics which constitute friendship quality. Researchers have consistently found a high negative correlation between friendship quality and children's loneliness (e.g., Hoza, Bukowski, & Beery, 2000; Nangle, Erdley, Newman, Mason, & Carpenter, 2003; Parker & Asher, 1993; Sun, Zhou, Fan, & Ke, 2009). Specifically, studies have indicated that companionship and support from friends were conducive to lessening or eliminating children's loneliness (Parker & Asher, 1993; Uruk & Demir, 2003). Moreover, longitudinal research has shown that friendship quality in middle childhood significantly and negatively predicted loneliness two years later (Zhou, Zhao, Sun, & Ding, 2006).

Nonetheless, relatively few studies have examined the link between the quantity of friends children have and children's loneliness, and the findings of these studies were inconsistent. Ladd, Kochenderfer, and Coleman (1997) found that although the quantity of friends did not predict children's loneliness as they made the transition to kindergarten, it positively predicted other aspects of children's adjustment (such as academic readiness and school involvement). Parker and colleagues (Parker & Asher, 1993; Parker & Seal, 1996) have shown that in middle childhood having one friend is predictive of lower loneliness, but they did not explore the effect of having more than one friend. Recent research in middle childhood samples has suggested that the number of friends is important in predicting loneliness. With a sample of third– through sixth–grade students, Nangle et al. (2003) demonstrated that the number of both best friends and good friends was important in predicting children's positive adjustment, including a lower level of loneliness. The number of early mutual friends at the ages of 9–10 has been shown to significantly and negatively predict loneliness two years later (Zhou et al., 2006).

Although research consistently suggests that both friendship processes and acceptance by peers can predict children's feelings of loneliness, these variables as typically measured (e.g., by friendship nominations and social preference scores) still leave much of the variance in loneliness unexplained (Asher & Paquette, 2003). One possible explanation is that not all friends are equally important in protecting a child from being lonely. Indeed, a child's friendships and social status in the peer group are embedded in a larger social network of peer relations and experiences (Gifford–Smith & Brownell, 2003). The peer group is a network in which most children connect with several other children by friendship. However, each of these friends has a distinct influence on the child, and one reason for this variation is that a child's friends may vary in social status. For example, friends with many connections in the peer group might provide more information or resources to the child than more isolated friends could. Thus one way to assess a child's connectedness in the peer group is to take into account not just the number of friends, but also the number of "friends of friends." Even if a child has few direct friends, a high number of indirect friends might reduce the risk of loneliness. A useful index of the quantity of indirect friends is called proximity prestige, a measure that developed from social network analysis.

In recent years, social network analysis, a method used to study the structure and characteristics of social networks (Scott, 2000), has been identified as an appropriate method for studying the peer context of child and adolescent social behaviors and emotions, such as aggression (Faris & Ennett, 2010), substance use (Ennett et al., 2006; Mercken, Steglich, Sinclair, Holliday, & Moore, 2012; Mundt, 2011), prosocial behaviors (Gest, Graham-Bermann, & Hartup, 2001), depression (Okamoto et al., 2011), and loneliness (Chamberlain, Kasari, & Rotheram-Fuller, 2007). This method generates a visual depiction of the complex inter-relations in a social network. Each actor (e.g., a child) is represented by one node, and the relationships (e.g., friendships) between any two actors are represented by lines. Some actors are linked directly, and some are linked indirectly through relationships with other actors. In the visual representation of these relationships, the distance between two actors is one unit if the actors are linked directly, two units if they are linked indirectly through one other actor, and so on. The shorter the distance is between actors, the closer the relationships are between them. An actor's influence domain is the number or percentage of all actors linked to him or her directly and indirectly in a closed group. However, the intensity of the actor's influence varies depending on the closeness of the relationship. The farther away that neighbors are from an actor, the smaller the actor's influence on them.

Using this type of information about a social network,

Lin (1976) first proposed a measure of prestige called proximity prestige. According to Wassermann and Faust (1994), the proximity prestige of an actor is the proportion of all other actors directly or indirectly connected to the actor in a closed group (influence domain) divided by the average distance these other actors are from the actor (the mean of influence intensity). This means that when relationships are defined by friend nominations, then the relationships are weighted differently depending on whether the link is direct (one child is nominated by another) or indirect (one child is nominated by another, who in turn is nominated by someone else). In a friend nominations network, the influence domain is the quantity of direct and indirect friends. The influence intensity represents the closeness of these relationships. Therefore, proximity prestige is a composite representation of the quantity and the closeness of children's direct and indirect friends. Meanwhile, this index, which is calculated based on friend nominations by others, also reflects children's objective friendship status in the peer group.

In social network analysis, one widely used index is centrality, which is the proportion of nominations one actor receives from other actors in a group. The actors who receive many positive nominations are considered to be prominent and to occupy a central position in the peer group. However, the centrality index makes no distinction in terms of the popularity of the nominators. For example, two actors who receive the same number of nominations, but who differ in how active they are in the larger network, would have the same centrality index score. Proximity prestige extends the concept of centrality by taking the popularity of the nominators into consideration.

It is important to distinguish proximity prestige from other concepts of peer relationships (e.g., peer acceptance, social preference), especially because the number of direct and indirect friends in the social network may highly coincide with the number of "like most" nominations, the basis of peer acceptance and social preference scores. But naming a specific child as a friend involves affirmation of a special dyadic relationship, whereas peer acceptance and social preference are unilateral constructs representing the general affective inclination of the group toward an individual. Even though these three constructs (peer acceptance, social preference, and proximity prestige) all lie at the group level of peer experience, proximity prestige is based on the social network of friendships and thus represents greater social complexity than the other two indexes.

Friendship quality, social preference (a measure of acceptance by the group) and proximity prestige lie at the dyadic and group level of peer relations, and they reflect different facets of children's experience with peers. Therefore, the first goal of the present study was to test whether variables representing peer relationships at different levels of social complexity were associated with children's loneliness, and it was hypothesized that the proximity prestige index would provide extra explanatory power above that of friendship quality and social preference in predicting children's loneliness.

1.2 Self-perceived social competence and loneliness

Measures of peer relations are often objective measures of social status, which may not correspond with children's subjective experience of loneliness. According to cognitive processes theory, it is a child's cognitive appraisal of peer relations that gives rise to the feeling of loneliness (Terrell-Deutsch, 1999). Self-perceived social competence is the internal cognitive estimation of one's own social competence. Children who perceive higher social competence are more likely to be satisfied with their peer relations, and less likely to feel lonely. Sun et al. (2009) discovered that self-perceived social competence was more predictive of loneliness than was objective social status. In addition, children's social success may lead to

positive social evaluations of their own competencies (Cole, 1991); in turn, positive self-perception may further lead to less loneliness. It is supposed that self-perception of social competence is a mediator of the link between peer relations and loneliness. Studies with third- to sixth-grade children have shown that self-perceived social competence mediates the association between direct peer relations (peer acceptance and friendship quality) and loneliness (Sun et al., 2009; Zhou et al., 2005). However, these studies did not take indirect peer relations into consideration. Therefore, the second goal of the present study was to test whether proximity prestige (a reflection of both direct and indirect friendships and objective friendship status) in the peer network is associated with children's loneliness through the mediator of self-perceived social competence.

1.3 Peer relations, self-perception, and loneliness

Previous studies have extensively explored different levels of peer relations as predictors of children's loneliness; however, the studies have tended to concentrate on each predictor independently rather than integratively. Based on an extensive critical review, Gifford-Smith and Brownell (2003) emphasized that children's peer relations are multifaceted, intersecting and overlapping systems, and it is necessary and valuable to attend to multiple aspects of the peer experiences simultaneously from an integrative perspective. Hinde (1987) also emphasized that events and processes at each level of peer experience are constrained and influenced by events and processes at other levels (Rubin et al., 2006). Additionally, researchers have pointed out that peer relations constitute just one facet of the contexts in which peer interactions occur and develop. These contexts determine the particular skills, perceptions, and cognitions that will be most effective for children's successful functioning in the peer group (Brownell & Gifford-Smith, 2003; Sheridan, Buhs, & Warnes, 2003). Thus, the third goal of the present research was to test a conceptual model in which peer experience at multiple levels of social complexity (namely friendship quality, social preference, proximity prestige, and self-perceived social competence) jointly predicts children's loneliness; in this model, children's self-perceived social competence is also tested as a mediator, representing a possible underlying mechanism in the association between peer relations and children's loneliness.

It was noted that the level of social interaction was not represented in the conceptual model. Social interation encompassed different kinds of behaviors and behavioral tendencies (Rubin et al., 2006). The theoretical and substantial evidences showed that social behaviors were the antecedents or predictors of peer relations in childhood (Chen, Rubin, & Sun, 1992; Ladd, Buhs, & Troop, 2002). According to the concept and theory of loneliness, it was highly emphasized that loneliness was the response to the lack of social connection (peer relations). The present study mainly focused on the link between peer relations and the feeling of lonely in the theoretical framework of loneliness. Thus, the antecedents of peer relations (e.g., the factors of social interaction) were not included in the analyses.

Both the social needs theory and the cognitive processes approach of loneliness indicate that the satisfaction or perceived satisfaction of interpersonal relationships could lower or eliminate feeling lonely (Terrell-Deutsch, 1999). The interpersonal relationships mentioned by the theories are the positive relations what individuals want to obtain (contrary to negative relations, such as peer victimization). Most of the research on children's loneliness has focused on the great contributions of two kinds of positive peer relations, acceptance by peer group and the friendship (Asher & Paquette, 2003; Margalit, 2010). The research has consistently found that children who are better accepted by peers or have a high-quality friendship with a best friend report experiencing less loneliness than other children. On the contrary, as one facet of friendship, the quantity of friends children has been occasionally examined its relation with loneliness, and the results are not consistent. Meanwhile, the research has not taken indirect friendships and the statuses of friends into consideration. Therefore, the index of proximity prestige is introduced to represent the number of indirect friends and the statuses of friends. The positive peer relations in the peer group can be represented by the total of proximity prestige (composite representation of the quantity and the closeness of children's direct and indirect friends), friendship quality (the relationship with best friend) and social preference (the affective inclination of the group toward an individual). The present study aimed to explore the simultaneous influence of the three kinds of peer relationships (friendship quality, social preference, and proximity prestige) on children's loneliness, especially the unique contribution of proximity prestige on loneliness.

1.4 Culture and loneliness

Most research on children's loneliness has been conducted in Western cultures, expecially North American, which are typified by an individualistic orientation. Individualist cultures value individualization, autonomy, and privacy. Asian countries including China are collectivistic cultures that prioritize relational bonds and group cohesion (Oyserman, Coon, & Kemmelmeier, 2002). The Western cultural orientation of individualism has been found to be associated with higher loneliness among adults, even in Western societies (Rokach & Bauer, 2004; Seepersad, Choi, & Shin, 2008). Multinational studies of the loneliness of 3rd to 6th grade children conducted in Canada, Brazil, China, and Southern Italy revealed no mean differences in self-reported loneliness across these four samples. However, the overall patterns of relations between social behaviors (such as aggression and shyness-sensitivity) and loneliness differed across samples. For example, shyness is viewed as maladaptive in Western cultures (Fox, Henderson, Marshall, Nichols,

& Ghera, 2005), and indeed, results of the multinational study showed that shyness was positively associated with loneliness in Canadian, Brazilian and Italian children, but not associated with loneliness in Chinese children. These results suggested that the nature of children's loneliness may be affected by the broad socio-cultural context (Chen, He, De Oliveira, et al., 2004).

In contemporary China, one major mission of schooling is to help children develop collectivistic ideologies (Chen, 2000). Children are encouraged to participate in group activities, develop positive attitudes toward the group, and learn social skills that facilitate group functioning. The emphasis on social connections results in an interdependent sense of self (Markus & Kitayama, 1991). In this context, social alienation and loneliness are more likely to be viewed as maladaptive or problematic (Chen, 2000). Chinese children's self-perceived social competence is also likely to be influenced by cultural context. Self-evaluation has been encouraged traditionally in Chinese culture (e.g., Luo, 1996). In Chinese schools, children are regularly required to engage in self-evaluation, in the belief that children who are conscious of their strengths and weaknesses are likely to perform and regulate their behavior better according to social norms and values. A particular emphasis is placed on self-awareness of the negative aspects of one's social and moral character (Chen, He, & Li, 2004). Therefore, self-perceived social competence would be expected to have close links with peer relations and feelings of loneliness in Chinese children.

1.5 Study hypotheses

Researchers consider peer relations to be a multidimensional social network structure (Gifford-Smith & Brownell, 2003; Margalit, 2010), and many studies have explored the effects of one or two facets of peer relations on children's loneliness. However, little research has examined multiple facets of peer relations in relation to children's loneliness, and the underlying mechanisms of these associations. The goal of the current study was to explore multiple indexes of peer relations as predictors of loneliness in middle childhood, and to test children's self-perceived social competence as a mediator of these links in the context of Chinese culture. Based on the empirical research and associated conceptual models to date, a new conceptual model was developed and tested in the current study (see Figure 1). It was hypothesized that: (a) friendship quality, social preference, proximity prestige, and self-perceived social competence would be negatively correlated with children's loneliness; (b) proximity prestige would explain unique variance in loneliness scores; (c) self-perceived social competence would mediate the association between peer relations and loneliness in both separate and integrated analyses.

2 Method

2.1 Participants

Data for the present research were collected from 509 third- to sixth-grade students of eight classes in a public ordinary elementary school in Wuhan, a big city in central China, at the end of the fall semester, December, 2008. The school had six classes at each grade level, and two classes of each grade were randomly selected for participation. The sample was composed of 233 girls and 276 boys (roughly equal numbers of girls and boys in each class). Almost all children fell in the age range of 8 to 12, with 125 third-graders, 127 fourth-graders, 129 fifth-graders, and 128 sixth-graders (62 to 66 students in each class). Four children were absent from school on the day of the survey, so the overall participation rate was 99.2%. The participation rates for six classes were 100%, and that for the other two classes were 95% and 98%. By teachers' report most of the students (>95%) were Han, the majority ethnic group in China, with a small number of students being from

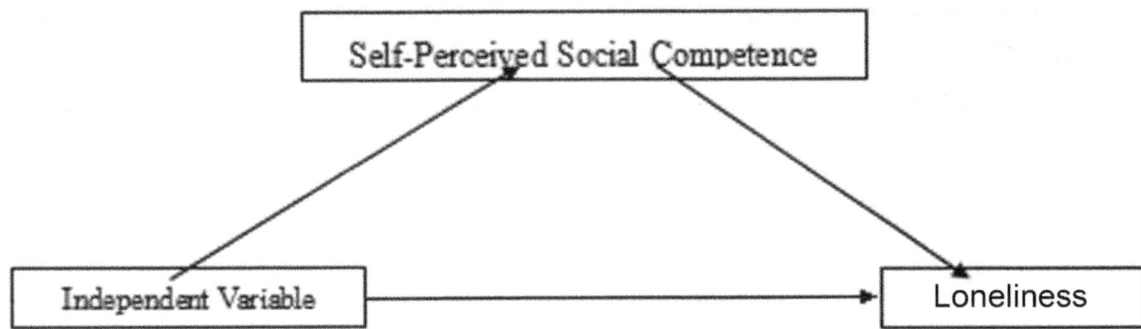

Fig. 1. The mediation model of peer variables, self-perceived social competence and loneliness. The independent variables were friendship quality, social preference, and proximity prestige.

minority groups. The students were primarily of middle socioeconomic status (> 80%), the majority of parents received college or university education (> 85%), and most (>95%) of the children were the only child in the family, an effect of China's 'one-child-per-family' policy implemented in the late 1970s. The sample was representative of ordinary elementary school children in urban China.

2.2 Measures

All the measures used in this study were originally English versions, which have been widely used in Western countries. Previous researchers translated these measures into Chinese language to apply to Chinese children, and found that the Chinese versions achieved the standards of psychometrics. The measures were commonly used in the research of Chinese children. The information about reliability and validity of each measure in prior research of Chinese children are reported in the descriptions of each measure.

2.2.1 Peer nominations for "like most" and "like least"

Researchers provided each child a printed name list of all children in the class, and children were asked to circle the names of classmates whom they "liked most" and "liked least" with no limit on the nomination number. Compared with the limited nomination method, the unlimited-choice procedure could provide a greater range, more normal-like distribution of values on the nomination items, and make the nominations more reliable

(Marks, Babcock, Cillessen, & Crick, 2012; Terry, 2000). Marks et al. (2012) demonstrated that the Cronbach's α s of unlimited liked-most nomination was approximately .75 with a 95% participation rate. The raw number of the children rated on the liked-most nomination was 9.73 (SD=6.92) on average, and 6.90 (SD=8.33) on the liked-least nomination. Then, the numbers of positive (like most) and negative (like least) nominations were calculated for each student and standardized within classrooms. This sociometric method has been widely used to assess children's peer acceptance and rejection in a variety of cultures including Chinese (e.g., Chen, Zappulla, Coco, et al., 2004; Zhou et al., 2006). Following Coie, Dodge, and Coppotelli's (1982) procedure, an index of social preference was formed to indicate the child's relative likeability in the group. This score was generated from the difference between the positive nomination score and the negative nomination score; higher scores indicate that compared to others in the class, the child is relatively more liked by peers. This index has been extensively used to reflect the peer relations of third- to sixth-grade children in China (e.g., Chen, Zappulla, Coco, et al., 2004; Zhou et al., 2005). One research with a sample of 3rd to 5th grade Chinese children showed that the test – retest correlation of social preference was .65 over a one-year interval (Zhou, Sun, Xiang, & Liu, 2007). Social preference was moderately and positively associated with sociability (r= .35) and self-perceived social competence (r=0.23), and negatively associated with aggression (r=−.46) in Chinese children (Chen, Zappulla,Coco, et al., 2004).

2.2.2 Friendship nominations

Participants were provided with a name list of all students in the class, and were asked to circle the names of classmates who were their friends, with no restriction on the number of nominations that could be made. Each child received 9.74 friendship nominations on average (*SD*=6.13). This procedure has been widely used across cultures, including Chinese culture, to identify children's friends (e.g., Chen et al., 1992; Parker & Asher, 1993; Zhou et al., 2006). Marks et al. (2012) demonstrated that unlimited friendship nominations were more reliable than 'top 3' friend nominations (Cronbach's α s were .60 and .38, respectively) when the participation rate was 95%. A previous study found that the quantity of good friends was positively correlated with children's popularity (r=0.65, .72), and negatively correlated with children's loneliness (r =−.39, −54) (Nangle et al., 2003).

In the current study, the key information was the number of "friend" nominations that each child received (rather than gave). A direct nomination occurred when a classmate nominated the child as a friend. An *indirect* nomination occurred when the classmate who had nominated the child had in turn been nominated as a friend by someone else in the class.

In mathematical terms, these connections can be described according to distance. The path distance of a direct nomination is 1; when the nomination is one step removed (that is, there is an indirect nomination), the path distance increases by 1. For example, child A names child B as a friend (a direct nomination with a path distance of 1), and child B names child C as a friend (another direct nomination, with a path distance of 1), but child A does not name child C as a friend. Child A and child C are indirect friends, and the path distance between them is 2. A student's proximity prestige is the ratio of the proportion

of all other classmates who directly or indirectly nominated the student to the mean path distance of these nominations. The formulation is as following (Wassermann & Faust, 1994):

$$P_P(n_i) = \frac{I_i/(g-1)}{\sum d(n_j,n_i)/I_i}$$

$P_{p(ni)}$ represents the prestige of actor I; I_i represents the total number of actors who directly or indirectly nominate actor i; grepresents the total number of actors in the group; $d_{(nj, ni)}$ represents the distance between the nominator j and the actor i. If no one nominates actor i, $P_{p(ni)}$= 0; if all actors in the group directly nominate actor i, $P_{p(ni)}$= 1.

2.2.3 Loneliness

Children's loneliness was assessed by the modified version of the Loneliness and Social Dissatisfaction Questionnaire (Asher & Wheeler, 1985). This self-report measure provides a five-point scale of 1 (*not at all true*) to 5 (*always true*) to rate 24 items, 16 of which assess feelings of loneliness and social dissatisfaction at school (e.g., "I have nobody to talk to in class;" "I am lonely at school;" "I do not have any friends in class"), and eight of which are filler items about personal interests (e.g., "I like to read"). The average score of the 16 items was used to represent loneliness, with higher the scores representing higher self-reported loneliness. Zhou et al. (2006) revised this measure for use with Chinese children, and found the 16 items of loneliness loaded on one factor which did not related to the 8 filler items, the Cronbach's alpha coefficient was .92. In a sample of Chinese children from grade 3 to grade 6, Chen, He, De Oliveira, et al. (2004) reported an alpha reliability coefficient of .95, and loneliness was moderately and negatively associated with social preference (r =−.31) and sociability (r =−.27). The test−retest correlation was .42 over a one−year period in another sample of Chinese children from grade 3 to grade 5 (Zhou et al., 2007). The internal consistency (Cronbach's alpha) was .92 in the present study.

2.2.4 *Self-perceived social competence*

Harter's (1982) Perceived Competence Scale for Children (PCSC) consists of four subscales representing children's self-perceived competence in four domains: social, cognitive, sport skill and global competence. Stigler, Smith, and Mao (1985) revised this scale for use with Chinese children and showed that the four-factor structure identified in U.S. samples of children in grades 3-9 (Harter, 1982) was replicated in a sample of Chinese fifth-grade children. For the social subscale (self-perceived social competence), in prior research on samples of Chinese children, children's self-perceived social competence was moderately and positively correlated with their sociability (.29) and social preference (.23) (Chen, He, & Li, 2004) and negatively related to victimization (−.19) (Zhou et al., 2006). The Cronbach's alpha coefficient was reported to be .67 and .64 (Chen, He, & Li, 2004; Stigler et al., 1985). Test-retest correlation was .48 over a one-year period in a sample of Chinese children in grades 3-5 (Zhou et al., 2007). The self-perceived social competence subscale was used in the present study. This subscale contains six items, with each item depicting two kinds of kids in two statements (e.g., "Some kids find it hard to make friends;" "Other kids find it's pretty easy to make friends"). Children were asked first to decide whether they were more like the child depicted on the left or the one on the right, and then to decide whether that part of the statement was "really true for me" or only "sort of true for me." This procedure of narrowing the choice gradually is helpful to allow even young children to produce meaningful ratings on a 4-point scale (Stigler et al., 1985). The average score of the 6 items was calculated as the subscale score, and the higher the score, the more positive the children perceived their social competence to be. In the present sample, Cronbach's alpha for the subscale was .56.

2.2.5 *Friendship quality*

Friendship quality was measured using an abbreviated version of the Friendship Quality Questionnaire (FQQ) (Parker & Asher, 1993). The abbreviated version of the FQQ consisted of the three items that showed the highest factor loadings on each dimension of the FQQ: Validation and Caring (e.g., "He or she tells me I am good at things"), Companionship and Recreation (e.g., "We always pick each other as partners for things"), Conflict and Betrayal (e.g., "We argue a lot"), Help and Guidance (e.g., "We help each other with schoolwork a lot"), Conflict Resolution (e.g., "We make up easily when we have a fight"), and Intimate Exchange (e.g., "We always tell each other our problems") as reported by Parker and Asher (1993), for a total of 18 items (Nangle et al., 2003). Zhou et al. (2006) revised the abbreviated version of the FQQ for use with Chinese children, and confirmed this measure's six-factor structure in the sample of Chinese children in grades 3-6, with an alpha reliability coefficient of .83; scores on this measure were moderately and positively correlated with peer acceptance (.26, .27) and number of mutual friends (.29, .36). Test-retest reliability was .40 over a one-year period in a sample of Chinese children in grades 3-5 (Zhou et al., 2007). In the present study, the children were asked to nominate their very best friend and assess their relationship using this questionnaire. Children responded to these items on a 5-point scale ranging from 0 (*not at all true*) to 4 (*really true*). After reversing the item scores on the dimension of Conflict and Betrayal, the scores of the 18 items were averaged, with higher scores indicating better friendship quality. In the present sample the Cronbach's alpha for the scale was .85.

2.3 Procedure

The survey was conducted through a passive parental consent process, and was overseen by the office of the Research Provost at the researchers' institution. The children were also told that they could drop out if they did not want to participate before the survey. All measures were collected in two group sessions in each classroom, and each session lasted approximately 40 minutes. The order of presentation of the surveys in the group sessions was counterbalanced by classroom. Surveys were administered by two trained graduate students in developmental psychology. These graduate students had received one-hour-training which including how to conduct the survey to children, how to cope with the problems they might meet, the points of caution and the special requirements for researchers. These researchers told the purpose of the investigation, assured the children that their answers were confidential, gave instructions on each measure and provided individual assistance when children had difficulties completing the survey. Researchers checked the questionnaires when they were turned in and asked children to complete any missing items. By using a Filemaker 6.0 template, all data were entered and cross-checked by two groups of trained graduate students in developmental psychology, those who had got half-an-hour-training about the form of the Filemaker template, especially how to enter the data of nominations.

2.4 Method of statistical analysis

First, descriptive statistics and correlations among all variables were generated. T-tests and F-tests were then performed to examine gender and grade differences on the study variables. Then, bias-corrected bootstrap analyses were employed to test whether self-perceived social competence mediated the effect of each peer variable (friendship quality, social preference, proximity prestige) on children's loneliness, using Mplus 7.0 statistical software (Muthén & Muthén, 1998-2010). Bootstrap analysis is a nonparametric approach that directly tests the significance of mediation effects, makes no assumptions about the distribution of variables or the sampling distribution of the statistic, and can be applied to smaller samples with greater confidence than is possible with other methods (Fairchild & McQuillin, 2010). The present study used a 99% confidence interval and 1000 bootstrap

samples to evaluate the magnitude and significance of indirect effects. An effect is considered to be statistically significant if the confidence interval does not include zero.

Finally, structural equation model (SEM) analyses were conducted to test an integrative mediation model in which peer relationship variables (friendship quality, social preference, and proximity prestige) predicted children's loneliness both directly and indirectly through children's self-perceived social competence. Path analyses were conducted with Mplus 7.0 using maximum likelihood estimation of model parameters, and bias-corrected bootstrap analyses were also used to test the significance of the mediation effects (Muthén & Muthén, 1998-2010). A just-identified model was first analyzed, including all direct and indirect pathways from friendship quality, social preference and proximity prestige to self-perceived social competence and loneliness; this model assumed that there were correlations among friendship quality, social preference and proximity prestige. Then, based on the modification indices, a trimmed model was analyzed excluding the non-significant (p > .01) pathways from the initial saturated model. Lastly the Monte Carlo simulation was used to verify whether modification model via modification indices would generalize. The Monte Carlo simulation is increasingly popular in evaluating statistical estimators for structural equation models (see ntroduction in Paxton, Curran, Bollen, Kirby, & Chen, 2001). The present study specified and analyzed the hypothesized integrative model and the modified model using the same simulated data set (10000 replications for N=509), and compared the modelfit statistics and indices of the two models from the Monte Carlo simulation.

2.4.1 Missing data

Social preference and proximity prestige for the children in two classes were calculated with 95% and 98% participation rates. According to Cillessen and Marks (2011), at least 60－70% were necessary to obtain reliable peer nomination data for social acceptance or preference. Thus, the class participation rates in the present research were sufficient to achieve reliable peer nomination data. Missing information on other variables was as following: loneliness (8 children, 1.6%), friendship quality (5 children, < 1%), and self-perceived social competence (26 children, 5.1%). In order to account for the missing data, all the analyses were completed using maximum likelihood estimation (ML) which is considered a state-of-the-art missing data technique and produces less biased estimates than traditional techniques (listwise deletion and single imputation methods; see Baraldi & Enders, 2010).

3 Results

3.1 Descriptive statistics

Descriptive statistics for the observed variables are presented by gender and grade in Table 1. Compared to boys, girls reported lower loneliness, higher perceived social competence and friendship quality; girls also had higher social preference scores and higher proximity prestige. The only significant grade difference was in proximity prestige. The proximity prestige of sixth graders was higher than that for the other grades, and the proximity prestige of fourth and fifth graders was higher than that of third graders, but there was no significant difference between fourth and fifth graders' proximity prestige. Fig.2 shows students' proximity prestige within one class. In this figure, students (called actors in the language of social network analysis) are situated within the network as a function of direct and indirect friendship nominations. The students who received more friendship nominations (direct and indirect) had higher proximity prestige (e.g., v2, v3). In addition, some students received few nominations, but the nominators themselves had many nominations, so that the students indirectly had comparatively high proximity prestige. For example, v30 and v35 each had one direct nomination, but their proximity prestige differed because the proximity prestige of their nominators differed: The child who nominated v30 (i.e., v16) received more nominations than the child who nominated v35 (i.e., v44).

3.2 Correlations among peer relations, self-perceived social competence, and loneliness

Table 2 shows that, as expected, loneliness was negatively and significantly associated with self-perceived social competence (the hypothesized mediator) and with friendship quality, social preference, and proximity prestige (the hypothesized predictors of children's self-reported loneliness). Self-perceived social competence was positively and significantly correlated with the three indicators of peer relations (friendship quality, social preference, and proximity prestige). This pattern of correlations indicates that it would be appropriate to test the proposed mediation model.

3.3 Mediating effect of self-perceived social competence

Table 3 presents the results of the bootstrap analyses for each of the separate mediation models. The 99% bias-

Table 1 Means and standard deviations of the research variables by gender and grade

Variable	Gender				T	Grade								F
	Boys		Girls			Third		Fourth		Fifth		Sixth		
	n = 276		n = 233			n = 125		n = 127		n = 129		n = 128		
	M	(SD)	M	(SD)		M	(SD)	M	(SD)	M	(SD)	M	(SD)	
Loneliness	1.88	(0.78)	1.56	(0.60)	5.17***	1.76	(0.70)	1.70	(0.73)	1.73	(0.76)	1.74	(0.70)	0.16
Self-perceived Social Competence	2.78	(0.54)	2.93	(0.50)	-3.37***	2.85	(0.53)	2.85	(0.57)	2.86	(0.52)	2.84	(0.49)	0.05
Friendship quality	2.89	(0.75)	3.11	(0.73)	-3.34***	2.95	(0.63)	2.86	(0.89)	3.07	(0.75)	3.06	(0.68)	2.24
Social preference	-0.53	(1.78)	0.63	(1.40)	-8.20***	0.00ᵃ	(1.65)	0.00ᵃ	(1.76)	0.00ᵃ	(1.72)	0.00ᵃ	(1.75)	0.00
Proximity prestige	0.41	(0.12)	0.44	(0.08)	-3.90***	0.35	(0.11)	0.43	(0.10)	0.42	(0.08)	0.49	(0.09)	45.07***

Note: N = 509. ᵃ Values are Z scores. *** p < .001.

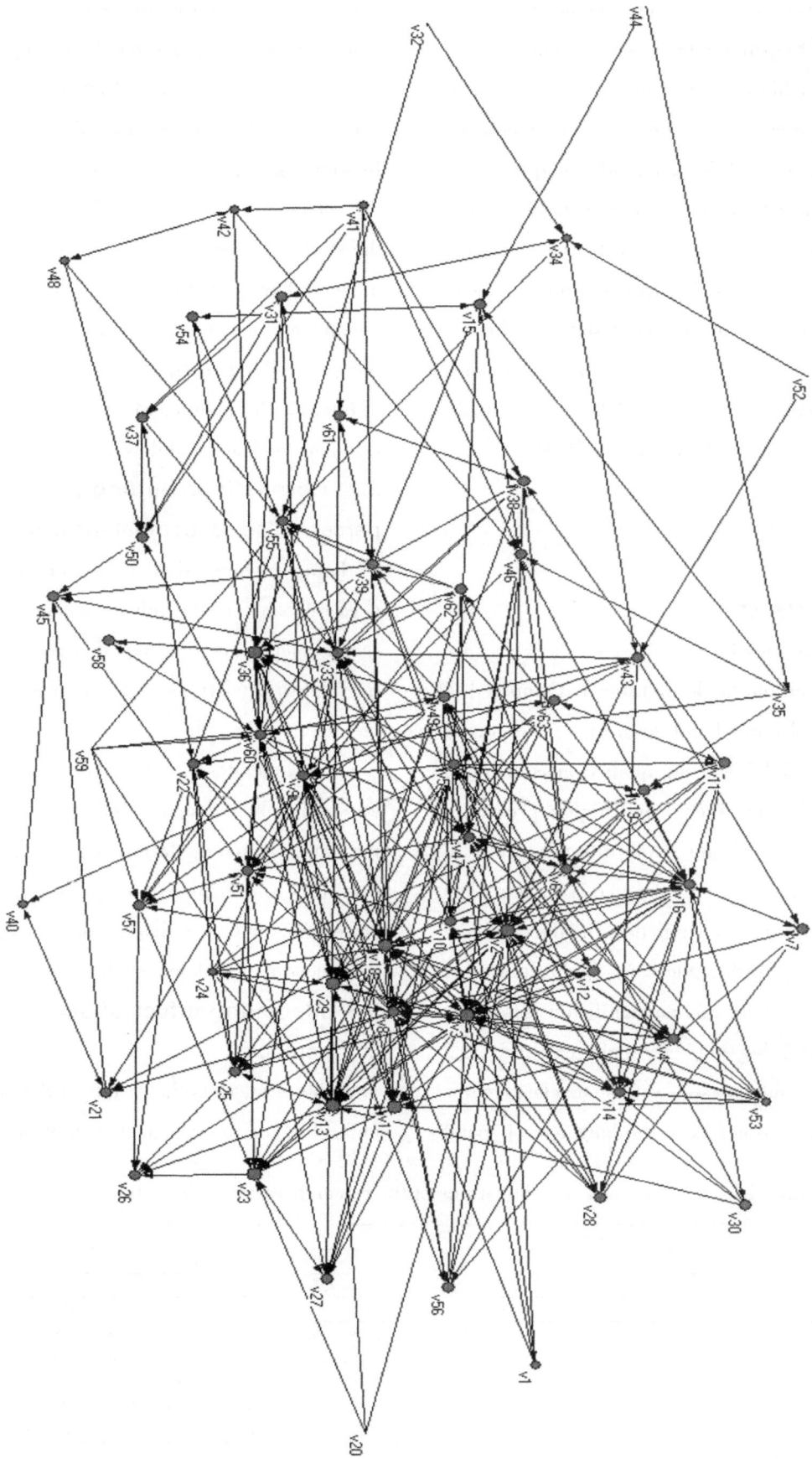

Fig.2. The sociogram of one class (class size = 63), depicting each child's proximity prestige. Each node denotes one child in the class; the line denotes one child (sender) who nominates another child (receiver) as a friend. The size of the nodes denotes the children's proximity prestige, and the bigger the node is, the more prestigious the child is. The sociogram was drawn with Pajek.

Table 2. Correlations among the research variables

Variables	1	2	3	4	5
1. Loneliness					
2. Self-perceived Social Competence	−.69**				
3. Friendship quality	−.41**	.34**			
4. Social preference	−.44**	.29**	.31**		
5. Proximity prestige	−.37**	.23**	.26**	.60**	
Mean	1.73	2.85	2.99	0.00ª	0.42
Standard deviation	0.72	0.53	0.75	1.72	0.10

Note. N = 509. ªValues are Z scores. ** $p < .01$.

corrected bootstrap confidence interval excluded zero for each of the three peer variables. In regard to model fit, the three models are all just-identified which have df= 0, CFI=1, TLI=1.000, $RMSEA$=0, $SRMR$=0. The results suggested that self-perceived social competence was the mediator of the relations between friendship quality and loneliness, social preference and loneliness, and proximity prestige and loneliness, respectively.

3.4 Path model linking peer relations with loneliness

The hypothesized integrative model was a just-identified model, the information for model fit as following: df=0, CFI=1, TLI=1.000, $RMSEA$=0, $SRMR$=0. There was no significant pathway from proximity prestige to self-perceived social competence (.06, p=0.24>.01). The 99% bias-corrected bootstrap confidence intervals (99% BC CI) for the indirect effect of friendship quality or social preference did not contain zero (−.240, −.083; −.084, −.009, respectively). However, there was zero in the 99% BC CI for the indirect effect of proximity prestige (−.949, .376). The indirect effect of proximity prestige was not significant. The insignificant pathway was then excluded and the modified model was set. The result showed that the modified model was acceptable and a good fit to the data: χ^2 (1)=1.39 (df=0.24), CFI =1.00, TLI=0.99 and RMSEA =0.03 (90% confidence interval values ranging from .00 to.13). Monte Carlo simulations also gave support for the modified model over the hypothesized integrative model. The result for the hypothesized integrative model from the Monte Carlo simulation showed that: for the information

criteria (AIC, BIC, $ABIC$), the values observed in the Monte Carlo replications were close to the theoretical values; for χ^2, $RMSEA$, and $SRMR$ the observed values were not close to the theoretical values. Meanwhile, for the pathway from proximity to self-perceived social competence, there was only 77.8% of 10000 replications for which the parameter estimate was significantly different from zero at α = .05, and it was smaller than the cut-off point value of .80 often used (Cohen, 1988). In regard to the modified model, the observed values were all close to the theoretical values for the model fit statistics and indices from the Monte Carlo simulation, and the lowest percentage of 10000 replications for which the parameter estimate was significantly different from zero at α =0.05 was 92.4% (larger than the cut-off point value of .80). The Monte Carlo simulation for the modified model in present study provided good results, and demonstrated that the modifications made via modification indices would generalize. Fig.3 presents the final integrative model of the SEM analyses.

The hypotheses were partially supported. In the modified model, there was a direct link from proximity prestige to loneliness, and a lack of a direct link between proximity prestige and self-perceived social competence. This means that self-perceived social competence did not act as a mediator between proximity prestige and loneliness when the influences of social preference and friendship quality were taken into account. There were direct links between friendship quality and loneliness, and between social preference and loneliness. Meanwhile,

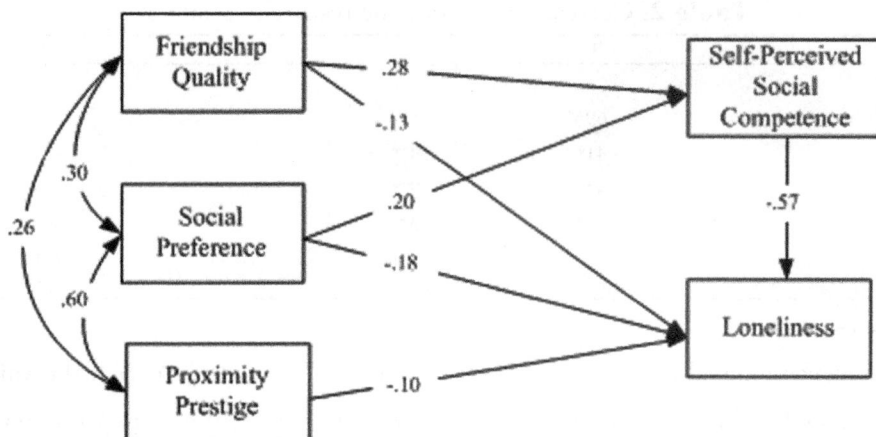

Fig. 3 Path model showing standardized path coefficients for associations among peer variables, self-perceived social competence and children s loneliness. Data were also analyzed using MLR in Mplus to correct for non-normality and the results did not change. N=509.

the effects of friendship quality on loneliness and of social preference on loneliness were transmitted through the intermediary predictor variable of self-perceived social competence, with the mediated (indirect) pathways having effects of −.16 and −.12 (p<.01), respectively. The 99% BC CI for the mediation (indirect) effect with friendship quality was (−.225, −.088), and that for social preference was (−0.183, −.049). These mediating effects accounted for 55% of the total effects of friendship quality on loneliness (standardized path coefficient of −.29), and 38% of the total effects of social preference on loneliness (−.29). The peer relationship variables (friendship quality, social preference, and proximity prestige) and self-perceived social competence together accounted for 56% of the variance in loneliness.

4 Discussion

According to the cognitive processes theory, loneliness could be influenced by self-cognition of social relationships, and peer relations provide the basis for self-cognition. Differing from past studies on loneliness in middle childhood, the present study assessed peer relations at multiple levels of social complexity, and introduced a measure of prestige in the social network. It was hypothesized that self-perceived social competence (self-cognition) would mediate the associations between peer relations (friendship quality, peer acceptance and proximity prestige) and children's loneliness, and mixed support for this hypothesis was found. The three

Table 3. The Confidence Mediated Effects of Self-Perceived Social Competence and the Corresponding Bias-corrected Bootstrap Confidence Intervals for Peer Variables as Predictors of Loneliness

Independent Variable	Estimate	SE	99% BC CI a	R^2_{med} [b]
Friendship quality	−.21**	.03	(−.28, −.14)	.13
Social preference	−.17**	.03	(−.24, −.10)	.13
Proximity prestige	−.13**	.04	(−.23, −.04)	.09

Note. Data set contains 4 cases with missing on all three variables (friendship quality, self-perceived social competence and loneliness). These cases were not included in the analysis of the mediation model with friendship quality (N = 505). All cases were included in the analyses of the other two mediation models (N = 509).

[a] BC CI represents bias-corrected confidence intervals. [b]R_{med} represents one effect size measure for mediated effect. R^2_{med} is computed by using squared bivariate correlations and the overall model R^2 from a model where Y is predicted from both X and M, as following (Fairchild & McQuillin, 2010): . $R^2_{med} = r^2_{Mr} - (R^2_{r,Mx} - r^2_{xr})$. r^2_{Mr} is the squared correlation between the outcome (Y) and the mediator (M), $R^2_{Y,Mx}$ is the overall model R^2 from the regression equation where Y is predicted from X and M. r^2_{xr} is the squared correlation between the outcome and the independent.

** p < .01 (significant mediation effect).

measures of peer relations were each significantly and negatively associated with loneliness, and each made a unique contribution to the variance in loneliness scores. Specifically, friendship quality made the highest contribution, in line with previous studies (Nangle et al., 2003; Sun et al., 2009), suggesting that a high quality friendship may be an important protective factor for children's loneliness. Tests of an integrative model showed that when the relations among all variables were taken into consideration, friendship quality and social preference were both associated with loneliness through the mediating effect of children's self-perceived social competence. However, self-perceived social competence did not mediate the relation between proximity prestige and loneliness.

4.1 Proximity prestige and children's loneliness

Previous studies have extensively explored the possible influence of friendships and acceptance by the peer group on children's loneliness. The present study also examined the association between loneliness and a more comprehensive index of peer relationships at the group level, namely proximity prestige, which reflects the quantity and the closeness of children's friendships in the peer social network. Proximity prestige was significantly correlated with loneliness, and in the integrative model it made a unique contribution to the variance in loneliness scores. But it was also highly associated with the other measures of peer relations, namely social preference (r =0.60) and friendship quality (r=0.25). Previous research also has found moderate correlations between network centrality (proximity prestige being one extension of network centrality) and peer acceptance (r=0.49; see Gest et al., 2001). What is unique about the proximity prestige measure is that it takes into account not only children's direct relations with each other (as do the measures of friendship quality and social preference), but also the indirect relations among children. Therefore, the significant

path coefficient from proximity prestige to loneliness in the integrative model suggests that indirect friends were conducive to a reduction in the experience of loneliness. Although a single indirect nomination contributes little to a child's score for proximity prestige, many "distant nominations" may contribute as much as one "close nomination" (de Nooy, Mrvar, & Batagelj, 2011). Being nominated by one friend with many friends has a direct benefit to the child but also increases the child's number of indirect friends, thus promotes the child's prestige (see Fig.2).

4.2 The mediating rffect of delf-perceived social competence

In the present study, children's subjective cognition about social competence, namely self-perceived social competence, made a greater contribution to loneliness scores than did other aspects of peer relations. This finding is consistent with the results of previous studies (Cheng & Furnham, 2002; Zhou et al., 2005) and suggests that subjective cognition about one's own competence plays a crucial role in the experience of loneliness. This result provides substantial support for cognitive processes theory as it relates to children's loneliness.

Previous studies also have found that cognitive factors mediated the relation between peer relationships and emotion adaption (Hymel et al., 2002). For example, Boivin, Hymel and Bukowski (1995) found that the association between negative peer relations and depression was mediated by the experience of loneliness and cognition about the social environment. The present study tested the mediation effect of self-perceived social competence in the relations between peer variables (friendship quality, social preference, and proximity prestige) and loneliness. The three peer variables lie at the dyadic and group levels of peer relations. When considered individually, each variable not only had a direct effect on the experience of loneliness, but also imposed an indirect effect through the individual's cognition. Using an integrative perspective, the simultaneous influence of the peer variables and self-perceived social competence on loneliness was explored. When all variables were included in the model, friendship quality and social preference still had indirect effects on loneliness, in line with previous studies (Sun et al., 2009; Zhou et al., 2005). However, the indirect effect of proximity prestige was no longer apparent. The results suggest that there is a mediational pathway between some aspects of peer relations and children's loneliness. Specifically, it is possible that the poorer friendship quality or the lower social preference that children have, the more negatively they perceive their social competence, resulting in greater loneliness. These results are consistent with the cognitive processes perspective of loneliness, which holds that loneliness is the distressing feeling accompanying the perception that one's social needs are not being met by the quantity or especially the quality of one's social relationships, and that the experience of loneliness depends on the perception of deficiencies in one's own social relations (Hawkley & Cacioppo, 2010; Peplau & Perlman, 1982; Terrell-Deutsch, 1999).

There are two possible reasons that proximity prestige did not have an indirect influence on loneliness through the mediator of self-perceived social competence in the integrative model. First, the prestige index takes into account a different type of information of which the child may not be aware, namely the social connectedness of the child's friends and the friends' friends. It is likely difficult for children to assess the extent of indirect relationships in the social network and the impact of remote relationships on their own social standing. This would be even more difficult over time, as children's social circles may have changed, for instance, their social connections become more complex (a pattern seen in this study, in which proximity prestige scores increased from grades 3 to 6), and some children may transfer to or from the classes.

Second, prestige is measured based on friendships identified by others; as the reflection of a more objective measure of friendship status, children may be unaware of their prestige in the peer group. The results of the current study suggest that, although the indirect and the objective relationships in the peer network are out of children's perception, they can influence how lonely a child feels, independent of the effects of other aspects of peer relations in which the child is more directly involved.

It should be noted that the present study examined perceived social competence only as the mediator between peer relationships and loneliness based on cognitive processes theory of loneliness. Another possibility is that perceived social competence is itself an important predictor of loneliness, and that the link between perceived social competence and loneliness may be mediated by variables of peer experience. For example, perhaps as potential indicative of actual social competence, self-perceived social competence could influence the extent of children's success in peer relations, which could in turn negatively predict loneliness. One longitudinal study with a sample of Chinese children found that self-perception of social competence had positive and unique contribution to the prediction of social preference two years later (Chen, He, & Li, 2004). Social preference may mediate the association between self-perceived social competence and loneliness. Furthermore, longitudinal research is needed to clarify the relations among children's perceived competence, peer relationships and loneliness.

4.3 The integrative perspective on the research of peer relations

Considering children's peer relations are multifaceted, intersecting and overlapping systems, Gifford-Smith and Brownell (2003) suggested that researchers could be better informed of children's psychosocial development by attending to multiple aspects of the peer experiences simultaneously from an integative perspective. Based on this perspective, the present study conceived and tested one conceptaul model of the links among peer experiences and children's emotional adjustment (loneliness). In this model, three kinds of peer relationships at the dyadic and group levels of peer experiences (friendship quality, social preference, and proximity presitge) were presumed to have simultaneous impacts on children loneliness through the mediator of individuals' cognition (self-perceived social competence). The pathways of the integrative model were compared with those of single peer relationship model. It was found that the indirect effect of proximity prestige on loneliness which was significant in the single peer relationship model was no longer significant in the integrative model. This result implied that the conclusions drawn at single levels of analysis may be not adaptive to the situations with multiple levels. Rubin et al. (2006) noted about children's peer experiences that: the analyses in each level of peer relations—interactions, relationships, groups—are scientifically legitimate and raise interesting questions. However, it has not always clearly demonstrated the important ways in which processes at one level are influenced by those at other levels. Sometimes there were limits for the conclusions drawn at the single levels of analysis. The integrative model of the present study took into consideration the correlations among the three different kinds of peer relationships, which are closer to children's real peer relations.

4.4 Strengths, limitations, and future research

The current study contributes to the literature on the relations between peer experiences and children's loneliness in four distinct ways. First, a social network index was introduced to assess the effect of indirect friends, a topic to which relatively less attention has been afforded in previous research. Proximity prestige reflects the quantity and the closeness of direct and indirect friendships (de Nooy et al., 2005), and it also demonstrates the objective status of a child's friends. Second, based on the cognitive processes theory of loneliness, the present research explored multidimensional aspects of peer relations (friendship quality, social preference, and proximity prestige) in relation to children's loneliness, and an individual cognitive factor (self-perceived social competence) as a possible underlying mechanism responsible for these effects. The mechanism revealed by this research might be suggestive for formulating measures to prevent children from loneliness. Third, under the framework of an integrative and comprehensive perspective of peer experiences, the current study tested the simultaneous influence of the multiple aspects of peer relations on children's loneliness with individual's cognition as a mediator, and compared results to the pathways of the single peer relationship model. The pathways of the integrative model were different from that of the single peer relationship model. The results indicated that it was valuable and necessary to explore the integrative effects of the multiple aspects of peer experiences on children's social development and adjustment. Fourth, much of the previous work on the effects of peer experience on children's loneliness has been conducted within Western cultures, and the current study adds perspectives based on a representative sample from China, a country typified by collectivist culture. The meaning and influence of social interactions for Chinese children may be related to their cognitions of peer relations and loneliness in ways that would be less common in the West. Cross-culture research is a promising avenue for future work.

Several issues should be noted. First, the present study was conducted in a sample of Chinese children. Chinese culture places great value on relational bonds and group cohesion, and the self-evaluation that has been traditionally encouraged places more stress on social connections than the self-evaluation seen in Western cultures (Heine, 2001; Markus & Kitayama, 1991). The neuroimaging research also provided evidence that the "Western self" was different from the Chinese self at a neural level (Zhu, Zhang, Fan, & Han, 2007). In addition, the children of this study were from an urban area of China. Thus, more investigation is needed to determine whether the results of the present study are generalizable to rural areas of China and to other countries and cultures, where most peer research has been conducted.

Second, due to the cross-sectional nature of the data, causal direction cannot be determined in the present study. The results can provide a foundation for future studies but cannot explain the causal mechanisms amongst peer relationships, children's cognition, and feeling of loneliness. Future research with longitudinal data and integrative perspectives is needed to study the dynamics of the relations among peer relationships, individual cognition and loneliness.

Third, the results indicated that there were significant gender differences on all five study variables. Boys felt lonelier than girls, had less positive self-perceived social competence, and had poorer peer relationships (lower friendship quality, social preference and proximity prestige). The analyses in this study focused on the general patterns of relations among the variables, without taking gender differences into consideration. Theory and research suggest that, compared to boys, girls are more involved in close relationships, receive greater emotional provisions in their friendships, more likely to adopt connection-oriented goals and pay greater concerns to social evaluation (for a review, see Rose & Rudolph, 2006). There may be gender differences in the patterns of relations among peer relationships, self-cognition (e.g., self-perceived social competence), and children's loneliness. Research on this topic would be benefit to provide process-oriented interpretation of gender differences in loneliness.

Another limitation with the analyses was that the present study did not take the nested nature of the data into consideration. The children were nested in eight classes and four grades, and there were significant grade differences in proximity prestige. Children's friendship networks may become more complex and closer with their increasing grades. Future studies could apply the analysis of multi-level models to provide more information about the effects of the class or grade.

The internal consistency of the Harter's Perceived Social Competence subscale was low (.56) in this study. Hoyle and Kenny (1999) showed that low reliability in the measure of the mediator leaded to underestimation of true mediation effect when analyzing observed variables. The simulation studies found that the bias associated with the mediation effect was approximately equal to the product of the reliabilities of two composite measures (the measures of the independent variable and the mediator; see Ledgerwood & Shrout, 2011). For instance, when the reliabilities of the two measures both were moderate (.70), the estimated effect was approximately half (.49) the size of the effect used to generate the data. It was supposed that the independent variable was measured without error, and the reliability of the measure of the mediator was .56, the estimated effect was approximately half (.56) the size of the real effect. The simulation studies also confirmed that the analysis of observed variables generally produces acceptable Type I error rates when there is in fact no real indirect effect (i.e., when pathway from independent variable to mediator and/or pathway from mediator to dependent variable is set to 0 in the simulation; see Ledgerwood & Shrout, 2011). Therefore, although the mediator was measured with low reliability in the present study, the results discovered in this study were credible. In addition, the reliability of the Harter's Perceived Social Competence subscale varied across studies in different cultural contexts. Other research in China with the same age group also found low reliability, with Cronbach's alphas of .67 and .64 (Chen, He, & Li, 2004; Stigler et al., 1985). This subscale has also been found to have low reliability in Brazil (.50) and Italy (.57) samples (Chen, Zappulla,Coco, et al., 2004). However, the subscale has shown good reliability in North American samples (e.g., .80 with Canadian children and .75 to .84 with U.S. children; see Chen, Zappulla, Coco, et al., 2004; Harter, 1982). This range of values suggests that the internal consistency of the subscale was influenced by translation or cultural differences. Therefore, more efforts are needed to explore influences on the reliability of the Harter's Perceived Social Competence subscale, especially those due to cultural factors.

Finally, the test–retest correlations were moderate for the measures of loneliness (.42), self–perceived social competence (.48), and friendship quality (.40) over a one–year interval. The temporal stability of the measures decreases along with the interval increases. During the one–year interval, the classroom rosters may be changed across the school years. Meanwhile, the children in middle childhood may experience dramatically changes in peer relations (such as, the expansion of peer circle, increase in the number of friends) and cognitive abilities. These factors are all likely to lower the temporal stability of the measures. Future research would be informative to provide the test–retest correlations of the measures over shorter time intervals.

4.5 Implications for practice

Results from the present study may have important implications for enriching practical work in this domain. Hawkley and Cacioppo (2010) summarized interventions for loneliness, and concluded that there were four main types of interventions: (1) enhancing social skills, (2) providing social support, (3) increasing opportunities for social interaction, and (4) addressing maladaptive social cognition. A meta–analysis of loneliness reduction interventions revealed that among the randomized studies, interventions that addressed maladaptive social cognition had a larger mean effect size compared to interventions that addressed social support, social skills, and opportunities for social intervention (Masi et al., 2011). The current study also found that a cognitive factor, self–perceived social competence, made a significant contribution to the variance in children's loneliness, and direct peer relations (friendship quality and social preference) were related to loneliness partially through self–perceived social competence.

As indicated by the current findings and the results of other research, it may be effective to prevent or diminish children's loneliness by changing maladaptive social cognitions such as sensitivity to social threats, preferential attention to negative social information, and negative social expectations. Meanwhile, in light of the cognitive level of children, it might be better to offer lonely children more opportunities for social interaction along with attempts to change their social cognition. A high quality friendship with a best friend, having friends who themselves have many friends, and acceptance by the peer group all appear beneficial for preventing the feelings of loneliness. Therefore, a three–tiered model is proposed to ameliorate social/emotional/behavioral concerns in order to prevent children's loneliness, and the improvements can be implemented in the classroom. Firstly, the teachers need to construct friendly atmosphere of peer relations, and encourage students to help each other. Secondly, based the effects of peer relations found in the present study, some peer–mediated interventions can be recommended. (a) In light of the finding that indirect friends might also make contributions to buffering children from feelings of loneliness, the teachers could guide highly prestigious children to befriend lonely classmates. This measure is beneficial to increase the friends of the lonely children, and also a way to directly and indirectly provide lonely children with more opportunities for peer interaction and potentially improved peer relations. (b) More class activities involving all classmates can be organized to increase peer interaction and peer acceptance for lonely children. (c) Some social skills can be taught to children for making and boosting their friendships, such as learning how to join in an activity or game, communicating (e.g., talking and listening), cooperating (e.g., taking turns and sharing materials), validating or supporting (e.g., giving attention or help), resolving conflicts (Oden & Asher, 1977). Finally, individual help should be provided to lonely children for changing their maladaptive social cognitions. For instance, cognitive behavioral therapy (CBT), which targets maladaptive social cognitions is an effective approach for reducing loneliness. The lonely children can be taught to identify automatic negative thoughts and look for disconfirming evidence, to decrease biased cognitions, and/or to reframe perceptions of loneliness and personal control (McWhirter, 1990).

4.6 Conclusions

The present study provided valuable insights on the role of peer relations in understanding children's loneliness. Unlike previous research on loneliness, this study assessed peer relations at multiple levels of social complexity. A key finding was that children's self-perceived social competence predicted children's loneliness both directly and as a mediator of associations between other peer variables and loneliness. A unique contribution of this literature was the use of a proximity prestige index, which documented that "friends of friends" may act as a buffer against loneliness. Together, the results suggest possible pathways by which peer relations, assessed both subjectively and objectively, might contribute to children's feelings of loneliness, thus laying the groundwork for identifying and intervening with lonely children.

References

Asher, S. R., & Paquette, J. A. (2003). Loneliness and peer relations in childhood. *Current Directions in Psychological Science, 12*,75–78. http://dx.doi.org/10.1111/1467-8721.01233

Asher, S. R., & Wheeler, V. A. (1985). Children's loneliness: A comparison of rejected and neglected peer status. *Journal of Consulting and Clinical Psychology, 53*, 500–505. http://dx.doi.org/10.1037/0022-006X.53.4.500

Baraldi, A. N., & Enders, C. K. (2010). An introduction to modern missing data analyses. *Journal of School Psychology, 48*, 5–37. http://dx.doi.org/10.1016/j.jsp.2009.10.001

Benner, A. D. (2011). Latino adolescents' loneliness, academic performance, and the buffering nature of friendships. *Journal of Youth and Adolescence, 40*, 556–567. http://dx.doi.org/10.1007/s10964-010-9561-2

Berguno, G., Leroux, P., McAinsh, K., & Shaikh, S. (2004). Children's experience of loneliness at school and its relation to bullying and the quality of teacher interventions. *The Qualitative Report, 9*, 483–499.

Boivin, M., Hymel, S., & Bukowski, W. M. (1995). The roles of social withdrawal, peer rejection, and victimization by peers in predicting loneliness and depressed mood in childhood. *Development and Psychopathology, 7*, 765–785. http://dx.doi.org/10.1017/S0954579400006830

Brownell, C. A., & Gifford-Smith, M. E. (2003). Context and development in children's school-based peer relations: Implications for research and practice. *Journal of School Psychology, 41*, 305–310. http://dx.doi.org/10.1016/s0022-4405(03)00052-9

Cillessen, A. H. N., & Marks, P. E. L. (2011). Conceptualizing and measuring popularity. In A. H. N. Cillessen, D. Schwartz, & L. Mayeux (Eds.), *Popularity in the peer system* (pp. 25 - 56). New York: Guilford Press.

Chamberlain, B., Kasari, C., & Rotheram-Fuller, E. (2007). Involvement or isolation? The social networks of children with autism in regular classrooms. *Journal of Autism and Developmental Disorders, 37,* 230–242. http://dx.doi.org/10.1007/s10803-006-0164-4

Chen, X. (2000). Growing up in a collectivistic culture: Socialization and socio-emotional development in Chinese children. *International perspectives on human development.* In A. L. Comunian & U. P. Gielen (Eds.), International perspectives on human development (pp. 331- 353). Lengerich, Germany: Pabst Science Publishers.

Chen, X., He, Y., & Li, D. (2004). Self-perceptions of social competence and self-worth in Chinese children: Relations with social and school performance. *Social Development, ,* 570–589. http://dx.doi.org/10.1111/j.1467-9507.2004.00284.x

Chen, X., He, Y., De Oliveira, A. M., Coco, A. L., Zappulla, C., Kaspar, V., DeSouza, A. (2004). Loneliness and social adaptation in Brazilian, Canadian, Chinese and Italian children: A multi-national comparative study. *Journal of Child Psychology and Psychiatry, 45,* 1373–1384. http://dx.doi.org/10.1111/j.1469-7610.2004.00329.x

Chen, X., Rubin, K. H., & Sun, Y. (1992). Social reputation and peer relationships in Chinese and Canadian children: A cross-cultural study. *Child Development, 63,* 1336–1343. http://dx.doi.org/10.1111/j.1467-8624.1992.tb01698.x

Chen, X., Zappulla, C., Coco, A.L., Schneider, B., Kaspar, V., De Oliveira, A. M.,Bergeron, N. (2004). Self-perceptions of competence in Brazilian, Canadian, Chinese and Italian children: Relations with social and school adjustment. *International Journal of Behavioral Development, 28,* 129–138. http://dx.doi.org/10.1080/01650250344000334

Cheng, H., & Furnham, A. (2002). Personality, peer relations, and self-confidence as predictors of happiness and loneliness. *Journal of Adolescence, 25,* 327–339. http://dx.doi.org/ 10.1006/jado.2002.0475

Cohen, J. (1988). *Statistical power analysis for the behavioral sciences.* New Jersey: Erlbaum.

Coie, J. D., Dodge, K. A., & Coppotelli, H. (1982). Dimensions and types of social status: A cross-age perspective. *Developmental Psychology, 18,* 557–570. http://dx.doi.org/10.1037/0012-1649.18.4.557

Cole, D. A. (1991). Change in self-perceived competence as a function of peer and teacher evaluation. *Developmental Psychology, 27,* 682–688. http://dx.doi.org/10.1037/0012-1649.27.4.682

de Nooy, W., Mrvar, A., & Batagelj, V. (2011). *Exploratory social network analysis with Pajek* (2nd ed.). New York: Cambridge University Press.

Ennett, S. T., Bauman, K. E., Hussong, A., Faris, R., Foshee, V. A., Cai, L., & DuRant, R. H. (2006). The peer context of adolescent substance use: Findings from social network analysis. *Journal of Research on Adolescence, 16,* 159–186. http://dx.doi.org/10.1111/j.1532-7795.2006.00127.x

Fairchild, A. J., & McQuillin, S. D. (2010). Evaluating mediation and moderation effects in school psychology: A presentation of methods and review of current practice. *Journal of School Psychology, 48,* 53–84. http://dx.doi.org/10.1016/j.jsp.2009.09.001

Faris, R., & Ennett, S. (2012). Adolescent aggression: The role of peer group status motives, peer aggression, and group characteristics. *Social Networks, 34,* 371–378. http://dx.doi.org/10.1016/j.socnet.2010.06.003

Fox, N. A., Henderson, H. A., Marshall, P. J., Nichols, K. E., & Ghera, M. M. (2005). Behavioral inhibition: Linking biology and behavior within a developmental framework. *Annual Review of Psychology, 56,* 235–262. http://dx.doi.org/10.1146/annurev.psych.55.090902.141532

Gest, S. D., Graham-Bermann, S. A., & Hartup, W. W. (2001). Peer experience: Common and unique features of number of friendships, social network centrality, and sociometric status. *Social Development, 10*, 23-40. http://dx.doi.org/10.1111/1467-9507.00146

Gifford-Smith, M. E., & Brownell, C. A. (2003). Childhood peer relationships: Social acceptance, friendships, and peer networks. *Journal of School Psychology, 41*, 235-284. http://dx.doi.org/10.1016/S0022-4405(03)00048-7

Harris, R. A., Qualter, P., & Robinson, S. J. (2013). Loneliness trajectories from middle childhood to pre-adolescence: Impact on perceived health and sleep disturbance. *Journal of Adolescence, 36*, 1295-1304. http://dx.doi.org/10.1016/j.adolescence.2012.12.009

Harter, S. (1982). The perceived competence scale for children. *Child Development, 53*, 87-97. http://dx.doi.org/10.1111/j.1467-8624.1982.tb01295.x

Hawkley, L. C., & Cacioppo, J. T. (2010). Loneliness matters: A theoretical and empirical review of consequences and mechanisms. *Annals of Behavioral Medicine, 40*, 218-227. http://dx.doi.org/10.1007/s12160-010-9210-8

Heine, S. J. (2001). Self as cultural product: An examination of East Asian and North American selves. *Journal of Personality, 69*, 881-905. http://dx.doi.org/10.1111/1467-6494.696168

Heinrich, L. M., & Gullone, E. (2006). The clinical significance of loneliness: a literature review. *Clinical Psychology Review, 26*, 695-718. http://dx.doi.org/10.1016/j.cpr.2006.04.002

Hinde, R. A. (1987). Individuals, relationships and culture: *Links between ethology and the social sciences*. New York: Cambridge University Press.

Hornby, A. S. (2010). *Oxford advanced learner's dictionary* (8th ed.). Oxford: Oxford University Press.

Hoyle, R. H., & Kenny, D. A. (1999). Sample size, reliability, and tests of statistical mediation. In R. H. Hoyle (Ed.), *Statistical strategies for small sample research* (pp. 195-222). Thousand Oaks, CA: Sage.

Hoza, B., Bukowski, W. M., & Beery, S. (2000). Assessing peer network and dyadic loneliness. *Journal of Clinical Child Psychology, 29*, 119-128. http://dx.doi.org/10.1207/S15374424jccp2901_12

Hymel, S., Vaillancourt, T., McDougall, P., & Renshaw, P. D. (2002). Peer acceptance and rejection in childhood. In P. K. Smith & C. H. Hart (Eds.), *Blackwell handbook of childhood social development* (pp. 265-284). Malden: Blackwell.

Jobe-Shields, L., Cohen, R., & Parra, G. R. (2011). Patterns of change in children's loneliness: Trajectories from third through fifth grades. *Merrill-Palmer Quarterly, 57*, 25-47.

Ladd, G. W. (1999). Peer relationships and social competence during early and middle childhood. *Annual Review of Psychology, 50*, 333-359. doi: 10.1146/annurev.psych.50.1.333

Ladd, G. W., Buhs, E. S., & Troop, W. (2002). Children's interpersonal skills and relationships in school settings: Adaptive significance and implications for school-based prevention and intervention programs. In K. Smith & C. H. Hart (Eds.), *Blackwell handbook of childhood social development* (pp. 394-415). Malden: Blackwell.

Ladd, G. W., Kochenderfer, B. J., & Coleman, C. C. (1997). Classroom peer acceptance, friendship, and victimization: Destinct relation systems that contribute uniquely to children's school adjustment? *Child Development, 68*, 1181-1197. http://dx.doi.org/10.2307/1132300

Ledgerwood, A., & Shrout, P. E. (2011). The trade-off between accuracy and precision in latent variable models of mediation processes. *Journal of personality and social psychology, 101*, 1174-1188. http://dx.doi.

org/10.1037/a0024776

Lin, N. (1976). *Foundations of Social Research*. New York: McGraw-Hill.

Luo, G. (1996). Chinese traditional social and moral ideas and rules. Beijing, China: *The University of Chinese People Press*.

Margalit, M. (2010). *Lonely children and adolescents*. New York: Springer.

Marks, P. E. L., Babcock, B., Cillessen, A. H. N., & Crick, N. R. (2012). The effects of participation rate on the internal reliability of peer nomination measures. *Social Development, 22*, 609–622. http://dx.doi.org/10.1111/j.1467-9507.2012.00661.x

Markus, H. R., & Kitayama, S. (1991). Culture and the self: Implications for cognition, emotion, and motivation. *Psychological Review, 98*, 224–253. http://dx.doi.org/10.1037/0033-295X.98.2.224

Masi, C. M., Chen, H. Y., Hawkley, L. C., & Cacioppo, J. T. (2011). A meta-analysis of interventions to reduce loneliness. *Personality and Social Psychology Review, 15*, 219–266. doi: http://dx.doi.org/10.1177/1088868310377394

McWhirter, B. T. (1990). Loneliness: A review of current literature, with implications for counseling and research. *Journal of Counseling & Development, 68*, 417–422. http://dx.doi.org/10.1002/j.1556-6676.1990.tb02521.x

Mercken, L., Steglich, C., Sinclair, P., Holliday, J., & Moore, L. (2012). A longitudinal social network analysis of peer influence, peer selection, and smoking behavior among adolescents in British schools. *Health Psychology, 31*, 450–459. http://dx.doi.org/10.1037/a0026876

Mouratidis, A. A., & Sideridis, G. D. (2009). On social achievement goals: Their relations with peer acceptance, classroom belongingness, and perceptions of loneliness. *The Journal of Experimental Education, 77*, 285–308. http://dx.doi.org/10.3200/JEXE.77.3.285–308

Mundt, M. P. (2011). The impact of peer social networks on adolescent alcohol use initiation. *Academic Pediatrics, 11*, 414–421. http://dx.doi.org/10.1016/j.acap.2011.05.005

Muthén, L. K. , & Muthén, B. O. (1998–2012). *Mplus user´s guide* (7th ed.). Los Angeles, CA: Muthén & Muthén.

Nangle, D. W., Erdley, C. A., Newman, J. E., Mason, C. A., & Carpenter, E. M. (2003). Popularity, friendship quantity, and friendship quality: Interactive influences on children's loneliness and depression. *Journal of Clinical Child and Adolescent Psychology, 32*, 546–555. http://dx.doi.org/10.1207/S15374424JCCP3204_7

Oden, S., & Asher, S. R. (1977). Coaching children in social skills for friendship making. *Child Development, 48*, 495–506. http://dx.doi.org/10.2307/1128645

Okamoto, J., Johnson, C. A., Leventhal, A., Milam, J., Pentz, M. A., Schwartz, D., & Valente, T. W. (2011). Social network status and depression among adolescents: An examination of social network influences and depressive symptoms in a Chinese sample. *Research in Human Development, 8*, 67–88. http://dx.doi.org/10.1080/15427609.2011.549711

Oyserman, D., Coon, H. M., & Kemmelmeier, M. (2002). Rethinking individualism and collectivism: Evaluation of theoretical assumptions and meta-analyses. *Psychological Bulletin, 128*, 3–72. http://dx.doi.org/10.1037/0033-2909.128.1.3

Parker, J. G., & Asher, S. R. (1993). Friendship and friendship quality in middle childhood: Links with peer group acceptance and feelings of loneliness and social dissatisfaction. *Developmental Psychology, 29*, 611–621. http://dx.doi.org/10.1037/0012-1649.29.4.611

Parker, J. G., & Seal, J. (1996). Forming, losing, renewing, and replacing friendships: Applying temporal parameters to the assessment of children's friendship experiences. *Child Development, 67*, 2248–2268. http://dx.doi.org/10.1111/j.1467-8624.1996.tb01855.x

Paxton, P., Curran, P. J., Bollen, K. A., Kirby, J., & Chen, F. (2001). Monte Carlo experiments: Design and implementation. *Structural Equation Modeling, 8*, 287–312. http://dx.doi.org/10.1207/S15328007SEM0802_7

Peplau, L. A., & Perlman, D. (1982). Perspectives on loneliness. In L. A. Peplau & D. Perlman (Eds.), Loneliness: *A sourcebook of current theory, research and therapy* (pp. 1–18). New York: Wiley.

Qualter, P., Brown, S. L., Munn, P., & Rotenberg, K. J. (2010). Childhood loneliness as a predictor of adolescent depressive symptoms: An 8–year longitudinal study. *European Child & Adolescent Psychiatry, 19*, 493–501. http://dx.doi.org/10.1007/s00787-009-0059-y

Rokach, A., & Bauer, N.. (2004). Age, culture, and loneliness among Czechs and Canadians. *Current Psychology, 23*, 3–23. http://dx.doi.org/10.1007/s12144-004-1005-2

Rose, A. J., & Rudolph, K. D. (2006). A review of sex differences in peer relationship processes: Potential trade-offs for the emotional and behavioral development of girls and boys. *Psychological Bulletin, 132,* 98–131. http://dx.doi.org/10.1037/0033-2909.132.1.98

Rotenberg, K. J. (1999). Childhood and adolescent loneliness: An introduction. In K. J. Rotenberg & H. Shelley (Eds.), *Loneliness in childhood and adolescence* (pp. 3–8). Cambridge, UK: Cambridge University Press. http://dx.doi.org/10.1017/CBO9780511551888.001

Rubin, K. H., Bukowski, W. M., & Parker, J. G. (2006). Peer interactions, relationships, and groups. In N. Eisenberg (Ed.), Handbook of child psychology, *Vol.3: Social, emotional, and personality development* (6th ed., pp. 571–645). Hoboken, NJ: Wiley. http://dx.doi.

org/10.1002/9780470147658.chpsy0310

Scott, J. (2000). Social network analysis: *A handbook.* London: Sage Publications Ltd.

Seepersad, S., Choi, M. K., & Shin, N. (2008). How does culture influence the degree of romantic loneliness and closeness? The Journal of Psychology: *Interdisciplinary and Applied, 142*, 209–220. http://dx.doi.org/10.3200/JRLP.142.2.209-220

Sheridan, S. M., Buhs, E. S., & Warnes, E. D. (2003). Childhood peer relationships in context. *Journal of School Psychology, 41*, 285–292. http://dx.doi.org/10.1016/s0022-4405(03)00049-9

Shin, Y. (2007). Peer relationships, social behaviours, academic performance and loneliness in Korean primary school children. *School Psychology International, 28*, 220–236. http://dx.doi.org/10.1177/0143034307078103

Sletta, O., Val?s, H., Skaalvik, E., & S?bstad, F. (1996). Peer relations, loneliness, and self-perceptions in school-aged children. *British Journal of Educational Psychology, 66*, 431–445. http://dx.doi.org/10.1111/j.2044-8279.1996.tb01210.x

Stigler, J. W., Smith, S., & Mao, L-W. (1985). The self-perception of competence by Chinese children. *Child Development, 56*, 1259–1270. http://dx.doi.org/10.2307/1130241

Sun, X., Zhou, Z., Fan, C., & Ke, S. (2009). Loneliness in middle childhood and its relation to multi-level peer experience. *Psychological Science (China), 32*, 567–570.

Terrell-Deutsch, B. (1999). The conceptualization and measurement of childhood loneliness. In K. J. Rotenberg & S. Hymel (Eds.), *Loneliness in childhood and adolescence* (pp. 11–33). New York: Cambridge University Press. http://dx.doi.org/10.1017/CBO9780511551888.002

Terry, R. (2000). Recent advances in measurement theory

and the use of sociometric techniques. *New Directions for Child and Adolescent Development, 2000*(88), 27–53. http://dx.doi.org/10.1002/cd.23220008805

Uruk, A. C., & Demir, A. (2003). The role of peers and families in predicting the loneliness level of adolescents. The Journal of Psychology: *Interdisciplinary and Applied, 137,* 179–193. http://dx.doi.org/10.1080/00223980309600607

van Dulmen, M. H. M., & Goossens, L. (2013). Loneliness trajectories. *Journal of Adolescence, 36,* 1247–1249. http://dx.doi.org/10.1016/j.adolescence.2013.08.001

Vanhalst, J., Goossens, L., Luyckx, K., Scholte, R. H.J., & Engels, R. C.M.E. (2012). The development of loneliness from mid– to late adolescence: Trajectory classes, personality traits, and psychosocial functioning. *Journal of Adolescence, 36,* 1305–1312. http://dx.doi.org/10.1016/j.adolescence.2012.04.002

Wassermann, S., & Faust, K. (1994). Social network analysis: *Methods and applications.* New York: Cambridge University Press.

Weiss, R. S. (1973). Loneliness: *The experience of emotional and social isolation.* Cambridge, MA: MIT Press.

Yu, G., Zhang, Y., & Yan, R. (2005). Loneliness, peer acceptance, and family functioning of Chinese children with learning disabilities: Characteristics and relationships. *Psychology in the Schools, 42,* 325–331. http://dx.doi.org/10.1002/pits.20083

Zhou, Z., Sun, X., Xiang, Y., & Liu, J. (2007). Peer interaction and loneliness in middle childhood: A cross–lagged analysis. *Psychological Science (China), 30,* 1309–1313.

Zhou, Z., Sun, X., Zhao, D., & Hsueh, Y. (2005). The test of the mediator variable between peer relationship and loneliness in middle childhood. *Acta Psychologica Sinica (China), 37,* 776–783.

Zhou, Z., Zhao, D., Sun, X., & Ding, X. (2006). Children's experiences with peers and loneliness: A two–year longitudinal study. *Acta Psychologica Sinica (China), 38,* 743–750.

Zhu, Y., Zhang, L., Fan, J., & Han, S. (2007). Neural basis of cultural influence on self–representation. *Neuroimage, 34,* 1310–1316. http://dx.doi.org/10.1016/j.neuroimage.2006.08.047

心理辅导研究所团队成员

江光荣，博士，教授，华中师范大学心理辅导研究所所长，主要研究领域为心理咨询与治疗的过程与效果、青少年自杀与自伤行为、国民心理健康素养与心理求助行为、学校心理健康教育。

陶嵘，博士，副教授，华中师范大学心理辅导研究所副所长，主要研究领域为情绪异常的心理生理机制及心理治疗，团体心理治疗、中国文化下的心理治疗、梦的工作等。

吴才智，博士，副教授，主要研究领域为心理健康教育、心理辅导的组织与管理、危机干预与自杀预防。

夏勉，博士，副教授，主要研究领域为中国人的求助行为与助人行为、心理咨询与治疗。

孙启武，博士，副教授，主要研究领域为基干依恋理论的心理病理与治疗改变机制、创伤心理学、心理咨询与治疗中的文化维度、心理咨询与治疗中的评估与诊断及咨询伦理等。

朱旭，博士，讲师，主要研究领域为心理咨询的过程与效果、咨询关系、团体治疗。

于丽霞，博士，讲师，主要研究领域为心理咨询与治疗、青少年自杀与自伤行为。

段文婷，硕士，讲师，主要研究领域为心理健康教育、心理咨询与治疗、大学生求助决策。

心理学报 , 2009, 14(3), 259‒266.

暴力电子游戏的短期脱敏效应：两种接触方式比较 *

郭晓丽[1,2]，江光荣[1]，朱　旭[1]

（[1] 华中师范大学心理学院，人的发展与心理健康湖北省重点实验室，武汉 430079 ）

（[2] 华中科技大学学生工作处，武汉 430074)

摘　要　比较主、被动接触暴力电子游戏的脱敏效应，以 44 名男性大学生为被试，利用生物反馈仪测量被试主动参与游戏或被动观看游戏录像前后，及随后观看暴力视频过程中皮电与心率的变化（脱敏效应的生理指标）。结果表明 :(1) 暴力电子游戏可以产生脱敏效应。接触游戏 15 分钟后，暴力游戏组观看暴力视频过程中皮电的增加值明显小于非暴力游戏组 ;(2) 游戏的接触方式对于脱敏效应的程度无显著影响，但主动参与组对于游戏内容知觉到更高的愉快与更低的沮丧。

关键词　暴力电子游戏 ; 脱敏效应 ; 接触方式

分类号　R395

1　问题提出

人们相信娱乐媒体中的暴力对于现实生活中的攻击行为有不可推卸的责任。美国 3 个科学委员会的调查——外科综合委员会 (SGC) 报告，国家心理卫生研究院 (NIMH)10 年跟踪，以及美国心理学会暴力与青少年委员会调查——断定观看暴力可以增加攻击行为 (Mullin & Daniel, 1995)。

日常生活中，人们通过电影、电视等基于屏幕的媒体接触到的暴力远远大于在真实生活中接触到的暴力。过去，电视与电影是主要的基于屏幕的媒体，接触方式主要是观看，但是当游戏出现后，人们接触暴力的方式就从"观看"发展到"参与"了。在过去 30 年中，电子游戏产业发展迅速，伴随着游戏产业的暴发式增长，公众对于接触电子游戏的负面效应

的担忧也与日俱增。担忧的原因之一是现有游戏中暴力内容的普遍性 (VanMierlo & VandenBulck, 2004)。一项对于游戏内容的分析表明，89% 的游戏含有部分暴力内容，50% 的游戏含有对于其他人物的严重暴力内容 (Dill & Dill, 1998)。这个百分比足以证明"暴力"是游戏的一个普遍特点。另一个原因是大量研究揭露接触暴力媒体的确有着消极效应。一项元分析研究发现接触暴力电子游戏导致攻击行为、攻击情感、攻击认知与生理唤起的增加，以及亲社会行为的减少 (Anderson & Bushman 2001)。美国心理学会暴力与青少年委员会还指出，攻击行为的增加不是观看暴力的唯一消极后果，观看暴力还可能会增加对于暴力的脱敏，导致对于暴力麻木的态度，从而降低暴力发生时采取行动帮助受害者的可能性 (Mullin & Daniel, 1995)。

* 本文获人的发展与心理健康湖北省重点实验室资助，资助项目为"基于网络平台的学校心理健康教育服务系统建设研究"(0601)。

通讯作者 : 江光荣 , E-mail:grjiang@mail.ccnu.edu.cn

脱敏 (desensitization) 一词的概念源于临床心理学中的系统脱敏治疗，指反复面对一个会导致焦虑、恐惧的刺激时，焦虑、恐惧等负面情绪反应逐渐消退的现象。暴力脱敏则特指持续暴露于暴力刺激时的情绪反应钝化现象。这种脱敏在生理上通常表现为皮肤电、心率的降低或血流量的减少。在有关媒体暴力脱敏的研究中，以上3种指标被公认为是反映脱敏的有效生理指标。脱敏并非简单的生理变化，它伴随着一系列认知与行为改变。暴力脱敏效应不仅可以导致暴力行为的增加，还有可能减少助人行为。暴力脱敏效应还可能是导致攻击人格形成的重要途径之一 (Carnagey, Anderson, & Bushman,2007)。

Anderson 和 Bushman(2002) 提出了一般攻击模型 (General Aggression Model) 来解释暴力媒体对于攻击行为的影响机制。根据一般攻击模型，个体变量（例如：敌意特质、对暴力的态度）和情境变量（例如：媒体暴力）交互影响个体的内部状态，认知、情感与生理唤起彼此作用，整体上影响个体对攻击行为的解释，并进一步影响后续的决策过程与行为表现 (Anderson, Carnagey, & Eubanks,2003;Gentile & Anderson, 2003;Kirsh, 2003)。一般攻击模型指出，暴力电子游戏对于攻击行为可以产生短期与长期效应 (Anderson, 2004; Carnagey & Anderson, 2005; Gentile & Lynch,2004)。在短期效应中，游戏作为情境变量导致攻击认知、生理唤起水平的提高，以及攻击情感的增加。在长期效应中，游戏通过促进形成攻击信念和攻击态度，促进产生攻击图式、攻击行为与攻击期望，以及降低个体对攻击的敏感性来影响攻击行为。相应的，这些因素也促进了个体攻击性人格的发展 (Anderson & Griffiths, 2004)。

一般攻击模型提供了一个有用的社会——认知框架来理解脱敏过程。人类对于暴力内容的原始反应是害怕与焦虑，当暴力刺激在积极情绪反应伴随下重复呈现时，这种原始的焦虑反应会由于抗条件作用 (counterconditioning) 而减弱。暴力脱敏发生的表现之一是个体重复接触媒体暴力后，在接触真实暴力时生理唤起水平降低。一旦脱敏发生，真实暴力就不再能引发原始的恐惧与焦虑。个体在认知上也发生相应的改变。例如，可能更少注意到攻击事件，知觉到较少或较轻的伤害，对于暴力受害者的同情降低等。

国外早在六七十年代就已经有人开始研究电视或电影暴力内容的脱敏效应。早先关于暴力媒体潜在脱敏效应的研究以被试的皮肤电为指标，发现被试对于电影中流血场面的观看降低了之后对于类似场面的生理唤起，表现出脱敏反应 (Cline & Croft,1973;Thomas & Horton, 1977)。同时，相关研究的结果也表明，过去对暴力影视的接触与所呈现的暴力视频的脱敏效应有正相关 (Funk & Baldacci,2004;Funk & Bushman, 2003)。

2000 年之后，开始出现少量针对暴力电子游戏脱敏效应的研究。相关研究的结果表明过去对于游戏暴力的接触与低同情、高亲暴力态度有关。而实验研究的结果表明暴力电子游戏不但可以导致对于媒体暴力的脱敏，对于真实暴力同样可以表现出脱敏。以往反映脱敏效应的指标为问卷或皮电、心率等传统生理指标，最近的研究开始尝试利用事件相关电位分析脱敏发生的过程 (Bartholow, Bushman, & Sestir, 2006)。

暴力媒体的负面效应也开始引起国内研究者的重视，目前国内的相关文献一部分集中于暴力媒体影响机制的理论探讨（辛自强，池丽萍，2004; 郑宏明，孙延军，1997)，少量的实证研究则关注媒体暴力对内隐攻击认知的影响（陈美芬，陈舜蓬，2005; 崔丽娟，胡海龙，吴明证，2006)。虽有一些类似的提法（张镇，刘月霞，张建新，2006)，但没有看到暴力媒体情绪脱敏效应的实证研究。

与传统的影视媒体相比，电子游戏作为一种新型媒体具有新的特点：参与性与互动性。游戏接触者会与虚拟的对手发生互动，做出暴力行为，而不仅仅是被动地观看暴力，因此在游戏过程中注意与情感卷入的水平可能更高，对于暴力角色有更多的认同，暴力行为也得到更有力的强化，因此玩暴力游戏对于他们的负面影响可能比观看暴力影视更为严重。

这种"主动参与"与"被动接受"的区别已经有不少研究者提及 (Anderson, Berkowitz, & Donnerstein,

2003; Anderson & Bushman, 2002;Anderson & Carnagey, 2004;Griffths, 1999),但到目前为止,还缺乏有力的实证比较。这里可能存在一个技术上的困难,即保证"主动参与暴力游戏"与"被动观看暴力游戏"除自变量外无关变量(如画面内容、暴力程度、图像质量等)的等值性。

综观国内外对于暴力媒体的研究,可以发现存在这样一些特点及问题:(1) 相关研究设计较多(探索暴力媒体的暴露量与另一些变量如攻击认知、情感与行为的相关),缺乏对改变机制(如脱敏效应)的实验研究;(2) 脱敏效应的研究多集中于暴力影视,针对暴力电子游戏的研究较少;(3) 有关暴力电子游戏由于主动参与的特点而有更严重脱敏效应的看法仅仅是理论假设,未经实证检验。

本研究聚焦于暴力电子游戏所导致的短期心理脱敏效应,并加入"暴露方式"这一新的自变量("主动参与"对"被动接受"),以实验研究设计检验有关暴力游戏脱敏效应的以下假设:

H1: 暴力电子游戏能够产生脱敏效应———接触游戏 15 分钟之后,暴力游戏组观看视频过程中生理指标的增加值显著低于非暴力游戏组。

H2: 主动参与暴力游戏的脱敏效应程度大于被动接受暴力游戏———主动参与组观看视频过程中生理指标的增加值显著低于被动接受组。

假设一着眼于暴力电子游戏的暴力内容所产生的即时效应,如果得到验证则说明暴力电子游戏的心理后效符合一般攻击模型,与已有的暴力影视的效应有共同的理论依据。假设二则基于暴力接触方式,探索"主动参与式"与"被动接受式"暴露于暴力媒体心理效应的区别。

2　研究方法

2.1　被试

通过广告招募大学生被试,为控制性别变量,被

试全部为男性。要求被试有一定电脑使用的经验。共有 44 名学生参加实验,年龄从 19 到 26 岁,平均年龄为 21 岁。被试被随机分配到各实验组。实验结束后付给被试一定的报酬。

2.2　测量问卷

2.2.1　Buss-Perry 量表　Buss-Perry 量表 (Buss-Perry Scale) 包含 4 个组成部分:身体攻击分量表,言语攻击分量表,愤怒分量表与敌意分量表 (Buss & Perry, 1992)。本研究使用了其中的身体攻击 (Physical Aggression) 分量表测量被试的身体攻击性水平。该分量表包含 9 个条目,7 点记分,α 系数为有影响,[1] 因此该量表得分作为协变量进入方差分析。

2.2.2　游戏使用习惯问卷　游戏使用习惯问卷 (Video Game Questionnaire) 要求被试列出他们最喜欢的五种游戏,然后就每种游戏的内容与画面的暴力程度以及他们玩游戏的频率评定等级,用以反映被试过去对暴力游戏的接触程度 (Anderson & Dill, 2000)。被试的游戏使用习惯可能会影响自变量的效应,因此所得结果作为协变量进入方差分析。

2.2.3　游戏评定量表　游戏评定量表 (Video Game Rating Sheet) 要求被试对所接触游戏的画面暴力水平、内容暴力水平、动作快慢水平、游戏激动水平、游戏沮丧水平、游戏愉快水平、游戏难度水平进行评定 (Anderson & Dill, 1986)。该量表为 7 点记分,α 系数为 0.78。被试在该量表上的得分反映被试对于游戏的感知,并作为自变量操作有效性的证据。具体就本研究设计来说,如果暴力接触组和非暴力接触组对游戏画面暴力水平和内容暴力水平的评价有差别且前者高于后者,说明对媒体暴力程度的差别性操作有效;如果主、被动接触组在画面暴力水平和内容暴力水平的评价上无差异,说明对媒体暴力程度的等值性操作有效。

2.3　实验材料

2.3.1　暴力电子游戏　游戏的选择遵循以下标准:难度适中(指经过 5 分钟的练习之后,有一定电脑使用经验的人都能够顺利玩游戏)、画面逼真、游戏人物为人类、暴力为主要特征(而非恐怖或色情)、游

---

[1]　被试的身体攻击水平可能对自变量的效应有影响。

戏进度与难度可控。根据以上标准，通过试玩比较，认为《喋血街头 II》(单机版角色扮演动作射击类游戏)最符合以上标准，将其用作暴力游戏的实验材料。

2.3.2 非暴力电子游戏 非暴力游戏的选择关键在于避免游戏中有任何形式的暴力行为，同时考虑到动态抓屏软件的运行效果，最后选定一款弹球休闲游戏作为实验材料。以往的研究中也曾使用过这类游戏 (Anderson, Carnagey, & Flanagan, 2007;Carnagey, Anderson, & Bushman, 2004)。

2.3.3 暴力视频 选取电影《杀死比尔 I》10 分钟的片段，该片断包含十分血腥的暴力镜头。该影片定级为 R 级。该视频资料呈现给所有被试，作为引发情绪唤起的暴力刺激源。

2.4 实验设备

2.4.1 生物反馈仪 该套多参数生物反馈仪由加拿大 Thought 公司生产，品牌为 ProCompInfiniti。

2.4.2 动态抓屏软件 利用虚拟视频软件 V3.0.2(CamtasiaStudioV3.0.2) 录制主动组被试玩游戏的全过程，保存为视频文档作为被动组的实验材料。

2.5 实验设计

实验采用 2×2 完全随机设计，包括两个自变量：游戏暴力程度 (暴力游戏 VS 非暴力游戏)、游戏接触

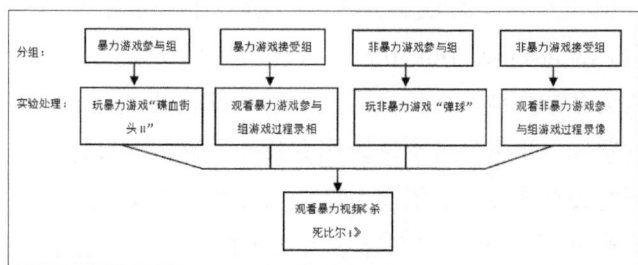

图 1 实验设计

方式 (主动参与游戏 VS 被动接受游戏录像)。因变量为被试的皮电与心率值。具体实验设计见图 1。

游戏接触方式是在以往研究的基础上新加入的变量，分为主动参与组与被动接受组。主动参与组是指被试自己亲身玩游戏，被动接受组是被试观看主动参与组游戏的录像。即一位主动参与组的被试在玩游戏时，将其游戏过程用软件 (CamtasiaStudioV3.0.2) 录制并保存下来，而这段游戏的录像就作为被动接受组

某位被试的实验材料。这样安排是为了使主动组与被动组的实验材料在画面内容与质量上等值，从而实现二组接触的暴力程度等值。经比较，两组被试对于各自游戏暴力水平的主观评定结果无差异 (详见 3.1)。

2.6 实验程序

被试个别进行实验，被告知研究目的是评估不同类型的媒体。具体程序包括以下几步：

（1）签订协议书、填写问卷：协议书中告知被试实验内容与注意事项，之后被试完成《身体攻击分量表》与《游戏使用习惯问卷》。实验结束后，被试完成《游戏评价问卷》。

（2）测量基线：进行 5 分钟心率与皮电的基线测量。

（3）接触游戏：将被试随机分配到四种实验条件中，接触游戏的时间为 15 分钟 (Anderson, Carnagey, & Flanagan, 2007)。为控制被试对于游戏的熟练程度，游戏的难度都设置为最低。

（4）后测：所有被试再次进行 5 分钟心率与皮电的测量。

（5）观看视频：所有被试观看《杀死比尔 I》的 10 分钟片段。在观看过程中，持续测量被试的心率与皮电。

2.7 统计方法

采用 SPSS11.5 进行描述性统计与方差分析。

3 研究结果

3.1 游戏评价问卷结果

2×2 随机方差分析的结果表明，对于游戏画面暴力水平的评定，游戏暴力程度的主效应显著，$F(1, 40)=21.20$, $p<0.001$；对于游戏内容暴力水平的评定，游戏暴力程度的主效应显著，$F(1, 40)=28.47$, $p<0.001$；对于游戏的沮丧水平，游戏暴力程度的主效应显著，$F(1, 40)=13.24$, $p<0.05$，接触方式的主效应显著，$F(1, 40)=13.24$, $p<0.05$；对于游戏的愉快程度，游戏暴力程度的主效应显著，$F(1, 40)=4.32$, $p<0.05$，接触方式的主效应显著，$F(1, 40)=16.45$, $p<0.001$；对于游戏的难度水

表1 各实验组游戏评价问卷项目的平均数与标准差

问卷项目	暴力游戏参与组		暴力游戏接受组		非暴力游戏参与组		非暴力游戏接受组	
	M	SD	M	SD	M	SD	M	SD
画面暴力水平	5.45	1.70	5.27	1.74	2.82	2.27	2.55	1.97
内容暴力水平	5.45	1.29	5.45	1.70	2.64	2.11	2.55	1.92
动作快慢水平	4.55	1.64	3.82	1.78	3.82	1.66	3.00	1.48
游戏激动水平	3.36	1.57	3.36	1.50	3.91	1.76	2.91	1.64
游戏沮丧水平	3.45	1.70	5.36	1.36	2.00	1.10	3.45	1.86
游戏愉快水平	3.55	1.64	2.00	1.41	4.85	1.60	2.64	1.43
游戏难度水平	1.82	1.25	3.64	1.21	2.00	1.00	2.36	1.29

注：每组被试均为11人。

平，接触方式的主效应显著，$F(1, 40)=9.23$, $p<0.05$。所有交互作用均不显著。

以上结果表明，被试对暴力游戏的暴力与血腥程度的评价均显著高于非暴力游戏，说明实验对于暴力程度这一自变量的操作是有效的。除此之外，暴力游戏与非暴力游戏在沮丧程度与愉快程度上也有显著差异。在后续的分析中，将这两个评分作为协变量进行方差分析，结果显示无显著效应。

对于另一个自变量———游戏的接触方式，发现对于游戏画面与内容暴力水平的评定，主动参与组与被动接受组无显著差异，这说明对实验材料在暴力程度上等值性的操作是有效的。结果显示主动参与组与被动接受组在游戏难度、愉快程度与沮丧程度三个维度上均有显著差异。主动参与组认为游戏更简单，感受到更低的沮丧和更高的愉快。这说明对于相同的游戏内容，不同的接触方式会使得被试的感受有所区别，主动参与者对游戏过程有更积极的感受。

3.2 皮电结果

3.2.1 描述性统计 以基线值作为因变量，以接触方式和暴力程度作为自变量，进行 2 × 2 随机方差分析，主效应和交互效应均不显著，说明四个组别的基线值在统计上无显著差异。非暴力游戏接受组的标准差较高，则可能与样本较小或该组实验刺激的单调

性有关，个别被试有可能没有集中注意于实验刺激，而是受到其它心理活动的影响，造成数据的不稳定。

3.2.2 皮电值随时间的变化趋势 以皮电为因变量，进行 2(游戏暴力程度：暴力，非暴力)×2(接触方式：主动参与，被动接受)×3(测量时间：基线阶段，后测阶段，观看视频阶段) 三因素混合设计方差分析。游戏暴力程度与接触方式为组间变量，测量时间为组内变量。结果表明，三向交互作用不显著，测量时间与游戏暴力程度的两向交互作用显著，$F(2, 40)=4.54$, $p<0.05$,测量时间的主效应显著，$F(2, 40)=14.68$, $p<0.001$。

由于测量时间与暴力程度的两向交互作用显著，进一步考查其简单效应，结果表明：暴力情境下，游戏后的皮电值显著低于基线阶段与视频观看阶段，而基线阶段与视频观看阶段无显著差异 (M 基线 =6.09,M 后测 =4.85, M 观看视频 =6.90);非暴力情境下，视频观看阶段的皮电值显著高于基线阶段与游戏后阶段，而游戏前后无显著差异 (M 基线 =4.89,M 后测 =5.77, M 观看视频 =7.33)。

3.2.3 暴力电子游戏的脱敏效应 用视频观看阶段与基线阶段皮电值的差值作为因变量———脱敏发生的指标，以身体攻击性分量表与游戏使用习惯问卷得分作为协变量，进行 2(游戏暴力程度：暴力，非暴力)×2(接触方式：主动参与，被动接受)随机方差分析。

表2 皮电值的平均值与标准差

实验处理	基线		后测		观看视频	
	M	SD	M	SD	M	SD
暴力游戏参与组	6.04	3.53	5.24	3.11	6.93	3.66
非暴力游戏参与组	3.75	1.69	4.04	2.50	5.42	2.96
暴力游戏接受组	6.14	3.55	4.47	2.71	6.87	2.85
非暴力游戏接受组	6.04	8.82	7.49	12.86	9.23	12.13

结果表明个体身体攻击水平与游戏使用习惯均没有任何效应显著，即这两个个体变量对于自变量的效应均无影响。游戏暴力程度与接触方式无交互作用，$F(1, 40)=1.11$，$p>0.05$。游戏暴力程度的主效应显著 ($M_{暴力}=0.81$，$M_{非暴力}=2.43$，$F(1, 40)=4.19$，$p<0.05$)，说明暴力组在观看暴力视频的过程中皮电的增加值明显小于非暴力组，表现出更低的生理唤起。H1得到验证，即暴力电子游戏可以产生脱敏效应。游戏接触方式的主效应不显著 ($M_{主动}=1.28$，$M_{被动}=1.97$，$F(1, 40)=0.75$，$p>0.05$)，主动参与组与被动接受组在观看视频的过程中皮电的增加值没有差异，H2没有得到验证，接触方式对于脱敏效应的影响不显著。

看阶段，而基线阶段与视频观看阶段无显著差异；非暴力情境下，视频观看阶段的皮电值显著高于基线阶段与游戏后阶段，而游戏前后无显著差异。根据一般攻击模型，被试基线阶段无生理唤起，皮电值应较低；接触游戏刺激后，无论是否为暴力游戏，皮电值都应有所增加。但进入暴力视频观看阶段，皮电值的具体变化则无法预测，唯一能假设的是非暴力游戏组的生理唤起应高于暴力游戏组。本研究发现在暴力情境下，游戏后的皮电值反而下降，而前人有研究发现皮电值随时间无显著变化 (Carnagey, Anderson, & Bushman, 2007)，这些结果均在一定程度上与一般攻击模型相

表3 心率的平均值与标准差

实验处理	基线		后测		观看视频	
	M	SD	M	SD	M	SD
暴力游戏参与组	83.44	8.80	78.86	8.55	77.38	7.37
非暴力游戏参与组	79.72	11.63	80.20	12.10	77.97	14.43
暴力游戏接受组	75.74	12.16	74.86	9.06	73.09	7.84
非暴力游戏接受组	79.84	8.30	77.46	13.15	74.50	9.18

3.3 心率结果

3.3.1 描述性统计 见表3。

3.3.2 方差分析结果 以心率为因变量，进行2(游戏暴力程度：暴力，非暴力)×2(接触方式：主动参与，被动接受)×3(测量时间：基线，后测，观看视频)三因素混合设计方差分析。结果表明，只有测量时间的主效应显著，心率在基线阶段、游戏后与视频阶段呈递减的趋势 ($M_{基线}=79.68$, $M_{后测}=77.85$, $M_{观看视频}=75.73$, $F=8.44$, $p<0.05$)。这一变化趋势与以往的研究结果正好相反 (Anderson & Bushman, 2001; Carnagey, Anderson, & Bushman, 2007)，与一般攻击模型也有矛盾之处。

4 讨论

4.1 皮电值随时间的变化趋势

以皮电值作为因变量的三因素混合设计的方差分析表明测量时间与游戏暴力程度有交互作用——暴力情境下，游戏后的皮电值显著低于基线阶段与视频观

违。考虑其中原因，一个可能是一般攻击模型不完全适用于暴力电子游戏；但也可能与生理指标的测量误差有关。本实验所用的生物反馈仪对于环境温度、湿度的变化较为敏感，有可能由此造成测量误差。

4.2 个体变量的调节作用

有关媒体暴力的文献中，经常受到关注的一个主题是个体对于媒体暴力易感性的差异。根据一般攻击模型，个体变量与情境变量交互影响个体的内部状态，包括认知过程、情绪变化与生理唤起。本研究将身体攻击性与游戏使用习惯这两个个体差异变量作为控制变量纳入了研究。结果并未发现这两个个体变量对于暴力游戏的脱敏效应有调节作用。近期其他人的一些研究也得到了类似的结果 (Bartholow, Bushman, & Sestir, 2007; Carnagey, Anderson, & Bushman, 2006)，提示游戏的脱敏效应有可能在一定程度上独立于个体变量。

4.3 游戏的接触方式

本研究加入了一个新的自变量：游戏的接触方式——主动参与游戏或被动观看游戏录像，以此类比

游戏媒体与传统影视媒体，从而尝试比较这两种不同类型媒体暴力脱敏效应的差异。这样操作实际上假定参与性和互动性是这两种媒体接触方式的根本区别，所以实验要使参与性和互动性之外的其他方面，例如游戏内容、画面激动程度、特别是暴力程度等保持等值。本研究采取用主动组画面录像给被动组观看的办法，来保证这种等值性。被试对于游戏的评价结果也支持了此实验操作的有效性。

本研究的结果并不支持假设二，游戏的接触方式的不同并没有导致脱敏效应的差异。这个结果可以有两种解释，一是主、被动接触方式的确不影响脱敏程度。另一可能是，实验处理中，暴力接触时间长度不够。实验处理为 15 分钟，可能这个时间太短，主、被动接触在情感、注意卷入上的不同还未能积累到产生差异效应的程度。

5　结论与建议

5.1　结论

由本实验结果得出以下结论：

（1）暴力电子游戏能够产生短期脱敏效应。在接触游戏 15 分钟后，与非暴力游戏组相比，暴力游戏组观看暴力视频过程中生理唤起的水平更低。

（2）游戏的接触方式（主动参与游戏和被动接受游戏录像）对于脱敏效应没有显著影响；但对于内容完全等值的游戏，主动组知觉到更低的沮丧与更高的愉快程度。

5.2　建议

对于本研究所得出的主动参与与被动接受暴力电子游戏，其脱敏效应没有不同的这一结果，应持谨慎态度。后继的研究在实验处理中对于暴力接触时间的把握、样本量的设定以及非暴力游戏的选择方面需要有所改进，再检查两种接触方式的脱敏效应是否有差别。另外，本研究中用作检验脱敏效应的实验材料为一段暴力影片，在以后的研究中可使用包含真实暴力内容的实验材料来更为有效地检验脱敏效应。

参考文献

Anderson, C. A., Berkowitz, L., & Donnerstein, E. (2003). The influence of media violence on youth. *Psychological Science in Public Interest, 4*, 81 - 106.

Anderson, C. A., & Bushman, B. J. (2001). Effects of violent video games on aggressive behavior, aggressive cognition, aggressive affect, physiological arousal, and prosocial aehavior: A meta-analytic review of the scientific literature. *Psychological Science, 12*, 353 - 359.

Anderson, C. A., & Bushman, B. J. (2002). Media violence and the American public revisited. *American Psychologist, 57*, 448 - 450.

Anderson, C. A., & Dill, K. E. (1986). Affect of the game player: Short-term consequences of playing aggressive video games. *Personality and Social Psychology Bulletin, 12*, 390 - 402.

Anderson, C. A., & Dill, K. E. (2000). Video games and aggressive thoughts, feelings, and behavior in the laboratory and in life. *Journal of Personality and Social Psychology, 78*, 772 - 790.

Anderson, C. A., & Griffiths, M. (2004). Cotemporary issues in adolescent video game playing: Brief overview and introduction to the special issue. *Journal of Adolescence, 27*, 1 - 3.

Anderson, C. A. (2004). An update on the effects of playing violent video games. *Journal of Adolescence, 27*, 113 - 122.

Anderson, C. A., Carnagey, N. L., & Eubanks, J. (2003). Exposure to violent media: The effects of songs with violent lyrics on aggressive thoughts and feelings. *Journal of Personality and Social Psychology, 84*, 960 - 971.

Anderson, C. A., Carnagey, N. L., & Flanagan, M. (2004). Violent video games: Specific effects of violent content on aggressive thoughts and behavior. *Advances in Experimental Social Psychology, 36,* 199 – 246.

Bartholow, B. D., Bushman, B. J., & Sestir, M. A. (2006). Chronic violent video game exposure and desensitization to violence: Behavioral and event-related brain potential data. *Journal of Experimental Social Psychology, 42,* 532 – 539.

Bushman, B. J., & Anderson, C. A. (2002). Violent video games and hostile expectation: A test of the General Aggression Model. *Personality and Social Psychology, 28,* 1679 – 1686.

Buss, A. H., & Perry, M. P. (1992). The aggression question. *Journal of Personality and Social Psychology, 63,* 452 – 459.

Carnagey, N. L., & Anderson, C. A. (2005). The effects of reward and punishment in violent video games on aggressive affect, cognition, and behavior. *Psychological Science, 16,* 882 – 889.

Carnagey, N. L., Anderson, C. A., & Bushman, B. J. (2007). The effect of video game violence on physiological desensitization to real-life violence. *Journal of Experimental Social Psychology, 43,* 489 – 496.

Chen, M. F., & Chen, Y. P. (2005). The effect of aggressive games on individual implicit aggressiveness. *Psychological Science, 28,* 458 – 460.

[陈美芬 , 陈舜蓬 . (2005). 攻击性网络游戏对个体内隐攻击性的影响 . *心理科学 , 28,* 458 – 460.]

Cline, V. B., & Croft, R. G. (1973). Desensitization of children to television violence. *Journal of Personality and Social Psychology, 27,* 360 – 365.

Cui, L. J., Hu, H. L., & Wu, M. Z. (2006). A study on the implicit aggressiveness of gaming addicts. *Psychological Science, 29,* 570 – 573.

[崔丽娟 , 胡海龙 , 吴明证 . (2006). 网络游戏成瘾者的内隐攻击性研究 . *心理科学 , 29,* 570 – 573.]

Dill, K. E., & Dill, J. C. (1998). Video game violence: A review of the empirical literature. *Aggression and Violent Behavior, 3,* 407 – 428.

Funk, J. B., & Buchman, D. D. (2003). Playing violent video games, desensitization, and normal evaluation in children. *Applied Developmental Psychology, 24,* 413 – 436.

Funk, J. B., & Baldacci, H. B. (2004). Violence exposure in real-life, video games, television, movies, and the internet: Is there desensitization? *Journal of Adolescence, 27,* 23 – 29.

Gentile, D. A., & Anderson, C. A. Violent video game: The newest media violent hazard. Retrieved December 20, 2008, from http://www.psychology.iastate. edu/~dgentile/106027_07.pdf

Gentile, D. A., & Lynch, P. J. (2004). The effect of violent video game habits on adolescent hostility, aggressive behavior, and school performance. *Journal of Adolescence, 27,* 5 – 22.

Griffths, M. (1999). Violent video games and aggression: A review of the literature. *Aggression and Violent Behavior, 4,* 203 – 212.

Kirsh, S. J. (2003). The effects of violent video games on adolescents: The overlooked influence of development. *Aggression and Violent Behavior, 8,* 377 – 389.

Mullin, C. A., & Daniel, L. (1995). Desensitization and resensitization to violence against women: Effects of exposure to sexually violent films on judgments of domestic of domestic violence victims. *Journal of Personality and Social Psychology, 69,* 449 – 459.

Thomas, M. H., & Horton, R. W. (1977). Desensitization to portrayals of real-life aggression as a function of exposure to television violence. *Journal of Personality and Social Psychology, 35,* 450 – 458.

Van Mierlo, J., & Van den Bulck, J. (2004). Benchmarking the cultivation approach to video game effects: A comparison of the correlates of TV viewing and game play. *Journal of Adolescence, 27,* 97 – 111.

Xin, Z. Q., & Chi, L. P. (2004). The mechanism of violence

in the virtual on the aggressive behaviors of children. *Journal of the Chinese Society of Education*, (5), 41 – 44.

[辛自强 , 池丽萍 . (2004). 虚拟世界的暴力对儿童攻击行为的影响机制 . *中国教育学刊* , (5), 41 – 44.]

Zhang, Z., Liu, Y. X., & Zhang, J. X. (2006). The theory model of the effect of violent media on the youths aggressiveness. *Chinese Journal School Health, 27*, 915 – 916.

[张镇 , 刘月霞 , 张建新 . (2006). 媒体暴力影响青少年攻击行为的理论模型 . *中国学校卫生 , 27*, 915 – 916.]

Zheng, H. M., & Sun, Y. J. (1997). Violent video games and aggression. *Advances in Psychologicial Science, 14,* 266 – 272.

[郑宏明 , 孙延军 . (1997). 暴力电子游戏对攻击行为及相关变量的影响 . *心理科学进展 , 14,* 266 – 272.]

Short-term desensitizing effects of violent video games: Comparison between two exposure ways

GUO Xiaoli[1,2] , JIANG Guangrong[1] , ZHU Xu[1]

([1]School of Psychology, Huazhong Normal University, Hubei human beings development and mental health key laboratory, Wuhan, 430079, China)

([2]Student Work Department, Huazhong University of Science and Technology, Wuhan, 430074, China)

Abstract　This study examined the desensitizing effect of active or passive exposures to violent video games on male college students' physiological arousal when viewing a violent film. The study employed a 2 (active or passive exposure) x 2 (violent or nonviolent video game) factorial design. Half of the forty-four participants were randomly assigned to either playing a violent video game or watching the records of someone else playing the violent game for 15 minutes, and the other half assigned to playing or watching a nonviolent video game. Then all the participants were presented with a 10-minute long violent film segment while their heart rate (HR) and galvanic skin response (GSR) were being recorded. The result showed that participants who previously played or viewed a violent video game had lower GSR while viewing the violent film than those who previously played or watched a nonviolent video game. This result demonstrated a physiological desensitization effect of exposure to violent video games on physiological arousal toward violence. However, the way of exposure, active or passive, to violent video games failed to show any influence on the degree of desensitization, although actively-playing group reported more enjoyment and less frustration than did passively-viewing group. Results were interpreted and discussed using the General Aggression model.

Keywords　violent video games; desensitization; exposure ways

心理学报 , 2014, 46(7), 960 – 975.

当事人对领悟的看法：质化分析

胡姝婧 [1,2]，江光荣 [2]，鲁艳桦 [2]，张莎莎 [2]，陈锐娟 [2]，于丽霞 [2]，杜　睿 [2]

([1] 海军工程大学理学院 , 武汉 430033)

([2] 青少年网络心理与行为教育部重点实验室 , 华中师范大学心理学院 , 湖北省人的发展与心理健康重点实验室 , 武汉 430079)

摘　要　为了从当事人的视角理解领悟 , 采用协商一致的质化研究方法对 15 位当事人的访谈结果进行分析。结果发现 7 个与领悟有关的域 : 领悟的内容 , 领悟的效果 , 影响领悟产生的因素 , 评估领悟质量的依据 , 领悟出现时的反应 , 领悟的来源和阻碍领悟发挥作用的因素。形成了领悟的概念界定 : 领悟是对自己和他人 (主要是自己) 的新认识 , 对自己的认识内容包括 , 自己的问题模式 , 心理困扰或问题模式的原因、影响和解决办法 , 以及自己内在的心理状态。

关键词　心理治疗 ; 领悟 ; 质化研究

分类号　R395

1　前言

心理咨询中有一个变量从咨询诞生时就受到重视 , 但直到最近才获得了越来越多的研究关注 , 这就是领悟 (insight)。弗洛伊德十分重视领悟 , 追求领悟是精神分析的标志性特征。其他咨询取向对领悟重要性的认识在早期有差异 , 但发展至今 , 领悟的重要性已经得到了比较一致的认同。在咨询理论越来越趋向于整合的今天 , 在较有影响力的跨理论咨询模型中 , 领悟也被提到了相当的高度。如 Hill(2009) 提出的三阶段咨询模型 , 探索—领悟—行动 , 领悟就是关键的第二阶段。

除了理论对领悟的重视之外 , 研究也一再证实了领悟在咨询中的重要性。对咨询会谈中重要事件类型的研究 , 代表性的是 Elliott 的助益事件 (helpfulevents) 的研究 , 研究和 Mahrer 的好的时刻 (good moments) 研究。Elliott, James, Reimschuessel, Cislo 和 Sack(1985) 通过访谈当事人在会谈结束后的体验 , 得到当事人所认为的会谈中的助益事件 , 8 类助益事件领悟是其中之一。Mahrer 和 Nadler (1986) 根据理论和前人的研究 , 归纳出 11 个会谈中好的时刻 , 领悟也是其中之一。此后其他各种对重要事件的研究 , 领悟的重要性都一再被证实 : 如格式塔治疗中好的时刻 (Boulet, Soulière, Sterner, & Nadler, 1992), 抑郁症治疗中当事人对治疗获益的觉知 (Gershefski,Arnkoff, Glass, & Elkin, 1996), 团体咨询中的重要事件 (Moreno, Fuhriman, & Hileman, 1995), 当事人在咨询中的获益体验 (Paulson, Truscott, & Stuart,1999), 当事人认为的咨询中的重要内容 (Levitt,Butler, & Hill, 2006), 等等 , 这些研究结果无一例外都包含领悟。

通讯作者 : 江光荣 , E-mail: grjiang@yeah.net

理论和实证研究都表明领悟是咨询中一个十分重要的因素，然而对领悟重要性的认识，却与对领悟的研究不成正比，专门的研究还很缺乏。Gibbons, Crits-Christoph, Barber 和 Schamberger(2007) 在对领悟的相关实证研究进行文献综述后感叹道："鉴于领悟在理论文献中的重要性，过去40年中，在这一概念的操作化和与治疗效果关系的检验上所付出的努力如此之少，是很令人惊讶的"。

心理咨询中的领悟指的是什么，这是研究的出发点。以下将总结西方主流咨询取向对领悟的看法，对领悟的跨理论界定，以及国内学者对领悟的理解，来梳理现有的对领悟基本概念的认识。

1.1 主流咨询取向的观点

主流咨询取向对领悟的认识各不相同。在《InsightinPsychotherapy》(Castonguay&Hill,2007) 一书中，主流的心理咨询取向对各自所认为的领悟进行了界定和诠释，概括如下。

精神分析是领悟概念最早出现也最重视领悟的取向。(1) 弗洛伊德的着作中很少直接提及领悟一词，但是其理论构想和实践中贯穿着领悟。在他那里，领悟是对潜意识动机和防御机制的发掘，对痛苦真相的寻找，对创伤性经历与当前心理痛苦关系的了解。(2) 在自我心理学中，领悟有两种涵义：一种是指内观的过程及其发现，另一种是承认自己的问题，而这种承认预示着治疗的成功。领悟既可以是治疗的手段也可以是最终目标，当其作为目标时，是将以前无意识的驱力、愿望、幻想、冲突和其他非理性的斗争整合到现实自我中。这两种理论都把领悟看作心理改变的原因。(3) 在关系理论中，领悟被视为心理改变的结果，改变发生的证据。它是在安全的治疗关系中努力澄清当事人的困扰后所产生的结果。当领悟发生时，治疗双方会察觉自己在咨访互动中表现着当事人生活中的主题。(Messer &McWilliams, 2007) 总体来看，在心理动力学取向中，领悟是无意识的冲突、驱力、愿望、动机等的意识化，它既是治疗的手段也是治疗的目标，既可以是心理改变的原因也可能是改变的结果。

在体验疗法中（包括当事人中心、格式塔、过程-体验和某些存在疗法），领悟通常被等同为觉察 (awareness) 和元觉察 (meta-awareness)。觉察是指明确地关注当下体验的某一方面，而元觉察是对感知事物、信息加工或建构个人体验的方式的特殊觉察。(1) 当事人中心疗法中，罗杰斯把领悟描述为"当事人达到的一种体验"，以联结和接纳的方式，是一种感觉到的，而不是理智上的体验。在罗杰斯那里，领悟似乎可以和觉察、感觉到的体验、符号化等互换。(2) 强调体感聚焦 (focusing) 的体验治疗，把领悟视为在当下的觉察过程中解释和创造新涵义的产物。(3) 存在疗法中的领悟是存在性的领悟，通常是在面对终极关怀（死亡、孤独、无意义和自由）时获得的觉察，是以情感为基础的对生命和生活的看法。(4) 格式塔疗法也认为领悟就是觉察和元觉察，是发现一个人的体验和行为，以及行为的方式。(Pascual-Leone & Greenberg, 2007) 可以看出，体验疗法十分强调领悟中的体验成分，领悟是内在体验和体验方式的意识化，领悟发生时必定伴有体验。

与前两种取向相比，认知-行为疗法最初对领悟重要性的强调最弱。但随着实践中领悟不断地偶然出现，以及对治疗产生的促进作用，逐渐引起了理论家的重视；与此同时，实证研究的结论也加速了该流派对领悟的接纳。(1) Ellis 在其理-情行为治疗中区分了理性领悟 (intellectual insight) 和情绪性领悟 (emotional insight)，认为后者造成的信念和行为改变的强度要超过前者。(2) 在其他的认知疗法中，领悟的涵义与认知改变、认知重构、理性重构、认知调整、理性再评价、发现非理性等近似。Beck 认为，认知改变过程由对自己想法的觉察、识别不适当的想法和用更适当的想法替换组成，领悟则包含在识别非理性的自动化想法和觉察替代性的认知中。Meichenbaum 认为认知重构是行为改变的关键，它既是治疗改变的手段也是改变的目的。认知重构反映着图式的改变，图式改变与单纯的理性领悟不同，它包含着心理机能的多重维度。(3) 从图式理论的观点来看，领悟是自我和他人图式的改变。在此过程中，个体有意识地觉察到两个或多个图式的联结，而该联结是之前不存在或以特殊方式联结在一起的。图式作为在长时记忆中储存的心理表征，其表征水平与储存位置在外显或内隐的记忆系统有关，理性领悟是在外显水平上建立图式间新的联结，而情绪性领悟则需整合外显和内隐的表征 (Holtforth et al., 2007)。综上所述，认知 - 行为疗法重视领悟中的认知成分，同时也强调认知改变时所伴随的情绪体验，它在心理表征的层面上进一步阐明了领悟的实质。

总体来看，心理动力学把领悟看作联结的形成，联结的内容包括过去与现在、内在冲突与外在表现、依恋关系与移情，等等，当两两间内在的联系贯通时，领悟就发生了。体验疗法把领悟等同于觉察和元觉察，当内在的体验意识化时，领悟就出现了。认知 - 行为疗法认为领悟是认知重构或图式改变，发现以前的非理性信念或图式，用新的理性信念或新图式取代，这就是领悟。而图式改变的含义亦即在图式间建立新的联结。由此看来，心理动力学和认知 - 行为疗法都认为领悟的实质是建立新联结，但是对联结的内容、联结形成方式等的看法有所不同。

Pascual-Leone 和 Greenberg (2007) 从领悟加工方式的角度抽取了两个维度，来对不同取向中的领悟进行比较，见图 1。这两个维度是抽象程度和加工类型。

抽象是指提取跨情境的具体的稳定因素并内化的过程，跨越时间和空间。抽象程度越高，归纳的范围越宽，抽象的来源越广。加工类型是指情感和认知加工的相对分量，加工一种体验既可以采用对知觉和情绪即时化的方式，也可以通过概念和理性思考的方式。

该模型认为主流取向中的领悟在两个维度上都有各自的位置，相应地表现出一些典型特征：(1) 大多数体验疗法中的领悟主要是觉察，抽象水平最低，聚焦于此时此地，加工方式是知觉 - 情绪 (如 "我现在感到对父亲很愤怒"); (2) 存在疗法和某些体验疗法中的领悟主要是体验性的元觉察，抽象水平比觉察高，加工方式既有知觉 - 情绪，也有概念 - 理性，但以知觉 - 情绪为主 (如 "我现在有种把整个世界都看作我的对立面的感觉"); (3) 认知疗法中的领悟主要是理性的元觉察，抽象水平比体验 - 存在性领悟更高，加工方式既有知觉 - 情绪，也有概念 - 理性，但以概念 - 理性为主 (如 "我现在意识到我是害怕失败所以不愿意尝试"); (4) 心理动力学中的领悟主要是概念联结，抽象水平最高，跨越时间和空间，加工方式是概念 - 理性 (如 "现在我明白，从小就缺乏安全感，所以我一直不敢亲近任何人")。这四类领悟，抽象程度依次递增，加工类型逐渐从知觉 - 情绪向概念 - 理性转移，一起共同构成了完整的领悟概念。

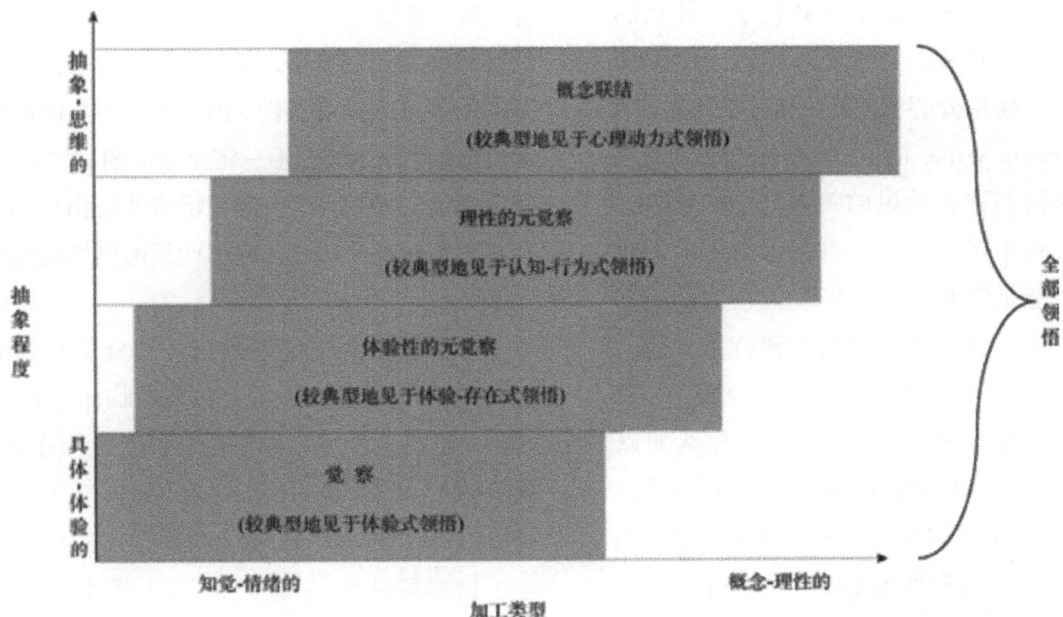

图 1　领悟的二维模型 (Pascual-Leone & Greenberg, 2007)

1.2 跨理论的观点

APA 心理学词典 (VandenBos, 2006) 对领悟的定义是：在心理治疗中，对自己或他人的情绪、认知或行为困难的潜在来源的觉知。Hill 等 (2007) 曾与 30 位不同取向的临床心理学家进行讨论，最后对领悟的概念得出了较为一致的意见：领悟是包含新联结的有意识的意义转变（即"这个与那个相联系"或某些因果的感觉）。

在跨理论的实证研究中，研究者对领悟给出了不同的界定。Elliott (1985) 在对会谈中"助益事件" (helpful events) 的研究中，对领悟的界定为：当事人描述认识到一些和自己有关的新东西，包括获得认知领悟，看到与自我或人际关系中的自我的一些新的联结。Mahrer 和 Nadler (1986) 在对会谈中"好的时刻" (good moments) 的研究中，对领悟的界定为：当事人表达或陈述一个重要的领悟–理解，有 3 个特点：(1) 表达情绪唤起的感受；(2) 在看待（认识和/或建构）自己和自己的世界的方式上表现出确实的改变；(3) 对自己的生活和个人/人际行为具有重要含义。Hill (1992) 在对会谈中的当事人行为进行分类的《当事人行为系统》(Client Behavior System) 中，对领悟的界定是：当事人表达出对自己的理解，可以明确说出行为、想法或感受的模式或原因；领悟通常包含一个"啊哈"的体验，当事人以一种新的方式知觉自己和这个世界；当事人承担适当的责任而不是责怪他人、使用外界强加的"应该"或合理化。Gelso, Kivlighan, Wine, Jones 和 Friedman (1997) 在研究领悟、移情与咨询效果的关系时，对领悟的界定为：当事人对正在探索的材料表达出正确理解的程度。理解可能是针对关系、当事人在咨询室外的机能状况、或者当事人的心理动力和行为方面。Gelso 和 Harbin (2007) 在精神分析理论的基础上，对领悟给出了一个较宽泛的定义，将领悟界定为对以下内容的理解和觉察：(1) 潜在的感受，想法和行为；(2) 它们之间的内在联系以及与早期事件的关系;(3) 内部事件（想法和感受）和外部事件（行动）的关系。他们强调领悟最基本的特点是对以前无意识的感受、想法和行动的觉察和理解。它并不需要直接和个人遥远的过去有关（如童年早期），也不需要反映无意识的加工，尽管往往会这样。

虽然表达各异，但仔细比较可以发现，不同学者对领悟的认识是大同、小异。大同表现为对领悟本质特征的认识是一致的，都认为领悟是当事人对自己有了一些新的认识和理解，表述为"认识到一些和自己有关的新东西"、"在看待（认识和/或建构）自己和自己的世界的方式上表现出确实的改变"、"表达出对自己的理解"、"表达出正确理解"，"对以前无意识的感受、想法和行动的觉察和理解"，都是在说当事人对自己的新理解。小异主要表现在两个方面：其一，对领悟内容的认识不尽相同，换句话说，对当事人对自己的什么有了新的认识和理解看法不同，有的认为是"自我或人际关系中的自我"、有的是"自己和自己的世界"、有的认为是"行为、想法或感受的模式或原因"、还有的是"关系、在咨询室外的机能状况、或者心理动力和行为"以及"以前无意识的感受、想法和行动"。对这些内容进行总结，可以归纳为两部分，一部分是自己的心理机能，包括认知、情绪、行为倾向和它们之间的相互影响；一部分是自己与他人的关系。其二，对领悟特征的认识不尽相同。有的界定中表述了领悟的特征，有的没有，如"领悟通常包含一个'啊哈'的体验"、"表达情绪唤起的感受"，等等。这些有关特征的表达在不同的界定中没有共性，因此有可能反映的是研究者对领悟的个人看法，不一定是领悟的必要特征。如"表达情绪唤起的感受"，会在情绪性领悟中出现，但不会在理性领悟中出现，但理性领悟仍然是领悟的一种类型。

理性领悟和情绪性领悟的区分是对领悟的一种比较重要的划分，得到了大多数临床心理学家的认同。心理动力学早先并没有对领悟进行分类，直到自我心理学家 James Strachey 在 1934 年提出"突变解释" (mutative interpretation) 的概念，这一概念强调整合了情感和认知的解释，要大大优于只有认知的解释。在此之后，大多数心理动力学家就区分了理性领悟和情绪性领悟（引自 Messer &McWilliams, 2007）。Albert Ellis 最早在理 – 情行为治疗中将领悟区分为理性领悟和情绪性领悟。两种领悟中当事人都能认识到错误的信念、自我挫败的行为，也都会体验到改变信念和行为的愿望。但是，两种领悟在影响程度上存在差别，在影响行为类型的数量、影响效力和承诺等方面，情绪性领悟都要优于理性领悟（Ellis, 1963）。值得注意的是，心理动力学和理 – 情行为疗法所说的情绪性领悟，尽管使用的是同一个词，但二者的内涵并不一样，心理动力学认为的情绪性领悟是"伴随着理解的宣泄过程"（Gelso et al., 1997），而理 – 情行为疗法认为的情绪性领悟是"伴随着理性领悟的确信感"（Ellis,2001）。Wachtel (1997) 对理性领悟和情绪性领悟的区分进行了综述，将二者概括为：理性领悟可以看作一个认知过程，帮助当事人掌握内在冲突的因果关系；情绪性领悟包含情感，当事人不但在理智上掌握了某种内部事件，而且体验到与那些事件有关的感受，这些感受是之前难以获得或没有体验过的。对如此区分的临床心理学家来说，情绪性领悟对促进行为改变更有效。Gelso 和 Harbin (2007) 对"情绪性领悟"这一名称提出了意见，认为这一命名并不恰当，因为容易和情绪宣泄相混淆，而且尽管其含义里包含认知成分，但这一命名本身却并没有体现出认知元素。因此，他们建议将其改为"综合性领悟"（integrative insight），它是认知理解和情绪体验的综合。当事人获得综合性领悟时，可以在认知上掌握自己的冲突和问题的原因，同时能体验到之前没有觉察和联系到这一认知理解的感受。

以上的文献回顾显示，目前获得较多认同的研究结论有：(1) 领悟的实质是建立新联结。尽管主流理论取向对领悟的界定不一致，但基本都认同领悟的实质是建立新联结，这一观点也得到了大多数临床心理学家的认可。(2) 领悟是当事人对自己的新认识和理解。对领悟的跨理论界定基本都持这一观点，新认识和理解的内容包括两部分，一是自己的心理机能，包括认知、情绪、行为倾向和它们之间的相互影响，二是自己与他人的关系。(3) 领悟有两种类型，理性领

悟和情绪性领悟。这一区分得到了大多数临床心理学家的认同。临床观察表明，在促进行为改变上，情绪性领悟比理性领悟更有效。

1.3 国内学者的观点

国内学者对领悟的关注不多，且大多是在个案报告中提及。最有影响力的研究当属钟友彬教授的认识领悟疗法。该疗法是在精神分析的理论基础上，植根于中国本土的一种心理动力疗法，因此，它也秉承了精神分析的理论精髓，强调领悟的重要性。钟友彬认为，领悟的本质是要"暗示病人认识并厌弃那些过时的或幼稚的感情和行为模式而代之以较为成熟的、健康的行为模式"（钟友彬，1985)，而其本土性就表现在病人领悟的内容，是"结合了中国的历史文化背景和病人的生活经验的"（钟友彬,1985)。还有一些学者用中国古诗词来描述他们对心理治疗中出现领悟的理解，如胡岚和周和玲 (2006) 认为领悟的出现会经过三个境界。第一境界是"昨夜西风凋碧树。独上高楼，望尽天涯路"，意指咨询师清楚地看到患者的许多问题，对患者的问题及相关因素尽收眼底，理清来龙去脉。第二境界是"衣带渐宽终不悔，为伊消得人憔悴"，意指咨询师在准备领悟到来的过程中，冥思苦想，呕心沥血，仔细研究患者症状轻重程度，认真分析患者问题的因果关系，通盘考虑与患者问题有关的因素，然后慎重选择有针对性的理论和技巧，用于最好解决的问题。第三境界是"众里寻他千百度，蓦然回首，那人却在，灯火阑珊处"，意指灵感到来，领悟出现。咨询师达到前两个境界，为第三境界的到来做好准备，而第三境界才是心理治疗的真正结果，要达到这一境界，咨询师要善于用心倾听，分析综合，联想推理，奇思妙想，经过漫长的准备，才能引导患者看到自己症状的根源及因果关系，最终引导患者领悟，达到"那人却在，灯火阑珊处"的境界。此外，还有研究者在分析比较西方主流理论取向对领悟看法的基础上，提出了领悟的跨理论界定："当事人在治疗师的辅助下采取各种方式，对自己和自己的世界形成新的觉知，表现为在个人意义系统中建立新的联结"（胡姝婧，江

光荣，2010）。

尽管该领域的研究已经取得了一些成果，但总体来看还存在一些问题：(1) 缺乏跨理论的研究。当今心理治疗的大趋势是走向整合，因此，跨理论治疗中的领悟研究就显得愈发需要。但目前这类研究一方面还比较少，另一方面也缺乏统一的概念界定，相应地也缺乏在概念界定基础上编制的专门的测量工具，在这些基本问题得到解决之后，才能对跨理论治疗中领悟的作用大小、作用机制等问题进行更深入的研究。(2) 缺乏对领悟的系统分类。对领悟进行分类，有助于深化对领悟现象的认识，也有利于考察不同类型领悟在咨询中所起的作用。目前只有理性领悟和情绪性领悟这一种区分得到了比较多的认同。这种区分来自于临床实践，在临床上具有一定的价值，但在研究中往往很难区分，也难以进一步研究 (Gelso & Harbin, 2007)。(3) 缺乏当事人视角的研究。现有研究几乎都是站在咨询师或临床研究者的角度来对领悟进行界定，但这一视角和当事人的是否一致？当咨询师致力于促进当事人获得领悟时（注意是咨询师认为的领悟），当事人是否真的从这样的领悟中获益，还是他们认为的领悟与此不同？现有研究无法回答这样的问题。(4) 缺乏本土的研究。从前面的文献回顾可以看出，现有的实证研究都是西方的，中西方当事人在不同的文化背景中，在不同的咨询氛围中，对领悟的认识是否会有所不同，钟友彬的研究从实务层面提供了一些证据支持，但还缺乏实证研究的归纳提炼。

基于以上分析，该研究将以国内的跨理论咨询为背景，对中国当事人眼中的领悟进行质化研究。质化研究的意义在于：首先，通过质化研究对领悟进行概念化，它为之后的测量工具编制和其他量化研究提供了基础。其次，有助于全面了解跨理论咨询中出现的领悟，对当事人在领悟前、中、后的体验有清晰的认识。再次，质化研究可以发现一些重要的问题，以后用量化研究的方法进一步检验和分析。最后，质化研究将以领悟的亲身体验者——当事人为访谈对象，它有助于我们从另一个视角丰富现有的认识。

2 研究方法

2.1 受访者

15 位当事人接受了访谈，其中男性 1 人，女性 14 人，年龄在 19 至 27 岁之间 ($M=22.67, SD=2.09$)。其中 14 人为某大学的学生，接受咨询时均为在校生，在该校心理咨询中心进行咨询，另有 1 人已工作，在社会心理咨询机构接受咨询。咨询次数最少 8 次，最多 56 次 ($M=12.73, Md=9, SD=12.34$)。参加研究时 13 人已经结案，2 人尚未结案。咨询的问题基本为发展性问题，包括自我探索与个人成长、人际关系、学习压力、亲密关系、人生规划等。15 位咨询师为这些当事人提供咨询，没有单一取向的咨询师，都是以某一种取向为主，同时兼用其他取向作为辅助，咨询取向主要是以人为中心疗法、认知疗法和心理动力疗法。

2.2 数据收集

在征得该校心理咨询中心负责人的书面同意后，获得了该中心部分接受心理咨询的当事人名单。从中筛选出会谈次数在 8 次及以上的心理咨询个案。通过手机短信告知这些当事人研究的相关信息，邀请其参与研究。将研究的访谈提纲通过电子邮件发送给同意参与研究的当事人，使其了解访谈内容并有所准备。然后和当事人约定好时间进行访谈。访谈形式有面谈、网络访谈和电话访谈，根据现实条件和当事人的意愿进行选择。访谈前向当事人介绍研究目的、访谈录音、文字记录、保密等相关事宜，征得当事人同意并签署知情同意书后开始访谈。访谈一般在 1 个小时左右。访谈结束后向当事人表示感谢，对面谈的当事人赠送了小礼品。通过这种方式获得了在该中心咨询的 14 位当事人的访谈资料。另有一位当事人在社会心理咨询机构接受了咨询，由他人推荐参与研究，通过相同的流程也对其进行了访谈。这样一共获得 15 份访谈材料。将这 15 份材料全部转成逐字稿，略去了可识别当事人和咨询师身份的信息，共计 11 万余字。

2.3 访谈提纲

访谈提纲由三部分组成。第一部分为当事人的基本信息，包括年龄、参加咨询的时间、咨询次数、是否结案等。第二部分是当事人咨询的主要问题。第三部分是咨询中所获领悟的相关问题，主要包括三个方面，一是对领悟的界定，认为领悟是什么；二是领悟的内容及对其质量评价；三是按照时间顺序，领悟出现前、中、后咨询的相关情况，包括咨访互动、当事人的想法感受、咨询的变化、当事人及其生活的变化等。该访谈为半结构式访谈，访谈问题均为开放式问题，访谈内容基本固定，但访谈过程中可根据当事人的回答适当增加或调整问题，以使答案更清晰。访谈提纲中的问题主要由本文第一作者拟定，在问题基本形成后，请两位咨询方向的博士生进行了审阅，根据其提出的意见进行了修改。

2.4 资料分析

对访谈资料的分析采用协商一致的质化研究方法 (Consensual Qualitative Research, CQR)。该方法由 Hill 等提出 (Hill, Thompson, & Williams, 1997;Hill et al., 2005)，是一种在心理咨询的研究中应用较广泛的质化研究方法。该方法与其他质化研究方法最大的区别在于，访谈资料的分析者不是一个人，而是一个小组；对访谈资料的分析要在组员间达成一致；并且会对结果的代表性进行检查。这样的分析过程有助于克服传统质化研究中存在的一些问题，如单一研究者在分析时可能存在偏见，研究结果难以被重复，研究过程达不到科学研究要求的规范和严谨等 (Hill et al., 1997)。

CQR 要求受访者至少 8 至 15 人，研究小组由 3 至 5 人组成，另外包括 1 至 2 名审核者。研究步骤主要是三步: (1) 将开放式问题的回答划入不同的域 (domain)，即不同的话题范围。(2) 将每个域中的内容提炼核心观点 (core idea)，即写摘要或简短的总结。(3) 交叉分析，将所有个案的同一个域的核心观点放在一起，对其进行归类 (category)，并对类别的代表性予以评定。每一步都必须通过研究小组的讨论，在组员间达成一致。审核者对每一步的结果予以审查，提出修改意见，供研究小组参考。

2.4.1 研究小组

本研究小组共 6 名成员，均为咨询方向的博士生或硕士生，其中博士生 4 人，硕士生 2 人，均为女性。所有组员事先对 CQR 的操作规程进行了学习和讨论，并阅读了采用 CQR 方法所做研究的相关文献。在研究开始前，按照 CQR 的要求，为了避免个人偏见和期待对研究过程的影响，每位组员先按照访谈提纲，根据自己的认识陈述对领悟的看法。主要有：领悟是对感受的澄清，了解自己的感受是什么及感受的来源；领悟是对带入咨询中的问题、在咨询中涉及的问题的新认识；领悟是对解决问题的方法途径的认识；领悟是对自己或自己的人际交往的新认识；领悟是将无意识意识化。组员表达了自己的看法后，被提醒尽量搁置这些看法，忠实数据本身。在此后的数据分析过程中，组员间时常会相互提醒其分析是否受自己看法的影响。

2.4.2 研究流程

按照 CQR 的操作要求，研究主要包括三步，划域，提取核心观点和交叉分析。具体操作如下。

（1）划域。阅读每一份访谈材料，将当事人的回答中和领悟有关的内容，划入相应的域中，并对每个域给予命名。组员先独立地对个案进行划域，然后一起讨论，逐渐确定下每个域的含义、命名以及逐字稿的划分方法。用确定下来的域再去划分之后的个案，并增补前面个案中未出现的域，或根据新出现的内容对域或域名进行修改调整。最终，每个域的含义和命名，以及对每一个个案的划域都在小组成员间达成一致。所有个案的划域结束后，组员再对每个个案独立进行检查，看是否需要修改，然后在小组中讨论，达成一致。

（2）提取核心观点。对每一个个案的每一个域里的内容，组员先独立地提取核心观点，即概括中心思想。核心观点要忠实原意，能用原文表达时最好用原文。然后对每个核心观点的内容和表述进行讨论，在小组中达成一致。在全部个案讨论结束后，组员独立检查每个个案，结合上下文考察核心观点的表述是否准确，并对同一个个案重复的核心观点予以合并。修改的结果经小组讨论，再次达成一致。

（3）交叉分析。以域为分析单位，将不同个案的同一个域的核心观点全部放在一起。组员先独立地对每一个域的核心观点进行分类，按照某一个分类标准将一个域里的核心观点分成几类。然后在小组中讨论，对分类标准、类别的命名以及核心观点的归类达成统一。先对15个个案中的13个进行分析，保留2个个案用于检查分类的稳定性，包括类别是否适用，类别的代表性是否会发生变化。结果这2个个案并没有改变原有结果，因而认为结果是稳定的。

（4）审核。将研究结果送给4位咨询方向的博士生审核，审核内容包括域的划分和命名是否准确，分类标准及其命名是否恰当，核心观点的归类是否合适等。研究小组对审核的结果进行讨论，确定是否需要修改和如何修改。

（5）类别的代表性评定。根据最终的结果对类别的代表性进行评定。按照 Hill 等（2005）的标准，如果类别适用于所有的个案或只有一个个案不符合，则该类别是普遍的（general），在本研究中是15或14个个案；如果适用于一半以上的个案，则该类别是典型的（typical），在本研究中是8至13个个案；如果适用于至少2个个案，则该类别是变异的（variant），本研究中是2至7个个案；如果只适用于一个个案，则将该类别放入杂类中，在结果中不予报告。

3 结果

和领悟有关的内容被划分成7个域：(1) 领悟的内容；(2) 领悟的效果；(3) 影响领悟产生的因素；(4) 评估领悟质量的依据；(5) 领悟出现时的反应；(6) 领悟的来源；(7) 阻碍领悟发挥作用的因素。表1呈现了所有的域，每个域里的类别（包括母类和子类），以及类别的代表性评定。

3.1 领悟的内容

当事人在咨询中获得的领悟主要是围绕自己的心理困扰或问题模式，认识到自身存在的问题模式，心理困扰或问题模式的原因、造成的影响和解决办法。此外，还有少数领悟是对他人的理解或对自己内在心理状态的觉察。

问题模式是当事人惯常采用的会造成不良影响的应对方式，它常常被运用却往往没有自觉。在本研究中，80% 的当事人对自己的问题模式有所领悟。当事人领悟的问题模式可以归纳为四种：（1）达到外界标准，限制真我表达。当事人认识到自己为了达到外界的要求，得到外界的肯定和接纳，而隐藏了真实的自我，把自己变成外界希望的样子。如当事人领悟到，"自己很多时候表现的不是真我，其实自己是特立独行和清高的，但做出来的样子是大家喜欢的、合群的"，"我一直通过达到外在的标准来获得价值感，因为无法自己肯定自己的价值，所以要通过控制别人，和有性格缺陷的异性交往来获得价值感"。（2）被动回避倾向。当事人认识到自己在生活中存在被动回避面对问题的倾向，如"我还像一个没有长大的小孩，生活中总是被动应对，没有主动操控，遇到困难会逃避面对，减少和外界的接触"，"我以前过于被动，总是等待机会或回避困难"。（3）不当的防御方式。当个体的需求没有得到满足而出现负面情绪，或潜意识中预感到负面情绪将要出现时，不直接面对和处理，而采用一些防御方式。如"在情感上不接受不好的经历，总会合理地解释它，从而不能客观地看待问题"，"我通过别人（姑姑、男友）来发泄自己的负面情绪。"（4）不良的人际交往模式。除了上述三种对自己问题模式的认识，还有一种对人际交往中自己问题模式的认识，它将焦点从个体自身转移到个体与他人的互动中。如"我对朋友太依赖，我让朋友很窒息"，"虽然我很想和别人建立好的关系，但是没有发自内心地想要去信任或者尊重别人"。

近3/4的个案对自己的心理困扰或问题模式的形成原因有所领悟，可以概括为三种：（1）完美主义倾向。因为追求完美导致对自己和他人的不满意、不接纳，如"一系列问题表现（如不自信、对自己不满意等）的原因是太追求完美，对自己要求太高"，"我的道德标准很高，我认为应该达到的道德要求，别人达不到时我会很困扰。这其实是要求别人和我一样。"（2）未获满足的积极关注的需要。被关注、肯定和爱的需要没有得到充分的满足，为了获得他人的积极

表 1　领悟的 CQR 研究结果

域	类别（母类 / 子类）	代表性评定
1. 领悟的内容	1.1 问题模式	12T
	1.1.1 达到外界标准，限制真我表达	4V
	1.1.2 被动回避倾向	4V
	1.1.3 不当的防御方式	3V
	1.1.4 不良的人际交往模式	5V
	1.2 心理困扰或问题模式的原因	11T
	1.2.1 完美主义倾向	4V
	1.2.2 未获满足的积极关注的需要	8T
	1.2.3 人际互动中的不良倾向	4V
	1.3 心理困扰或问题模式的影响	5V
	1.3.1 对自己的影响	2V
	1.3.2 对人际交往的影响	3V
	1.4 心理困扰或问题模式的解决办法	10T
	1.4.1 降低完美主义倾向	3V
	1.4.2 提高对负性事件的接纳度	2V
	1.4.3 增强主动性	3V
	1.4.4 改变人际交往的观念	4V
	1.4.5 调整人际界限	4V
	1.5 对他人的理解	2V
	1.6 觉察	3V
2. 领悟的效果	2.1 自己的变化	15G
	2.1.1 态度	12T
	2.1.1.1 接纳自己	6V
	2.1.1.2 成为自己	4V
	2.1.1.3 趋于成熟	2V
	2.1.1.4 接纳他人	5V
	2.1.2 情绪感受	8T
	2.1.2.1 平静轻松	5V
	2.1.2.2 开朗积极	3V
	2.1.2.3 存在感、清晰感、掌控感	5V
	2.1.3 行为	6V
	2.1.3.1 人际行为的变化	5V
	2.1.3.2 个体行为的变化	3V
	2.1.4 自我认识与调节	11T
	2.1.4.1 自我理解和觉察	8T
	2.1.4.2 自我调整	10T
	2.1.5 整体变化	5V
	2.2 人际关系的变化	5V
	2.3 咨询的变化	8T
	2.3.1 更信任喜欢咨询师	4V
	2.3.2 咨询更清晰流畅	4V
	2.3.3 考虑结束咨询	3V
	2.3.4 无变化	5V
	2.4 其他（改变的状态）	4V
3. 影响领悟产生的因素	3.1 咨询师方面的促进因素	15G
	3.1.1 倾听	4V
	3.1.2 反馈	11T
	3.1.3 解释	5V
	3.1.4 即时化	2V
	3.1.5 接纳、非指导的态度	5V
	3.2 当事人方面的促进因素	10T

续表 1

域	类别（母类 / 子类）	代表性评定
	3.2.1 内省思考	7V
	3.2.2 信任咨询师	2V
	3.2.3 阅读书籍	3V
	3.2.4 咨询中尝试行动	2V
	3.2.5 咨询中的状态	3V
	3.3 阻碍因素	2V
4. 评估领悟质量的依据	4.1 领悟对当事人及其生活的影响力	13T
	4.2 领悟的特性（重要性、深刻性、完善性）	5V
	4.3 领悟时的感受强度	3V
5. 领悟出现时的反应	5.1 豁然开朗	7V
	5.2 希望感	5V
	5.3 惊喜震撼意外	5V
	5.4 释放放松	4V
	5.5 感动	2V
	5.6 与个人问题相关的感受	2V
6. 领悟的来源	6.1 咨询师	11T
	6.2 当事人	7V
	6.3 咨访双方	5V
7. 阻碍领悟发挥作用的因素	7.1 旧有习惯的惯性	4V
	7.2 现实因素的制约	4V
	7.3 改变动力的不足	2V

注：代表性评定中，G 代表 general（普遍），T 代表 typical（典型），V 代表 variant（变异）；数字代表个案数；如"7V"表示该类别在 7 个

关注，而形成问题模式或产生心理困扰，如"为了在重男轻女的家庭环境里得到更多关注，让自己向男性性别角色认同"，"小时候受冷落的自己希望通过更完美、更乖、学习更好来引起别人的注意和爱，但又达不到，所以自卑。"（3）人际互动个案中出现，属于变异的。中的不良倾向。人际互动中不良的心理或行为倾向，导致人际交往中出现心理困扰或产生问题模式，如"只关注自己的紧张情绪，没把精力投入到对方身上，是导致我人际交往困难的原因"，"和父亲关系不好的一个原因是我总是用教训的口吻和他说话"。

1/3 的个案对于心理困扰或问题模式对自己造成的影响有所领悟。有两种影响：一种是对自己的心态和生活造成的影响，如"自卑让我在做很多事情之前就把结果想得很糟糕，所以不敢做"，"超越界限使我做事收效不好"；另一种是对人际交往造成的影响，如"表现得像小孩子能逗大家喜欢，但不一定能获得尊重"，"被动使我不能被别人理解"。2/3 的个案对如何解除自己的心理困扰或问题模式有所领

悟。可概括为 5 种方式：(1) 降低完美主义倾向。调整自己对于完美的认识，接纳自己的不完美，如"世界不完美，我也不可能做到完美"，"人本身就是不完美的，正是这种不完美，才让人不断努力不断追求，所以要接纳自己的不完美"。(2) 提高对负性事件的接纳度。在认知上引导自己接纳消极负面的事件，如"要客观地看待父亲对我的培养方式，不能为了不破坏父亲的形象，就找借口来掩饰"，"只有跟过去和解了才能更好地活在当下"。(3) 增强主动性。调整心态，积极主动地面对生活，如"我要用积极的而非退缩的方式处理问题"，"生活中，困难和问题会时时存在，主动解决会带来好心情、自信心"。(4) 改变人际交往的观念。对人际交往的方式增加了一些新的认识，或者改变了原有不正确的观念，如"人际交往中表达和理解同样重要，学会表达能获得主动权。适时适度的表达会得到真正的理解"，"我值得父母关心，不一定要很优秀才能得到父母的爱，对他们对我的关爱不用内疚。父母关心子女是其价值的一种体现"。(5) 调整人际界限。从认识上或行动意向上调整人际距离，

保持适当的人际界限，如"我和弟弟是两个独立的个体，不应要求弟弟走和自己一样的路"，"自己应该独立了，不应再依赖别人的支持"。

绝大部分当事人的领悟属于以上四种类型，是对自身存在的问题的新认识。但也有个别当事人对他人的心理状态或行为方式有所领悟，获得了新的理解，本研究中有两个个案谈到这种领悟，如一个当事人认识到"即将高考的弟弟压力很大，除了自身的压力和父母的期望，还有超越姐姐和父亲的压力，所以用表面看起来漫不经心、对高考不屑的方式来应对压力"，还有一个当事人谈到"站在爷爷的角度去考虑问题，我了解了他那样做的原因"。另外有3位当事人获得的领悟可以概括为觉察，是对自己内在心理状态的体认，如"其实我对父亲是有不满的"，"有些情绪和事情其实是对自己有影响的"，"我的力量增强了，敢于面对情敌了"。

3.2 领悟的效果

在获得领悟后，当事人自身、其人际关系和咨询都会发生一些变化，将这些变化统称为领悟的效果。

所有的当事人自身都有变化，可归纳为五个方面：(1) 态度的变化。80% 的个案在对待自己和他人的态度上有所变化。对于自己态度的变化，一是要接纳自己，如"要接受自己，扬长避短"，"比较能接受自己的不足了"；二是要成为自己，做自己而不再被外界所塑造，如"不想再为别人而活，想要坚持自己的路"，"要让本真的自我力量增强，不受外界的打压"；三是要让自己的为人处世更成熟，如"觉得自己处事应更成熟、大度，不能还像小孩一样任性"，"不再采用旧的行为模式，要展现自己成熟的一面"。对于他人态度的变化主要是接纳他人，对他人更理解、包容和关爱，如"对他人更包容"，"觉得父亲挺不容易的，感激父亲，想跟父亲走得更近"。(2) 情绪感受的变化。60% 的个案在情绪感受上有所变化。主要表现在三个方面：一是感到平静轻松，以前的消极情绪消除或缓解，如"烦躁降低"，"内心平静了很多"；二是更加开朗积极，心情更豁然，对待生活更积极，如"豁然很多"，"认识到生活有希望，困难可以解决，

不再过于畏难了"；三是体验到存在感、清晰感或掌控感，在生活中体验到真我的存在，对自己的认识更清晰，更能掌控自己的问题，如"日常生活中体验到真实的自己的存在"，"思想理顺了，通透了"，"比以前更有能力处理情绪问题，不再困在情绪问题里不知所措"。(3) 行为的变化。近一半的当事人在行为上有变化，有两种变化：一种是在人际交往方面，主动与外界接触沟通，如"敢和父母说出内心真实的感受了"，"跟外界的接触更多"；另一种是当事人自身行为的变化，如对他人的抱怨减少，或为了解决自己的问题进行一些有针对性的行为练习，如"上课主动发言锻炼自己，主动与同学结组参加每周活动，以提高自信心"，"在生活中尝试拒绝"。(4) 自我认识与调节。近 90% 的个案在自我认识与调节方面出现变化。在自我认识上，对自我的理解和觉察能力增强，对与自身问题相关的认知、情绪和行为的理解和觉察力提高，如"对自己认识更清晰，知道想法产生的原因和如何应对"，"能从生活细节中觉察自己退缩的模式"；在自我调节上，面对自己的问题和会引发自己问题的情境时，能够从心态上进行调整，更好地应对问题，如"不再理睬以前让自己很纠结的小事情，平静地做事，试着依靠自己而非掌控环境来获得安全感"，"在看到他人缺点时，会回想起领悟的内容，从而更倾向于包容他人"。(5) 整体变化。1/3 的个案谈到了自己整体性的变化，如"自卑降低"，"更有力量生活，生活更积极主动，心情好，更能承担责任"。

1/3 的个案的人际关系出现一些变化，大部分是积极的变化，如"朋友关系更和谐，朋友更多，自己很开心"，"和同学、男友的关系改善"；但也有个别个案出现消极的转变，如"和父母的关系有些纠结"，或者是没有变化，如"想过能否根据领悟改善人际关系，但觉得难就放弃了，所以没有变化"。

80% 的个案在领悟后咨询会出现一些变化，可概括为四个方面：(1) 更信任喜欢咨询师。在领悟后对咨询师更信任喜欢，对咨询师更开放，暴露更多更重要的个人信息，如"信任咨询师，告诉她从未说过的对自己影响很大的事情"，"更信任咨询师，暴露更

重要的事"。(2) 咨询更清晰流畅。领悟提供了线索，使接下去的探索更清晰流畅，如"咨询有了重点和线索，自己的思路更清晰"，"在以后的咨询中涉及人际方面时，会谈到领悟的内容"。(3) 考虑结束咨询。在领悟后感到问题基本解决，咨询可别个案出现消极的转变，如"和父母的关系有些纠结"，或者是没有变化，如"想过能否根据领悟改善人际关系，但觉得难就放弃了，所以没有变化"。80% 的个案在领悟后咨询会出现一些变化，可概括为四个方面：(1) 更信任喜欢咨询师。在领悟后对咨询师更信任喜欢，对咨询师更开放，暴露更多更重要的个人信息，如"信任咨询师，告诉她从未说过的对自己影响很大的事情"，"更信任咨询师，暴露更重要的事"。(2) 咨询更清晰流畅。领悟提供了线索，使接下去的探索更清晰流畅，如"咨询有了重点和线索，自己的思路更清晰"，"在以后的咨询中涉及人际方面时，会谈到领悟的内容"。(3) 考虑结束咨询。在领悟后感到问题基本解决，咨询可以结束，如"知道如何面对问题，愿意自己应对，咨询中发现的新问题少了，可以考虑结束咨询"，"清楚了自己的问题，不再迷茫，没有要咨询的困惑了"。(4) 咨询没有变化。也有个案表示领悟后咨询没有变化。

有 4 位当事人谈到了在领悟后自己改变的状态，表达了较为一致的信息：领悟后的改变不是很快很容易发生的，而是慢慢的一点点的改变，需要时间和努力，如"让自己更清晰，但距离实际的生活改变还有很长的距离"，"明确了改变的大方向，但具体到生活中的小事不是很容易改变，会尝试"。

3.3 影响领悟产生的因素

影响领悟产生的因素包括促进因素和阻碍因素两部分。促进领悟产生的因素可以分为两个方面，一是咨询师方面，一是当事人方面。所有的当事人都谈到了咨询师方面的促进因素，包括五种：(1) 倾听。近 1/3 的当事人表示咨询师的倾听促进了他们的自我探索，并获得领悟。(2) 反馈。近 3/4 的当事人获得领悟是经由咨询师的反馈，咨询师陈述他们对于当事人问题的看法。(3) 解释。1/3 的当事人通过咨询师给予

解释获得领悟。(4) 即时化，有 2 位当事人通过咨询师在咨询过程中做即时化的反应获得领悟。(5) 接纳、非指导的态度。1/3 的当事人感到咨询师对自己的接纳促进了其对自己的接纳，在这样放松、没有限制的环境中可以充分地自我探索，并获得领悟。1/3 的当事人感到咨询师对自己的接纳促进了其对自己的接纳，在这样放松、没有限制的环境中可以充分地自我探索，并获得领悟。2/3 的当事人谈到了自己方面的促进因素，包括五种；(1) 内省思考。近一半的当事人谈到自己的内省思考对领悟形成的作用，他们会通过各种方式，包括对咨询师反馈的思考、和身边的人讨论以及自己的反省，来获得领悟。(2) 信任咨询师。1 位当事人谈到对咨询师的信任会促进自己充分的自我表露，从而获得领悟；另 1 位当事人谈到对咨询师的信任使自己愿意接受咨询师的分析和反馈，从而获得领悟。(3) 阅读书籍。3 位当事人在咨询过程中看书受到了启发获得领悟。(4) 咨询中尝试行动。2 位当事人在咨询过程中，于日常生活中尝试了一些与以前不同的行为，进而获得了领悟。(5) 咨询中的状态。3 位当事人谈到自己在咨询中无拘无束地表达的这样一种状态，以及迫切地想摆脱困扰的状态，促进了领悟的出现。还有 2 位当事人谈到了阻碍领悟产生的因素，主要是在咨询关系方面，咨询双方互不接纳，或者联结不够紧密，当事人不能够完全开放。

3.4 评估领悟质量的依据

当事人对领悟质量的评估主要是从三个方面：(1) 领悟的影响力。近 90% 的当事人认为领悟对自己和生活的影响力的大小，是评估领悟质量的最重要的依据。如果领悟能够对自己和生活产生积极的影响，当事人对领悟质量的评分较高，反之则较低。这是从领悟效果的角度对领悟的评估。(2) 领悟的特性。1/3 的个案从领悟的重要性、深刻性、完善性的角度对领悟质量进行评估，领悟内容对自己的重要程度，接近核心问题的程度，以及完善全面的程度，是他们评估领悟质量时的依据。(3) 领悟时的感受强度。3 个个案从领悟时自己的感受强度来评估领悟质量，感受包括两种，一种是和问题相关的个人化的情绪感受，另

一种是领悟出现时感到的惊讶（震撼）。当情绪感受强烈的时候，当事人倾向于对领悟质量评高分。

3.5 领悟出现时的反应

在领悟出现时，当事人会有一些感受和反应，可概括为6种：(1) 豁然开朗。近一半的当事人谈到这一感受，如"打开一条通路的感觉，拨云见日，豁然开朗"，"眼前一亮，找到出口的感觉"。(2) 希望感。领悟后1/3的当事人的希望感增强，感到生活有了希望，有了力量和信心去应对问题。如"突然有了信心，觉得一切都不可怕了"，"感到有力量，有信心应对问题"。(3) 惊喜震撼意外。1/3的当事人对领悟感到惊喜、震撼或意外，如"对自己表达的想法和表达时思路的清晰感到震惊"，"感到意外"。(4) 释放放松。4位当事人谈到自己在领悟后哭了，尽情地宣泄，感到放松，如"尽情地哭和宣泄，感觉很放松"，"一下哭出来，压抑的情绪顿时释放"。(5) 感动。2位当事人谈到对于咨询师对自己的关注和帮助感到感动，如"对咨询师的帮助很感动"，"为咨询师对我的关注感动"。(6) 与个人问题相关的感受。2位当事人谈到领悟后对自己的问题产生了一些个人化的感受，如"觉得委屈，心疼以前的自己"，"羞愧"。

3.6 领悟的来源

研究还考察了领悟的来源：来源于咨询师的有近3/4的个案，咨询师对当事人反馈后，当事人一下子感到说得很对、很准，从而有所领悟；来源于当事人自身的有近一半的个案，他们可能是在咨询探索的过程中自己逐渐领悟，也可能是在生活中通过内省、看书、行动获得领悟；来源于咨访双方的有1/3的个案，当事人在咨询师的引导下逐渐获得领悟，或者是在咨询师反馈后当事人在会谈中或会谈后继续思考，确认咨询师反馈的正确性后再获得领悟。

3.7 阻碍领悟发挥作用的因素

本研究中，有一些个案谈到在咨询中虽然有领悟，但是领悟并没有对自己造成很大的影响，自己还是没有多大的改变，或是改变起来感到很困难。总体来说，有三个因素阻碍了领悟发挥作用：(1) 旧有习惯的惯性。4个个案认为旧有的习惯或模式已经养成，形成了一种惯性，尽管它会对自己造成不良的影响，但一方面习惯的影响力很大，改变起来很困难，另一方面维持惯性很舒服，而如果要改变，会感到不舒服。(2) 现实因素的制约。4个个案认为现实因素或外部环境限制了自己的改变，如面临现实生活中的压力，没有精力考虑如何改变；旧有的人际关系已经形成，难以突破等。(3) 改变动力的不足。2个个案谈到改变动力不足，如生活忙碌了就不把问题放在心上；以后会离开现在的人际环境，所以不用为了改善现在的人际环境而去改变，等等。

4　讨论

4.1 领悟的界定

从最一般的意义上说，心理咨询是一个解决问题的过程，当事人带着自己的问题而来，在咨询师的协助下解决。在咨询中要解决的问题既可以是当事人自诉的心理困扰，也可以是潜伏在心理困扰之下，悄悄发挥作用的更根本的问题。当对这些问题进行工作时，当事人会获得一些新的认识，这些新认识包括自身存在的问题模式，造成自己的心理困扰或者问题模式的原因，心理困扰或者问题模式对自己造成的影响，以及消除心理困扰或问题模式的办法，这四个方面共同构成了当事人在咨询中获得的领悟的主要类型。除此之外，当事人也可能会对他人有所领悟，更能理解他人，或者是对自己内心的状态有所体察。因此，若从这一研究结果来界定咨询中的领悟，可以将其定义为：领悟是对自己和他人（主要是自己）的新认识，对自己的认识内容包括，自己的问题模式，心理困扰或问题模式的原因、影响和解决办法，以及自己内在的心理状态。

将这一定义和前人从咨询师角度的跨理论定义进行比较，可以发现一些异同。共同点在于，都认为领悟是当事人对自己的新认识，这是对领悟本质特征的最基本认识。而区别主要有两点：其一，对领悟内容的认识层面不同。前人的界定中对内容的描述有的是在比较抽象宏观的层面，如"自我或人际关系中的自我"(Elliott, 1985)，"自己和自己的世界"(Mahrer & Nadler, 1986)，这样的表述无疑是正确的，但因为过于宏观，所以提供的信息量不大。有的是在比较微观的层面从心理机能的角度来描述领悟内容，如"行为、想法或感受的模式或原因"(Hill, 1992)，"关系、在咨询室外的机能状况、或者心理动力和行为"(Gelso et al., 1997)，"以前无意识的感受、想法和行动"(Gelso & Harbin, 2007)。而本研究可以说是在二者之间从中观的层面对领悟内容的描述，围绕心理困扰或问题模式来建构领悟。这与当事人的身份及其对领悟的直观感知是一致的，他们报告的是自己知觉到的领悟内容，因而是与其咨询问题密切相关的，自己的问题是什么，是如何产生的，对自己造成了什么影响，以及如何解决，这就是他们所领悟的主要内容。而咨询师受过心理学专业训练，因此在看待领悟的内容时难免刻上心理学的烙印。双方视角不同，对领悟内容的解读也就存在差异。这一结果提示咨询师在咨询中可以多一个视角，尝试从当事人的角度来看待和促进领悟。其二，对领悟内容的认识不尽相同。尽管认识层面不同，但当事人和咨询师认为的领悟内容绝大部分还是一致的，这主要是指问题模式、原因、影响和觉察。除此之外，当事人认为的领悟内容还包括心理困扰或问题模式的解决办法，前人的界定中没有这一部分。这说明当事人和临床心理学家在这一点上有分歧。按照经典精神分析的观点，领悟后症状会自然消失，但有些持跨理论观点的学者对此提出了异议，认为领悟后仍然需要行动的跟进，才能消除症状 (Hill, 2009)。本研究表明，对有些当事人来说，获得领悟并不一定能使症状消除，他们需要在理解自己问题的基础上，继续探索并对如何解决问题有所领悟，然后依此行动或做出调整，才能有所改善。此外，当事人的领悟还包含对他人的新认识，尽管绝大多数领悟是对自己的新认识，但当对他人有新的认识和理解时，当事人也认为是领悟。而前人的界定中，除了 APA 心理学词典 (VandenBos, 2006) 的定义中涉及了他人，其他都只包括对自己的认识。

4.2 领悟的来源、出现时的反应和作用

领悟既可以来自于咨询师的反馈，也可以来自于当事人自身的思考，还可以来自于咨访双方的合作，在本研究中，分别是 11 个、7 个和 5 个个案。从这个结果来看，咨询师是最主要的领悟来源，当事人完全通过自己获得的领悟相对少一些，但差别不是很大。完全通过一种来源获得领悟的个案有 8 个，通过两种或三种来源获得领悟的个案有 7 个。采用哪种方式使当事人获得领悟，在几种传统的主流咨询取向中表现出比较大的差异，精神分析和认知疗法中主要的领悟来源是咨询师，体验疗法中主要的来源是当事人自己。这也说明领悟可以有不同的来源。在跨理论取向的治疗中，领悟究竟来自咨询师还是当事人也许并不太重要，既然是领悟，那说明它们最终都是被消化吸收而成为了当事人自己的东西，尽管殊途却是同归，"白猫黑猫，抓住老鼠就是好猫"的理论也许在这里同样适用。但从另一方面来说，不同来源的领悟在质量上是否存在差异，对咨询过程和效果的影响是否有所不同，这些问题还有待研究检验。

在领悟出现时，当事人相应地会有一些感受和反应，西方的研究中将主要的反应称为"Aha"，类似于本研究中的豁然开朗，都是指一种突然通了、亮了、清楚了、明白了的感受。除了这一种常与领悟伴随出现的感受外，本研究表明，还会有一些其他的感受或反应。如果从理性领悟和情绪性领悟的角度来看这几种反应，可以发现，豁然开朗、希望感、惊喜震撼意外、感动是两种领悟都可能导致的反应；而释放放松、与个人问题相关的感受却只和情绪性领悟有关。由此看来，可以将这些反应分为两类，一类是由领悟导致的积极情绪反应，一类是由领悟引起的具有个人意义的情绪反应，前者可以由理性领悟或情绪性领悟引发，而后者的出现才表明获得的是情绪性领悟。这样，可以通过了解当事人在领悟出现时的反应来判断获得的是理性领悟还是情绪性领悟。当然，这倚赖于当事人的自我报告，如果当事人在咨询中没有明确的言语说明或非言语表示（如哭泣），咨询师和观察员很难凭观察来判断其是否获得了情绪性领悟。这或许

可以为理性领悟和情绪性领悟的测量提供一点启示。

领悟与咨询效果的关系一直是研究者关心的一个问题，领悟对咨询效果起多大作用，是如何起作用的，对这些问题的回答有助于深化对咨询机制的理解。本研究虽然不能从量上回答这些问题，但研究结果反映出领悟的确会导致当事人发生改变，而这些改变中有一部分是咨询效果本身，如态度、情绪感受和行为的变化，还有一部分是导致咨询产生效果的中间变量，如加深了自我认识，增强了自我调节的能力。此外，结果显示领悟会促进工作同盟质量的提升，如当事人更信任喜欢咨询师——增强了情感联结，咨询更清晰流畅——增强了目标和任务的一致性。这一结果支持 Wampold, Imel, Bhati 和 Johnson-Jennings (2007) 提出的理论模型，他们认为领悟会促进工作同盟的进一步巩固，而工作同盟与咨询效果的关系已得到大量研究的确证，因此领悟会通过促进工作同盟来促进咨询效果的实现。以后的研究可以采用定量的方式进一步检验这些假设，并考察领悟的影响力大小。

虽然大多数当事人认为领悟对自己有积极的影响，但也有个案谈到虽然有领悟，但改变并不容易。这说明，领悟并不一定能直接改善当事人的症状或改变其生活，仍然需要在对自己的问题有所认识后，付出努力才能产生切实的改变。如果因为惯性、现实因素、缺乏动力就不付出努力和行动，也就不会有很大的改变。这与 Hill (2009) 领悟并不一定能直接导致行动的看法是一致的，也正因为如此，她将行动放在领悟之后，作为咨询的最后一个阶段，而不是在领悟之后就结束咨询。

4.3 启示

对于临床工作，该研究提示：(1) 咨询师应该重视促进当事人在咨询过程中获得领悟，因为领悟既可以给当事人带来积极的改变，也有助于巩固工作同盟，使咨访双方更好地合作，以实现咨询目标。但也应注意，并非领悟后咨询就可以结束，而应在领悟的基础上进一步陪同当事人制定改变计划，并监督其付诸实施，否则有可能当事人在回归现实环境后因为较大的困难而不改变。(2) 咨询师可以转换视角，从当事人的角度来看待和促进领悟，按照问题解决的思路，思考当事人心理困扰的前因后果，进行概念化，然后在咨询中和当事人讨论，进行验证，使当事人获得领悟。(3) 促进领悟形成的因素很多，咨询师既可以适时采用合适的咨询技术，也可以促动当事人调用自身的因素。同时应注意不良的咨询关系会阻碍领悟形成，因此在咨询关系欠佳时，咨询师应首先致力于修复咨询关系，而不应急于促成当事人的领悟，否则有可能会适得其反。

对于后续的研究工作，该研究提供了概念化支持，可以在此基础上编制领悟测量工具。还可以根据本研究的一些发现，采用定量的方式进一步研究，如理性领悟与情绪性领悟的区分，及其与后续咨询过程和效果的关系；不同的促进领悟形成的因素对领悟的影响力大小；领悟与咨询效果的关系，以及工作同盟、自我调节等在其间所起的作用；等等。

从文化角度审视该研究的结果，笔者认为，总体来说并未显示出十分强烈的文化特异性。尤其是在领悟的内容方面，大多数内容与精神分析或人本主义理论的基本观点是一致的，如对防御方式、价值条件的领悟等。导致这一状况的原因，既可能是因为这些理论观点的确是具有普适性的，也有可能是由于该研究中的咨询师，都受过这些理论取向的训练，因此在咨询中解释或引导时都会受到理论的影响，从而当事人的领悟也都有明显的理论特色。正如钟友彬 (1985) 指出的："病人的领悟内容和医生的理论观点有密切关系，显然医生的解释起了重要作用"。Wampold 等 (2007) 也认为，领悟的真实性对于当事人接受解释或解释发挥作用并不必要，不存在一种解释比另一种解释更接近真实，解释的真实性也与效果无关，重要的是一个解释是否被当事人接受，而不在于解释是否精准。因此，领悟的内容是本土还是舶来也许并不是那么重要，关键在于当事人是否能够接受。咨询师在帮助当事人获得领悟时，既无需和理论牵强附会，也不必刻意强调本土性而抵制西方理论，而应该贴近实际，灵活处置。

该结果中有一处是否体现出文化特异性尚值得

探讨：影响领悟产生的因素中，来自咨询师方面的促进因素，其中占比最大的是咨询师的反馈，15 个个案中有 11 个获得领悟是由于有咨询师的反馈。这说明，很多当事人会通过咨询师陈述对其问题的看法而获得领悟，尽管也有当事人从咨询师的倾听和接纳、非指导的态度中获得领悟，但当事人也并不排斥咨询师直接表达自己的观点，只要恰当，他们也愿意接受。许多论述心理咨询本土化问题的文章都提到，中国当事人相较西方当事人，更愿意接受咨询师的权威角色（如吴垠，2011；孔德生,2007）。咨询师在反馈时，的确是扮演着权威角色的，而当事人获得了领悟，说明他们认同了咨询师的反馈，也就是接受了咨询师的权威。如果中国当事人的确更乐意咨询师扮演权威，而咨询师直接反馈也的确能使当事人获得领悟，且这样的领悟与当事人自己探索获得的领悟在效果上并无差异，那么这种方式可能是更适合于中国当事人的方式。但这诸多的前提条件是否能满足，还需要更多的研究检验。

4.4 局限与展望

本研究呈现了跨理论心理咨询中当事人对领悟的基本认识，但研究存在一些局限，以后可以改进。首先是样本问题，访谈对象中绝大多数为女性，接受咨询时都是大学生，均为非临床样本，大多在一个咨询中心接受咨询，咨询的都是发展性问题，除了一个个案有 56 次会谈之外，其余均为短程个案，这虽然符合 Hill 提出的样本同质性的要求，但同样因此而减少了代表性。在与此不同的样本中，结果是否同样适用，还需要研究检验。其次，研究小组的成员均为女性，尽管在结果审查时有男性参与，但仍然不确定完全同性的成员组成是否会对结果造成实质性的影响。最后，该研究对领悟内容的分类，可进一步通过量化研究，如聚类分析或因素分析来检验其可靠性。

参考文献

Boulet, D. B., Soulière, M. D., Sterner, I., & Nadler, W. P.(1992). Development of a category system of goodmoments in Gestalt therapy. *Psychotherapy, 29,* 554 - 563.

Castonguay, L. G., & Hill, C. E. (Eds.) (2007). *Insight inpsychotherapy*. Washington, DC: American PsychologicalAssociation.

Elliott, R. (1985). Helpful and nonhelpful events in briefcounseling interviews: An empirical taxonomy. *Journal ofCounseling Psychology, 32,* 307 - 322.

Elliott, R., James, E., Reimschuessel, C., Cislo, D., & Sack, N.(1985). Significant events and the analysis of immediatetherapeutic impacts. *Psychotherapy, 22,* 620 - 630.

Ellis, A. (1963). Toward a more precise definition of "emotional" and "intellectual" insight. *PsychologicalReports, 13,* 125 - 126.

Ellis, A. (2001). "Intellectual" and "emotional" insightr evisited. *NYS Psychologist, 13,* 2 - 6.

Gelso, C. J., & Harbin, J. (2007). Insight, action, and thetherapeutic relationship. In L. G. Castonguay & C. E. Hill(Eds.), *Insight in psychotherapy* (pp. 293 - 311). Washington, DC: American Psychology Association.

Gelso, C. J., Kivlighan, D. M., Wine, B., Jones, A., &Friedman, S. C. (1997). Transference, insight, and thecourse of time-limited therapy. *Journal of CounselingPsychology, 44,* 209 - 217.

Gershefski, J. J., Arnkoff, D. B., Glass, C. R., & Elkin, I.(1996). Clients' perceptions of treatment for depression: I.helpful aspects. *Psychotherapy Research, 6,* 233 - 247.

Gibbons, M. B. C., Crits-Christoph, P., Barber, J. P., &Schamberger, M. (2007). Insight in psychotherapy: Areview of empirical literature. In L. G. Castonguay & C. E.Hill (Eds.), *Insight in psychotherapy* (p. 159). Washington,DC: American Psychology Association.

Hill, C. E. (1992). An overview of four measures developed totest the Hill Process Model: Therapist intentions, therapistresponse modes, client reactions, and client behaviors.*Journal of Counseling & Development, 70,* 728‑739.

Hill, C. E. (2009). Helping skills: *Facilitating exploration,insight, and action* (3rd ed.). Washington, DC: AmericanPsychology Association.

Hill, C. E., Castonguay, L. G., Angus, L., Arnkoff, D. B.,Barber, J. P., Bohart, A. C., ··· Wampold, B. E. (2007).Insight in psychotherapy: Definitions, processes,consequences, and research directions. In L. G. Castonguay& C. E. Hill (Eds.), *Insight in psychotherapy* (p. 442).Washington, DC: American Psychology Association.

Hill, C. E., Knox, S., Thompson, B. J., Williams, E. N., Hess,S. A., & Ladany, N. (2005). Consensual qualitativeresearch: An update. *Journal of Counseling Psychology, 52,*196‑205.

Hill, C. E., Thompson, B. J., & Williams, E. N. (1997). Aguide to conduct Consensual Qualitative Research. *TheCounseling Psychologist, 25,* 517‑572.

Holtforth, M.G., Castonguay, L. G., Boswell, J. F., Wilson, L.A., Kakouros, A. A., & Borkovec, T. D. (2007). Insight incognitive‑behavioral therapy. In L. G. Castonguay, & C. E.Hill (Eds.), *Insight in psychotherapy* (pp. 57‑80).Washington, DC: American Psychology Association. Hu, L., & Zhou,

H. L. (2006). The comprehension inpsychological consultation and treatment. *Medicine &Philosophy (Humanistic & Social Medicine Edition), 27,*41‑43.

[胡岚 , 周和玲 . (2006). 心理咨询与治疗中的领悟 . *医学与哲学 (人文社会医学版),27,* 41‑43.]

Hu, S. J., & Jiang, G. R. (2010). Definition of "insight" inpsychotherapy. *Advances in Psychological Science, 18,*1489‑1495.

[胡姝婧 , 江光荣 . (2010). 心理咨询中的 "领悟" 概念辨析 . *心理科学进展, 18,* 1489‑1495.]

Kong, D. S. (2007). Exploration and practice for localizationof psychological counseling in China.

Academic Exchange,(12), 31‑35.

[孔德生 . (2007). 我国心理咨询本土化的探索与实践 . *学术交流 , (12),* 31‑35.]

Levitt, H., Butler, M., & Hill, T. (2006). What clients findhelpful in psychotherapy: developing principles forfacilitating moment‑to‑moment change. *Journal of Counseling Psychology, 53,* 314‑324.

Mahrer, A. R., & Nadler, W. P. (1986). Good moments inpsychotherapy: A preliminary review, a list, and somepromising research avenues. *Journal of Consulting andClinical Psychology, 52,* 10‑15.

Messer, S. B., & McWilliams, N. (2007). Insight inpsychodynamic therapy: Theory and assessment. In L. G.Castonguay & C. E. Hill (Eds.), *Insight in psychotherapy* (pp.9‑29). Washington, DC: American Psychology Association.

Moreno, J. K., Fuhriman, A., & Hileman, E. (1995).Significant events in a psychodynamic psychotherapygroup for eating disorders. *Group, 19,* 56‑62.

Pascual‑Leone, A., & Greenberg, L. S. (2007). Insight andawareness in experiential therapy. In L. G. Castonguay, & C.E. Hill (Eds.), *Insight in psychotherapy* (pp. 31‑56).Washington, DC: American Psychology Association.

Paulson, B. L., Truscott, D., & Stuart, J. (1999). Clients'perceptions of helpful experiences in counseling. *Journalof Counseling Psychology, 46,* 317‑324.

VandenBos, G. R. (2006). *APA Dictionary of Psychology* (pp.484‑485). Washington, DC: American PsychologicalAssociation.

Wachtel, P. L. (1997). *Psychoanalysis, behavior therapy, andthe relational world.* Washington, DC: AmericanPsychological Association.

Wampold, B. E., Imel, Z., Bhati, K. S., & Johnson‑Jennings,M. (2007). Insight as a common factor.

In L. G. Castonguay,& C. E. Hill (Eds.), *Insight in psychotherapy* (pp. 119 – 139).Washington, DC: American Psychology Association.

Wu, Y. (2011). The tension between the Chinese "relationship–oriented" social psychology and thelocalization of psychological counseling. *Social SciencesJournal of Universities in Shanxi, 23*, 91 – 93.

[吴垠 . (2011). 中国人 "关系取向" 的社会心理特征与心理咨询本土化之间的张力 . *山西高等学校社会科学学报* , 23,91 – 93.]

[Zhong, Y. B. (1985). "Insight" in psychodynamic therapy. *Journal of International Psychiatry, (3)*, 137 – 140.

[钟友彬 . (1985). 论动力学心理疗法中的 "领悟" . 国外医学 . *精神病学分册* , *(3)*, 137 – 140.]

Chinese psychotherapy clients' perspectives on insight: A qualitative examination

HU Shujing[1,2], JIANG Guangrong[2], LU Yanhua[2], ZHANG Shasha[2], CHEN Ruijuan[2], YU Lixia[2], DU Rui[2]

([1] College of Science, Naval University of Engineering, Wuhan 430033, China)

([2] Key Laboratory of Adolescent Cyberpsychology and Behavior (Ministry of Education); School of Psychology, Central China Normal University; Key Laboratory of Human Development and Mental Health of Hubei Province, Wuhan 430079, China)

Abstract　Insight is a crucial phenomenon in counseling with its importance been confirmed by counseling theory andempirical studies. However, there is a lack of specific studies on this topic, and most researchers focused on psychodynamic area whereas little attention has been laid on pantheoretical counseling. There is no unified definition of insight available at the moment. Psychodynamic therapy regards insight as the formation of connections, e.g. between past and present, inner conflict and external performance, attachment relationships and transference. Experiential therapy equates insight to awareness and meta-awareness. Cognitive - behavioral therapy views insight as a cognitive restructuring or change of schema; insight occurs when previous irrational beliefs or schemas are recognized and replaced by new, rational beliefs or schemas. Meanwhile, different tresearchers also gave different operational definitions in cross-theory empirical research. As a result, the lack of a unified definition of insight, in either theoretical or empirical research, has made it impossible to integrate different results and prevented the development of a specialized and reliable measurement of insight. In addition, nearly all definitions were proposed by researchers or clinicians. Whether or not clients have a different tperspective is unknown. Moreover, all the definitions currently in use were developed by western researchers focusing on western psychotherapy. Would Chinese psychotherapy present a different perspective? This study sought to delineate insight from Chinese clients' perspective via a qualitative approach.

Fourteen clients counseled in a university counseling center and a client counseled in a social counseling institution participated in the study. They were interviewed with a semi-structured protocol, either face-to-face, by phone, or through internet, about their opinions and experiences of insight during counseling. Their responses were analyzed by a research team using Consensual Qualitative Research (CQR) method.

Analysis revealed seven key issues: contents of insight, effects of insight, factors influencing the appearance of insight, basis for evaluating insight quality, feelings when insight appears, source of insight, and factors that hinder insight from having an effect. Based on interviewees' responses, a definition of "insight" from the client perspective is proposed: "insight is a new understanding of oneself and others (mainly oneself), the contents of which include problematic patterns, their reasons, their effects, solutions for psychological distress or problematic patterns, and awareness of one's internal mental state."

Results suggested that most contents of insight from client and counselor's perspectives are in concordance. However, the description level is different and some contents which haven't been thought as insight before was confirmed by clients. It reminds counselors that they could change their angle of view in order to understand insight from client's point. The current study also provides the foundation of future study's development of scal efor insight.

Keywords psychotherapy; insight; qualitative research

心理学报，2007, 39(5), 892－900.

归因、自我效能和社会容认度对心理求助行为的影响

夏　勉，江光荣

(华中师范大学心理学院，武汉 430079)

摘　要　在江光荣心理求助行为的"阶段－决策模型"框架下，探查处于第三阶段的被试的求助行为，侧重研究(1)对心理问题的归因、(2)作为心理咨询当事人的自我效能和(3)心理求助行为的社会容认度这三个变量对实际求助行为的影响，同时探讨变量之间的作用机制。结果表明，这三个变量对求助行为均有预测作用。其中心理求助行为的社会容认度除了对求助行为有直接作用外，还以作为心理咨询当事人的自我效能为中介变量间接影响求助行为。整个模型对求助行为解释的变异量达到 26%。

关键词　归因，自我效能，社会容认度，心理求助行为。

分类号　B849:R395

1　导言

心理求助是指个体在遇到心理困扰或障碍的时候，向个人之外的力量寻求帮助以达到解决困扰的过程。现有的研究表明，许多需要心理帮助的人并不寻求专业援助。而东方人 (包括中国人) 较之西方人对求助的态度更为消极，以致不少人最后出现较严重的心理疾病甚至精神崩溃。费立鹏等 (2002) 的调查发现，在中国的自杀死亡者中，仅有 17% 的人死前曾寻求过专业心理帮助。事实上，即使在心理咨询服务很容易获得的地方例如大学，人们也不愿意向心理咨询专业人员求助。因此，为了更好地帮助那些有潜在治疗需要的人，就必须了解他们不愿意求助的原因。

以往的研究从人口统计学因素、社会文化因素、心理因素三个方面考察妨碍个体求助的原因。在人口统计学方面，研究者发现一个非常突出的现象是男性较女性对心理求助持更消极的态度，并发现这跟主流文化中男性性别角色观念有直接联系。在社会和文化方面，研究者对受文化影响，与求助有关的偏见、信念和价值观等因素进行了较广泛的调查，并发现其中一些因素对当事人的求助行为有影响。从个体的心理因素探讨原因的研究相对较少。已有的研究探讨了当事人的情绪管理能力，归因方式和控制感，对治疗的恐惧、自我表露和自我隐藏、相互依赖和自我依赖、自我效能等因素与求助的关系。

在上述三类研究中，对人口统计学因素和社会文化因素研究比较多，而对心理因素的研究的较少。然而较近的研究表明，社会和文化的变量合在一起，对实际求助行为的解释比例仍然在一个较低的水平。这提示我们应该关注心理变量所起的作用。另外，国

通讯作者：江光荣，E－m ail:grjiang@m ai.l ccnu. edu. Cn

内外的大部分研究对求助态度和求助行为没有做出恰当的区分，大部分研究考察的是求助态度，极少考察求助行为，其原因主要是观察样本难以获得。已有研究表明，尽管态度与行为可能会存在某种程度的相关，但这种相关受态度的强度、态度的稳定性、态度的可获得性、态度的醒目程度、情境压力等一系列因素的影响，因而实际相关并不能高到可以用态度代替行为进行研究的程度。如果能够直接研究实际求助行为的话，显然更有科学价值。

为了深入考察当事人求助前的心理决策过程，本研究以江光荣 (2006) 提出的心理求助行为的阶段 - 决策模型作为理论基础。该阶段模型认为求助行为包括先后相继的三个阶段：（1）心理问题自我觉察阶段——当事人是否察觉自己有心理问题；（2）自助评估阶段——当事人认为自己能不能有效地解决问题；（3）他助评估阶段——当事人决定是否求助于个人之外的力量，及向何种力量求助。本研究聚焦于第三阶段的当事人，即意识到自己有心理问题且不能自己有效解决的大学生，试图探索影响他们向专业机构求助的动机性因素。

江光荣，王铭的研究初步发现自我效能是影响个体求助的一个变量。余晓敏，江光荣进一步将自我效能分为解决心理问题的自我效能和作为心理咨询当事人的自我效能，并发现这两种自我效能对求助态度都有显著影响。但在这两项研究中，都没有考虑求助的阶段因素。从上述"阶段 - 决策模型"来考虑，个体解决个人心理问题的自我效能可能主要在第二阶段起作用，而作为心理咨询当事人的自我效能则主要影响处于第三阶段（他助评估阶段）的当事人。故本研究只考察作为心理咨询当事人的自我效能。

国外有少量研究涉及归因对求助行为的影响。这些研究大部分以维纳的归因理论为基础，考察归因的三个维度（稳定性、部位、可控性）与求助行为的关系，其中研究较多的是部位和可控性。结果发现，可控性归因对求助态度有影响，部位与求助行为的研究则得到了不一致的结果。有些研究发现，如果个体把造成心理问题的原因归于内部因素时，则倾向于不

求助，当归因于外部因素时，则倾向于求助。另外一些研究则发现，内部 - 外部归因对求助行为没有显著影响。在"阶段 - 决策模型"中，对心理问题的归因也被认为是一个影响求助的因素，但这个因素在各阶段起作用的情形却不易推测，本研究拟探索归因的两个维度（部位和可控性）在第三阶段起作用的情况。

此前的研究者一直比较重视社会文化因素对求助行为的影响，然而实际调查结果却发现社会文化对求助行为的解释比例比较低。江光荣认为，这是因为所有外在的因素都要通过个体的内在表征起作用，因此实际上对个体产生影响的是个体知觉到的社会文化。其他研究者也提到过类似的观点，如 Fishbe in 和 A jzen 认为主观社会规范是影响行为意向的重要变量。据此，江光荣用一个新的概念——"社会容认度"——来表征个体知觉到的社会文化信念。这个概念是指在个体知觉中，公众对某种行为的接受——排斥程度。这样就将社会文化变量转化为个体的心理变量。"阶段 - 决策模型"假设：心理求助行为的社会容认度是在他助评估中起作用的一个重要变量，本研究也将探讨这个变量的影响力。

本研究除了考察上述三个变量对求助行为的影响，还准备考察变量之间的作用方式或机制。有研究提示，归因会对自我调节过程中的自我效能因素产生一定的作用，这是因为人们对其行为的自我归因，能产生一种情感体验，这种情感体验会影响后继行为。但归因与社会容认度是否存在关系我们还无法推测。至于社会容认度跟自我效能的关系，班杜拉曾指出社会文化会影响自我效能，因为文化中对于特定行为的价值规范以支持、鼓励或约束的

方式影响着文化中的个体。在经验积累的基础上，会影响到个体的自我效能感。心理求助行为的社会容认度体现着文化中有关求助行为的价值规范。如果个体知觉到的心理求助行为的社会容认度较高，那么知觉到社会偏见如羞耻、丢面子等就比较低，这一高一低的经验，最终会使当事人对心理咨询专业服务及专业人员的信任程度和接受程度提高，对求助的有效性，对可预见的与咨询有关的压力也都会持比较积

极的态度。而这些都是当事人自我效能的核心成分。

因此本研究假设归因和社会容认度会影响作为心理咨询当事人的自我效能，即归因和社会容认度除直接影响求助行为外，还以自我效能为中介变量影响求助行为。

2　研究程序与方法

2.1　取样

本研究以实际求助行为为因变量，以克服此前的研究多以求助态度或求助意图为因变量的局限。为此选择了两组被试，一组是意识到自己有心理困扰且不能自己有效解决，正在向专业机构求助的被试（以下称为"主动求助组"）。一组是同样意识到自己有心理困扰且不能自己有效解决，但没有向专业人员求助的被试（以下称为"未求助组"）。两组样本均取自武汉市7所高校。

以往的研究发现，求助经验会影响个体求助的意愿。为了控制求助经验的影响，曾经向专业人员求助过的被试不纳入取样范围。确定未求助组被试的程序如下：被试在回答问卷的主体部分前须回答三个问题：①您是否曾经向心理咨询师求助过？②"在最近一段时间内（三个月），您在多大程度上感到自己有过心理困扰？"（以 Likert 7 点量表由低到高记分）；"如果没有求助，您觉得您自己是否能有效解决。"同时满足三个条件（第一题回答"否"，第二题得分 >5，第三题回答"不能自己有效解决"）的学生，被作为未求助组被试。

具体取样过程如下：将 180 份问卷发放到六所高校的心理咨询中心，请经过教育部《大学生心理健康评定量表》筛查有问题的学生填答＊，这些学生均为大一学生，回收 150 份有效问卷，有效回收率 83.3%，其中符合未求助组条件者 65 份。为了取得其它年级学生的样本，另将 720 份问卷发放到七所高校的公共课堂，当堂发放，当堂回收，共回收有效问卷 700 份，有效回收率 97.2%，其中符合未求助组条件者 89 份。这样未求助组共获得有效问卷 154 份。

主动求助组被试以到 7 所高校的心理咨询中心主动求助且属首次求助的学生构成。为避免会谈经验的影响，问卷请被试在第一次咨询前填答。共发放问卷 180 份，回收有效问卷 151 份，有效回收率为 76.6%。

研究的最终样本量为 305，其中未求助组 154，主动求助组 151。被试的具体资料见表 1。

表 1　被试的基本资料 (N =305)

组别	性别		来源			年级					专业				
	男	女	大城市	中小城市	乡镇农村	大一	大二	大三	大四	研究生	文科	理科	工科	农科	医科
求助组	69	82	17	47	87	40	48	34	23	6	60	37	33	4	14
未求助组	82	72	19	40	95	42	73	32	7	0	27	54	41	8	24
总 体	151	154	36	87	192	82	121	66	30	6	87	91	74	12	38

2.2　测量工具

2.2.1　作为心理咨询当事人的自我效能　作为心理咨询当事人的自我效能包括以下一些信念：相信自己能够坚持完成咨询，能够有效地与咨询员交流自己的问题，能够处理心理咨询中的突发事件，并相信自己能从咨询中获益，等等。

此变量由问卷测量。问卷由余晓敏，江光荣 (2004) 编制的《作为心理咨询当事人的自我效能问卷》修订而来。原问卷包括 10 个项目，探索性因素分析结果表明，这 10 个项目较好地聚合在单一的维度上，可以解释的变异量为 41.68%，内部一致性信度为 0.84。在本研究中，用验证性因素分析进行修订，修订后的问卷由 8 个项目组成，有较好的结构效度，问卷的内部一致性信度为 0.81。自我效能问卷验证性因素分析的结果为：$\chi^2=74.20$, $df=20$, $\chi^2/df=3.71$, RMSEA=0.094, CFI=0.92, IFI=0.92, NNFI=0.89, GFI=0.94。项目举例："我相信自己能够自如地和心理咨询员讨论我的问题"。采用 5 点 Likert 量表记分，分值越高代表自我效能越高。

2.2.2　归因　对心理问题的归因只包括部位、可控性两个维度。由自编问卷测量，每个维度包括一个测量问题。部位维度："假如这段时间内，您经历了一些心理困扰，您觉得您的心理困扰在多大程度上是自己的原因造成的？"；可控性维度："假如这段时间内，您经历了一些心理困扰，您觉得造成您心理困扰的原因在多大程度上是自己可以控制的？"。问卷用6点Likert量表记分。

2.2.3 心理求助行为的社会容认度 心理求助行为的社会容认度是指在被试心目中，公众对于自己寻求心理帮助的接受——排斥程度。变量亦采用自编问卷测量。问卷采用语义差别量表的形式，题干是"假如这段时间内，您经历了一些心理困扰，并且想向心理咨询员求助，那么您觉得"您的求助行为"在社会上大多数人的眼里是——"。评价部分由14对双极形容词组成。如"明智的 – 愚蠢的"，"积极的 – 消极的"。两极间距离从 –3 到 3，共分为 7 级。级数越高表示对心理求助行为的接受程度越高。问卷的 α 信度系数为 0.91。

试测分析结果表明，该问卷为单一维度，单个维度可解释的变异量为 49.23%，对正式施测结果进行验证性因素分析，χ^2=287.72, df=77, χ^2/df=3.71, RMSEA=0.094, CFI=0.91, IFI=0.91, NNFI=0.90, GFI=0.88, 结果表明单一维度的结构可以接受。

2.3 数据处理

本研究采用 SPSS 11.0 和 LISREL 8.30 进行统计分析，主要的分析方法为相关分析、Logistic 回归和结构方程模型。

3 研究结果

本研究的预测变量为：作为心理咨询当事人的自我效能、对心理问题的归因、心理求助行为的社会容认度。结果变量为求助行为。求助行为是二分变量，即求助和未求助。未求助组和主动求助组在各个预测变量上的平均值和标准差见表2，预测变量和结果变量之间的相关矩阵见表3。

3.1 预测变量对结果变量的 Logistic 回归结果

预测变量对结果变量的 Logistic 回归结果见表4。

由表4可知，可控性归因、心理求助行为的社会容认度、作为心理咨询当事人的自我效能这三个变量对求助行为影响都达到显著水平，而内部 – 外部归因没有进入最优回归方程。变量间的关联是：个体知觉中心理问题的原因越是可控的，就越倾向于不求助；个体作为心理咨询当事人的自我效能越高，越有可能去求助。个体知觉到的心理求助行为的社会容认度越高，越可能求助。从 Ward 统计值来看，作为心理咨询

表 2 未求助组和主动求助组的平均数和标准差

| 组别 | 心理问题归因 | | | | 求助行为的社会容认度 | | 作为心理咨询当事人的自我效能 | |
| | 内部—外部归因 | | 可控性归因 | | | | | |
	M	SD	M	SD	M	SD	M	SD
未求助组	4.09	0.94	3.83	1.06	4.92	0.93	4.96	0.79
主动求助组	4.04	1.15	3.61	1.15	5.33	0.89	5.62	0.61
总体	4.07	1.03	3.74	1.10	5.09	0.93	5.24	0.79

注：① ** P <0.01；* P <0.05；
② 求助行为与其他变量的相关为点二列相关，其余为积差相关。

表 3 三个预测变量以及结果变量之间的相关矩阵 (N =305)

变量	M	SD	1	2	3	4	5
1 内部—外部归因	4.07	1.03	1.00				
2 可控性归因	3.74	1.10	0.058	1.00			
3 心理求助行为的社会容认度	5.09	0.93	0.054	0.025	1.00		
4 作为心理咨询当事人的自我效能	5.24	0.79	-0.139*	0.079	0.165**	1.00	
5 求助行为	0.50	0.50	-0.026	-0.096	0.217**	0.413**	1.00

表 4　预测变量对结果变量的 Logistic 回归结果

变量	B	SE	Wald	df	Sig.	Exp(B)
可控性归因	−0.380	0.122	9.637	1	0.002	0.684
社会容认度	0.475	0.150	9.967	1	0.002	1.608
自我效能	1.442	0.218	43.937	1	0.000	4.231
确定系数	Cox & Snell R Square			Nagelkerke R Square		
	0.239			0.319		

注：①进入回归方程的方法为 Forw ard S tepw ise (C ond itional)　②B 为非标准化回归系数

表 5　心理求助行为的结构方程模型拟合指数

	χ^2	df	χ^2/df	RMSEA	CFI	IFI	NNFI	R^2
虚模型	3090.72	276	11.20					
模型	560.19	248	2.26	0.064	0.90	0.90	0.89	0.26

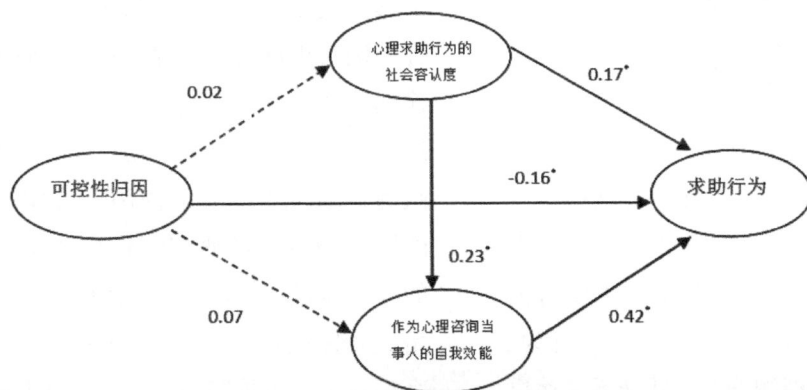

图 1　心理求助行为的结构方程模型

注：①路径系数均为标准化解②＊表示路径系数显著

当事人的自我效能作用最大，其次是心理求助行为的社会容认度，再次是可控性归因。由于自我效能、社会容认度、可控性归因的单位不同，分别为 5 级计分，7 级计分和 6 级计分，因此不再进行风险比的比较。

3.2　结构方程模型分析及中介效应检验

根据导言中的讨论，将四个变量之间的关系构建如图 1 的模型。该模型包括三个相互关联的假设：可控性归因影响求助行为（包括直接效应和通过社会容认度的中介效应以及通过作为心理咨询当事人的自我效能的中介效应）；社会容认度影响求助行为（包括直接效应和通过自我效能的中介效应）；作为心理咨询当事人的自我效能影响求助行为。结构方程的分析结果见图 1。

模型的拟合情况见表 5。

研究结果发现，可控性归因到求助行为这条路径是显著的，但是通过社会容认度和自我效能对求助行为的影响不显著。心理求助行为的社会容认度到求助行为这条路径以及社会容认度通过自我效能到求助行为这条路径都是显著的。即社会容认度除了直接影响求助行为以外，还以自我效能为中介变量对求助行为产生影响。用 LISRL 8. 30 计算中介效应，发现心理求助行为的社会容认度通过"作为心理咨询当事人的自我效能"对求助行为的中介效应是 0.05，标准误是 0.01，Z 值是 3.16。社会容认度对求助行为的总效应是 0.13，标准误是 0.03，Z 值是 4.46。因为 Z>1.96(p<0.05)，所以无论是社会容认度对求助行为的总效应还是通过自我效能的中介效应都是显著的，中介效应占总效应的比是 0.38。

作为心理咨询当事人的自我效能对求助行为的影响也是显著的，自我效能不但是一个重要的预测变量，而且在模型中起中介作用。总体来说模型是可以接受的，整个模型解释了求助行为变异的 26%。

4 讨论

4.1 研究实际求助行为的必要性

本研究在研究设计上的一个显著进步是研究个体实际的求助行为。由于取样及观察上的困难，以往的研究多考察求助意愿或求助态度。社会心理学的知识告诉我们，态度与行为的关系非常复杂，由态度并不能准确地预测行为。在心理求助领域，求助态度和求助行为的研究也发现了同样的情况。如 Kelly 和 A chte r 的研究发现，与低自我隐藏的人相比，高自我隐藏的人求助态度更消极，但实际求助的可能性却更高。又如在对治疗恐惧的研究中，Kushne r 和 Sher 发现，治疗恐惧水平的增加会引发消极的求助态度。但是 Segal, Hodges 和 Hardiman 对已经求助的被试进行研究，发现反而是那些有较高治疗恐惧的个体实际上选择了向专业人员求助。由于求助态度和求助行为并不完全一致，而我们的目标通常是想要预测行为，所以直接以实际求助行为为观测变量，其结果显然更有说服力，实践价值更大。

4.2 预测变量对求助行为的影响

本研究以认知动机理论为基础，考察了自我效能、归因和社会容认度对求助行为的影响。下面分别讨论研究得到的结果。

自我效能是指人们关于自己实现特定领域内的行为目标所需能力的信心或信念。班杜拉认为自我效能会影响个体对于未来环境和行为的选择。一般而言，个体会选择自认为他能加以有效应对的环境，而回避自感无法控制的环境。余晓敏、江光荣在此前的研究中已经发现，作为心理咨询当事人的自我效能对求助意愿有明显的预测作用。但在该项研究中，没有考虑求助阶段因素，样本中包括了处于不同求助阶段的被试。如果依阶段 – 决策模型，从逻辑上推测，这会低估自我效能对处于第三阶段被试求助行为的预测力。另外，该研究是以求助意愿为被预测变量，根据社会心理学的知识，实际求助行为会显著低于求助意愿。本研究一方面将求助阶段因素纳入设计考虑，另一方面改用实际求助行为为被预测变量。在此基础上比较前一研究和本研究的结果，发现当事人自我效能与实际求助行为的相关 (r=0.413) 略高于该自我效能与求助意愿的相关 (r=0.376)。可以推测，前一研究中自我效能的预测力的确被低估了，至少对处于第三阶段的被试是这样。无论如何，这两项研究表明，无论是对求助态度还是求助行为，自我效能都是一个非常重要的预测变量。

前人有研究表明可控性归因与求助意愿有关，而部位与求助意愿的关系不一致。本研究进一步考察这两个变量对实际求助行为的影响。结果发现，可控性归因对求助行为有显著影响，而内部 – 外部归因的作用不显著。由于考察归因与求助关系的研究积累还比较少，现在还无法推测本研究中内外归因的作用跟此前研究不一致的原因。可控性归因对求助行为的影响与以往的研究一致。个体知觉到的心理问题的原因越是可控的，就越倾向于不求助。对此的解释是：如果个体认为导致心理问题的原因是可控的，那么相对于不可控归因来说，当事人更可能产生负性反应，如自责、容易知觉到别人对自己有不良反应、掩饰行为等。作为掩饰行为的一部分，当事人可能会延迟求助或不求助。

本研究的一个尝试是解决作为群体现象的文化规范如何在个体层面上发生作用，影响个体的求助行为这个问题。这个尝试是借助提出求助行为的社会容认度这个概念实现的。文献表明，一个文化中关于心理问题的信念、该文化所赞许的心理问题表达方式和解决方式，会影响个体在遭遇心理困难时寻求专业帮助的态度。在中国文化中，患有精神障碍及因心理问题向外人求助，通常被视为羞耻的事。有关中国人性格的研究表明，中国人的一个典型心理特征是他人取向，表现为对他人的意见、褒贬特别敏感，总希望在他人心目中留下良好印象，在行为上努力与别人相一致，并尽量避免他人的责罚、讥笑、拒绝、尴尬及冲突。因此，中国人对心理求助会非常忌讳和回避。然而实际研究发现，虽然中国人的确比西方人较少利用心理卫生服务，但文化偏见所起的作用并不如预想的那么大。问题在于，文化规范并非对同一文化中的个体都是划一地起作用。总是对有的个体影响大，有的影响小。这也许是此前的研究发现文化偏见对于求助变量的解释力并不很大的原因。Sue 提出在考察与文化相关的行为时，在实践和研究中都应关注某一文化群体内的个体差异。江光荣提出的社会容认度的概念（在个体知觉中，公众对某种行为的接受——排斥程度）较好地解决了这一问题。社会容认度代表着个体对社会文化规范的内在表征，是一种个体现象。这样就可以从个体水平上考察文化变量对行为的影响。

研究结果表明，心理求助行为的社会容认度对求助行为有显著预测作用。个体知觉到的心理求助行为的社会容认度越高，越倾向于求助。这还仅仅是心理求助行为的社会容认度的影响，可以设想，如果把心理问题（罹患心理障碍）的社会容认度加进来，对求助行为的预测力还会增加。

4.3 预测变量之间的关系

有关预测变量之间关系的假设部分得到支持。结构方程模型分析表明，心理求助行为的社会容认度影响作为当事人的自我效能。进一步分析发现，社会容认度通过自我效能对求助行为产生影响的中介效应占社会容认度总效应的比例为 38%，这个比例是相当高的。也就是说，如果个体知觉到的社会上大多数人对求助行为越接受，那么作为心理咨询当事人的自我效能也就比较高，个体就越倾向于求助。这个结果在实践上有一定意义：心理健康教育工作可以通过宣传，提高公众求助行为的社会容认度，从而提高潜在求助者的当事人自我效能，进而提高求助率。

但心理问题归因跟当事人自我效能之间关系的假设没有得到支持，结构方程分析表明二者之间不存在关联。分析原因，可能在于心理问题的归因跟作为心理咨询当事人的自我效能在逻辑上的不对应。按照阶段 – 决策模型，心理问题的归因跟求助的关系可能主要发生在第二阶段——自助评估阶段。在这个阶段，心理问题的归因可能影响着当事人独立处理个人心理问题的自我效能。但这需要另行研究。

4.4 局限

本研究以阶段 – 决策模型为理论基础，考察了在第三个阶段影响个体求助行为的三个认知性动机变量——社会容认度、自我效能和归因，这三个变量可以解释求助行为 26% 的变异。检讨整个研究设计和过程，研究者觉得这个解释比例很可能有所低估。例如问题的严重程度这个变量的控制，就不是很精确，本研究的被试在问题严重程度这个变量上，基本是以"自觉存在心理问题且这问题不能自己解决"为入选标准。相信如果引入一个较为客观精确的严重程度测量作为控制变量，上述三个预测变量的效应会有所提高。需要说明的是，本研究只选取了三个动机变量考察其预测作用。实际上在第三阶段影响个体求助行为的还有许多别的因素，如性格特点、时空便利性、支付能力等，所以我们并不预期本研究的三个变量能解释大部分求助行为的变异。另外，本研究所用的归因问卷每个维度只用一个项目测量，从测量的角度来说可能会对问卷的信度有所影响。最后，本研究所选样本是大学生，因此研究结果的可推广性会受到限制。

5 结论

（1）心理问题可控性归因对求助行为有预测作用。个体知觉到的造成自己心理困扰的原因越是可控的，就越倾向于不求助。

（2）心理求助行为的社会容认度对求助行为有预测作用，包括直接作用及以"作为心理咨询当事人的自我效能"为中介变量的间接作用。个体知觉到的心理求助行为的社会容认度越高，越倾向于求助。

（3）作为心理咨询当事人的自我效能对求助行为有预测作用。作为心理咨询当事人的自我效能越高，当事人越倾向于求助。

参考文献

Andrews, G., Hall, W., Teesson, M., & Henderson, S. (1999). *The Mental Health of Austra lians*. National

Survey of Mental Health & Well- being Report No 2. Canberra: Mental Health Branch, Common wealth Department of Aged Care.

Bryan, S. K. K., & Michael, M. O. (2003). A sian Cultural Values, Attitudes toward Seeking P rofessional Psychological Help, and Willingness to See a Counselor. *The Counseling Psychologist, 31,* 343–361.

Cheatham, H. E., Shelton, T. O., & Ray, W. J. (1987). Race, sex, causal attribution, and help–seeking behavior. *Journal of College Student Personnel, 28*(6), 559–568.

Ciarrochi, J. V., & Deane, F. P. (2001). Emotional competence and willingness to seek help from professional and nonprofessional sources . *British Journal of Guidance & Counseling, 29*(2), 233–246.

Ciarrochi, J. V., Deane, F. P., & Wislon, C. J. (2002). Adolescentswho need help themost are the least likely to seek it :the relationship between low emotional competence and low intention to seek help. *British Journal of Guidance & Counseling, 30*(2), 173–188

Corrigan, P. W., Backs– Edwards, A., Green, A., Lickey– Diwan, S., Penn, D. (2001). Prejudice, socia l distance, and familiarity with mental illness. *Schizophrenia Bulletin, 27,* 219–225.

Flisher, A. J., De, & Beer, J. P. (2002). Characteristics of students receiving counseling services at the university of Cape Town, South Africa. *British Journal of Guidance & Counseling, 30*(3), 299–310.

Kung, W. W. (2004). Cultural and practical barriers to seeking mental health treatment for Chinese Americans. *Journal of Community Psychology, 32*(1), 27–43.

Guo Benyu. (2003). *The new advance of current psycho logy.* Jinan: Shandong education press, 6–7.

[郭本禹 . *当代心理学的新进展* . 济南：山东教育出版社 , 2003 . 6 - 7.]

Jiang G. R., & Wang, M. (2003). The research about college students' help–seeking behavior. (In Chinese) *Chinese journal of clinical psychology, 11*(3), 180–184.

[江光荣，王铭 . (2003). 大学生求助行为研究 . *中国临床心理学杂志，11*(3), 180–184.]

Jiang G. R., & Xia, M. (2006). Psychological help–seeking behavior: the research review and phase –decision model (In Chinese). *Advances in Psychological Science, 14*(6), 888–894.

[江光荣，夏勉 . (2006). 心理求助行为：研究现状及阶段 - 决策模型 . *心理科学进展，14*(6), 888–894]

Kelly, A. E., & Achter, J. A. (1995). self–conceal ment and attitude toward counseling in university students. *Journal of counseling psychology, 42,* 40–46.

Kung, W. W. (2004). Cultural and practical barriers to seeking mental health treatment for Chinese Americans. *Journal of Community Psychology, 32*(1), 27–43.

Kushner, M. G., & Sher, K. L. (1989). Fear of psychological treatment and its relation tomental health service avoidance. *Professional Psychology Research and Practice, 20,* 251–257.

Kushner, M. G., & Sher, K. L. (1991). The relation of treat ment fearfulness and psychological service utilization:an overview. *Professional Psychology Research and Practice, 22,* 196–203.

Martini, T. S. (1996). A ttributions and the stigma of illiteracy: understanding help seeking in low literate adults, *Canadian Journal of Behavioural Science, 28*(2), 121–129.

Matsuoka, J. K., Breaux, C., & Ryujin, D. H. (1997). National utilization of mental health services by Asian Americans/Pacific Islanders. *Journal of Community Psychology, 25*(2), 141–145.

Mei, J. R., & Sui, Y. J. (1998). The graduates' help – seeking tendency (In Chinese). *Chinese journal of clinical psychology, 6* (4), 210–215.

[梅锦荣，隋玉杰 . (1998). 大学生的求助倾向 . *中国临*

床心理学杂志, 6(4):210 ～ 215]

Nadler, A. (1991). Help-seeking behavior: Psycholog ical costs and instrumental benefits. In: M S Clark (Ed.), *Review of persona lity and social psychology, 12*, 290-311.

Phillips, M. R., Yang, G., Zhang, Y., Wang L., Ji, H., & Zhou M. (2002). Risk factors for suicide in China:a national case-control psychological autopsy study. *The Lancet, 360*, 1728-1736.

Pugh, J. S. (2002). Help seeking and personality among college students. *The Sciences & Engineering, 62*(7 - B), 3387.

Schohn, M. (1991). *Antecedents and consequences of attributions for causes of psychopathology.* D issertation abstract international, B 52 /02, 1082.

Segal, S. S., Hodges, J. Q., & Hardiman, E. R. (2002). Factors in decisions to seek help from self-help and co-located community mental health agencies . *American journal of orthopsychiatry, 72* (2), 241-249.

Senior, V., Weinman, J., & Marteau, T. M. (2002). The influence of perceived control over causes and responses to health threats: A vignette study. *British Journal of Health Psychology, 7*, 203-211.

Sue, S.(1999). Science, ethnicity, and bias: where have we gone wrong! *American psychologist, 54*, 1070-1077.

Taylor, S. E., Replau, L. A., & Sears, D. O. (2004). Translate in Chinese by Xie X F, Xie D M, Zhang Y L. *Social psychology.* Beijing:Beijing university press, 2004. 171-178.

[Taylor, S. E., Replau, L. A., & Sears, D. O. 著 (2004). 谢晓非, 谢冬梅, 张怡玲等译. *社会心理学*. 北京 : 北京大学出版社, 171-178.]

Tsang, W. H., Tam, P. K. C., Chan, F., et al. (2003). Stigmatizing attitudes towards individualswithmental illness in Hong Kong: Implications for their recovery. *Journal of Community Psychology, 31* (4), 383-397.

Uba, L. (1994). *Asian Americans: personality patterns, identity, and mental health.* New York: Guilford.

Vogel, D. L., & Wester, S. R. (2003). To seek help or not to seek help: the risks of self-disclosure. *Journal of counseling psychology, 50*(3), 351-361.

Yang, K. S. (1993). *The social orientation of Chinese: the viewpoint of social interaction.* In Yang Kuo -Shu, Yu, A. B. The psychology and behaviorof Chinese——the theory and method. Taiwan: Guiguan press, 87-142.

[杨国枢. (1993). *中国人的社会取向：社会互动的观点*. 见：杨国枢, 余安邦主编. 中国人的心理与行为——理念及方法篇. 台湾：桂冠图书公司, 1993. 87-142.]

Yeh, C. J. (2002). Taiwanese students' gender, age, interdependent and independent self - construal, and collective self-esteem as predictors of professional psycho logical help-seeking attitudes . *Cultural diversity and ethnicm inority psychology, 8*(1), 19-29.

Yu, X. M., & Jiang, G. R. (2006). The relationship of locus ofmental health, self-efficacy and help seeking of college students. *Clinical Journal School Health, 27*(5), 444-445.

Zhang, A. Q. (1999). Motivation theory:the motive research in 21st century. Wuhan: CentralChina Nor malUniversity Press, 133-134.

[张爱卿. (1999). 动机论：迈向二十一世纪的动机心理学研究. 武汉：华中师范大学出版社, 133-134.]

Relationship among attribution, self-efficacy, perceived social acceptance, and help-seeking behavior

XIA Mian, JIANG Guangrong

(Schoo l of Psychology, Centra l China Norma l University, Wuhan 430079, China)

Abstract Chinese people hesitate to seek professional help and hold negative attitudes toward help seeking. Earlier studies have examined various demographic, social–cultural and personal factors in the attempt of identifying the barriers. However, recent studies demonstrate that social–cultural variables can only explain a relatively small amount of variance in actual help–seeking behavior. Therefore, we should pay as much attention to personal factors in understanding help seeking behaviors. To understand the individual's decision–making process before help–seeking behavior, Jiang (2005) developed a Phases–Decision–Making Model (PDM). This model illustrates three phases involved in help–seeking decision: 1) the perception of having psychological problems; 2) Self–help evaluation; and 3) Other–help evaluation. To study the actual help–seeking behavior in this study, we employed this model and examined the barriers to help–seeking at the third phase. We studied those who were aware that they had psychological problems, recognized that they could not solve them effectively by themselves, and were prepared to seek help from others. In order to understand factors that may lead to help–seeking behavior, we examined attribution of psychological problems, self–efficacy of being a counseling client (the belief in one's ability of being a "good" client), and perceived social acceptance of help–seeking. Three hundred and five college students (154 female, 151 male) who were aware of having psychological problems and realized that they couldn't solve them effectively by themselves (as identified by a survey) participated in the study, with the first group being those actually sought professional help (HS) and the second never did (NHS). All the participants were enrolled from seven universities in Wuhan, China, with 39.7% sophomores, 26.9% freshmen, 21.6% juniors, 9.8% seniors, and 2% graduate students. We used questionnaires to measure the three variables. The questionnaires are The Questionnaire of Self–efficacy of being Counseling Client (Yu, Jiang, 2004), The Questionnaire of the Perceived Social Acceptance and the Questionnaire of attribution of the mental problem. The result showed that the self–efficacy of being counseling client was a significant predictor of help–seeking behavior. Higher self–efficacy led to more help seeking. Controllability predicted help seeking in that if students perceived their psychological problems as controllable they would be less likely to seek help. The locus of attribution failed to differentiate HS and NHS students. Perceived social acceptance of help–seeking predicted help–seeking behavior positively. Those who perceived higher social acceptance of help–seeking would perceive higher self–efficacy of being counseling client, higher self–efficacy of being counseling client would be positively associated with help–seeking. The self–efficacy of being counseling client and perceived social acceptance were important motivational factors that predicting help–seeking behavior in the third phase of PDM. This study supported the PDM on some extent.

Keywords attribution; self–efficacy; perceived social acceptance; psychological help–seeking

心理学报, 2013, 45(3), 320 - 315.

自伤青少年的冲动性 *

于丽霞 [1,2], 凌　霄 [2,3], 江光荣 [1,2]

([1] 青少年网络心理与行为教育部重点实验室)

([2] 湖北省人的发展与心理健康重点实验室 , 华中师范大学心理学院 , 武汉 430079)

([3] 武汉外国语学校 , 武汉 430022)

摘　要　以自我报告、行为学和脑电为指标 , 检验自伤青少年的冲动性。研究 1, 对 820 名普通中学生和 72 名工读生进行问卷调查 , 探讨自伤行为与情绪调节困难、冲动性的关系。结果表明 , 冲动性能够预测自伤行为 , 且预测效应量大于情绪调节困难。研究 2, 采用 Go/Nogo 范式的 ERPs 实验 , 检验自伤组与对照组冲动控制的行为学与脑电差异。结果表明 , 自伤组 Nogo 正确反应的 N2 波幅显著高于对照组 , N2 潜伏期在部分电极点处高于对照组。脑电地形图显示两者的脑电差异主要体现在前额叶区。结论 : 自伤青少年的冲动性高于同龄普通青少年。

关键词　非自杀性自伤 ; 自伤 ; 冲动性 ; ERPs; N2

分类号　R395;B844

1　问题提出

自我伤害 (下文简称自伤) 行为指在没有明确自杀意图的情况下 , 个体故意、重复地改变或伤害自己的身体组织 , 如用利器割伤 / 划伤、打火机烧伤、以头撞墙等 ; 这种行为不为社会所认可 , 且不具致死性或致死性较低 (Gratz, 2001)。自伤行为与很多心理问题或精神障碍相关 , 如边缘型人格障碍 (BPD)、摄食障碍、抑郁、焦虑、药物滥用、创伤后应激障碍、分离障碍 (Gratz & Gunderson, 2006;Haw, Hawton, Houston, & Townsend, 2001; Sho etal., 2009; Svirko & Hawton, 2007; Zlotnick, Mattia,& Zimmerman, 1999) 等。

虽然自伤的目的不在自杀 , 但自伤当事人的自杀风险远高于普通人群 (Hawton,Zahl, & Weatherall, 2003; Sinclair, Hawton, & Gray,2010), 因此危险性极大。自伤的相关研究正受到越来越多的研究者和临床工作者的重视 , "为什么要自伤"成为研究者最关切的问题。围绕这个问题 , 目前主要在两方面展开研究 , 一是自伤的功能 , 一是自伤的影响因素。

自伤具有多种功能 , 包括情绪管理、对抗分离感、对抗自杀、人际影响、自我惩罚等 (Klonsky, 2007;Messer & Fremouw, 2008; Suyemoto, 1998), 其中首要功能是情绪管理 (Klonsky, 2007; 郑莺 , 2006)。情绪管理指个体知觉、理解、接受自己的情绪体验以及

---

* 国家科技科技支撑计划项目 (2009BA177B02) 资助。

** 通讯作者 : 江光荣 , E-mail:grjiang@mail.ccnu.edu.cn

灵活地运用策略做出合适的行为，以上任何一种能力的缺失均称为情绪管理障碍 (Gratz, 2007)。情绪管理障碍是自伤者的核心问题，具体表现为情绪感受的脆弱性 (情绪易唤起、强度高、难平复) 和情绪管理能力的缺乏 (Chapman, Gratz, & Brown, 2006;Gratz, 2007;冯玉, 2008; 于丽霞，江光荣，吴才智, 2011)。研究发现，当事人在自伤前通常会经历强烈的、难以自控的负性情绪体验，比如愤怒、悲伤、受挫、内疚等，自伤能够帮助当事人在短时间内迅速释放情绪和恢复平静 (Klonsky, 2009)。因此自伤是一种适应不良的应对方式，目的是帮助自伤者调节强烈的负性体验。Chapman (2006) 将情绪管理障碍具体分为情绪表达不能、情绪调节困难和高情绪强度三个成分。冯玉 (2008) 以青少年为被试，检验了三个成分对自伤的预测作用，结果情绪表达不能和情绪调节困难进入了回归方程，共解释了自伤行为总变异的 18.5%，其中情绪调节困难的解释力最高，解释了总变异的 12.1%。冯玉的研究同时发现情绪调节困难在早期创伤经验对自伤的影响中起中介作用。另一项以大学生为被试的研究表明，情绪调节困难能将 64% 的自伤者与非自伤者区分开，准确率达 80% (Gratz & Chapman, 2007)。由此可见情绪调节困难是自伤的重要影响因素，本文将选取情绪调节困难作为自伤的预测因素之一。

自伤的影响因素既有来自后天的社会环境因素，如早期创伤性经验 (Gratz & Chapman, 2007;Klonsky & Moyer, 2008; Yates, Carlson, & Egeland,2008)，也有来自先天的人格 或气质特点，如冲动性。不少研究发现，自伤者从考虑到正式实施自伤的时间间隔通常不到 5 分钟 (Nock & Prinstein, 2005; 郑莺, 2006)，且常同时伴有其他冲动性行为，如暴食、酒精 / 物质滥用、病理性赌博等。并且，根据 DSM-IV 的诊断标准，自伤满足冲动 - 控制障碍的三个核心特点 (Dell' Osso, Altamura, Allen, Marazziti,& Hollander, 2006)：(1) 无法控制伤害自己或他人的冲动性行为；(2) 行为发生之前情绪唤起或紧张感不断加剧；(3) 行为之后有愉悦感或释放感。因此冲动性被认为是影响自伤的核心因素之一 (Herpertz,Sass, & Favazza, 1997)，甚至有研究者

认为自伤应属于一种冲动 - 控制障碍 (转引自 Glenn & Klonsky,2010)。Miller 等 (2003) 对有自杀未遂和自伤经历的 BPD 患者的研究发现，控制 BPD 的效应之后，冲动性仍然可以将自伤者与自杀者区分开，而控制冲动性的效应后，BPD 的区分效应则不显著。这些证据都表明冲动性是自伤的重要风险因素。

冲动性是一种人格特质，有较强的生物学基础，主要表现为个体对于来自内部或外部的刺激产生迅速、无计划的反应倾向，而不考虑对自己或他人造成的负面后果 (Moeller, Barratt, Dougherty,Schmitz, & Swann, 2001)。冲动性与冲动 - 控制障碍、边缘型人格障碍 (Links, Heslegrave, & Reekum,1999)、注意缺陷多动障碍 (Dougherty et al., 2004)、摄食障碍 (Fischer, Smith, & Anderson, 2003)、攻击 (Barratt, 1994; Berkowitz, 1974)、物质滥用 (Jentsch& Taylor, 1999)、自杀 (Baca-Garcia et al., 2005;Swann et al., 2005) 等多种心理病理行为有关联，是研究者给予较多关注的心理病理行为的影响因素。但冲动性的概念和结构多年来仍然没有获得统一的认识。学界都认同冲动性并非单一结构，而是包含有多种独立成分的复合结构 (Evenden, 1999)，但具体包含哪些成分研究者们则莫衷一是 (Whiteside& Lynam, 2001)。由于测量工具依赖于特定的理论框架 (Evenden, 1999)，理论上界定的不统一也导致了冲动性测量工具的多样化 (Gerbing, Ahadi, &Patton, 1987)。当前发展出的冲动性测量方法中，仅自评工具就达数十种，其中最为普遍使用的是第 11版 Barratt 冲动性量表 (BIS-11, Stanford et al.,2009)，另有多个行为学实验范式 (如行为抑制 / 冲动控制、冒险决策、延迟折扣等)。近年来神经生物学实验 (Carver, Johnson, & Joormann, 2008; Cools,Roberts, & Robbins, 2008) 和脑电实验 (Ramautar,Kok, & Ridderinkhof, 2004; Ruchsow et al., 2008a) 也成为检验冲动性的重要方法。在测量过程中，由于自我报告受个体主观因素的影响，研究者通常建议将自我报告与行为学、神经生物学和 / 或神经心理学等客观指标相结合进行综合评定 (周亮，何晓燕，肖水源, 2006)。

虽然冲动性是自伤的重要影响因素之一，但对自

伤者冲动性研究的结论并不一致。首先是不同研究结论之间存在冲突。如，Evan 等 (1996) 对住院自伤当事人冲动性的研究发现，重复自伤者的冲动性高于一次自伤者，后者又高于普通人群。但 Herpertz 等 (1997) 研究发现自伤组与对照组只在冲动性的某些方面 (如计划维度) 存在差异。也有研究发现冲动性与不同自伤者的自伤程度存在相关，但自伤组与对照组的冲动性水平差异不显著 (Simeon et al., 1992)。Hawton 等 (2002) 以 6020 名 15~16 岁的中学生为样本，发现仅女性被试自伤组的冲动性水平高于对照组，而男性被试的两组差异不显著。其次是不同测量方法的结果之间存在冲突。自伤组自我报告的冲动性水平明显高于正常组 (Glenn &Klonsky, 2010; Herpertz et al., 1997; Janis & Nock, 2009; Mc Closkey, Look, Chen, Pajoumand, &Berman, 2012)，但行为学实验结果则显示两组在实验任务中的表现均无显著差异 (Glenn & Klonsky, 2010; Janis & Nock, 2009; Mc Closkey et al., 2012)。Glenn 和 Klonsky (2010) 采用 UPPS 冲动性量表 (Whiteside & Lynam, 2001) 和停止信号任务 (stopsignaltask) 探讨冲动性与自伤的关系，结果表明自伤组的 UPPS 测量结果显著高于对照组，而两组在停止信号任务中表现一致。Janis 和 Nock (2009) 分别以青少年和成人两个样本为被试，发现两个样本自伤组自我报告的冲动性水平均高于对照组，但两组在不同实验范式 (行为去抑制实验、冒险决策实验、延迟折扣) 下的任务表现无差异。当前对自伤者冲动性研究结论的不一致表明，了解冲动性与自伤的关系还需要更多的研究证据，多个指标比单一指标的测量结果更可靠。

除了传统的自评和行为学指标，事件相关电位 (ERPs) 研究是当前被较多用于检验冲动性的方法。ERPs 实验通常通过反应抑制范式的 Go/Nogo 任务刺激诱发脑部额区 N2 负波来检验冲动性水平 (Falkenstein, Hoormann, & Hohnsbein, 1999) 。Go/Nogo 范式给被试呈现两种刺激，要求被试对一种刺激 (Go) 做反应而对另一刺激 (Nogo) 不做反应。每个刺激出现后 200 ~ 350 ms, 在脑部额叶区会诱发出一个 ERPs 的主要成分—— N2 负波。由于 Go 刺激的呈现频率

高于 Nogo 刺激，Go 反应成为优势反应，因此抑制对 Nogo 刺激的反应需要更多的神经加工，耗费更多的神经能量。Nogo-N2 因此被认为反映了神经加工的早期反应抑制功能 (Folstein &Van Petten, 2008; Jodo & Kayama, 1992; van Boxtel, van der Molen, Jennings, & Brunia, 2001), 而反应抑制功能损伤的一个重要表现就是冲动性 (Newman, 1987)。二者的对应关系是，冲动性较高者的冲动抑制能力较差，其 Nogo 刺激下的反应 (包括正确和错误反应) 诱发的 N2 波幅较之正常组更小、潜伏期更短 (Falkenstein et al., 1999); 在 Nogo 刺激正确反应情况下，冲动性高者由于成功抑制一个冲动反应需要耗费更多的神经能量，其 N2 幅值更高、潜伏期更长。

当前已有多个病理性人群冲动性的 ERPs 研究的 N2 证据，但不同人群的研究结论很不一致。Sunohara 等 (1999) 发现 ADHD 儿童的冲动性水平高于对照组，表现为 Nogo 反应的 N2 潜伏期比正常组更短。Kiehl 等 (2000) 对精神分裂症和精神病患者冲动性的研究发现，病患组在 Go/Nogo 刺激中的 N2 效应 (N2d, 即 Go 和 Nogo 刺激反应的 N2 波幅的差值) 显著小于正常组，表明病患组的冲动控制能力受损。Yang 等 (2009) 对海洛因成瘾者的研究发现，成瘾组与控制组在 Go/Nogo 的任务中的 N2 效应差异 (N2d) 显著，但两组的 Nogo 反应的 N2 幅值差异不显著。Ruchsow 等 (2008a) 对 BPD 冲动性的研究发现，BPD 组与正常组的 Nogo 反应 N2 幅值、潜伏期差异均不显著。Munro 等 (2007) 以犯罪人员为样本，发现犯罪组与正常组的任务反应时和准确率的行为学指标差异显著，但 Nogo 反应的 N2 指标无差异。Ruchsow 等 (2008b) 对正常人群高低冲动组的 ERPs 研究发现两组的 Nogo 反应的 N2 指标不存在差异。Luijten 等 (2011) 对吸烟成瘾者的反应抑制能力研究发现，吸烟组比控制组的反应抑制能力更差，表现为吸烟组的 Nogo 反应的 N2 波幅明显降低。对于自伤冲动性的研究当前多采用问卷或行为学实验的方法，还没有来自神经科学方面的证据。

本文将采用 ERPs 实验的 N2 指标结合自我报告和行为学结果共同检验自伤青少年的冲动性。青少年

是自伤的高发群体，国外流行学调查发现自伤在普通青少年中的发生率约为 14%~56% (Bjarehed& Lundh, 2008; Hilt, Cha, & Nolen- Hoeksema,2008)，国内青少年自伤的发生率在 40% 以上 (冯玉 ,2008; 郑莺 , 2006)，因此本文以青少年为研究对象。情绪管理障碍是自伤者的核心问题，但从情绪管理的过程来看，实施自伤行为的直接原因在于对激发的冲动或行为表达的调节，即冲动控制。因此本文将进行两个研究：研究 1 将比较情绪调节困难和冲动性对自伤行为的预测力；在研究 1 结论的基础上，研究 2 将采用 ERPs 的 N2 指标，并结合行为学指标 (反应时、错误率)，比较自伤青少年和普通青少年冲动性的差异。

2 研究 1

2.1 研究目的

研究 1 的目的是检验情绪调节困难和冲动性与自伤行为的关系。研究假设：(1) 情绪调节困难和自我报告的冲动性水平对自伤行为有显著预测作用 ;(2) 自我报告冲动性水平对自伤的预测作用大于情绪调节困难。

2.2 研究方法与程序

2.2.1 被试

采用整群抽样法，向武汉市某普通中学和某工读学校学生集体发放问卷 920 份，有效收回 892 份，回收率为 96.96%。其中，普通中学学生 820 名，男生 430 名，女生 390 名，平均年龄 13.64 岁 (SD=0.99); 工读学校学生 72 名，男生 53 名，女生 19 名，平均年龄 14.28 岁 (SD=0.95)。892 份问卷中有 9 份多题漏答，予以剔除，实际参与统计分析共 883 人。

2.2.2 测量工具

（1）青少年自我伤害行为问卷

该问卷由郑莺 (2006) 编制、冯玉 (2008) 修订，根据自伤史的频次和对身体的平均伤害程度的乘积来综合评估自伤水平 (下文的自伤水平即指频次与对身体平均伤害程度的乘积)。频次和平均伤害程度分别采用 0~3 的四级 (0 次、1 次、2~4 次、5 次以上) 和 0~4 的五级 (无、轻、中、重、极重) 计分。前期研究发现，该问卷内部一致性信度为 0.85，并具有理想的区分效度、效标效度和聚合效度 (冯玉 , 2008)。

（2）情绪调节困难量表 (Difficulties in Emotion Regulation Scale, DERS)

原问卷由 Gratz 和 Roemer(2004) 编制，本文采用冯玉 (2008) 翻译修订的中文版。DERS 由 5 个维度，共 31 个项目组成，5 个维度分别为：（1）难以意识到自己的情绪反应；（2）不接纳自己的情绪反应;（3）缺乏有效的情绪调节策略；（4）当体验到消极情绪时，难以控制自己的冲动反应；（5）当体验到消极情绪时，难以进行有预定目标的行为。被试根据五级评分直接报告项目描述是否符合自己，5 个维度共解释总变异的 50.76%。总量表和分量表的内部一致性信度系数均在 0.70 以上。本研究采用 DERS 总分作为预测指标。

（3）Barratt 冲动性量表 (Barratt Impulsiveness Scale, BIS-11)

BIS 由 Barratt 于 1959 年初次编制后便不断得到重新修订，经由 Patton 等 (1995) 修订而成的最新版 BIS-11 是当前使用最为广泛的冲动性自评工具 (Stanford et al., 2009)。BIS-11 共 30 个项目，从三个维度评估冲动性人格：注意冲动性 (attentionalimpulsiveness)、运动冲动性 (motor impulsiveness) 和无计划冲动性 (no-planning impulsiveness)。问卷采用 Likert5 级评分 (不是、极少、有时、经常、总是)，总分在 30~150 分之间变化，得分越高，冲动性越强。来自中国大学生样本和普通社区样本的研究表明，中文版 BIS 的结构与原问卷基本一致 (杨会芹等 , 2007)，总量表和各分量表的内部一致性信度系数在 0.56~0.76 之间，重测信度在 0.67~0.85 之间，条目 - 总分相关有统计学显著性意义，适合于中国群体冲动性水平的研究 (周亮，肖水源，何晓燕，厉洁，刘慧铭，2006)。本文采用由北京心理危机研究与干预中心翻译修订的 BIS-11 中文版问卷 (安静 , 2008)，以总分为预测指标。

由于对同一批被试同时采用了三个问卷共 89 个条目的测量，有可能存在因社会称许性、被试作答偏向和习惯作答等因素导致的共同方法变异 (Spector, 2006)。本文在问卷施测过程中采用了匿名作答以尽量减少共同方法变异的产生，同时采用单一方法潜变量法进行检验 (Podsakoff, MacKenzie,Lee, & Podsakoff, 2003; 熊红星，张璟，叶宝娟，郑雪，孙配贞，2012; 周浩，龙立荣，2004)。将共同方法作为单独一个潜变量纳入结构方程模型，分析结果 (见表1) 表明，当加入方法因子后，模型的卡方量变化显著 ($\triangle \chi^2 = 1508.87$, $\triangle df=97$, $\triangle \chi^2/\triangle df=15.56$)，说明各变量之间存在一定程度的共同方法变异。由于 $\triangle \chi^2$ 同时受到样本量的影响，在比较两个模型时还应参考其他拟合指标的变化情况 (温忠麟，侯杰泰，马什赫伯特，2004)。从表1 结果可见，纳入方法因子后模型的拟合指数改善不明显，两个模型均与数据有非常好的拟合，这说明各变量之间不存在严重的共同方法变异问题。

2.2.3 分析与统计

采用 SPSS 17.0 做描述性分析，用 Mplus 6.11 做相关分析、logistic 回归分析和结构方程模型。

2.3 结果

2.3.1 青少年自伤行为的发生率

分别统计普通中学生与工读学校学生的自伤率，两个群体自伤的基本情况见表2。由描述性分析结果可见，从自伤水平看，普通中学生中从未实施自伤行为的比例为 74.39%；单从自伤频次看，从未实施过自伤的比例为 64.02%。工读生中，从自伤水平看，得分为 0 分的比例为 11.11%；单从自伤频次看，从未实施自伤的比例仅为 6.94%。

2.3.2 青少年自伤行为与情绪调节困难、冲动性的相关分析

将自伤频次和自伤水平作为分类变量，将被试分别分为四组 (0 次、1 次、2~4 次、5 次以上; 0 分、1 分、2~10 分、10 分以上)，其中自伤水平为人为分组，方差分析表明，四组被试在情绪调节困难、冲动性总分和各维度得分的组别差异显著，表明分组有区分效应。做自伤与情绪调节困难和冲动性的总分与各维度的相关分析。根据表3 和表4 的结果可见，自伤频次和自伤水平与情绪调节困难、冲动性总分均呈中度正相关。

2.3.3 自伤行为对情绪调节困难、冲动性的 logistic 回归分析

表 1　模型比较 ($n=883$)

模型	χ^2	df	χ^2/df	RMSEA	90%C.I.	CFI	TLI
原模型	8284.69	4553	1.82	0.030	0.029–0.032	0.87	0.87
纳入方法因子模型	6775.82	4456	1.52	0.024	0.023–0.025	0.92	0.92

表 2　样本青少年自伤基本情况

自伤水平	普通中学生 ($n=820$)		工读生 ($n=72$)		自伤频次	普通中学生 ($n=820$)		工读生 ($n=72$)	
	n	%	n	%		n	%	n	%
0分	610	74.39	8	11.11	0次	525	64.02	5	6.94
1分	51	6.22	3	4.17	1次	63	7.68	4	5.56
2~10分	115	14.02	21	29.17	2~4次	92	11.22	8	11.11
10分以上	44	5.37	40	55.56	5次以上	140	17.07	55	76.39

表 3　自伤行为与 DERS 及各维度的相关分析 ($n=883$)

自伤行为	情绪意识困难	情绪接纳困难	缺乏调节策略	冲动控制困难	目标执行困难	DERS 总分
自伤频次	0.17**	0.19**	0.37**	0.34**	0.31**	0.39**
自伤水平	0.16**	0.20**	0.36**	0.36**	0.33**	0.38**

表 4　自伤行为与 BIS 及各维度的相关分析 ($n=883$)

自伤行为	注意冲动性	运动冲动性	无计划冲动性	BIS 总分
自伤频次	0.31**	0.43**	0.38**	0.45**
自伤水平	0.32**	0.43**	0.36**	0.44**

将自伤频次和自伤水平作为分类变量，以 WLSMV 法评估情绪调节困难、冲动性总分对自伤频次及自伤水平的预测效应。Logistic 回归结果 (见表 5) 显示，冲动性和情绪调节困难共同解释自伤频次的 22.42%，自伤水平的 21.65%。其中，冲动性的解释力较高，能解释自伤频次总变异的 14.60%，自伤水平总变异的 14.24%；情绪调节困难能解释自伤频次总变异的 7.82%，自伤水平总变异的 7.41%。

2.4 讨论

2.4.1 青少年自伤行为的发生率

本结果发现，以自伤频次和自伤水平为指标，分别有 35.98% 和 25.61% 的普通中学生至少实施过一次以上的自伤行为，均低于国内前期调查结果 (冯玉，2008; 郑莺，2006)。郑莺 (2006) 曾采用分层随机抽样方法，从武汉市七个中心城区按初高中生和年级比例共随机抽取了 13 所中学、共 1283 名普通中学生为样本，以自伤频次为指标的调查结果发现，中学生的自伤率高达 57.4%。冯玉 (2008) 曾以三所普通中学的 340 名中学生为样本、以自伤水平为指标，结果发现

有自伤史的比例为 45.6%。不同调查结果的差异可能与样本的代表性有关。本研究样本来自一所重点初中，取样单一，样本可能不具代表性。

本研究发现工读学校学生的自伤率高达 88.89%，这一结论基本与中西方以往调查结果一致。冯玉 (2008) 以 115 名少年犯为样本结果发现自伤的发生率为 83.5%。Nock 等 (2006) 发现自伤在病理性问题青少年中的发生率超过 80%(Nock &Prinstein, 2004)。由此可见，问题青少年是发生自伤行为的高危群体。

2.4.2 青少年自伤行为与情绪调节困难、冲动性的关系

相关分析和 logistic 回归分析结果表明，青少年的自伤频次、自伤水平与情绪调节困难、冲动性总分均有密切关联，这与前期研究结果一致 (Janis& Nock, 2009; 冯玉，2008)。其中，自我报告的冲动性对自伤行为总变异的解释量 (14.24%) 比情绪调节困难的解释量 (7.41%) 更大。在冯玉 (2008) 前期研究中，情绪调节困难能解释自伤行为 12.1% 的总变异，本研究发现情绪调节困难对自伤的解释力低于自我报告的冲

表 5　自伤行为对情绪调节困难、冲动性人格的 logistic 回归分析 (n=883)

自伤行为	预测变量	Estimate	SE	StdYX Estimate	P	R^2
自伤频次	冲动性	0.02	0.003	0.33	<0.001***	0.224
	情绪调节困难	0.01	0.003	0.20	<0.001***	
自伤水平	冲动性	0.02	0.003	0.33	<0.001***	0.216
	情绪调节困难	0.01	0.003	0.19	<0.001***	

动性水平。这一结论的差异可能源于情绪调节困难与冲动性存在高相关。尽管多重共线性检验表明情绪调节困难与冲动性不存在共线性问题 (容忍度 =0.66)，但两者相关达 0.58，这表明两者存在较高的共变关系，情绪调节困难对自伤的效应量一部分被冲动性的效应量所解释。这说明冲动性是自伤的重要危险因子。

3 研究 2

3.1 研究目的

研究 1 结果发现冲动性是影响自伤的重要因素之一，其对自伤的预测作用超过情绪调节困难。研究 2 采用基于 Go/Nogo 范式的 ERPs 实验，比较检验自

伤青少年冲动性的行为学及脑电特征。研究假设为自伤组冲动性水平高于对照组，表现为：(1) 在行为学指标上，自伤组在 Go/Nogo 实验任务中的按键反应时比对照组更短，或 / 和错误率更高；(2) 在脑电指标上，自伤被试成功抑制一个冲动行为应需要耗费更多神经能量，表现为 Nogo 刺激成功抑制 (不按键) 的 N2 (下文统称 Nogo-N2, 即 Nogo 刺激正确反应的 N2) 幅值更大、潜伏期延迟。

3.2 研究程序与方法

3.2.1 被试

被试均来自研究 1。比较检验研究 1 中不同自伤水平被试的情绪调节困难和冲动性水平，方差分析结果发现自伤的组别效应显著，得分在 10 分以上 (即

自伤频次 ≥ 5 次、伤害程度中度以上) 自伤者的情绪调节困难和冲动性的总分及分量表得分均显著高于其他三组 (0 分、1 分、2~10 分,p<0.001)。为增加区分效应,选取自伤得分在 10 分以上和 0 分者分别为研究 2 的自伤组和对照组。

作者向研究 1 中符合条件的学生发出书面及电话邀请,对所有受邀被试再次进行自伤问卷及访谈筛查。自伤组问卷筛查标准满足:(1) 过去曾自伤 5 次以上;(2) 自伤程度为中度及以上。对照组问卷筛查标准满足自伤频次为 0。访谈内容为问卷填写的真实性。最终确定自伤组 19 人,其中男生 10 人,女生 9 人。对照组 15 人,其中男生 9 人,女生 6 人。被试的自伤及自我报告冲动性水平见表 6。所有被试均为右利手,视力或矫正视力正常,智力正常。通过对陪同前来的家长或教师及被试本人的访谈,排除其他精神疾病及

表 6　被试的自伤程度与冲动性水平

变量	自伤组 (n=19)	对照组 (n=15)
年龄 (M ± SD)	14.11 ± 0.94	13.88 ± 0.62
自伤频次 (M ± SD)	17.95 ± 9.25	0
自伤频次 × 程度 (M ± SD)	29.58 ± 25.23	0
冲动性 (M ± SD)	33.65 ± 2.89	16.23 ± 1.39

严重躯体疾病、24 小时内服用镇静药物或精神活性药物。被试在正式实验之前签署知情同意书,实验之后给予适当报酬。

3.2.2　实验刺激材料

用 Stim 软件编制、呈现刺激序列。刺激为单个正立和倒立的正三角图形 (边长 6.5 cm),以黑色背景白色线条呈现于显示器中央 (亮度 60 cd/m2)。刺激序列包括 300 个刺激,倒立三角形 200 个 (66.67%),正立三角形 100 个 (33.33%)。刺激随机呈现,正立三角连续出现次数小于 3 次。视距 120cm,垂直视角 2.8°,水平视角 2.9°。刺激呈现时间 100 ms,刺激间隔 (ITI) 在 650~750 ms 之间随机 (见图 1)。

3.2.3　ERPs 记录

使用 Neuroscan 系统记录脑电,采样率 500 Hz,记录带通 0.05 Hz ~ 100 Hz,双侧乳突参考,头皮和电极之间阻抗小于 5kΩ。被试佩戴 10-20 系统 64 导电极帽,双眼外眦记录水平眼电 (HEOG),左眉上和眶下记录垂直眼电 (VEOG)。同时记录连续 EEG 与行为学数据 (见图 2)。

3.2.4　程序和任务

被试舒适坐位,双眼平视屏幕中心。被试被告知这是一个按键游戏,要求他 / 她在倒立三角形呈现时尽快尽准确地按键 (Go 刺激),对正立三角不反应 (Nogo 刺激),游戏时长 3 分钟,他 / 她在游戏中的表现决定最终获得的报酬额度。这一处理用以激发被试的行为动机,并引起对行为结果的预期。之后刺激快速呈现 (100 ms),限速反应 600 ms。其中 Go 刺激出现概率为 66.7%,按键反应成为优势反应;Nogo 刺激出现概率为 33.3%,不按键为抑制性反应。正式实验前,16 人的预实验 (大学生 10 人,初中生 6 人;其中自伤初中生 4 人,对照组 2 人) 表明 Nogo 刺激错误反应的反应时比 Go 刺激的正确反应时短,Nogo 刺激错误率高,证明实验任务能够成功诱发被试的冲动反应 (Falkenstein, Hohnsbein,Hoormann, & Blanke, 1991)。

正式实验流程为:首先是 50 个刺激的练习环节,确认被试明白作业任务后,开始正式实验。按键左右手在被试者中交叉平衡。整个实验持续 45 ~60 min。

3.2.5　ERPs 数据分析和统计

采用 Scan 4.3 离线分析数据。自伤组 4 人,对照组 2 人 (男女各半) 用以预实验参数调整,不计入正式实验数据分析。正式实验中,自伤组 3 人,对照组 1 人的数据因脑电信噪比低 (帽子参考电极出现故障) 被剔除。因此,最终进入统计分析的有效被试为 24 人,自伤组 12 人,对照组 12 人。

具体数据处理过程为:合并行为数据,相关法去除眼电,分析时程为刺激前 100 ms (用于基线校正) 至刺激后 600 ms。基线校正后波幅超过 ± 100 μV 的在叠加中被剔除。0.1 Hz ~ 16 Hz 无相移数字滤波后对反应正确的 EEG 进行分类叠加,得到 Go、Nogo 刺激产生的两类 ERPs 成分。组内变量包括:任务类型 (Go/Nogo) 和电极位置 (F3/Fz/F4/FC3/FCz/FC4/C3/Cz/C4) 等。组间变量为自伤 / 非自伤。

使用 SPSS 17.0 进行统计分析。(1) 行为学分析为组别 2 (自伤组 / 对照组) × 任务类型 2 (Go 刺激正确反应 /Nogo 刺激错误反应) 重复测量的方差分

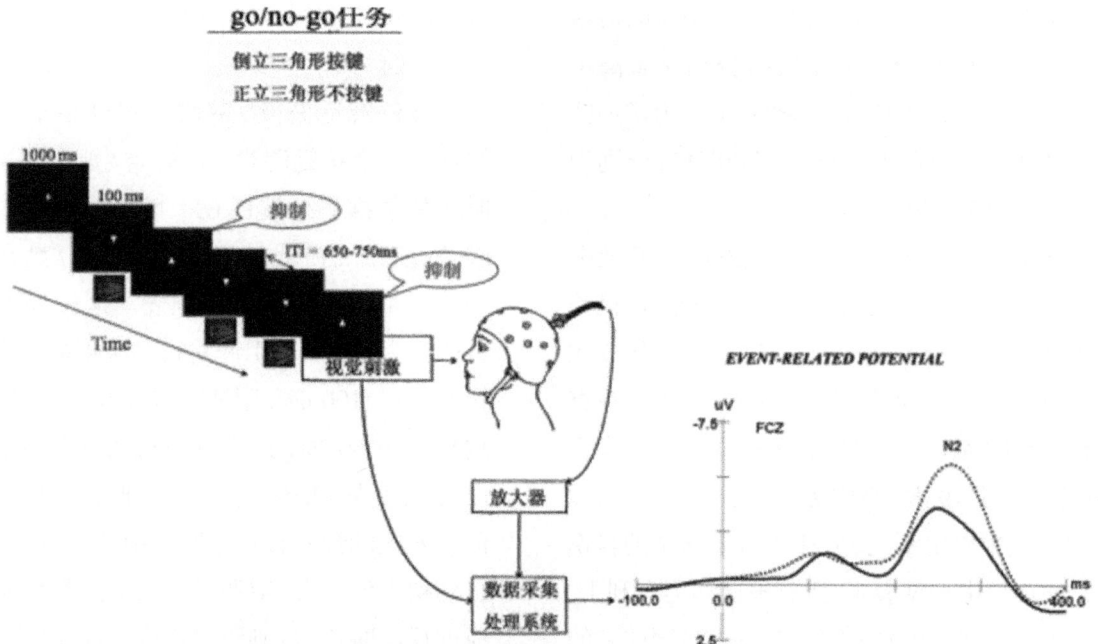

图 1　实验流程及数据采集分析示意图

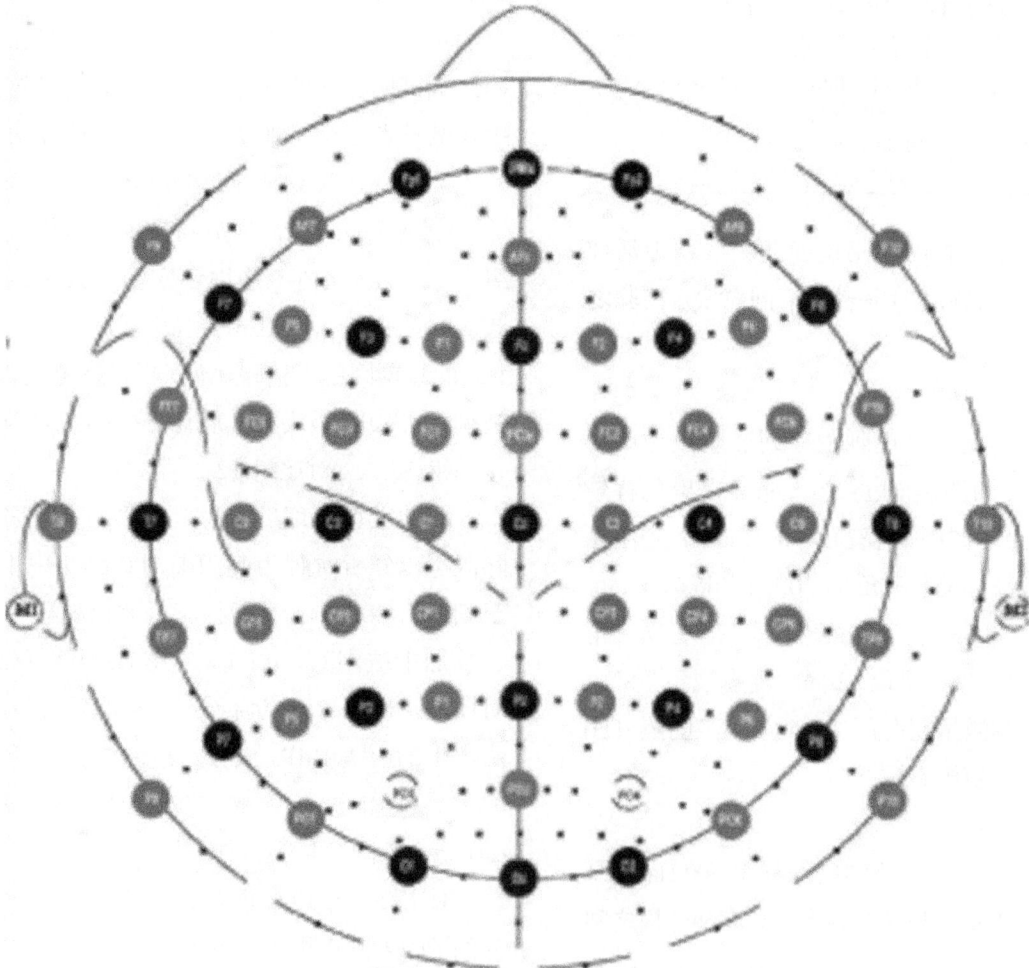

图 2　64 导电极位置分布图

析，因变量为按键反应的反应时与错误率。(2) 脑电分析为组别 2（自伤组 / 对照组）× 电极位置 9（9 个电极点）重复测量的方差分析，因变量为 Nogo-N2 幅值和潜伏期。并绘制地形图。当自由度大于 1 时，用 Greenhouse-Geisser 法校正 p 值。

3.3　结果

3.3.1　行为学指标——反应时与错误率的组间差异比较

分别比较两组被试在不同任务中的反应时与错误率（见表 7）。对两组按键反应 (Go 刺激正确反应与 Nogo 刺激错误反应) 的反应时进行重复测量的方差分析，结果发现任务与组别的交互作用不显著，$F_{(1, 22)} = 0.12$, $p > 0.05$。任务的主效应显著，$F_{(1, 22)} = 426.61$, $p < 0.001$，即 Nogo 刺激的错误反应时比 Go 刺激的正确反应时短。组别主效应不显著，$F_{(1, 22)} = 0.92$, $p > 0.05$，两组的按键反应时差异不显著。

对两组按键反应的错误率进行重复测量的方差分析，任务与组别的交互作用不显著，$F_{(1, 22)} = 0.04$, $p > 0.05$。组别主效应显著，$F_{(1, 22)} = 7.42$, $p < 0.05$，自伤组 Go 刺激与 Nogo 刺激按键的错误率都显著高于对照组。

3.3.2　Nogo-N2 效应的有效性检验

分别对 N2 幅值和潜伏期进行 2 (Go 刺激按键 / Nogo 刺激抑制按键) × 9 (电极位置) × 2 (自伤 / 非自伤) 重复测量的方差分析以检验实验任务是否成功诱发冲动控制反应。结果发现，任务对 N2 潜伏期的

表 7　两组按键反应的反应时、错误率 (M ± SD)

组别	Go 刺激正确反应		Nogo 刺激错误反应	
	反应时 (ms)	错误率 (%)	反应时 (ms)	错误率 (%)
自伤组 (n=12)	254.96 ± 18.01	37.50 ± 8.85	212.29 ± 20.36	26.22 ± 11.69
对照组 (n=12)	260.98 ± 15.97	27.83 ± 15.70	219.69 ± 16.60	17.89 ± 7.07

表 8　不同组别 Nogo-N2 幅值、潜伏期差异比较

脑区			潜伏期 (ms)	幅值 (μV)
额区	F3(左)	自伤组	259.00 ± 31.34	-7.74 ± 4.15
		对照组	254.50 ± 49.46	-3.89 ± 2.29
		$F_{(1, 22)}$	0.07	7.93*
	Fz(中)	自伤组	270.33 ± 19.91	-8.99 ± 4.30
		对照组	258.00 ± 37.67	-4.17 ± 2.00
		$F_{(1, 22)}$	1.01	12.34**
	F4(右)	自伤组	265.17 ± 34.09	-8.37 ± 4.52
		对照组	241.50 ± 39.38	-3.96 ± 2.24
		$F_{(1, 22)}$	2.48	9.18**
额中央区	FC3(左)	自伤组	264.00 ± 22.12	-6.61 ± 3.03
		对照组	247.17 ± 36.95	-2.72 ± 1.64
		$F_{(1, 22)}$	1.83	15.26**
	FCz(中)	自伤组	271.50 ± 9.84	-8.73 ± 2.86
		对照组	254.17 ± 26.81	-3.64 ± 1.37
		$F_{(1, 22)}$	6.75*	30.79***
	FC4(右)	自伤组	269.33 ± 15.76	-7.69 ± 3.75
		对照组	242.50 ± 33.07	-3.22 ± 1.59
		$F_{(1, 22)}$	6.44*	14.47**
中央区	C3(左)	自伤组	269.67 ± 40.80	-3.91 ± 3.62
		对照组	251.83 ± 49.99	-1.36 ± 1.08
		$F_{(1, 22)}$	0.92	5.47*
	Cz(中)	自伤组	269.50 ± 16.25	-1.85 ± 1.68
		对照组	251.67 ± 27.38	-0.43 ± 0.43
		$F_{(1, 22)}$	7.12*	8.04*
	C4(右)	自伤组	281.00 ± 33.56	-3.47 ± 2.34
		对照组	240.50 ± 43.48	-1.57 ± 0.98
		$F_{(1, 22)}$	6.53*	6.75*

注：*$p < 0.05$；*$p < 0.01$；**$p < 0.001$；

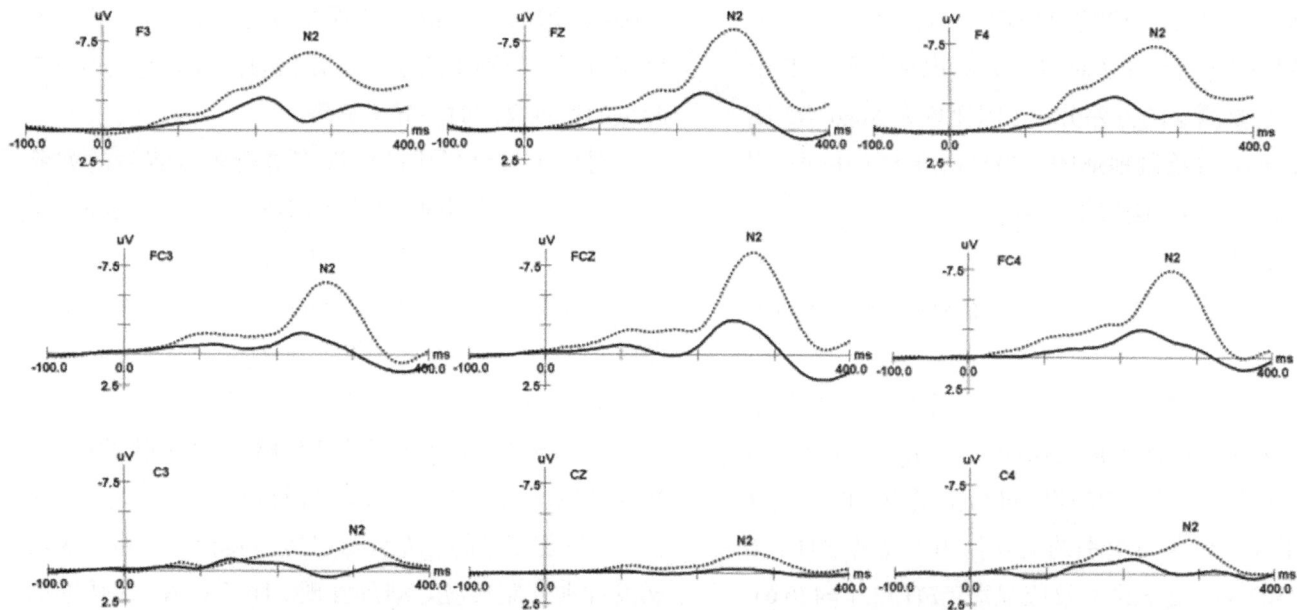

图 3　自伤组与对照组 Nogo-N2 幅值、潜伏期

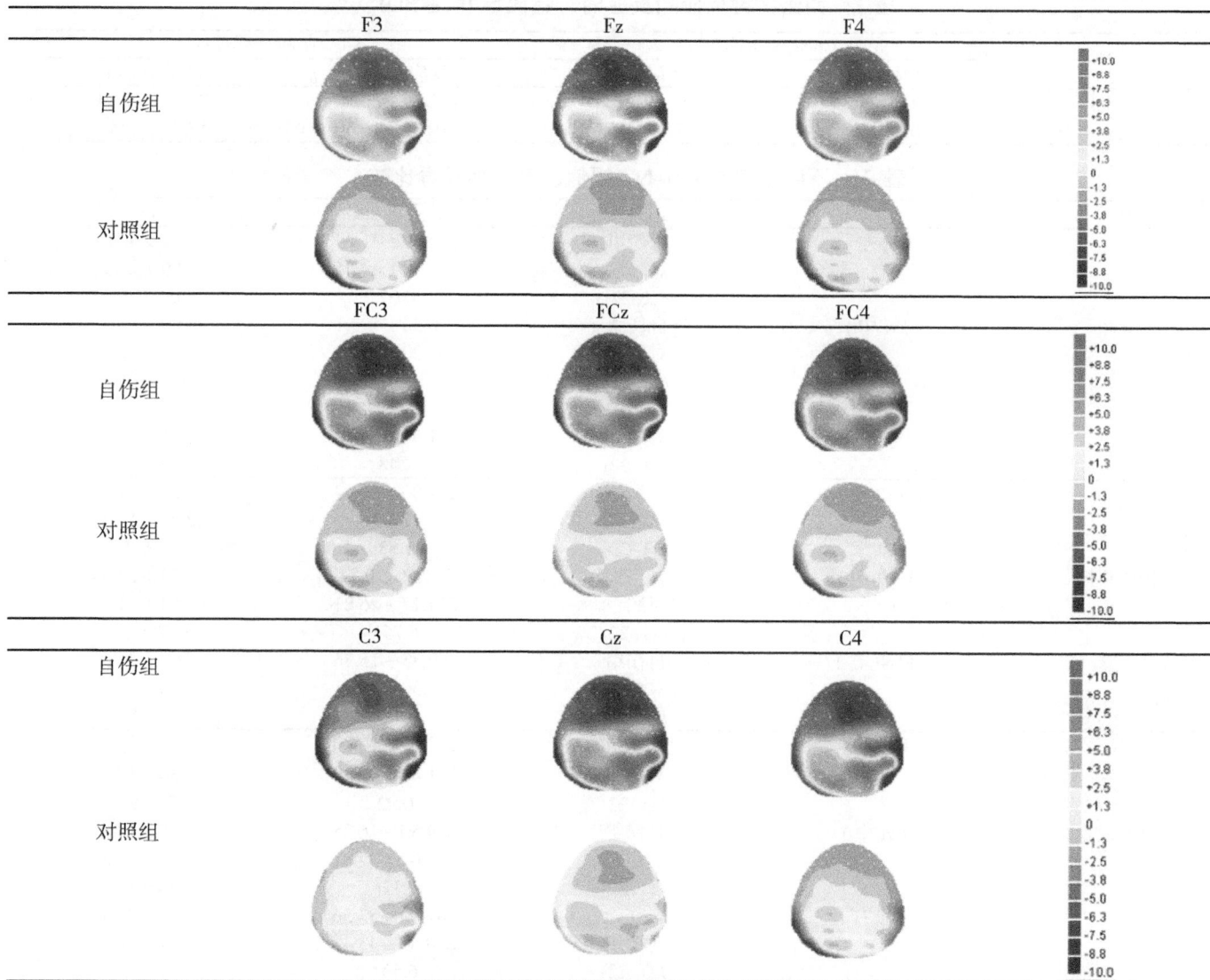

图 4　自伤组与对照组 Nogo-N2 的地形图比较 (270ms)

主效应边缘显著，F(1, 22) =4.16, p=0.053,Nogo- N2 的潜伏期比 Go-N2 延迟较为显著。任务对 N2 幅值的主效应显著，F(1, 22) =23.19,$p<0.001$,Nogo-N2 幅值显著高于 Go-N2。表明实验程序成功诱发了被试的冲动控制反应。

3.3.3 Nogo-N2 幅值、潜伏期的组间差异比较

（1）Nogo-N2 幅值

对两组在不同电极位置的 Nogo-N2 幅值进行 2(组别) × 9 (电极位置) 的重复测量方差分析。组别与电极位置的交互效应显著，F(8, 176) =4.51, $p<0.01$。组别主效应显著，F(1, 22) =14.66, $p=0.001$，自伤组的 Nogo-N2 幅值显著高于对照组，$p<0.001~0.05$。电极位置主效应显著，F(8, 176)=40.28, $p<0.001$。

（2）Nogo-N2 潜伏期

对两组在不同电极位置的 Nogo-N2 潜伏期进行 2 (组别) × 9 (电极位置) 的重复测量方差分析。电极位置与组别的交互效应不显著，F(8, 176) =1.07,$p>0.05$。电极位置主效应不显著，F(8, 176) =0.81,$p>0.05$。组别主效应显著，F(1, 22) =4.34, $p<0.05$, 自伤组 Nogo-N2 潜伏期明显迟于对照组。进一步检验发现，在 FCz、FC4、Cz、C4 电极位置，自伤组的 Nogo-N2 潜伏期迟于对照组，$p<0.05$，而其他电极位置没有差异，$p>0.05$ (见表 8 和图 3)。

3.3.3 Nogo-N2 脑电地形图

进一步绘制脑电二维地形图 (刺激开始 270 ms 左右) 比较两组的脑电差异 (见图 4)。红色轴为正电压，标尺正波幅值。蓝色轴为负电压，标尺负波幅值，蓝色调越深，负电压越大。实验任务诱发了 N2 负波，地形图的额区附近均以蓝色调为主。由图可见，成功抑制 Nogo 刺激引起的脑电活动主要集中在额区 (F3/Fz/F4) 附近。在 9 个电极点处，自伤组成功抑制 Nogo 刺激的电压都比对照组更大 (蓝色更深)。与上述方差分析结果一致。

3.4 讨论

3.4.1 行为学指标—— 反应时与错误率

研究 2 在研究 1 的基础上，采用 Go/Nogo 实验范式的 ERPs 方法，检验自伤组与对照组冲动控制的行为学和脑电差异。本研究中 Nogo 刺激错误反应的反应时比 Go 刺激正确反应的反应时短，与以往同类研究一致 (Ciesielski, Harris, & Cofer, 2004;Gehring & Willoughby, 2002; 王振宏, 2005)。Falkenstein 等 (1991) 研究认为，错误反应的反应时比正确反应的反应时短，表明出错是由快速决策或提前引发反应造成的，是一种冲动反应。本研究中 Nogo 错误反应的反应时比 Go 正确反应的反应时短了 40 多毫秒，且效应显著，这表明实验程序成功诱发了被试的冲动反应。

自伤组与非自伤组按键反应的反应时差异不显著，但自伤组错误率明显高于非自伤组。反应时与错误率是冲动控制的两个行为学指标 (Menon,Adleman, White, Glover, & Reiss, 2001; Ruchsow etal., 2008b; Saunders et al., 2008), 反应时越短或 / 且错误率越高，表明冲动控制能力越差、冲动性越高。Janis 等 (2009) 分别以 64 名自伤青少年和 30 名普通青少年为被试，采用冲动控制实验范式 (Conner' sContinuous Performance Test, CPT) 比较了两组的冲动性水平，结果显示两组被试按键错误率没有显著差异。Glenn 和 Klonsky (2010) 以 82 名大学低年级有过自伤史的学生和 86 名对照组为被试，采用反应控制实验范式 (Stop-Signal Task, SST) 发现两组被试在反应时上差异不显著。两个行为学实验结果均与"自伤组的冲动性更高"的假设不符。本研究中两组在按键错误率上差异显著，证实了研究假设，但反应时与错误率同样作为冲动控制实验范式的行为学指标，在本研究中并未得到一致的结论。根据速度 - 准确率之间的权衡关系可知，当反应速度增加，错误率也会相应增长，表面看来，反应时与错误率的同步变化应该能够同步反映行为特点。但研究 (Prinzmetal, McCool, & Park, 2005; Santee &Egeth, 1982) 表明两者结论的不一致并非偶然，反应时和准确 (或错误) 率并非一定同步表征行为，两者可能来自不同的加工机制。Prinzmetal 等 (2005) 指出，人的注意同时包括自主注意和非自主注意两部分，他们研究发现，在实验室任务中，自主注意同时影响个体的反应时和准确率，而非自主注意仅影响反应时。

由此 Prinzmental 等提出实验室任务中的反应时和准确率结论不一致是正常情况。同时，Santee 和 Egeth (1982) 的研究也表明，当以速视仪呈现刺激（即刺激短时呈现）时，准确率易受由目标刺激与非目标刺激早期引起的感知觉因素的干扰，反应时则更易受认知加工后期引起的行为反应因素的干扰。而在刺激长时呈现条件下，两者均易受知觉后期加工的影响。这说明反应时与准确率有可能是两个独立的成分，并不能总是同步反映相同的知觉加工，或者说两个指标有可能分别体现了冲动反应的不同方面。本研究中自伤组的按键错误率高于对照组，表明自伤青少年较同龄普通人群的冲动反应程度更高。Go/Nogo 任务是检验停止一个正在进行的反应（或反应类型）的能力（趋近行为管理能力），冲动反应程度高的个体在抑制趋近行为反应时有更高的错误率 (Newman, 1987)，是由于缺乏管理趋近行为的能力 (Lansbergen, 2007)。因此从本研究结论可以推论为，与普通青少年相比，自伤青少年缺乏趋近行为管理能力，在本研究实验任务中可能具体表现为两组在自主注意和早期感知觉特点上存在差异。

3.4.2 脑电指标—— Nogo-N2 幅值、潜伏期与地形图

Go/Nogo 实验范式中，冲动性高者成功抑制一个优势反应需要耗费更多的神经能量，表现为抑制 Nogo 刺激比反应 Go 刺激的 N2 幅值更高、潜伏期更长。本研究结果发现，Nogo-N2 的潜伏期比 Go-N2 延迟较为显著，且 Nogo-N2 幅值显著高于 Go-N2，表明实验程序成功诱发了被试的冲动控制反应。与前文行为学检验结果一致。

对 Nogo-N2 幅值的重复测量方差分析结果显示，自伤组的 Nogo-N2 幅值在所有电极点上均显著高于对照组 (p<0.001~0.05)，自伤组的 Nogo-N2 潜伏期在 FCz、FC4、Cz、C4 电极位置明显迟于对照组。前期有部分研究已显示 N2 可能是反映冲动控制水平的电生理指标。Falkenstein 等 (1999) 发现视觉刺激 ERPs 任务中，错误率高被试的 Nogo 刺激（同时包括正确和错误）反应 N2 幅值更小、潜伏期提前；Sunohara 等

(1999) 发现 ADHD 患者的冲动性高于对照组，表现为 Nogo 反应的 N2 潜伏期提前。本研究中 Nogo-N2 幅值和潜伏期的组间差异显著表明自伤青少年比普通青少年成功抑制优势反应需要更多的神经能量，也有理由进一步推断这一结果有可能反映了两组冲动控制能力的差异：自伤组的冲动控制能力更差。该结果同时也为 N2 可能是冲动控制的电生理指标提供了证据。

就两组 Nogo-N2 幅值、潜伏期差异的脑区位置看，两组差异在额中央区 (FC3/FCz/FC4) 最大；从左/中/右脑区分布来看，中间脑区 (Fz/FCz/Cz) 的差异最大。就脑电地形图来看，成功抑制 Nogo 刺激的脑电活动主要集中在额区附近，自伤组成功抑制 Nogo 刺激的额区电压比对照组更大（蓝色更深）。综合电极位置和脑区地形图的结果可以发现组间差异主要表征在前额皮层 (Prefrontal Cortex, PFC)。前期研究表明 PFC 与行为抑制能力有关。PFC 是认知控制的主要负责脑区，在情绪加工、识别刺激物的情绪意义及调节情绪反应和行为表现中具有重要作用 (Davidson, Pizzagalli, Nitschke, & Kalin, 2003;Ochsner & Gross, 2005)。反应抑制的 fMRI 研究也表明，背外侧和腹外侧 PFC 及眶额皮层 (Orbitofrontal Cortex, OFC) 与行为抑制能力密切相关 (Berlin, Rolls, & Iversen, 2005; Horn, Dolan,Elliott, Deakin, & Woodruff, 2003; Liddle, Kiehl, &Smith, 2001)。本研究中发现组间差异主要表征在前额皮层，进一步表明前额皮层与冲动控制能力的加工有关。Nogo-N2 的幅值增高、潜伏期延迟、分布面积增大也说明，与普通青少年相比，自伤青少年抑制一个优势反应需要的神经活动过程可能有所不同，两者可能存在脑机制的差异。进一步确认该结论还需要更多来自 ERPs、fMRI 研究的佐证。

4 总讨论

本文通过两个研究，结合自我报告、行为学和脑电指标，检验了自伤青少年的冲动性。结果发现自我报告的冲动性能够显著预测自伤水平；自伤组在实

验任务中的按键反应错误率高于对照组；与对照组相比，自伤组在实验任务中成功抑制优势反应诱发的 N2 负波幅值增高、N2 潜伏期延迟、前额区脑电活动增强。自我报告、行为学指标和脑电指标结果一致表明自伤青少年比同龄普通青少年的冲动控制能力更差。

当前为数不多的关于自伤者冲动性水平的研究多来自自我报告和实验室冲动控制实验的行为学的结果。自我报告的结论与研究假设比较一致，自伤组自我报告的冲动性水平显著高于普通组，但自我报告法通常被认为不能准确获得诸如动机、自我控制等需要高级认知参与的行为的信息，从而影响报告结果的可靠性 (Nisbett & Wilson, 1977;Rachlin, 1995)。而多个采用不同研究范式的行为学实验结果均与研究假设和自我报告结果不符，即两组在实验任务中的成绩并无显著差异。行为学指标与自我报告结果的不一致，一方面与冲动性的概念和结构的多样性有关 (Evenden, 1999)，不同测量工具和测量方式有可能测量的是冲动性的不同方面，研究者不应期望自我报告与行为学测量有一致的结果 (Gerbing et al., 1987)；另一方面也表明自伤冲动性的研究需要更多的证据。本研究发现自伤组按键反应的错误率明显高于对照组、相关脑电成分差异显著，为自伤青少年的冲动性提供了新的行为学和脑电证据。

在自伤研究领域，本研究首次采用 ERPs 方法检验自伤群体冲动性，并提供了 N2 指标的证据。在研究冲动控制的脑电实验中，除 N2 之外，P3 是反映冲动控制的另一个脑电成分。P3 是刺激出现之后 300ms 左右出现的脑电正波，研究者检验被试冲动控制能力时通常会同时参考 N2 和 P3 两个指标的结果（如，Ramautar, Kok, & Ridderinkhof, 2004;Ruchsow et al., 2008a)。N2 和 P3 是否反映了冲动控制加工一直存在很多争议 (Bruin, Wijers, & VanStaveren, 2001; Donkers & Van Boxtel, 2004;Falkenstein et al., 1999; Smith, Johnstone, & Barry,2007)，甚至有人提出 P3 而非 N2, 才是冲动控制的指标 (Smith et al., 2007)。但在一些同时考察 N2 和 P3 两个指标的研究中，两者的

结论也不尽一致。比如，Ruchsow 等 (2008a) 对 BPD 患者冲动控制的研究中，BPD 组与对照组在 P3 指标上差异显著，而 N2 指标显示两组不存在差异。而在 Falkenstein (1999) 对反应抑制高低两组脑电成分比较中，两组在 N2 指标上差异显著，但 P3 成分却未出现组间差异。尽管 N2 和 P3 成分的意义表征仍存在争论，但对于自伤群体冲动性水平的后期研究仍然可以同时结合 N2 和 P3 两个指标进行检验。此外，后期研究还可以采用功能性核磁共振 (fMRI) 来提供更多的研究证据。

本研究证实了自伤者冲动控制能力受损，可以帮助临床工作者有针对性的对自伤个案进行冲动控制能力训练（如 Ainslie, 1975), 以减少自伤行为的发生。但在本研究中冲动性对自伤总变异的解释量不足 15%, 与情绪调节困难一起的解释量也不足 30%, 这一结果可能与自伤行为本身的复杂性有关。首先是自伤的影响因素众多。除冲动性外，早期创伤经验（如受虐待、受忽视等）、情绪管理障碍和某些生物学因素等都是自伤的重要危险因子（见江光荣，于丽霞，郑莺，冯玉，凌霄，2011)，且各个影响因素之间交互作用复杂，这使得单一因素对自伤的解释力非常有限。一个类似的例子是，早期研究曾强调早期创伤经验尤其是性虐待对自伤的影响，但元分析结果 (Klonsky & Moyer, 2008) 却发现性虐待对自伤行为的解释量尚不足 5%。这也提示着在研究和临床干预中，研究者和临床工作者应综合多种因素来理解自伤行为的发生。其次是自伤群体的异质性。从临床案例的观察来看，自伤群体的异质性很高，看似同样的行为实则包含着不同的亚类型。不同自伤亚类背后可能有着非常不同的影响因素。但目前的实际研究均据当前自伤行为的定义（如 Gratz, 2001) 来选取被试，并未区分不同亚类群体，这使得对某一亚类群体有效的影响因素的解释力被其他亚类所稀释，从而对自伤行为的预测力降低。

5　结论

（1）自伤青少年自我报告的冲动性能显著预测

自伤水平。

（2）自伤青少年的冲动控制能力弱于普通青少年，表现为 Go/Nogo 实验任务中按键反应的错误率更高、成功抑制按键反应产生的 N2 幅值更高、部分电极点处的 N2 潜伏期延迟。

（2）冲动控制能力加工主要表征在前额叶区。

参考文献

Ainslie, G. (1975). Specious reward: A behavioral theory of impulsiveness and impulse control. *Psychological Bulletin, 82*(4), 463–496.

An J. (2008). *The study on the relationship of impulsiveness with suicidal ideation*. Unpublished master's thesis, China University of Geosciences.

[安静. (2008). *冲动性人格特征与自杀意念关系的研究*. 硕士学位论文, 中国地质大学.]

Baca-Garcia, E., Diaz-Sastre, C., Garcí a Resa, E., Blasco, H., Braquehais Conesa, D., Oquendo, M. A., & De Leon, J. (2005). Suicide attempts and impulsivity. *European Archives of Psychiatry and Clinical Neuroscience, 255*(2), 152–156.

Barratt, E. S. (1994). Impulsiveness and aggression. Violence and Mental Disorder: *Developments in Risk Assessment, 10*, 61–79.

Berkowitz, L. (1974). Some determinants of impulsive aggression: Role of mediated associations with reinforcements for aggression. *Psychological Review, 81*(2), 165–176.

Berlin, H. A., Rolls, E. T., & Iversen, S. D. (2005). Borderline personality disorder, impulsivity, and the orbitofrontal cortex. *American Journal of Psychiatry, 162*(12), 2360–2373.

Bjarehed, J., & Lundh, L. -G. (2008). Deliberate self-harm in 14-year-old adolescents: How frequent is it, and how is it associated with psychopathology, relationship variables, and styles of emotional regulation. *Cognitive Behaviour Therapy, 37*(1), 26–37.

Bruin, K., Wijers, A., & Van Staveren, A. (2001). Response priming in a go/nogo task: Do we have to explain the go/nogo N2 effect in terms of response activation instead of inhibition? *Clinical Neurophysiology, 112*(9), 1660–1671.

Carver, C. S., Johnson, S. L., & Joormann, J. (2008). Serotonergic function, two-mode models of self-regulation, and vulnerability to depression: What depression has in common with impulsive aggression. *Psychological Bulletin, 134*(6), 912–943.

Chapman, A. L., Gratz, K. L., & Brown, M. Z. (2006). Solving the puzzle of deliberate self-harm: The experiential avoidance model. *Behaviour Research and Therapy, 44*(3), 371–394.

Ciesielski, K. T., Harris, R. J., & Cofer, L. F. (2004). Posterior brain ERP patterns related to the go/no-go task in children. *Psychophysiology, 41*(6), 882–892.

Cools, R., Roberts, A. C., & Robbins, T. W. (2008). Serotoninergic regulation of emotional and behavioural control processes. *Trends in Cognitive Sciences, 12*(1), 31–40.

Davidson, R. J., Pizzagalli, D., Nitschke, J. B., & Kalin, N. H. (2003). Parsing the subcomponents of emotion and disorders of emotion: Perspectives from affective neuroscience. In R. J. Davidson, K. R. Scherer, & H. H. Goldsmith (Eds.), *Handbook of affective sciences* (pp. 8–24). New York: Oxford University Press.

Dell' Osso, B., Altamura, A. C., Allen, A., Marazziti, D., & Hollander, E. (2006). Epidemiologic and clinical updates on impulse control disorders: A critical review. *European Archives of Psychiatry and Clinical Neuroscience, 256*(8), 464–475.

Donkers, F. C. L., & Van Boxtel, G. J. M. (2004). The N2 in go/no-go tasks reflects conflict monitoring not response inhibition. *Brain and Cognition, 56*(2), 165–176.

Dougherty, D. M., Mathias, C. W., Marsh, D. M., Papageorgiou, T. D., Swann, A. C., & Moeller, F. G. (2004). Laboratory measured behavioral impulsivity relates to suicide attempt history. *Suicide and Life-Threatening Behavior, 34*(4), 374-385.

Evans, J., Platts, H., & Liebenau, A. (1996). Impulsiveness and deliberate self-harm: A comparison of 'first-timers' and 'repeaters'. *Acta Psychiatrica Scandinavica, 93*(5), 378-380.

Evenden, J. L. (1999). Varieties of impulsivity. *Psychopharmacology, 146*(4), 348-361.

Falkenstein, M., Hohnsbein, J., Hoormann, J., & Blanke, L. (1991). Effects of crossmodal divided attention on late ERP components. II. Error processing in choice reaction tasks. *Electroencephalography and Clinical Neurophysiology, 78*(6), 447-455.

Falkenstein, M., Hoormann, J., & Hohnsbein, J. (1999). ERP components in Go/Nogo tasks and their relation to inhibition. *Acta Psychologica, 101*(2-3), 267-291.

Feng Y. (2008). *The relation of adolescents self-harm behaviors, individual emotion characteristic and family environment factors*. Unpublished master's thesis, Central China Normal University.

[冯玉. (2008). *青少年自我伤害行为与个体情绪因素和家庭环境因素的关系*. 硕士学位论文, 华中师范大学.]

Fischer, S., Smith, G. T., & Anderson, K. G. (2003). Clarifying the role of impulsivity in bulimia nervosa. *International Journal of Eating Disorders, 33*(4), 406-411.

Folstein, J. R., & Van Petten, C. (2008). Influence of cognitive control and mismatch on the N2 component of the ERP: A review. *Psychophysiology, 45*(1), 152-170.

Gehring, W. J., & Willoughby, A. R. (2002). The medial frontal cortex and the rapid processing of monetary gains and losses. *Science, 295*(5563), 2279-2282.

Gerbing, D. W., Ahadi, S. A., & Patton, J. H. (1987). Toward a conceptualization of impulsivity: Components across the behavioral and self-report domains. *Multivariate Behavioral Research, 22*(3), 357-379.

Glenn, C. R., & Klonsky, E. D. (2010). A multimethod analysis of impulsivity in nonsuicidal self-injury. *Personality Disorders: Theory, Research, and Treatment, 1*(1), 67-75.

Gratz, K. L. (2001). Measurement of deliberate self-harm: Preliminary data on the Deliberate Self-Harm Inventory. *Journal of Psychopathology and Behavioral Assessment, 23*(4), 253-263.

Gratz, K. L. (2007). Targeting emotion dysregulation in the treatment of self-injury. *Journal of Clinical Psychology, 63*(11), 1091-1103.

Gratz, K. L., & Chapman, A. L. (2007). The role of emotional responding and childhood maltreatment in the development and maintenance of deliberate self-harm among male undergraduates. *Psychology of Men & Masculinity, 8*(1), 1-14.

Gratz, K. L., & Gunderson, J. G. (2006). Preliminary data on an acceptance-based emotion regulation group intervention for deliberate self-harm among women with borderline personality disorder. *Behavior Therapy, 37*(1), 25-35.

Gratz, K. L., & Roemer, L. (2004). Multidimensional assessment of emotion regulation and dysregulation: Development, factor structure, and initial validation of the difficulties in emotion regulation scale. *Journal of Psychopathology and Behavioral Assessment, 26*(1), 41-54.

Haw, C., Hawton, K., Houston, K., & Townsend, E. (2001). Psychiatric and personality disorders in deliberate self-harm patients. *British Journal of Psychiatry, 178*, 48-54.

Hawton, K., Rodham, K., Evans, E., & Weatherall, R. (2002). Deliberate self harm in adolescents: Self report survey in schools in England. *BMJ: British Medical*

Journal, 325(7374), 1207–1211.

Hawton, K., Zahl, D., & Weatherall, R. (2003). Suicide following deliberate self-harm: Long-term follow-up of patients who presented to a general hospital. *British Journal of Psychiatry, 182*(6), 537–542.

Herpertz, S., Sass, H., & Favazza, A. (1997). Impulsivity in self-mutilative behavior: Psychometric and biological findings. *Journal of Psychiatric Research, 31*(4), 451–465.

Hilt, L. M., Cha, C. B., & Nolen-Hoeksema, S. (2008). Nonsuicidal self-injury in young adolescent girls: Moderators of the distress-function relationship. *Journal of Consulting and Clinical Psychology, 76*(1), 63–71.

Hilt, L. M., Nock, M. K., Lloyd-Richardson, E. E., & Prinstein, M. J. (2008). Longitudinal study of nonsuicidal self-injury among young adolescents: Rates, correlates, and preliminary test of an interpersonal model. *The Journal of Early Adolescence, 28*(3), 455–469.

Horn, N., Dolan, M., Elliott, R., Deakin, J. F. M., & Woodruff, P. W. R. (2003). Response inhibition and impulsivity: An fMRI study. *Neuropsychologia, 41*(14), 1959–1966.

Janis, I. B., & Nock, M. K. (2009). Are self-injurers impulsive? Results from two behavioral laboratory studies. *Psychiatry Research, 169*(3), 261–267.

Jentsch, J. D., & Taylor, J. R. (1999). Impulsivity resulting from frontostriatal dysfunction in drug abuse: Implications for the control of behavior by reward-related stimuli. *Psychopharmacology, 146*(4), 373–390.

Jiang, G. R., Yu L. X., Zheng, Y., Feng Y., & Ling X. (2011). The current status, problems and recommendations on non-suicidal self-injury in China. *Advances in Psychological Science, 19*(6), 861–873.

[江光荣，于丽霞，郑莺，冯玉，凌霄. (2011). 自伤行为研究：现状、问题与建议. *心理科学进展，19*(6), 861–873.]

Jodo, E., & Kayama, Y. (1992). Relation of a negative ERP component to response inhibition in a Go/No-go task. *Electroencephalography and Clinical Neurophysiology, 82*(6), 477–482.

Kiehl, K. A., Smith, A. M., Hare, R. D., & Liddle, P. F. (2000). An event-related potential investigation of response inhibition in schizophrenia and psychopathy. *Biological Psychiatry, 48*(3), 210–221.

Klonsky, E. D. (2007). The functions of deliberate self-injury: A review of the evidence. *Clinical Psychology Review, 27*(2), 226–239.

Klonsky, E. D. (2009). The functions of self-injury in young adults who cut themselves: Clarifying the evidence for affect-regulation. *Psychiatry Research, 166*(2), 260–268.

Klonsky, E. D., & Moyer, A. (2008). Childhood sexual abuse and non-suicidal self-injury: Meta-analysis. *British Journal of Psychiatry, 192*(3), 166–170.

Lansbergen, M. (2007). *Impulsivity: A deficiency of inhibitory control?* Unpublished doctorial dissertation, Utrecht University.

Liddle, P. F., Kiehl, K. A., & Smith, A. M. (2001). Event-related fMRI study of response inhibition. *Human Brain Mapping, 12*(2), 100–109.

Links, P. S., Heslegrave, R., & Reekum, R. (1999). Impulsivity: Core aspect of borderline personality disorder. *Journal of Personality Disorders, 13*(1), 1–9.

Luijten, M., Littel, M., & Franken, I. H. A. (2011). Deficits in inhibitory control in smokers during a Go/Nogo task: An investigation using event-related brain potentials. *PLoS ONE, 6*(4), e18898.

Mc Closkey, M. S., Look, A. E., Chen, E. Y., Pajoumand, G., & Berman, M. E. (2012). Nonsuicidal self-injury: Relationship to behavioral and self-rating measures of impulsivity and self-aggression. *Suicide and Life-*

Threatening Behavior, 42(2), 197–209.

Menon, V., Adleman, N., White, C., Glover, G., & Reiss, A. (2001). Error–related brain activation during a Go/Nogo response inhibition task. *Human Brain Mapping, 12*(3), 131–143.

Messer, J., & Fremouw, W. (2008). A critical review of explanatory models for self–mutilating behaviors in adolescents. *Clinical Psychology Review, 28*(1), 162–178.

Miller, J., Flory, K., Lynam, D., & Leukefeld, C. (2003). A test of the four–factor model of impulsivity–related traits. *Personality and Individual Differences, 34*(8), 1403–1418.

Moeller, F. G., Barratt, E. S., Dougherty, D. M., Schmitz, J. M., & Swann, A. C. (2001). Psychiatric aspects of impulsivity. *American Journal of Psychiatry, 158*(11), 1783–1793.

Munro, G. E. S., Dywan, J., Harris, G. T., McKee, S., Unsal, A., & Segalowitz, S. J. (2007). Response inhibition in psychopathy: The frontal N2 and P3. *Neuroscience Letters, 418*(2), 149–153.

Newman, J. P. (1987). Reaction to punishment in extraverts and psychopaths: Implications for the impulsive behavior of disinhibited individuals. *Journal of Research in Personality, 21*(4), 464–480.

Nisbett, R. E., & Wilson, T. D. (1977). Telling more than we can know: Verbal reports on mental processes. *Psychological Review, 84*(3), 231–259.

Nock, M. K., Joiner, T. E., Jr., Gordon, K. H., Lloyd–Richardson, E., & Prinstein, M. J. (2006). Non–suicidal self–injury among adolescents: Diagnostic correlates and relation to suicide attempts. *Psychiatry Research, 144*(1), 65–72.

Nock, M. K., & Prinstein, M. J. (2004). A functional approach to the assessment of self–mutilative behavior. *Journal of Consulting and Clinical Psychology, 72*(5), 885–890.

Nock, M. K., & Prinstein, M. J. (2005). Contextual features and behavioral functions of self–mutilation among adolescents. *Journal of Abnormal Psychology, 114*(1), 140–146.

Ochsner, K. N., & Gross, J. J. (2005). The cognitive control of emotion. *Trends in Cognitive Sciences, 9*(5), 242–249.

Patton, J. H., Stanford, M. S., & Barratt, E. S. (1995). Factor structure of the Barratt impulsiveness scale. *Journal of Clinical Psychology, 51*(6), 768–774.

Podsakoff, P. M., MacKenzie, S. B., Lee, J. Y., & Podsakoff, N. P. (2003). Common method biases in behavioral research: A critical review of the literature and recommended remedies. *Journal of Applied Psychology, 88*(5), 879–903.

Prinzmetal, W., McCool, C., & Park, S. (2005). Attention: Reaction time and accuracy reveal different mechanisms. *Journal of Experimental Psychology: General, 134*(1), 73–92.

Rachlin, H. (1995). Self–control: Beyond commitment. *Behavioral and Brain Sciences, 18*(1), 109–121.

Ramautar, J., Kok, A., & Ridderinkhof, K. (2004). Effects of stop–signal probability in the stop–signal paradigm: The N2/P3 complex further validated. *Brain and Cognition, 56*(2), 234–252.

Ruchsow, M., Groen, G., Kiefer, M., Buchheim, A., Walter, H., Martius, P., ···Falkenstein, M. (2008a). Response inhibition in borderline personality disorder: Event–related potentials in a Go/Nogo task. *Journal of Neural Transmission, 115*(1), 127–133.

Ruchsow, M., Groen, G., Kiefer, M., Hermle, L., Spitzer, M., & Falkenstein, M. (2008b). Impulsiveness and ERP components in a Go/Nogo task. *Journal of Neural Transmission, 115*(6), 909–915.

Santee, J. L., & Egeth, H. E. (1982). Do reaction time and accuracy measure the same aspects of letter recognition? Journal of Experimental Psychology:

Human Perception and Performance, 8(4), 489–501.

Saunders, B., Farag, N., Vincent, A. S., Collins Jr, F. L., Sorocco, K. H., & Lovallo, W. R. (2008). Impulsive errors on a Go/Nogo reaction time task: Disinhibitory traits in relation to a family history of alcoholism. *Alcoholism: Clinical and Experimental Research, 32*(5), 888–894.

Sho, N., Oiji, A., Konno, C., Toyohara, K., Minami, T., Arai, T., & Seike, Y. (2009). Relationship of intentional self-harm using sharp objects with depressive and dissociative tendencies in pre-adolescence-adolescence. *Psychiatry and Clinical Neurosciences, 63*(3), 410–416.

Simeon, D., Stanley, B., Frances, A., Mann, J. J., Winchel, R., & Stanley, M. (1992). Self-mutilation in personality disorders: Psychological and biological correlates. *American Journal of Psychiatry, 149*(2), 221–226.

Sinclair, J. M. A., Hawton, K., & Gray, A. (2010). Six year follow-up of a clinical sample of self-harm patients. *Journal of Affective Disorders, 121*(3), 247–252.

Smith, J. L., Johnstone, S. J., & Barry, R. J. (2007). Response priming in the Go/Nogo task: The N2 reflects neither inhibition nor conflict. *Clinical Neurophysiology, 118*(2), 343–355.

Spector, P. E. (2006). Method variance in organizational research. *Organizational Research Methods, 9*(2), 221–232.

Stanford, M. S., Mathias, C. W., Dougherty, D. M., Lake, S. L., Anderson, N. E., & Patton, J. H. (2009). Fifty years of the Barratt Impulsiveness Scale: An update and review. *Personality and Individual Differences, 47*(5), 385–395.

Sunohara, G. A., Malone, M. A., Rovet, J., Humphries, T., Roberts, W., & Taylor, M. J. (1999). Effect of methylphenidate on attention in children with attention deficit hyperactivity disorder (ADHD): ERP evidence. *Neuropsychopharmacology, 21*(2), 218–228.

Suyemoto, K. L. (1998). The functions of self-mutilation. *Clinical Psychology Review, 18*(5), 531–554.

Svirko, E., & Hawton, K. (2007). Self-injurious behavior and eating disorders: The extent and nature of the association. *Suicide and Life-Threatening Behavior, 37*(4), 409–421.

Swann, A. C., Dougherty, D. M., Pazzaglia, P. J., Pham, M., Steinberg, J. L., & Moeller, F. G. (2005). Increased impulsivity associated with severity of suicide attempt history in patients with bipolar disorder. *American Journal of Psychiatry, 162*(9), 1680–1687.

van Boxtel, G. J. M., van der Molen, M. W., Jennings, J. R., & Brunia, C. H. M. (2001). A psychophysiological analysis of inhibitory motor control in the stop-signal paradigm. *Biological Psychology, 58*(3), 229–262.

Wang, Z. H. (2005). *Affective style and aggressive behavior in adolescents.* Unpublished doctoral dissertation. Capital Normal University.

[王振宏 . (2005). 青少年情感风格与攻击行为 . 博士论文 , 首都师范大学 .]

Wen, Z. L., Hau, K. T., & Marsh, H. W. (2004). Structural equation model testing: Cutoff criteria for goodness of fit indices and chi square test. *Acta Psychologica Sinica, 36,* 186–194.

[温忠麟 , 侯杰泰 , 马什赫伯特 . (2004). 结构方程模型检验 : 拟合指数与卡方准则 . *心理学报 , 36,* 186–194.]

Whiteside, S. P., & Lynam, D. R. (2001). The five factor model and impulsivity: Using a structural model of personality to understand impulsivity. *Personality and Individual Differences, 30*(4), 669–689.

Xiong, H. X., Zhang, J., Ye, B. J., Zheng, X., & Sun, P. Z. (2012). Common method variance effects and the models of statistical approaches for controlling it. *Advances in Psychological Science, 20*(5), 757–769.

[熊红星 , 张璟 , 叶宝娟 , 郑雪 , 孙配贞 . (2012). 共同方法变异的影响及其统计控制途径的模型分析 . *心*

理科学进展, *20*(5), 757–769.]

Yang, B., Yang, S. Y., Zhao, L., Yin, L. H., Liu, X., & An, S. S. (2009). Event–related potentials in a Go/Nogo task of abnormal response inhibition in heroin addicts. *Science in China Series C: Life Sciences, 52*(8), 780–788.

Yang, H. Q., Yao, S. Q., Zhu, X. Z., Auerbach, R. P., Abela John, R. Z., & Tong, X. (2007). The Chinese version of the Barratt Impulsiveness Scale 11th version (BIS–11) in college students: Its reliability and validity. *Chinese Mental Health Journal, 21*(4), 223–225.

[杨会芹, 姚树桥, 朱熊兆, Auerbach, R. P., Abela John, R. Z., & Tong, X. (2007). Barratt 冲动量表中文版在 209 名大三学生中的试用. *中国心理卫生杂志, 21*(4), 223–225.]

Yates, T. M., Carlson, E. A., & Egeland, B. (2008). A prospective study of child maltreatment and self–injurious behavior in a community sample. *Development and Psychopathology, 20*(2), 651–671.

Yu, L. X., Jiang, G. R., & Wu, C. Z. (2011). Psychological assessment and treatment of self–injury. *Chinese Mental Health Journal, 25*(12), 937–941.

[于丽霞, 江光荣, 吴才智. (2011). 自伤行为的心理学评估与治疗. *中国心理卫生杂志, 25*(12), 937–941.]

Zheng, Y. (2006). *Epidemiologic investigation of self–mutilation behavior among adolescents in Wuhan and its functional model.* Unpublished master's thesis, Central China Normal University.

[郑莺. (2006). *武汉市中学生自我伤害行为流行学调查及其功能模型.* 硕士学位论文, 华中师范大学.]

Zhou, H., & Long, L. R. (2004). Statistical remedies for common method biases. *Advances in Psychological Science, 12*(6), 942–950.

[周浩, 龙立荣. (2004). 共同方法偏差的统计检验与控制方法木. *心理科学进展, 12*(6), 942–950.]

Zhou, L., He, X. Y., & Xiao, S. Y. (2006). Measurement methods of impulsiveness. *Chinese Journal of Clinical Psychology, 14*(5), 455–457.

[周亮, 何晓燕, 肖水源. (2006). 冲动性测量的方法学问题. *中国临床心理学杂志, 14*(5), 455–457.]

Zhou, L., Xiao, S. Y., He, X. Y., Li, J., & Liu, H. M. (2006). Reliability and validity of Chinese version of Barratt Impulsiveness Scale–11. *Chinese Journal of Clinical Psychology, 14*(4), 342–344.

[周亮, 肖水源, 何晓燕, 厉洁, 刘慧铭. (2006). BIS–11 中文版的信度与效度检验. *中国临床心理学杂志, 14*(4), 342–344.]

Zlotnick, C., Mattia, J. I., & Zimmerman, M. (1999). Clinical correlates of self–mutilation in a sample of general psychiatric patients. *Journal of Nervous and Mental Disease, 187*(5), 296–301.

Impulsivity in non-suicidal self-injurious adolescents in China

YU Lixia[1,2]; LING Xiao[2,3]; JIANG Guangrong[1,2]

([1] Key Laboratory of Adolescent Cyberpsychology and Behavior (CCNU), Ministry of Education, Wuhan 430079, China)

([2] Key Laboratory of Human Development and Mental Health of Hubei Province, School of Psychology,

Central China Normal University, Wuhan 430079, China)

([3] Wuhan Foreign Languages School, Wuhan 430022, China)

Abstract Impulsivity has been proposed as an important risk factor in Non-Suicidal Self-Injury (NSSI). Yet, research outcomes on the relationship of impulsivity to NSSI have been mixed. The present study clarifies this relationship using event-related potentials (ERPs), along with self-reports and behavioral measures. Study 1 aimed to detect the prediction of emotion dysregulation and impulsivity to NSSI. 820 local common high school students and 72 counterparts with problematic behaviors were surveyed, and then the relation among NSSI, difficulties of emotion regulation (DER) and impulsivity were investigated by self-report measurements. Regression analysis results indicated that both DER and impulsivity could well predict NSSI, and contribution of impulsivity was much bigger than that of DER. In Study 2, a Go/Nogo paradigm was adopted to test the impulsivity of the injurers using behavioural measures and Nogo-N2 of ERPs. Participants were 12 confirmed self-injurious adolescents and 12 typical school middle students chosen from Study 1. The group differences (injurers vs controls) in behavior (response time and false alarm) and ERPs index (N2 amplitude and latency in successful Nogo trials) were analyzed in detail. Results disclosed that the NSSI group's probability of false alarm was higher than the control group's probability of false alarm in both Go and Nogo trials. In ERPs experiment, the NSSI group's N2 amplitudes were significantly higher than the controls in correct Nogo trials, and NSSI group's N2 latencies were clearly more delayed than the controls' in correct Nogo trial. Results from Nogo-N2 amplitudes and latencies combining with the topographic maps showed that impulse processes occurred in prefrontal cortex mainly. According to the results from self-reports, behavioural measures and Nogo-N2 of ERPs, it can be concluded that self-injurious adolescents possessed stronger impulsivity; and they needed much more neural energy to fulfill an impulse inhibition; moreover, they were insensitive to Nogo stimuli. The present study is the first to examin Nogo-N2 in NSSI, and provides further evidence for impaired response inhibition in NSSI.

Keywords non-suicidal self-injury; NSSI; impulsivity; ERPs; N2

心理学报，2011, 43(4), 420－431.

当事人眼里的工作同盟：质的分析

朱　旭，江光荣

（华中师范大学心理学院，湖北省人的发展与心理健康重点实验室，武汉 430079）

摘　要　为了解中国当事人如何看待咨询中的工作同盟，对工作同盟做初步的本土概念化，采用协商一致的质的研究方法对来自 1 所大学心理咨询中心的 20 名当事人的访谈结果进行了分析。结果发现，与工作同盟相关的域有 6 个，分别是情感联结、任务、投入、合作模式、发展变化、影响因素。前三个域可以看作是工作同盟的构成要素，而后三个域则是对其外部特征的描述。对每个域的含义及其与现有理论和研究的关系进行了讨论。

关键词　咨询关系；工作同盟；质的分析

分类号　R395

1　前言

工作同盟成为近三十年来治疗研究中最受关注 的 变 量 (Orlinsky, Ronnestad, & Willutzki, 2004)。原因可能有三，一是工作同盟的概念虽然源于精神分析，但现在已被看作是各种疗法所共有的治疗要 素 (Bordin, 1979; Gelso & Carter, 1985; Horvath & Bedi, 2002)。二是工作同盟与治疗效果之间的联系被大量研究所证实，效果量虽然不大（r = 0.21 至 0.26），但却非常稳定 (Horvath & Bedi, 2002; Horvath & Symonds, 1991; Martin, Garske, & Davis, 2000)。最后，相对于以往咨询理论对咨询师或当事人单方面因素的关注 (e.g., Rogers, 1957; Strong, 1968)，工作同盟强调咨询师与当事人相互合作的概念为理解咨询关系提供了一个更为全面的视角。

早期，Freud 认为积极移情中有一部分是有意识的、可以接受的（unobjectionable positive transference），可以利用这部分移情帮助分析治疗取得成功 (1912/1958, p. 105)。后来，Freud 提到分析师和当事人相对正常的自我结盟而对抗其自体中不受控制部分的必要性 (1937/1964, p. 235)。Sterba (1934) 提出了"自我同盟（ego alliance）"的概念，认为分析师与当事人自我中的理性部分的同盟可以使当事人能够自我观察，从而从分析师的解释中获益。Zetzel (1956) 在对移情进行讨论时使用"治疗同盟（therapeutic alliance）"的概念来指当事人自我中健康的部分与分析师联合而完成治疗任务的能力。Greenson (1965, 1967) 首先使用了"工作同盟（working alliance）"一词，并使其成为精神分析中广泛使用的概念。Greenson 将工作同盟定义为当事人与分析师之间相对非神经症的、理性的、中立的关系，这种关系使得当事人有可能与分析师合作，并在分析情境中有目的地工作。工作同盟既包括当事人对治疗师的情绪感受，也包括当事人的动机与工作能力。

Bordin (1979) 整合不同的理论观点，认为工作同

通讯作者：江光荣，E-mail: grjiang@mail.ccnu.edu.cn

盟是咨询师与当事人之间一种相互协商、相互合作的治疗关系，包括三个成分，即咨询师与当事人对咨询目标的一致看法（goal agreement），对如何实现目标所涉及的一系列任务达成共识（task agreement）及相互之间的情感联结（bond）。Bordin 的工作同盟概念强调咨询师与当事人之间的相互性与合作，而非单方面的作用。而且，Bordin 明确指出工作同盟是存在于所有疗法之中的共同要素，工作同盟的强度决定了治疗效果。Bordin 的工作同盟理论一经提出就受到了广泛关注，激起了大量有关工作同盟的研究。

虽然工作同盟的研究已经持续了几十年，理论家们对工作同盟的概念仍存在着诸多分歧。工作同盟与其他关系成分的区分，工作同盟与治疗技术的关系等都是争论的焦点。虽然理论家们对工作同盟的定义各有侧重，不过多数人认为咨询师与当事人之间有目的、有意识的合作是工作同盟的核心 (Hatcher & Barends, 1996; Horvath & Bedi, 2002)。至于这种合作具体包括哪些成分，或者说是工作同盟的结构如何，则不同的学者有不同的看法。其中，影响最大的是前面介绍的 Bordin 关于工作同盟三成分的论述。

从工作同盟概念的发展过程可以看出，目前关于工作同盟的知识主要来自理论家，代表的是专业人士或咨询师对合作过程的看法，而对于合作的另一方——当事人的观点知之甚少。在研究中，实际上是假定当事人与咨询师的看法相同，用同样的维度去评价咨询中的合作过程。例如，根据 Bordin 的理论假设编制的《工作同盟问卷》（Working Alliance Inventory，WAI）(Horvath & Greenberg, 1989) 的当事人版本与咨询师版本具有相同的结构，只是项目表述上有所差异。但是，这一假定受到了挑战。元分析的结果表明咨询师与当事人对工作同盟评价的相关并不高（r = 0.36）(Tryon, Blackwell, & Hammel, 2007)，说明当事人与咨询师对工作同盟的看法很可能存在差异。Bedi (2006) 认为研究者是在以专家的概念框架来理解当事人，而许多在当事人看来重要的同盟要素可能没有被触及。

少数几个研究采用质的方法探索了当事人对工作同盟的看法。Bachelor (1995) 对 34 位当事人在咨询的三个阶段提供的 66 个对工作同盟的描述进行了现象学分析。结果发现有三类不同的同盟，一类是抚育型（nurturant）（46%），强调的是咨询师的助长态度；一类是领悟型（insight-oriented）（39%），看重增进自我理解；一类是合作型（collaborative）（15%），体现为当事人的积极参与。另外，Bedi 等 (Bedi, 2006; Bedi, Davis, & Williams, 2005) 分别使用关键事件技术（critical incident technique, CIT）和多变量概念绘图（multivariate concept-mapping, MVCM）的方法研究了当事人对咨询早期有助于建立同盟的外显言行活动的知觉。结果发现，咨询环境、身体语言、非言语姿势、情感支持、坦诚、确认、指导与挑战、教育、推荐相关材料、当事人的个人责任和会谈管理等均被当事人认为有助于同盟的建立。上述研究者认为所得结果与已有工作同盟的文献与理论并不吻合。例如，当事人提到的一些变量，如咨询师的友好、幽默、建议、当事人的自我理解等，现有文献并未充分说明；而被研究者看重的概念，如合作、相互性，在当事人那里则显得不是那么重要。

导致质的研究结果与理论不符的部分原因可能来自研究的目的与方法。Bedi 等关注的是有助于建立同盟的具体的、可观察的言行活动，所得结果自然会包括咨询环境、非言语反应等这些通常不被认为是同盟构成要素的内容，而合作与相互性这些较为抽象的元素则可能被分解为具体的事件。另外，上述研究者为了使当事人明白研究的内容，均使用"治疗关系"或"工作关系"来指代工作同盟。将工作同盟与咨询关系等同看待与通常对工作同盟的理解并不一致，这也可能导致了研究结果与理论的差异。

目前研究关注的都是当事人眼里良好的同盟关系，而对不良的同盟关系没有涉及。Bedi (2006) 认为好的同盟与不好的同盟并不一定是一个维度，即两者的区别可能不仅只是在一些因素上存在量的差别，而还可能有结构上的不同。显然，此假设对工作同盟的测量有着重要的意义，但要对此假设进行验证必须同时对积极和消极的工作同盟进行探索性的研究。

此外，当我们对源于西方的概念进行研究时，还应注意概念的对等性（construct equivalence），即这些概念在不同的文化中是否具有相同的涵义，如概念的结构与维度是否一致（梁觉，周帆，2010）。虽然工作同盟在国外有较长的研究历史，但在国内却少有研究者涉及，采用质的方法对其进行本土的概念化是后续研究的基础。

鉴于以上原因，本研究将紧扣工作同盟的定义，即咨询中当事人与咨询师之间的合作关系，对中国当事人眼里的工作同盟进行质的研究，如合作关系的特点、发展变化和影响因素等。为此，我们选择协商一致的质的研究方法（Consensual Qualitative Research, CQR）(Hill et al., 2005; Hill, Thompson, & Williams, 1997) 来开展研究。在咨询的过程研究领域，CQR 是一种被广泛使用的质的研究方法 (Orlinsky, et al., 2004, p. 309)，其最大的特点是数据的分析是由一个小组来完成，所有的决定均需小组成员协商达成一致。这样既保证了对复杂现象进行研究时视角的多样性，又减少了单一研究者容易产生偏差的影响。

2 方法

2.1 参与者

2.1.1 当事人　来自一所大学心理咨询中心的 20 位当事人参与了研究。3 人为在校研究生，其余为本科生。年龄在 19 到 29 岁之间（$M=22.05$，$SD=2.44$）。3 人为男性。7 人已经结束咨询，结束的时间均在半年之内，其余当事人的咨询仍在继续。在参与研究时，咨询的会谈次数为 8 至 27 次（$M=11.10$，$Mdn=9$，$SD=4.53$）。结束咨询和未结束咨询的当事人在年龄和会谈次数上没有显著差异。其中 4 人的主诉问题为抑郁，其他均为发展性问题，主要包括人际关系、自我认识等。

2.1.2 研究小组　所有的访谈工作由 1 位心理咨询方向的博士生（本文的第一作者，男，30 岁）完成。此外，研究小组还包括 5 位心理咨询方向的硕士生（1 男 4 女，年龄均为 20 多岁）。部分成员参加了 Clara Hill 举办的介绍 CQR 方法的工作坊（2009 年 3 月，武汉）。所有人被要求阅读有关介绍 CQR 方法的两篇文献及数篇使用 CQR 方法的研究文献。然后大家一起对 CQR 的研究方法和程序进行讨论（6 小时），特别强调了鼓励表达和建立平等的交流氛围的重要性。

在研究开始之前，结合访谈提纲，所有组员一起讨论了各自对咨询中的工作同盟的体验与看法，包括个人的偏见（bias）与期望（expectation）。大家均认为，与西方相比，将咨询关系看作是合作显得较为陌生，中国的当事人可能更倾向于将这种合作看作是对咨询师的配合。大家也一致同意情感因素在良好合作关系中的重要性。大家也承认，对工作同盟或合作关系的理解会受到相关理论的影响，如 Bordin 的工作同盟理论。在此提供这些信息，作为读者理解研究结果的背景。组员表达各自的看法后，被提醒尽可能搁置这些期望，而忠实数据本身。在此后的数据分析过程中，组员们继续交流各自对工作同盟的看法。

2.2 访谈提纲

访谈提纲的第一部分询问当事人的一些基本信息，如年龄、咨询是否结束、疗程及主诉问题等。第二部分询问当事人对咨询过程中合作或合作关系的看法，比如好的合作关系是个什么样子，哪些因素比较重要等，要求尽可能详细地说明和举例。如果当事人认为与咨询师的合作关系并不好，则让其对自己的经历进行描述。第三部分让当事人描述合作关系的变化过程。最后，让当事人对其与咨询师之间的合作关系做一个 5 级的评价（1 = 非常不好，3 = 一般，5 = 非常好）。访谈提纲为半结构式，对每个当事人访谈的内容基本一致，但具体问法则根据当事人的回答进行调整。在访谈提纲的开发过程中，对 3 位有过当事人经历的硕士研究生及 3 位当事人进行了预访谈，根据反馈，对访谈提纲进行了反复修改。

2.3 数据收集

2.3.1 招募当事人　通过两种方式招募当事人。第一种方式是在一所大学咨询中心的接待室放置广告，对研究进行介绍，有兴趣参与的当事人留下联系方式，随后研究者与其联系。结果发现有兴趣参与的当事人大多刚开始咨询，预访谈发现他们能够提供的信息有限。为了找到对咨询有更多了解的当事人，考虑采用第二种招募方式。在征得咨询中心负责人的批准后，通过该咨询中心的当事人信息数据库获取本学年咨询会谈在 8 次及 8 次以上当事人的名单及联系方式（包括结束咨询和未结束咨询的）。先向当事人发送短信简单介绍研究，然后在随后的一周里再一一进行电话确认。愿意接受访谈的当事人付给象征性的报酬（10 元）。事实上，许多参与研究的当事人并未领取报酬。选取当事人时特别排除了那些有可能与研究者认识的当事人，如同院系的学生。

2.3.2 访谈 所有的访谈由一名研究者完成，采用电话访谈的形式。先对研究目的及保密、录音等事项进行介绍，若当事人同意，则开始正式访谈。访谈大约持续半小时左右。访谈结束后对当事人表示感谢，邀请其对访谈过程提供一些反馈，如有哪些地方不明白或可以改进。询问当事人是否愿意留下邮箱地址，以便可以接收访谈逐字稿和研究结果。最后访谈者记下此次访谈的备忘，如访谈的感觉等，作为后续分析材料的背景信息。

2.3.3 转录 访谈录音由几位组员转录为逐字稿。其后访谈者又对照录音对所有逐字稿进行了核对，以保证转录的准确性，并将所有身份信息移除。每个个案使用一个代号，并将访谈备忘附在每个个案的后面，作为背景信息。访谈逐字稿约为 10 万字。

2.4 数据分析

采用协商一致的质的研究方法（Consensual Qualitative Research, CQR）(Hill, et al., 2005; Hill, et al., 1997) 对数据进行分析。按照 Hill 等 (2005) 的说法，CQR 主要是建构主义取向，在其中加入了一些后实证主义的元素。其分析方法大致可以分为以下三步：首先，将文本中所有与合作相关的信息划分为几个域（domain，即话题范围）；然后，将每个个案中各个域的信息精练成核心观点（core idea，相当于写摘要，尽量忠实原意）；最后，将所有个案中同一个域的核心观点放在一起做交叉分析（cross analysis），

找出其中的共同主题，聚成不同的类别（category），形成研究结果。所有的数据分析由一个研究小组来完成，整个分析过程均需小组成员共同协商，达成一致，并且邀请研究小组之外的人来做审核员（auditor），以避免整个小组出现重大偏差。在本研究中，每次讨论至少有 4 名研究小组成员参加，整个数据分析过程花费 160 余小时。

2.4.1 域编码 按访谈提纲和逐字稿的内容将所有与合作相关的内容划分为几个域（即话题范围）。小组成员先各自将每个个案中的意义单元划分到一个或多个域中，然后大家一起讨论，直到所有材料的划分达成一致。在分析的过程中，由于新信息的出现或对材料理解的加深，可能需要对原来的域进行修改，如合并或增加新的域。最后，按照最终确定的域对所有数据重新划分一遍，使得前后编码一致。

2.4.2 提炼核心观点 每个组员先各自对每个个案每个域中的内容进行概括、提炼，写成核心观点。核心观点尽可能用当事人的原话，保持原意，其形式往往是形成几个不同含义的句子。然后所有组员对每个核心观点的内容及文字表述进行讨论，直到达成一致。

2.4.3 交叉分析 首先对 18 个个案的数据进行交叉分析（留下两个个案做稳定性检查）。每个组员各自对所有个案同一个域中的核心观点进行检查，提出分类。然后所有组员一起对分类及其名称，还有每个核心观点的归属进行讨论，直到全部达成一致。

2.4.4 稳定性检查 剩余的两个个案被加入，检查初步交叉分析所得的类别是否也适用于新的个案，各个类别的代表性是否会发生变化。结果发现，新增个案并没有改变原有结果，因此所得的结果被认为是稳定的。

2.4.5 审核 审核分为内部审核和外部审核。在得出初步结果后，组员们又一起对照原始的逐字稿重新对每个个案进行回顾，以防止一些背景信息被忽略而影响对抽取出来的信息的理解（因为一些类别的划分是建立在对整个个案的理解之上的），同时对所得结果进行检查。当组员达成一致后，这一步的

结果被发给 4 位心理咨询方向的博士生，邀请其对结果进行审核。审核的内容包括是否还有其他划分框架，域和类别的名称是否合适，核心观点的归属是否准确等。研究小组对所有的反馈意见再次进行讨论，以决定是否修改结果。

2.4.6　当事人反馈　将每个当事人的访谈逐字稿及研究结果（域、类别、核心观点及相应说明）通过电子邮件发给当事人，让其对身份的保密性进行确认，并对研究结果提供一些反馈，为研究提供证明效度（testimonial validity）(Stiles, 1993)。在本研究中没有当事人给予回复。

2.4.7　类别的代表性　按照 Hill 等 (2005) 的标准，可以对结果中各个类别的代表性进行划分。若此类结果能代表所有个案或只有一个个案例外，在本研究中即为 19 或 20 个个案，则该类别被视作是普遍的（general）；如果能代表一半以上的个案，在本研究中为 11 至 18 个个案，则为典型的（typical）；若能代表 4 至 10 个个案则为变异的（variant）；代表 2 至 3 个个案为少有的（rare）。所有只包括一个个案的结果，放在"其他"类中，不予报告。

3　结果

在 20 名当事人中，对合作关系持积极评价的有 12 人（当事人自己的评分为 4、4.5 或 5 分），认为合作关系一般的有 6 人（评分为 3 或 3.5 分），认为合作关系不好的有 2 人（评分为 2 分）。结束咨询和未结束咨询的当事人在对合作关系的评价上没有差异。为了检验积极与消极的同盟关系是否在结构上存在不同，可以看不同性质的同盟关系能否使用同一个框架进行描述。结果发现，积极的和消极的同盟关系都可以用一些共同的类别来描述，两者正好从正反两面提供证据。和同盟相关的内容被分为情感联结、任务、投入、合作模式、发展变化和影响因素 6 个域。所有域、类及其下子类的结果呈现在表 1 中，表中还包括每个类别的代表性及核心观点举例。文中则按域来呈现结果，对每个域的含义及其中的类别（以典型类别为主）进行介绍。在每个域中，先呈现积极的同盟关系，再呈现消极的同盟关系的结果。

3.1　情感联结

情感联结是咨询师与当事人之间的情绪感受。其中，当事人对咨询师的信任是一个典型的结果。如果当事人对咨询师很信任则会"在咨询师面前表露最真实的自己"，而"如果信任打破了对当事人则会有毁灭性的影响"。同盟关系出现问题的当事人也往往是由于没有对咨询师产生充分的信任，如"当事人很少信任别人，因为不想让咨询师太为难，才有什么话尽可能与咨询师说，但过几天回想会觉得不舒服，时间长了就觉得怯场，有种怕了咨询师的感觉"。其他被当事人认为良好的工作同盟所包含的情感还有坦率真诚、关注关心、轻松自然、接纳、依赖和理解。

3.2　任务

任务是合作过程中咨询师与当事人各自或双方要做的一些事情和活动，包括围绕着问题解决的倾诉 – 了解、探索 – 分析和指导 – 执行，还有对整个咨询的过程调控。在刚开始咨询时，当事人往往"很主动地说自己的情况、经历和思想状态"，认为"信息提供得越充分，感受表达得越真实，咨询师就越能在了解的基础上提出有帮助的建议"。此后，咨询师"具体分析问题，让当事人明白了一些以前没有意识到的问题，恍然大悟"，或是"在当事人说出一些困惑的东西时，咨询师及时给出有针对性的提示，能够说出当事人隐隐约约的感觉，让当事人意识到是哪方面出了问题"。而在有问题的同盟里，"都是当事人一个人在讲，咨询师没有针对当事人的问题做分析，提出一些剖析性的见解或建议，当事人不知道该怎么做，当事人要知道自己怎么了，然后才能自救"。最后，咨询师还会"给当事人建议、方法或布置一些作业"，而当事人则会"努力去做"、"尽力完成"。工作同盟不佳的当事人要么觉得"咨询师讲当事人知道的大道理，但不给具体的方法指引，这让当事人不知道如何做"，要么"刚开始时更听从咨询师的建议，到后来咨询师布置的任务有

表 1 当事人眼里的工作同盟：CQR 的结果

域、类及子类	代表性	核心观点举例
情感联结		
当事人的信任	T	
咨询师安全可靠	T	当事人信任咨询师，说问题时不会有保留。
咨询师的能力	R	当事人相信咨询师能帮助自己解决问题。
坦率真诚	V	当事人很坦诚，完全暴露自己的问题；咨询师与当事人的交流很坦率。
咨询师的关注关心	V	咨询师对当事人表示关注，发自内心地想去帮助当事人，而不是像完成任务一样。
轻松自然	V	当事人和咨询师像好朋友，很自然很放松。
咨询师的接纳	V	咨询师不会对当事人讲的内容进行批评，觉得当事人不对。
当事人的依赖	V	咨询师请假，当事人体验到了对咨询的依赖，觉得咨询是重要的心理支柱和情感出口。
咨询师的理解	V	当事人有被理解的感觉，咨询师非常容易听懂当事人说什么，能身临其境地感受当事人的感受。
任务		
倾诉－了解	T	
信息收集	T	当事人提供一些有用的信息给咨询师。
情感宣泄	V	早期当事人急于倾诉，咨询师起到安抚作用。
探索－分析	T	
咨询师的分析与启发	T	咨询师对当事人的问题进行分析、总结，让当事人挺受启发，能看清自己。
当事人的反思与反馈	V	当事人会去思考、反省咨询师指出来的一些问题，并慢慢接受。
指导－执行	T	
咨询师指导	T	
具体建议	T	咨询师给当事人一些指导，如让当事人练习自信地讲话。
一般方法	V	咨询师教当事人一些做事情的方法，例如做选择之前分析各个选择的后果，让当事人感觉很贴心。
当事人执行	V	当事人按照咨询师提出的建议去做，例如去调节自己的心情，看有没有效果。
过程调控	T	
结构化	V	咨询师应该在咨询前做些介绍，增加当事人对咨询活动的认识，减少当事人表述自己时的顾虑。
目标计划	V	咨询师与当事人建立共同目标，然后讨论具体问题。
反馈调整	V	咨询师始终提醒当事人把自己当作咨询师这样一个角色来理解，不能依赖自己，后来当事人就反复提醒自己不能依赖咨询师。
投入		
积极主动	V	当事人愿意说，而且有意识地往点子上说，如果做不到，至少要勤开口，不要有太多的疑虑和隐瞒。
克服困难	V	当事人不想与外界联系，但当事人每周能坚持咨询，很努力地让自己从那个情绪当中走出来。
认真专注	V	当事人和咨询师要有认真的心态，当事人要将咨询当作重要的事情来做。
合作模式		
咨询师主导	V	好的合作就是咨询师对当事人进行指导，当事人去做，而不会有质疑。
相互合作	V	合作就是咨询师与当事人建立共同目标，然后共同努力。
当事人主导	R	咨询中当事人是主导，咨询师是辅助作用，咨询师帮助当事人探索，当事人要主动探索，并做一些努力和调整。
发展变化		
稳定	T	合作关系没什么变化。
上升	V	合作关系一开始就比较好，是一个上升的过程，但上升的幅度不大。
倒 U 型	R	当事人觉得开始时比较生疏，不愿意讲所有的事情；中间时最亲密、关系最好，什么都愿意讲；再到后来反而没什么可讲的，关系又拉远了。
影响因素		
咨询师的专业能力和经验	V	咨询师要有专业素养，有很多方法和技巧，有很硬的专业功底，有丰富的经验。
当事人的个人特点	V	当事人是一个感性的人，做事情凭感觉，如果对咨询师感觉好，会更愿意配合。
当事人的改变动机	V	当事人一直希望能够得到帮助，所以一直很配合。
当事人对咨询的期望	V	当事人对咨询不要抱太高期望，咨询师不能帮当事人解决所有问题。
当事人的咨询经历	V	当事人曾经做过心理咨询，所以知道咨询师不会给倾向性的内容，也知道不用怕咨询师，不用顾忌得罪咨询师，怎么想就怎么说。
效果	V	当事人对咨询师很满意，感觉咨询师对自己有帮助，当事人也很愿意配合咨询师，配合出来的结果对当事人也比较有帮助。
当事人对咨询的信念	R	当事人坚信心理咨询有效，相信咨询师的实力能帮助自己，即积极的心理作用。
当事人对咨询的接纳	R	当事人对咨询活动的认识比较客观，不认为自己是有病，就是对现实不太满意，做一个积极的调整，所以没有很大的戒备心理。
性别	R	咨询师与当事人都是女性，当事人能自如、随意地说。
匹配	R	咨询师与当事人感觉比较对，就可以讲很多。

注：$N=20$；T=typical，11–18 个个案；V=variant，4–10 个个案；R=rare，2–3 个个案。

时会拖过去不做"。当然，以上三个阶段的划分是相对的，事实上三者往往是交叉或循环进行的，如"当事人回去实践咨询师布置的任务，然后跟咨询师分享，咨询师再帮当事人分析"。

与此同时，咨询师还需对咨询的过程进行调控，以使各项任务活动能够顺利进行。比如，咨询师"在刚开始时引导当事人说要建立信任的关系，把咨询比喻成走迷宫的过程"，"反复强调良好关系的重要性，让当事人对咨询引起重视，帮当事人树立积极的态度，让当事人投入到咨询过程中"，与当事人"建立共同目标，然后讨论具体问题"。而身处不良同盟关系中的当事人则抱怨"咨询师没有给当事人任何明示或暗示，告诉当事人如何配合，所以当事人不知如何合作，如不知道在咨询中说什么"，或是"咨询师跟当事人说什么都要说，每次基本上都是当事人在说，当事人不明白咨询的主要问题，感觉糊涂"，而且"当事人跟咨询师提出一些自己的问题和想法，表达自己的不满，咨询师却没有做出相应的改变"。

3.3　投入

投入是合作关系中咨询师与当事人的付出程度，具体包括积极主动、克服困难和认真专注三个方面。双方投入才能保证合作最终产生成效，实现目标。一些当事人在咨询中会"特别积极主动地探索解决方式，多想多思考，表述自己特别真实的想法，形象地描述自己的感觉"，在咨询之外也会"刻意寻求别人的帮助，改变自己不好的生活习惯，充实自己的生活，搜集相关书籍和资料，来调节自己的情绪"。而且，"当事人刚开始的时候觉得进展不大，但当事人挺沉得住气，给自己积极的暗示，反复提醒自己要相信咨询师，应该继续去做没有做到的事情"。当然，"当事人和咨询师的主动和投入都很重要"，咨询师的投入主要体现为"咨询师认真负责"。不过，也有一些当事人"一直很被动，咨询的时候常有大段的沉默"，"平时很忙，练习咨询师教给当事人的方法次数不多"。

3.4　合作模式

合作模式是根据当事人对咨询师的依赖程度和在咨询过程中的参与程度来划分的，可以分为咨询师主导、相互合作和当事人主导三种类型。一些当事人认为"咨询更多依靠咨询师，当事人就想着把自己的问题说出来让咨询师解决"，或者是"比较好的合作就是咨询师给很好的方法指导当事人，给建议让当事人去做"，当事人"希望咨询师帮自己解决问题，没有思考过怎样把咨询做得更好，很少想自己要在里面做什么，只会跟着咨询师的思路走"。而其他一些当事人则认为"合作就是咨询师与当事人建立共同目标，然后共同努力"，"双方都需要积极主动"。少数当事人认为"咨询中当事人是主导，咨询师起辅助作用，咨询师帮助当事人探索，当事人要主动探索，并做一些努力和调整"。当然，这三种合作模式还会随着咨询过程而有所变化，如"刚开始当事人向咨询师讲自己的问题，时间由当事人来掌握，合作关系的重点在咨询师，获取当事人的信赖；后来一起互动，时间由两人掌握，合作关系重点落到当事人身上，要当事人行动与思考"。

3.5　发展变化

同盟关系在整个咨询过程中典型的发展变化是保持稳定，如"合作关系没什么变化"；也有上升的，如"关系越来越熟悉，越来越亲切，越来越好，对咨询师的信任感逐步增加"；少数出现倒 U 型，如"当事人觉得开始时比较生疏，不愿意讲所有的事情；中间时最亲密、关系最好，什么都愿意讲；再到后来反而没什么可讲的，关系又拉远了"。

3.6　影响因素

对同盟关系有影响的因素包括咨询师的专业能力和经验，如"咨询师要有专业素养，有很多方法和技巧，有很硬的专业功底，有丰富的经验"；当事人的个人特点，如"当事人的性格，例如防备心和表达水平，会影响当事人在咨询中表述的充分程度"；当事人的改变动机，如"当事人将咨询作为最后的希望，需要得到咨询师的救助，非常想通过咨询帮助自己成长，愿意全力配合"；当事人对咨询的期望，如"当事人对咨询不要抱太高的期望，咨询师不能帮当事人解决所有问题"；当事人的咨询经历，如"当事人之前做过咨询，让当事人对咨询有足够的信任，是后面求助咨询的基础"；效果，如"当事人对咨询师很满意，感觉咨询师对自己有帮助，当事人也很愿意配合咨询师，配合出来的结果对当事人也比较有帮助"。少数当事人还提到当事人对咨询的信念、当事人对咨询的接纳、性别和当事人与咨询师的匹配对同盟关系也会产生影响。

4 讨论

本研究通过质的方法对当事人眼里的同盟关系进行分析，得出了与工作同盟相关的 6 个域，分别是情感联结、任务、投入、合作模式、发展变化和影响因素。若将咨询看作是一个咨询师与当事人相互合作而达成目标的过程 (Hatcher & Barends, 2006)，则前三个域可以看作是这个合作过程的构成要素，而后三个域则是对其外部特性的描述。

4.1 情感联结、任务和投入

假如将合作作为工作同盟的核心内涵，则情感联结可以看作是合作的情感基础，即愿不愿意合作。在情感基础上，任务则是合作的内容，即如何合作。而为了保证合作能达到预期目标，双方的投入不可缺少。投入是合作的动力，即花多大力气合作。这三者构成了良好合作的三个成分。Hougaard (1994) 认为同盟关系包含两个方面的内容，一个是与社会情感相关的个人关系；一个是和任务相关的合作关系。显然，情感联结属于个人关系，而任务与投入则属于合作关系。

情感联结域与 Bordin 工作同盟理论中的情感联结因素基本对应。Bordin 指出各种治疗关系都需要基本水平的信任，而若要探索更加隐秘的内在体验时，则需要更深的信任关系 (1979, p. 254)。同样，情感联结域中最突出的也是当事人对咨询师的信任。情感联结域中的坦率真诚、关注关心、接纳和理解与 Rogers (1957) 强调的助长条件一致。这些助长条件既促使了当事人的改变，也促进了当事人与咨询师的合作，成为良好合作的情感基础。

咨询师与当事人在合作过程中需要完成的任务大部分指向问题解决，也包括了对咨询过程进行调控的要求。许多当事人提到的咨询师的活动通常可以被归为治疗技术，如指导、结构化、建议、解释、倾听等。其中，当事人最为看重的是咨询师的指导，如建议、教育、布置任务等，反映了中国当事人对咨询师指导性及咨询过程结构化的要求。不过，在国外有关工作同盟的质的研究中，国外当事人也同样看重这些带有较强咨询师指导性的活动，如效果聚焦的技术和策略 (Bedi, et al., 2005)、教育 (Bedi, 2006)、咨询师指导当事人的能力、促进当事人理解的干预、建议 (Bachelor, 1995) 等。这说明工作同盟中的合作任务可能具有一定程度的文化普遍性。另外，Bordin (1979, 1994) 工作同盟理论中的任务类似于各个流派的治疗策略与技术，如自由联想、空椅子技术等，而本研究结果中的任务则不局限于某种特定治疗取向。

投入本来应该包括咨询师与当事人双方，但考虑到咨询师往往较为"认真负责"，所以理论家们大多关注的是当事人的投入。本研究投入域中的内容也大都是针对当事人而言。Frieswyk 等主张将同盟定义为当事人在治疗过程中的合作 (Frieswyk, et al., 1986; Frieswyk, Colson, & Allen, 1984)，实际上就是当事人积极投入到治疗任务中的程度，如当事人在会谈中积极主动，谈论重要的主题，坦诚地提供信息和表达感受，好好利用治疗师的帮助，将治疗收获应用于生活，积极改变等 (Allen, Newsom, Gabbard, & Coyne, 1984)。Gaston (1990, 1991) 认为应该对治疗同盟与工作同盟的概念进行区分，前者指当事人对咨询师的情感依附，后者侧重当事人在同盟中的工作能力。这些能力包括当事人自我揭示重要的信息，自我观察，探索自己的问题，体验情绪，对咨询师的意见反应积极，加深对重要主题的探索，为解决问题有目的地工作等。从中可以看出，虽然 Gaston 称之为当事人的"工作能力"，但其内容与这里所说的当事人的投入其实非常相似。另外，Horvath 和 Bedi (2002) 认为工作同盟不仅指咨询师与当事人有共同的目标，还包括双方承担各自的责任，共同为完成各项目标任务而努力。这其中的"承担责任"与"努力"也与本研究所说的投入的含义相吻合。

Bachelor (1995) 发现在当事人的描述中有三类不同的同盟，一类是抚育型（nurturant），强调的是咨询师的助长态度；一类是领悟型（insight-oriented），看重增进自我理解；一类是合作型（collaborative），体现为当事人的积极参与。这三类同盟的特点恰好与本研究中情感联结、任务和投入三个合作成分一一对应。事实上，抚育型的同盟除了看重咨询师的态度，同样也包括咨询师的指导和当事人充分的自我揭示；领悟型的同盟除了看重咨询师的澄清，也少不了信任和与咨询师的合作；合作型的同盟除了看重当事人的参与，也强调咨询关系与咨询师的反应技术。这说明 Bachelor (1995) 发现的三类同盟中其实均包含着与本研究所得结果类似的几个共同成分，只不过当事人对这些成分的重要性及在同盟中的作用理解不一，所以形成了不同的同盟类型。同样的现象两个研究采用了不同的分析方法，一个是类型学，一个是因素论，所得结果却可以相互补充。

4.2 合作模式、发展变化和影响因素

根据当事人对咨询师的依赖程度和在咨询过程中的参与程度将工作同盟分为咨询师主导、相互合作和当事人主导三种合作模式。大部分的当事人属于前两种合作模式。这说明当事人在同盟关系中仍较多依赖咨询师，或至少认为"合作是一方对另一方的响应，合作关系需要咨询师引导——提供一个大的框架，然后在具体问题上纠正和帮助——这样当事人才会更积极地配合"。国外的研究也一致发现同盟关系的质量更多取决于咨询师，当事人更多将责任放到咨询师身上，在同盟理论中被强调的合作并没有被当事人过多提及 (Bachelor, 1995; Bedi, 2006; Bedi, et al., 2005)。在 Bachelor (1995) 的研究中，合作型的同盟仅占15%（本研究中与之含义相近的当事人主导的合作模式占10%），而且当事人更多认为合作是咨询师与当事人各出各的力，而不是同盟理论中通常提到的双方的协商与一致。显然，当事人理解的合作仍然少不了咨询师的积极参与和指导。

在本研究中，工作同盟有三种发展模式，稳定、上升与倒 U 型。前两种变化趋势在研究中均有发现 (de Roten et al., 2004; Kivlighan & Shaughnessy, 2000; Kramer, Roten, Beretta, Michel, & Despland, 2008; Stevens, Muran, Safran, Gorman, & Winston, 2007; Stiles et al., 2004)，而倒 U 型的发展模式则与已有的理论假设相反。Gelso 和 Carter (1994) 认为工作同盟的发展是"U"型（高 - 低 - 高）变化模式。他们认为特别是在疗程较短的治疗中，由于当事人对于治疗的乐观主义，开始阶段会有较高水平的工作同盟。随后，当事人往往会经历失望、沮丧等负面情绪，工作同盟的水平也随之下降。但在成功的治疗中，在最后阶段当事人的反应变得更加现实，工作同盟也会恢复到一个较高的水平。倒 U 型的发展模式则正好与之相反。开始当事人与咨询师的同盟关系不断加深，中间双方密切合作完成咨询的各项任务，待各项任务告一段落，合作关系又逐渐疏远。可能的情况是，若咨询目标达到，咨询随即结束，则最后逐渐疏远的阶段不会持续太长。若咨询还有下一步的任务，则这段倒 U 型的关系就只是整个同盟关系发展过程中的一个片段，疏远阶段既可能代表暂时的调整，如商讨下一步的目标，也可能是关系出现了破裂，需要进一步的处理。此外，也不排除咨询师对咨询进程把握不当的可能。不过，此发展模式的代表性并不高（为 rare）。

本研究呈现了 10 个对同盟质量有影响的因素。其中一些在其他质的研究里也被当事人提及，如效果 (Bachelor, 1995)、咨询师的个人特点以及当事人之前的咨询经历等 (Bedi, et al., 2005)。另一些则出现在检验同盟关系影响因素的研究里，如当事人与咨询师的依恋风格、咨询师的经验、咨询师与当事人的匹配 (Horvath, 2001)、当事人的期望 (Patterson, Uhlin, & Anderson, 2008)、动机 (Rumpold et al., 2005)、性别 (Wintersteen, Mensinger, & Diamond, 2005) 等。咨询师的各类反应技术也是常被研究者关注的同盟关系的影响因素 (e.g., Ackerman & Hilsenroth, 2001, 2003)。不过在本研究中，咨询师的反应技术被视作合作过程中的任务活动，被放在了任务域中。

4.3 局限

采用质的方法对中国当事人眼里的工作同盟进行描述，在得到一些有意义的结果的同时，也要认识到本研究的局限。首先，本研究对结果的解释有赖于对工作同盟概念的理解。例如，投入到底是一个当事人变量，或是同盟的构成要素，还是良好同盟的结果 (Horvath & Bedi, 2002)？对咨询的信心是同盟本身，还是同盟的影响因素 (Hatcher & Barends, 1996)？工作同盟是咨询关系的一种，还是一个看待咨询过程的视角 (Hatcher & Barends, 2006)？对此类问题不同研究者可能有不同答案，以致对研究的结果也可能有着不同的解释。

在谈到 CQR 的取样问题时，Hill 等 (1997, 2005) 建议研究者应该从同质的总体中随机选取那些对所研究的问题有着丰富知识的被试，而且最好是近来有过相关的经历，即基于标准的取样。据此，本研究将取样标准定为咨询会谈达到 8 次及以上的当事人，而且咨询经历发生在半年内。但事后检讨，发现在努力满足这个标准的同时，却对另一问题考虑不周：会谈次数达到 8 次及以上的当事人中，消极工作同盟的比例很可能比较小，进而使得研究结果对这类同盟关系的代表性不足。另外，研究中所有当事人都来自大学咨询中心，均为在校学生，且绝大数为女性，这虽然符合 Hill 提出的同质性要求，但也同样可能因此而减少了样本的代表性。不过，代表性问题是质的研究方法的固有特性，其根本解决办法可能需要采取量的研究范式。

此外，不清楚当事人确切的诊断信息，而当事人的症状与问题类型显然会影响其建立同盟的能力 (Gaston, 1991)。比如，边缘型人格障碍的当事人建立同盟关系会特别困难。因此，对这些当事人而言，工作同盟的建立本身就是一个关键的效果变量 (Frieswyk, et al., 1986)。本研究也没有收集有关咨询师的信息，所得结果可能与某些特定的咨询师变量有着密切的关系，比如咨询师的治疗取向、经验，甚至性别等。这些限制使得本研究的结果只能被看作是对工作同盟本土概念化的一个初步探索。

4.4 对临床的启示

当事人来寻求帮助时，大都指望咨询师帮其解决问题，对咨询师较为依赖。当事人的这种想法是可以理解的，原因可能有二。一是当事人找到咨询师时，大多已经非常脆弱，无力或不愿再承担更多的责任。当事人"将咨询作为最后的希望，需要得到咨询师的救助"，而国外的当事人同样觉得"不能处理这么多的责任了"(Bedi, 2006, p. 32)。另外一个原因是当事人对咨询过程不了解，不知道该做些什么，如何与咨询师合作。于是，当事人根据自己的日常经验，如到医院看病，开始向咨询师寻求问题的解决之道。因此，在所有研究里（包括本研究）当事人都强调了咨询师指导的重要性。

事实上，咨询师可以利用各种同盟管理或调控策略来调动当事人，以建立更好的同盟关系。在本研究中，咨询过程的调控也同样是任务域中的重要内容。调控的内容既包括咨询结构上的设置，如结构化、目标任务等，也包括咨询关系的反馈与调整。也就是说，咨询师一方面要告诉当事人在咨询中做什么，如何做，对咨询过程保持一定的结构化，同时，又要鼓励当事人积极参与，主动反馈，保持一定的开放性与灵活性。Bachelor (1995) 发现当当事人可以对咨询师的行为或咨询效果进行讨论或评论时，可以增强同盟关系。也许在此之前，当事人并不知道自己是可以参与到咨询过程之中，而且是被咨询师所鼓励和看重的。同样，当事人也希望更多地参与咨询过程，如"整个咨询都是咨询师说，当事人回答，当事人充当一个小孩子，听咨询师说应该怎么办，感觉不是特别好，希望更积极地加入咨询过程，且能提出一些自己的想法"。咨询师若不够开放或缺乏灵活性与敏感，则常常会抑制当事人的积极性，如"当事人跟咨询师提出一些自己的问题和想法，表达自己的不满，但咨询师没有做出相应的改变"。咨询师能够察觉同盟关系中的问题并与当事人进行协商，既能促进同盟的质量，也会对当事人有治疗性的帮助 (Hill & Knox, 2009)。

一些理论家认为当事人会随着咨询的进程越来

越投入，如 Luborsky (1976) 的二阶段／类型同盟理论认为当事人与咨询师建立了信任关系之后则会与咨询师共同努力，共度难关。但 Bachelor (1995) 发现合作型的同盟比例并没有随着咨询的进程而提高，而 Bedi 等 (2005, p. 317) 更是认为即使使用合适的角色导入策略，当事人还是会有意或无意地将建立或增强同盟的责任放在咨询师身上。看来，无论何时咨询师身上的担子都轻不了。

4.5 对研究的启示

本研究为工作同盟这一治疗过程领域的重要变量的本土概念化提供了初步证据。结果表明，中国当事人对工作同盟的看法与西方工作同盟的理论与研究基本吻合。这在一定程度上证明了已有工作同盟理论和结构的跨文化适用性，为后续开展本土的工作同盟研究奠定了基础。例如，可以在此基础上修订与编制相应的工作同盟的测量工具。另外，Bedi (2006) 认为积极与消极的同盟关系在结构上可能不同。为检验此假设，本研究同时对积极的、一般的和消极的工作同盟进行了探索性的分析。结果发现，积极的和消极的同盟关系都可以用一些共同的类别来描述，意味两者有共同的结构。这个结果显然对工作同盟的测量有着重要的意义。这使得目前的测量工具不仅能告诉我们工作同盟有多好，而且还可以让我们知道工作同盟有多差。研究结果表明，合作过程中咨询师的指导与对咨询过程的调控对工作同盟有着重要的意义，而这些变量目前并没有得到充分地研究。这些变量可以成为下一步的研究方向。

致谢: 作为研究小组的成员，梁佳、王曼、朱文臻、赵丽和赵金在数据分析的过程中付出了大量精力，在此表示衷心的感谢。

参考文献

梁觉，周帆. (2010). 跨文化研究方法的回顾及展望. 心理学报, 42, 41–47.

Ackerman, S. J., & Hilsenroth, M. J. (2001). A review of therapist characteristics and techniques negatively impacting the therapeutic alliance. *Psychotherapy: Theory, Research, Practice, Training, 38*, 171–185.

Ackerman, S. J., & Hilsenroth, M. J. (2003). A review of therapist charcteristics and techniques positively impacting the therapeutic alliance. *Clinical Psychology Review, 23*, 1–33.

Allen, J. G., Newsom, G. E., Gabbard, G. O., & Coyne, L. (1984). Scales to assess the therapeutic alliance from a psychoanalytic perspective. *Bulletin of the Menninger Clinic, 48*, 383–400.

Bachelor, A. (1995). Clients' perception of the therapeutic alliance: A qualitative analysis. *Journal of Counseling Psychology, 42*, 323–337.

Bedi, R. P. (2006). Concept mapping the client's perspective on counseling alliance formation. *Journal of Counseling Psychology, 53*, 26–35.

Bedi, R. P., Davis, M. D., & Williams, M. (2005). Critical Incidents in the Formation of the Therapeutic Alliance from the Client's Perspective. *Psychotherapy: Theory, Research, Practice, Training, 42*, 311–323.

Bordin, E. S. (1979). The generalizability of the psychoanalytic concept of the working alliance. Psychotherapy: Theory, *Research & Practice, 16*, 252–260.

Bordin, E. S. (1994). Theory and research on the therapeutic working alliance: New directions. In A. O. Horvath & L. S. Greenberg (Eds.), The working alliance: *Theory, research, and practice* (pp. 13–37). New York: John Wiley & Sons.

de Roten, Y., Fischer, M., Drapeau, M., Beretta, V., Kramer, U., Favre, N., et al. (2004). Is one assessment enough? Patterns of helping alliance development and outcome. *Clinical Psychology & Psychotherapy, 11*, 324–331.

Freud, S. (1958). The dynamics of transference. In J. Strachey (Ed. & Trans.), *The standard edition of the* complete psychological works of Sigmund Freud (Vol. 12, pp. 99–108). London: Hogarth Press. (Original work published 1912)

Freud, S. (1964). Analysis terminable and interminable. In J. Strachey (Ed. & Trans.), *The standard edition of the complete psychological works of Sigmund Freud* (Vol. 23, pp. 209–253). London: Hogarth Press. (Original work published 1937)

Frieswyk, S. H., Allen, J. G., Colson, D. B., Coyne, L., Gabbard, G. O., Horwitz, L., et al. (1986). Therapeutic alliance: Its place as a process and outcome variable in dynamic psychotherapy research. *Journal of Consulting and Clinical Psychology, 54,* 32–38.

Frieswyk, S. H., Colson, D. B., & Allen, J. G. (1984). Conceptualizing the therapeutic alliance from a psychoanalytic perspective. *Psychotherapy: Theory, Research, Practice, Training, 21,* 460–464.

Gaston, L. (1990). The concept of the alliance and its role in psychotherapy: Theoretical and empirical considerations. *Psychotherapy: Theory, Research, Practice, Training, 27,* 143–153.

Gaston, L. (1991). Reliability and criterion–related validity of the California Psychotherapy Alliance Scales–Patient version. *Psychological Assessment: A Journal of Consulting and Clinical Psychology, 3,* 68–74.

Gelso, C. J., & Carter, J. A. (1985). The relationship in counseling and psychotherapy: Components, consequences, and theoretical antecedents. *The Counseling Psychologist, 13,* 155–243.

Gelso, C. J., & Carter, J. A. (1994). Components of the psychotherapy relationship: Their interaction and unfolding during treatment. *Journal of Counseling Psychology, 41,* 296–306.

Greenson, R. R. (1965). The working alliance and the transference neurosis. *The Psychoanalytic Quarterly, 34,* 155–179.

Greenson, R. R. (1967). *The technique and practice of psychoanalysis.* New York: International Universities Press.

Hatcher, R. L., & Barends, A. W. (1996). Patients' view of the alliance in psychotherapy: Exploratory factor analysis of three alliance measures. *Journal of Consulting and Clinical Psychology, 64,* 1326–1336.

Hatcher, R. L., & Barends, A. W. (2006). How a return to theory could help alliance research. Psychotherapy: Theory, Research, Practice, Training, 43, 292–299.

Hill, C. E., & Knox, S. (2009). *Processing the therapeutic relationship. Psychotherapy Research, 19,* 13–29.

Hill, C. E., Knox, S., Thompson, B. J., Williams, E. N., Hess, S. A., & Ladany, N. (2005). Consensual Qualitative Research: An Update. *Journal of Counseling Psychology, 52,* 196–205.

Hill, C. E., Thompson, B. J., & Williams, E. N. (1997). A guide to conducting consensual qualitative research. *The Counseling Psychologist, 25,* 517–572.

Horvath, A. O. (2001). The alliance. *Psychotherapy: Theory, Research, Practice, Training, 38,* 365–372.

Horvath, A. O., & Bedi, R. P. (2002). The alliance. In J. C. Norcross (Ed.), *Psychotherapy relationships that work: Therapist contributions and responsiveness to patients* (pp. 37–69). New York: Oxford University Press.

Horvath, A. O., & Greenberg, L. S. (1989). Development and validation of the Working Alliance Inventory. *Journal of Counseling Psychology, 36,* 223–233.

Horvath, A. O., & Symonds, B. D. (1991). Relation between working alliance and outcome in psychotherapy: A meta–analysis. *Journal of Counseling Psychology, 38,* 139–149.

Hougaard, E. (1994). The therapeutic alliance: A conceptual analysis. *Scandinavian Journal of*

Psychology, 35, 67–85.

Kivlighan, D. M., Jr., & Shaughnessy, P. (2000). Patterns of working alliance development: A typology of client's working alliance ratings. *Journal of Counseling Psychology, 47*, 362–371.

Kramer, U., Roten, Y. d., Beretta, V., Michel, L., & Despland, J.-N. (2008). Patient's and therapist's views of early alliance building in dynamic psychotherapy: Patterns and relation to outcome. *Journal of Counseling Psychology, 55*, 89–95.

Leung, K., & Zhou, F. (2010). Cross-cultural research methods: Review and prospect. *Acta Psychologica Sinica, 42*, 41–47.

Luborsky, L. (1976). Helping alliances in psychotherapy. In J. L. Claghorn (Ed.), *Successful psychotherapy* (pp. 92–116). New York: Brunner/Mazel.

Martin, D. J., Garske, J. P., & Davis, M. K. (2000). Relation of the therapeutic alliance with outcome and other variables: A meta-analytic review. *Journal of Consulting and Clinical Psychology, 68*, 438–450.

Orlinsky, D. E., Ronnestad, M. H., & Willutzki, U. (2004). Fifty years of psychotherapy process-outcome research: Continuity and change. In M. J. Lambert (Ed.), *Bergin and Garfield′s handbook of psychotherapy and behavior change* (5th ed., pp. 307–389). New York: Wiley.

Patterson, C. L., Uhlin, B., & Anderson, T. (2008). Clients' pretreatment counseling expectations as predictors of the working alliance. *Journal of Counseling Psychology, 55*, 528–534.

Rogers, C. R. (1957). The necessary and sufficient conditions of therapeutic personality change. *Journal of Consulting Psychology, 21*, 95–103.

Rumpold, G., Doering, S., Smrekar, U., Schubert, C., Koza, R., Schatz, D. S., et al. (2005). Changes in motivation and the therapeutic alliance during a pretherapy diagnostic and motivation-enhancing phase among psychotherapy outpatients. *Psychotherapy Research. Special Issue: The Therapeutic Relationship, 15*, 117–127.

Sterba, R. (1934). The fate of the ego in analytic therapy. *The International Journal of Psychoanalysis, 15*, 117–126.

Stevens, C. L., Muran, J. C., Safran, J. D., Gorman, B. S., & Winston, A. (2007). Levels and patterns of the therapeutic alliance in brief psychotherapy. *American Journal of Psychotherapy, 61*, 109–129.

Stiles, W. B. (1993). Quality control in qualitative research. *Clinical Psychology Review, 13*, 593–618.

Stiles, W. B., Glick, M. J., Osatuke, K., Hardy, G. E., Shapiro, D. A., Agnew-Davies, R., et al. (2004). Patterns of Alliance Development and the Rupture-Repair Hypothesis: Are Productive Relationships U-Shaped or V-Shaped? *Journal of Counseling Psychology, 51*, 81–92.

Strong, S. R. (1968). Counseling: An Interpersonal Influence Process. *Journal of Counseling Psychology, 15*, 215–224.

Tryon, G. S., Blackwell, S. C., & Hammel, E. F. (2007). A meta-analytic examination of client-therapist perspectives of the working alliance. *Psychotherapy Research, 17*, 629–642.

Wintersteen, M. B., Mensinger, J. L., & Diamond, G. S. (2005). Do Gender and Racial Differences Between Patient and Therapist Affect Therapeutic Alliance and Treatment Retention in Adolescents? *Professional Psychology: Research and Practice, 36*, 400–408.

Zetzel, E. R. (1956). Current concepts of transference. *The International Journal of Psychoanalysis, 37*, 369–375.

The working alliance in clients' eyes: A qualitative analysis

ZHU Xu; JIANG Guangrong

(School of Psychology, Huazhong Normal University, Wuhan 430079, China)

Abstract　The working alliance has been a focus of psychotherapy research for several decades in the western literature. Although theorists differ somewhat in their conceptualizations of the alliance, most of them agree that the core of this construct is the collaboration between therapist and client that emphasizes the contributions of both participants. Most of our knowledge in this field derives from theorists' hypotheses, and clients' perspective is often neglected. In fact, it was often assumed that clients use the same conceptual dimensions as therapists do to rate the collaborative processes in psychotherapy. This assumption, however, has been challenged by reported low correlation between client–therapist alliance ratings and by results of qualitative studies that show divergence between clients' and therapists' perspectives. Moreover, the literature regarding clients' actual experiences of the alliance is sparse. In an attempt to examine clients' perspectives in viewing working alliance and the cultural adaptability of western alliance theories to Chinese culture, the present study sought to delineate the alliance from Chinese clients' perspective via a qualitative approach and addressed the concordance between clients' perceptions and theoretician–derived views of the alliance. Twenty clients at a university counseling center consented to participate in the study. Clients were interviewed by phone about their experiences of being collaborative with their therapists and their opinions about what is important to this collaborative process in counseling. Data were analyzed by a research team using Consensual Qualitative Research (CQR) (Hill et al., 2005; Hill, Thompson, & Williams, 1997). CQR requires that research team members reach consensus about the classification and the meaning of data through three steps, namely, domain coding, core ideas, and cross subject analysis. Six domains with regard to the collaborative process in counseling emerged from the data: bond, task, engagement, collaboration pattern, development, and influencing factors. Bond is the emotional and personal aspect of the client–therapist relationship, which serves as the emotional base for collaborative work. Task refers to the activities and things client and therapist need to do during the collaborative process, which is the content of the collaborative work. Engagement implies the degree to which client and therapist devote themselves to the process, lending an impetus to the collaboration. These three domains can be viewed as components of working alliance and the other three as descriptions of major features of the alliance.

The study results suggest that the working alliance is perceived by Chinese clients to include bond, task, and engagement. This finding is comparable to current theories of alliance in the western literature and advances conceptualizations of the alliance especially when culture differences are concerned.

Keywords　therapeutic relationship; working alliance; qulitative analysis

管理心理研究所团队成员

马红宇，博士，教授，管理心理研究所所长，主要研究领域为工作生活平衡与组织文化建设、新生代职业心理与管理、胜任特征模型构建与应用、网络心理与应急管理等。

洪建中，特聘教授，主要研究领域为心理与知识管理，包括文化－历史活动理论在知识管理中的应用、知识管理中心理与文化机制、合作性学习与知识创新、心理与知识管理的跨文化研究、网络环境下知识管理的心理研究等。

徐富明，博士，教授，主要研究领域为经济心理与行为决策、教师职业心理学、组织行为和人力资源管理等。

王忠军，博士，副教授，管理心理研究所副所长，主要研究领域为员工发展与职业健康，包括新生代职业心理、工作动机与激励、上下级关系、情绪劳动与情感管理、职场压力与健康、职业生涯成功等。

刘亚，博士，副教授，主要研究领域为组织行为学与人力资源管理、中国文化与中国式管理、大学生职业决策等。

谢员，博士，讲师，主要研究领域为组织行为学与人力资源管理、生涯心理辅导与咨询、员工心理援助计划 EAP，包括压力管理、员工使命感、野心家型员工、生涯适应、生涯决策风格、教师职业发展与心理健康等主题。

唐汉瑛，博士，讲师，主要研究领域为组织行为与人力资源管理，包括反生产行为、工作－生活平衡、谦卑领导、胜任特征建模与应用等

心理学报，2014, 46(4), 540 - 551.

边界弹性与工作 - 家庭冲突、增益的关系：基于人 - 环境匹配的视角 *

马红宇 [1,2]，申传刚 [1,2]，杨 璟 [3,4]，唐汉瑛 [1,2]，谢菊兰 [1,2]

([1] 青少年网络心理与行为教育部重点实验室)

([2] 华中师范大学心理学院暨湖北省人的发展与心理健康重点实验室，武汉 430079)

([3] 中国科学院心理研究所，北京 100101)

([4] 中国科学院大学，北京 100039)

摘 要 本研究从人 - 环境匹配理论的视角探讨工作和家庭边界弹性能力和边界弹性意愿对个体工作 - 家庭冲突和工作 - 家庭增益的交互影响。通过问卷法共获得 494 份有效数据，基于多项式回归分析和反应曲面分析的结果表明：工作弹性能力与工作弹性意愿的匹配对工作→家庭冲突有显著的负向效应，对工作→家庭增益无显著影响；家庭弹性能力与家庭弹性意愿的匹配对家庭→工作冲突、家庭→工作增益有显著的负向效应。

关键词 人 - 环境匹配；边界弹性能力；边界弹性意愿；工作 - 家庭冲突；工作家庭增益

分类号 B849:C93

1 问题提出

工作和家庭是现今成年人生活中的两个重要组成部分，工作角色和家庭角色是个体最主要的社会角色，它们之间相互影响，密不可分。目前关于工作和家庭角色之间关系的研究主要基于两种视角：一种是消极视角，认为由于个体有限的时间、高水平的压力和竞争性的行为期望而引发角色之间的相互冲突，即工作 - 家庭冲突 (work-family conflict, WFC) (Eby, Casper, Lockwood, Bordeaux & Brinley, 2005; Greenhaus & Beutell, 1985)；另一种是积极视角，认为个体可以从工作和家庭角色的投入中获得有意义的资源 (如自

尊、经济收入等)，继而提升个体在相对角色领域的表现，即工作 - 家庭增益 (work-family enrichment, WFE) (Greenhaus & Powell, 2006)。这两种不同的视角为个体处理工作和家庭之间的关系提供了新的思路，即个体在履行工作和家庭角色时如何尽可能在减少角色间冲突的同时增加角色间的积极溢出。

边界理论的提出为我们回答上述问题提供了可能。根据边界理论，个体会围绕其所在的工作领域或家庭领域建立起不同的角色边界，个体每天投身于工作和家庭领域并进行着跨边界的角色转变活动 (Ashforth, Kreiner, & Fugate, 2000; Clark, 2000; Kreiner, 2006)。不同的个体，由于价值观、对工作和家庭所

---

* 国家自然科学基金青年项目（31200795）、国家自然科学基金重大研究项目（91324201）、华中师范大学中央高校基本科研业务费专项资金科研项目（CCNU13A05047）、华中师范大学优秀博士学位论文培育计划资助项目（2013YBYB12）资助

通讯作者：马红宇，E-mail: mahy@mail.ccnu.edu.cn 申传刚，E-mail: psychshen@126.com

持有的分割 - 整合偏好不同，在面对另一领域的角色需求时，个体是否愿意进行相应角色转变的意愿程度有所不同，这种是否愿意从一个角色向另外一个角色转变的程度称之为边界弹性意愿，包括工作弹性意愿 (work flexibility-willingness, WFW) 和家庭弹性意愿 (family flexibility-willingness, FFW) (Matthews & Barnes-Farrell, 2010; Matthews, Bernes-Farrell, & Bulger, 2010)。另一方面，由于组织政策、家庭责任、管理者和家人支持等外部环境因素的影响，环境对其进行角色转换所提供条件或资源也不相同 (Ashforth et al., 2000; Clark, 2000; Kossek, Lautsch, & Eaton, 2004)。研究者将个体对其能否离开所在领域去满足另外一个领域需求的外部环境特征的认知性评估称之为边界弹性能力，包含着工作弹性能力 (work flexibility-ability, WFA) 和家庭弹性能力 (family flexibility-ability, FFA) (Matthews & Barnes-Farrell, 2010; Matthews et al., 2010)。当个体身处一个领域，而另一领域有需求需要其进行角色转变时，除了会受到组织政策、家庭责任等环境因素的影响之外，还会受到个人的角色转变意愿的影响。这提示我们，个体对其所处的环境所能提供的角色转变的条件或资源的认知性评估与个体进行角色转变的意愿之间的匹配可能是影响个体工作 - 家庭结果变量的重要决定性因素。因此，运用人 - 环境匹配 (person-environment fit, PE fit) (Edwards, 2008; Kristof-Brown, Zimmerman, & Johnson, 2005) 理论来分析边界弹性意愿 (个人需求) 和边界弹性能力 (环境资源) 之间的匹配对工作 - 家庭冲突和工作 - 家庭增益的影响将增进我们对工作 - 家庭界面的理解。

尽管工作 - 家庭界面的研究已经相当丰富，但是以往的研究仍然存在一些不足。首先，尽管人 - 环境匹配一直都是学者们关注的核心理论构念 (Edwards, 2008)，但是纵观国内外的研究，多数研究仍聚焦在情境需求对工作 - 家庭界面的影响上 (Allen et al., 2012; Ford, Heinen, & Langkamer, 2007; Greenhaus, Ziegert, & Allen, 2012; McNall, Masuda, & Nicklin, 2010; Odle-Dusseau, Britt, & Greene-Shortridge, 2012; Straub, 2012)，只有少数研究运用人 - 环境匹配的构念探讨组织环

境和个人需求之间的匹配对工作 - 家庭界面重要结果变量的影响，但是这些研究却忽视了家庭作为另一类重要的环境变量与个人需求之间的匹配 (Chen, Powell, & Greenhaus, 2009; Kreiner, 2006; 马丽，徐枞巍，2011)；其次，尽管个体可能同时体验到工作 - 家庭冲突和工作 - 家庭增益，但是多数研究仍然将冲突和增益作为独立的现象来进行分析，少有研究将它们放在同一框架下进行探讨 (Maertz & Boyar, 2011; Powell & Greenhaus, 2006)。因此，本研究的目的是为了针对上述不足，拟以人 - 环境匹配为基础，以边界弹性为切入点，检验工作和家庭领域的边界弹性能力和个体的边界弹性意愿之间的匹配对工作 - 家庭冲突和工作 - 家庭增益的影响，以期更为全面地揭示冲突和增益发生的机制，为工作 - 家庭界面的组织实践和个人应对提供更为有效的意见和建议。

2　假设提出

2.1　弹性能力和弹性意愿之间匹配和工作 - 家庭冲突之间的关系

工作 - 家庭冲突指来自工作与家庭双方的需要在某些方面出现难以调和的矛盾时，个体所经历的一种角色交互冲突，包含着工作→家庭冲突 (work-family conflict, WFC) 和家庭→工作冲突 (family-work conflict, FWC) (Greenhaus & Beutell, 1985)。工作和家庭领域的相关变量对两种不同方向冲突的产生有着不同的效应量。研究表明工作领域的变量与工作→家庭冲突有着更强的相关，而家庭领域的变量与家庭→工作冲突有着更强的相关 (Byron, 2005; Eby et al., 2005)。尽管有元分析研究表明工作领域的变量与家庭→工作冲突有显著的相关以及家庭领域的变量与工作→家庭冲突有显著的相关，但是这些效应很微弱，工作领域的变量与家庭→工作冲突的相关在 -0.04~0.10 之间，家庭领域的变量与工作→家庭的相关在 -0.04~0.14 之间 (Michel, Mitchelson, Kotrba, LeBreton, & Baltes, 2009; Ng & Feldman, 2008)，所以本研究只分析工作弹性能力与工作弹性意愿之间的匹配对工作→家庭冲

突的影响，以及家庭弹性能力与家庭弹性意愿之间的匹配对家庭→工作冲突的影响。

人－环境匹配理论认为，当环境不能提供个体需求的条件或资源时，其需求将得不到满足，进而会导致紧张、消极情绪的产生以及冲突的体验（Edwards, 2008; Edwards & Rothbard, 1999; Jansen & Kristof-Brown, 2006)。基于此，在工作－家庭领域，一个工作边界弹性意愿高的个体，可能希望组织能够提供相应的资源（如弹性工作时间或弹性工作地点）使其能迅速进行角色的转变，当个体知觉到组织提供的资源使其能够从认知或行为上进行角色转变时，就会与个体转变角色的意愿相符，从而使得个体验到较少的工作→家庭冲突；相反，当个体知觉到组织提供的资源不能使其顺利进行角色转变去满足家庭领域的需求时，即工作弹性能力低于其工作弹性意愿时，个体会体验到较高的工作→家庭冲突。另外，在实际的情境中，还可能会出现此种情况，即一个以工作为中心的个体，其工作弹性意愿可能会很低，倾向于工作时专注于工作任务，尽可能避免工作角色的中断，而此时若组织强制执行弹性工作政策（弹性工作时间或弹性工作地点），反而可能使其面临着更多的角色转变，进而引起紧张、体验到较高的冲突。这一点已经得到了相关研究的支持。已有研究发现对那些希望保持工作和家庭高分割的个体而言，当组织提供更多的整合策略时（如日托中心，弹性工作时间、地点等），他们会有着更为频繁的角色转变，会导致他们的工作和家庭角色模糊，进而使得他们有着更低的满意感、组织承诺和更高的工作→家庭冲突（Desrochers, Hilton, & Larwood, 2005; Rothbard, Phillips, & Dumans, 2005)。所以，当个体知觉到工作边界弹性能力高于工作弹性意愿或低于工作弹性意愿时，均会存在一定的冲突，只有当个体知觉到其工作弹性能力与其工作弹性意愿相匹配时，工作→家庭冲突的体验才最低。同理，对个体的家庭弹性意愿而言，如果个体知觉到家庭弹性能力与其家庭弹性意愿正好匹配时，个体会有较低的家庭→工作冲突体验，而知觉到家庭弹性能力高于

或低于家庭弹性意愿时，个体均会有着较高的家庭→工作冲突体验。因此，我们提出以下假设：

H1：工作弹性能力和工作弹性意愿之间的匹配负向影响工作→家庭冲突。即当工作弹性能力朝向工作弹性意愿逐渐增加的时候，个体所体验到的工作→家庭冲突会逐渐降低，但是当工作弹性能力超过工作弹性意愿继续增加的时候，个体所体验到的工作→家庭冲突反而会逐渐上升。

H2：家庭弹性能力与家庭弹性意愿之间的匹配负向影响家庭→工作冲突。即当家庭弹性能力朝向家庭弹性意愿逐渐增加的时候，个体所体验到的家庭→工作冲突将会降低，但是当家庭弹性能力增加超过了家庭弹性意愿继续增加的时候，个体所体验到的家庭→工作冲突反而会逐渐上升。

2.2 弹性能力和弹性意愿之间的匹配和工作－家庭增益之间的关系

关于工作－家庭冲突产生的原因，研究者多用稀缺假说来解释，认为工作－家庭冲突的产生是由于个体的时间和精力等资源的有限而相互竞争所导致的（Rothbard, 2001)。然而也有学者认为多重角色除了可能会导致角色紧张外，还可能给个体提供获得满意感和自我增益（self-enrichment）的机会（Siber, 1974)。在积极心理学兴起之后，这一观点得以继续发展，研究者们认为个体在某一角色表现中所获得的收益可能大于其角色投入所造成的损失，一个角色所产生的资源（如知识，技能和能力的发展、社会资本、弹性等）会增加另外一个角色的表现，即工作－家庭增益（Greenhaus & Powell, 2006; Carlson, Kacmar, Wayne, & Grzywacz, 2006)。与工作－家庭冲突类似，工作－家庭增益也包括工作→家庭增益（work-family enrichment, WFE）和家庭→工作增益（family-work enrichment, FWE），并且工作→家庭增益与工作相关的变量有着更强的相关，家庭→工作增益与家庭领域的变量有着更强的相关（McNall, Nicklin, & Masuda, 2010)。所以，与工作－家庭冲突的分析类似，本研究只分析工作弹性能力与工作弹性意愿之间的匹配对工作→家庭增益的影响，以及

家庭弹性能力与家庭弹性意愿之间的匹配对家庭→工作增益的影响。

人–环境匹配理论认为，环境的供给或奖赏与个体的需求之间的匹配是以个体的需求是否得到满足为基础的，个体需求的满足影响着个体的态度和行为，当环境的供给与个人的需求之间达到匹配时，会增加个体的满意感、幸福感等积极的体验 (Edwards, 2008; Jansen & Kristof-Brown, 2006)。当个体知觉到组织环境所提供的资源正好与其角色转变意愿相一致时，个体可能会体验到工作满足感等积极情绪，而这种积极情绪可能会使个体更愿意履行家庭角色，进而增加家庭领域的积极体验。当个体知觉到组织环境提供的资源不能使其顺利进行角色转变时，即工作弹性能力低于其工作弹性意愿时，个体的满足感等积极情绪会减少，冲突、紧张等消极情绪会增加，进而不利于家庭角色的履行；同样，当个体知觉到组织所提供的条件或资源使得其工作和家庭领域的整合超出了其期望时，就会导致个体的工作满意感和组织承诺的降低 (Rothbard et al., 2005)，也不利于个体家庭角色的履行。同理，当个体知觉到家庭弹性能力和其家庭弹性意愿一致时，个体同样会体验到幸福、满足等积极情绪，这种积极情绪同样可能会使得个体有着更多的精力和更积极的情绪去履行工作角色，进而增加工作领域的积极体验，使得个体体验到较高家庭→工作的增益，当家庭弹性能力高于或低于家庭弹性意愿时，个体会体验到较低的家庭→工作增益。基于上述分析，我们提出以下假设：

H3：工作弹性能力和工作弹性意愿之间的匹配正向影响工作→家庭增益，即当工作弹性能力朝向工作弹性意愿逐渐增加的时候，个体所体验到的工作→家庭增益会逐渐上升，但是当工作弹性能力超过工作弹性意愿继续增加的时候，工作→家庭增益反而会逐渐降低

H4：家庭弹性能力与家庭弹性意愿之间的匹配正向影响家庭→工作增益。即当家庭弹性能力朝向家庭弹性意愿逐渐增加的时候，个体所体验到的家庭→工作增益将会逐渐上升，但是当家庭弹性能力超过

了家庭弹性意愿继续增加的时候，家庭→工作增益反而会逐渐降低

3　研究方法

3.1　研究对象

本研究采用问卷调查法获取研究数据，对上海某高校的 MBA 已婚学员及湖北、河南、上海、江苏、江西等多家企事业单位的已婚员工发放了问卷。参与调查的员工所在的企业涉及制造、IT、金融、电子等多个行业。问卷采用被调查者自评的方式，总共发放 525 份问卷，在删除无效问卷之后，最终回收有效问卷 494 份，有效回收率为 94.1%。其中男性 261 人，女性 230 人，被试年龄介于 24~53 岁之间，平均年龄 35.39 岁，样本中 348 人有 18 岁以下的小孩需要照顾。

3.2　研究工具

3.2.1　工作家庭边界弹性量表

结合中国的实际情况，对 Matthews 和 Barnes-Farrell (2010) 编制的工作家庭边界弹性量表进行了修订，形成了符合我国实际情况的工作家庭边界弹性量表，共计 16 个条目。该量表包含着工作弹性能力，工作弹性意愿，家庭弹性能力和家庭弹性意愿 4 个维度。问卷采用 Likert 5 点设计，1 表示完全不符合，5 表示完全符合。本研究中 4 个维度的内部一致性系数分别为 0.82, 0.78, 0.74, 0.77, 整个问卷的内部一致性系数为 0.82。对本研究中的工作家庭边界弹性量表进行验证性因素分析，结果表明 χ^2/df=3.23<4, GFI=0.92, CFI=0.92, IFI=0.92, NFI=0.88, RSMEA=0.067, 说明工作家庭边界弹性量表具有良好的结构效度。

3.2.2　工作–家庭冲突量表

采用 Netemeyer 和 Boles (1996) 编制的工作–家庭冲突问卷。该量表包含工作→家庭冲突和家庭→工作冲突两个维度，每个维度 5 个条目。问卷采用 Likert 5 点设计，1 表示完全不符合，5 表示完全符合。该问卷的有效性在国内也得到了验证 (李永鑫，赵娜，2009)。本研究中两个维度的内部一致性系数分别为 0.87, 0.87, 整个问卷的内部一致性系数为 0.89。

3.2.3 工作 - 家庭增益量表

采用 Wayne, Musisca 和 Fleeson (2004) 编制的工作家庭增益量表。该量表包含工作→家庭增益和家庭→工作增益两个维度，每个维度 4 个条目。问卷采用 Likert 5 点设计，1 表示完全不符合，5 表示完全符合。该问卷的有效性在国内也得到了验证（陈恒盼，2008）。本研究中两个维度的内部一致性系数分别为0.72, 0.77, 整个问卷的内部一致性系数为 0.74。

3.2.4 控制变量的测量

本研究中，我们将年龄、工龄、工作时间、性别、是否有孩子和配偶的工作状态作为控制变量。其中年龄、工龄、工作时间等是连续测量，男性赋值为 1, 女性赋值为 2; 没有小孩赋值为 0, 有小孩赋值为 1; 配偶没有全职工作赋值为 0, 有工作赋值为 1。

3.3 统计分析

本研究采用 SPSS 18.0 统计软件包和 AMOS 18.0统计软件包进行统计分析。首先使用验证性因素分析来验证问卷的效度，然后运用 SPSS 18.0 统计软件

包进行描述性统计，并分析弹性能力与弹性意愿之间的匹配对工作 - 家庭结果变量的影响。

4 研究结果

4.1 验证性因素分析及共同方法偏差检验

本研究使用 AMOS 18.0 软件对工作弹性能力、工作弹性意愿、家庭弹性能力、家庭弹性意愿、工作→家庭冲突、家庭→工作冲突、工作→家庭增益、家庭→工作增益 8 个构念进行验证性因素分析，并将拟合指数与另外几个模型进行比较。验证性因素分析的具体结果见表 1。研究结果表明，八因子模型各拟合指标均达到了推荐的标准（温忠麟，侯杰泰，马什赫伯特，2004), 且明显的优于其他备选模型，证明了这 8 个变量确实是 8 个不同的构念。同时，本研究采用了 Harman 的单因素因子分析来进一步的检验了共同方法偏差，结果表明未旋转时，共生成 8 个因子，解释了 64.83% 的变异，第一个因子解释了 18.16%

表 1　验证性因子分析结果 (*N*=494)

模型	χ^2	*df*	χ^2/df	RMSEA	GFI	CFI	IFI	NFI
单因素模型	5679.73	527	10.59	0.141	0.45	0.29	0.30	0.28
四因素模型 1	3545.21	521	6.80	0.11	0.60	0.59	0.59	0.55
四因素模型 2	3047.88	521	5.85	0.10	0.66	0.65	0.65	0.61
六因素模型 1	2311.88	512	4.52	0.09	0.71	0.75	0.75	0.71
六因素模型 2	2022.56	512	3.95	0.08	0.78	0.79	0.79	0.74
八因素模型	1276.66	499	2.56	0.056	0.86	0.89	0.89	0.84

注：单因素模型：指所有项目负荷在一个因子上；四因素模型 1：工作弹性能力 + 家庭弹性能力、工作弹性意愿 + 家庭弹性意愿、工作 - 家庭冲突、工作 - 家庭增益；四因素模型 2：工作弹性能力 + 工作弹性意愿、家庭弹性能力 + 家庭弹性意愿、工作 - 家庭冲突、工作 - 家庭增益；六因素模型 1：工作弹性能力 + 家庭弹性能力、工作弹性意愿 + 家庭弹性意愿、工作→家庭冲突、家庭→家庭冲突、工作→家庭增益、家庭→工作增益；六因素模型 2：工作弹性能力 + 工作弹性意愿、家庭弹性能力 + 家庭弹性意愿、工作→家庭冲突、家庭→家庭冲突、工作→家庭增益、家庭→工作增益；八因素模型：项目负荷在各自的理论维度上

表 2　各研究变量的平均数、标准差与相关矩阵 (*N*=494)

变量	*M*	*SD*	1	2	3	4	5	6	7	8
1.WFA	3.39	0.82	1.00							
2.WFW	3.16	0.75	0.45**	1.00						
3.FFA	3.83	0.69	0.36**	0.08	1.00					
4.FFW	3.61	0.76	0.17**	0.12*	0.51**	1.00				
5.WFC	2.68	0.88	−0.23**	0.05	−0.09*	0.14*	1.00			
6.FWC	2.20	0.81	−0.13**	0.18**	−0.29**	−0.07	0.55**	1.00		
7.WFE	3.18	0.73	0.16**	0.12*	0.23**	0.19**	−0.04	0.03	1.00	
8.FWE	3.87	0.61	0.15**	0.14**	0.36**	0.31**	−0.01	−0.22**	0.28**	1.00

注：** *p*<0.01, * *p*<0.05 其中 WFA 代表工作弹性能力，WFW 代表工作弹性意愿，FFA 代表家庭弹性能力，FFW 代表家庭弹性意愿，WFC 代表工作→家庭冲突，FWC 代表家庭→工作冲突，WFE 代表工作→家庭增益，FWE 代表家庭→工作增益

的方差变异，远小于 Harrison, McLaughlin 和 Coalter (1996) 推荐的 50% 的判断标准，表明本研究中共同方法偏差并不严重。

4.2 描述性统计分析

表 2 呈现了本研究中所有变量的均值、标准差及变量之间相关系数的结果。从表 2 可以得知工作弹性能力与工作→家庭冲突呈显著的负相关，与工作→家庭增益呈显著的正相关；家庭弹性能力与家庭→工作冲突呈显著的负相关，与家庭→工作增益呈显著的正相关，家庭弹性意愿与家庭→工作增益呈显著正相关。

4.3 假设检验

以往关于人 – 环境匹配的研究中，关于两个构念之间匹配程度的衡量主要采用差异分数 (difference scores)，即通过个体在同一属性量表分别评估自己的愿望 (或能力) 和环境的回报 (或要求)，然后计算两个分数的差值，具体的形式有代数差 (X–Y)、差的绝对值 (｜ X–Y ｜)、差的平方 (X–Y)^2。但是它们只能反映匹配的整体特征，无法区分超过或不及情况

下对结果变量的影响，并且存在着容易丢失信息、对差异性来源不敏感、无法反映单个变量的贡献率等缺点。相关研究者指出人、环境和结果变量之间的关系应被看作 3 个方面，提出应该使用多项式回归 (polynomial regression) 来产生一个三维反应面来代表 P 和 E 的整合关系对结果变量的影响 (Edward & Parry, 1993; Edward & Harrison, 1993; Edward & Rothbard, 1999, 2000)。基于此，我们沿用了 Edwards 所提出来的多项式回归来分析边界弹性能力与边界弹性意愿之间的匹配对工作家庭界面的影响。具体而言，每一个结果变量分别对弹性能力 (A)、个体的弹性意愿 (W)、能力和意愿的交互 (A*W)、弹性能力的平方 (A^2)、弹性意愿的平方 (W^2) 进行回归，回归方程中的 5 个项目即包含了变量线性匹配对结果变量的影响，同时也包含着曲线匹配效应。具体的表达方程式如下：$Z = b_0 + b_1X + b_2Y + b_3Y^2 + b_4XY + b_5X^2 + e$。其中 Z 代表工作→家庭冲突，家庭→工作冲突，工作→家庭增益和家庭→工作增益；X 代表弹性意愿分数；Y 代表弹性能力分数。为了避免高的共线

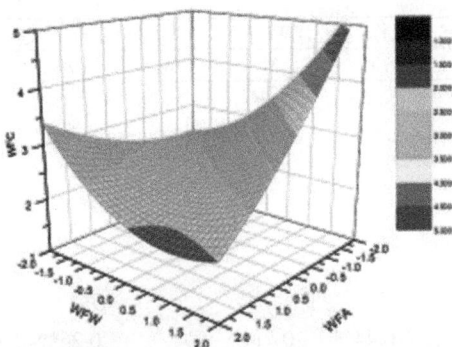

图 1　WFA 与 WFW 匹配对 WFC 的影响

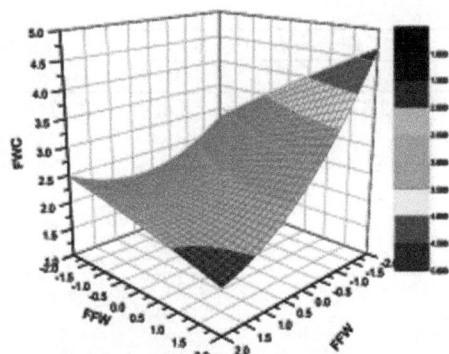

图 2　FFA 与 FFW 匹配对 FWC 的影响

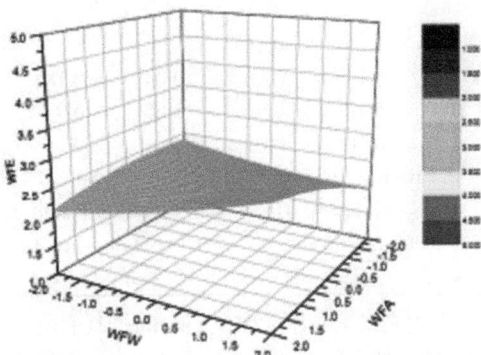

图 3　WFA – WFW 匹配对 WFE 的影响

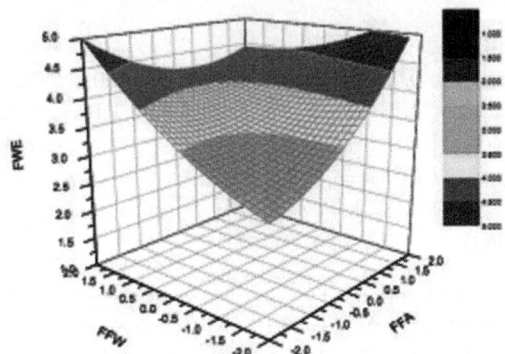

图 4　FFA 与 FFW 匹配对 FWE 的影响

性，我们将所有自变量进行了中心化处理。

在具体的分析过程中，我们均采用了多层回归的方法，将控制变量放在第一层 (M_1)；将边界弹性能力和边界弹性意愿放在第二层，检验边界弹性能力和边界弹性意愿的主效应 (M_2)；将边界弹性能力的平方、边界弹性意愿的平方、以及弹性意愿和弹性能力的交互项纳入到回归方程第三层 (M_3)。在纳入了变量的高阶项之后，R^2 显著地增加，或者方程中的高阶项系数显著，即说明匹配和结果变量之间存在着非线性关系，此时，相关的具体结果可以通过反应曲面方法 (response surface methodology) 呈现出三维图形。在三维图形中，在"弹性意愿 = 弹性能力"对角线曲面的形状表明当个体的弹性能力与其弹性意愿相匹配时结果变量的情形。"弹性意愿 = − 弹性能力"对角线的曲面的形状表明当个体的弹性能力与弹性意愿不匹配时结果变量的变化情况，而所有的假设均是通过此曲面的发展趋势来进行检验的。先计算 $x_1 = b_1-b_2$，$x_2 = b_3-b_4+b_5$，进而对 x_1 和 x_2 进行显著性检验，如果 x_1 显著，则表明沿着"弹性意愿 = − 弹性能力"对角线方向有线性关系，如果 x_2 显著，则表明沿着"弹性意愿 = − 弹性能力"对角线方向有一个曲面 (Edwars & Parry, 1993)，此时可以通过计算曲面的第一主轴和第二主轴来判断曲面的发展趋势。同时，本研究还计算出 $a_1 = b_1+b_2$，$a_2 = b_3+b_4+b_5$ 来进一步的分析"弹性意愿 = 弹性能力"对角线曲面的发展趋势。当 $a_1>0$ 且显著时表明高弹性意愿和高弹性能力的匹配比低弹性意愿和低弹性能力的匹配有着更为积极的结果，当 $a_1<0$ 且显著时，表明高弹性意愿和高弹性能力的匹配比低弹性意愿和低弹性能力的匹配有着更为消极的结果。当 a_2 显著时，表明"弹性意愿 = 弹性能力"对角线方向有一个曲面。

本研究详细的结果见表 3，除了工作→家庭增益的二次项回归方程中的二次项系数没有达到显著性以外，其他的回归方程中均有二次项系数达到显著。我们对表 2 中的方程运用 origin 8.0 绘图软件绘制出了三维图。具体图形见图 1、图 2、图 3 和图 4。在

表 3　二次项回归方程结果 ($N=494$)

变量	工作→家庭冲突 (WFC)			工作→家庭增益 (WFE)			家庭→工作冲突 (FWC)			家庭→工作增益 (FWE)		
	M_1	M_2	M_3	M_1	M_2	M_3	M_1	M_2	M_3	M_1	M_2	M_3
常数项	2.37	2.37	2.42	2.47	2.42	2.40	2.73	2.39	2.46	3.25	3.62	3.66
控制变量	--	--	--	--	--	--	--	--	--	--	--	--
WFA		−0.35**	−0.34**		0.11*	0.10*						
WFW		0.23**	0.23**		0.05	0.06						
WFA²			0.07			−0.03						
WFW²			0.17**			0.02						
WFA*WFW			−0.29**			0.08						
FFA								−0.44**	−0.41**		0.26**	0.26**
FFW								0.14*	0.14*		0.12**	0.15**
FFA²									0.13*			0.12*
FFW²									−0.01			0.07+
FFA*FFW									−0.16*			−0.20**
R^2	0.06	0.14	0.18	0.03	0.05	0.06	0.05	0.15	0.16	0.04	0.18	0.21
F	2.18	5.11	5.44	1.16	1.78	1.62	1.86	5.42	4.91	1.66	6.97	6.80
△R^2		0.08	0.04		0.02	0.01		0.10	0.01		0.14	0.03
A=W 对角线	$a_1=b_1+b_2=-0.11$			$a_1=b_1+b_2=0.16*$			$a_1=b_1+b_1=-0.27**$			$a_1=b_1+b_2=0.41**$		
	$a_2=b_3+b_4+b_5=-0.05$			$a_2=b_3+b_4+b_5=0.07$			$a_2=b_3+b_4+b_5=-0.04$			$a_2=b_3+b_4+b_5=-0.01$		
A=−W 对角线	$x_1=b_1-b_2=0.57**$			$x_1=b_1-b_2=-0.04$			$x_1=b_1-b_2=0.55**$			$x_1=b_1-b_2=-0.11*$		
	$x_2=b_3-b_4+b_5=0.53**$			$x_2=b_3-b_4+b_5=-0.09$			$x_2=b_3-b_4+b_5=0.28**$			$x_2=b_3-b_4+b_5=0.39**$		

注：** $p<0.01$，* $p<0.05$+ $p<0.10$，表中 a1 和 a2 代表着"弹性能力 = 弹性意愿"对角线的斜率和曲率，x1 和 x2 代表着"弹性能力 = − 弹性意愿"的对角线的斜率和曲率，b1 代表弹性意愿的系数；b2 代表着弹性能力的系数，b3 代表这弹性能力平方的系数，b4 代表着弹性能力 * 弹性意愿的系数，b5 代表着弹性意愿平方的系数

图形中，在弹性能力＝弹性意愿对角线的右边区域表示弹性能力低于弹性意愿，在对角线的左侧则表示弹性能力高于弹性意愿的情况。

4.3.1 假设 1 的检验

根据表 3 中所呈现的工作→家庭冲突的二次项回归方程，以及图 1 所呈现的三维反应曲面，我们发现当工作弹性能力朝向工作弹性意愿逐渐增加时（从 WFW=−WFA 对角线最右端向最左端移动的时候的反应曲面），工作→家庭冲突不断下降；当工作弹性能力超过了工作弹性意愿继续增加时（即从反面曲面的中心移向最左端时），工作→家庭冲突呈上升的趋势。斜率的分析证实了图 1 的结果，在 "WFW= −WFA" 对角线的形状有一个显著的曲线形状 (x_2=b_3−b_4+b_5= 0.53, p<0.01)，第一主轴的斜率为 −1.4，第二主轴的斜率为 0.72。进一步的分析发现沿着 "WFW=WFA" 对角线的趋于一条直线，斜率分析证实了这一结果 (a_1= b_1+b_2 =−0.11, p>0.10; a_2= b_3+b_4+b_5 =−0.05, p>0.10)，说明高工作弹性能力 (WFA) 和高工作弹性意愿 (WFW) 匹配情况下的工作→家庭冲突与低工作弹性能力和低工作弹性意愿匹配情况下的工作→家庭冲突没有显著的差异。假设 1 得到验证。

4.3.2 假设 2 的检验

根据表 3 中所呈现的家庭→工作冲突的二次项回归方程，以及图 2 所呈现的反应曲面，我们可以发现当家庭弹性能力朝向家庭弹性意愿增加时（从 FFW=−FFA 对角线最右端向最左端移动的反应曲面），家庭→工作冲突逐渐降低；当家庭弹性能力超过了家庭弹性意愿继续增加时（即从反应曲面的中心移向最左端的时候），家庭→工作冲突有一个缓慢上升的趋势。斜率的分析证实了此结果，在 "FFW=−FFA" 对角线的形状有一个显著的曲线形状 (x_2=b_3−b_4+b_5= 0.28, p<0.01)，第一主轴的斜率为 −0.44，第二主轴的斜率为 2.18。假设 2 得到验证。进一步的分析发现沿着 "FFW=FFA" 的对角线上，家庭→工作冲突的发展趋势是一条直线 (a_1=b_1+b_2=−0.27, p<0.01; a_2= b_3+b_4+b_5=−0.04, p>0.10)，且高家庭弹性意愿和高家庭弹性能力匹配情境下家庭→工作冲突比低家庭弹性意愿和低家庭弹性能力匹配情境下的家庭→工作冲突更低。

4.3.3 假设 3 的检验

根据表 3 中所呈现的工作→家庭增益的二次项回归方程，我们发现二次项的系数均不显著，且 x_1=b_1−b_2=0.04, p>0.05; x_2= b_3−b_4+b_5=0.09, p>0.05，表明沿着 "弹性意愿 ＝ 弹性能力" 对角线，整个反应曲面基本上一条直线，此时，工作弹性能力的系数显著，说明工作弹性能力对工作→家庭增益的主效应显著，即随着工作弹性能力的增加，工作→家庭增益也随之增加，这一点在图 3 中也得到证实，如图 3 所示，工作→家庭增益的反应面基本上是一个平面。假设 3 并未得到验证。

4.3.4 假设 4 的检验

根据表 3 中所呈现的家庭→工作增益的二次项回归方程，以及图 4 所呈现的反应曲面，我们发现当家庭弹性能力朝向家庭弹性意愿增加时（从 FFW=−FFA 对角线最右端向最左端移动的反应曲面），此时我们发现家庭→工作增益并非如果我们所期望的那样逐渐增加，而是逐渐降低，当家庭弹性能力超过了家庭弹性意愿继续增加时（即从反应曲面的中心移向最左端时），家庭→工作增益 (FWE) 同样也并非如我们所期望的那样逐渐降低，反而逐渐上升。假设 4 未得到验证。进一步的分析发现沿着 "FFW=FFA" 的对角线上，家庭→工作增益的发展趋势是一条直线 (a_1=b_1+b_2=0.41, p<0.01; a_2=b_3+b_4+b_5=−0.01, p>0.10)，且说明高家庭弹性意愿和高家庭弹性能力匹配情境下的家庭→工作增益比低家庭弹性意愿和低家庭弹性能力匹配情境下的家庭→工作增益更高。

5　讨论

5.1　边界弹性能力和边界弹性意愿的匹配对工作 – 家庭冲突的影响

本研究的结果表明工作（或家庭）弹性能力和工作（或家庭）弹性意愿之间的匹配有助于降低员工的工作→家庭冲突（家庭→工作冲突）。从图 1 和图 2

中我们可以看出，当工作（或家庭）弹性意愿很高，而其工作（或家庭）弹性能力很低的时候，此时个体会体验到最为强烈的工作→家庭冲突（家庭→工作冲突）；而当个体的弹性能力趋近其弹性意愿时，工作→家庭冲突（家庭→工作冲突）呈下降的趋势，当弹性能力超过弹性意愿进一步增加时，其冲突的体验又会有所上升。此结果表明，双向的工作－家庭冲突不仅受外部环境资源的影响，还受到个体特征的影响，并且它们还交互作用共同决定了冲突的发生，因此，这一结果比较直观地揭示了工作－家庭冲突的发生机制。具体来说，当个体某一领域的弹性意愿很高，而个体知觉到其所在的环境不能提供进行相应的角色转变的资源时，如组织政策不允许，领导不支持或家人反对等，就会导致个体心理紧张以及冲突体验的产生；而当个体在某一领域的弹性意愿很低，此时个体可能倾向于在此领域将工作和家庭分割开来，不愿意将工作和家庭角色进行高度整合，而此时若个体知觉到此领域有着较高的弹性能力，即环境提供的资源会促进领域整合时，可能会导致个体产生较高的角色模糊，进而使个体工作－家庭冲突的体验增强（Glavin & Schieman, 2012）。

5.2 边界弹性能力和边界弹性意愿的匹配对工作－家庭增益的影响

本研究结果表明工作弹性能力和工作弹性意愿之间的匹配对工作→家庭增益的作用不显著，只有工作弹性能力能够显著预测个体的工作→家庭增益；而家庭弹性能力和家庭弹性意愿的匹配对家庭→工作增益的影响与假设完全相反，即当家庭弹性能力逐渐向家庭弹性意愿增加的时候，家庭→工作增益呈下降趋势，当家庭弹性能力超过家庭弹性意愿继续增加时，家庭→工作增益逐渐上升，在家庭弹性能力和家庭弹性意愿匹配的情况下，只有当个体的家庭弹性能力和意愿均很低的时候，家庭→工作增益最低。通过图4不难发现，只有当个体的家庭弹性能力和家庭弹性意愿均很低的时候，家庭→工作增益最低，随着家庭弹性能力或家庭弹性意愿单方面增加或两者同时增加时，家庭→工作增益也随之增加。这说明了家庭

弹性能力和家庭弹性意愿均能显著地增加个体的家庭→工作增益。究其原因，可能是在我们国家人们普遍持有一种工作优先的行为规范（张勉，李海，魏钧，杨百寅，2011），认为工作不仅是个人的事情，而是提升家庭整体利益和荣耀的手段，勤奋工作是一种有家庭责任感的表现（Aryee, Field, & Luk, 1999; Wang, Lawler, Walumbwa, & Kan, 2004; 李晔，2003）。个体无论在认知上，还是在行为上均可能聚焦在如何更好地履行工作角色上，即使组织提供再多的资源让其能够更容易转变工作角色，个体通常也不太关注家庭角色的履行，因此工作→家庭增益变化不大。反而，只要个体持工作优先规范的原则，或家人给予支持能够让其顺利转变家庭角色去更好地履行工作职责时，均会出现较高的家庭→工作增益，而只有个体在认知和行为上均不以工作优先，且没有足够的资源去转变角色时，才会出现较低的家庭→工作增益。为了更清楚地揭示工作－家庭增益的发生机制，将来的研究应充分考虑我国工作优先的规范对员工工作－家庭增益的影响，可以通过将深度访谈等质化研究的方法和问卷调查等量化研究方法进行有机的结合，对上述的推论进行进一步的验证。

5.3 理论和实践意义

本研究的理论贡献主要表现为：(1) 从人－环境匹配的视角来探索了工作和家庭领域的边界弹性能力和弹性意愿对工作－家庭界面结果变量的交互影响，弥补了以往研究多关注情境需求而忽视个体特征同样在其工作－家庭管理中发挥着重要作用的不足；而且，同时考虑工作和家庭环境与个人需求的匹配对双向的工作－家庭冲突和增益的影响，发现了个体在工作家庭领域的弹性意愿和其知觉到的弹性能力对家庭→工作冲突和增益的交互影响，扩充了以往研究仅关注工作环境和个人需求之间匹配的研究模式，并通过多项式回归和三维反应曲面将工作－家庭冲突和增益的发展趋势清晰地展示出来，证实了个体在实际的生活中，工作－家庭冲突和工作家庭增益同时存在的假设。通过比较图1和图3，当工作弹性能力朝向工作弹性意愿逐渐增加并进一步超过工作弹性意

愿时，工作→家庭冲突先下降而后呈逐渐上升趋势，而工作→家庭增益的变化趋势却始终不显著；通过比较图 2 和图 4，当家庭弹性能力朝向家庭弹性意愿逐渐增加并进一步超过家庭弹性意愿时，家庭→工作冲突和家庭→工作增益有着同样的变化趋势，均为先逐渐降低而后逐渐升高。以上结果表明，对低家庭弹性能力和高家庭弹性意愿的个体而言，同时有着高的家庭→工作冲突和家庭→工作增益，这一发现为我们更清晰地理解工作－家庭平衡提供了帮助；(2) 本研究为前期研究成果中不一致的结论提供了可能的解释。Kossek 和 Nichol (1992) 研究表明弹性的工作制度、日托中心等家庭友好政策会导致员工工作－家庭冲突显著减少，而 Solomon (1994) 研究却发现家庭友好政策对员工的工作－家庭冲突的影响不大，Desrochers 等 (2005) 研究发现家庭友好政策会增加员工的角色模糊，进而导致员工的工作－家庭冲突增加。基于本文人－环境匹配视角的研究结果，导致上述研究结果不一致的原因可能是组织在实施相关的家庭友好政策时，忽略了在工作－家庭界面中与组织政策共同起作用的个人因素，特别是个体对工作和家庭所持的价值观、态度和期望的影响作用。

本研究结果为组织管理实践提供了一定的指导。研究发现个体工作弹性能力与工作弹性意愿的匹配会有效地降低其工作→家庭冲突的体验，而不匹配则会导致个体工作→家庭冲突，特别是当个体的弹性意愿很高而其感知到的工作弹性能力很低的时候尤为明显。这提示组织在制定工作和家庭的相关政策制度时，要切实考虑到不同员工的实际需求，做到组织提供的资源和员工个人需求相匹配，而不能"一刀切"式地强制执行某一固定的工作－家庭政策。其次，对于组织内部的员工，要积极宣传组织相关的工作－家庭政策，通过社会化的过程改变员工对工作和家庭的偏好，进而改变其意愿。再次，组织在招聘新进员工时，要尽量录用与组织所能提供的工作家庭边界管理文化相匹配的员工。另外，本研究结果也为促进员工的家庭边界管理提供了途径。通过比较图 2 和图 4，我们不难发现当个体在家庭弹性能

力很低，而家庭弹性意愿很高的情况下，个体会同时体验到较高的家庭→工作增益和家庭→工作冲突；但是当家庭弹能力愿很高而家庭弹性意愿很低的情况下，个体所体验到的家庭→工作增益很高、家庭→工作冲突的增幅却很小；并且当家庭弹性能力和家庭弹性意愿均很高时的家庭→工作冲突要显著低于家庭弹性能力和家庭弹性意愿均很低时的家庭→工作冲突，而家庭→工作增益则相反。这说明了家庭弹性能力在缓解家庭→工作冲突和增加家庭→工作增益上具有独立的效能，所以，家庭成员应该尽可能的给员工提供生活上的支持，减少其在家庭中的负担，当工作有需求时，使其能够顺利及时地进行角色上的转变，这样才能够在有效降低其冲突体验的同时，增加其增益的体验。

5.4 局限与展望

本研究的局限主要表现在以下几方面。首先，本研究的各变量均采用员工自评的方式来进行，对于工作→家庭冲突和工作→家庭增益而言，个体的感受可能与家人存在着一定的差异，对家庭→工作冲突和家庭→工作增益而言，个体的感受也可能与上司或同事的感受存在一定的差异，所以在未来的研究中应该采用多源评价的方式来全面地测量个体的边界弹性能力和工作－家庭界面的结果变量。其次，本研究对工作和家庭环境的测量采用了个体弹性能力这一主观的评估，可能存在同一环境中的不同个体对其所处的环境的感知有所不同的情况，所以在将来的研究中应该检验组织的实际工作－家庭政策、主管和家人实际的行为等与个人弹性意愿的匹配对个体工作－家庭界面的影响。再次，本研究采用了横切面研究，不能进行因果关系的推断，而且工作－家庭冲突和工作－家庭增益在实际的生活中并非一成不变，而是随着不同的时间段有着不同的冲突和增益水平 (Maertz, Jr & Boyar, 2011)，所以在将来的研究中，可以采用纵向追踪或者是日志的研究方法对这些变量的因果关系进行进一步的探讨，以更为清晰、全面地揭示工作－家庭冲突和增益的发生机制。

参考文献

Allen, T. D., Johnson, R. C., Saboe, K. N., Cho, E., Dumani, S., & Evans, S. (2012). Dispositional variables and work-family conflict: A meta-analysis. *Journal of Vocational Behavior, 80,* 17-26.

Ashforth, B. E., Kreiner, G. E., & Fugate, M. (2000). All in a day's work: Boundaries and micor role transitions. *Academy of Management Review, 25,* 472-491.

Aryee, S., Field, D., & Luk, V. (1999). A cross-cultural test of a model of the work-family interface. *Journal of Management, 25,* 491-511.

Byron, K. (2005). A meta-analytic review of work-family conflict and its antecedents. *Journal of Vocational Behavior, 67,* 169-198.

Carlson, D. S., Kacmar, K. M., Wayne, J. H., & Grzywacz, J. G. (2006). Measuring the positive side of the work-family interface: Development and validation of a work-family enrichment scale. *Journal of Vocational Behavior, 68,* 131-164.

Chen, H. P. (2008). The relationship among core self-evaluations and emloyee's satisfaction, turnover intentions: The mediating role of work-family facilitation. Unublished master's thesis, Zhejiang University.

[陈恒盼 . (2008). 核心自我评价与员工满意感、离职意向的关系研究：工作家庭促进的中介作用 . 硕士学位论文 , 浙江大学 .]

Chen, Z., Powell, G. N., & Greenhaus, J. H. (2009). Work-to-family conflict, positive spillover, and boundary management: A person environment fit approach. *Journal of Vocational Behavior, 74,* 82-93.

Clark, S. C. (2000). Work/Family border theory: A new theory of work/family balance. *Human Relations, 53,* 747-770.

Desrochers, S., Hilton, J. M., & Larwood, L. (2005). reliminary validation of the work-family integration-blurring scale. *Journal of Family Issues, 26,* 442-466.

Eby, L. T., Caser, W. J., Lockwood, A., Bordeaux, C., & Brinley, A. (2005). Work and family research in IO/OB: Content analysis and review of the literature (1980-2002). *Journal of Vocational Behavior, 66,* 124-197.

Edwards, J. R. (2008). Person environment fit in organizations: An assessment of theoretical progress. *The Academy of Management Annals, 2,* 167-230.

Edwards, J. R., & Harrison, R. V. (1993). Job demands and worker health: Three dimensional reexamination of the relationship between person environment fit and strain. *Journal of Alied sychology, 78,* 628-648.

Edwards, J. R., & Parry, M. E. (1993). On the use of olynomial regression equations as an alternative to difference scores in organizational research. *Academy of Management Journal, 36,* 1577-1613.

Edwards, J. R., & Rothbard, N. P. (1999). Work and family stress and well-being: An examination of person environment fit in the work and family domains. *Organizational Behavior and Human Decision processes, 77,* 85-129.

Edwards, J. R., & Rothbard, N. P. (2000). Mcchanisms linking work and family: Clarifying the relationship between work and family constructs. *Academy of Management Review, 25,* 178-198.

Ford, M. T., Heinen, B. A., & Langkamer, C. L. (2007). Work and family satisfaction and conflict: A meta-analysis of cross domain relations. *Journal of Alied sychology, 92,* 57‐80.

Glavin, P., & Schieman, S. (2012). Work-family role blurring and work-family conflict: The moderating influence of job resources and job demands. *Work and Occu ations, 39,* 71-98.

Greenhaus, J. H., & Beutell, N. J. (1985). Sources of conflict between work and family roles. *Academy of Management Review, 10,* 76-88.

Greenhaus, J. H., & Powell, G. N. (2006). When work and family allies: A theory of work–family enrichment. *Academy of Management Review, 31*, 72–91.

Greenhaus, J. H., Ziegert, J. C., & Allen, T. D. (2012). When family supportive supervision matters: Relations between multiple source of support and work–family balance. *Journal of Vocational Behavior, 80*, 266–275.

Harrison, D. A., McLaughlin, M. E., & Coalter, T. M. (1996). Context, cognition, and common method variance: psychometric and verbal protocol evidence. *Organizational Behavior and Human Decision processes, 68*, 246–261.

Jansen, K. J., & Kristof, B. A. (2006). Toward a multidimensional theory of person environment fit. *Journal of Management Issues, 18*, 193–212.

Kossek, E. E., Lautsch, B., & Eaton, S. C. (2004). Flexibility enactment theory: Relationships between type, boundaries, control and work family effectiveness. In E. E. Kossek & S. J. Lambert (Eds.), *Work and life integration: Organizational, cultural and individual persectives* (pp243–261). Mahwah, NJ: Erlbaum.

Kossek, E. E., & Nichol, V. (1992). The effects of on site child care on employee attitudes and performance. *personnel psychology, 45*, 485–509.

Kreiner, G. E. (2006). Consequences of work home segmentation or integration: A person environment fit perspective. *Journal of Organizational Behavior, 27*, 485–507.

Kristof-Brown, A. L., Zimmerman, R. D., & Johnson, E. C. (2005). Consequences of individual's fit at work: A meta-analysis of personpjob, personporganization, personpgroup, and personpsupervisor fit. *personnel psychology, 58*, 281–342.

Li, Y. (2003). Empirical research of Work–family conflict in China. *Ergonomics, 9*, 14–17.

[李晔. (2003). 工作－家庭冲突的影响因素研究. 人类工效学, 9, 14–17.]

Li, Y. X., & Zhao, N. (2009). Structure and measurement of work–family support and its moderation effect. *Acta psycholigica Sinica, 41*, 863–874.

[李永鑫, 赵娜. (2009). 工作－家庭支持的结构与测量及其调节作用. 心理学报, 41, 863–874.]

Ma, L., & Xu, C. W. (2011). Boundary management and work–family interface: From personpenvironment fit perspective. *Nankai Business Review, 14*, 41–47.

[马丽, 徐枞巍. (2011). 基于个人－环境匹配理论的边界管理与工作家庭界面研究. 南开管理评论, 14, 41–47.]

Maertz, Jr. C. –., & Boyar, S. (2011). Work–family conflict, enrichment, and balance under "levels" and "Episodes" approaches. *Journal of Management, 37*, 68–98.

Matthews, R. A., & Barnes-Farrell, J. L. (2010). Development and initial evaluation of an enhanced measure of boundary flexibility for the work and family domains. *Journal of Occupational Health psychology, 15*, 330–346.

Matthews, R. A., Barnes-Farrell, J. L., & Bulger, C. A. (2010). Advancing measurement of work and family domain boundary characteristics. *Journal of Vocational Behavior, 77*, 447–460.

McNall, L. A., Masuda, A. D., & Nicklin, J. M. (2010). Flexible work arrangements, job satisfaction, and turnover intentions: The mediating role of work–to–family enrichment. *The Journal of psychology, 144*, 61–81.

McNall, L. A., Nicklin, J. M., & Masuda, A. D. (2010). A meta-analytic review of the consequences associated with work–family enrichment. *Journal of Business psychology, 25*, 381–396.

Michel, J. S. Mitchelson, J. K., Kotrba, L. M., LeBreton, J. M., & Baltes, B. B. (2009). A comparative test of workpfamily conflict models and critical examination of work-family linkages. *Journal of Vocational Behavior, 74*, 199-218.

Netemeyer, R. G., & Boles, J. S. (1996). Develo-ment and validation of work-family conflict and family-work conflict scales. *Journal of Applied psychology, 81*, 400-410.

Ng, T., & Feldman, D. (2008). Long work hours: A social identity perspective on meta-analysis data. *Journal of Organizational Behavior, 29*, 853-880.

Odle-Dusseau, H. N., Britt, T. W., & Greene-Shortridge, T. M. (2012). Organizational workpfamily resource as predictors of job performance and attitudes: The process of work pfamily conflict and enrichment. *Journal of Occupational health psychology, 17*, 28-40.

Owell, G. N., & Greenhaus, J. H. (2006). Is the opposite of positive negative? Untangling the complex relationship between work-family enrichment and conflict. *Career Development International, 11*, 650‐659.

Rothbard, N. P. (2001). Enriching or depleting? The dynamics of engagement in work and family roles. Administrative Science Quarterly, 46, 655-684.

Rothbard, N. P., Philips, K. W., & Dumas, T. L. (2005). Managing multiple roles: Work-family policies and individuals' desires for segmentation. *Organization Science, 16*, 243‐258.

Siber, S. D. (1974). Toward a theory of role accumulation. *American Sociological Review, 39*, 567-578.

Solomon, C. (1994). Work/family's failing grade: Why today's initiatives are not enough. *personnel Journal,* 72, 72-87.

Straub, C. (2012). Antecedents and organizational consequences of family supportive supervisor behavior: A multilevel conceptual framework for research. *Human Resource Management Review, 22*, 15‐26.

Wang, P., Lawler, J. J., Walumbwa, F. O., & Kan, S. (2004). Work-family conflict and job withdrawal intentions: The moderating effect of cultureal differences. *International Journal of Stress Management, 11*, 392-414.

Wayne, J. H., Musisca, N., & Fleeson, W. (2004). Considering the role of personality in the work-family experience: Relationships of the big fve to work‐family confict and facilitation. *Journal of Vocational Behavior, 64*, 108-130.

Wen, Z. L., Hau, K. T., & Herbert, W. M. (2004). Structural equation model testing: cutoff criteria for goodness of fit indices and Chi-square test. *Acta sycholigica Sinica, 36*, 186-194.

[温忠麟，侯杰泰，马什赫伯特. (2004). 结构方程模型检验：拟合指数与卡方准则. *心理学报, 36*, 186-194.]

Zhang, M., Li, H., Wei, J., & Yang, B. Y. (2011). Cross-over effects or direct effects? The mechanism linking work-family conflict with outcomes. *Acta sychologica Sinica, 43*, 573-588.

[张勉，李海，魏钧，杨百寅. (2011). 交叉影响还是直接影响？工作‐家庭冲突的影响机制. *心理学报, 43*, 573-588.]

Boundary flexibility and work-family interface: From person-environment fit perspective

MA Hongyu[1,2]; SHEN Chuangang[1,2]; YANG Jing[3,4]; TANG Hanying[1,2]; XIE Julan[1,2]

([1]Key Laboratory of Adolescent Cyberpsychology and Behaivor (CCNU), Ministry of Education, Wuhan 430079, China)

([2]School of Psychology, Central China Normal University, and Hubei Human Development of Mental Health Key Laboratory, Wuhan 430079, China)

([3]Institute of Psychology, Chinese Academy of Sciences, Beijing, 100101, China)

([4]University of Chinese Academy of Sciences, Beijing 100039, China)

Abstract Work and family are two important domains in an individual's life. How to balance work and family domains have become an increasingly compelling and pressing issue for both the organizational scholars interested in theoretical advances, and for human resources practitioners seeking to promote the employee's daily life. Individuals negotiate the boundaries between work and family in their daily activities. There are differences between the individual's preference and the resource that provided by the organization in boundary management of work and family domain. In our study, using a person–environment (PE) fit theoretical base, we explored how the interaction between an individual's boundary flexibility willingness and the perceived flexibility ability of the domain boundary affects work–family conflict and work – family enrichment. Specifically, we predict that the fit of domain boundary flexibility–ability and individual's domain boundary flexibility–willingness would be associated with lower work–family conflict and higher work–family enrichment. Data were collected from a sample of 494 fulltime married employees from different industries. The questionnaire for employee included work–family boundary flexibility scale, work–family conflict and work–family enrichment. Among the major measures, the 16–items boundary flexibility scale was adopted from Matthews and Barnes–Farrell (2010), WFC was measured via 10 items that was adopted from Netemeyer and Boles (1996), the eight item WFE scale was adopted from Wayne, Musisca and Fleeson (2004). Results show that the Cronbach's alpha coefficients for the above measures range from 0.72 to 0.89. Polynomial regression and response surface methodology were utilized to examine the proposed hypotheses. In line with the predictions, results of polynomial regression and response surface methodology demonstrate that work–to–family conflict decreased as work flexibility–ability (WFA) approached work flexibility–willingness (WFW), and increased as WFA exceeded WFW, family–to–work conflict decreased as family flexibility–ability (FFA) approached family flexibility–willingness (FFW), and increased as FFA surpassed FFW. The results also showed that the fit of WFA and WFW has no effect on work–to–family enrichment, and the fit of FFA and FFW has the significant effect on family–to–work enrichment, but it is opposite to the hypotheses. Specifically, family–to–work enrichment decreased as FFA approaching FFW, and increased as FFA exceeded FFW. The present study extends to our understanding the mechanism of the process of the work–family conflict and work–family enrichment happens. Finally, the theoretical and managerial implications of the findings, limitations and future research directions were also discussed.

Keywords person–environment fit; boundary flexibility–ability, boundary flexibility–willingness; work–family conflict; work–family enrichment.

心理学报 , 2012, 44(12), 1677－1686.

上司不当督导与下属绩效：反馈寻求行为和学习目标定向的作用 *

申传刚 [1,2]，马红宇 [1,2]，杨　璟 [3,4]，刘腾飞 [1,2]

([1] 青少年网络心理与行为教育部重点实验室)

([2] 华中师范大学心理学院暨湖北省人的发展与心理健康重点实验室 , 武汉 430079)

([3] 中国科学院心理研究所 , 北京 100101)

([4] 中国科学院研究生院 , 北京 100039)

摘　要　本研究从下属反馈管理行为的视角来探索领导与下属的社会交换过程。具体为探讨下属的反馈寻求行为在上司不当督导与下属绩效之间的中介作用，下属的学习目标定向对上述过程中的调节作用。通过问卷法获得 306 名下属与上司的对偶数据，基于层级回归和 Bootstrap 分析的结果表明：上司不当督导不仅直接影响下属的绩效，还能通过抑制下属的反馈寻求行为间接地影响员工的绩效；下属的学习目标定向调节着上司不当督导与下属的反馈寻求行为的关系，当下属的学习目标定向越低，上司不当督导对反馈寻求行为的抑制作用更加明显。

关键词　上司不当督导；员工绩效；反馈寻求行为；学习目标定向

分类号　BB49：C93

1 问题提出

领导力一直都是组织行为学研究的热点，以往关于领导力的研究多是从积极的视角来研究有效的领导行为。而领导在与下属互动的过程中，为了影响下属的行为，除了通过建立愿景、个别关怀等积极领导行为以外，还可能会经常做出一些非常细微、很难捕捉并且容易被社会所接受的负面领导行为，如奚落挖苦、控制资源等。相对于前面的积极领导行为，

后面被称为不当督导 (abusive supervision) 的消极领导行为尚未得到深入的研究。

Tepper（2000）针对组织中上述的奚落挖苦、控制资源等负性领导行为，首次提出了不当督导的概念：下属感知到的被上司反复表现出来的具有敌意性的言语和非言语行为，但不包括身体上的接触。此概念提出之后，相关学者将此概念与职场欺负(workplace bullying) 等其他一些类似的职场负性行为进行了相应的区分（Tepper, 2007）。职场欺负指一

---

　* 国家自然科学基金青年项目 (31200795); 全国教科规划课题 (BBA090066); 华中师范大学国家教师教育理论创新与实践研究重大招标项目 (985ZB0303); 青少年网络心理与行为教育部重点实验室开放课题 (2012B04) 资助。

通讯作者 : 马红宇 , E-mail: mahy@mail.ccnu.edu.cn

个个体长期遭受难以反抗的负性行为，这些行为来源于一人或多人，如侮辱，贬低、戏弄、孤立某人（Einarsen, 1999）。从概念和测量内容上看，它们之间存在着一定的重叠，均指个体反复地暴露于职场中的敌对行为情境中，均包含着被戏弄、隐私被侵犯、受侮辱、承担费力不讨好的工作、工作成绩得不到肯定等方面（吴隆增，刘军，刘刚，2009；李永鑫，聂光辉，李艺敏 等，2011）。但它们之间也存在着一些差异。首先，不当督导行为发起者是上司，是一对一的行为，而欺负行为的发起者，可以是上司、同事或下属，可以是一对一的行为，也可以是多对一的行为；其次，欺负行为包含着行为的消极结果，但是不当督导并未预测行为的结果；再次，上司不当督导不包含对个体身体接触的行为，如打耳光，推拉等，而职场欺负行为包含着身体伤害的行为。

作为负性领导的典型代表，自不当督导的概念提出以后，得到了国内外学者的积极关注。一系列研究结果表明，上司不当督导会影响众多工作结果变量，如工作满意度、情绪耗竭、心理紧张、自尊、建言行为、抵抗、攻击、偷窃、酗酒、组织公民行为和绩效等（Tepper, 2000; Tepper, Duffy & Shaw, 2001; Duffy, 2002; Inness, Barling & Turner, 2005; Burton & Hoolber, 2006; Bamberger & Bacharach, 2006; Harvey et al, 2007; Mitchell & Ambrose, 2007; Harris, Kacmar & Zivnuska, 2007; Tepper, et al. 2006, 2008; Tsung-Yu Wu & Changya Hu, 2009; 吴宗祐，2008；李锐，凌文辁，柳士顺，2009；刘军，吴隆增，林雨，2009；吴隆增等 2009）。其中，有关上司不当督导与绩效之间的关系得到了相关学者的重视。上司不当督导对下属绩效的负向影响已经在中西方文化下都得到了验证（Harris, et al. 2007; 吴隆增等，2009）。而有关不当督导对下属绩效的影响是如何产生以及如何有效应对上司不当督导提升下属绩效的实证研究还相对较少。所以，探讨上司不当督导影响下属绩效的作用机制，更全面地揭示有效应对上司不当督导、改善个人绩效的方法，是本研究的出发点。

社会交换理论和互惠准则（Blau, 1964）是用来解释上司和下属互动的重要理论之一。对员工而言，上司是其重要的交换对象之一。在员工与上司的交换过程中，主管提供下属完成任务的资源，下属以好的绩效回报主管。下属可能为了个人职业的发展，期望能够与主管形成良好的互惠关系，以及能够从主管那里得到更多的支持和资源，他们可能会在完成任务的过程中，可能会寻求如何满足他们主管要求、期望的反馈。而主管的这些反馈信息主要是关于下属绩效或者表现，下属根据此反馈来判断其行为是否正确与核实，减少关于工作和自我的不确定，这种行为即反馈寻求行为（feedback seeking behavior）（Ashford & Cummings, 1983）。从工具性的视角来看，一些研究表明反馈对员工增强他们的绩效来说确实是一个相当有价值的资源（Asfhord & Cumming, 1983; VandeWalle; VandeWalle, Ganesan, Challagalla & Brown, 2000, ）。已有研究表明，与那些不经常做出反馈寻求行为的员工相比，经常做出反馈寻求行为的员工的目标达成情况会更好，上司对其给予的绩效评价也更高（Deshon, Kozlowski, Schmidt, Milner & Wiechmann, 2004; Locke & Latham, 2004; Northcraft & Ashfrod, 1990），所以，不难推测员工主动向主管寻求反馈的行为会促进主管和员工之间的社会交换过程。而作为员工最重要、最有价值的交换对象和反馈源的上司，他们的领导方式反过来又影响着员工向其寻求反馈的行为（VandeWalle, et al. 2000; Madzar, 2001; Levy, Cober, Miller, 2002）。研究表明，当下属在与上司互动的过程中，知觉到上司对其表现出信任、尊重、关心、支持、鼓励时，他们会更频繁地向其上司询问关于工作绩效的信息（Chen, Lam & Zhong, 2007; Lam, Huang & Snape, 2007）；相反，当下属知觉寻求反馈所付出的代价大时，会抑制个体的反馈寻求频率（Ashford, 1986）。因此，不难推测，当下属在面对一个经常做出不信任、不支持、甚至辱骂、贬损、不理不睬、冷嘲热讽等不当督导行为的上司时，会减少向其寻求反馈的频率，但是减少了寻求反馈的频率，也同时让他们丧失了获得一些提升个人绩效的有价值信息的机会，不利于其绩效的提高。基于以上的分析，我们提出以下假设：

H1：上司不当督导通过影响下属的反馈寻求行为间接影响下属的绩效

领导的有效性一直都是领导力研究的焦点。领导力通过影响下属得以体现，下属对领导的接受或拒绝，决定了领导的效能，这充分的体现了领导和下属的对立统一的关系，在研究领导对下属的过程中，要充分的考虑下属的特性（Fiedler, 1967; House & Mitchell, 1974; Jago & Vroom, 1980）。根据Kerr 和 Jermeir（1978）的替代领导理论（substitute for leadership theory），下属的行为除了受到正式的层级领导的影响之外，还会受到个体、群体、任务和组织因素的影响，此理论强调了个体、任务和组织方面的因素会抵消（neutralize）或替代（substitute）领导行为，从而降低了领导行为对下属的影响。在本研究中，就下属的反馈寻求行为本身而言，除了受领导这一外界的因素的影响之外，员工自身的人格特质——学习目标定向——也是影响其寻求反馈信息的重要影响因素（VandeWalle & Cummings, 1997; VandeWalle, 2003）。学习目标定向的个体将绩效反馈视为激发完成当前任务和将来任务所需能力的一种手段，是提升任务掌控力的诊断性信息（Dweck & Leggett, 1988; VandeWalle, 2003）。高学习目标定向型的个体更多地关注反馈信息对个人能力提升和发展的价值，而较少地关注做出反馈寻求行为所付出的代价（VandeWalle & Cummings, 1997）。所以，根据Kerr 和 Jermeir（1978）的替代理论的思想，我们推测，当主管展现出不当督导等负性行为时，下属的学习目标定向会起到一定的替代作用，即下属的学习目标定向这一人格特质会削弱不当督导对下属反馈寻求行为的影响。具体表现为对于高学习目标定向的个体而言，当面对一个经常做出不当督导的上司时，即使面临着受辱和丢面子的风险，他们对能力提升和任务控制的期望以及对反馈和困难所持有的信念仍会驱使他们向上司寻求绩效的反馈。相反，对于低学习目标定向的个体而言，他们本身就不会频繁地主动向上司寻求反馈，如果再面对一个经常贬损、

辱骂下属的不当督导类型的上司，这加大了寻求反馈的代价，会导致他们更少做出寻求反馈的行为。基于以上分析，提出以下假设：

H2：下属学习目标定向在上司不当督导与反馈寻求行为之间起调节作用。下属学习目标定向越低，不当督导对其反馈寻求行为的抑制作用越明显，进而对下属的绩效产生更大的负面效应。

综上所述，具体的研究框架如图1所示。

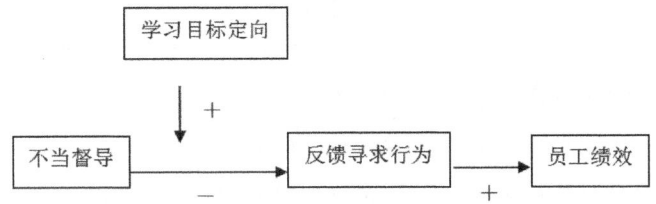

图1 研究框架

2 研究方法

2.1 被试与调查程序

本研究采用上司–下属的二元对偶研究设计。调查对象来自湖北、广州、厦门、宁波的7家企业中的不同工作部门的上司及其直接下属。调查问卷分为员工问卷（问卷A）和上司问卷（问卷B），问卷A包括员工对上司的领导行为方式（不当督导），反馈寻求行为，和自我学习目标定向的评价，问卷B包括上司对下属一般任务绩效的评价。为了尽量消除被试的疑虑和保证问卷的隐匿性，员工问卷装在带有双面胶的信封内，指导语中强调该问卷调查以不记名的方式进行，并提醒被试填写完毕后将信封进行封口后，交还给调查小组成员或人力资源部门负责人。为了保证数据的配对，我们采用了相应的编码系统，以配对上司评定与下属的回答。上司调查采取方便抽样的原则确定参与调查上司140位，再通过随机的方式各选择3-5名直接下属参与调查。

此调查共发放上司问卷和员工问卷各500份，回收上司问卷451份，员工问卷439份，回收率分别为90.2%和87.8%，剔除无效问卷后，最终得到上司–下属匹配数据306份。除去未填写项目后，调查样本中

员工的平均年龄为 33.5 岁（SD=8.53），平均工龄为 8.52 年（SD=6.75），其中男性占 35.7%，女性 65.3%。

2.2 研究工具

2.2.1 不当督导 采用 Aryee（2007）研究中所使用的 10 条目单维结构量表。此量表是 Ayree 等人根据中国文化情境，从 Tepper（2000）的原始量表中选出的 10 个条目组成的量表。该 10 条目量表的有效性在李锐，凌文辁，柳士顺（2009）的研究中得到了再次验证。该量表采用该量表采用 5 级计分，1 代表我不记得他（她）对我做出过这种行为，5 代表他（她）经常的对我做出这种行为。在本研究中，该量表的内部一致性系数为 0.91。经过验证性因素分析，所得结果为 χ^2/df=4.86<5，拟合指标 GFI=0.89，CFI=0.91，NFI=0.90，IFI=0.92，RMSEA=0.11。该结果表明不当督导的单维度结构拟合良好。

2.2.2 反馈寻求行为 关于反馈寻求行为的测量工具，主要有 Ashford 和 Cummings（1983）年编制的反馈寻求量表或其后来修订版本，Moss 等（2003）从印象管理的角度编制的反馈寻求量表，以及日本学者 Yanagizawa（2008）编制的关于目标达成情况的反馈寻求量表。Ashford 和 Cummings（1983）所编制的反馈寻求量表主要是测量员工的于反馈寻求行为的数量、努力程度等方面，Moss 等（2003）所编制的量表主要是测量个体做出反馈寻求行为的动机和意愿。因为本研究中的目的是为了探索不当督导对下属绩效的影响，为了突出反馈源以及员工寻求反馈的目的，所以我们采用日本学者 Yanagizawa（2008）所编制的反馈寻求量表，并在测量的过程中将反馈源进行了相应的确定。该量表由员工自评，共 6 个条目，例如，"我向上司寻求在处理工作问题时我的判断是否正确的信息"，"我向上司寻求关于我完成的工作有多好的信息"等。该自评问卷采用 6 点计分，1 代表从不，6 代表非常频繁。该问卷为单维结构，本研究中该量表的内部一致性系数为 0.88。

2.2.3 学习目标定向 采用 Vandewalle（1997）开发的工作领域成就目标定向量表中的学习目标定向分量表。该量表包含 5 个条目，如"我愿意选择一个能从中学习很多东西且具有挑战性的工作任务"、"我经常寻找机会来丰富自己的知识和提高自己的技能"等。该问卷采用 7 点计分。1 代表非常不同意，7 代表非常同意。该问卷的有效性在国内也得到验证（王雁飞，凌文辁，朱瑜，2004)。本研究中该量表的内部一致性系数为 0.75。

2.2.4 员工绩效 上司评价绩效问卷采用 Tsui 等（1997）编制的绩效评估问卷。该问卷测量的是员工的一般绩效问卷，而不是具体的某个行业的绩效问卷。该问卷包含 6 个条目，单维结构，6 个条目分别从员工工作的数量、质量和效率等方面与平均水平的比较来测量员工的基本任务绩效。例如，"该员工的工作数量高于平均水平"，"该员工的工作质量远高于平均水平"。该问卷测量采用 7 点计分，1 代表非常不同意，7 代表非常同意。该问卷的单维性和有效性得到了相关学者的验证（Aryee, Budhar & Chen, 2002; 韦慧民，龙立荣，2009）。本研究中，该量表的内部一致性系数为 0.80。

2.3 统计分析

本研究采用 SPSS15.0 和 Amos7.0 进行统计分析。首先使用验证性因素分析来验证问卷的效度，然后运用逐步回归来分析不当督导对员工绩效的影响，以及反馈寻求行为的中介作用，最后运用 Edwards 和 Lambert（2007）提出的"总效应调节效应"来分析学习目标定向的调节作用。

3 数据分析和结果

3.1 验证性因素分析结果

3.3.1 员工测量量表验证性因素分析结果

运用 Amos7.0 对员工调查数据进行了验证性因素分析，比较了单因素模型（三个变量同属于一个因素），两因素模型（学习目标定性与反馈寻求行为同属于一个因素），三因素模型（不当督导，学习目标定向，反馈寻求行）从而确认研究中员工调查变量（不当督导，学习目标定性、反馈寻求行为）

表1 验证性因素分析结果（N=306）

模型	χ^2	df	χ^2/df	RMSEA	GFI	CFI	IFI	NFI
单因素模型	1366.91	189	7.23	0.14	0.55	0.60	0.61	0.57
两因素模型	545.55	188	2.90	0.08	0.84	0.88	0.88	0.83
三因素模型	436.59	186	2.34	0.07	0.88	0.92	0.92	0.86

表2 各研究变量的平均数、标准差与相关矩阵（N=306）

	M	SD	1	2	3	4
1. 不当督导	2.11	0.87	1.00			
2. 反馈寻求	3.59	0.95	−0.38**	1.00		
3 学习目标定向	5.19	0.92	−0.31**	0.57**		1.00
4. 员工绩效	5.17	0.68	−0.33**	0.35**	0.36**	1.00

注：** $p<0.01$，* $p<0.05$

表3 层级回归结果：员工反馈寻求行为的中介（N=306）

变量	Step1 绩效 β	反馈寻求 β	Step2 反馈寻求 β	β	Step3 绩效 β	β
控制变量				0		
年龄	0.02	−0.03	0.03	−0.02	0.02	−0.03
员工性别	0.16	0.10	0.09	0.04	0.16	0.09
上司性别	−0.06	0.02	0.03	0.11	−0.06	−0.01
工龄	−0.03	−0.10	0.32**	0.25**	−0.03	−0.17
与上司工作年限	−0.10	−0.01	−0.30**	−0.21**	−0.10	0.05
自变量						
不当督导		−0.33**		−0.32**		−0.24**
中介变量						
反馈寻求行为						0.30**
R^2	0.30	0.13	0.13	0.22	0.03	0.20
F	1.66	7.09**	8.63**	14.46**	1.66	10.53**
ΔR^2		0.10		0.09		0.17

注：** $p<0.01$，* $p<0.05$

的区分效度。结果显示，三因素对数据的拟合程度最好，卡方与自由度之比小于4，RMSEA为0.07，低于0.080，而GFI，CFI，IFI，NFI等介于0.86和0.92之间，接近于0.90，说明三因素模型可以接受，所以，员工测量量表测量了三个不同的因素，具有较高的区分效度（见表1）。

3.2 各研究变量的描述性统计分析

运用这些配对数据，对不当督导、反馈寻求行为、反馈逃避行为、员工绩效进行描述性统计和相关分析，其结果见表2。从结果中可以看出，不当督导与反馈寻求、员工绩效呈显著的负相关，反馈寻求行为与员工绩效呈显著的正相关关系，这为后面的中介效应检验提供了良好的基础。

3.3 研究假设的检验

3.3.1 反馈寻求行为的中介效应检验

采用层级回归对假设1进行检验。检验员工反馈寻求行为的中介效应时，遵循Baron和Kenny（1986）提出的中介效应检验步骤，各步骤标准化的回归系数和方程检验结果见表3。

第一步检验了上司不当督导对员工绩效的显著影响。由表3结果可知，在控制了人口学变量之后，不当督导对员工绩效有显著的影响（$\beta = -0.35$，$P<0.01$）。第二步检验了上司不当督导对员工反馈寻求行为的影响。结果显示，上司不当督导对员工的反馈寻求行为达到显著影响（$\beta = -0.33$，$P<0.01$）。第三步将不当督导和反馈寻求行为同时纳入回归方程

来解释员工绩效时，检验不当督导的效应消失（完全中介效应）或减弱（部分中介效应）。而结果表明，将不当督导和反馈寻求行为同时纳入方程来预测员工绩效时，反馈寻求行为的效应显著（β =0.29，$P<0.01$），不当督导的效应有所减弱，但是依然显著（β =-0.25, $P<0.01$）。可以得出，反馈寻求行为在上司不当督导和员工绩效之间起部分中介作用。进一步的 sobel 检验表明反馈寻求行为在不当督导和员工绩效之间的间接效应显著（$Z=3.87, p<0.01$），假设 1 得到检验。

3.3.2 学习目标定向的调节效应检验

本研究调节效应的分析采用 Edwards 和 Lambert（2007）所提出的总效应调节模型（Total effect moderation model）的方法，将调节效应和中介效应纳入到同一个分析框架中进行分析。因为在上述的中介效应分析中，反馈寻求行为起部分中介作用，所以，在总效应调节的过程中，上司不当督导→反馈寻求行为，反馈寻求行为→员工绩效，以及上司不当督导→员工绩效三条路径都有可能受到调节变量学习目标定向的影响。因此，采用此方法将直接效应（上司不当督导→员工绩效）和间接效应（上司不当督导→反馈寻求行为→员工绩效）结合起来进行调节分析，克服了以往研究中将中介效应和调节效应进行分开的弊端，以便从更完整的角度来探讨员工学习目标定向对整个中介模型的调节效应（Edwards & Lambert, 2007; 李锐，凌文辁，柳士顺，2009）。根据

Edwards 和 Lambert（2007）所提出的分析程序，将建立以下两个回归方程：（1）回归方程一：$FSB=a_{05}+a_{X5}AS+a_{Z5}LGO+A_{xz5}(AS×LGO)+e_{m5}$；（2）回归方程二：$Pf=b_{020}+B_{X20}AS+b_{M20}FSB+b_{Z20}LGO+B_{XZ20}(AS×LGO)+B_{MZ20}(FSB×LGO)+e_{Y20}$。其中 AS、FSB、LGO、Pf 分别代表不当督导，反馈寻求行为，学习目标定向和绩效。根据上述两个回归方程，得出或计算出下列系数或效应：（1）第一阶段：由不当督导到反馈寻求行为的回归系数（根据回归方程一得到）；（2）第二阶段，由反馈寻求行为到员工绩效的回归系数（根据回归方程二得到）；（3）直接效应：由不当督导到员工绩效的回归系数（根据回归方程二）；（4）间接效应：由第一阶段和第二阶段的回归系数相乘得到；（5）总效应：直接效应和间接效应相加得到；（6）差异：高学习目标定向个体（1 个标准差之上）情况下的系数或效应减去低学习目标定向个体（1 个标准差之下）情况下的系数或效应所得的差。

因为上面所得的系数或效应（路径系数、直接效应、间接效应、总效应等）都是从回归方程中得到的，其中单一路径（第一阶段、第二阶段以及直接效应）系数的显著性检验遵循简单斜率检验流程，高、低学习目标定向下第一阶段、第二阶段和直接效应的差异的显著性检验等同于单一系数的显著性检验，对于间接效应和总效应以及其差异的显著性检验使用拔靴法（bootstrap），以本研究中的 306 个样本为"母本"，采用有放回的抽样方式从母本中随机抽取 306

表 4 学习目标定向调节效应分析结果

阶段与效应	低学习目标定向	高学习目标定向	差异
阶段			
第一阶段：			
不当督导→反馈寻求行为	−0.34**	−0.12	−0.22**
第二阶段：			
反馈寻求行为→员工绩效	0.20**	0.01	0.19**
效应			
直接效应	−0.16**	−0.14*	−0.02
间接效应	−0.07**	0	−0.07**
总效应	−0.23**	−0.14**	−0.09**

注：（1）表格中的数字为回归系数（第一阶段、第二阶段及直接效应），以及运用这些回归系数计算而得到的数值（间接效应、总效应及差异）。低学习目标定向是将学习目标定向的均值减一个标准差，高学习目标定向是将学习目标定向的均值加一个标准差；（2）** $P<0.01$，* $P<0.05$

个样本，共抽 1000 组样本，根据抽取的 1000 个样本计算出单纯路径系数、间接效应、总效应的估计值。然后根据这 1000 组估算值，推导出"偏差校正置信区间"，最后根据这些置信区间来确定各路径系数、间接效应、总效应及差异的显著性。分析结果见表 4。

从表 4 中可以得出，对于低学习目标定向个体而言，反馈寻求行为对上司不当督导与员工绩效之间的关系具有中介作用，而对高学习目标定向的个体而言，反馈寻求行为对上司不当督导与员工绩效之间的关系中介作用不显著。在低学习目标定向情况下，不当督导对反馈寻求的负向效应为较强（-0.34，$p<0.01$）；在学习目标定向较高的情况下，不当督导对反馈寻求的负向效应较弱，且变得不显著（-0.12，$p>0.05$），两者的差异显著（-0.22，$P<0.01$）；相应地，通过反馈寻求行为传递的间接效应的差异也同样达到了显著水平（-0.07，$p<0.01$）；在学习目标定向高、低两种情况下，不当督导对员工绩效的负向效应都达到了显著，但是差异并不显著；在低学习目标定向情况下，模型的总效应较高，（-0.24，$p<0.01$)，在高学习目标定向情况下，模型总效应较弱（-0.13，$p<0.01$），并且差异也达到了显著 (-0.11，$p<0.01$)。因此 H3 得到验证。

4　讨论

4.1　不当督导对员工绩效的影响

本研究结果表明，上司不当督导既直接地影响着员工的绩效，也通过抑制员工的反馈寻求行为间接地影响员工绩效。此结果与 Harris 等（2007）在西方情景下的研究结论一致，与吴隆增等（2009）在中国情境下所得结果也基本相同，这说明了在中西方的企业中，上司的不当督导都会对下属绩效产生消极的影响。同时，本研究结果与 Chen 等（2007）的研究结果在某种程度上具有一致性。即员工的反馈寻求行为在领导与下属社会交换的过程中扮演着重要的角色。本研究结果表明上司不当督导会通过抑制下属的反馈寻求行为进而影响下属的绩效。这一结果让我们从新的视角来理解上司与下属之间的社会交换过程。根据 Blau（1964）的社会交换原则以及 Cropanzano 和 Mitchell（2005）的消极互惠原则的观点当个体受到他人的消极对待时，也同样会以的消极的方式来对待他人，即所谓的以眼还眼，以牙还牙。因此，当下属遭到了来自上司的奚落挖苦、控制资源等不当对待行为时，他们会以同样的方式给予上司消极的回报。但是，在我国文化背景下，上下级之间权距较大，等级制度深严，"上尊下卑"的思想严重，上级控制着下级的各种资源和命运（龙立荣，刘亚，2004），直接与上司发生冲突可能会给自己带来"厄运"，所以，个体消极回报上司时可能会有更多的顾虑，不敢轻易地以上司表现出来的不当对待的方式反过来对待上司，而会通过直接地降低个人绩效这种比较"隐蔽"的方式来报复上司，或者通过降低个人的反馈寻求行为这样一种更为"隐蔽"的方式，不主动地去寻求获得提升个人绩效的一些有价值的信息和方法，进而影响个人绩效的完成，以达到报复主管的目的。本研究结果表明下属的反馈寻求行为在上司不当督导和下属绩效之间起部分中介作用，这说明了在上司不当督导对下属的绩效影响的过程中，并非只有通过影响下属的反馈寻求行为才能进一步的影响下属的绩效，它一方面可能直接地影响了下属的绩效（Harris et al., 2007），另一方面，可能通过影响下属的其他方面进而影响下属绩效，如通过影响下属对主管的信任（吴隆增，等 2009）。

4.2　学习目标定向的调节作用

本研究采用了 Edwards 和 Lambert（2007）的分析方法和程序，分析了下属的学习目标定向是否会调节不当督导对下属绩效的影响效应。研究结果表明，下属的学习目标定向可以调节不当督导和反馈寻求行为之间的关系，并能进一步的调节相应的中介链效应（"不当督导→反馈寻求行为→员工绩效"）。对低学习目标定向的下属而言，不当督导与反馈寻求行为之间的负向关系越强，进而对下属的绩效产生不利影响；对高学习目标定向的下属而言，不当督导与反馈寻求行为之间的负向关系越弱，但是并未进一步的影响其绩效。

首先，此结果证实了下属的个体特质对上司领导过程的影响。即上司不当督导对下属的影响过程受下属学习目标定向的影响。这可能是由于学习目标定向在某种程度上决定着员工的工作态度、工作动机、努力程度以及影响着下属对工作中困难和失败的态度。学习目标定向型的个体在面对困难时会做出控制导向（mastery-oriented）的反应模式（Dweck，1986）。因此，高学习目标定向的员工可能会将上司的贬损行为理解为自己的绩效仍未"达标"，会削弱上司不当督导的消极效应。相反，对低学习目标定向的个体而言，他们可能认为完成工作仅仅是为了与领导进行交换，他们关注的焦点在于领导如何对待他们，而不是如何去提升自身的能力和技能，因此，他们不会太主动地去向上司寻求关于自己工作绩效的一些反馈，再加上要面对一个对自我形象等构成威胁的不当对待型上司的时候，他们更不会主动向上司询问关于他们绩效的相关信息。

另外，对于学习目标定向的不同水平下，反馈寻求行为的中介作用的差异。这可能是因为下属的学习目标定向水平不同，反馈寻求行为对绩效的影响不同（具体见表 4 第二阶段）。前人研究表明，学习目标定向与自我效能感、工作目标的设定水平呈正相关（VandeWalle，William，& John，2001），因此，有可能低学习目标定向的下属对工作的控制感和目标设定比较低，所以，当他们从上司那里寻求反馈时，会获得一定提升绩效的有用信息，有利于其绩效的提高；而对于高学习目标定向的下属而言，他们本身对工作有着很高的控制感，通过主动寻求获得的反馈信息对其控制感的增强不是很大，对其绩效的影响也不是很大。

总的来说，本研究结果证实了 Howell 和 Shamir（2005）的观点：尽管上司和下属在权力上不平衡，但是在形成他们相互的关系的过程中均扮演着重要的角色，并且共同决定着组织的结果变量，即下属在上司的领导过程中同样起着相当关键的作用。同时，本研究证实了替代领导理在负性领导情境中的存在，为我们更为全面地理解不当督导的作用机制提供了一个新的视角。

4.3 研究意义与局限

本研究采取了上司和下属互评的方式收集数据，克服了员工单一数据所造成的共同方法偏差，使得本研究的结果可信度更高。另外，本研究结果为组织的管理实践提供了一定的指导。本研究结果显示上司的不当督导会抑制员工的反馈寻求行为，并进一步的影响员工的工作绩效，并且员工个人的学习目标定向会影响上司的领导效能。因此，为了让员工更加积极主动的寻求关于他们绩效的反馈信息，消除顾虑，进一步提高他们的工作绩效，组织和管理者应从以下两个方面来进行：一方面组织管理者应该充分认识到不当督导的消极影响，并应在实际的管理中减少和杜绝不当督导现象的发生，例如可以从制度，文化氛围等方面对上司的领导方式进行约束，还可以通过对上司进行培训，让他们知道如何人性化地去管理他们的下属以及哪些领导行为可能会导致下属产生不当对待的知觉；另一方面，应该积极地出台相应的制度和奖惩措施来支持员工主动寻求反馈的愿望和行动，构建员工帮助计划平台，帮助员工形成应对组织负性事件的技巧，以及在将来再次受到领导不当对待时，如何更好地保护自己。

本研究还存在一些局限：（1）本研究虽然采用的是上司-下属对偶设计，在一定程度上控制了共同方法偏差，但是由于测量内容本身的敏感性，社会称许性在所难免，所以，在将来的研究中，可以尝试同时从下属和第三方来收集上司的不当督导水平，然后将所搜集数据进行整合作为不当督导的水平来探讨不当督导对下属和第三方的影响及其机制，同时也可以运用深度访谈等质化研究的方式，进一步的探索不当督导对下属绩效的影响过程；（2）本研究采用的横切面研究，不能进行因果关系的推断，所以在将来的研究中，可以采用纵向研究，在收集自变量，中介变量和因变量的数据时，有一定的时间跨度或者是采取实验室模拟实验来对这些变量的因果关系进行进一步的探讨。

5 结论

本研究结果表明：（1）上司的不当督导不仅对员工的绩效有直接的消极影响，还通过抑制员工的反馈寻求行为间接对员工的绩效产生影响；（2）员工的学习目标定向对不当督导与员工反馈寻求行为的关系，以及"不当督导→员工反馈寻求行为→员工工作绩效"中介效应链和不当督导对员工绩效的总效应均具有调节的作用，当下属的学习目标定向越低时，上述关系及效应也越强。

参考文献

Aryee, S., Budhwar, P. S., & Chen, Z. X. (2002). Trust as mediator of the relationship between organizational justice and work outcomes: Test of a social exchange model. *Journal of Origanizational Behavior, 23*, 267–285.

Aryee, S., Chen, Z. X., Sun, L., & Debrah, Y. D. (2007). Antecedents and outcomes of abusive supervision: Test of a trickle–down model. *Journal of Applied Psychology, 92*, 191–201,

Ashford, S. J. (1986). Feedback–seeking in individual adaptation: A resource perspective. *Academy of management Journal, 29*, 465–487.

Ashford, S. J., & Cummings, L.L. (1983). Feedback as an individual resource: personal strategies of creating information. *Organizational Behavior and Human Performance, 32*, 370–398.

Baron, R. M., & Kenny, D. A.(1986). The moderator–mediator variable distinction in social psychological research: Conceptual, strategic, and statistical considerations. *Journal of Personality and Social Psychology, 51*, 1173–1182.

Bamberger, P. A., & Bacharach, S. B. (2006). Abusive supervision and subordinate problem drinking: Taking resistance, stress, and subordinate personality into account. *Human relations, 59*, 1–30.

Blau, P. M. (1964). *Exchange and power in social life*. New York: John Wiley.

Burton. P. J. & Hoolber, M. J.(2006), Subordinate self–esteem and Abusive supervision. *Journal of Managerial Issues, 3*, 340–355.

Chen, Z. G., Lam, W. & Zhong, J, A. (2007). Leader–Member exchange and member performance: A new look at individual–level negative feedback–seeking behavior and team–level empowerment climate. *Journal of Applied Psychology. 92*, 202–212.

Cropanzano, R, & Mitchell, M. S. (2005). Social exchang theory: An interdisciplinary review. *Journal of Management, 31*, 1–27.

Deshon, R., Kozlowski, S. W. J., Schmidt, A. M., Milner, K. R., & Wiechmann, D. (2004). A multiple–goal, multilevel model of feedback effects on the regulation of individual and team performance. *Journal of Applied Psychology, 89*, 1035–1056.

Duffy, M. K., Ganster, D., & Pagon, M., C. (2002). Social undermining in the workplace. *Academy of Management Journal, 45*, 331–351.

Dweck, C. S. (1986). Motivational processes affecting learning. *American psychologist, 41*, 1040–1048.

Dweck, C. S., & Leggett, E. L. (1988). A social–cognitive approach to motivation and personality. *Psychological Review, 95*, 256–273.

Edwards, J. R. & Lambert, L. S. (2007). Methods for intergrating moderation and mediation: A general analytical framework using moderated path analysis. *Psychological Methods, 12*, 1–11.

Einarsen, S. (1999). The nature and causes of bullying at work. *International Journal of Manpower, 20*, 16–27.

Fiedler, F. E. (1967). *A theory of leadership effectiveness*. New York: Academic Press.

Harris, K. J., Kacmar, K. M., & Zivnuska, S. (2007). An

investigation of abusive supervision as a predictor of performance and the meaning of work as a moderator of the relationship. *Leadership quarterly, 18,* 252–263.

Harvey, P., Stoner, J. Hochwarter, W., & Kacmar,C. (2007). Coping with abusive bosses: The neutralizing effects of ingratiation and positive affect on negative emplyee outcomes. *Leadership Quarterly, 18,* 264–280.

House, R.J., & Mitchell, T. R. (1974). Path – goal theory of leadership. *Journal of Contemporary Business, 5,* 81–97.

Howell, J. M., & Shamir, B. (2005). The role of followers in the charismatic leadership process: relationships and their consequences. *Academy of Management Review, 30,* 96–112.

Inness, M., Barling, J., & Turner, N. (2005). Understanding supervisor–targeted aggression: A within–person, between–jobs design. *Journal of Applied Psychology, 90,* 731–739.

Jago, A. G., & Vroom, V. H. (1980). An evaluation of two alternatives to the Vroom/Yetton Normative Model. *Academy of Management Journal, 23,* 347–355.

Kerr, S., & Jermeir, J. (1978). Substitutes for leadership: their meaning and measurement. *Organizational Behavior and Human Performance, 22,* 375– 403.

Lam, W., Huang, X., & Snape, E.(2007). Feedback–seeking behavior and LMX: do supervisor–attributed motives matter? *Academy of Management Journal. 50,* 348–363.

Levy, P. E., Cober, R. T. & Miller, T. (2002). The effect of transformational and Transactional leadership perception on Feedback– seeking intentions. *Journal of Applied Psychology, 32,* 1703–1720.

Li, R., Ling, W. Q. & Liu, S. S. (2009). The Mechanisms of How Abusive Supervision Impacts on Subordinates' Voice Behavior. *Acta Psychologica Sinica, 12,* 1189–1202.

[李锐，凌文辁，柳士顺. (2009): 上司不当督导对下属建言行为的影响及其作用机制. *心理学报，41,* 1189–1202.]

Li, Y. X., Nie, G. H., Li, Y. M., Wang, M. H., & Zhao, G. X. (2011). The structure and measurement of workplace bullying in china. *Journal of Psychological Science, 34,* 1201–1208

[李永鑫，聂光辉，李艺敏，王明辉，赵国祥. (2011). 工作场所欺负的内容结构与测量. *心理科学，34,* 1201–1208.]

Liu, J., Wu, L. Z., & Lin, Y. (2009).Coping with Abusive Supervision: Mechanisms of Ingratiation and Political Skill. *Nankai Business Review, 2,* 52–58.

[刘军，吴隆增，林雨. (2009). 应对辱虐管理：下属逢迎与政治技能的作用机制研究. *南开管理评论，2,* 52–58.]

Locke, E. A. & Latham, G. P. (2004). What should we do about motivation theory? Six reconmmendations for the twenty–first century. *Academy of Management Review, 29,* 388–403.

Madzar, S. (2001). Subordinate's information inquiry: Exploring the effect of perceived leadership style and individual difference. *Journal of Occupational Psychology, 74,* 221–232.

Mitchell, M. S., & Ambrose, M. L. (2007). Abusive supervision and workplace deviance and the moderating effects of negative reciprocity beliefs. *Journal of Applied Psychology. 92,* 1159–1168.

Moss, S. E., Valenzi, E. R., & Taggart, W. (2003). Are you hiding from your boss? The development of a taxonomy and instrument to assess the feedback management behaviors of good and bad. *Journal of Management, 29,* 487–510.

Northcraft, G. B., & Ashford, S. J. (1990). The preservation of self in everyday life: The effects of performance expectations and feedback context on feedback inquiry. *Organizational Behavior and Human Decision Processes, 47,* 42–64.

Tepper, B. J., (2000). Consequences of abusive supervision. *Academy of Management Journal, 43*, 178–190.

Tepper, B. J., Duffy, M. K., & Shaw, J. D., (2001). Personality moderators of the relationships between abusive supervision and subordinates' resistance. *Journal of Applied Psychology, 86*, 974–983.

Tepper, B. J., Duffy, M. K., Henle, C. A., & Lambert, L. S. (2006). Procedural injustice, victim precipitation, and abusive supervision. *Personnel psychology, 59*, 101–123.

Teeper, B. J., Henle, A. C., Lambert, S. L. Duffy, M. K. & Giacalone, A. R. (2008). Abusive supervision and subordinates' organization deviance. *Journal of Applied Psychology. 93*, 721–732.

Tsui A. S., Pearce, J. L. Porter, L. W. Tripoli, A. M. (1997). Alternative approaches to the employee– organization relationship: does investment in employees pay off? *Academy of Management Journal, 40*, 109–121.

Tsung–Yu Wu. (2008). Abusive Supervision and Emotional Exhaustion: The Mediating Effects of Subordinate Justice Perception and Emotional Labor. *Chinese Journal of Psychology, 2*, 201–221.

[吴宗祐 . (2008). 由不当督导到情绪耗竭：部属正义知觉与情绪劳动的中介效果。*中华心理学刊, 2*, 201–221.]

Tsung–Yu Wu, Changya Hu (2009) Abusive supervision and employee emotional exhaustion: dispositional antecedents and boundaries. *Group & Organizational Management 34*, 143–169.

VandeWalle, D. (1997). Development and validation of a work domain goal orientation instrument. *Educational and Psychological Measurement,56*, 995–1015.

VandeWalle, D. (2003). A goal orientation model of feedback seeking behavior. *Human Resource Management Review,13*, 581–604.

VandeWalle, D., & Cummings, L. L..(1997). A test of the influence of goal orientation on the feedback–seeking process. *Journal of Applied Psychology, 82*, 390–400.

VandeWalle, D., Ganesan, S., Challagalla, G. N., & Brown, S. P. (2000). An integrated model of feedback–seeking behavior: Disposition,context and cognition. *Journal of Applied Psychology, 85*, 996–1003.

VandeWalle, D., & William, L. C. & John, W. S. (2001). The role of goal orientation following performance feedback. *Journal of Appilied Psychology, 86*, 629–640.

Wang, Y. F., Ling, W. Q., & Zhu, Y. (2004). The relationship among achievement goal orientation, self-efficacy and feedback–seeking behavior. *Psychological Science, 27*, 31–33.

[王雁飞 , 凌文辁 , 朱瑜 . (2004). 成就目标定向、自我效能与反馈寻求行为的关系 . *心理科学, 27*, 31–33.]

Wei, H. M., Long, L. R. (2009). Effects of cognition– and affect–base trust in supervisors on task performance and OCB. *Acta Psychologica Sinica, 41*, 86–94.

[韦慧民 , 龙立荣 . (2009). 上司认知信任和情感信任对员工行为及绩效的影响 . *心理学报, 41*, 86–94.]

Wu, L. Z., Liu, J., & Liu, G. (2009). Abusive Supervision and Employee Performance: Mechanisms of Traditionality and Trust. *Acta Psychologica Sinica, 6*, 510–518.

[吴隆增 , 刘军 , 刘刚 . (2009). 辱虐管理与员工表现：传统性与信任的作用 . *心理学报, 41*, 510–518.]

Yanagizawa, S. (2008). Effect of goal difficulty and feedback seeking on goal attainment and learning. *Japanese Psychological Research. 50*, 137–144.

Abusive supervision and employee' performance: Mechanisms of FSB and learning goral orientation

SHEN Chuangang[1,2], MA Hongyu[1,2], YANG Jing[3,4], LIU Tengfei[1,2]

([1] Key Laboratory of Adolescent Cyberpsychology and Behavior (CCNU), Ministry of Education, Wuhan 430079, China)

([2] School of Psychology, Central China Normal University, and Hubei Human Development and Mental Health Key Laboratory, Wuhan 430079, China)

([3] Institute of Psychology, Chinese Academy of Sciences, Beijing 100101, China)

([4] Graduate School, Chinese Academy of Sciences, Beijing 100039, China)

Abstract The literature on abusive supervision has consistently demonstrated the negative relationship between member perception of supervisor's abusive behavior and member performance. The process through which relationship supervisor's abusive behavior influences subordinates' performance, however, is still not fully understood. The present study provides a mechanism for the process. Specifically, we predict that the feedback seeking behavior(FSB) of members mediates these relationships, and learning goal orientation moderates the relationship between abusive supervision and FSB. In order to avoid the common method variance problem, two sources of survey were administrated. Data was from a total of 306 matched supervisor-subordinate dyads in 7 enterprises located in Hubei, Zhejiang, Xiamen. Two structured questionnaires were employed as the research instrument for this study. one consisted of three scales designed to measure abusive supervision, FSB and learning goal orientation, Among the major measures, the 15-items abusive supervision was adopted from Tepper(2000) study; FSB was measured via 6 items that was adopted from Saori Yanagizawa(2008) study; the five item learning goal orientation scale was adopted from Vandewalle & Cummings(1997) study. The other questionnaire was consisted of one scale which was adopted from Tusi et al(1997) study to measure supervisor-rated subordinates' performance. Results show that the Cronbach's alpha coefficients for above measures were from 0.75 to 0.94. Hierarchical regression and the total effect moderation model were utilized to analyze the date for testing the hypotheses proposed. In line with predictions, results of hierarchical regression demonstrate that abusive supervision had negative influence on FSB, supervisor-rated performance, and FSB played partially mediated role in the relationship between abusive supervision and supervisor-rated performance. Specifically, the negative effect of abusive supervision on subordinates' performance was partially mediated by subordinates' FSB. In addition, results of total effect moderation model analysis reveal that subordinates' learning goal orientation moderated the relationship between abusive supervision and FSB. Abusive supervision was more strongly related to FSB when subordinates' learning goal orientation was low.The present study extends our understanding of social exchange between supervisor and subordinate in the link between abusive supervision and subordinate's performance. Finally, the theoretical and managerial implications of the findings, limitations and future research directions were discussed.

Keywords abusive supervision; performance; feedback seeking behavior; learning goal orientation

心理学报，2015, 47(11), 1379 - 1394.

仕途"天花板"：公务员职业生涯高原结构、测量与效果 *

王忠军[1]，龙立荣[2]，刘丽丹[3]，黄小华[4]，贾文文[1]，李　璐[1]，马红宇[1]

([1] 青少年网络心理与行为教育部重点实验室，华中师范大学心理学院暨湖北省人的发展与心理健康重点实验室，武汉 430079)

([2] 华中科技大学管理学院，武汉 430074)

([3] 湖北中医药大学人文学院，武汉 430065)

([4] 江西省委统战部，南昌 330006)

摘　要　中国公共组织中基层公务员的职业生涯高原是值得关注的重要问题。首先通过深度访谈和开放式问卷调查，归纳出公务员职业生涯高原的典型特征与表现，然后通过问卷调查获得三批处级以下公务员的数据。基于样本一 (n=279) 的探索性因素分析发现，公务员职业生涯高原为两维度的结构：升迁停滞、职位边缘化。基于样本二 (n=517) 的验证性因素分析证实了公务员职业生涯高原两维度结构模型，公务员职业生涯高原测量工具有较理想的信度和效度。基于样本三 (n=520)、样本四 (n=230) 的研究表明，职业生涯高原对公务员的组织承诺、职业倦怠、工作退缩行为均有不同程度的消极的影响，但相对于升迁停滞，公务员职业边缘化对于上述效果变量的影响更为显著。

关键词　公务员；职业生涯高原；升迁停滞；职位边缘化

分类号　B849：C93

1　问题的提出

公务员是我国社会中的一个特殊群体，指"我国各级党政机关、人大政协、法院检察院，以及承担行政职能的事业单位和群众团体的人员"（依据《公务员法》）。与其他职业群体相比，公务员拥有较多的社会、文化和经济资源，掌握较大的社会公共权力，承担公共事务的管理责任 (董鑫，2007)。职业生涯高原 (career plateau) 指的是个体在当前组织中职业生涯发展出现停滞的一种现象 (Smith–Ruig, 2009; Miles, Gordon, & Storlie, 2013)。即在职业生涯的某一阶段，个体进一步① "晋升" (Ference, Stoner, & Warren, 1977)、② "工作流动" (Veiga, 1981)、③ "承担更大或更多责任" (Feldman & Weitz, 1988)、④ "学习新知识与新技能" (Lee, 2003) 的可能性很小。鉴于职业生涯高原对个人和组织效能的消极影响，近年来，中国企业员工的职业生涯高原问题受到较多的关注 (谢宝国，龙立荣，2008; 余琛，2006; 李华，2006; 王竹青，

---

* 国家自然科学基金重点项目 (71232001)、教育部人文社科青年基金项目 (14YJC630084)、湖北省社会科学基金项目 (2012143)、华中师范大学中央高校基本科研业务费项目 (CCNU14Z02015) 资助。

张慧，2007)，也有少量针对教师职业生涯高原的研究（寇冬泉，2007；高峰，2011；惠善康，曹健，2010)，但针对中国公共组织中公务员群体的职业生涯高原现象，尚未有系统的研究。

1.1 公务员仕途的"天花板"现象

公务员群体中是否存在职业生涯高原现象？从我国公务员管理体制的历史与现状来看，答案应是肯定的。近10年来，中国公务员职业生涯发展所依托的制度基础是2006年颁布实施的《中华人民共和国公务员法》，该法对公务员录用、分类管理、考核、职务任免、升降、交流与退出做出系统规定。与中国公务员法相配套的法规还有《党政领导干部选拔任用工作条例》、《公务员职务与级别管理规定》等。以上法规构成了中国公务员群体职业生涯发展的基本政治与制度背景。但没有任何制度是万能的，或完美无缺。许多研究者认为，我国公务员群体的职业生涯发展，存在许多制度性的瓶颈制约，比如：①晋升渠道单一，发展空间狭小；②人力资源配置不够优化；③官本位现象严重，个人目标与组织目标脱节；④内外流动不畅，人才难尽其用；⑤培训重形式轻实效；⑥"强者流失、弱者易留"，"逆淘汰"现象严重（梁文懋，杨龙兴，2006；梁丽芝，郑凤娇，2007)。直到今天，这些问题都没有得到很好地解决。

公务员管理体制的瓶颈，制度变革的滞后效应，加上公务员队伍年轻化、知识化的发展趋势，加剧了公务员职业发展的难题，比如职业路径单一、晋升通道狭窄、晋升空间有限、工作流动困难（张文勤，2006；张再生，2005；杜兴洋，田进，2011)。由于我国各类党政机关为高耸的"金字塔"式组织结构，我们推测，大量有潜质的公务员的职业生涯发展会过早地停滞不前，他们在不同人生阶段，都长期陷入职业生涯高原状态，也被媒体称为官员或干部仕途的"天花板现象"，即在干部成长过程中，大多数人达到一定级别后，晋升空间便会越来越小，从而在不同阶段遇到自身仕途的"天花板"的状况（杜凤娇，2009)。例如，人民论坛杂志社曾联合多家网络媒体和研究机构进行了一项"干部成长天花板"的大型调查（受调查人数总计8311人)，结果显示，64%的被调查者认为县处级"天花板"干部最多，尤其是年龄在45-55岁之间，而全国绝大部分公务员都在处级以下，在县乡两级公务员中，更多的人只能在"科员"与"办事员"这"两级台阶"上走完"仕途"（杜凤娇，2009)。但遗憾的是，学术界尚未对这一重要的现象开展过系统的研究。

1.2 公务员职业生涯高原的结构

公务员的职业生涯高原是否有别于企业员工？这是值得研究的问题。从前文所述，职业生涯高原概念的内涵十分丰富，导致有关职业生涯高原概念结构的研究长期存在较大的争议。比如职业生涯高原概念先后出现：①单维观（指晋升的可能性小）(Tremblay & Roger, 1993, 2004)；②二维观（层级高原、工作内容高原）(Milliman, 1992)；③三维观（结构高原、内容高原、生活高原）(Bardwick, 1986)；④四维观（结构高原、内容高原、个人选择和工作技能）(Joseph, 1996)。其中，结构高原和层级高原均指在组织中难以晋升；内容高原指在工作中难以获得新知识和技能；生活高原指个体所感受到在所有生活领域的低成就感或停滞感。国内研究者谢宝国、龙立荣和赵一君(2008)通过实证研究，提炼出三维结构：层级高原、内容高原和中心化高原。其中，中心化高原指很难被赋予更多的责任，向组织中心转移的可能性很小。但以上都是针对企业员工的研究结果。

根据《公务员法》，中国公务员分为领导职务（包括乡科级、县处级、市厅级、省部级、国家级）和非领导职务（包括科员与办事员、正副主任科员、正副调研员、正副巡视员）两大类别或系列，但两类公务员之间存在交错流动的现象（施康，2006)。概括而言，主要有两种职业生涯运动形式：第一，跨越职务、职级向上流动，即"晋升"。值得注意的是，向上流动可能会有两种结果：一是"被重用"，地位上升、资源更多、职权和职责扩大、待遇更好；二是职责与权力没有明显变化，甚至下降或缩小，即被"边缘化"、"明升暗降"。第二，跨越公共部门职能边界平行或横向流动，即"交流"。《公务员法》规定，公务员需要在不同地区、部门、职务进行交流、挂职锻炼、调任或转任。相对于企业员工在组织内部的转岗、轮岗而言，公务员的交流范围更大、形式较多、内涵更丰富，并伴随着职权、职责的变化。并且，交流有一定"趋高性"，往往成为晋升的前提，没有交流的机会，也表明晋升希望渺茫 (施康，2006; 刘欣，2014)。同样地，交流可能会给公务员带来或"起"或"落"的感受。

鉴于公共组织的职位设计与公务员职业生涯运动轨迹的复杂性，我们推测，公务员职业生涯高原的概念结构也有别于企业员工。比如，晋升与交流的停滞与机会缺失可能是公务员职业生涯高原的核心，并伴随着权力资源的缺乏，以及被组织"冷落"或"边缘化"的感受。但以上推测还有待实证研究的检验。

1.3　公务员职业生涯高原的测量

对职业生涯高原的实证与干预研究都必须建立在对这一概念的有效测量的基础上。20 世纪 90 年代以前，研究者主要采用从他人判断的角度对职业高原进行测量，并提出了一些客观测量指标，如年龄、任职时间、两次晋升的间隔时间等 (Slocum, Jr, Cron, Hanson, & Rawlings, 1985; Evans & Gilbert, 1984)。其中任职时间是一个相当重要的指标，如 Tremblay 和 Roger (1993) 通过不同样本的测量，均发现如果员工的任职时间达到或超过 5 年，便可判定其达到职业生涯高原期。虽然使用任职时间、年龄、晋升间隔期等指标较为客观，但无法解释诸如为什么有些员工在同一职位干了很长时间，却依然能保持较高的工作热情和不俗的绩效等现象。

事实上，员工的工作态度和行为的基础，是基于他们对客观现实的主观认知，而非客观现实本身。因此上世纪 90 年代以后，学者们更多地从个人感知的角度，对职业生涯高原进行主观测量。Chao (1990) 就明确指出采用主观知觉测量比用客观指标更为恰当，因为个体对其职业生涯发展的主观评估，决定了其对当前工作的反应，Chao 因此使用了包含两个条目的量表来测量知觉的结构高原。随后，采用主观测量方法的研究逐渐成为主流。比如 Joseph (1996) 和 Milliman (1992) 各自开发了一些条目来测量结构高原和内容高原。谢宝国等 (2008) 也基于中国企业员工开发了一套包含 16 个条目的主观测量工具。作为一项探索性的研究，我们对公务员职业生涯高原也将采用主观测量的方式，同时结合一些客观指标进行交互验证。

综上所述，探究基层公务员职业生涯高原现象，是深入了解中国公务员队伍职业生涯发展现状，揭示公务员职业心理与行为规律的一个独特视角。虽然学术界对职业生涯高原现象的研究已有了一些积累，但这些研究主要基于营利性的企业组织和员工样本，缺乏对非营利性的公共管理组织与公务员队伍的研究。且在以往文献中，有关职业生涯高原的理论与概念结构并不统一，缺乏统整的、权威的理论或结构模型作为实证研究的基础。另一方面，国内外文献中缺乏对公职人员职业生涯高原的理论与实证研究，由于文化与体制的差异，国外的研究结果也很难被直接适用于中国公务员对象。

考虑到中国文化中的高权力距离、"官本位"的传统思想，以及公共组织的"金字塔式"层级结构，"职位"与"权力"可能是中国公务员职业发展特殊"生态环境"中最为核心的两大要素，而这一点与企业员工也有较大差异 (如薪酬回报、知识经验、工作技能、可雇佣性对于企业员工职业发展尤其重要)。因此，本研究拟采用"归纳式"研究取向，即从现象、经验和数据出发，"自下而上"地建构公务员职业生涯高原的理论结构。由于处级及其以下级别的公务员是职业生涯高原多发的阶段，且在人数上也占中国公务员队伍的绝大多数 (杜凤娇，2009), 本研究主要以处级以下的基层公务员为研究对象，拟初步探索公务员职业生涯高原的内涵结构与测量，进一步考察职业生涯高原与公务员职业心理与行为的关系。

2　研究1：公务员职业生涯高原的质性研究及结构要素的探索

2.1　研究方法

2.1.1　公务员职业生涯高原的访谈、开放式问卷调查及问卷的初步编制

第一步：深度访谈。首先，研究人员对职业生涯

高原的相关文献进行了分析，制定了访谈提纲。访谈提纲分为两个部分：第一部分为受访者的基本信息，比如性别、年龄、工龄、学历、单位、职务、职责、职级等；第二部分为公务员职业生涯高原的表现，比如要求受访者回答以下问题：在党政机关里工作，陷入职业发展停滞状态的公务员是否普遍，哪些人员表现比较突出，有何征兆，请举例说明等。访谈是半结构化的深度访谈，以个别访谈的方式进行，每名访谈对象的访谈时间为 1 个小时左右。

考虑到访谈对象身份与访谈内容的特殊性，同时为了充分地收集信息，首先研究人员确定选取 16 名处级以下公务员（领导职务、非领导职务各 8 名）作为访谈对象，其中，领导职务中有 2 名正处、2 名副处、2 名正科、2 名副科；非领导职务中有 2 名调研员、2 名副调研员、2 名主任科员、1 名副主任科员和 1 名科员。接下来，在江西省南昌市和新余市的市委组织部门的支持和帮助下，研究人员在两市公务员名册上，每间隔一定数量的页码选取一页，然后基于该页名单中的个人信息与研究需要，从中选取 1 人作为访谈对象，并通过电话与其约定访谈时间，这样直到选取所需的 16 名对象为止。在访谈对象选取时，注意涵盖了不同性别、年龄、职务、级别和组织。在性别上，男性 10 名，女性 6 名；访谈对象的年龄在 35 ～ 47 岁之间；其工作单位包含林业局、农业局、审计局、经贸委、民政局、教育局、科技局、统战部、人大、司法局等多家单位。

第二步：开放式问卷调查。为了在更大范围内搜集公务员职业生涯高原的信息，在江西省南昌市某干部培训班上进行了开放式问卷调查。调查内容与访谈研究相同，比如公务员陷入职业生涯发展停滞状态的具体信号、征兆、表现等，要求调查对象对每一问题至少列出 3 项信息或事例。被试是来自江西省不同地市的公务员，总共发放 55 份问卷，回收有效问卷 43 份。其中，男性占 71%，女性占 29%；平均工龄 24 年；专科占 11%，本科占 76%，研究生以上占 13%；正科级占 23%，副处级占 68%，正处级占 9%。

第三步：编码、条目归类与汇总。首先由 3 名

研究人员独立对访谈和开放式问卷调查所得资料进行编码，然后集中讨论编码结果，在达成一致的基础上将所有条目输入计算机，进行筛选工作，然后归类、汇总。筛选标准为：(1) 被试的描述必须有清楚的含义；(2) 对频次和重要性排序，选取出现频次大于 3 的描述。根据以上标准，最后得到 30 个有关公务员职业生涯高原的描述性陈述。

第四步：问卷条目的初步编制。在确定了公务员职业生涯高原所包含的特征之后，由 2 名人力资源管理专家参照国内外职业生涯高原测量问卷的描述，并根据前一研究阶段所汇总的公务员职业生涯高原现象描述，经分析讨论后确定了公务员职业生涯高原测量问卷的 20 个初始条目。为了确保条目的内容效度，另请 8 名研究人员对这 20 个初始条目进行了评定与修改。在综合考虑了问卷的内容效度，条目的单维性、重要性、文字表述清晰和简洁性，以及是否适合公务员对象这几个方面的因素后，最终缩减为 14 个条目构成公务员职业生涯高原预试问卷。采用 Likert 5 点等级量表计分，1 表示"完全不符合"，5 表示"完全符合"。

第五步：试测。在江西省南昌市某公务员学习班上对初编问卷进行了试测。试测时，共发放 50 份问卷，回收有效问卷 41 份。在样本中，男性占 68.3%，女性占 31.7%；正科级占 4.9%，副处级占 90.2%，正处级占 4.9%；大专占 4.9%，本科占 61.0%，研究生及其以上占 34.1%。在填答问卷的同时，要求调查对象对条目中描述不合适、表述不恰当或理解不清的地方进行标记。问卷填写完毕后，研究人员随机对其中 6 名公务员进行了回访，征求他们对问卷的意见。最后综合上述信息对部分条目的用词表达进行了调整，并将包含 14 个条目的职业生涯高原问卷用于下一阶段的大样本问卷调查。

2.1.2 大样本问卷调查

大样本问卷调查被试一部分是来自江西省南昌市的一个公务员培训班学员，学员是来自该省不同地市的公务员，另一部分来自江西省新余市党校培训班的学员，其学员是来自该市所辖各区、县、乡的公务员。

问卷由研究人员在培训期间集中发放，并当场回收。

本次调查共发放 320 份问卷，回收有效问卷 279 份，有效回收率为 87.19%。在有效样本中，男性占 82.4%，女性占 17.2%；已婚占 96.8%，未婚占 0.7%；中专及其以下占 0.4%，大专占 17.6%，本科占 60.9%，研究生及其以上占 20.4%；乡镇占 16.8%，县（市、区）占 20.8%，设区市占 40.5%，省直占 19.0%，其他地区占 1.4%；正乡科级占 30.1%，副县处级占 49.8%，正县处级占 18.6%，其他职级占 0.4%；领导职务占 89.9%，非领导职务占 7.6%；30 岁（含）以下占 1.9%，31～35 岁占 7.6%，36～40 岁占 30.5%，41～45 岁占 41.8%，46～50 岁占 12.4%，51～55 岁占 4.4%，55 岁以上占 1.5%；调查对象的平均工龄为 22.65 年，标准差为 6.07。

2.2 探索性因素分析结果

为了寻求公务员职业生涯高原的概念结构，采用 SPSS16.0 统计软件对 14 个问卷条目进行探索性因素分析，提取因子的方法为主成分分析法 (principal components analysis)，转轴的方法是极大方差法，选取特征根大于等于 1 作为保留因子的标准。第一次探索性因素分析后删除了因素负荷过低的条目"调任至外单位任职的希望不大"（因素负荷小于 0.35）；第二次探索性因素分析后删除了因素负荷过低的条目"在单位内部调动工作的可能性不大"（因素负荷小于 0.35），以及存在双重负荷的条目"我不想继续升职"；第三次探索性因素分析后删除了存在双重负荷的条目"上

级领导不关注我"；第四次因素分析的结果比较理想，10 个测量条目呈现出比较清晰的两因素结构，各测量条目的因素负荷在 0.62~0.81 之间，解释的方差总量为 60.61%。具体见表 1。

因素一包括 6 个条目，反映公务员知觉到在现任职位上得不到上级和组织的重视与关怀，缺乏权力和影响力，很难获得更大的职权，缺乏成就感，并倍感失落的一种状态，体现的是在目前职位与权力体系中被"边缘化"后的感受。因此，我们将该因素命名为"职位边缘化"。因素二包括 4 个条目，反映公务员知觉到的其职务、级别、工作岗位在公职系统中难以得到进一步晋升、交流的状况。因此，我们将该因素命名为"升迁停滞"。

经探索性因素分析所得两因素间具有中等程度的相关 ($r = 0.51$, $p < 0.01$)，职位边缘化、升迁停滞分别与总分的相关为 0.86、0.88 ($p < 0.01$)，说明两因素既共同反映了相同构念的内容，又具有一定程度的独立性。职位边缘化、升迁停滞的内部一致性信度系数 (Cronbach α) 分别为 0.85 和 0.81，总问卷的内部一致性系数为 0.87，表明职业生涯高原问卷的信度较好。

3　研究 2：公务员职业生涯高原结构的验证及信效度检验

3.1 研究方法

选取江西省南昌、九江、新余、景德镇、萍乡、宜春、赣州共 7 个地市的研究样本，采用本研究所开发的公务员职业生涯高原问卷，将 10 个测量条目在不同维度上进行交叉排列，以 Likert 5 点等级量表来计分。通过各地公务员培训与学习班、党校、机关单位来搜集数据，样本覆盖面宽、来源广泛，具有代表性，基本涵盖了该省各地区、层级、单位的公务员。采用集中发放问卷，当场回收，研究人员进行现场指导，并强调本调查的学术目的和匿名性。总共发放 1250 份问卷，回收有效问卷 1037 份，有效回收率为 82.96%。由于本次问卷调查所搜集的样本量较大，我们通过计算机将其随机分为两半，一半样本 (n=517)

表 1　公务员职业生涯高原结构的探索性因素分析

测量条目	因素 1：职位边缘化	因素 2：升迁停滞
Q1：不被上级领导关怀和重视	0.81	0.22
Q2：已很少得到组织的关怀和重视	0.80	0.15
Q3：在工作中缺乏成就感	0.77	0.23
Q4：在单位里经常感到很失落	0.74	0.14
Q5：很难获得更大的职权	0.66	0.19
Q6：所在的职位缺乏权力和资源	0.62	0.20
Q7：继续升迁调动的可能性不大	0.13	0.78
Q8：对得到上级的提拔不抱希望	0.37	0.78
Q9：职务级别很难得到提升	0.33	0.77
Q10：晋升与交流的愿望不再强烈	0.09	0.73
解释的方差（变异量）（共计 60.61%）	35.07%	25.54%
内部一致性 Cronbach α 系数	0.85	0.81

注：表中数据为各测量条目的因素负荷值。

用于验证性因素分析，另一半样本 (n=520) 用于考察职业生涯高原与效果变量的关系。

在样本一 (n=517) 中，男性占 67.4%，女性占 32.6%；已婚占 88.3%，未婚占 11.7%；中专及其以下占 5.7%，大专占 32.0%，本科占 53.9%，研究生及其以上占 8.4%；乡镇占 10.5%，县区市占 17.6%，设区市占 56.7%，省直占 13.4%，其他地区占 1.8%；副乡科级占 19.4%，正乡科级占 33.3%，副县处级占 19.2%，正县处级占 11.7%，其他职级占 16.4%；领导职务占 70.3%，非领导职务占 29.7%；30 岁（含）以下占 18.1%，31～35 岁占 14.2%，36～40 岁占 21.5%，41～45 岁占 21.7%，46～50 岁占 10.4%，51～55 岁占 11.4%，55 岁以上占 2.8%；调查对象的平均工龄为 19.89 年，标准差为 10.003。采用 SPSS16.0 和 LISREL8.30 统计软件对数据进行统计分析。

3.2 研究结果

3.2.1 验证性因素分析

采用验证性因素分析 (CFA) 来验证公务员职业生涯高原两因素结构。验证性因素分析技术的关键在于通过比较多个模型之间的优劣，来确定最佳匹配模型。从前面的探索性因素分析 (EFA) 结果可知，公务员职业生涯高原是一个两因素的结构，包括升迁停滞和职位边缘化。但这两个因素之间具有显著的相关，那么，公务员职业生涯高原是否为单维度结构，即单因素模型？本研究对单因素模型和双因素模型进行比较。选择的拟合指数包括 χ^2、df、χ^2/df、RMSEA、SRMR、IFI、CFI、GFI。验证性因素分析结果见表 2、表 3 所示。

从表 2 所呈现的模型拟合指数的结果来看，单因素模型的 χ^2/df 值超过了 5，RMSEA 值超过了 0.08，SRMR 值超过了 0.06，说明单因素模型的拟合情况较差。相对而言，双因素模型的各项拟合指数明显较好，达到可接受的临界水平，并且与单因素模型比较，$\Delta\chi^2$ 为 128.45、Δdf=1，达到 0.001 的显著水平，我们接受拟合效果更好的双因素模型，这一结果也说明我们编制的公务员职业生涯高原问卷具有较好的结构效度。

3.2.2 信度分析

对样本二 (n=517) 的数据进行信度分析。其中，升迁停滞、职位边缘化两个分量表的内部一致性 Cronbach α 信度系数分别为 0.82、0.85，总问卷的内部一致性系数为 0.89，所有的信度系数均高于 0.70 的推荐值。这表明，公务员职业生涯高原问卷的信度质量比较理想。同时也说明，研究 2 和研究 1 中，公务员职业生涯高原问卷信度比较稳定。

3.2.3 效标检验

根据《公务员法》和其他相关法规，如《党政领导干部选拔任用工作条例》、《公务员职务与级别管理规定》，我国选任制和委任制公务员大多具有工作任期制度。选任制公务员每一职务任期大约 5 年，在任期届满后，如果不能晋升，便可能体会到"仕途停滞"感，而委任制公务员如果在同一职位或职级上连续工作满 5 年以上，如果无法改变，也同样可能存在"仕途停滞"感。在本研究的访谈和开放式问卷调查中，接受调查的公务员也普遍认为，如果在某一职级或职务连续工作满 5 年，还得不到提拔升迁的话，晋升的希望就比较小。在以往文献中，Slocum 等 (1985)、Savery (1989)、Tremblay 和 Roger (1993) 也通过不同的企业样本进行测量均发现，如果企业员工在当前职位上的任职时间达到或超过 5 年，便可判定其达到了职业生涯高原阶段。综上考虑，我们选择了三个效标。

表 2　公务员职业生涯高原的验证性因素分析结果

模型	χ^2	df	χ^2/df	RMSEA	SRMR	IFI	CFI	GFI
虚模型	4979.02	45	110.65					
单因素模型	283.51	35	8.10	0.12	0.06	0.95	0.95	0.89
双因素模型	155.06	34	4.56	0.08	0.04	0.98	0.98	0.94

表 3　双因素结构模型中各潜变量在外源变量及误差上的负荷

升迁停滞			职位边缘化		
测量项目	负荷	误差负荷	测量项目	负荷	误差负荷
T1	0.72	0.49	Q1	0.76	0.43
T2	0.76	0.42	Q2	0.77	0.41
T3	0.70	0.52	Q3	0.67	0.55
T4	0.76	0.43	Q4	0.63	0.60
			Q5	0.73	0.46
			Q6	0.64	0.58

首先，在问卷中编制了一个题目，要求调查对象根据实际情况评估自己"在未来5年内晋升的可能性"。该题采用5级计分，1表示"完全可能晋升"，5表示"完全不可能晋升"。通过对样本数据进行相关分析，结果发现，公务员的"升迁停滞"与其感知的"未来5年内晋升的可能性"显著负相关($r=-0.57$, $p<0.01$)，公务员的"职位边缘化"与其感知的"未来5年内晋升的可能性"显著负相关($r=-0.28$, $p<0.01$)。

其次，考虑到中国公务员管理体制的特殊性，如公务员的职级与职务既有联系又有区别，职务指公务员所具有的岗位头衔称谓，主要体现工作能力和职责大小，如科长、处长、县长、局长、厅长等；职级指一定职务层次所对应的级别，主要体现资历，如县长所对应的职级一般是县处级正职，与正处级、调研员可能为相同级别。职务与职级不一定具有"同步性"，比如存在"高职低就"和"低职高就"的情况（施康，2006；刘欣，2014），我们采用独立样本t检验方法进行差异分析，结果发现：(1) 任目前职级5年以上的公务员，其升迁停滞得分($M=3.18$, $SD=0.96$)显著高于5年以内者($M=2.85$, $SD=0.99$)($t=3.79$, $p<0.01$)，其职位边缘化得分($M=2.85$, $SD=0.90$)也高于5年以内者($M=2.70$, $SD=0.87$)(边缘显著, $t=1.77$, $p<0.10$)。(2) 任目前职务5年以上公务员，其升迁停滞得分($M=3.24$, $SD=0.97$)显著高于5年以内者($M=2.85$, $SD=0.98$)($t=4.17$, $p<0.01$)，其职位边缘化得分($M=2.90$, $SD=0.92$)也显著高于5年以内者($M=2.69$, $SD=0.86$)($t=2.49$, $p<0.05$)。以上结果表明，公务员职业生涯高原测量问卷，对公务员晋升和工作任期具有较好的区分度，如果5年没有得到升迁，公务员容易出现职业生涯高原状态。同时，上述结果表明公务员感知的升迁停滞与职位边缘化程度有差异，这也在一定程度上说明公务员职业生涯高原双因素结构的合理性。

最后，职业成功(career success)是个体职业生涯发展最重要的结果，指的是个体在工作经历中逐渐积累和获得的积极的心理感受（主观职业成功），以及与工作相关的成就（客观职业成功）(Seibert, Grant, & Kraimer, 1999)。在当代职业生涯管理领域中，个体主观的或心理成功感日益受到学者们的重视。职业满意度(career satisfaction)是个体从自身的角度，采用内在的标准对其职业生涯发展状况的总体性评价与感受，是衡量个体主观职业成功的最重要的、普遍采用的指标(Heslin, 2005)。理论上，职业生涯高原是公务员知觉到的一种严重的负性职业发展状态，与职业成功应是相反的结果。

为了检验上述假定，同时考虑到问卷的长度问题，我们在问卷中加入一道题目，请公务员评价"对目前职业发展现状的总体满意度"，采用Likert 5点计分，1表示"非常不满意"，5表示"非常满意"。该题目来源于Greenhuas、Parasuraman和Wormley (1990)编制的职业满意度问卷，该指标目前被学术界广泛用于测量主观职业成功。我们假设：公务员职业生涯高原与自评的职业满意度负相关。对公务员职业生涯高原和职业满意度进行相关分析结果表明，公务员的升迁停滞与职业满意度呈现显著的负相关($r=-0.41$, $p<0.01$)，公务员的职位边缘化与职业满意度同样呈现显著的负相关($r=-0.54$, $p<0.01$)。该结果验证了我们的假设，说明当公务员处于职业生涯高原状态时，对自身职业生涯发展满意度的评价也不高。

3.2.4 公务员职业生涯高原在人口学因素上的特征

进一步通过独立样本t检验、单因素方差分析和相关分析，探讨公务员职业生涯高原在人口统计学因素上的特征。结果表明：(1) 公务员的升迁停滞在性别($M_{男}=2.99$, $SD_{男}=0.98$, $M_{女}=2.94$, $SD_{女}=0.97$, $t=0.56$, $p>0.05$)上不存在显著差异，公务员的职位边缘化在性别($M_{男}=2.75$, $SD_{男}=0.86$, $M_{女}=2.78$, $SD_{女}=0.92$, $t=0.30$, $p>0.05$)上也不存在显著差异；(2) 公务员的升迁停滞在地区因素上不存在显著差异($F_{(4,486)}=0.51$, $p>0.05$)，职位边缘化在地区因素上差异显著($F_{(4,478)}=3.36$, $p<0.05$)，经事后多重比较发现，县区市公务员($M=2.69$, $SD=0.90$)的职位边缘化显著高于乡镇公务员($M=2.69$, $SD=0.90$)；(3) 已婚公务员的升迁停滞($M=2.99$, $SD=0.96$)与未婚公务员($M=2.78$, $SD=1.03$)的差异不显著($t=1.52$, $p>0.05$)，已婚公务员的职位边缘化($M=2.72$, $SD=0.84$)与未婚公务员

(M=2.87, SD=0.97) 的差异不显著 (t=1.07, p>0.05)；(4) 领导职务公务员的升迁停滞 (M=2.97, SD=0.96) 与非领导职务公务员 (M=3.01, SD=1.02) 的差异不显著 (t=0.37, p>0.05)，但是，非领导职务公务员的职位边缘化 (M=3.01, SD=0.88) 显著地高于领导职务公务员 (M=2.67, SD=0.87)(t=3.92, p<0.01)；(5) 最后，公务员的升迁停滞与其年龄显著的正相关 (r=0.32, p<0.01)，与其工作年限显著地正相关 (r=0.28, p<0.01)，与其教育程度显著的负相关 (r=-0.22, p<0.01)，公务员的职位边缘化与其年龄的相关不显著 (r=0.08, p>0.05)，与其工作年限的相关也不显著 (r=0.01, p>0.05)，但与其教育程度显著的负相关 (r=-0.20, p<0.01)。以上结果不仅能初步反映中国基层公务员的职业生涯高原现状，也为本研究开发的公务员职业生涯高原问卷的有效性提供了更多的证据。

4 研究3：职业生涯高原与公务员职业心理与行为的关系

职业生涯发展是人们获得成就感和价值感的重要来源。而较早进入职业生涯的停滞期，往往会使他们发生机能失调，产生挫败感，进而降低工作积极性和组织的效能。正是由于职业生涯高原所引发的一系列问题，使得该现象成为人力资源管理研究的重要内容。以往研究证实，职业生涯高原会对企业员工的工作满意度、组织承诺、组织认同、工作绩效、缺勤、工作投入、工作压力、离职等一系列工作态度与行为变量产生消极影响 (Lemire, Saba, & Gagnon, 1999; McCleese, Eby, Scharlau, & Hoffman, 2007; Ji-hyun & Jinkook, 2008; 余琛，2006；李华，2006；谢宝国，龙立荣，2008；白光林，凌文辁，李国昊，2011)。

但值得注意的是，以上研究主要针对的是企业员工，职业生涯高原与公务员职业心理与行为关系的研究尚是空白。更重要的是，从中国公务员的历史与现状来看，与企业员工的高流动性不同，公务员队伍的离职率极低，这是不争的事实。例如，在我们的前期访谈中，管理部门和访谈对象普遍反映，即使职位多

年无法升迁，也很少有公务员会选择离职。此外，公务员的工作绩效的标准、绩效考核与评价的机制问题一直是困扰中国公共管理的一个难题，这一点也与企业员工有着明显的差异 (王骚，2011)。因此，作为一项探索性的研究，为了揭示职业生涯高原与公务员心理与行为的关系，同时检验公务员职业生涯高原问卷的预测效度，在综合考虑以上因素后，我们拟选择组织承诺、职业倦怠与工作退缩行为作为公务员职业生涯高原的效果变量。

4.1 研究假设

组织承诺 (organizational commitment) 反映个人对所属组织的参与、忠诚和认同 (Meyer & Allen, 1984)。高组织承诺的员工会主动增加工作投入，创造更高的工作绩效，从而提升组织绩效。公务员对政府的组织承诺也是保持公务员队伍稳定、提高行政效率的重要手段，提高公务员对政府的组织承诺，意义深远 (安世民，蔺全录，2007)。总所周知，中国公务员的职业生涯运动轨迹与发展结果受组织与体制因素的影响较大，且公务员向组织外部流动、主动离职的机会也较少 (梁文懋，杨龙兴，2006；梁丽芝，郑凤娇，2007)。一方面，根据归因理论 (attribution theory) 和人类成就归因中的"自我服务偏差"倾向 (self-service bias) (Weiner,1985) 来推测，公务员很可能将高原状态归因于与组织、政策、体制等相关的外因，并因此而降低对组织的承诺。另一方面，根据社会交换理论 (Blau, 1964)，长期处于职业生涯高原状态的公务员，也容易产生受到组织不公正对待的感受，会降低自己回报组织的义务感，可能表现为降低对组织的承诺。例如，针对企业员工样本的研究证实，职业生涯高原与情感承诺存在负相关关系 (Ji-hyun & Jinkook, 2008; 谢宝国，龙立荣，2008)，但对继续承诺和规范承诺的影响还缺乏研究。根据组织承诺的三维度理论，情感承诺强调对组织的情感依恋，规范承诺强调留在当前组织的义务感知，情感承诺与规范承诺间一般存在显著的正相关 (Allen & Meyer, 1990)，可推测职业生涯高原与公务员对组织的情感承诺、规范承诺间存在正相关关系。而继续承诺强调对离职的成本感知，是一种

功利性的结果，由于公共组织在中国社会的先天"优越性"，职业发展越"停滞"，越可能强化公务员对于离开公共组织所带来利益损失的感知，而不得不继续留在组织中。因此，我们提出如下假设：

假设 1：职业生涯高原与公务员的组织承诺相关。具体而言，升迁停滞、职位边缘化分别与其对组织的情感承诺负相关，与继续承诺正相关，与规范承诺负相关。

从职场压力的视角来看，职业生涯高原也是一种与组织和工作相关的工作压力源，职业倦怠则是目前被广泛关注的一种职业压力症状。职业倦怠 (occupational burnout)，也称工作倦怠、工作耗竭、职业枯竭等，指一种情绪耗竭、犬儒主义及个人专业效能降低的现象，是对工作中长久的情绪和压力的持续反应，表现为身体上的耗竭与长期性的疲倦，感觉无助与无望，进而对工作、生活或其他人产生负面的观念及态度 (Maslach & Jackson, 1981)。其中，情绪耗竭是指个体情绪与情感处于极度疲劳状态，乃至无法应付工作之需要；犬儒主义又称为讥诮态度、情感疏离等，是指个体以消极、否定或麻木不仁的态度对待工作，以应对耗竭，这是对工作本身的态度；专业效能包含对过去和现在的成就感的满足，可评估个体未来继续努力工作的可能 (Maslach, Jackson, & Leiter, 1996)。

对公务员而言，升迁的停滞，以及职位的边缘化，会在一定程度上导致公务员体验到不同程度的职业发展压力。一般而言，人们都带着期望到工作中，每个人都有成长的动机，希望在组织中不断成长。对于那些具有较高期待的公务员来说，如果期望没有得到满足，在工作中没有获得认可与提升，同时受体制的限制，又没有办法进行灵活的职业发展规划，人生中又缺乏其他的机会时，很容易产生压力，增加焦虑、恐惧等不良情绪。在以往文献中，"工作压力→工作倦怠"的模型已得到广泛支持 (Schaufeli & Taris, 2014)。根据资源保存理论 (the conservation of resources, COR) (Hobfoil, 1989) 来推测，公务员的升迁停滞与职位边缘化，会导致个体资源的实际的、潜在

的损失，因而感知到压力和威胁，工作资源的相对缺乏，会进一步对身心反应产生负面影响，引发工作倦怠感。这一推论在其他样本的研究中也得到初步的证实，如研究发现，企业员工和教师的职业生涯高原与工作倦怠正相关 (陈子彤，金元媛，李娟，2011；赵寅汝，2012)。因此，我们提出以下假设：

假设 2：公务员职业生涯高原与职业倦怠相关。具体而言，升迁停滞、职位边缘化分别与情绪耗竭正相关，与犬儒主义正相关，与专业效能负相关。

工作退缩行为 (Withdrawal behavior) 是组织成员在维系组织运作与工作角色关系的前提下，逃避工作角色和工作任务的一种行为，例如不合理的迟到、早退，找借口逃离工作、擅离岗位、缺勤、假装生病等 (Hanisch & Hulin, 1990)。由于这些行为背后的原因与机制存在某些共性，研究者统称之为退缩行为。人们的行为会受到情绪和认知的影响，当员工处于消极情绪时，更容易引发工作退缩行为 (Judge, Scott, & Ilies, 2006)。如研究发现，退缩行为可能是源于工作满意度的下降，以"闹情绪"方式表现出来，也可能是个体面临角色冲突和工作压力的一种被动性应对方式 (Griffeth, Gaertner & Sager, 1999; Taris, Schreurs, & van Iersel-van Silfhout, 2001)。

如同前面的分析，从归因的角度来看，公务员很可能将职业生涯高原归因于与公共管理组织、政策、体制等相关的外因。根据社会交换理论，处于职业生涯高原状态的公务员，会体验到组织不公正对待的感受，但在公共组织中，公开的报复或离职行为一般较为少见，公务员更容易做到的是减少对工作的投入，减少在组织中的主动参与行为，即更多地表现出退缩行为。中国文化强调以和为贵、遇事忍为先，因此在长期面对职业发展停滞时，基层公务员更可能出现消极情绪和负面认知，增加公务员身心双方面回避工作的倾向，例如消极怠工、拒绝帮助、"出工不出力"、"在其位不谋其政"等。作为国家公共管理系统中的公务员，退缩行为不仅会给个人和组织带来消极影响，还有可能给社会和民众带来长久的负面效应。因此，我们提出以下假设：

假设 3：公务员职业生涯高原与工作退缩行为正相关。具体而言，公务员升迁停滞与工作退缩行为正相关、职位边缘化与工作退缩行为正相关。

4.2 方法

4.2.1 样本

（1）样本一来自江西省南昌、九江、新余、景德镇、萍乡、宜春、赣州共 7 个地市的共 1037 名处级（含）以下的基层公务员中随机分配的另一半样本（n = 520），问卷调查方法和数据收集过程与研究 2 相同，该样本量也相对较大，基本覆盖了各级、各类公务员。在该样本中，我们同时测量与收集了公务员职业生涯高原与组织承诺的数据，我们基于该样本来检验假设 1。其中，男性占 67.1%，女性占 32.9%；已婚占 90.2%，未婚占 9.8%；中专及其以下占 2.2%，大专占 31.1%，本科占 59.9%，研究生及其以上占 6.8%；乡镇占 8.6%，县区市占 13.5%，设区市占 61.3%，省直占 15.3%，其他地区占 1.2%；副乡科级占 20.7%，正乡科级占 32.1%，副县处级占 18.6%，正县处级占 11.8%，其他职级占 16.8%；领导职务占 69.7%，非领导职务占 30.3%；30 岁以下占 18.4%，31~35 岁占 17.9%，36~40 岁占 16.3%，41~45 岁占 22.7%，46~50 岁占 11.7%，51~55 岁占 10.5%，55 岁以上占 2.5%；调查对象的平均工龄为 19.51 年，标准差为 9.64。

（2）样本二来自湖北省某厅级单位及其下属的 2 家单位的公务员，研究人员对不同部门的公务员进行现场问卷调查，并当场收回问卷。问卷内容主要涉及公务员职业生涯高原、职业倦怠和工作退缩行为。共发放问卷 250 份，有效回收问卷 230 份，有效回收率为 92%。样本二集中于某一特定的机关系统，主要为处级以下公务员，虽然样本量小于样本一，但被试在性别、学历、年龄、工龄与职级上分布比较均匀，样本代表性较强，我们基于该样本的数据来检验假设 2 和假设 3。其中，男性占 57.8%，女性占 42.2%；学历为中专及以下占 2.6%，大专占 8.3%，本科占 48.7%，研究生及以上占 40.4%；在年龄上，25 岁以下占 4.3%，2630 岁占 22.2%，3135 岁占 22.2%，3640 岁占 16.5%，4145 岁占 12.6%，46 岁以上占 22.2%；工龄为 5 年以下者占 29.1%，610 年者占 17.4%，1115 年者占 13%，

1620 年者占 12.2%，21 年以上占 28.3%；办事员占 28.7%，科员占 43.9%，副科级占 11.7%，正科级占 10%，副处级占 4.8%，正处级占 0.9%。

4.2.2 测量工具

职业生涯高原。采用本研究所编制的公务员职业生涯高原问卷，包含"升迁停滞"（4 个测量项目）和"职位边缘化"（6 个测量项目）两个维度，将 10 个测量项目在不同维度上进行交叉排列，以 Likert 5 点计分，1 表示"完全不符合"，5 表示"完全符合"。在样本一中，升迁停滞、职位边缘化两个分量表的内部一致性 α 系数分别为 0.83 和 0.85，总问卷的内部一致性系数为 0.89；在样本二中，升迁停滞、职位边缘化两个分量表的内部一致性 α 系数分别为 0.91 和 0.89，总问卷的内部一致性系数为 0.92。这表明，本研究开发的公务员职业生涯高原问卷的信度较为稳定。

组织承诺。采用 Allen 和 Meyer（1990）编制的三维度组织承诺问卷（即情感承诺、继续承诺和规范承诺），该问卷被国内外的众多研究普遍使用。采用 Likert 5 点等级量表计分，1 表示"完全不同意"；5 表示"完全同意"。在本研究的样本一中，情感承诺、继续承诺和规范承诺的 α 系数分别为 0.76、0.79 和 0.80，总问卷的 α 系数为 0.89。

职业倦怠。采用 Maslach 等开发的 MBI-GS（第三版）量表（Maslach Burnout Inventory-General Survey），该量表是国际上应用最广的职业倦怠测量工具，适用更广泛职业领域的员工职业倦怠水平的测评（Maslach et al., 1996）。该量表包括三个维度，分别是情绪耗竭、犬儒主义、专业效能，共 16 个测量项目，采用 Likert 5 点计分，1 表示"非常不同意"，5 表示"非常同意"。在本研究的样本二中，情绪耗竭、犬儒主义、专业效能的的 α 系数分别为 0.81、0.72 和 0.76，总问卷的 α 系数为 0.87。

工作退缩行为。采用 Lehman 和 Simpson（1992）编制的退缩行为问卷，共 4 个测量项目，如"我不想上班"、"我对目前的工作不愿全力以赴"，采用 Linkert 4 点记分，1 表示"从不如此"，4 表示"始终如此"。在本研究的样本二中，该问卷的内部一致性

α 系数为 0.74。

4.2.3 变量测量的验证性因素分析

对样本一中的变量构建了一个五因素的测量模型（升迁停滞、职位边缘化、情感承诺、继续承诺、规范承诺），基于样本一的数据 (n=520) 进行验证性因素分析，五因素模型的各项拟合指数如下：χ^2= 955.75、df=265、χ^2/df=3.61、RMSEA=0.07、SRMR=0.06、IFI=0.95、CFI=0.94、NFI=0.93、NNFI=0.94。模型拟合的结果比较理想，这说明在样本一中变量测量之间具有较好的区分效度。

同样，我们对样本二中的变量构建了一个六因素的测量模型（升迁停滞、职位边缘化、情绪耗竭、犬儒主义、专业效能、工作退缩行为），基于样本二的数据 (n=230) 进行验证性因素分析，各项拟合指数如下：χ^2=869.68、df=390、χ^2/df=2.23、RMSEA=0.07、SRMR=0.07、IFI=0.96、CFI=0.96、NFI=0.93、NNFI=0.96。模型拟合的结果同样比较理想，这说明在样本二中变量测量之间也具有较好的区分效度。

4.2.4 共同方法偏差的控制与检验

由于本研究采用问卷调查法，所有的问卷题目均由被试本人填答，因此测量中可能存在共同方法偏差 (common method bias) 问题。在程序控制上：(1) 本研究在数据收集过程中强调匿名性、保密性以及数据仅限于科学研究的说明；(2) 在不同变量的测量项目上采用不同的计分方式，例如有的测量问卷采用 Likert

5 点计分，有的则采用 Likert 4 点计分；(3) 对不同问卷采用不同的反应方式，如有的采用同意程度，有的采用符合程度，有的采用行为频次。由于程序控制只能对共同方法偏差起到部分的修正作用，我们采用统计方法进一步检验共同方法偏差效应。

首先根据 Podsakoff, MacKenzie, Lee 和 Podsakoff (2003) 的建议，我们进行了 Harman 单因子检验，也就是同时对所有变量的测量项目进行未旋转的主成分因素分析。如果得到了多个因子，且第一个因子解释的变异量没有超过40%，则表明共同方法变异问题并不严重。本研究中，样本一的未旋转的主成分因素分析结果表明，有 4 个因子的特征根值大于 1，且第一个因子解释的变异量只有 24.17%；样本二有 6 个因子的特征根值大于 1，第一个因子解释的变异量只有 27.07%。这一结果初步说明本研究中的共同方法偏差问题并不严重。

考虑到单因子方法检验共同方法偏差的局限性，进一步通过控制非可测潜在因子影响法检验共同方法偏差 (Anderson & Williams, 1992)。即将共同方法偏差作为一个潜变量进入结构方程模型，允许所有的标识变量在该方法潜变量上负载，通过比较含有共同方法偏差潜变量与不含方法潜变量的两种情况下的拟合程度，来检验共同方法偏差效应。通过比较样本一中两模型的 χ^2 和 df 的差值可知，当在五因素模型中加入一个共同方法变异因子后，模型的卡方值并没有

表 4　变量的描述性统计与相关分析结果

变量 (n = 520)	M	SD	1	2	3	4	
1. 升迁停滞	3.02	0.98	(0.84)				
2. 职位边缘化	2.77	0.91	0.64**	(0.86)			
3. 情感承诺	3.14	0.82	−0.03	−0.23**	(0.76)		
4. 继续承诺	2.75	0.83	0.09*	0.16**	0.45**	(0.78)	
5. 规范承诺	2.76	0.87	0.04	−0.03	0.69**	0.64**	(0.81)

变量 (n = 230)	M	SD	1	2	3	4	5	
1. 升迁停滞	3.39	1.13	(0.91)					
2. 职位边缘化	2.83	0.94	0.67**	(0.89)				
3. 情绪耗竭	2.92	0.91	0.54**	0.60**	(0.81)			
4. 犬儒主义	2.86	0.82	0.51**	0.62**	0.67**	(0.72)		
5. 专业效能	3.78	0.66	−0.27**	−0.50**	−0.44**	−0.52**	(0.76)	
6. 工作退缩行为	1.96	0.57	0.39**	0.56**	0.60**	0.56**	−0.53**	(0.74)

注：括号中的数据为 Cronbach α 系数；**p < 0.01，*p < 0.05。

表5 公务员职业生涯高原与各效果变量的层级回归分析

自变量	组织承诺 (n=520)			职业倦怠与工作退缩行为 (n=230)			
	情感承诺	继续承诺	规范承诺	情绪耗竭	犬儒主义	专业效能	工作退缩行为
第一步：控制变量							
性别	−0.09*	−0.07	−0.04	0.17*	0.06	−0.11	0.25**
年龄	0.15*	0.06	0.13*	0.16*	0.17*	0.10	−0.04
教育程度	−0.02	−0.10*	−0.10*	−0.11	−0.15*	0.04	−0.09
职务	−0.18**	−0.21**	−0.23**	0.05	−0.04	0.17*	0.07
第二步：预测变量							
升迁停滞	0.11	0.01	0.05	0.22*	0.19*	−0.05	0.09
职位边缘化	−0.37**	0.15*	−0.11*	0.48**	0.52**	−0.46**	0.52**
$Adj. R^2$	0.14	0.07	0.06	0.43	0.44	0.25	0.35
$\triangle R^2$	0.10	0.02	0.01	0.35	0.38	0.22	0.30

注：$^\uparrow p < 0.10$，$*p < 0.05$，$**p < 0.01$。

得到显著的改变（$\Delta df=25$，$\Delta \chi^2=35.63$）。当在样本二的六因素模型中加入一个共同方法变异因子后，模型的卡方值也没有显著的改变（$\Delta df=30$，$\Delta \chi^2=41.02$）。此外，样本一与样本二在纳入共同方法因子前后两模型的 RMSEA、CFI、IFI、NFI、NNFI 等主要拟合指标的改变在 0.010.02 之间，其改变并不显著。因此可判定，加入共同方法偏差因子后，模型并未得到十分明显的改善，即共同方法偏差对模型中的变量关系并未产生显著影响，基于本研究的数据得出的职业生涯高原与各效果变量的关系是可信的。

4.3 结果

表4呈现的是公务员职业生涯高原、组织承诺、职业倦怠、工作退缩行为的描述性统计及相关分析结果。

以职业生涯高原为自变量，分别以组织承诺的三个维度（情感承诺、继续承诺、规范承诺）、职业倦怠的三个维度（情绪耗竭、犬儒主义、专业效能）和工作退缩行为作为因变量，采用层级回归(Hierarchical Regression)分析方法探讨职业生涯高原与各结果变量之间的关系。控制变量包含性别（1为男，0为女）、年龄、教育程度和职务（1为领导职务，0为非领导职务）。表5为层级回归分析的汇总结果。

层级回归分析的结果表明，控制人口学变量后，公务员的升迁停滞与组织情感承诺（β =0.11，$p>0.05$）、继续承诺（β =0.01，$p>0.05$）、规范承诺（β =0.05，$p>0.05$）的相关均不显著；公务员的职位边缘化与组织情感承诺显著地负相关（β =−0.37，

$p<0.01$），与继续承诺显著地正相关（β =0.15，$p<0.05$），与规范承诺的负相关边缘显著（β =−0.11，$p<0.10$），本研究的假设1得到部分支持。其次，公务员的升迁停滞与情绪耗竭显著地正相关（β =0.22，$p<0.01$），与犬儒主义同样显著地正相关（β =0.19，$p<0.05$），但与专业效能的相关不显著（β =−0.05，$p>0.05$）；公务员的职位边缘化与情绪耗竭显著地正相关（β =0.48，$p<0.01$），与犬儒主义同样显著地正相关（β =0.52，$p<0.01$），与专业效能显著地负相关（β =−0.46，$p<0.01$），本研究的假设2基本得到支持。最后，公务员的升迁停滞与工作退缩行为的相关不显著（β =0.09，$p>0.05$），但职位边缘化与工作退缩行为具有显著的正相关（β =0.52，$p< 0.01$），假设3得到部分支持。此外，在上述结果中不难发现，公务员升迁停滞虽然能在一定程度上引发职业倦怠（如与情绪耗竭、犬儒主义正相关），但相对而言，职位边缘化却是一种更为严重的职业生涯高原状态，其消极影响更为严重。

5 讨论

5.1 公务员职业生涯高原现象及概念结构

某一构念的结构模型一般通过两条途径来建构：一是理论驱动型，即在文献分析与理论的基础上进行建构；二是数据驱动型，主要通过因素分析来实现。鉴于以往针对企业员工职业生涯高原的研究中，缺乏统一、权威的理论基础及结构模型，加上受文化与制度的约束，中国公务员职业生涯发展轨迹比较特殊，

本研究主要基于经验数据"自下而上"地建构公务员职业生涯高原的结构模型。本研究的"升迁停滞"维度主要是从外显的、可见的"流动"角度来刻画公务员的职业生涯高原，不仅包含针对企业员工研究所提出的层级高原 (Milliman, 1992; Bardwick, 1986; 谢宝国等, 2008)、结构高原 (Joseph, 1996) 以及针对中小学教师研究中的层级高原 (寇冬泉, 2007) 所表达的晋升或向上流动的含义，即职业生涯纵向运动的"晋升"，还包含中国公务员职业生涯发展中较为特殊的横向交流、调动与转任等职务的变化，即横向运动的"交流"。无论"升"还是"迁"对公务员而言都有着同等的意义。在中国现行公务员管理体制下，也唯有"升迁"二字最能概括公务员职业生涯运动的特殊、丰富的含义。

本研究的"职位边缘化"维度主要是从内隐的、不可见的"权力"角度来刻画公务员的职业生涯高原，具有更强烈的主观心理体验色彩，体现了公务员职业生涯高原的特殊性，与谢宝国等人 (2008) 提出的"中心化高原"也有区别。在企业中，员工即便没有获得职位晋升，也可能被组织赋予更大的工作职权与责任，或者被纳入到核心的决策团队中。但在党政机关等公共组织中，组织结构为典型的"金字塔"式科层结构，职务、级别往往与特定的权力、责任和待遇挂钩，等级森严，不容错乱，故有"在其位，谋其政"，"不在其位，则不谋其政也"的说法，因而在公务员中难以出现类似于企业员工的中心化高原。对于公务员而言，职务级别的升迁往往伴随着权力资源与社会影响力的变化，二者如影随形。换言之，公务员在升迁停滞的同时，容易感知到所拥有权力和资源的相对静止或减少，即影响力下降。更进一步，在遭遇升迁停滞与影响下降后，公务员会感受到不被上级和组织重视，由此产生低成就感与失落感，"门前冷落车马稀"正是其后果的写照。

不过，在本研究中并没有发现企业员工中广泛存在的内容高原维度 (Milliman, 1992; Bardwick, 1986; Joseph, 1996; 谢宝国等, 2008)。究其原因，可能与组织背景的差异有关。目前，企业组织中技术更新速度

快，员工的知识技能与学习能力是适应竞争、取得职业成功的关键要素。但在公共组织中，官僚行政文化仍占主导地位，公务员"官本位"思想比较严重，很多公务员在设定目标的过程中，无疑会把"晋升"、"当领导"作为实现人生价值的唯一目的或途径 (杜兴洋, 田进, 2011)。在影响公务员职业生涯发展和工作胜任的因素中，人际关系、政治背景、思想品德往往是极为重要的因素 (何胤, 2005; 候奕斌, 2007)。因此，除了少量专业技术性职务类型的公务员外，公共组织对员工知识和技能的要求总体上不如企业组织高，这在本研究前期访谈中也有相同的发现。因而，内容高原在公务员群体中并不凸显，也是容易理解的。

5.2 公务员职业生涯高原的现状特征

本研究发现，在中国基层 (处级以下) 公务员群体中，存在较为普遍的职业生涯高原现象。当陷入职业生涯高原状态时，最直接的表现就是职务、职级升迁调动的停滞 (升迁停滞)，更严重的是公务员所拥有的权力、资源减少，个人影响力下降，不被领导和组织重视、发展无望所导致的心理上的困扰 (职位边缘化)。具体而言，我们通过 t 检验发现，基层公务员的职业生涯高原在性别、婚姻上不存在显著差异。相关分析表明，随着年龄的增长，公务员职业生涯高原状况越严重，这一特征在两个维度上具有一致性，其中以 45 岁以上的公务员的职业生涯高原最为明显。并且，由于工龄与年龄具有较高的相关 ($r = 0.92$, $p < 0.01$)，所以在工龄上也基本呈现与年龄一致的情况。此外，本研究还发现，职业生涯高原与教育程度显著地负相关，其基本趋势是教育程度越高，职业生涯高原状况越好，这与当前干部选拔中对"年轻化"、"知识化"的要求是一致的。

中国是一个具有两千多年封建官僚传统的国家，自古以来就有"官者，管也"的文化传统。在职务上，本研究发现相对于领导职务，非领导职务公务员感知的职业生涯高原更为严重。由于领导职务的权力相对较大，地位较高，在某种程度上来讲，许多长期居于非领导职务系列的公务员觉得自己"有职无权"，或"无职无权"，处于边缘化状态，缺乏影响力，个人成

就感低便是必然的结果。我国政府公务员领导职务与非领导职务的划分虽然为公务员提供了"两个职业发展通道"，但由于在非领导职务中，只有极少数的公务员能晋升到上一级领导职务，专业技术人员很难沿着非领导职务的系列向上攀升（宋斌，2010）。而高层次的非领导职务（如调研员）往往成了给担任领导职务的公务员以高一级职务待遇的"虚职"，这就直接造成领导职务的"通道"吸引力依然大于非领导职务，这种"官本位"的制度设计思路使得公职人员都想方设法往领导职务的目标奋斗（施康，2006）。

5.3 职业生涯高原与公务员职业心理与行为的关系

本研究通过层级回归分析，在控制了重要的人口学变量后，考察了职业生涯高原与公务员的多种职业心理与行为效果变量（如组织承诺、职业倦怠、退缩行为）的关系。首先，众多研究证实企业员工的职业生涯高原与情感承诺的负向关系（Ji-hyun & Jinkook, 2008; Lemire et al., 1999; Chay, Argee, & Chew, 1995; Allen, Poteet, & Russell, 1998; 谢宝国，龙立荣，2008）。本研究同样证实，职位边缘化对公务员的组织情感承诺产生显著的消极影响。与我们的假设一致，本研究还发现，职位边缘化会在一定程度上削弱公务员对组织的规范承诺，同时也会增强公务员对组织的继续承诺。但本研究同时发现，升迁停滞与公务员的组织情感承诺、继续承诺和规范承诺的相关均不显著。这说明，对于组织承诺，职业生涯高原不同维度的预测作用存在差异，职位边缘化是更为显著的预测源。

与企业员工不同，公务员的收入稳定，有良好的社会保障和福利待遇，工作环境和条件较好，职业地位和职业声望较高，使得公务员成为一个热门职业。对公务员而言，放弃工作的损失较大，而重新就业的风险和难度也较大，这对年龄较大、工龄较长、学历较低的公务员更是如此。因此，即便遭遇职业生涯高原，但继续承诺却越发增强，这可能也是公务员队伍低离职率的一个原因。另一方面，在公共组织中，公务员群体大多直接受共产党与政府多年的培养和教育，文化素质、思想政治觉悟普遍较高，受制度的管束也较多，随着年龄与工龄的增长，其忠诚于组织的规范承诺应该不低。但本研究证实，职业生涯高原不仅降低公务员对组织的情感承诺，增强继续承诺，也会在一定程度上削弱公务员的规范承诺。

本研究还表明，职业生涯高原不仅会导致公务员的职业倦怠，还会促发工作退缩行为。例如，升迁停滞会造成公务员在工作中表现出更多的情绪耗竭和犬儒主义的现象，这显然会对工作效率带来消极影响；而职位边缘化会导致公务员出现更多的情绪耗竭、犬儒主义、退缩行为，还会降低专业效能感。有意思的是，本研究再次发现，相对于升迁停滞，职位边缘化对各效果变量（组织承诺、职业倦怠、工作退缩行为）均有更加显著的预测作用，这一研究结果再次说明公务员职务升迁的复杂性。这也说明，对于公务员而言，职位边缘化是一种更应值得关注和重视的职业生涯高原状态，对公务员工作态度与行为的影响更为消极。

目前，国内缺乏公务员职业生涯高原及其后果的研究，本研究基于公务员群体的研究发现既有特殊性，也进一步丰富了职业生涯高原的研究。由于组织承诺、职业倦怠、工作退缩行为也可能与公务员的工作投入、工作绩效以及众多工作偏差行为相关联，那么，公务员职业生涯高原是否能引发"心理偏差"、"不平衡感"以及职业价值的扭曲，是否能导致当前中国社会普遍关注的官员"腐败行为"、"不作为"或者"过度行为"呢？这是未来值得研究的问题，本研究给我们带来一些重要的启示，未来可从不同理论视角（例如归因、公平、社会交换、社会比较等）进行深入探究。

5.4 研究贡献与展望

本研究采用层层递进的深度访谈、多阶段大样本问卷调查的方法，通过"自下而上"地探索并验证了公务员职业生涯高原的双因素理论结构，即体现"工作流动"的"升迁停滞"，体现"权力影响"的"职位边缘化"，这一理论结构提炼出中国公共组织中公务员职业生涯发展停滞现象的本质与特色，有助于我们深入理解中国公务员职业生涯发展规律，并做好公共组织成员的职业生涯管理。本研究开发出公务员职业生涯高原问卷，具有良好的信度和效度，符合心理测量学标准，可应用于未来研究，以及公职人员职业生涯高原的咨询、诊断与干预。此外，本研究还考察了公务员职业生涯高原对一系列重要的职业心理与行为结果的影响，有助于认识职业生涯高原对公务员个人和公共组织所造成的负面影响。但作为一项探索性的研究，本研究对公务员职业生涯高原现象的研究还并不深入。未来的研究可在此基础上，进一步探讨公务员职业生涯高原对一些严重的工作偏差行为（如贪污腐败或行政不作为等）的影响机制，可从个人层面、组织层面、政策层面考察引致公务员职业生涯高原的前因以及公务员的心理反应，还可针对公务员职业生涯高原现象开展一些咨询、诊断、管理对策及干预研究。

参考文献

Allen, N. J., & Meyer, J. P. (1990). The measurement and antecedents of affective, continuance and normative commitment to the organization. *Journal of Occupational Psychology, 63*(1), 1‐18.

Allen, T. D., Poteet, M. L., & Russell, J. E. A. (1998). Attitudes of managers who are more or less career plateaued. *The Career Development Quarterly, 47*(2), 159‐172.

An, S. M., & Lin, Q. L. (2007). View of civil servants' organizational commitment system to the government. *Chinese Public Administration,* (10), 76‐78.

[安世民, 蔺全录. (2007). 论公务员对政府的组织承诺制度. *中国行政管理,* (10), 76‐78.]

Anderson, S. E, & Williams, L. J. (1992). Assumptions about unmeasured variables with studies of reciprocal relationships: The case of employee attitudes. *Journal of Applied Psychology, 77*(5), 638‐650.

Bai, G. L., Lin, W. Q., & Li, G. H. (2011). Research on the relationship among career plateau structure dimensions, job satisfaction and resignation intention. *Science & Technology Progress and Policy, 28*(3), 144‐148.

[白光林, 凌文辁, 李国昊. (2011). 职业高原结构维度与工作满意度、离职倾向的关系研究. *科技进步与对策, 28*(3), 144‐148.]

Bardwick, J. M. (1986). *The plateauing trap: How to avoid it in your career and your life.* New York: American Management Association.

Blau, P. M. (1964). *Exchange and power in social life.* New York: Wiley.

Chao, G. T. (1990). Exploration of the conceptualization and measurement of career plateau: A comparative analysis. *Journal of Management, 16*(1), 181‐193.

Chay, Y. W., Aryee, S., & Chew, I. (1995). Career plateauing: Reactions and moderators among managerial and professional employees. *International Journal of Human Resource Management, 6*(1), 61‐78.

Chen, Z. T., Jin, Y. Y., & Li, J. (2011). Empirical study on the relationship between career plateau and job burnout of knowledge worker. *Journal of Wuhan Textile University, 24*(2), 31‐33.

[陈子彤, 金元媛, 李娟. (2011). 知识型员工职业高原与工作倦怠关系的实证研究. *武汉纺织大学学报, 24*(2), 31‐33.]

Dong, X. (2007). An analysis of the intension and extension of civil servant concept. *Qi Lu Journal,* (1), 148‐150.

[董鑫. (2007). 浅析公务员概念的内涵与外延. *齐鲁学刊,* (1), 148‐150.]

Du, F. J. (2009). The ceiling of official career: A survey of cadres' career development. *People,s Tribune,* (23), 16‐19.

[杜凤娇. (2009). 仕途"天花板"：突不破的"围城"——干部成长"天花板"调查. *人民论坛,* (23), 16‐19.]

Du, X. Y., & Tian, J. (2011). The study of civil servant's career development path based on competence. *Chinese Public Administration,* (11), 105‐109.

[杜兴洋，田进 . (2011). 基于公务员胜任力的职业发展路径研究——以湖北省为例 . *中国行政管理，*(11), 105 - 109.]

Evans, M. G., & Gilbert, E. (1984). Plateaued managers: their need gratifications and their effort–performance expectations. *Journal of Management Studies, 21*(1), 99 - 108.

Feldman, D. C., & Weitz, B. A. (1988). Career plateaus in the sales force: Understanding and removing blockages to employee growth. *Journal of Personal Selling & Sales Management, 8*(3), 23 - 32.

Ference, T. P., Stoner, J. A. F., & Warren, E. K. (1977). Managing the career plateau. *Academy of Management Review, 2*(4), 602 - 612.

Gao, F. (2011). Research on the situation of the career plateau of the youth P. E. Teacher in college: Take the college P. E. Teachers in Beijing as example. *Journal of Beijing Sport University, 34*(7), 102 - 105.

[高峰 . (2011). 高校青年体育教师职业生涯高原现状的研究——以北京市高校青年体育教师为例 . *北京体育大学学报，34*(7), 102 - 105.]

Greenhaus, J. H., Parasuraman, S. & Wormley, W. M. (1990). Effects of race on organizational experiences, job performance evaluations, and career outcomes. *Academy of Management Journal, 33*(1), 64 - 86.

Griffeth, R. W., Gaertner, S., & Sager, J. K. (1999). Taxonomic model of withdrawal behaviors: The adaptive response model. *Human Resource Management, 9*(4), 577 - 590.

Hanisch, K. A., & Hulin, C. L. (1990). Job attitudes and organizational withdrawal: An examination of retirement and other voluntary withdrawal behaviors. *Journal of Vocational Behavior, 37*(1), 60 - 78.

He, Y. (2005). *Research on the achievement motivation & attribution of primary level leader* (Unpublished master's thesis). Suzhou University, Suzhou.

[何胤 . (2005). *基层党政干部成就动机及其归因的研究（硕士学位论文）. 苏州大学，苏州 .]

Heslin, P. A. (2005). Conceptualizing and evaluating career success. *Journal of Organizational Behavior, 26*(2), 113 - 136.

Hobfoil, S. E. (1989). Conservation of resources: A new attempt at conceptualizing stress. *American Psychologist, 44*(3), 513 - 524.

Hou, Y. B. (2007). *The study on the civil servants of section chief's competency and relative factors* Unpublished doctorial dissertation. Jinan University, Guangzhou.

[候奕斌 . (2007). *科级公务员胜任特征及相关因素研究（博士学位论文）. 暨南大学，广州 .]

Hui, S. K., & Cao, J. (2010). Analysis on factors affecting the shaping of teacher's career plateau in elementary and middle school. *Psychological Development and Education, 26*(5), 527 - 533.

[惠善康，曹健 . (2010). 中小学教师职业生涯高原现象的特征及相关因素 . *心理发展与教育，26*(5), 527 - 533.]

Ji-hyun, J., & Jinkook, T. (2008). The effects of perceived career plateau on employees' attitudes: Moderating effects of career motivation and perceived supervisor support with Korean employees. *Journal of Career Development, 35*(2), 187 - 201.

Joseph, J. (1996). An exploratory look at the plateauism construct. The Journal of Psychology: *Interdisciplinary and Applied, 130*(3), 237 - 244.

Judge, T. A., Scott, B. A., & Ilies, R. (2006). Hostility, job attitudes, and workplace deviance: Test of a multilevel model. *Journal of Applied Psychology, 91*(1), 126 - 138.

Kou, D. Q. (2007). *A study on the structure and characteristics of teacher's career plateau and its relationship to the job effect* (Unpublished doctorial dissertation). Southwest University, Chongqing.

[寇冬泉 . (2007). *教师职业生涯高原的结构、特点及

其与工作效果的关系（博士学位论文）. 西南大学，重庆.]

Lee, B. C. P. (2003). Going beyond career plateau: Using professional plateau to account for work outcomes. *Journal of Management Development, 22*(6), 538 – 551.

Lehman, W. E., & Simpson, D. D. (1992). Employee substance use and on–the–job behaviors. *Journal of Applied Psychology, 77*(3), 309 – 321.

Lemire, L., Saba, T., & Gagnon, Y. C. (1999). Managing career plateauing in the Quebec Public Sector. *Public Personnel Management, 28*(3), 375 – 391.

Li, H. (2006). *A study of relationship between career plateau, job satisfaction, organizational commitment and turnover intention of business managers* (Unpublished doctorial dissertation). Chongqing University, Chongqing.

[李华. (2006). *企业管理人员职业高原与工作满意度、组织承诺及离职倾向关系研究*（博士学位论文）. 重庆大学，重庆.]

Liang, L. Z., & Zheng, F. J. (2007). The development of China's professional technical civil servants career path. *Seeker, (2)*, 73 – 75, 78.

[梁丽芝，郑凤娇. (2007). 中国专业技术类公务员职业的发展路径. *求索, (2)*, 73 – 75, 78.]

Liang, W. M., & Yang, L. X. (2006). Opinions on the construction of civil servants' career planning support system in China. *Jiangxi Social Sciences, (8)*, 129 – 132.

[梁文懋，杨龙兴. (2006). 我国公务员职业生涯规划支持体系建设刍议. *江西社会科学, (8)*, 129 – 132.]

Liu, X. (2014). *Research on the horizontal flowing of civil servants of leading posts in China* (Unpublished master's thesis). South China University of Technology, Guangzhou.

[刘欣. (2014). *我国领导职务公务员横向流动问题研究*（硕士学位论文）. 华南理工大学，广州.]

Maslach, C., & Jackson, S. E. (1981). The measurement of experienced burnout. *Journal of Occupational Behavior, 2*(2), 99 – 113.

Maslach, C., Jackson, S. E., & Leiter, M. P. (1996). *Maslach Burnout Inventory: Manual* (3rd ed.). Palo Alto, CA: Consulting Psychologist Press.

McCleese, C. S., Eby, L. T., Scharlau, E. A., & Hoffman, B. H. (2007). Hierarchical, job content, and double plateaus: A mixed–method study of stress, depression and coping responses. *Journal of Vocational Behavior, 71*(2), 282 – 299.

Meyer, J. P., & Allen, N. J. (1984). Testing the "side-bet theory" of organizational commitment: Some methodological considerations. *Journal of Applied Psychology, 69*(3), 372 – 378.

Miles, S., Gordon, J., & Storlie, C. (2013). Job satisfaction, perceived career plateau, and the perception of promotability: A correlational study. *The Journal of International Management Studies, 8*(1), 1 – 9.

Milliman, J. F. (1992). *Causes, consequences and moderating factors of career plateauing* (Unpublished doctorial dissertation). University of Southern California.

Podsakoff, P. M., MacKenzie, S. B., Lee, J. –Y., & Podsakoff, N. P. (2003). Common method biases in behavioral research: A critical review of the literature and recommended remedies. *Journal of Applied Psychology, 88*(5), 879 – 903.

Savery, L. K. (1989). Comparing plateaued and non-plateaued employees. *Journal of Managerial Psychology, 4*(1), 12 – 16.

Schaufeli, W. B., & Taris, T. W. (2014). A critical review of the job demands–resources model: Implications for improving work and health. In W. B. Schaufeli, & T. W. Taris (Eds.), *Bridging occupational, organizational and public health* (pp. 43 – 68). Netherlands: Springer.

Seibert, S. E., Grant, J. M., & Kraimer, M. L. (1999).

Proactive personality and career success. *Journal of Applied Psychology, 84*(3), 416 - 427.

Shi, K. (2006). *Research on relationship and integration among civil servant´s recruitment, management and exit mechanism in China* (Unpublished doctorial dissertation). Nanjing Agricultural University, Nanjing.

[施康 . (2006). *我国公务员录用、管理与退出机制的关系及整合研究* (博士学位论文). 南京农业大学，南京 .]

Slocum, J. W., Jr., Cron, W. L., Hanson, R. W., & Rawlings, S. (1985). Business strategy and the management of plateaued employees. *Academy of Management Journal, 28*(1), 133 - 154.

Smith-Ruig, T. (2009). Exploring career plateau as a multi-faceted phenomenon: Understanding the types of career plateaux experienced by accounting professionals. *British Journal of Management, 20*(4), 610 - 622.

Song, B. (2010). Research on career development channels of public officials and SWOT model. *Chinese Public Administration,* (12), 69 - 72.

[宋斌 . (2010). 公职人员职业发展通道和 SWOT 模型初探 . *中国行政管理*, (12), 69 - 72.]

Taris, T. W., Schreurs, P. J. G., & van Iersel-van Silfhout, I. J. (2001). Job stress, job strain, and psychological withdrawal among Dutch university staff: Towards a dual process model for the effects of occupational stress. Work & Stress: An International Journal of Work, *Health & Organisations, 15*(4), 283 - 296.

Tremblay, M., & Roger, A. (1993). Individual, familial, and organizational determinants of career plateau: An empirical study of the determinants of objective and subjective career plateau in a population of Canadian managers. *Group & Organization Management, 18*(4), 411 - 435.

Tremblay, M., & Roger, A. (2004). Career plateauing reactions: The moderating role of job scope, role

ambiguity and participation among Canadian managers. *The International Journal of Human Resource Management, 15*(6), 996 - 1017.

Veiga, J. F. (1981). Plateaued versus non-plateaued managers: Career patterns, attitudes, and path potential. *Academy of Management Journal, 24*(3), 566 - 578.

Wang, S. (2011). Problems with civil servant performance appraisal and counter measures. *Journal of Shandong University* (Philosophy and Social Science), (1), 25 - 31.

[王骚 . (2011). 公务员绩效考核中的问题与对策分析 . *山东大学学报* (哲学社会科学版), (1), 25 - 31.]

Wang, Z. Q., & Zhang, H. (2007). Research into relationship between career plateau and job satisfaction of enterprise employees. *Journal of Chongqing Technology and Business University (Social Sciences Edition), 24*(6), 56 - 59.

[王竹青 , 张慧 . (2007). 企业员工的"职业高原"和工作满意度的关系研究 . *重庆工商大学学报 (社会科学版), 24*(6), 56 - 59.]

Weiner, B. (1985). An attributional theory of achievement motivation and emotion. *Psychological Review, 92*(4), 548 - 573.

Xie, B. G., & Long, L. R. (2008). The effects of career plateau on job satisfaction, organizational commitment and turnover intentions. *Acta Psychologica Sinica, 40*(8), 927 - 938.

[谢宝国 , 龙立荣 . (2008). 职业生涯高原对员工工作满意度、组织承诺、离职意愿的影响 . *心理学报, 40*(8), 927 - 938.]

Xie, B. G., Long, L. R., & Zhao, Y. J. (2008). Development of career plateau questionnaire and research on its validity and reliability. *Chinese Journal of Clinical Psychology, 16*(4), 344 - 347.

[谢宝国 , 龙立荣 , 赵一君 . (2008). 职业高原问卷的编制及信效度研究 . *中国临床心理学杂志, 16*(4), 344 - 347.]

Yu, C. (2006). Research on construct of career plateau and the relationship between it and job performance of knowledge employees. *Studies in Science of Science, 24*(6), 929‑933.

[余琛 . (2006). 知识型员工的职业高原与工作绩效的关系研究 . *科学学研究* , 24(6), 929‑933.]

Zhang, W. Q. (2006). *On the career management of common civil servants in China government* (Unpublished master's thesis). Northeastern University, Shenyang.

[张文勤 . (2006). *论我国普通公务员职业发展管理* (硕士学位论文). 东北大学 , 沈阳 .]

Zhang, Z. S. (2005). International comparison and reference research of civil servant's career development management. *Journal of US‑China Public Administration, 2*(5), 1‑6.

[张再生 . (2005). 公务员职业生涯发展管理的国际比较与借鉴研究 . *美中公共管理* , 2(5), 1‑6.]

Zhao, Y. R. (2012). *The relationship of middle school teachers' career plateau, job burnout and turnover intention* (Unpublished master's thesis). Shandong Normal University, Jinan.

[赵寅汝 . (2012). *中学教师职业生涯高原、职业倦怠和离职倾向的关系* (硕士学位论文). 山东师范大学 , 济南 .]

The career plateau of Chinese public servants: Construct, measurement and its psychological and behavioral influence

WANG Zhongjun[1]; LONG Lirong[2]; LIU Lidan[3]; HUANG Xiaohua[4]; JIA Wenwen[1]; LI Lu[1]; MA Hongyu[1]

([1] Key Laboratory of Adolescent Cyberpsychology and Behavior, Ministry of Education; School of Psychology, Central China Normal University; Key Laboratory of Human Development and Mental Health of Hubei Province, Wuhan 430079, China)

([2] School of Management, Huazhong University of Science and Technology, Wuhan 430074, China)

([3] School of Humanities, Hubei University of Chinese Medicine, Wuhan 430065, China)

([4] United Front Work Department, CPC Jiangxi Provincial Committee, Nanchang 330006, China)

Abstract　The career plateau is a serious problem for Chinese public servants, especially for those who at the lower level of public organizations, which occurs when an individual has limited vertical and horizontal career mobility. Since career plateau is of great influence on individual career development as well as organizational efficiency, this study investigated the career plateau phenomenon of Chinese public servants. Firstly, the typical features of Chinese public servants' career plateau were found with the content analysis of data collected from 16 public servants, using structured interviews and open-ended questionnaires for other 43 public servants from various public organizations. Meanwhile, the initial items for Public Servants Career Plateau Questionnaire (PSCPQ) were also established. Secondly, data collected from a sample of 279 public servants of diverse organizations was used to explore the CSCPQ's conceptual construct, reliability and validity by exploratory factor analysis (EFA). Thirdly, conducting Confirmatory Factor Analysis (CFA) for data from another larger sample of 517 public servants, the validity and reliability of the PSCPQ was confirmed. Eventually, this study investigated the relationship between career plateau and civil servants' occupational psychological and behavioral outcomes, such as organizational commitment, occupational burnout and withdrawal behavior in workplace by hierarchical multiple regression analysis for data of 520 public servants and another sample (n=230) from three public organizations. The research results indicated the career plateau of public servants was a two-dimension structure in Chinese public organizations context. The dimensions include the stagnation of promotion and the marginalization of position, which are completely different from the constructs of employees' career plateau of enterprises found in previous studies. The Cronbach α coefficients of the two dimensions are all above 0.80 and show a steady and acceptable status in different samples. The Cronbach α coefficients of the entire PSCPQ is also above 0.87 in different sample. These results showed that PSCPQ developed in this study had good psychometric reliability and high validity. Furthermore, the results showed that after controlling demographic variables, the stagnation of promotion of public servants was significantly positive related to emotional burnout and cynicism, and was not related to organizational commitment, professional efficiency as well as withdrawal behavior at work. After controlling demographic variables, the positional marginalization of public servants was significantly negative related to organizational affective commitment, normative commitment and professional efficiency, and was positive related to organizational continuance commitment, emotional burnout, cynicism as well as withdrawal behavior. These results would be of enlightenment to the management practices and career development for public servants in Chinese public organizations. Finally, implications and directions for future research are discussed.

Keywords public servants; career plateau; stagnation of promotion; marginalization of position

心理学报，2011，43(7)，798－809.

组织中主管－下属关系的运作机制与效果 *

王忠军[1]，龙立荣[2]，刘丽丹[3]

（[1] 华中师范大学心理学院暨湖北省人的发展与心理健康重点实验室，武汉 430079）

（[2] 华中科技大学管理学院，武汉 430074）

（[3] 湖北中医药大学社会科学部，武汉 430065）

摘　要　基于社会交换的理论视角，以下属关系投入－主管资源回报的概念架构来展现组织中主管与下属关系互动的实质，对主管－下属关系的运作效果与机制进行跨层次的实证研究。通过问卷法获得 54 个工作群体的 426 名下属与主管的对偶数据，基于 HLM 分析的结果表明：下属在工作之余对主管的私人关系投入不仅能直接获得主管的工具性资源回报与情感性资源回报，还能通过领导－成员交换 (LMX) 间接地获得主管的工具性与情感性资源回报，而在工作群体内基于私人关系进行人力资源管理决策的特征对主管与下属之间的关系互动与关系质量也存在一定程度的影响。

关键词　主管－下属关系；关系投入；资源回报；领导－成员交换；关系导向人力资源管理

分类号　B849：C93

1　问题的提出

在中国社会，关系现象充斥于人们的日常生活、经济活动以及组织行为之中。关系管理作为中国式管理的核心，备受企业实践者与组织研究者的关注。随着现代化的发展，许多组织在管理中力图淡化关系的影响，却无法动摇传统儒家文化和价值观的社会基础，正是这种文化与价值塑造了中国的组织管理行为。许多研究者认为关系的作用在未来中国以及东亚国家及其组织中将一直持续下去 (Lovett, Simmons, & Kali, 1999; Millington, Eberhardt, & Wilkinson, 2005; Yang, 2002)。在组织的各类关系中，最重要和吸引人的是上下级之间的关系，如主管－下属关系。与西方不同，中国员工普遍重视与领导、同事建立并维持良好的私人关系，而处理并维护好与下属的关系也是管理者有效管理下属的关键要素 (Law, Wong, Wang, & Wang, 2000)。因此，基于中国社会文化情境，探究组织中主管－下属关系运作的机制与效能，对于理解组织中的关系现象，丰富关系管理有重要的理论与实践意义。

1.1　文献回顾

由中国传统社会文化所孕育的关系 (guanxi) 概念一直是华人学者进行本土心理学建构，并循此了解中国人心理与行为的核心概念，学界从概念层面对关系的文化意涵、定义、类型特征、互动法则等进行了广泛探讨（周丽芳，2002）。在管理学领域，研究者发

---

　*　国家自然科学基金资助（70671046），教育部人文社科青年基金资助（10YJC630267），华中师范大学人文社科丹桂计划资助（09DG003）

现中国人的关系展现方式及其结果与西方的人际关系 (interpersonal relationship) 有很大差异，关系在中国人的商业活动、企业管理与组织行为中扮演不可言喻的重要角色 (Xin & Pearce, 1996)。因此，由关系概念来透视组织管理成为中国式管理研究的最佳进路。

不过，对组织内部的关系研究存在多元层次与视角，如对偶层次 (dyad)、三方关系 (triads)、关系网络 (networks)。就对偶关系而言，大体有两类：一是上下级之间，如下属与主管、员工与领导，属于垂直型对偶关系；二是同级之间，如同事关系，高管团队中两两关系，属于水平型对偶关系。目前组织行为学的研究大多集中于前者。

对于主管 – 下属关系，早期研究集中于关系基础 (guanxi basis) 及其效能。即探讨主管与下属之间既定的特殊性关系连带（如血缘关系、九同关系等）对主管与下属的关系品质（如亲信、友谊、认知性信任与情感性信任）以及主管对下属的绩效评估的影响 (Farh, Tsui, Xin, & Cheng, 1998; Tsui & Farh, 1997; Xin, Farh, Cheng, & Tsui, 1998)。在上述研究基础上，Tsui、Farh 和 Xin(2000) 根据关系分类与互动原则的架构，提出了一个华人组织中关系与效能的概念模型，即"关系基础（家人、熟人、生人）→关系形式（义务、友谊、认同）→直接效果（人际信任、人际喜好、忠诚、偏私）→间接效果（职业生涯与事业成功）"。显然，关系基础的研究视角承袭的是费孝通"差序格局"的思想，关系大多被定义为一种"特殊性社会连带"(King, 1985; Yang, 1986)。在此定义下，关系经常被操作为一种二分变量，即要么存在某种类型的关系基础，要么不存在，并且不同的关系基础具有不同义务规范，会受到不同对待。但总体来看，关系基础与相关效果变量间的关系不是很稳定，其原因有二：其一，由关系基础到关系成分（情感、义务、工具等）的推论不够清晰，例如，即使过去有同宗或同学的关系基础，并不必然会出现某种形式的关系成分（周丽芳，2002）。其二，更重要的是，在现代组织中具有关系基础的对偶双方或网络成员的比例相当低，即组织成员之间"沾亲带故"的现象并不普遍，一定

程度上影响了研究结果的可靠性。例如在 Farh 等人 (1998) 的研究中，在全部 560 组主管与下属配对样本中没有发现师生关系，过去曾是上下级关系的仅占 0.04%，而同学、亲戚、同姓、同乡、同事、邻里关系出现的频次均在 2.10% 至 3.40% 之间。

因此，近期对于主管 – 下属关系的研究则逐渐转向下属与主管的私人关系质量 (guanxi quality) 及其对个体的积极影响。在该类研究中，主管 – 下属关系被定义为"主管与下属在工作时间之外通过非工作相关的行为活动而建立的私人关系质量" (Law et al., 2000)，"关系"是在互相满足关系双方个人目标的过程中所建立起来的，并为工具性目的服务。此类研究结果表明，主管与下属的私人关系质量能影响主管的管理决策，如对下属的晋升、奖酬分配、工作安排等 (Law et al., 2000)；能预测下属对主管的满意度以及对组织的情感承诺 (Wong, Tinsley, Law, & Mobley, 2003)；能让下属从主管处获得更多的关系性报酬 (guanxi payoff)，如奖酬分配、晋升机会和任务安排等，并且提升下属知觉的程序公平感 (Chen, Friedman, Yu, & Sun, 2008)；对员工的职业生涯发展也有着积极的影响（刘军，宋继文，吴隆增，2008）。

从关系质量视角的研究引出的思考是：下属要想与主管拥有良好的私人关系质量，就必然要有建立、维持和运作关系的行为活动。换言之，有了建立私人关系行为的投入，才可能具备良好的关系质量。而在以往研究中，对主管与下属关系质量的操作与测量也大多是基于双方的关系互动行为。事实上，关系概念的复杂性和丰富性也正是体现在关系的互动方面，即不具备特殊性关系连带的个体之间如何建立和发展关系，并会因此带来什么样的后果。因此，从关系行为(guanxi behavior)的视角来探讨主管 – 下属关系的运作机制及其效果是一个新的、重要的研究视角，而社会交换理论能为该视角提供理论解释。

1.2 社会交换视角的主管 – 下属关系

中国人的关系行为本质上是一种社会交换行为的观点早已得到学界的认同 (King, 1985; Hwang, 1987)，对于组织中主管 – 下属关系也不例外。根

据社会交换理论 (Blau, 1964; Foa & Foa, 1980)，本研究将组织中主管－下属关系 (supervisor-subordinate guanxi, SSG) 定义为："组织中下属通过工作范围之外的互动行为与其主管建立的非正式、特殊性社会交换关系"。由此定义引出的问题是：下属与主管关系交换的内容是什么？换言之，下属投入什么，主管相应地回报什么？

从下属方面来看，要建立、维持或经营与主管的特殊性私人关系，必然要付出一些成本，比如时间、金钱、情绪乃至机会成本，即关系投入的行为。比如 Law 等人 (2000) 通过问卷调查，搜集下属与其主管建立良好私人关系的各种行为活动，最后确立了六种最具代表性和有效性的行为。Wong 等人 (2003) 研究发现主管与下属关系互动行为主要表现在如下五个方面：社会活动、经济支持、优先照顾、节日庆祝和情绪支持。本研究提出"关系投入"(guanxi input, GI) 的概念，并将其界定为"为与主管建立良好的私人关系，下属在工作范围之外对其主管进行的各种时间、经济与情感的投入行为"。

而从主管方面来看，也会相应地给予下属各种资源回报。在权力距离较大的中国组织中，不同组织层级的资源是不平衡的。研究发现，中国企业的决策权力一般更多地集中于中高层，管理人员在员工选拔、薪酬和雇佣等人事决策上有着更大的影响力 (Wang & Heller, 1993)。下属与主管所拥有的资源差距在组织的中高层与基层之间表现得尤为明显（王忠军，龙立荣，2009）。从社会资本（social capital）的观点来看，对于主管所拥有的资源，下属只有通过与主管建立较强的社会连带才能得以借用。因此，下属对主管的关系投入行为也才有了一个最基本的动因和条件。

根据社会资源理论 (Lin, 2001)，社会行动者主要有两类行动：①工具性行动（寻找和获得额外有价值资源），以及②情感性行动（寻找情感和支持的行动），并且情感性行动往往比工具性行动更重要。而社会交换理论也认为，人类的社会交换行为不仅仅是工具性资源交换，还有情感性资源交换。因此，本研究认为，主管给予下属的回报主要有两种：一是工具性资源

回报，亦简称工具性回报 (instrumental output, IO)，本研究将其界定为"主管基于私人关系给予下属直接的、客观性物质利益或好处，如晋升机会、任务安排、奖金分配、绩效考评、工作支持等"。这一概念类似于其他研究者所谓的"关系性报酬" (Friedman, Chi, & Liu, 2006; Chen et al., 2008)。二是情感性资源回报，亦简称情感性回报 (affective output, AO)，本研究将其界定为"主管基于私人关系给予下属以间接的、主观性精神利益或好处，如接纳、友善、信任、认可、鼓励、关怀、宽容等"。 显然，以往中国人关系研究常将焦点置于工具性利益与义务性情感，而较少关注自我表露、内心交流、情感性支持等真实情感或情绪层面（周丽芳，2002）。从下属方面来看，下属与主管建立和维持关系，不仅仅是想获取主管的工具性资源，更想博取情感性回报。而从主管方面来看，工具性资源具有客观性和有限性，并且需要在不同下属之间来平衡，而情感性资源具有主观性和丰富性，在组织环境中，管理者对下属往往一手运用工具性资源，一手运用情感性资源，交互运作，以更好地管理和驾驭下属。因此，探究情感性回报是极为重要的，但以往的研究很少涉及。

根据以上分析，在差序格局、关系取向以及特殊主义的中国社会文化背景下，组织中主管－下属关系的实质表现为下属在工作之余对其主管进行关系投入，主管相应给予下属不同程度的工具性、情感性资源回报的互动过程。

1.3 研究假设

1.3.1 下属关系投入与主管资源回报

中国人常言，投之以桃，报之以李，来而不往非礼也。根据社会交换理论，若主管能感知到下属对其关系行为的投入，基于以下三个原因：①回报的要求、②信任、③互惠原则，下属的投入终会有回报。譬如 Zhang 和 Yang(1998) 研究发现中国企业管理者对于奖金分配的决策不仅受公平原则的影响，还受到关系的影响。Law 等人 (2000) 在对中国大陆的 189 对主管－下属的对偶关系研究发现，关系会影响主管的管理决策，与主管拥有良好的私人关系的下属

能获得更多的晋升机会、更多的奖金分配和更好的工作安排。Zhou 和 Martocchio(2001) 的研究报告中国管理者会给予那些与其拥有良好关系的下属更多的非货币性报酬。Chen 和 Tjosvold(2007) 的研究也发现中国企业员工与管理者的个人关系能带来更好的工作安排以及晋升机会。Chen 等人 (2008) 的研究也发现，良好的主管－下属关系能换来更好的关系性报酬。不过正如前文的分析，以往研究大多关注关系所带来的工具性资源回报，而很少关注情感性资源回报。而由社会交换理论不难推出，下属对主管的私人关系投入行为也能获得主管的情感性资源回报，因为下属与其主管建立良好私人关系的各种行为中存在许多情感性投入的成分，这在 Law 等人 (2000) 的研究中已有所展现。对于中国人而言，直接的、赤裸裸的物质利益交换往往让人难以接受，而最有效的方式则是在利益交换的过程中渗透情感的投入与交换。比如在 Wong 等人 (2003) 研究中所发现的主管与下属关系互动中也无一不展现了情感互动的成分（如社会活动、优先照顾、节日庆祝和情绪支持）。因此，本研究提出如下假设：

假设 1-1：下属的关系投入正向影响主管的工具性回报。

假设 1-2：下属的关系投入正向影响主管的情感性回报。

1.3.2 领导－成员交换的中介作用

现有文献大多认为主管－下属关系与领导－成员交换关系 (Leader-member exchange, LMX) 是彼此独立的概念 (Law et al., 2000)。其主要的区别是：LMX 是建立在工作职责上的正式的工作关系，被定义为领导与成员彼此之间在工作上展现出信任、忠诚、情感、贡献与责任的行为 (Graen & Uhl-Bien, 1995)；而主管－下属关系反映的是工作范围之外的、非正式的私人关系 (Wong et al., 2003)。Hui 和 Graen(1997) 曾深入比较过关系概念与 LMX 概念的区别。基于此，以往的研究大多单独或并行地考察主管－下属关系和 LMX 各自的作用机制，并加以比较（比如 Law et al., 2000；Chen & Tjosvold, 2007），却很少有研究探讨二者之

间的关系。那么，LMX 在主管与下属的私人关系交换过程中起着什么样的作用呢？以往的研究发现，下属影响主管的行为，如与主管结盟、相互交换、逢迎主管等会影响 LMX 的质量 (Deluga, 1994)。而 LMX 也会进一步影响员工的晋升、工作安排和薪酬 (Wakabayashi, 1988; Chen & Tjosvold, 2007)。根据中国社会背景来看，关系在一定程度上体现了中国人所谓"做人"的一面。与西方不同，中国人是极其重视"做人"的，这直接来源于"会做人"的好处以及"不会做人"的坏处。因此，下属与主管在工作范畴之外发展出的具有强烈的"组织规定外"、"私人情感"色彩的关系会渗透到正常工作中，从而在组织制度内发挥作用（刘军等，2008）。由此推论，下属在工作之余对主管的私人关系投入行为，可能会对彼此在工作场所中的 LMX 关系质量产生一定程度的积极影响，而良好的 LMX 关系质量也会进一步为下属带来各种情感性和工具性资源回报。基于以上分析，本研究提出如下假设：

假设 2-1：下属的关系投入通过 LMX 间接地影响主管的工具性回报。

假设 2-2：下属的关系投入通过 LMX 间接地影响主管的情感性回报。

1.3.3 关系导向人力资源管理的调节作用

主管与下属的关系交换与运作虽然是个体间的互动行为，但却是嵌入在群体或组织的背景中，受到群体或组织特征的约束，而这种嵌入性在个体层面的研究中往往被忽视。在中国组织中，关系的作用会渗透进组织的各项管理决策之中，成为"正式法制支持的替代品" (Xin & Pearce, 1996)。Chen, Chen 和 Xin(2004) 因此提出"关系导向人力资源管理"（Guanxi-based human resources management practice, GHRM）概念，指的是人力资源管理决策中以私人关系为基础的总体状况。需要说明的是，该概念既适用于组织层面，也适用于群体层面，本研究中采取的是群体层面的概念。在注重制度规范和公平正义的组织环境下，关系导向人力资源管理被认为具有众多负面性，比如破坏程序公平 (Chen et al., 2008)，降低

员工对组织管理的信任 (Chen, Chen & Xin, 2004)，损害员工的角色内和角色外绩效 (Hsu & Wang, 2007)。总之，人力资源管理决策的关系导向越强，说明"人治"气氛越浓厚，基于私人关系的弹性操作空间越大，同时也意味着制度规范性越差，在这样的环境下，主管根据私人关系而给予下属差别化的特殊对待的现象会拥有"制度合法性"的背景，并得以强化。因此，本研究提出如下假设：

假设 3-1：工作群体的关系导向人力资源管理对下属关系投入与主管工具性回报的关系具有正向调节作用。

假设 3-2：工作群体的关系导向人力资源管理对下属关系投入与主管情感性回报的关系具有正向调节作用。

2 研究方法

2.1 被试与调查程序

由于本研究属于跨层次的研究，涉及个体与群体两个层面的数据搜集，因此，问卷调查均以工作群体 (work group) 为抽样单位，并在每个工作群体（部门）内，采取了主管与下属的二元对偶研究设计。调查包含两份问卷，分别由部门内的员工及其直接主管填写。在内容上，下属问卷包括自评的任务绩效、主管资源回报、LMX、关系导向人力资源管理；部门主管则仅需填答不同下属的关系投入。为了保证问卷的隐匿性以及数据的主管－下属配对，采用了一个编码系统，以匹配主管评定与下属回答。之所以采用上述研究设计，主要出于以下考虑：

第一，主管对于下属的关系投入行为一般会有感知，并会据此相应地给予下属各种资源回报；同样，下属对于主管给予的资源回报也会有直接的感知，其感知的结果会进一步影响双方的后续互动。第二，一名主管需要评价多名下属，主管的负担会较重，因此仅要求主管填答项目数量相对较少的下属关系投入问卷。出于同样的考虑，任务绩效也由下属自评。第三，主管与下属"错位式"的互评可在一定

程度上降低问卷项目敏感性带来的心理压力。总之，上述研究设计既考虑了研究需要，又兼顾了可行性，从不同来源获取数据。

被试来自湖北、江西、北京、上海、广东地区8家企业中的不同工作群体（工作部门）。在调查程序上，首先，研究人员与企业人力资源部门一起确定了调查的部门，主要是企业中层部门。判定群体的依据有：①不同员工属于同一工作部门；②不同员工拥有一个共同的直接主管；③他们长时期在一起工作。然后，研究人员在企业助手的带领和协助下，进行现场调查。最后，由研究人员当场收回问卷，回收问卷后进行主管与下属问卷的配对组合。以现场调查方式所获得的样本占全部有效样本的81.50%，这在一定程度上能保证数据的质量，而少量委托调查则给受托者及其单位的人力资源部门提供了指导语。

研究者对有效数据进行了筛选：①首先剔除了空白过多、反应倾向过于明显的问卷；②然后剔除了下属人数过少的群体样本（少于5人）。最终回收了 54 个有效群体样本，总共包含 426 份有效个体问卷，平均每个群体包含 8 人，人数最少的群体有 6 人，人数最多的群体有 13 人。主管－下属匹配后的有效填答率为 82%，其中有 83.10% 的被试与其直接主管保持的上下级关系年限在 1 年以上。在有效样本中，国有企业 40.80%，民营企业 55.50%，外资企业 3.70%；男性占 55.70%，女性占 44.30%；25 岁及以下占 20.50%，26~30 岁占 13.90%，31~35 岁占 14.90%，36~40 岁占 20.80%，40 岁以上者占 30%；管理岗位占 34.70%，生产岗位占 16.70%，技术岗位占 22.90%，销售岗位占 8.50%，行政后勤占 17.20%。

2.2 研究工具

2.2.1 下属关系投入

下属关系投入采用经本研究修订过的单维度问卷。修订过程如下：采用与 Law 等人 (2000) 编制关系问卷相同的方法，首先通过对来自不同企业的 27 名员工进行开放式问卷调查，搜集员工在工作之余与其直接主管建立并保持良好私人关系的行为，结果

发现大部分行为与 Law 等人 (2000) 的研究相似，但也有少部分行为包含在 Chen 等人 (2008) 和 Wong 等人 (2003) 的问卷项目之中。因此，本研究在综合以上相关问卷项目的基础上，初步编制了包含 9 个项目的下属关系投入问卷，为避免被试填答问卷时的"趋中性"，采取 Likert 6 点计分，1 表示"非常不符合"，6 表示"非常符合"。接下来，对来自江西 5 家企业的共 260 名员工进行初试，获得 211 名员工的有效数据，其中 99.50% 的员工与其主管保持了 1 年以上的上下级关系。进行探索性因素分析 (EFA) 后发现下属关系投入问卷为单维度，解释的变异量为 65.15%。

由于正式施测中的关系投入问卷是由主管来填答，所以我们从主管的角度对问卷项目的人称进行了相应修改，比如："该职工在平时会打电话或上门拜访我"、"该职工总是主动地与我交流他（她）的想法、问题、需要和感受"。对正式施测的样本（n=426）数据进行验证性因素分析 (CFA)，下属关系投入问卷的单维度模型的各项拟合指数均达到或接近临界值，具体如下：x^2=109.89，df=27，RMSEA=0.07，SRMR=0.05，IFI=0.91，CFI=0.91，NFI=0.90，NNFI=0.90，这表明问卷具有较好的结构效度。下属关系投入问卷的 Cronbach α 信度系数为 0.93，符合测量学的标准。此外，本研究还表明该问卷具有较好的效标效度。

2.2.2 主管资源回报

主管资源回报问卷采用本研究自编的问卷。编制过程如下：首先对来自不同企业的 27 名员工进行开放式问卷调查，请被试列举"下属与直接主管建立良好私人关系后，主管会给下属带来哪些利益或好处"。对开放式问卷调查的资料进行内容分析后，初步编制了包含 15 个项目的主管资源回报问卷，为避免被试填答问卷时的"趋中性"，采取 Likert 6 点计分，1 表示"完全不同意"，6 表示"完全同意"。

接下来，通过对初试样本（n=211）的有效数据进行探索性因素分析 (EFA)，结果表明主管资源回报问卷具有十分清晰的两维度结构：一是工具性回报，包含 6 个项目，比如："他（她）会尽量给我安排我期望的工作岗位"、"他（她）会想方设法提拔我"；二是情感性回报，包含 7 个项目，比如："在生活中，他（她）很关心照顾我"、"他（她）会与我分享他（她）的经验、想法和感受"。两维度解释的变异量为 68.09%。

最后，利用正式施测的样本（n=426）数据进行验证性因素分析 (CFA)，主管资源回报的两维度模型的各项拟合指数均达到临界值，具体如下：x^2=231.68，df=64，RMSEA=0.06，SRMR=0.43，IFI=0.92，CFI=0.92，NFI=0.91，NNFI=0.90，这表明问卷具有较好的结构效度。信度分析表明，工具性回报和情感性回报的 Cronbach α 系数分别为 0.90 和 0.92，总问卷的 α 系数为 0.94，说明问卷的信度质量较好。此外，本研究还表明问卷具有较好的效标效度。

2.2.3 领导 - 成员交换

对于如何测量 LMX，学界还存在争议，主要源于其结构是单维的还是多维的差异。但 Liden 和 Maslyn(1998) 提出 LMX 的维度不一定需要得到一个确定的模式，而是需要与不同的考察目的和结果变量挂钩。Schriesheim，Castro 和 Coglister(1998) 运用元分析 (meta-analysis) 技术，检验了各种量表的内部一致性，结果表明 Graen 和 Uhl-bien(1995) 研制的 7 个项目量表具有最高的信度和效度，简称 LMX-7。由于本研究主要关心组织中领导 - 成员交换关系的质量，而非不同方面的交换内容。因此，本研究采用被广泛应用的 LMX-7 量表，并采用 Likert 6 级计分，1 表示"完全不同意"，6 表示"完全同意"。在本研究中，该问卷的 α 系数为 0.92。

2.2.4 关系导向人力资源管理

本研究中关系导向人力资源管理是一个群体层次的变量，而对其测量是通过对群体中的个体的测量来完成的，这里面就有一个指称迁移问题。根据 Chan（1998）所提出的"指称迁移共识模型"（referent-shift consensus model），在测量时，所用的项目不是群体中单个成员的行为描述，而必须把所有成员作为一个整体来看待，以整体为出发点来描述群体成员的行为。在本研究中，关系导向人力资源管理采用 Chen，Chen 和 Xin(2004) 开发的量表，原量表共 5 个项目，α 系数为 0.93。所有项目均以"在我所工作的部门内"为指称。此量表在 Chen 等人 (2008) 的一项研究中被修订为 4 个项目，α 系数为 0.88。上述两个版本的量表均为英文，本研究对其进行翻译和回译后，综合了两个版本的项目，获得了 7 个项目，其中的一个项目"在我所工作的部门内，培训发展机会的获得依靠与主管的关系"为本研究新加入的一个项目。采用 Likert 6 点计分，1 表示"非常不符合"，6 表示"非常符合"。

对关系导向人力资源管理问卷 7 个项目进行了探索性因素分析，结果得到单一因素，解释的变异量为 70.90%，问卷的 Cronbach α 信度系数为 0.93，这一结果同样表明关系导向人力资源管理问卷具有较好的信度和结构效度。此外，我们运用方差分析检验了关系导向人力资源管理在企业性质上的差异，结果发现不同性质的企业之间存在显著地差异，其中关系导向人力资源管理在国有企业的表现程度最高，民营企业次之，外资企业的关系导向最低（$M_{国有企业}$=3.53，$M_{民营企业}$=3.03，$M_{外资企业}$=2.54，F=11.18，$p < 0.001$），该结果与实际情况基本相符，也说明关系导向人力资源管理问卷具有较好的同时效度。

本研究中，关系导向人力资源管理用于在群体层次上代表群体的人力资源管理决策特征与氛围。对应于每一个工作群体，其关系导向应当是唯一的，所以有必要将群体中个体提供的评估数据汇聚到群体层次。$ICC(1)$、$ICC(2)$ 和 R_{wg} 是三个最常用的用于判断个体数据汇聚是否可靠的指标，本研究同时考察三者。为了判断的一致性，我们先通过方差分析 (ANOVA) 进行了组间差异性检验，结果组内相似性高于组间相似性，即不同工作群体之间存在显著的差异 $F(53,372) = 4.81, p < 0.001$。计算得到 54 个工作群体的 R_{wg} 值在 0.27~0.96 之间，尽管少数群体的 R_{wg} 值较低，但均值为 0.75，高于 0.70 的标准 (Dixon & Cunningham, 2006)。同时，本研究计算所得 $ICC(1)$ 为 0.32，在 James (1982) 推荐的 0 到 0.5 的临界值范围之内，这表明变量在各群体中有充足的内部同质性；$ICC(2)$ 为 0.79，大于 Klein 等人 (2000) 推荐的临界值 0.70，这表明采用个体的平均数作为群体变量的指标的可信度较高。总之，以上结果均一致表明，可以用群体中个体知觉到的关系导向人力资源管理数据的平均数作为群体层面变量的观察值。

除了以上关键变量外，本研究还控制了可能会影响主管给予下属资源回报的一个重要变量，即员工的任务绩效。任务绩效采用员工自评式问卷，包含 4 个项目，来源于 Williams 和 Anderson(1991) 编制的任务绩效量表，这 4 个项目分别是："和同事相比，我的工作成绩比较优秀"、"我的领导对我的工作成绩比较满意"、"同事对我的工作成绩评价比较高"、"我的工作成绩经常受到单位的表扬"。本研究对以上 4 个项目进行探索性因素分析，结果得到单一因素，解释的变异量为 67.98%，问卷的 α 系数为 0.84。

2.3 统计方法

采用 SPSS 11.5 进行描述统计、相关分析、探索性因素分析和信度分析，采用 LISREL 8.30 进行验证性因素分析，采用多层线性模型 HLM 6.02 对研究假设进行检验。

表 1 各研究变量的平均数、标准差与相关矩阵

变量（n=426）	M	SD	1	2	3	4	5
1. 任务绩效	4.15	0.82					
2. 关系投入	3.09	1.10	0.36**				
3. 工具性回报	3.47	1.05	0.40**	0.59**			
4. 情感性回报	3.96	0.99	0.36**	0.50**	0.69**		
5. LMX	3.75	0.98	0.43**	0.58**	0.74**	0.80**	
6. 关系导向人力资源管理	3.34	1.19	0.07	−0.06	−0.04	−0.24**	−0.18**

注：关系导向人力资源管理为个体层面的数据；**$p < 0.01$,*$p < 0.05$。

表 2　下属关系投入对主管资源回报的影响

因变量	模型	参数估计							
		γ_{00}	γ_{10}	γ_{20}	σ^2	τ_{00}	τ_{11}	τ_{22}	作用
工具性回报	**M1**：零模型　L1：$IO_{ij}=B_{0j}+r_{ij}$　　L2：$B_{0j}=\gamma_{00}+\mu_{0j}$	3.47***			0.88	0.24***			
	M2：任务绩效→工具性回报　L1：$IO_{ij}=B_{0j}+B_{1j}(TP_{1ij})+r_{ij}$　L2：$B_{0j}=\gamma_{00}+\mu_{0j}$　　$B_{1j}=\gamma_{10}+\mu_{1j}$	3.47***	0.47***		0.69	0.26***	0.14**		0.22
	M3：关系投入、任务绩效→工具性回报　L1：$IO_{ij}=B_{0j}+B_{1j}(GI_{1ij})+B_{2j}(TP_{2ij})+r_{ij}$　L2：$B_{0j}=\gamma_{00}+\mu_{0j}$　　$B_{1j}=\gamma_{10}+\mu_{1j}$　　$B_{2j}=\gamma_{20}+\mu_{2j}$	3.47***	0.43***	0.27**	0.48	0.29***	0.06**	0.10**	0.30
情感性回报	**M1**：零模型　L1：$AO_{ij}=B_{0j}+r_{ij}$　　L2：$B_{0j}=\gamma_{00}+\mu_{0j}$	3.95***			0.69	0.30***			
	M2：任务绩效→情感性回报　L1：$AO_{ij}=B_{0j}+B_{1j}(TP_{1ij})+r_{ij}$　L2：$B_{0j}=\gamma_{00}+\mu_{0j}$　　$B_{1j}=\gamma_{10}+\mu_{1j}$	3.95***	0.34***		0.51	0.32***	0.22**		0.26
	M3：关系投入、任务绩效→情感性回报　L1：$AO_{ij}=B_{0j}+B_{1j}(GI_{1ij})+B_{2j}(TP_{2ij})+r_{ij}$　L2：$B_{0j}=\gamma_{00}+\mu_{0j}$　　$B1j=\gamma_{10}+\mu_{1j}$　　$B2j=\gamma_{20}+\mu_{2j}$	3.95***	0.37***	0.18*	0.39	0.34***	0.02*	0.17***	0.24

注：① ***$p<0.001$，**$p<0.01$，*$p<0.05$。② σ^2 为水平 1 的残差；τ_{00} 为截矩残差，即 μ_{0j}；τ_{11} 和 τ_{22} 为斜率残差，即 μ_{1j} 和 μ_{2j}；③作用 =(原始残差－条件残差)/原始残差；④ IO 为工具性回报，AO 为情感性回报，TP 为任务绩效，GI 为关系投入。

3　研究结果

3.1　描述性统计及相关分析结果

表 1 呈现的是本研究中涉及的关键变量的描述性统计和相关分析结果。

3.2　下属关系投入对主管资源回报的影响

在运用多层线性模型 (HLM) 对假设进行验证时，本研究将下属的任务绩效作为一个关键的控制变量纳入 HLM 分析之中，表 2 为分析的结果。由表 2 可知，任务绩效对工具性回报具有显著的正向预测作用 (模型 M2，$\gamma_{10}=0.47$，$p<0.001$)，控制变量解释的方差为 0.22。在模型 M3 中，当同时纳入关系投入和任务绩效时，任务绩效对工具性回报仍具有显著的正向预测作用 ($\gamma_{20}=0.27$，$p<0.01$)，而关系投入对工具性回报同样具有更为显著的正向影响 ($\gamma_{10}=0.43$，$p<0.001$)，关系投入解释的方差为 0.30，这表明控制了任务绩效后，关系投入对工具性回报具有显著的正向作用，假设 1–1 得到验证。另外，由表 2 可知，任务绩效对情感性回报具有显著的正向预测作用 (模

型 M2，$\gamma_{10}=0.34$，$p<0.001$)，控制变量解释的方差为 0.26。在模型 M3 中，当纳入关系投入和任务绩效一起分析时，任务绩效对情感性回报仍具有显著的正向预测作用 ($\gamma_{20}=0.18$，$p<0.05$)，而关系投入对情感性回报同样具有更为显著的正向影响 ($\gamma_{10}=0.37$，$p<0.001$)，关系投入解释的方差为 0.24，这表明控制了任务绩效后，关系投入对情感性回报具有显著的正向作用，假设 1–2 得到验证。

3.3　LMX 的中介作用

在表 3 中，在模型 M1 中，关系投入对领导－成员交换 (LMX) 具有显著的正向作用 ($\gamma_{10}=0.46$，$p<0.001$)。在模型 M3 中，LMX 对工具性回报具有显著的正向作用 ($\gamma_{10}=0.83$，$p<0.001$)。而在模型 M4 中，当将关系投入和 LMX 作为预测变量一起纳入模型中时，LMX 对工具性回报具有显著的正向作用 ($\gamma_{10}=0.69$，$p<0.001$)，而关系投入虽然对工具性回报也具有显著的正向预测作用 ($\gamma_{20}=0.21$，$p<0.01$)，但是其影响系数要比模型 M2 中的系数 ($\gamma_{10}=0.50$，$p<0.001$) 明显降低。综合以上结果，据此可以推论，LMX 在关系投入与工具性回报之间起着部分中介的

表3 LMX 在下属关系投入对主管资源回报影响中的中介作用

模型	参数估计						
	γ_{00}	γ_{10}	γ_{20}	τ_{00}	τ_{11}	τ_{22}	σ^2
M1：关系投入→LMX	3.74^a	0.46^a		0.34^a	0.04^b		0.40
M2：关系投入→工具性回报	3.47^a	0.50^a		0.28^a	0.06^b		0.54
M3：LMX→工具性回报	3.47^a	0.83^a		0.30^a	0.06^b		0.40
M4：LMX、关系投入→工具性回报	3.47^a	0.69^a	0.21^a	0.31^a	0.14^a	0.06^a	0.33
M5：关系投入→情感性回报	3.95^a	0.42^a		0.33^a	0.04^b		0.45
M6：LMX→情感性回报	3.95^a	0.73^a		0.36^a	0.09^b		0.25
M7：LMX、关系投入→情感性回报	3.95^a	0.69^a	0.10^c	0.36^a	0.15^a	0.06^a	0.22

注：① a 为 $p < 0.001$，b 为 $p < 0.01$，c 为 $p < 0.05$。② σ^2 为水平1的残差；τ_{00} 为截矩残差，即 μ_{0j}；τ_{11} 和 τ_{22} 为斜率残差，即 μ_{1j} 和 μ_{2j}；③ IO 为工具性回报，AO 为情感性回报，GI 为关系投入，LMX 为领导–成员交换。

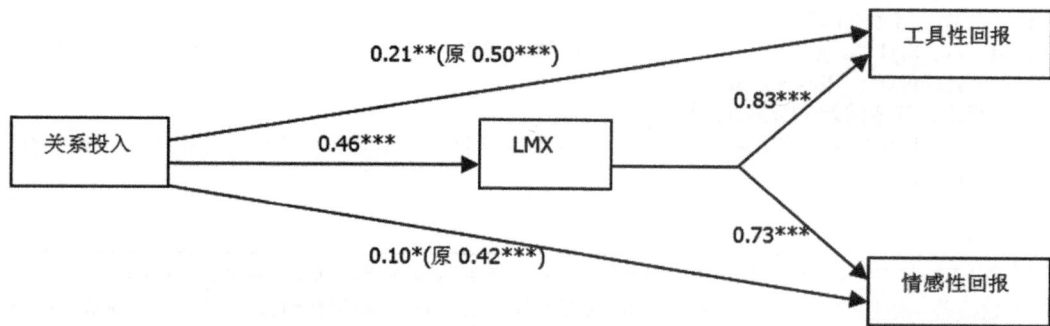

图1 主管–下属关系的社会交换模型的实证研究结果（基于 HLM 分析的路径图）

注：$***p < 0.001$，$**p < 0.01$，$*p < 0.05$；原：代表未加入中介变量时的影响系数

作用，假设 2-1 得到验证。

另外在模型 M6 中，LMX 对情感性回报具有显著的正向作用（$\gamma_{10}=0.73$，$p < 0.001$）。而在模型 M7 中，将关系投入和 LMX 作为预测变量一起纳入模型中时，LMX 对情感性回报具有显著的正向作用（$\gamma_{10}=0.69$，$p < 0.001$），而关系投入虽然对情感性回报也具有显著的正向预测作用（$\gamma_{20}=0.10$，$p < 0.01$），但是其影响系数要比模型 M5 中的系数（$\gamma_{10}=0.42$，$p < 0.001$）明显降低。综合以上结果，据此可以推论，LMX 在关系投入与情感性回报之间仍起着部分中介的作用。因此，本研究的假设 2-2 也得到验证。图1为基于上述研究结果而呈现的综合模型图。

3.4 关系导向人力资源管理的调节作用

由表3的结果可知，关系投入和 LMX 对工具性回报均具有显著的正向预测作用，但从表4的结果表明，关系导向人力资源管理对关系投入与工具性回报之间的关系系数（即斜率）不具有显著的调节

作用（$\gamma_{21}=-0.00$，$p > 0.05$），同样地，关系导向人力资源管理对 LMX 与工具性回报之间的关系系数的调节作用也不显著（$\gamma_{11}=0.13$，$p > 0.05$）。本研究的假设 3-1 没有得到验证。此外，在表3的结果中，关系投入和 LMX 对情感性回报均具有显著的正向预测作用，但在表4的结果中，关系导向人力资源管理对关系投入与情感性回报之间的关系系数（即斜率）不具有显著的调节作用（$\gamma_{21}=-0.01$，$p > 0.05$），同样，关系导向人力资源管理对 LMX 与情感性回报之间的关系系数的调节作用也不显著（$\gamma_{11}=0.05$，$p > 0.05$）。本研究的假设 3-2 也没有得到验证。

4 讨论

4.1 关系运作的个体效能

在现代企业组织中，工作绩效往往是极其重要的资源分配标准，比如很多企业实行绩效薪酬制度。

表 4　关系导向人力资源管理的调节作用模型

模型	参数估计									
	γ_{00}	γ_{01}	γ_{10}	γ_{11}	γ_{20}	γ_{21}	σ^2	τ_{00}	τ_{11}	τ_{22}
M1：对工具性回报的调节作用模型	3.32[a]	0.05	0.26	0.13	0.22	−0.00	0.33	0.32[a]	0.13[a]	0.06[a]
L1：$IO_{ij}=B_{0j}+B_{1j}(LMX_{1ij})+B_{2j}(GI_{2ij})+r_{ij}$										
L2：$B_{0j}=\gamma_{00}+\gamma_{01}(GXHRM_{ij})+\mu_{0j}$										
$B_{1j}=\gamma_{10}+\gamma_{11}(GXHRM_{ij})+\mu_{1j}$										
$B_{2j}=\gamma_{20}+\gamma_{21}(GXHRMij)+\mu_{2j}$										
M2：对情感性回报的调节作用模型	4.77[a]	−0.25	0.51[c]	0.05	0.13	−0.01	0.22	0.33[a]	0.15[a]	0.06[a]
L1：$AO_{ij}=B_{0j}+B_{1j}(LMX_{1ij})+B_{2j}(GI_{2ij})+r_{ij}$										
L2：$B_{0j}=\gamma_{00}+\gamma_{01}(GXHRM_{ij})+\mu_{0j}$										
$B_{1j}=\gamma_{10}+\gamma_{11}(GXHRM_{ij})+\mu_{1j}$										
$B_{2j}=\gamma_{20}+\gamma_{21}(GXHRM_{ij})+\mu_{2j}$										

注：①零模型见表 2，直接作用模型见表 3，IO 为工具性回报，GI 为关系投入，LMX 为领导－成员交换，GXHRM 为关系导向人力资源管理。② σ^2 为水平 1 的残差；τ_{00} 为截矩残差，即 μ_{0j}；τ_{11} 和 τ_{22} 为斜率残差，即 μ_{1j} 和 μ_{2j}；③ a 为 $p < 0.001$，b 为 $p < 0.01$，c 为 $p < 0.05$。

本研究的结果也证明，下属的任务绩效对主管的工具性资源回报和情感性资源回报均有显著的正向影响。但是当我们将下属对其主管的私人关系投入与任务绩效一起去预测主管的资源回报时，让人意外的是，关系投入的解释力明显大于任务绩效，说明下属在工作之余的关系投入行为对主管资源回报有着重要的影响力。这一研究结果与 Law 等人 (2000)、Chen 等人 (2008) 以及刘军等人 (2008) 的研究结论一致。正如 Warner(1993) 所指，尽管技术和规范在中国组织中已经变得更为必要，但关系的重要性在中国社会仍然占据主导地位。该研究结果也凸现了关系在当代中国组织资源分配中仍占据重要地位，也给本研究从关系行为和社会交换的理论视角来审视组织中主管－下属关系提供了实证支持。

从社会交换的观点来看，组织中的主管与下属之间的关系投入与资源回报应是一种基于人情的社会交换行为，并具有长期互动的性质，而非一次性的交换。由于主管与下属之间身份、位阶、职权以及占有资源上的差异，主管与下属之间发展的私人关系很难像生活中单纯的朋友关系（以情感支持与寄托为主）那样简单。在组织中，下属对主管的关系投入既需要满足情感性支持的需要，更隐含着工具性回报的期待，这一点在本研究中也能得到反映，比如工具性回报与情感性回报的相关为 0.69($p < 0.01$)，下属

关系投入对主管工具性资源回报的影响系数为 0.50($p < 0.01$)，对主管情感性资源回报的影响系数为 0.42($p < 0.01$)。总之，从本研究中可得到的启示与刘军等人 (2008) 的研究相同，即在中国组织中，除了工作上的努力与付出以外，发展与上级在生活上更为密切的私人关系更是下属不可忽视的。

4.2　关系运作的内在机制

主管－下属之间关系投入与资源回报的交换行为有着怎样的内在机制？本研究的结果表明：下属的关系投入除了对主管的工具性回报与情感性回报有着直接的影响外，还可以通过提升领导－成员交换（LMX）关系质量间接地影响主管的资源回报，即 LMX 起着部分中介的作用。在以往的研究中，往往将 LMX 看作是与关系（Guanxi）相平行的概念，前者代表正式的工作关系，后者代表着非正式的私人关系，并分别对主管的工具性回报有着积极影响（Law et al., 2000；Chen & Tjosvold, 2007），而很少有研究考察非正式的私人关系对正式的工作关系质量的影响。在国外的文献中，这一问题也许不太重要，因为在西方的组织中，二者之间往往是"泾渭分明"的。但在中国文化背景下，二者之间却可能有着剪不断的"千丝万缕"的关系。本研究发现，下属对主管的私人关系投入对 LMX 具有显著的正向作用，并且关系投入经由 LMX 的中介而获取主管的工具性

回报、情感性回报的假设也得到证实。这一结果说明，主管与下属之间在私底下建立与发展起来的、具有强烈"组织规定外"及"私人情感"色彩的关系会影响和渗透到正式的工作场所中，从而在组织制度的范畴内发挥作用（刘军等，2008）。正所谓："功夫尽在诗外"。根据社会资本（social capital）理论，下属在工作之余，对其主管的关系投入行为具有社会资本投资的性质，等同于个人社会资本的积累与运作。在本研究的结果也表明，在中国文化背景下，主管－下属之间的私人关系对工作范围内的领导－成员交换关系有着重要的影响，那些与领导搞好私人关系的下属更可能被领导视为工作领域中的"圈内人"，反之则有可能成为"圈外人"。这一研究结果对于 LMX 的研究也具有启示意义。

4.3 关系运作的制度强化

最后，本研究通过跨层次的研究设计，搜集不同层面（个体与群体）的变量数据，运用多层线性模型（HLM）的统计方法，检验了群体的关系导向人力资源管理对下属与主管的关系交换行为的调节作用。但结果发现关系导向人力资源管理的调节效应并不显著，本研究的假设没有得到验证。究其原因，可能有以下方面：其一，关系导向人力资源管理构念反映的是群体或组织中的人力资源管理决策依"私人关系"而论的程度，但由其测量项目可知，其中的各项人力资源管理决策更多地反映工具性资源分配，比如工作任务安排、奖金分配、薪水、晋升、考核、培训机会等，没有涉及情感性资源分配。换言之，关系导向人力资源管理的概念内涵存在局限性。其二，本研究定位于群体层次，但所属的组织样本仅 8 家企业，由于群体内又进行主管－下属的二元对偶设计，难度较大，最后仅筛选出 54 个有效群体，即群体与组织样本偏少，变异不大，今后的研究可适当增加组织或群体样本量。其三，由于中国企业组织中的各项决策权力更多地集中于中高层（Wang & Heller, 1993），而研究设计的难度，本研究的群体样本大部分为中层，也包含部分基层，部门主管能对其拥有的组织资源（主要是工具性资源）进行有效分配的职权

有限，导致关系导向人力资源管理的调节效应在中层与基层工作群体中难以展现，因此本研究的启示，今后有必要对更多国有企业和民营企业的高层群体进行探究。不过，本研究通过相关分析发现，关系导向人力资源管理与主管情感性资源回报 (r=-0.24, $p < 0.01$)、LMX(r=-0.18, $p < 0.01$) 均显著地负相关，这说明注重私人关系的人力资源决策可能会抑制上下级之间的情感性交换，并削弱上下级之间正式的工作关系质量，不过其原因和机制也还需进一步地探究。

4.4 贡献与局限

从理论贡献上来看，本研究基于社会交换的理论视角，以"下属关系投入－主管资源回报"的概念架构来展现组织中主管与下属的关系互动实质，将组织中关系的研究引向关系运作的层面，并在模型之中同时纳入工具性资源回报与情感性资源回报，拓展了主管－下属关系的研究空间，探究了主管－下属关系的运作效能与机制，验证了 LMX 的中介作用，对组织管理实践具有深刻的启示意义。

由于本研究属于横向研究设计，因此难以确证下属关系投入与主管资源回报之间的因果关系，并且在跨层次的研究中，也难于搜集到较大的群体样本。此外，本研究虽然采取主管－下属的对偶设计，试图克服共同方法变异的影响，但可能仍然无法有效解决测量的敏感性和社会称许性问题。未来还需要进一步探究主管与下属的不同方面的关系互动与交换及其效果差异，并考量对其他员工以及群体或组织可能产生的影响，包括积极的、消极的影响。

5 结论

本研究表明，尽管下属对主管的私人关系投入行为发生在组织规定的主管与下属工作交往范畴之外，但却能发展出一种带有特殊性、私人情谊的关系连带，不仅为下属获取主管的各种正式的和非正式的资源回报（工具性资源回报、情感性资源回报）带来积极影响，还能在组织制度的范畴之内发挥作

用，即通过促进和提升作为正式的工作关系的领导 -
成员交换关系质量（LMX），来间接地获取主管的各
类工具性与情感性的资源回报。最后，本研究也发
现工作群体的人力资源管理决策特征（如关系导向
人力资源管理）在一定程度也能制约主管与下属之
间的关系互动与关系质量，但其影响机制与效果还
需进一步地研究。

参考文献

Blau, P. M. (1964). *Exchange and power in social life.* Wiley, New York.

Chan, D.(1998). Functional relations among constructs in the same content domain at different levels of analysis: a typology of composition models. *Journal of Applied Psychology, 83*, 234–246.

Chen, C., Chen, Y., & Xin, K.(2004). Guanxi practices and trust in management: a procedural justice perspective. *Organization Science, 15*, 200–209.

Chen, N. Y.–f., & Tjosvold, D.(2007). Guanxi and leader member relationships between American managers and Chinese employees: open-minded dialogue as mediator. *Asia Pacific Journal of Management, 24*, 171–189.

Chen, Y., Friedman, R., Yu, E., & Sun, F.(2008, August). *Examining the positive and negative effects of Guanxi: a multi-level analysis of Guanxi and procedural justice.* Paper presented at Meeting of 3th International Association for Chinese Management Research, Guangzhou.

Chou, L. F.(2002). Guanxi and social network in Chinese organization, *Indigenous Psychological Research in Chinese Societies, 18*, 175–228.

[周丽芳 (2002). 华人组织中的关系与社会网络 . *本土心理学研究 , 18,* 175－228.]

Deluga, R. J.(1994). Supervisor trust building, leader-member exchange and organizational citizenship behavior. *Journal of occupational and Organizational Psychology, 67*, 315－327.

Dixon, M. A., & Cunningham, G. B.(2006). Data aggregation in multilevel analysis: a review of conceptual and statistical issues. *Measurement in Physical Education and Exercise Science. 10*, 85－107.

Farh, J. L., Tsui, A. S., Xin, K., & Cheng, B. S.(1998). The influence of relational demography and guanxi: the Chinese case. *Organization Science,9*, :471－487.

Foa, E. B., & Foa, U. G. (1980). Resource theory of social exchange. In K. J. Gergen, M. S. R. Greenberg, & H. Willis(Eds.), *Social exchange: advances in theory and research*(pp.99－131). Plenum, New York.

Friedman, R., Chi, S., & Liu, L. A. (2006). An expectancy model of Chinese-American differences in conflict avoiding. *Journal of International Business Studies, 37*, 76－91.

Graen, G. B., & Uhl-Bien, M.(1995). Relationship-based approach to leadership: Development of leader-member exchange(LMX) theory of leadership over 25 years: applying a multi-level multi-domain perspective. *Leadership Quarterly, 6*, 219－247.

Hsu, W., & Wang, An.(2007). *Downsides of guanxi practices in Chinese organizations.* Paper presented at 68th Academy of Management meeting, Philadelphia.

Hui, C., & Graen, G.(1997). Guanxi and professional leadership in contemporary Sino-American joint ventures in mainland China. *Leadership Quarterly, 8*, 451－465.

Hwang, K. K.(1987). Face and favor: the Chinese power game. *American Journal of Sociology, 92*, 944–974.

James, L. R.(1982). Aggregation bias in estimates of perceptual agreement. *Journal of Applied Psychology, 67*, 219–229.

King, Y. C. (1985). The individual and group in Confucianism: a relational perspective. In D. J. Munro(Ed.), *Individualism and holism: studies in Confucian and Taoist values.* Ann Arbor, MI: Center for Chinese Studies, University of Michigan.

Klein, K. J., Bliese, P. D., Kozlowski, S. W. J., Dansereau, F., Gavin, M. B., Griffin, M. A., et al.(2000). Multilevel analytical techniques. In K. J. Klein, & W. J. Kozlowski(Eds.), *Multilevel theory, research, and methods in organizations: foundations, extensions, and new directions*(pp. 512 - 553). San Francisco: Jossey-Bass.

Law, K. S., Wong, C. S., Wang, D. X., & Wang, L. H.(2000). Effect of supervisor-subordinate Guanxi on supervisory decisions in China: An empirical investigation. *International Journal of Human Resource Management, 11*, 751 - 765.

Liden, R. C., & Maslyn, J. M.(1998). Multidimensionality of leader-member exchange: an empirical assessment through scale development. *Journal of Management, 24*, 43 - 72.

Lin, N.(2001). Social capital: a theory of social structure and action. Cambridge: Cambridge University Press.

Liu, J., Song, J. W., & Wu, L. Z.(2008). Antecedents of employee career development: an examination of politics and Guanxi. *Acta Psychologica Sinica, 40*, 201 - 209.

[刘军，宋继文，吴隆增 (2008). 政治与关系视角的员工职业发展影响因素探讨 . *心理学报，40*, 201 - 209.]

Lovett, S., Simmons, L. C., & Kali, R.(1999). Guanxi versus the market: ethics and efficiency. *Journal of International Business Studies, 30*, 231 - 248.

Millington, A., Eberhardt, M., & Wilkinson, B.(2005). Gift giving, Guanxi and illicit payments in buyer-supplier relationships in China: analyzing the experience of UK companies. *Journal of Business Ethics, 57*, 255 - 268.

Schriesheim, C., Castro, S., & Coglister, C.(1998). Leader-member exchange(LMX) research : a comprehensive review of theory, measurement, and data-analytic praticices. *Leadership Quartly, 10*, 63 - 113.

Tsui, A. S., & Farh, J. L.(1997). Where Guanxi matters：

relational demography and Guanxi in the Chinese context. *Work and Occupations, 24*, 56 - 79.

Tsui, A. S., Farh, J. L., & Xin, K.(2000). Guanxi in the Chinese context. In J. T. Li, A. S. Tsui, & E.Weldon(Eds.), *Management and organizations in the Chinese context*. London: MacMillan.

Wang, Z. M., & Heller, F. A.(1993). Patterns of power distribution in managerial decision making in Chinese and British industrial organizations. *International Journal of Human Resource Management, 4*, 113 - 128.

Wang, Z. J., & Long, L. R.(2009). Mechanism of social capital on Chinese employees'career success. *Management Review, 21*, 30 - 39.

[王忠军，龙立荣 (2009). 员工的职业成功：社会资本的影响机制与解释效力 . *管理评论，21*, 30 - 39.]

Wakabayashi, M.(1988). Japanese management progress: mobility into middle management. *Journal of Applied Psychology, 73*, 217 - 227.

Warner, M.(1993). Human resource management "with Chinese characteristics" . *International Journal of Human Resource Management, 4*, 45 - 65.

Williams, L. J., & Anderson, S. E.(1991). Job satisfaction and organizational commitment as predictors of organizational citizenship and in-role behaviors. *Journal of Management, 17*, 601 - 618.

Wong, C., Tinsley, C., Law, K., & Mobley, W. H.(2003). Development and validation of a multidimensional measure of Guanxi. *Journal of Psychology Chinese Societies, 4*, 43 - 69.

Xin, K. R., & Pearce, J. L.(1996). Guanxi: connections as substitutes for structural support. *Academy of Management Journal, 36*, 1641 - 1658.

Xin, K. R., Farh, J. L., Cheng, B. S., & Tsui, A. S.(1998). *Guanxi in vertical dyads: evidence from Taiwan and the PRC*. Paper presented at the research conference Management and Organization in the Chinese Context.

Hong Kong University of Science and Technology, Hong Kong.

Yang, M. M.(2002). The resilience of Guanxi and its new deployments: a critique of some new Guanxi scholarship. *The China Quarterly, 170*, 459 - 476.

Yang, K. S.(1986). Chinese personality and its changes. In M. H. Bond(Ed.), *The Psychology of the Chinese People. Hong Kong*: Oxford University Press.

Zhang, Z. X., & Yang, C. F.(1998). Beyond distributive justice: the reasonableness norm in Chinese reward allocation. *Asian Journal of Social Psychology, 1*, 253 - 269.

Zhou, J., & Martocchio, J. J.(2001). Chinese and American managers' compensation award decisions: a comparative policy-capturing study. *Personnel Psychology, 54*, 115 - 145.

Operation mechanism and effects of supervisor-subordinate guanxi in Chinese organizations

WANG Zhongjun[1]; LONG Lirong[2]; LIU Lidan[3]

([1] School of Psychology, Central China Normal University, Hubei Province Key Lab for Human Development and Mental Health, Wuhan 430079, China)

([2] School of Management, Huazhong University of Science and Technology, Wuhan 430074, China)

([3] Department of Social Science, Hubei University of Chinese Medicine, Wuhan 430065, China)

Abstract Different from western society, Chinese employees attached much importance to developing good personal relationship with their leaders. So, the concept of "guanxi" and "guanxi management" were the most important aspects in Chinese management. In perspective of social exchange theory, this study enriched the concept of supervisor–subordinate guanxi, developed a social exchange model of supervisor–subordinate guanxi, and then investigated the mechanism of supervisor–subordinate guanxi operation and its effects on subordinate in Chinese organizations. By using questionnaire survey, the exploratory factor analysis (EFA) for data of 211 employees and confirmatory factor analysis (CFA) for data of 426 employees were implemented. The results showed that subordinate's guanxi input had only one dimension, and supervisor's resources output had two dimensions, including instrumental resources output and affective resources output. The study also showed that the subordinate guanxi input and supervisor resources output questionnaires had good reliability and high validity. By using questionnaire, Data was from a total of 426 matched supervisor–subordinate dyads in 54 work groups from different organizations. Hierarchical liner modeling (HLM) analysis was implemented. the results showed that after controlling task performance, subordinate's guanxi input had a positive effect on supervisor's instrumental resources output and affective resources output. Subordinate's guanxi input had a positive effect on leader–member exchange (LMX). The results also indicated that LMX partially mediated the relationship between subordinate's guanxi input and supervisor's resources output. Although our hypothesis that guanxi–based human resources management practice in work group had a positive moderating effect on the relationship between subordinate's guanxi input and supervisor's resources output was not tested, the results indicated that guanxi–based human resources management practices of work group were significantly relative to supervisor's affective resources output and LMX. The present study contributes to our understanding of the private guanxi operation behavior happened outside of work and its mechanism involved between supervisor and subordinate, as well as LMX in Chinese organizations. The results of this study will be of benefit to the guanxi management practices in organizations. Finally, the limitations in this study were discussed, and the future directions were also presented.

Keywords supervisor–subordinate guanxi; guanxi input; resources output; leader–member exchange (LMX); guanxi–based human resources management practice

Innovations in Education and Teaching International. (2015). Advance online publication.
doi:10.1080/14703297.2015.1060133.

Learning process and learning outcomes of video podcasts including the instructor and PPT slides: A Chinese case

Zhongling Pi , Jianzhong Hong*

(Key Laboratory of Adolescent Cyberpsychology and Behavior (Ministry of Education) ,

School of Psychology, Central China Normal University, Wuhan China)

Abstract Video podcasts have become one of the fastest developing trends in learning and teaching. The study explored the effect of the presenting mode of educational video podcasts on the learning process and learning outcomes. Prior to viewing a video podcast, the 94 Chinese undergraduates participating in the study completed a demographic questionnaire and prior knowledge test. The learning process was investigated by eye–tracking and the learning outcome by a learning test. The results revealed that the participants using the video podcast with both the instructor and PPT slides gained the best learning outcomes. It was noted that they allocated much more visual attention to the instructor than to the PPT slides. It was additionally found that the 22 min was the time at which the participants reached the peak of mental fatigue. The results of our study imply that the use of educational technology is culture bound.

Keywords video podcasts; learning process; learning outcomes; eye movements; visual attention; mental fatigue; culture

1 Introduction

With the rapid development and growing ubiquity of the Internet, video podcasts have become one of the fastest developing trends in learning and teaching. Video podcasts refer to video files that are distributed in a digital format through the Internet and accessed using personal computers or mobile devices (McGarr, 2009). They are generally used to record and transmit live lectures (Li, 2013).

Video podcasts are considered as having great promise for education, as they can present information through vivid visual and audio forms simultaneously (Homer, Plass, & Blake, 2008; Zhang, Zhou, Briggs, & Nunamaker, 2006), and the learning outcome of video podcasts has become a subject of considerable research interest (Copley, 2007; Homer et al., 2008; Zhang et al., 2006).

*Corresponding author. Email: jhong@mail.ccnu.edu.cn

The study was financially supported by Central China Normal University [grant number CCNU11C01005].

Answering the question whether video podcasts are effective for learning is far from straightforward. Previous studies have shown that the presenting mode of video podcasts influences the learning outcome (Chen & Wu, in press; Wang, Hao, & Lu, 2014). For example, Wang et al. (2014) compared the learning outcomes of two types of video podcasts in an investigation involving 60 Chinese undergraduates and graduates. The first type of video podcast included the instructor and PPT slides, whereas the second type of video podcast only included PPT slides. It was found that the learners viewing the first type of video podcast gained significantly higher scores. The findings of the study thus implied that the presenting mode of the video podcast affects the learning outcome and that the presence of the instructor in a video podcast enhances learning.

Some western studies, however, have produced different results (Homer et al, 2008; Kizilcec, Papadopoulos, & Sritanyaratana, 2014; Lyons, Reysen, & Pierce, 2012). For example, Homer, Plass, and Blake (2008) examined the effect on the learning outcomes of the presence of the instructor in video podcasts. Similar to Wang et al.'s study (2014), there were two different types of video podcast: the first included the instructor and PPT slides, and the second included only PPT slides. Twenty-six American undergraduates were assigned to view one of the two versions of the video podcasts. No significant differences were found between the two groups as regards the learning outcome. Similar results were obtained by Kizilcec et al. (2014).

A possible reason for the discrepancy between the studies is the fact that the participants in the reviewed studies were different. The participants in Wang et al.'s study (2014) were Chinese undergraduates and graduates, whereas in Homer et al.'s study (2008) and Kizilcec et al.'s study (2014) they were American undergraduates. The students lived in different cultures, which could

have affected the cognitive and learning process and the learning outcome. A growing number of studies have suggested that people from different cultures exhibit different cognitive processing styles (Nisbett & Miyamoto, 2005; Varnum, Grossmann, Kitayama, & Nisbett, 2010). Does this imply that culture may play an important role in video podcast learning?

Although previous studies on the topic have produced inconsistent results, the findings collectively imply that some presenting modes of video podcasts could be more appropriate and effective than others. Therefore, for researchers and educators, the key question to be investigated becomes not whether video podcasts are effective tools for learning, but rather to identify a presenting mode which optimizes the learning process and outcomes (Homer et al, 2008).

Understanding the learning process of video podcasts, in particular, viewing behaviors and mental fatigue of learners, is crucial to improving learning outcomes. Recently, Kizilcec et al. (2014) used an eye-tracking technique to explore viewing behavior while watching video podcasts. In the study, video overlays of the instructor were used such that the slides were presented alternately with and without the instructor overlay, with the form alternating approximately every 2 min. The study found that when the instructor was present, learners spent 41% of their time looking at the instructor and switched between the instructor and slides every 3.7 s.

Previous studies have suggested that mental fatigue seriously affects learning outcomes (Gonzalez, Best, Healy, Kole, & Bourne, 2011). Controlling the length of video podcasts thus becomes an important element in avoiding mental fatigue and thus ensuring a good learning outcome. Danforth, Schumacher, Cullen, and Ma (2012) explored the effect of video podcast length on nutrition knowledge learning. They presented nutrition knowledge through three formats. The first format focused on the ingredients of one dish, including nutrition and preparation information. The second was a quick recipe demonstration. The third was a long recipe demonstration that included all the steps involved in completion of the finished recipe. The three formats were included in video podcasts ranging from 2 to 11.5 min in length. After viewing the video podcasts, participants filled out a learning test. It was found that the participants viewing a 4–6 min video podcast gained the best learning scores. The results of the study thus implied that the learners experienced mental fatigue after 4–6 min of watching a learning video podcast.

Previous studies have compared the learning outcomes of different presentation modes, for example (Wang et al., 2014), but few studies have explored the viewing behavior and mental fatigue associated with video podcasts based on eye movements and within the specific context of China. Based on the above discussion, it becomes clear that we know relatively little about which presenting mode of video podcast is the most effective for learning, how learners process the learning content, how cultural context affects video podcast learning and when learners experience mental fatigue.

Our study focused on investigating which presenting mode of video podcast optimally enhances learning by comparing the learning outcomes of four frequently used modes of video podcast. The learning process of the video podcast was examined using eye-tracking, which is a promising tool for tracking the learning process because eye movements can be considered as reflecting in real time the attention and encoding processes and mental fatigue of learners during the entire learning period (Hyona, 2010; Kizilcec et al., 2014). Specifically, fixation data can be considered as an indicator representing learners' visual attention, saccade eye movement as reflecting visual attention transition (Kizilcec et al., 2014), and blink data, a classical indicator, can be used to assess learners' mental fatigue (Zhu, Wu, Wang, & Qi, 2008).

Based on the above literature review and analysis, the following hypotheses were formulated:

(1)Video podcasts including both the instructor and PPT slides best facilitate learning outcomes.

(2)Chinese learners using video podcasts including the instructor and PPT slides spend more than half the learning time visually fixated on the instructor.

(3)Learners' mental fatigue increases with time and reaches a peak at a certain point.

2 Method

2.1 Participants

Ninety-six undergraduate students (23 male and 73 female) were recruited from Central China Normal University (CCNU) via advertisement. Two of them were excluded from the study because of poor eye calibration. The study participants were aged 17.17 to 25.25 $years$ (M=20.25, SD=1.32). Using informal interviews, it was ensured that none of the participants was tired or knowledgeable on the topic of the video podcasts. The participants were from 10 study majors and no participant was a psychology major. All participants completed a prior knowledge test to assess their prior knowledge. The information gained was used to balance the groups and avoid the possibility of differences in prior knowledge introducing bias into the results. Participants all had normal or corrected-to-normal vision. They all gave written informed consent. On completion of the study, every participant received a small gift; they could choose a small item such as a delicate spoon and a pair of chopsticks, cleansing tissue, washing powder and candy.

2.2 Apparatus

Eye movements were registered by Eyelink 1000 eye tracker (SR Research Ltd., Canada). Prior to viewing the video podcasts, a 9-point calibration and validation procedure was applied to determine periods of fixation and eye reaction to stimuli presentation. The stimuli were presented on a 21-in. CRT monitor (NESOJXC FS210A) at a viewing distance of 60 cm. The resolution of the monitor was 1024 × 768 pixels and the refresh rate was 75 Hz. The participants listened to the auditory explanation via Philips headphones connected to the PC. Participants used a chin rest with a head-stop to stabilize head position during the experiment. Data were monitored for a 1000 Hz monocular sampling rate.

2.3 Materials

2.3.1 Stimuli

Four modes of 25 min video podcasts were presented on the topic of attachment by Dr Huang from the School of Psychology (CCNU). The modes used were (1) *the mode of PPT slides (PPT)*, in which only synchronized PPT slides were included; (2) *the mode of instructor without PPT slides (Instructor)*, in which only the instructor was included; (3) *the mode of instructor with PPT slides (Instructor and PPT)*, in which the instructor and synchronized PPT slides were included; and (4) *the mode of classroom (Classroom)*, in which the whole learning activity in the classroom was recorded, including the instructor, students and synchronized PPT slides.

The instructor's audio explanations were the same in all four modes. The difference between the four modes was the visual material. The PPT slides were simply to reinforce the learning points. Before the video podcasts were recorded, the instructor was asked to write a script and keep the learning content identical.

2.3.1 Measures

Demographic questionnaire. Participants were asked to report their gender, age, grade, QQ number (for the QQ social media network in China), e-mail address, study major, online learning experience and experience of studying psychology courses. The demographic data were collected as controlling factors in the study.

Prior knowledge test. The prior knowledge test included 16 multiple choice questions to test participants' knowledge of developmental psychology. The test drew on a popular Chinese developmental psychology textbook, *Developmental Psychology* (Lin, 2005). There were four options on each item, and every item had two or more right options. Only if all right options were selected, did the participant get 1 point on an item. The total score was 16. The test had high discrimination (t (60) =16.94, *p*< .001).

Learning test. This test included two parts. The first part, which tested learning recall, contained 15 multiple choice questions testing participants' comprehension of key concepts in the video podcast. The scoring method was the same as with the prior knowledge test. The second part, which tested learning transfer, contained 3 short-answer questions testing participants' ability to apply their knowledge to novel situations. The total score was 25. The test had high discrimination (*t* (60) = 20.79, *p*<.001).

2.4 Design

A between-subject design was used with the presenting modes of the video podcasts as the between-subject variable. The between-subject variable included 4 levels: PPT, Instructor, Instructor and PPT and Classroom. The between-subject design meant that the participants were randomly assigned to one of the four experimental condition groups.

The process of video podcast learning was analyzed by eye movement tracking. Specifically, it could be identified what information was processed and visually focused on based on data of the learners' visual fixation, such as fixation duration, fixation counts and mean fixation duration. The eye-mind hypothesis suggests that there is a positive link between learners' fixation and visual attention, i.e., the more time fixated on an item, the more visual attention is being paid (Just & Carpenter, 1984). Saccade, which is a rapid movement of an eye, can reflect transition of visual attention (Kizilcec et al., 2014). Blink duration, tracked by eye-tracking technique, is a highly sensitive measure for tracking fluctuating levels of mental fatigue. Mental fatigue describes the phenomenon that, over time, an individual engaged in cognitive activities gradually exhibits psychological loss. The longer the blink duration, the greater the mental fatigue; and the shorter the blink duration, the lighter the mental fatigue (Stern, Boyer, & Schroeder, 1994; Zhu et al, 2008).

2.5 Procedure

The study was carried out in a laboratory at the School of Psychology CCNU. After granting consent, all participants were first given the demographic questionnaire and prior knowledge test. They were then escorted to an eye-tracking room and randomly assigned to view, without pauses and in the absence of the experimenter, one of the four video podcasts. Before the video podcast started, participants were asked to follow a dot on the screen with their eyes for calibration and validation. Immediately after viewing the video podcast, participants filled out the learning test.

3 Results

3.1 Prior knowledge of different groups

One-way between-subjects analysis of variance was performed, with the scores of the prior knowledge test as the dependent variable and the presenting modes as the independent variable. No significant difference was found between the groups (F (3, 90) =1.36, p>.05).

3.2 Learning outcomes

The differences in learning test scores were compared between the different groups to assess the learning outcome

Table 1. Mean and standard deviations of the learning test scores.

Groups	N	M	SD
PPT	24	11.42	3.06
Instructor	22	8.52	3.80
Instructor and PPT	24	14.75	3.13
Classroom	24	11.25	2.73

of the different video podcasts by presenting mode. Table 1 shows the mean and standard deviation of the scores of the different groups. It was found that the participants in the Instructor and PPT group gained the highest scores on the learning test.

An analysis of covariance was performed with the scores of the prior knowledge test as the covariable, the presenting modes as the independent variable and the learning test scores as the dependent variable. This statistical method could exclude the effect of prior knowledge on the learning outcome. It was found that there was significant difference between the groups (F (3, 89) = 15.67, p < .001).

The results of the posthoc test (Bonferroni) are shown in Table 2. It was found that the learning test scores of the Instructor and PPT group were significantly higher than those of the other three groups. In addition, the Instructor group gained significantly lower scores than the PPT group and the Classroom group. There was no statistically significant difference between the PPT group and Classroom group. The results implied that video podcasts with Instructor and PPT optimally helped learners' learning.

3.3 Eye movements of the video podcast with Instructor and PPT

As the Instructor and PPT approach gave best results for learning, it is necessary to review the evidence from eye movements on how the participants viewed the video with this mode and how mental fatigue changed during the learning process.

3.3.1 Allocation of visual attention

In analysis of the Instruction and PPT mode, the video was divided into 2 areas of visual interest. These areas were the area of the instructor and that of the PPT slides. Table 3 shows the mean and standard deviation of the total fixation duration, fixation counts and mean fixation duration in the two areas. Looking at the percentage of

Table 2. Mean differences (*p*—value) between the learning test scores.

Groups	Instructor	Instructor and PPT	Classroom
PPT	3.24 (0.005) **	3.06 (0.007) **	0.15 (1.000)
Instructor		6.30 (0.000) ***	3.08 (0.008) **
Instructor and PPT			3.21 (0.004) **

**p<0.01; ***p<0.001.

Table 3. Mean and standard deviation of the total fixation duration, fixation counts and mean fixation duration for Instructor and PPT video podcasts.

Areas of interest	Total fixation duration (s)		Fixation counts		Mean fixation duration (ms)	
	M	SD	M	SD	M	SD
Instructor	691.04	19.00	2864.48	707.02	241.79	37.97
PPT slides	418.22	15.43	1113.62	355.36	378.09	94.20

total fixation duration on the areas of interest, it was found that the participants spent 62.30% of the time fixating on the instructor, and 37.70% fixating on the PPT slides, on average. The transition rate of visual attention between the areas, analyzed by saccade data, was every 24.59 s.

Paired samples *t*—tests were performed with the areas of interest. It was found that the total fixation duration in the area of the instructor was significantly longer than in that of the PPT slides ($t(23)=6.83$, $p< .001$). The fixation count in the area of the instructor was greater than in that of the PPT slides ($t(23)=14.92$, $p< .001$), and the mean fixation duration in the area of the instructor was shorter than in that of the PPT slides ($t(23)=7.53$, $p< .001$).

The above results indicate that the participants allocated much more visual attention to the instructor than to the PPT slides, which suggests that the instructor facilitated the enhanced learning outcome.

3.3.2 *Mental fatigue in the process of learning*

Mental fatigue was examined by blink duration, for which one minute was taken as the time unit and averaged blink duration per minute calculated. The data were sampled within two standard deviations.

There are three upturns in the participants' blink duration. The first is during the 10–13 min, with the participants' blink duration slowly rising from 567.20 to 1622.59 ms; the second is at the 22 min, where the participants' blink duration suddenly rises to 7881.79

ms, the peak value in this study; and the third upturn occurs at 25 min, where blink duration rises to 3984.37. The interval between the first and the second upturn was much longer than the interval between the second and the third upturn.

The results for mental fatigue therefore implied that the participants started to experience mental fatigue after 10 min of video podcast learning. The trend suggests that they were immediately aware that they felt tired and refreshed their mind; that is, they re—entered a normal learning mental status as found earlier in the learning period. However, they entered heavy mental fatigue status at the 22 min. At this point, although they were aware that they felt tired, they could no longer refresh their mind. In other words, they could no longer re—enter their normal learning mental status. The results for blink duration suggest that a video podcast with Instructor and PPT should not be longer than 21 min.

4 Discussion

The results of the study showed that of the four video podcast modes studied the Instructor and PPT mode most enhanced the learners' learning, which supported the first hypothesis that a video podcast with both the instructor and PPT slides best facilitates the learning outcome. Previous studies have shown that the presence of the instructor in video podcasts affects learners' engagement and cognitive load, which in turn influences the learning outcome. They suggest that presence of the instructor contributes social cues while simultaneously increases the cognitive load (Gunawardena, 1995; Homer et al., 2008; Paas & van Merriënboer, 1994). The cognitive load increases because the instructor's presence provides information beyond what is taught. Furthermore, the learners' attention is drawn away from the supporting material in the PPT slides (Pass & van Merrënboer, 1994). Learning engagement and cognitive load concurrently affect learning, and they need to be well balanced (Pi, 2014). The video podcast with PPT imposed the lowest cognitive load. However, it did not include any social cues, which suggests that it created the lowest learning engagement, which hindered effective learning. The other two presenting modes, i.e. Instructor mode and Classroom mode, can be considered as not being ideal in the sense that social presence and cognitive load are not well balanced.

The results for the learning outcomes in our study confirm the results of Wang et al. (2014) and are in contrast with the results of Homer, et al. (2008) and Kizilcec et al. (2014). One possible explanation for the difference is that cultural differences have affected the learners' perception of social cues, which further influenced the learning outcomes. The participants in Homer et al. (2008)and Kizilcec et al. (2014)were American students, and the participants in our study and in the study of Wang et al. (2014) were Chinese students. Cultural studies have indicated that Chinese are relation-oriented and live in complex social networks with prescribed role relations, whereas Americans tend to be task-oriented, live in less constraining social networks and pay much more attention to the focal object rather than background information (Hong, Heikkinen, & Blomqvist, 2010; Markus & Kitayama, 1991; Nisbett & Miyamoto, 2005). Thus, Chinese students can be expected to be

particularly sensitive to social cues and their cognitive outcomes are consequently more based on relationships than Western students. In contrast, American students may pay less attention to social cues but more attention to the content of the study. Therefore, the instructor in the video podcast has had little effect on the American participants. Comments by our participants during informal interviews after they had viewed the video podcasts provided further evidence of their preference for the presence of the instructor.

Additionally, the fixation data for eye movement in our study support the above explanation. The data showed that the learners spent much more time looking at the instructor than the PPT slides, which supported our second hypothesis that learners using video podcasts with Instructor and PPT spend more than half the learning time fixated on the instructor. In contrast to the results in this study, Kizilcec et al. (2014) found that their participants spent less time looking at the instructor than the PPT slides. It would appear that compared with American students, Chinese students spend more time on social cues. The results of our study on podcast learning imply that the use of educational technology is culture bound, despite today's global context of learning and teaching. However, this finding needs to be tested further with research focusing specifically on cross-cultural aspects.

The eye-tracking data revealed that in the podcast with Instructor and PPT the learners started to feel mental fatigue at the 10 min and reached the peak of mental fatigue at the 22 min. These results support our third hypothesis that learners' mental fatigue increases with time and reaches a peak at a certain point. The results for mental fatigue imply that if video podcasts can be divided into several sections, it would be best that the length of any section is kept below 10 min. If the video podcasts cannot easily be divided, the length should not exceed 22 min. The results, to some extent, are consistent with those of Danforth et al.

(2012), who found that 4–6 min video podcasts gave optimal learning outcomes. The length of the video podcasts in the study of Danforth et al. (2012)was, however, limited between 2 and 11.5 min, and hence nothing is known about changes in mental fatigue after 11.5 min.

To the best of our knowledge, this study is the first to systematically compare the learning outcome of four common presenting modes of video podcasts and explore via eye movement analysis the time when learners experience the onset of mental fatigue in the educational context of China. The focus of our study is the question of which presenting mode best facilitates learning. In the explanation of our major finding that the most effective presenting mode is Instructor and PPT, we assume that learners' engagement and cultural expectations play a critical role. We believe that Chinese sensitivity to social cues, which has been noted in other contexts, is equally applicable to the studied learning situation. At this stage, all conclusions on the mechanisms underlying the results of our study are inferred and thus need to be explored in more depth. Further research is also needed to investigate how social cues can enhance online teaching and learning in different cross–cultural contexts.

There are several limitations in this study. Firstly, in order to guarantee the accuracy of the eye movement data, the head of the participants was stabilized, which is a clear difference from natural learning situations and to a certain extent will affect the measured results. Secondly, mental fatigue was explored during the study of the psychological concept of *attachment*; a different learning subject might give different results. Despite these limitations, the study makes a meaningful contribution to discussion of video podcast learning by demonstrating the clear differences in learning effectiveness of the different presentation modes, by illustrating the use of quantitative visual behavior data in study of video podcast learning, and by presenting a rationale for the contradictory findings in the literature.

References

Chen, C. M., & Wu, C. H. (2015). Effects of different video lecture types on sustained attention, emotion, cognitive load, and learning performance. *Computers & Education, 80*, 108 – 121.

Copley, J. (2007). Audio and video podcasts of lectures for campus–based students: Production and evaluation of student use. *Innovations in Education and Teaching International, 44*, 387 – 399.

Danforth, S., Schumacher, J., Cullen, R., & Ma, Y. J. (2012). Evaluating format preferences and effectiveness of video podcasts related to nutrition education and recipe demonstrations. *Journal of Nutrition and Dietetics, 112*, A19.

Gonzalez, C., Best, B., Healy, A. F., Kole, J. A., & Bourne, L. E. J. (2011). A cognitive modeling account of simultaneous learning and fatigue effects. *Cognitive Systems Research, 12*, 19 – 32.

Gunawardena, C. N. (1995). Social presence theory and implications for interaction and collaborative learning in computer conferences. *International Journal of Educational Telecommunications, 1*, 147 – 166.

Homer, B. D., Plass, J. L., & Blake, L. (2008). The effects of video on cognitive load and social presence in multimedia–learning. *Computers in Human Behavior, 24*, 786 – 797.

Hong, J. Z., Heikkinen, J. & Blomqvist, K. (2010). Culture and knowledge co–creation in R&D collaboration between MNCs and Chinese universities. *Knowledge and Process Management, 17*, 62 – 73.

Hyona, J. (2010). The use of eye movements in the study of multimedia learning. *Learning and Instruction, 20*, 172 – 176.

Just, M. A., & Carpenter, P. A. (1984). Using eye fixations to study reading comprehension. In D. E. Kieras & M. A.

Just (Eds.), *New Methods in Reading Comprehension Research* (pp. 151 – 182). Hillsdale, NJ: Erlbaum.

Kizilcec, R. F., Papadopoulos, K., & Sritanyaratana, L. (2014). Showing face in video instruction: Effects on information retention, visual attention, and affect. *Proceedings of the SIGCHI Conference on Human Factors in Computing Systems* (pp. 2095–2102). New York: USA.

Li, S. K. (2013). Review of application of podcasts in teaching in China. *Software Guide, 12*, 195 – 197 (in Chinese).

Lin, C. D. (2005). *Developmental Psychology*. Hangzhou: Zhejiang Education Publishing House (in Chinese).

Lyons, A., Reysen, S., & Pierce, L. (2012). Video lecture format, student technological efficacy, and social presence in online courses. *Computers in Human Behavior, 28*, 181 – 186.

Markus, H. R., & Kitayama, S. (1991). Culture and the self: Implications for cognition, emotion, and motivation. *Psychological Review, 98*, 224 – 253.

McGarr, O. (2009). A review of podcasting in higher education: Its influence on the traditional lecture. *Australasian Journal of Educational Technology, 25*, 309 – 321.

Nisbett, R. E., & Miyamoto, Y. (2005). The influence of culture: Holistic versus analytic perception. *Trends in Cognitive Science, 9*, 467 – 473.

Paas, F., & van Merri nboer, J. J. G. (1994). Instructional control of cognitive load in the training of complex cognitive tasks. *Educational Psychology Review, 6*, 51 – 71.

Pi, Z. L. (2014). *The presenting formats of video podcasts and their effect on learning effectiveness and learning mechanism: An eye tracking study*. Unpublished master thesis. Wuhan: Central China Normal University (in Chinese).

Stern, J. A., Boyer, D., & Schroeder, D. (1994). Blink rate: A possible measure of fatigue. *The Journal of the Human Factors and Ergonomics Society, 36*, 285–297.

Varnum, M. E. W., Grossmann, I., Kitayama, S., & Nisbett, R. E. (2010).The origin of cultural differences in cognition. The social orientation hypothesis. *Current Directions in Psychological Science, 19*, 9 – 13.

Wang, J., Hao, Y. H., & Lu, J. L. (2014). The effect of presenting mode of teaching video on self-directed learning effectiveness: An experimental study. *E-education Research, 251*, 93 – 105 (in Chinese).

Zhang, D., Zhou, L., Briggs, R. O., & Nunamaker Jr, J. F. (2006). Instructional video in e-learning: Assessing the impact of interactive video on learning effectiveness. *Information & Management, 43*, 15 – 27.

Zhu, Z. H., Wu, X. J., Wang, L., & Qi, L. (2008). Detection method of driver fatigue based on blink duration. *Computer Engineering, 34*, 201 – 206.

附录：本书网络版目录

教育心理

因版面限制，本书纸质版未能完全收录各方向推荐的代表性论文，所有这些代表性论文均可在本书网络版中查阅全文，网址：http://psych.ccnu.edu.cn，也可根据此目录信息在《CNKI 中国学术期刊网络出版总库》检索到全文。本目录页码为网络版 PDF 文件页码。

脑与认知

人格与社会心理

管理心理